Mathematical Foundations
for
SIGNAL PROCESSING, COMMUNICATIONS, AND NETWORKING

Mathematical Foundations

for

SIGNAL PROCESSING, COMMUNICATIONS, AND NETWORKING

Edited by

Erchin Serpedin • Thomas Chen • Dinesh Rajan

CRC Press
Taylor & Francis Group
Boca Raton   London   New York

CRC Press is an imprint of the
Taylor & Francis Group, an **informa** business

CRC Press
Taylor & Francis Group
6000 Broken Sound Parkway NW, Suite 300
Boca Raton, FL 33487-2742

First issued in paperback 2017

© 2012 by Taylor & Francis Group, LLC
CRC Press is an imprint of Taylor & Francis Group, an Informa busine

No claim to original U.S. Government works

Version Date: 20111103

ISBN 13: 978-1-138-07216-9 (pbk)
ISBN 13: 978-1-4398-5513-3 (hbk)

Library of Congress Cataloging-in-Public

Mathematical foundations for signal processing, communicat
    editors, Erchin Serpedin, Thomas Chen, and Dinesh Rajan.
      p. cm.
    Includes bibliographical references and index.
    ISBN 978-1-4398-5513-3 (hardback)
     1. Telecommunication--Mathematics. I. Serpedin, Erchin,
III. Rajan, Dinesh.

    TK5102.5.M2966 2011
    621.38--dc23

Visit the Taylor & Francis Web site at
http://www.taylorandfrancis.com

and the CRC Press Web site at
http://www.crcpress.com

For Zara, Nisa, Aisha and Nesrin (ES)

For Robin and Kayla (TC)

In memory of Dilip Veeraraghavan (DR)

"*Mathematics is the queen of the sciences.*"

Carl Friedrich Gauss

Contents

3 Linear Algebra

Fatemeh Hamidi Sepehr and Erchin Serpedin

5 Numerical Analysis

Vivek Sarin

6 Combinatorics

Walter D. Wallis

7   Probability, Random Variables, and Stochastic Processes   205

Dinesh Rajan

8   Random Matrix Theory   245

Romain Couillet and Merouane Debbah

9 Large Deviations

Hongbin Li

10 Fundamentals of Estimation Theory

Yik-Chung Wu

11 Fundamentals of Detection Theory   369

Venugopal V. Veeravalli

14 Unconstrained and Constrained Optimization Problems   487

Shuguang Cui, Anthony Man-Cho So, and Rui Zhang

15 Linear Programming and Mixed Integer Programming   519

Bogdan Dumitrescu

16 Majorization Theory and Applications

Jiaheng Wang and Daniel Palomar

17 Queueing Theory

Thomas Chen

18 Network Optimization Techniques   627

Michał Pióro

19 Game Theory   691

Erik G. Larsson and Eduard Jorswieck

20 A Short Course on Frame Theory

Veniamin I. Morgenshtern and Helmut Bölcskei

List of Figures

List of Tables

Preface

The rationale behind this textbook is to provide all the necessary mathematical background to facilitate the training and education of students and specialists working in the interrelated fields of signal processing, telecommunications, and networking. Our intention was to create a self-contained textbook that enables mastering both the fundamental results in the areas of signal processing, telecommunications and networking as well as the more advanced results and recent research trends in these areas.

In our collective academic experience, students often begin their graduate education with widely varying undergraduate backgrounds in terms of needed subjects such as probability theory, stochastic processes, statistics, linear algebra, calculus, optimization techniques, game theory, and queueing theory. While some students are well prepared for advanced courses in signal processing, telecommunications, and networking, others are not as well prepared and must make extra remedial efforts. However, obtaining the necessary mathematical background is often difficult because these topics are usually dispersed across a large number of courses, where the emphasis is frequently put on topics different than signal processing, telecommunications, and networking.

We hope that this textbook will serve as a reference for graduate-level students seeking to reach a common standard level of preparedness before undertaking more advanced specialized studies. We believe that this book will also be useful for researchers, engineers, and scientists working in related areas in electrical engineering, computer science, bioinformatics, and system biology. We also believe that this textbook could be used for a one-semester (or two-semester) graduate-level course that will help students acquire all the necessary mathematical background to pursue more advanced graduate courses. Both introductory and more advanced graduate-level courses could be taught based on this textbook. This textbook is accompanied by a Solutions Manual, and a set of viewgraphs will be made available to interested instructors. This textbook could also be used as a reference for self-teaching by a motivated professional who has graduate-level math skills.

Each chapter of this 20-chapter textbook was contributed by different experts in the areas of signal processing, telecommunications, and networking. Collectively, the chapters comprehensively cover a broad range of mathematical techniques that

have been found to be necessary for a good underst
concepts and results from the fields of signal process
networking. Each chapter includes both mathemat
ples and homework exercises to illustrate the usage
concepts to solve different applications.

This textbook would not have been possible witho
port of contributing authors and our students. Ou
contributing authors, families, PhD student Mr. Ali
ciate Dr. Serhan Yarkan, who spent a huge amount o
correcting and merging together all the files and co
participation in this project are greatly appreciated.
our students: Amina Noor, Aitzaz Ahmad, and Sabit
support from the National Science Foundation (NSF)
appreciated.

Despite our efforts to minimize all sources of er
we believe that inconsistencies and errors might s
kindly asking our readers to email us their feedbac
Any feedback is welcomed to improve this book. '
maintained at the webpage of Dr. Serpedin at Texa
http://www.ece.tamu.edu/~serpedin/. Viewgrap
additional teaching materials will be posted at this '
terested in adopting this textbook can access the So
CRC Press.

Editors

Erchin Serpedin received (with highest distinction) a Diploma of Electrical Engineering from the University of Bucharest, Romania, in 1991. He received a specialization degree in signal processing and transmission of information from École Supérieure D'Électricité, Paris, France, in 1992, an M.Sc. degree from Georgia Institute of Technology, Atlanta, GA, in 1992, and a Ph.D. degree in Electrical Engineering from the University of Virginia, Charlottesville, in January 1999. In July 1999, he joined Texas A&M University (TAMU) in College Station, Wireless Communications Laboratory, as an assistant professor, and he currently holds the position of professor. His research interests lie in the areas of statistical signal processing, wireless communications and bioinformatics. He received the US NSF Career Award in 2001, CCCT 2004 Conference - Best Paper Award, TAMU Outstanding Faculty Award in 2004, National Research Council (NRC) Fellow Award in 2005, Texas Engineering and Experimentation Station (TEES) Fellow Award in 2005, and American Society for Engineering Education (ASEE) Fellow Award in 2006. He has served as an associate editor for 10 major journals including *Institute of Electrical and Electronics Engineers (IEEE) Communications Letters, IEEE Transactions on Signal Processing, IEEE Transactions on Communications, IEEE Transactions on Information Theory, IEEE Transactions on Wireless Communications, IEEE Signal Processing Letters*, and *European Association for Signal Processing (EURASIP) Journal on Advances in Signal Processing*. Dr. Serpedin served also as a guest editor for several journal special issues. He served as a technical co-chair of the Communications Theory Symposium at Globecom 2006 Conference, and the Wireless Access Track at the Vehicular Technology Conference (VTC) Fall 2006 Conference. Dr. Serpedin is the author of seventy journal papers, one hundred conference papers, two research monographs, and five book chapters.

Thomas Chen is a Professor in Networks at Swansea University in Swansea, Wales, UK. He received BS and MS degrees in Electrical Engineering from the Massachusetts Institute of Technology (MIT) in 1984, and a Ph.D. in Electrical Engineering from the University of California, Berkeley, in 1990. He worked on high-speed networks research at GTE Laboratories (now Verizon) in Waltham,

Massachusetts, before joining the Department of Ele
ern Methodist University, Dallas, as an associate ɪ
Swansea University in May 2008. He is co-authoɪ
House) and co-editor of *Broadband Mobile Multimeɛ
tions* (CRC Press). He is currently editor-in-chief
Electronics Engineers (IEEE) Network, senior editor
*Electronics Engineers (IEEE) Communications Mag
ternational Journal of Security and Networks*, associ
curity and Communication Networks, associate editoɾ
Digital Crime and Forensics, and technical editor fⱼ
Electronics Engineers (IEEE) Press. He was formɛ
of Electrical and Electronics Engineers (IEEE) Co
founding editor-in-chief of *Institute of Electrical and .*
Communications Surveys. He received the Institute
Engineers (IEEE) Communications Society's Fred ᴴ
1996.

Dinesh Rajan received the B.Tech. degree in Electɾ
Institute of Technology (IIT), Madras in 1997. He
degrees in Electrical and Computer Engineering in
from Rice University, Houston. He joined the Electɾ
at Southern Methodist University, Dallas, in August ᴢ
associate professor. His current research interests inɕ
wireless networks, information theory, and computati
NSF career award in 2006 for his work on applying infɛ
of mobile wireless networks. He also received the awaɾ
in the EE Department in 2006 and 2007. He was cɦ
the IEEE Communications and Vehicular Technologⱼ
chairman, he received the "best chapter award" in 20(
worldwide from the IEEE Vehicular Technology Sociɛ
Intern Award from Nokia Research Center in 1998.
member.

List of Contributors

Aitzaz Ahmad
Texas A&M University
College Station, USA

Helmut Bölcskei
ETH Zurich
Switzerland

Thomas Chen
Swansea University
Wales, UK

Romain Couillet
L'École Supérieure D'Électricité
(SUPELEC)
France

Shuguang Cui
Texas A&M University
College Station, USA

Merouane Debbah
L'École Supérieure D'Électricité
(SUPELEC)
France

Tolga M. Duman
Arizona State University
Tempe, USA

Bogdan Dumitrescu
Tampere University of Technology
Finland

Eduard Jorswieck
Technical University of Dresden
Germany

Erik G. Larsson
Linkoping University
Sweden

Hongbin Li
Stevens Institute of Technology
Hoboken, NJ, USA

Veniamin I. Morgenshtern
ETH Zurich
Switzerland

Daniel Palomar
Hong Kong University of Science and
Technology

Michał Pióro
Lund University
Sweden
Warsaw University of Technology
Poland

Khalid Qaraqe
Texas A&M University at Qatar

Dinesh Rajan
Southern Methodist University
Dallas, USA

Vivek Sarin
Texas A&M University
College Station, USA

Erchin Serpedin
Texas A&M University
College Station, USA

Anthony Man-Cho So
The Chinese University of Hong Kong
Hong Kong, China

Fatemeh Hamidi Sepehr
Texas A&M University
College Station, USA

Venugopal V. Veeravalli
University of Illinois at
    Urbana-Champaign
USA

Walter D
Southern I
Carbondal

Jiaheng
KTH Roya
Stockholm

Xiaodong
Columbia
New York,

Yik-Chun
University

Serhan Y
Texas A&
College St

Rui Zhan
National U

List of Acronyms

A/D	analog-to-digital
AnF	amplify-and-forward
AnNetC	analogue network coding
ARMA	autoregressive-moving average
ASEE	American Society for Engineering Education
AWGN	additive white Gaussian noise
BCH	Bose and Ray–Chaudhuri
BSC	binary symmetric channel
CD	compact disc
CDF	cumulative distribution function
CDMA	code division multiple access
CFO	carrier frequency offset
CGF	cumulant generating function
CLT	central limit theorem
CRLB	Cramér-Rao lower bound
CSCG	circular symmetric complex Gaussian
d.f.	distribution function
D/A	digital-to-analog
DFT	discrete Fourier transform
DNA	deoxyribonucleic acid
DTFT	discrete-time Fourier transform
DVD	digital versatile disc
e.s.d.	empirical spectral distribution
EURASIP	European Association for Signal Processing
FFT	fast Fourier transform
GLR	generalized likelihood ratio
GLRT	generalized likelihood ratio test
GSVD	generalized singular value decomposition
HMM	hidden Markov model
i.i.d.	independent and identically distributed
IEEE	Institute of Electrical and Electronics Engineers

IIT	Indian Institute of Technology
ISI	intersymbol interference
KKT	Karush-Kuhn-Tucker
l.s.d.	limit spectral distribution
LFP	linear fractional programming
LLN	law of large numbers
LMMSE	linear minimum mean-square error
LMP	locally most powerful
LP	linear programming
LRT	likelihood ratio test
LTI	linear time-invariant
MAP	maximum a posteriori
MCMC	Markov chain Monte Carlo
MFN	multicommodity flow networks
MIMO	multi-input multi-output
MIP	mixed integer programming
MMSE	minimum mean-square error
MSE	mean-square error
MVUE	minimum variance unbiased estimat
NASA	National Aeronautics and Space Ad
NRC	National Research Council
NSF	National Science Foundation
OFDM	orthogonal frequency division multip
OFDMA	orthogonal frequency division multip
ONB	orthonormal basis
PCA	principal component analysis
PDF	probability density function
PMF	probability mass function
PSD	power spectral density
PWM	position weight matrix
QPSK	quadrature phase shift keying
RLS	recursive least squares
ROC	receiver operating characteristics
RoC	region of convergence
SemiDP	semidefinite programming
SER	symbol error rate
SINR	signal-to-interference-plus-noise rati
SISO	single-input single-output
SMC	sequential Monte Carlo
SNR	signal-to-noise ratio
SVD	singular value decomposition
TAMU	Texas A&M University
TEES	Texas Engineering and Experimenta

TWRC	two-way relay channel
UMP	uniformly most powerful
VTC	vehicular technology conference
WH	Weyl–Heisenberg
WLLN	weak law of large numbers
WSS	wide-sense stationary

Notations and Symbols

$\mathcal{A}, \mathcal{B}, \ldots$	sets
$\{\cdot\}$	set
$\mathcal{A}\backslash\mathcal{B}$	set $\mathcal{A}$ without $\mathcal{B}$
$\forall$	for all (propositional logic)
$\cdot \wedge \cdot$	AND operator (propositional logic)
$\exists$	there exists (propositional logic)
$\exists!$	there exists only one (propositional logic)
$\lvert\cdot\rvert$	absolute value
i	$\sqrt{-1}$
$\triangleq$	definition
$\mathbb{R}, \mathbb{C}, \mathbb{Z}, \mathbb{N}$	real line, complex plane, set of all integers, set of natural numbers (including zero)
$\mathbb{R}^n, \mathbb{C}^n$	set of n-tuples with components in $\mathbb{R}, \mathbb{C}$
X	metric space X
$\mathcal{F}$	field $\mathcal{F}$
$\mathcal{V}$	vector space $\mathcal{V}$
$\dim(\mathcal{V})$	dimension of vector space $\mathcal{V}$
$\mathrm{span}(\mathcal{A})$	linear span of set of vectors
$\mathbf{a}, \mathbf{b}, \ldots$	vectors
$\mathbf{A}, \mathbf{B}, \ldots$	matrices
$\mathbf{a}^\mathsf{T}, \mathbf{A}^\mathsf{T}$	transpose of the vector $\mathbf{a}$ and the matrix $\mathbf{A}$
$a^*, \mathbf{a}^*, \mathbf{A}^*$	complex conjugate of the scalar a, element-wise complex conjugate of the vector $\mathbf{a}$, and the matrix $\mathbf{A}$
$\mathbf{a}^\dagger, \mathbf{A}^\dagger$	Hermitian transpose of the vector $\mathbf{a}$ and the matrix $\mathbf{A}$
$\mathbf{I}_N$	identity matrix of size $N \times N$
$\mathrm{rank}(\mathbf{A})$	rank of the matrix $\mathbf{A}$
$\lambda(\mathbf{A})$	eigenvalue of the matrix $\mathbf{A}$
$\lambda_{\min}(\mathbf{A}), \lambda_{\min}(\mathbb{A})$	smallest eigenvalue of the matrix $\mathbf{A}$, smallest spectral value of the self-adjoint operator $\mathbb{A}$
$\lambda_{\max}(\mathbf{A}), \lambda_{\max}(\mathbb{A})$	largest eigenvalue of the matrix $\mathbf{A}$, largest spectral value of the self-adjoint operator $\mathbb{A}$
$\mathrm{adj}(\mathbf{A})$	adjoint of the matrix $\mathbf{A}$

$\mathbf{A}^{-1}$	inverse of the matrix $\mathbf{A}$
$\star$	convolution operation
$\mathfrak{F}\{\cdot\}$	Fourier transform
$\mathfrak{F}^{-1}\{\cdot\}$	inverse Fourier transform
$\mathfrak{F}_c\{\cdot\}$	cosine transform (a.k.a. Fourie
$\mathfrak{F}_c^{-1}\{\cdot\}$	inverse cosine transform
$\oplus$	direct sum operation
$\mathcal{L}^2$	Hilbert space of complex-value
$\mathcal{L}^2(B)$	space of square-integrable func
$\mathcal{H}$	abstract Hilbert space
l^2	Hilbert space of square-summa
$\langle\mathbf{a}|\mathbf{b}\rangle$	inner product of the vectors $\mathbf{a}$
$\langle x|y\rangle$	depending on the context: inn
	Hilbert space $\mathcal{H}$ or inner prod
	and $y(t)$: $\langle x, y\rangle \triangleq \int_{-\infty}^{\infty} x(t)y^*($
$\|\mathbf{a}\|^2$	squared l^2-norm of the vector
$\|y\|^2$	depending on the context: squ
	Hilbert space $\mathcal{H}$ or squared $\mathcal{L}^2$
	function $y(t)$: $\|y\|^2 \triangleq \int_{-\infty}^{\infty} |y(t$
$\mathbb{I}_{\mathcal{H}}, \mathbb{I}_{l^2}, \mathbb{I}_{\mathcal{L}^2}, \mathbb{I}_{\mathcal{L}^2(B)}$	identity operator in the corres
$\mathbb{A}^*$	adjoint of operator $\mathbb{A}$
$\widehat{x}(f)$	Fourier transform of $x(t)$: $\widehat{x}(f$
$\widehat{x}_d(f)$	discrete-time Fourier transform
	of $x[k]$: $\widehat{x}_d(f) \triangleq \sum_{k=-\infty}^{\infty} x[k]e$
$\mathcal{R}(\mathbf{A}), \mathcal{R}(\mathbb{A})$	range space of the matrix $\mathbf{A}$, r
$\mathcal{I}\mathrm{m}(\mathbf{A})$	image of the matrix $\mathbf{A}$
$\mathcal{N}\mathrm{ull}(\mathbf{A})$	null space of the matrix $\mathbf{A}$
$\mathcal{K}\mathrm{er}(\mathbf{A})$	kernel of the matrix $\mathbf{A}$

Chapter 1 Introduction

The present textbook describes mathematical concepts and results that are of importance in the design, analysis, and optimization of signal processing algorithms, modern communication systems, and networks. This textbook is intended to offer a comprehensive overview of all the necessary concepts and results from linear algebra, numerical analysis, statistics, probability, stochastic processes, and optimization in order to give students and specialists easy access to the state-of-the-art literature and mastering of the key techniques and results in the above-mentioned areas.

This textbook consists of 20 chapters ordered in such a manner that all the necessary background and prerequisite information is presented before more advanced topics are addressed. At the same time, all the efforts have been taken to make each chapter self-contained and as much as possible independent with respect to the other chapters. Each chapter is accompanied by a set of homework exercises and presents pointers to further readings for additional topics and applications.

Chapter 2 of this textbook presents a brief overview of the fundamental transforms used for processing analog and discrete-time signals such as Fourier, Laplace, Z-transform, Hilbert, Hartley, discrete sine and cosine transforms, and sampling. The Sampling Theorem and aliasing errors introduced by sampling analog signals are also described.

Chapter 3 of the book provides a compressed summary of the broad field of linear algebra and matrix theory. Concepts such as vector space, vector norm, normed vector space, linear transformation, orthogonal projection, and Gramm-Schmidt orthogonalization are first described. This chapter also presents an overview of the most important concepts and factorizations from the field of matrix theory. Concepts such as condition number, eigenvalue, eigenvector, determinant, pseudoinverse, special types of matrices (circulant, Toeplitz, Hankel, Vandermonde, stochastic, positive and negative definite), condition number, and matrix factorizations (LU, singular value decomposition (SVD), QR, Cholesky) are described. This chapter ends by presenting various matrix operations such as Kronecker product, Hadamard product, dot (inner) product, direct sum and rules for differentiation of matrix and vector valued functions.

Chapter 4 of this textbook presents a compact description of the Galois field of

numbers from the viewpoint of coding and cryptograp
duces first the concepts of group, ring and fields, and
of finite (Galois) fields and potential applications of

Chapter 5 introduces basic concepts in numerica
point representation of numbers, condition number,
tion, determining the roots of polynomials, and num
roots of nonlinear equations, eigenvalues, and singula

Chapter 6 presents an incursion into the three ma
enumeration, graph theory, and design theory. Basic
as permutations, combinations, the principle of inclus
functions and recurrence relations are presented and
numerous examples and applications. Fundamental
field of graph theory such as path, cycle, tree, Eul
cycle are presented too. Finally, combinatorial desi
and analyzing experiments.

Chapter 7 covers the main concepts and results fr
random variables, and stochastic processes. The m
tributions and convergence laws for sequences of ra
The law of large numbers (LLN) and central limit t
plications in signal processing, telecommunications
presented. In the realm of stationary stochastic proc
on concepts such as power spectral density, correlatic
Markov model (HMM).

Chapter 8 is devoted to the study of matrix-value
cially to the study and applications of the asymptoti
ues of matrices whose entries as normally distributed
without limit. Applications in spectral analysis, sta
testing, and parameter estimation in large dimensior

Chapter 9 is dedicated to the large deviation the
rence of rare events. This chapter represents a sh
large deviation concepts and techniques including cc
function, Cramer's theorem, type analysis, and San
of the error exponents of several standard hypothesis
as an application of large deviation theory.

Chapter 10 and Chapter 11 cover the fundamen
tion theories, respectively. Chapter 10 presents first
bound and several approaches for finding minimum
such as the method based on Rao-Blackwell-Lehman
timation technique that exploits the concept of suffi
maximum likelihood estimator are then reviewed. V
estimation techniques such as recursive least squares (
constrained least squares and regularized least square
10 ends with the description of Bayesian estimation t
with a general formulation of a detection problem as

Three fundamental approaches for binary detection applications are considered: Bayesian, minimax, and Neyman-Pearson. The likelihood ratio test (LRT), uniformly most powerful (UMP), generalized likelihood ratio (GLR), and locally most powerful (LMP) detection scheme are also discussed.

Chapter 12 discusses the usage of Monte Carlo simulation techniques in solving various problems in the areas of signal processing, wireless communications, and bioinformatics. General features and some applications of Markov chain Monte Carlo (MCMC) and sequential Monte Carlo (SMC) techniques are presented.

Chapter 13 is dedicated to the general problem of statistical inference in graphs. The main focus is on factor graphs and message passing algorithms. Basic concepts regarding factor graphs and their construction are first presented. Common features of factor graphs with random fields and Bayesian networks are also described. The message passing algorithm, and in particular the sum-product and the max-product algorithms, for inference in factor graphs are presented. Various algorithms such as BCJR, Viterbi algorithm, Kalman filtering, and EM algorithm are described as instances of message passing in factor graphs. The chapter ends with an application of sum-product algorithm to the carrier frequency offset synchronization of an orthogonal frequency division multiplexing (OFDM) system.

Chapter 14 presents an overview of the basic mathematical approaches for solving constrained and unconstrained optimization problems. The chapter first reviews basic concepts in convex analysis, and then establishes optimality conditions for determining the solutions of unconstrained and constrained optimization problems. In particular, Karush-Kuhn-Tucker (KKT) conditions, Lagrangian duality and applications to several wireless communications systems designs are discussed.

Chapter 15 focuses on optimization problems that assume a linear criterion as well as linear constraints. Such optimization problems are efficiently solved in practice using linear programming and mixed integer programming techniques. This chapter provides an introduction into the basic theory and the main algorithmic approaches that fall under the umbrella of linear programming and mixed integer programming techniques. A few typical signal processing problems that could be resolved using standard linear programming and mixed integer programming techniques are also presented.

Chapter 16 provides a summary of basic results and applications of majorization theory. Basic concepts, such as Schur-convexity, multiplicative majorization, Schur inequality, T-transform, and stochastic majorization, are described. Numerous applications of majorization theory in the design of wireless communications transceivers are presented as well.

Chapter 17 presents basic concepts and results from queueing theory, with special emphasis on the simple M/M/1 queue and the intermediate M/G/1 queue.

Chapter 18 focuses on modeling and optimization techniques for communication network design and planning. The main focus of Chapter 18 is on optimization of the capacity of network resources and traffic routing. The link capacity and routing modeling are formulated in terms of multicommodity flow networks (MFN), and

integer programming techniques are proposed for opt
flow networks.

Chapter 19 presents the basic concepts and res
cooperative games, Nash equilibrium, cooperative g
information, repeated games, as well as a discussion
theory to signal processing, communications, and ne

Finally, Chapter 20 provides a short introductic
Applications of frame theory to the sampling of an
struction from discrete-time samples are presented.
such as wavelets and Weyl-Heisenberg frames are als

Chapter 2 Signal Processing Transforms

Serhan Yarkan[‡] and Khalid A. Qaraqe[♯]
‡Texas A&M University, College Station, USA
♯Texas A&M University at Qatar

2.1 Introduction

In a broader scope, a transformation in mathematics can broadly be defined as the operation which takes its input and "represents" it in a different form. Such a definition immediately implies preserving the essential characteristics of the input through several conservation rules or laws. In order to outline the complete input–output relationship in an abstract transformation, the input and output of the transformation should be characterized as well. Therefore, mathematically speaking, a transformation can be viewed as a special function (or a correspondence) whose input and output can be a single value or another function. In this context, signals are considered to be both input and output functions of the transformation of interest.[1]

Depending both on the transform and the input signal characteristics, different types of transforms exist in the literature. For instance, there are transforms which assume a deterministic input (e.g., Fourier transform), whereas there are transforms operating on a stochastic input (e.g., Karhunen–Loéve transform). A different way of classifying transforms is to consider the nature of the input signal such as continuous and discrete transforms, which assume continuous and discrete signals as input, respectively.

It is worth mentioning at this point that some of the transformations yield an output in a different domain — which is sometimes called "transform domain"— from that of the input. A very well-known example of this is a time–frequency transformation, which operates on a signal in the time domain and yields an output signal in the frequency domain or vice versa. However, such a domain change is not always observed for every transformation.

---

[1] Note that, depending on the type and characteristic of the transformation, it is not obligatory that a transformation is a bijection. As will be shown subsequently, there are some transformations which take a single value as their input and yield a set of values as their output.

2.2   Basic Transformations

When an independent variable is considered, several t
in the context given in Section 2.1. Basic transforma
any function $f(t)$, which can be imagined to be one

$$(\mathcal{A}.\textbf{I}) \quad f(t) \xrightarrow{\text{Scale}} f(a \cdot t), \quad \forall a \in \mathbb{R}$$

$$(\mathcal{A}.\textbf{II}) \quad f(t) \xrightarrow{\text{Reflection}} f(-t) \tag{Ref}$$

$$(\mathcal{A}.\textbf{III}) \quad f(t) \xrightarrow{\text{Shift}} f(t - t_0), \quad \forall t_0 \in \mathbb{R}$$

Note that the basic transformations listed above o
as in combinations. Such combinations help form the
even functions and periodic functions. Also, such con
appropriately applied, are very useful to simplify the
systems.

2.3   Fourier Series and Transfor

2.3.1   Basic Definitions

Fourier series and transform are two fundamental to
ence and mathematics. Before proceeding further, th

Definition 2.3.1 (Periodicity). A function is said
T if:

$$f(t) = f(t + T)$$

is satisfied for all t.

A direct consequence of the Definition 2.3.1 is the p
for all t and $n \in \mathbb{Z}$. This reasoning necessitates the f

Definition 2.3.2 (Fundamental Period). The fu
riodic function $f(\cdot)$ is the smallest positive value of $^{\gamma}$

It is noteworthy to state that if $f(\cdot)$ is constant,
is "undefined," since there is no "smallest" positive v
to be "aperiodic" if (2.4) does not hold.

2.3.2 Fourier Series

Theorem 2.3.1 (Fourier Series Expansion). *If a function $f(t)$ satisfying Dirichlet conditions is periodic with a fundamental period T_0, then it can be represented in the following Fourier series expansion:*

$$f(t) = \sum_{k=-\infty}^{\infty} a_k e^{jk\omega_0 t} \tag{2.5}$$

where $j = \sqrt{-1}$ and $T_0 = 2\pi/\omega_0$. In (2.5), the term obtained when $k = 1$ is called "fundamental mode" or "first harmonic" and the constant obtained when $k = 0$ is called "dc component" or "average value."

Note that, since (2.5) in Theorem 2.3.1 is an expansion, the convergence of its right-hand side expansion must be questioned. The convergence is guaranteed when $f(\cdot)$ has finite energy over one period. However, the Dirichlet conditions, if satisfied, offer a set of sufficient conditions for the existence of "equality" rather than a "convergence" except for the isolated values causing discontinuity [1].

Theorem 2.3.2 (Dirichlet Conditions). *Any function $f(t)$ is equal to its Fourier series representation at the values where the function $f(t)$ is continuous and converges to the mean of the discontinuity values (average of the left- and right-hand limits of $f(t)$ at values where $f(t)$ is discontinuous) if the following conditions are satisfied:*

- *Function $f(t)$ is bounded periodic,*

- *Function $f(t)$ has a finite number of discontinuities,*

- *Function $f(t)$ has a finite number of extrema,*

- *Function $f(t)$ is absolutely integrable $\int_{T_0} |f(t)|\, dt < \infty$.*

As stated earlier, the conditions given in Theorem 2.3.2 are "sufficient" to guarantee that the Fourier series expansion representation of $f(t)$ exists. This should not be interpreted as when conditions in Theorem 2.3.2 are not satisfied, Fourier series expansion representation does not exist. For the existence of Fourier series expansion of any function, "necessary" conditions are not yet known. However, in real-world applications, almost all of the signals encountered satisfy the conditions of Theorem 2.3.2 implying the existence of the Fourier series expansion. Therefore, from the practical perspective, seeking necessary conditions is not critical.

In conjunction with the statements above, it might be necessary to contemplate the behavior of Fourier series expansion at points where the function of interest exhibits discontinuities especially for a finite number of terms included (partial sums). Let one define $f_N(t)$ to be:

$$f_N(t) = \sum_{k=-N}^{N} a_k e^{(jk\omega_0 t)} \tag{2.6}$$

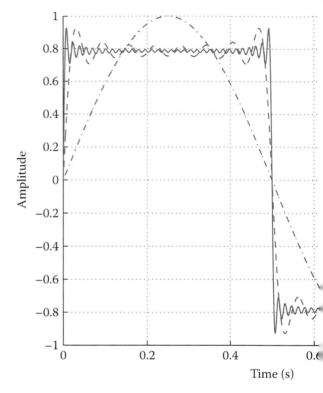

Figure 2.1: Gibbs phenomenon. The plot $\sum_{k=1}^{N_i} \frac{1}{2k-1} \sin\left(2\pi\left(2k-1\right) f_0 t\right)$ where the fundamer $\{N_i\}$ is $\{1, 8, 32\}$.

in parallel with (2.5). In this case, one might want t difference:

$$e(t) = f_N(t) - f(t)$$

i.e., $E_N = \int_{T_0} |e(t)|^2 \, dt$. It is clear that $\lim_{N \to \infty} E_N$ difference in terms of energy would be negligible as the increases [2]. However, at the values where disconti a strange phenomenon takes place and overshoot/un This is known as the "Gibbs phenomenon" in honor clarified Albert Michelson's concerns about a possibl harmonic analyzer calculating the "truncated" Four An illustration of the Gibbs phenomenon is plotted i

Gibbs phenomenon implies spectral growth. In if a time signal has discontinuities, its Fourier repres terms ($N \to \infty$) with respect to a signal that does

In practical applications, truncation due to finite support causes ringing (ripple) effects that are referred to as spectral leakage.

2.3.3 Fourier Transform

Periodic signals can be represented via Fourier series expansions as explained in Section 2.3.2. It is of interest to represent aperiodic signals in a similar way. The concept of Fourier series expansion can be utilized to represent aperiodic signals as well.

An aperiodic signal does not repeat itself over time. Mathematically speaking, one can view an aperiodic signal as a periodic signal that repeats itself with a period of infinity. It might be helpful to look at the same reasoning from the perspective of frequencies. Recall that Fourier series expansion actually yields a set of frequencies of discrete values each of which is a multiple integer of the fundamental frequency:

$$\omega_0 = \frac{2\pi}{T_0} \tag{2.8}$$

Then the difference between successive frequency components of the output of Fourier expansion is given by

$$\Delta\omega = \frac{2(k+1)\pi}{T_0} - \frac{2k\pi}{T_0} = \omega_0 \tag{2.9}$$

via (2.5). Therefore, in case the fundamental period T_0 gets larger, the gap between discrete frequency components becomes smaller, i.e., if $T_0 \to \infty$, then $\Delta\omega \to 0$. Of course, in the limiting case the summation in the Fourier series expansion converts into an integral that is referred to as the Fourier integral. Therefore, the Fourier transform integral can be introduced as:

$$F(j\omega) = \int_{-\infty}^{\infty} f(t)e^{-j\omega t}\, dt, \tag{2.10}$$

where $F(j\omega)$ is called the "Fourier transform of $f(t)$" and is equivalently denoted in terms of notation with $\mathfrak{F}\{f(t)\}$. Thus, $\mathfrak{F}\{f(t)\} = F(j\omega)$. The inverse Fourier transform is given by:

$$f(t) = \frac{1}{2\pi} \int_{-\infty}^{\infty} F(j\omega)e^{j\omega t}\, d\omega \tag{2.11}$$

and it can also be represented as $\mathfrak{F}^{-1}\{f(t)\}$. Both (2.10) and (2.11) are referred to as a Fourier transform pair. A direct consequence of both (2.10) and (2.11) are the following relationships:

$$\mathfrak{F}^{-1}\{\mathfrak{F}\{f(t)\}\} = f(t) \tag{2.12a}$$

$$\mathfrak{F}\{\mathfrak{F}^{-1}\{F(j\omega)\}\} = F(j\omega). \tag{2.12b}$$

Properties of Fourier Transform

Linearity.   If $F_1(j\omega) = \mathfrak{F}\{f_1(t)\}$ and $F_2(j\omega) = \mathfrak{F}\{$

$$\mathfrak{F}\{a_1 f_1(t) + a_2 f_2(t)\} = a_1 \mathfrak{F}\{f_1(t)\}$$
$$= a_1 F_1(j\omega) +$$

As a result of both (2.10) and (2.11), linearity applie

Symmetry.   If $F(j\omega) = \mathfrak{F}\{f(t)\}$ for $f(t) \in \mathbb{R}$, then

$$F(-j\omega) = F^*(j\omega).$$

It is important to state that even though $f(t) \in$
(2.10) can easily be extended to complex valued func
property since $f_z(t) = f_x(t) + j(f_y(t))$ for any $f_z(t)$
and $f_y(t) = \Im(f_z(t))$ with $\Re(\cdot)$ and $\Im(\cdot)$ denoting t
respectively.

Scaling.   If $F(j\omega) = \mathfrak{F}\{f(t)\}$, then

$$\mathfrak{F}\{f(at)\} = \frac{1}{|a|} F\left(\frac{\omega}{a}\right).$$

Shifting.   If $F(j\omega) = \mathfrak{F}\{f(t)\}$, then

$$\mathfrak{F}\{f(t - t_0)\} = e^{-j\omega t_0} F(j$$

Differentiation.   If $F(j\omega) = \mathfrak{F}\{f(t)\}$, then

$$\mathfrak{F}\left\{\frac{d}{dt} f(t)\right\} = j\omega F(j\omega)$$

Integration.   If $F(j\omega) = \mathfrak{F}\{f(t)\}$, then

$$\mathfrak{F}\left\{\int_{-\infty}^{t} f(x)\, dx\right\} = \frac{1}{j\omega} F(j\omega) + \pi$$

where $\delta(\cdot)$ denotes the Dirac delta function.

Duality.   If $F(j\omega) = g(\omega) = \mathfrak{F}\{f(t)\}$, then

$$\mathfrak{F}\{g(t)\} = 2\pi f(-\omega).$$

Even though duality seems to be an insignificant p
powerful, especially when considered along with the

Properties of Fourier Transform in Linear Systems

In the preceding subsection, very important fundamental properties of Fourier transform have been reviewed. However, the power of Fourier transform is better understood when linear systems are analyzed. As will be shown subsequently, the fundamental properties of Fourier transform will be of great help in the analysis of linear time–invariant (LTI) systems yielding very important results.

Parseval's Equality.   If $F_1(j\omega) = \mathfrak{F}\{f_1(t)\}$ and $F_2(j\omega) = \mathfrak{F}\{f_2(t)\}$, then

$$\int_{-\infty}^{\infty} f_1(t)f_2^*(t)\,dt = \frac{1}{2\pi} \int_{-\infty}^{\infty} F_1(j\omega)F_2^*(j\omega)\,d\omega \qquad (2.20)$$

A special case of Parseval's equality is obtained for $f_1(t) = f_2(t)$, which is known to be "Bessel's equality," or simply the energy conservation property.

Convolution.   It is known that the output of an LTI system that is fed by an input signal $x(t)$ is given by the convolution of the "impulse response" of the LTI system, say $h(t)$, with the input signal $x(t)$. Formally, this input–output relationship can be expressed as follows:

$$y(t) = h(t) \star x(t) = \int_{-\infty}^{\infty} x(\tau)h(t - \tau)\,d\tau, \qquad (2.21)$$

where $y(t)$ is the output of the LTI system at the time instant t. If the frequency domain representation of this output is desired, then the following very important result is obtained:

$$\mathfrak{F}\{y(t)\} = Y(j\omega) = H(j\omega)X(j\omega), \qquad (2.22)$$

where $H(j\omega)$ is known to be the "frequency response" of the LTI system of interest. As shown in (2.22), the Fourier transform allows one to investigate directly the output of LTI systems in the frequency domain. Note that, considering the duality property mentioned earlier, the convolution property implies a very important simplicity in the analysis. For instance, because of the convolution property of Fourier transform, ordering is not important in cascaded systems from the perspective of overall system response.

Modulation.   Through the use of duality, if convolution is applied in frequency domain, then the result in time domain should be characterized by multiplication. Formally, this is stated as:

$$y(t) = x_1(t)x_2(t) = \mathfrak{F}^{-1}\{X_1(j\omega) \star X_2(j\omega)\}. \qquad (2.23)$$

This property of Fourier transform is called "modulation" because the frequency domain convolution operation leads to a scaling of one of the signals with the

other signal in the time domain. Since multiplication
"amplitude modulation" (i.e., scale of the amplitude)
(2.23) is known as "modulation property."

2.4   Sampling

As stated in Section 2.1, a transformation is an oper
its input in another way. From this perspective, as
sampling can be considered to be a type of transforma
Moreover, sampling is the bridge between continuo
Considering the fact that dealing with discrete valu
and preferable in terms of saving memory and proc
at the heart of many engineering applications.

Before giving the details of the sampling theorem,
how sampling is mathematically established.

2.4.1   Impulse-Train Sampling

In order to collect samples from a continuous-time
mathematical object that allows to pick the value o
and to repeat this process in a periodic manner. Bo
by applying a Dirac delta function (impulse) in a pe
is the sequence of impulses (impulse train or samplin

$$p(t) = \sum_{n=-\infty}^{\infty} \delta(t - nT_s)$$

The period of $p(t)$ is T_s and it represents the sam
damental frequency of $p(t)$ is $\omega_s = \frac{2\pi}{T_s}$. Hence, th
expressed as:

$$f_p(t) = f(t)p(t).$$

Therefore, (2.25) can be rewritten as:

$$f_p(t) = \sum_{n=-\infty}^{\infty} f(nT_s)\delta(t - n$$

Note that (2.25) (and also (2.26)) represents a mult
between two sequences. Therefore, the output can be
of the Fourier transforms of $p(t)$ and $f(t)$. Since $\mathfrak{F}\{$
in frequency domain, the output is actually the sum
(see Exercise 2.12.9).

Note that the property mentioned above points ou
and their summation along the frequency axis. Thi

replicas of the Fourier transform overlapping in the frequency domain because if an overlap occurs in the frequency domain, the recovery of the signal becomes impossible. In order to guarantee that no overlap occurs in the frequency domain, impulses forming $\mathfrak{F}\{p(t)\}$ should be separated sufficiently apart from each other. This condition leads to the sampling theorem:

Theorem 2.4.1 (Sampling Theorem). *Let a bandlimited function $f(t)$ be defined as $F(j\omega) = 0$ for $|\omega| > \omega_B$. Then $f(t)$ can be uniquely represented by its samples $f(nT)$, $n \in \mathbb{Z}$, provided that:*

$$2\omega_B < \omega_s, \tag{2.27}$$

where

$$\omega_s = \frac{2\pi}{T}. \tag{2.28}$$

It is important to emphasize that reconstruction of the signal $f(t)$ in Theorem 2.4.1 is actually established by filtering the samples, which are sufficiently close to each other, with an ideal low-pass filter. This process is known to be "band-limited interpolation." However, due to practical considerations, it is difficult to implement an ideal low-pass filter. Therefore, different interpolation methods are applied in real-world applications.

2.4.2 Aliasing

One of the most important requirements in Theorem 2.4.1 is to sample the signal with a sufficiently high frequency such that the condition $2\omega_B < \omega_s$ is met. At this point, one might ask about why such a condition is imposed in the theorem.

In order to better understand what happens if the condition $2\omega_B < \omega_s$ is not satisfied, a frequency domain representation of a bandlimited baseband signal $f(t)$ is considered. Since the bandlimited signal in the frequency domain, i.e., $F(j\omega)$, has no frequency component beyond ω_B, after the sampling operation, the replicas of $F(j\omega)$ are formed by shifting them across the ω–axis at multiple integers of ω_s. This implies that, for instance, the first shifted replica of $F(j\omega)$ is centered at ω_s, which is the alias term $F(j(\omega - \omega_s))$. Therefore, the lowest frequency component of $F(j(\omega - \omega_s))$ is located at $\omega_L = \omega_s - \omega_B$. Now, in case the lowest frequency component of the first replica ($F(j(\omega - \omega_s))$), namely ω_L, overlaps with the highest frequency component of the original spectrum $F(j\omega)$, then information in the frequency domain is lost and the analog signal cannot be reconstructed perfectly from the samples of discrete-time signal. This phenomenon is referred to as "aliasing." Note that the highest frequency component of $F(j\omega)$ is denoted with ω_B by definition and it is also referred to as the signal bandwidth. Thus, in order not to have aliasing:

$$\omega_B < \omega_L \iff \omega_B < \omega_s - \omega_B \iff 2\omega_B < \omega_s \tag{2.29}$$

must be satisfied, as stated in Theorem 2.4.1.

2.5   Cosine and Sine Transform

Even though they are not as tractable as Fourier tran
forms exhibit very interesting and nice properties in
cessing such as spectral analysis of real sequences and
it is worth reviewing these transforms.

2.5.1   Cosine Transform

For any function $f(t) \in \mathbb{C}$ that is defined over $t \geq 0$, t
by:

$$\mathfrak{F}_c \left\{ f(t) \right\} = F_c(\omega) = \int_0^\infty f(t) \cos$$

where $\mathfrak{F}_c \left\{ \cdot \right\}$ represents the cosine transform operator
form of $f(t)$ [4]. Note that the integration begins fror
kernel is $\cos(\omega t)$. These details imply that $\mathfrak{F}_c \left\{ \cdot \right\}$ do
    The inverse cosine transform is given by:

$$\mathfrak{F}_c^{-1} \left\{ F_c(\omega) \right\} = \frac{2}{\pi} \int_0^\infty F_c(\omega) \cos$$

for $t \geq 0$.

Properties of Cosine Transform

Let $f(t)$ and $g(t)$ be defined as in Section 2.5.1. Som
of cosine transforms are listed below.

Linearity.   If $F_c(\omega) = \mathfrak{F}_c \left\{ f(t) \right\}$ and $G_c(\omega) = \mathfrak{F}_c \left\{ g \right.$

$$\mathfrak{F}_c \left\{ a_1 f(t) + a_2 g(t) \right\} = a_1 \mathfrak{F}_c \left\{ f(t) \right\} + a_2 \mathfrak{F}_c \left\{ g(t) \right\} =$$

for two arbitrary scalars a_1 and a_2.

Scaling.   If $F_c(\omega) = \mathfrak{F}_c \left\{ f(t) \right\}$, then

$$\mathfrak{F}_c \left\{ f(at) \right\} = \frac{1}{a} F_c \left(\frac{\omega}{a} \right)$$

Shifting. Since the cosine transform is defined over $t \geq 0$, shifting the signal to the left requires further investigation. If $f_E(t) = f(|t|)$, then:

$$\mathfrak{F}_c\{f(t+a) + f(|t-a|)\} = 2F_c \cos(a\omega). \tag{2.34}$$

With the same argument, if the shift is considered to be over the frequency domain, then:

$$F_c(\omega \pm a) = \mathfrak{F}_c\{f(t)\cos(\alpha t)\} \mp \mathfrak{F}_s\{f(t)\sin(\alpha t)\} \tag{2.35}$$

where $\mathfrak{F}_s\{\cdot\}$ is the sine transform in (2.45), which will be introduced subsequently in this chapter. If these properties are extended, the following relations are obtained:

$$\mathfrak{F}_c\{f(at)\cos(\alpha t)\} = \frac{1}{2a}\left(F_c\left(\frac{\omega+\alpha}{a}\right) + F_c\left(\frac{\omega-\alpha}{a}\right)\right), \tag{2.36}$$

and

$$\mathfrak{F}_c\{f(at)\sin(\alpha t)\} = \frac{1}{2a}\left(F_s\left(\frac{\omega+\alpha}{a}\right) - F_s\left(\frac{\omega-\alpha}{a}\right)\right), \tag{2.37}$$

where $F_s(\cdot)$ is the sine transform of $f(t)$.

Differentiation. As in shifting, due to the special structure of cosine transform, odd- and even-order differentiations exhibit different characteristics. The cosine transform of first-order derivative is given by:

$$\mathfrak{F}_c\left\{\frac{d}{dt}f(t)\right\} = \omega F_s(\omega) - f(0) - D\cos(\omega t_D) \tag{2.38}$$

where D is the magnitude of the discontinuity (discontinuity jump) at the time instant t_D given by $D = f(t_D^+) - f(t_D^-)$. Similarly, the second-order differentiation is given by:

$$\mathfrak{F}_c\left\{\frac{d^2}{dt^2}f(t)\right\} = -\omega^2 F_c(\omega) - f'(0) - \omega D\sin(\omega t_D) - D'\cos(\omega t_D) \tag{2.39}$$

where $D' = f'(t_D^+) - f'(t_D^-)$ [4]. It must be noted here that in order for all of the formulas given above to hold, all-order derivatives (including the zeroth-order derivative, i.e., the function itself) are assumed to vanish as $t \to \infty$.

When differentiation in frequency domain is considered, the odd-even structure is preserved. However, the odd-even structure manifests itself in a different way. The odd-order differentiation is given by:

$$\frac{d^{(2n+1)}}{dt^{(2n+1)}}F_c(\omega) = \mathfrak{F}_c\left\{(-1)^{n+1}t^{2n+1}f(t)\right\}, \tag{2.40}$$

whereas for the even-order differentiation is given by:

$$\frac{d^{(2n)}}{dt^{(2n)}}F_c(\omega) = \mathfrak{F}_c\left\{(-1)^n t^{2n}f(t)\right\}, \tag{2.41}$$

where $n \in \mathbb{Z}^+$. It must be emphasized that the existe.
a fact, which can only be guaranteed by imposing s
such as being piecewise continuous, etc.

Integration. Integration in the time domain for
via:

$$\mathfrak{F}_c \left\{ \int_t^\infty f(\tau)\, d\tau \right\} = \frac{1}{\omega} F_s(\omega$$

whereas the integration in the frequency domain is g

$$\mathfrak{F}_s^{-1} \left\{ \int_\omega^\infty f(\alpha)\, d\alpha \right\} = -\frac{1}{t} f$$

These results are valid under general conditions suc
lutely integrable.

Convolution. The cosine transform of the convo
transform and is expressed as:

$$\mathfrak{F}_c \left\{ \int_0^\infty f(\tau)\, [g(t+\tau) + g_o(t-\tau)]\, d\tau \right\}$$

where $g_o(\cdot)$ denotes the odd part of $g(t)$.

2.5.2   Sine Transform

For any function $f(t) \in \mathbb{C}$ that is defined over $t \geq 0$,
by:

$$\mathfrak{F}_s \{f(t)\} = F_s(\omega) = \int_0^\infty f(t) \sin$$

where $\mathfrak{F}_s \{\cdot\}$ represents the sine transform operator a
of $f(t)$, which assumes the domain $\omega \geq 0$. Note tha
zero and the transformation kernel is $\sin(\omega t)$. These
not necessarily exist.
    The inverse sine transform is then given by:

$$\mathfrak{F}_s^{-1} \{F_s(\omega)\} = \frac{2}{\pi} \int_0^\infty F_s(\omega) \sin ($$

and assumes the domain $t \geq 0$.

Properties of Sine Transform

Let $f(t)$ and $g(t)$ be defined as in Section 2.5.2. Then some of the important properties of sine transforms are listed below.

Linearity. If $F_s(\omega) = \mathfrak{F}_s\left\{f(t)\right\}$ and $G_s(\omega) = \mathfrak{F}_s\left\{g(t)\right\}$, then

$$
\begin{aligned}
\mathfrak{F}_s\left\{a_1 f(t) + a_2 g(t)\right\} &= a_1 \mathfrak{F}_s\left\{f(t)\right\} + a_2 \mathfrak{F}_s\left\{g(t)\right\} \\
&= a_1 F_s(\omega) + a_2 G_s(\omega)
\end{aligned}
\tag{2.47}
$$

for arbitrary scalars a_1 and a_2.

Scaling. If $F_s(\omega) = \mathfrak{F}_s\left\{f(t)\right\}$, then

$$
\mathfrak{F}_s\left\{f(at)\right\} = \frac{1}{a} F_s\left(\frac{\omega}{a}\right)
\tag{2.48}
$$

with $0 < a$.

Shifting. Since the sine transform is defined over $t \geq 0$, shifting the signal to the left requires further investigation. If $f_E(t) = f\left(|t|\right)$, then:

$$
\mathfrak{F}_s\left\{f(t+a) + f\left(|t-a|\right)\right\} = 2F_s \sin\left(a\omega\right).
\tag{2.49}
$$

With the same argument, if the shift is considered to be over the frequency domain, then:

$$
F_s\left(\omega + \alpha\right) = \mathfrak{F}_s\left\{f(t)\cos\left(\alpha t\right)\right\} + \mathfrak{F}_c\left\{f(t)\sin\left(\alpha t\right)\right\},
\tag{2.50}
$$

where $\mathfrak{F}_s\left\{\cdot\right\}$ is the sine transform. The following more general relationships hold:

$$
\mathfrak{F}_s\left\{f(at)\cos\left(\alpha t\right)\right\} = \frac{1}{2a}\left(F_s\left(\frac{\omega + \alpha}{a}\right) + F_s\left(\frac{\omega - \alpha}{a}\right)\right),
\tag{2.51}
$$

and

$$
\mathfrak{F}_s\left\{f(at)\sin\left(\alpha t\right)\right\} = \frac{1}{2a}\left(F_c\left(\frac{\omega + \alpha}{a}\right) + F_c\left(\frac{\omega - \alpha}{a}\right)\right),
\tag{2.52}
$$

where $F_s\left(\cdot\right)$ is the sine transform and $F_c\left(\cdot\right)$ is the cosine transform of $f(t)$, respectively.

Differentiation. As in shifting, due to the special structure of cosine transform, odd- and even-order differentiations present different characteristics. The sine transform of the first-order derivative is given by:

$$
\mathfrak{F}_s\left\{\frac{d}{dt}f(t)\right\} = -\omega F_c(\omega) + D\sin\left(\omega t_D\right),
\tag{2.53}
$$

where D is the magnitude of the discontinuity (dis
instant t_D: $D = f(t_D^+) - f(t_D^-)$. Similarly, the second
by:

$$\mathfrak{F}_s\left\{\frac{d^2}{dt^2}f(t)\right\} = -\omega^2 F_s(\omega) -$$

where $f(t)$ is assumed to be continuous to the first-
given above to hold, all-order differentiation (includ
the function itself) functions must vanish as $t \to$
order derivatives, the operational rule must be appli

When differentiation in the frequency domain is c
ture is preserved; however, such structures are repre
odd-order derivative is given by:

$$\frac{d^{(2n+1)}}{dt^{(2n+1)}}F_s(\omega) = \mathfrak{F}_c\left\{(-1)^n\, t^{2n}\right.$$

with the use of $\mathfrak{F}_c\{\cdot\}$, whereas the even-order differe

$$\frac{d^{(2n)}}{dt^{(2n)}}F_s(\omega) = \mathfrak{F}_s\left\{(-1)^n\, t^{2n}\right.$$

where $n \in \mathbb{Z}^+$.

Integration. The integration in the time domain

$$\mathfrak{F}_s\left\{\int_0^t f(\tau)\, d\tau\right\} = \frac{1}{\omega}F_c($$

whereas the integration in the frequency domain is

$$\mathfrak{F}_c^{-1}\left\{\int_\omega^\infty F_s(x)\, dx\right\} = \frac{1}{t}f$$

where these results hold under general conditions su
lutely integrable.

Convolution. The sine transform of the convolu
transform and is expressed as:

$$\mathfrak{F}_s\left\{\int_0^\infty g(\tau)\left[f(t+\tau) + f_o(t-\tau)\right] d\tau\right\}$$

where $f_o(\cdot)$ denotes the odd part of $f(t)$ and $G_c(\omega)$

2.6 Laplace Transform

In Section 2.3, recall that the Fourier transformation yields an output in the frequency domain by correlating the time domain signal $f(t)$ with the complex exponential $e^{j\omega t}$. Recall also that, because of this property, $\mathfrak{F}\{f(t)\}$ is represented as $F(j\omega)$. This is a direct consequence of the transform kernel employed. Since the Fourier transform employs a complex exponential of pure imaginary form (i.e., $e^{-j\omega t}$ in $\mathfrak{F}\{\cdot\}$), one might wonder what happens if the transform kernel is generalized in such a way that it employs a complex exponential of a general complex form $e^{st} = e^{(\sigma+j\omega)t}$ where s denotes the "complex frequency in generalized form" rather than being in pure imaginary form. It is easier to infer that e^{st} degenerates to the Fourier transform kernel $e^{j\omega t}$ when $\sigma = 0$.

In the light of the generalized transform kernel e^{st}, the Fourier transform integral in (2.10) can be expressed as:

$$F(\sigma + j\omega) = \int_{-\infty}^{\infty} f(t)e^{-(\sigma+j\omega)t}\, dt = \int_{-\infty}^{\infty} \underbrace{\left[f(t)e^{-\sigma t}\right]}_{g(t)} e^{-j\omega t}\, dt. \tag{2.60}$$

Note that (2.60) is regarded as the Laplace transform of signal $f(t)$. One can easily conclude that the Laplace transform of a function $f(t)$ is the Fourier transform of $g(t) = f(t)e^{-\sigma t}$. Thus, the Laplace transform is defined as:

$$F(s) = \int_{-\infty}^{\infty} f(t)e^{-st}\, dt, \tag{2.61}$$

where $s = \sigma + j\omega$. At this point, it is important to make the following observations.

First of all, although the lower limit of (2.61) is stretched back to negative infinity, the integration is often limited to the interval $[-\infty, 0)$ because of the multiplication of the real exponential $e^{-\sigma t}$. Therefore, sometimes it is said that Laplace transform deals with "causal" functions even though the function $f(t)$ can be defined all over the real axis. This is because of the behavior of the transform kernel.

Second, since (2.61) is an improper integral, its convergence must be carefully investigated:

$$\int_{-\infty}^{\infty} f(t)e^{-st}\, dt = \underbrace{\lim_{T\to 0} \int_{-\infty}^{T} f(t)e^{-st}\, dt}_{I^-} + \underbrace{\lim_{T\to\infty} \int_{0}^{T} f(t)e^{-st}\, dt}_{I^+}, \tag{2.62}$$

which forces one to consider the existence of both integrals on the right-hand-side of (2.62). Note that if a function $f(t)$ is defined on the interval $(-\infty, 0)$, then

(2.61) can only yield an output for I^- in (2.62). S
different functions might give the same algebraic ou
convergence (RoC) is introduced which consists of t
$s = \sigma + j\omega$, for which the Fourier transform of g
reason, (2.61) is referred to as the "bilateral" or "
and denoted with $\mathcal{L}_B\{\cdot\}$. If (2.61) is the bilateral
"unilateral" or "one-sided" Laplace transform is giv

$$F(s) = \int\limits_0^\infty f(t)e^{-st}\, dt,$$

where the convergence of the integral is assumed to
    The inverse Laplace transform can be found by ϵ
tween Fourier transform and Laplace transform and

$$f(t) = \frac{1}{2j\pi} \int\limits_{\sigma-j\infty}^{\sigma+j\infty} F(s)e^{st}$$

Computation of integral in (2.64) requires the use of
for the transforms that are of rational form, such
performed by using Cauchy's residue formula [5].

2.6.1   Properties of Laplace Transform

Since the Laplace transform can be considered as
transform, similar properties are observed for the La

Linearity.   If $F_1(s) = \mathcal{L}\{f_1(t)\}$ and $F_2(s) = \mathcal{L}\{f$
respectively, then:

$$\mathcal{L}\{a_1 f_1(t) + a_2 f_2(t)\} = a_1 \mathcal{L}\{f_1(t)\}$$
$$= a_1 F_1(s) + a_2$$

for arbitrary scalars a_1 and a_2 with a resulting RoC
it is worth mentioning that if $R_1 \cap R_2 = \emptyset$, then the
does not admit a Laplace transform.

Conjugation   If $F(s) = \mathcal{L}\{f(t)\}$, with a RoC R, t

$$\mathcal{L}\{f^*(t)\} = F^*(s^*)$$

with the same RoC R.

Scaling. If $F(s) = \mathcal{L}\{f(t)\}$ admits a RoC R, then:

$$\mathcal{L}\{f(at)\} = \frac{1}{|a|}F\left(\frac{s}{a}\right), \tag{2.67}$$

with a RoC R_1 satisfying $R_1 = \frac{R}{a}$.

Shifting. If $F(s) = \mathcal{L}\{f(t)\}$ has the RoC R, then:

$$\mathcal{L}\{f(t - t_0)\} = e^{-st_0}F(s), \tag{2.68}$$

with the same RoC R. Similar to the time domain shifting, the shift in the complex frequency plane is given by:

$$\mathcal{L}\{e^{s_0 t}f(t)\} = F(s - s_0) \tag{2.69}$$

with a RoC $R_1 = R + \sigma_0$ where $\sigma_0 = \Re(s_0)$.

Differentiation. If $F(s) = \mathcal{L}\{f(t)\}$ has the RoC R, then:

$$\mathcal{L}\left\{\frac{d}{dt}f(t)\right\} = sF(s) - f(0^+), \tag{2.70}$$

with a RoC including R. Similar to the differentiation in the time domain, differentiation in the complex frequency plane can be expressed as:

$$\mathcal{L}\{-tf(t)\} = \frac{d}{ds}F(s) \tag{2.71}$$

with the same RoC R.

Integration. If $F(s) = \mathcal{L}\{f(t)\}$, with a RoC R, then:

$$\mathcal{L}\left\{\int_{-\infty}^{t} f(x)\,dx\right\} = \frac{1}{s}F(s) + \frac{1}{s}\left[\int f(t)\,dt\right]_{t=0^+} \tag{2.72}$$

with a RoC including $R \cap \{\sigma > 0\}$.

Convolution. Although the convolution property was reviewed previously in the Section 2.3.3 dealing with the Fourier transform, for Laplace transform it will be expressed under a general framework. If $F_1(s) = \mathcal{L}\{f_1(t)\}$ and $F_2(s) = \mathcal{L}\{f_2(t)\}$ with RoCs R_1 and R_2, respectively, then:

$$\mathcal{L}\{f_1(t) \star f_2(t)\} = F_1(s)F_2(s) \tag{2.73}$$

with a resulting RoC containing $R_1 \cap R_2$. Here, it is worth mentioning that if $R_1 \cap R_2 = \emptyset$, then the linear convolution does not admit a Laplace transform.

2.7 Hartley Transform

In Section 2.3.3, the transform kernel is defined t
which implies a complex trigonometric series expansi
$\cos(\omega t) + j \sin(\omega t)$. It will be shown subsequently tha
drops j and yields a real-valued kernel rather than
that, especially when computational complexity is
real-valued functions seems to be an important altern

In the light of the discussion above, let the tra
Hartley transform as:

$$\text{cas}(\nu t) = \cos(\nu t) + \sin(\iota$$

where ν is the angular frequency and given by $\nu = 2$
Hertz. It is obvious that $\text{cas}(\cdot)$ can be represented i
forward trigonometric conversions such as:

$$\text{cas}(\nu t) = \sqrt{2}\cos\left(\nu t - \frac{\pi}{4}\right) = \sqrt{2}\,s$$

Thus, the Hartley transform is given by:

$$F_H(\nu) = \frac{1}{\sqrt{2\pi}} \int_{-\infty}^{\infty} f(t)\,\text{cas}(\nu$$

where $F_H(\nu)$ is called the "Hartley transform of $f(t)$"
as $\mathfrak{H}\{f(t)\}$.

As a direct consequence of (2.76), Hartley transfo
Therefore, the inverse Hartley transform is given by (
write the following:

$$f(t) = \mathfrak{H}\{\mathfrak{H}\{f(t)\}\}$$

since $\mathfrak{H}^{-1}\{\cdot\}$ is the same as $\mathfrak{H}\{\cdot\}$.

2.7.1 Properties of Hartley Transforn

Similar to the Fourier transform and cosine transfor
hibits the following properties:

Linearity. If $F_{H_1}(\nu) = \mathfrak{H}\{f_1(t)\}$ and $F_{H_2}(\nu) = \mathfrak{H}$

$$\mathfrak{H}\{a_1 f_1(t) + a_2 f_2(t)\} = a_1 \mathfrak{H}\{f_1(t)\}$$
$$= a_1 F_{H_1}(\nu) +$$

for some arbitrary scalars a_1 and a_2.

Scaling. If $F_H(\nu) = \mathfrak{H}\{f(t)\}$, then:

$$\mathfrak{H}\{f(at)\} = \frac{1}{|a|} F_H\left(\frac{\nu}{a}\right). \tag{2.79}$$

Shifting. If $F_H(\nu) = \mathfrak{H}\{f(t)\}$, then:

$$\mathfrak{H}\{f(t - t_0)\} = \cos(\nu t_0) F_H(\nu) + \sin(\nu t_0) F_H(-\nu). \tag{2.80}$$

Note that the transform domain shift can also be treated in the same way, since the Hartley transform is involuntary.

Differentiation. If $F_H(\nu) = \mathfrak{H}\{f(t)\}$, then:

$$\mathfrak{H}\left\{\frac{d^n}{dt^n} f(t)\right\} = \text{cas}'\left(\frac{n\pi}{2}\right) \nu^n H_F\left((-1)^n \nu\right) \tag{2.81}$$

where n stands for the order of differentiation.

Convolution. If $F_{H_1}(\nu) = \mathfrak{H}\{f_1(t)\}$ and $F_{H_2}(\nu) = \mathfrak{H}\{f_2(t)\}$, then:

$$\mathfrak{H}\{f_1(t) \star f_2(t)\} = \frac{1}{2}[F_{H_1}(\nu)F_{H_2}(\nu) + F_{H_1}(-\nu)F_{H_2}(\nu) + F_{H_1}(\nu)F_{H_2}(-\nu) \\ - F_{H_1}(-\nu)F_{H_2}(-\nu)] \tag{2.82}$$

Product. If $F_{H_1}(\nu) = \mathfrak{H}\{f_1(t)\}$ and $F_{H_2}(\nu) = \mathfrak{H}\{f_2(t)\}$, then:

$$\mathfrak{H}\{f_1(t)f_2(t)\} = \frac{1}{2}[F_{H_1}(\nu) \star F_{H_2}(\nu) + F_{H_1}(-\nu) \star F_{H_2}(\nu) + F_{H_1}(\nu) \star F_{H_2}(-\nu) \\ - F_{H_1}(-\nu) \star F_{H_2}(-\nu)] \tag{2.83}$$

2.8 Hilbert Transform

The Hilbert transform for a one-dimensional real-valued signal $f(t)$ is given by the following integral:

$$\mathcal{H}\{f(t)\} = \overset{\Delta}{\check{f}}(t) = \frac{-1}{\pi} \mathfrak{P} \int_{-\infty}^{\infty} \frac{f(\tau)}{\tau - t} d\tau \tag{2.84}$$

where $\mathcal{H}\{\cdot\}$ denotes the Hilbert transform operator, $\mathfrak{P}$ represents the Cauchy's principal value and it is defined as:

$$\mathfrak{P} = \int_a^b g(x)dx = \lim_{\epsilon \to 0+} \left(\int_a^{c-\epsilon} g(x)\, dx + \int_{c+\epsilon}^b g(x)\, dx \right) \tag{2.85}$$

when it exists for a complex-valued function $g(\cdot)$ def

The inverse Hilbert transform is defined in terms

$$\mathcal{H}^{-1}\left\{\overset{\Delta}{f}(t)\right\} = \frac{1}{\pi}\mathfrak{P}\int_{-\infty}^{\infty}\frac{\overset{\Delta}{f}(\tau}{\tau -}$$

By definition, Hilbert transform can also be defir
lution operation:

$$\mathcal{H}\left\{f(t)\right\} = \overset{\Delta}{f}(t) = f(t) \star$$

whereas the inverse Hilbert transform is expressed a

$$\mathcal{H}^{-1}\left\{\overset{\Delta}{f}(t)\right\} = -\overset{\Delta}{f}(t) \star \frac{1}{\pi}$$

From both (2.87) and (2.88), it follows that applyir
tively leads to the inverse Hilbert transform.

2.8.1   Properties of Hilbert Transform

There are many applications and properties for the
properties are reviewed below.

Linearity.   If $\overset{\Delta}{f_1}(t) = \mathcal{H}\left\{f_1(t)\right\}$ and $\overset{\Delta}{f_2}(t) = \mathcal{H}\left\{f_2(\right.$

$$\mathcal{H}\left\{a_1 f_1(t) + a_2 f_2(t)\right\} = a_1\mathcal{H}\left\{f_1(t)\right\}$$
$$= a_1\overset{\Delta}{f_1}(t) + a_2$$

for arbitrary scalars a_1 and a_2.

Scaling.   If $\overset{\Delta}{f}(t) = \mathcal{H}\left\{f(t)\right\}$, then:

$$\mathcal{H}\left\{f(at)\right\} = \overset{\Delta}{f}(at),$$

and:

$$\mathcal{H}\left\{f(-at)\right\} = -\overset{\Delta}{f}(-at)$$

for all $a > 0$.

Shifting.   If $\overset{\Delta}{f}(t) = \mathcal{H}\left\{f(t)\right\}$, then:

$$\mathcal{H}\left\{f(t - t_0)\right\} = \overset{\Delta}{f}(t - t_0$$

Differentiation. If $\overset{\triangle}{f}(t) = \mathcal{H}\{f(t)\}$, then:

$$\mathcal{H}\left\{\frac{d}{dt}f(t)\right\} = \frac{-1}{\pi t} \star \frac{d}{dt}\left(\overset{\triangle}{f}(t)\right). \tag{2.93}$$

Convolution. If $\overset{\triangle}{f_1}(t) = \mathcal{H}\{f_1(t)\}$ and $\overset{\triangle}{f_2}(t) = \mathcal{H}\{f_2(t)\}$, then:

$$\mathcal{H}\{f_1(t) \star f_2(t)\} = -\overset{\triangle}{f_1}(t) \star \overset{\triangle}{f_2}(t). \tag{2.94}$$

Also:

$$\mathcal{H}\left\{f_1(t) \star \overset{\triangle}{f_2}(t)\right\} = \overset{\triangle}{f_1}(t) \star f_2(t). \tag{2.95}$$

Product. If $\overset{\triangle}{f_1}(t) = \mathcal{H}\{f_1(t)\}$ and $\overset{\triangle}{f_2}(t) = \mathcal{H}\{f_2(t)\}$, then:

$$\mathcal{H}\{f_1(t)f_2(t)\} = f_1(t)\overset{\triangle}{f_2}(t) \tag{2.96}$$

where $f_1(t)$ and $f_2(t)$ present non-overlapping power spectra such as one being low-pass and the other being high-pass.

When formal analytic signals are of interest, several interesting properties of the Hilbert transform can be observed. Formally, an analytic signal is a complex-valued continuous-time function with a Fourier transform that vanishes for negative frequencies [6]. Therefore, for analytic signals one can write:

$$\mathcal{H}\{\phi_1(t)\phi_2(t)\} = \phi_1(t)\mathcal{H}\{\phi_2(t)\} \tag{2.97}$$

which is identical to:

$$\mathcal{H}\{\phi_1(t)\phi_2(t)\} = \mathcal{H}\{\phi_1(t)\}\phi_2(t) \tag{2.98}$$

where the analytic signals can be represented with:

$$\phi(t) = f(t) + j\mathcal{H}\{f(t)\} \tag{2.99}$$

2.9 Discrete-Time Fourier Transform

In parallel with the discussion of continuous-time Fourier transform, a similar transformation can be established for discrete-time signals. This idea leads to the discrete-time Fourier transform (DTFT).

If a signal in discrete time is represented as $f[n]$, then the DTFT of $f[n]$ is given by:

$$F\left(e^{j\omega}\right) = \sum_{n=-\infty}^{\infty} f[n]e^{-j\omega n}, \tag{2.100}$$

where $F\left(e^{j\omega}\right)$ is called the "discrete-time Fourier t
lently represented with $\mathfrak{F}_D\left\{f[n]\right\}$. Then, $\mathfrak{F}\left\{f(t)\right\} =$
given by:

$$f[n] = \frac{1}{2\pi} \int_0^{2\pi} F\left(e^{j\omega}\right) e^{j\omega n}$$

and can also be represented with $\mathfrak{F}_D^{-1}\left\{f[n]\right\}$. Both (
to as a Fourier transform pair. A direct consequenc
is:

$$\mathfrak{F}_D^{-1}\left\{\mathfrak{F}_D\left\{f[n]\right\}\right\} = f[n$$
$$\mathfrak{F}_D\left\{\mathfrak{F}_D^{-1}\left\{F\left(e^{j\omega}\right)\right\}\right\} = F\,($$

Properties of Discrete-Time Fourier Transfor

Linearity.   If $F_1\left(e^{j\omega}\right) = \mathfrak{F}_D\left\{f_1[n]\right\}$ and $F_2\left(e^{j\omega}\right) =$

$$\mathfrak{F}_D\left\{a_1 f_1[n] + a_2 f_2[n]\right\} = a_1\mathfrak{F}_D\left\{f_1[n]\right\} +$$
$$= a_1 F_1\left(e^{j\omega}\right) + a_2$$

for arbitrary scalars a_1 and a_2.
As a consequence of both (2.100) and (2.101), linear

$$\mathfrak{F}_D^{-1}\left\{a_1 f_1[n] + a_2 f_2[n]\right\} = a_1\mathfrak{F}_D^{-1}\left\{F_1\left(e^{j\omega}\right)\right\} + a_2\mathfrak{F}_D^{-1}$$

Symmetry.   If $F\left(e^{j\omega}\right) = \mathfrak{F}_D\left\{f[n]\right\}$ for $f[n] \in \mathbb{R}$, t

$$F\left(e^{j\omega}\right) = F^*\left(e^{-j\omega}\right)$$

It is important to state that even though $f[n] \in$
extended to complex-valued sequences due to the
$f_x[n] + jf_y[n]$, where $f_x[n] = \Re(f_z[n])$ and $f_y[n] =$
denote the real and imaginary parts, respectively.

Scaling.   In DTFT, scaling requires more attentior
main counterpart, since the discrete-time values are
$n \in \mathbb{Z}^+$. In the case $n \in \mathbb{Q}\setminus\mathbb{Z}$, $f[n]$ should be zero.
as:

$$f_{(a)}[n] = \begin{cases} f[n/a], & \text{if } n \text{ is a multipl} \\ 0, & \text{otherwise.} \end{cases}$$

If $F\left(e^{j\omega}\right) = \mathfrak{F}_D\left\{f[n]\right\}$, then

$$\mathfrak{F}_D\left\{f_{(a)}[n]\right\} = F\left(e^{ja\omega}\right)$$

Shifting. If $F\left(e^{j\omega}\right) = \mathfrak{F}_D\left\{f[n]\right\}$, then

$$\mathfrak{F}_D\left\{f[n - n_0]\right\} = e^{-j\omega n_0} F\left(e^{j\omega}\right). \tag{2.108}$$

Similarly,

$$\mathfrak{F}_D\left\{e^{j\omega_0 n} f[n]\right\} = F\left(e^{j(\omega - \omega_0)}\right). \tag{2.109}$$

Differencing. If $F\left(e^{j\omega}\right) = \mathfrak{F}_D\left\{f[n]\right\}$, then

$$\mathfrak{F}_D\left\{\mathfrak{D}^{(1)}\left\{f[n]\right\}\right\} = \left(1 - e^{-j\omega}\right) F\left(e^{j\omega}\right), \tag{2.110}$$

where $\mathfrak{D}^{(k)}\left\{\cdot\right\}$ is the difference operator of k–th order.

An interesting property occurs when differentiation (not differencing) is performed in the frequency domain. In this case:

$$\mathfrak{F}_D\left\{n f[n]\right\} = j \frac{d}{d\omega}\left(F\left(e^{j\omega}\right)\right). \tag{2.111}$$

Accumulation. If $F\left(e^{j\omega}\right) = \mathfrak{F}_D\left\{f[n]\right\}$, then

$$\mathfrak{F}_D\left\{\sum_{m=-\infty}^{n} f[m]\right\} = \frac{1}{e^{-j\omega}} F\left(e^{j\omega}\right) + \pi F\left(e^{j0}\right) \sum_{k=-\infty}^{\infty} \delta(\omega - 2\pi k). \tag{2.112}$$

Properties of Discrete-Time Fourier Transform in Linear Systems

In the preceding subsection, very important fundamental properties of DTFT have been reviewed. However, the power of DTFT is better understood when linear systems are analyzed. As will be shown subsequently, the fundamental properties mentioned above will be of help in the analysis of discrete-time LTI systems.

Parseval's Equality. If $F_1\left(e^{j\omega}\right) = \mathfrak{F}_D\left\{f_1[n]\right\}$ and $F_2\left(e^{j\omega}\right) = \mathfrak{F}_D\left\{f_2[n]\right\}$, then

$$\sum_{n=-\infty}^{\infty} f_1[n] f_2^*[n] = \frac{1}{2\pi} \int_{2\pi} F_1\left(e^{j\omega}\right) F_2^*\left(e^{j\omega}\right) d\omega \tag{2.113}$$

A special case of Parseval's equality is obtained for $f_1[n] = f_2[n]$, which is referred to as "Bessel's equality," or conservation of energy property.

Convolution. It is known that the output of a LTI
signal $x[n]$ is given by the convolution of the "impulse
say $h[n]$, with the input signal $x[n]$. Formally, this i
be expressed as follows:

$$y[n] = h[n] \star x[n] = \sum_{m=-\infty}^{\infty} x[m]$$

where $y[n]$ is the output of the LTI system at the
frequency domain representation of this output is c
very important result is obtained:

$$\mathfrak{F}_D\{y[n]\} = Y\left(e^{j\omega}\right) = H\left(e^{j\omega}\right)$$

where $H\left(e^{j\omega}\right)$ is called the "frequency response" of
shown in (2.115), DTFT allows one to investigate
the frequency domain as well. Note that the convol
important simplicity in the analysis of frequency dor
As an example, because of the convolution proper
important in cascaded systems from the perspective

Multiplication. Through the use of duality, if o
quency domain, then the corresponding time domain
of multiplication of original sequences. Formally, thi

$$y[n] = x_1[n]x_2[n] = \mathfrak{F}_D^{-1}\left\{X_1\left(e^{j\omega}\right)\right.$$

2.10 The Z-Transform

With the same reasoning used in Section 2.6, one m
the transform kernel is generalized for the discrete-
it employs a complex exponential of a general comp
generalized kernel, the Z-transform is defined as:

$$F(z) = \sum_{n=-\infty}^{\infty} f[n]z^{-n},$$

where $z = re^{j\omega}$. At this point, it is important to ma
In analogy with (2.117) which represents the bil
"unilateral" or "one-sided" Z-transform is defined as

$$F(z) = \sum_{n=0}^{\infty} f[n]z^{-n},$$

where the convergence of the sum is assumed.

The inverse Z-transform can be found by exploiting the relationship between the DTFT and the Z-transform, and assumes the following expression:

$$f[n] = \frac{1}{2j\pi} \oint F(z) z^{n-1}\, dz. \tag{2.119}$$

The integral (2.119) requires the use of contour integration and is calculated using Cauchy's residue formula.

2.10.1 Properties of Z-Transform

Since the Z-transform can be considered as an extended version of DTFT, similar properties can be observed for the Z-transform.

Linearity. If $F_1(z) = \mathfrak{Z}\{f_1[n]\}$ and $F_2(z) = \mathfrak{Z}\{f_2[n]\}$ with RoCs R_1 and R_2, respectively, then:

$$\begin{aligned}
\mathfrak{Z}\{a_1 f_1[n] + a_2 f_2[n]\} &= a_1 \mathfrak{Z}\{f_1[n]\} + a_2 \mathfrak{Z}\{f_2[n]\} \\
&= a_1 F_1(z) + a_2 F_2(z)
\end{aligned} \tag{2.120}$$

with a resulting RoC containing $R_1 \cap R_2$, for arbitrary scalars a_1 and a_2. Here, it is worth mentioning that if $R_1 \cap R_2 = \emptyset$, then the linear combination of interest does not admit a Z-transform.

Symmetry. If $F(z) = \mathfrak{Z}\{f[n]\}$ has the RoC R, then:

$$\mathfrak{Z}\{f^*[n]\} = F^*(z^*), \tag{2.121}$$

and admits the same RoC R.

Scaling. If $F(z) = \mathfrak{Z}\{f[n]\}$ has the RoC R, then:

$$\mathfrak{Z}\{z_0^n f[n]\} = F\left(\frac{z}{z_0}\right), \tag{2.122}$$

admits the RoC R_1 satisfying $R_1 = |z_0|\, R$. Here, note that the scaling is performed in the Z-domain.

Shifting. If $F(z) = \mathfrak{Z}\{f[n]\}$ has the RoC R, then:

$$\mathfrak{Z}\{f[n - n_0]\} = z^{-n_0} F(z), \tag{2.123}$$

with the same RoC R except for the possible appending or deletion of the origin or infinity. Similar to the time domain shifting, if the RoC for $F(z)$ includes the

unit circle, a shift in the complex frequency plane a
modulated sequence $e^{j\omega_0 n} f[n]$:

$$\mathcal{Z}\left\{e^{j\omega_0 n} f[n]\right\} = F(\omega - \omega$$

with the same RoC, which is referred to as the "free

Differentiation.   If $F(z) = \mathcal{Z}\{f[n]\}$ has the RoC

$$\mathcal{Z}\{nf[n]\} = -z\frac{d}{dz}\left(F(z)\right.$$

with the same RoC R. Similar to the differentiation
tiation in the complex frequency plane is given by:

$$\mathcal{Z}\left\{-tf(t)\right\} = \frac{d}{ds}F(s)$$

with the same RoC R.

Convolution.   If $F_1(z) = \mathcal{Z}\{f_1[n]\}$ and $F_2(z) = \mathcal{Z}$
respectively, then:

$$\mathcal{Z}\{f_1[n] \star f_2[n]\} = F_1(z)F_2$$

with a RoC containing $R_1 \cap R_2$. Here, it is worth men
then the linear convolution does not admit a Z-trans

Product.   If $F_1(z) = \mathcal{Z}\{f_1[n]\}$ and $F_2(z) = \mathcal{Z}\{f$
respectively, then:

$$\mathcal{Z}\{f_1[n]f_2[n]\} = \frac{1}{2\pi j} \oint F_1(z)F_2$$

with a resulting RoC containing $R_1 \cap R_2$. If $R_1 \cap$
signals does not admit a Z-transform.

2.11   Conclusion and Further R

Since transformations enable to look at things in diffe
pecially in the field of signal processing are somehow
Before applying any transformation to a specific pro
in mind that the suitability of the transformation c
hand is of utmost importance. Especially in enginee
on the system design and/or constraints, some of the
single transformation) become more appropriate und
fore, prominent characteristics of the transformation
so that the problem on hand can be tackled in a ver

Signal processing has a vast literature in terms of both applications and techniques. Transformations have a special place in the vast literature as well. In order to obtain a broad perspective, the readers might refer to [7]. There are even studies dedicated to specific transformations, which provide a very comprehensive perspective for each of them [8–11]. For a detailed state-of-the-art work which covers many signal processing transformations and explores the relationships between them, the readers might refer to [4].

Acknowledgments: This work was supported by QNRF–NPRP.

2.12   Exercises

Exercise 2.12.1 (Even-Odd Signals). Assume that $f(t)$ is an odd signal and $g(t)$ is an even signal. Show that $h(t) = f(t)g(t)$ is an odd signal.

Exercise 2.12.2 (Periodicity). Assume that $f(t) = g(kt)$ where $k \in \mathbb{R}/\{0\}$. If $f(t)$ is periodic with T, is $g(\cdot)$ periodic?

Exercise 2.12.3 (Basic Transformation). Assume that an LTI system is defined by its output, say $y(t)$, and by its input, say $x(t)$, through the following input–output relationship: $y(t) = \mathfrak{K}\{x(t)\}$, where $\mathfrak{K}\{\cdot\}$ denotes a specific sort of transformation. If another transformation such as $\mathfrak{K}^{-1}\{\cdot\}$ exists, it is said that the system of interest is "invertible" since $\mathfrak{K}^{-1}\{y(t)\}$ yields the original input, $x(t)$. In the light of this, decide whether a system characterized by the differentiation operator is invertible. Note that for such a system $\mathfrak{K}\{\cdot\} = \frac{d}{dt}(\cdot)$.

Exercise 2.12.4 (Linear Transformation and Fourier Series). Assume that $f(t)$ is a continuous–time periodic signal with a fundamental period T. Find the Fourier series coefficients for $g(t) = f(2t - 1)$.

Exercise 2.12.5 (Differentiation and Fourier Series). Assume that $f(t)$ is a continuous–time periodic signal with a fundamental period T. Find the Fourier series coefficients for $g(t) = \frac{d^2}{dt^2}(f(t))$.

Exercise 2.12.6 (Non-linear Transformation and Fourier Transform). Assume that $f(t) = e^{-3|t-1|}$. Find the Fourier transform of $f(t)$.

Exercise 2.12.7 (Inverse Fourier Transform). Calculate the inverse Fourier transform of $F(\omega) = \frac{1}{\omega}$.

Exercise 2.12.8 (Wiener-Khintchine Theorem). Show that $\mathfrak{F}\{\rho_{f,g}(\tau)\} = F(\omega)G^*(\omega)$, where $\rho_{f,g}$ denotes the cross-correlation of $f(t)$ and $g(t)$ in the delay domain τ.

Exercise 2.12.9 (Sampling). Assume that for a
and is denoted by $F(j\omega)$. Prove that

$$\Delta t \sum_{n=-\infty}^{\infty} f(t - n\Delta t) = \sum_{n=-\infty}^{\infty} F(n$$

Exercise 2.12.10 (Fourier Cosine Transforn
derivative of the cosine transform of a function $f(t)$.

Exercise 2.12.11 (Differentiation in Laplace T

Exercise 2.12.12 (Hartley Transform). Show t
forms an orthonormal basis on $(-\pi, \pi]$.

Exercise 2.12.13 (Autoconvolution Hilbert
$\rho_{ff}(t) = -\rho_{\substack{\Delta\Delta \\ \hat{f}\hat{f}}}$.

References

[1] A. V. Oppenheim, A. S. Willsky, and S. H.
Prentice Hall International, Upper Saddle Riv

[2] B. P. Lathi, *Linear Systems and Signals.*
York, 2005.

[3] J. Foster and F. B. Richards, "The Gibbs ph
approximation," *The American Mathematical*
47–49, 1991.

[4] A. D. Poularakis, *Transforms and Applicatio*
Poularakis, Ed., CRC Press, Boca Raton, FI

[5] R. W. Hamming, *Numerical Methods for Scie*
Publications, Mineola, NY, 1986.

[6] S. L. Marple, "Computing the discrete-time 'an
Transactions on Signal Processing, vol. 47, no.

[7] A. V. Oppenheim, R. W. Schafer, J. R. Buck
Processing. Prentice Hall, Englewood Cliffs,

[8] R. Bracewell, *The Fourier Transform and Its A*
Hill, New York, 2000.

[9] S. L. Hahn, *Hilbert Transforms in Signal Proce*
MA, 1996, vol. 2.

[10] N. Ahmed, T. Natarajan, and K. R. Rao, "Discrete cosine transfom," *IEEE Transactions on Computers*, vol. 100, no. 1, pp. 90–93, 1974.

[11] R. N. Bracewell, *The Hartley Transform.* Oxford University Press, New York, 1986.

Chapter 3 Linear Algebra

Fatemeh Hamidi Sepehr[‡] and Erchin Serpedin[‡]
[‡]Texas A&M University, College Station, USA

This chapter will first review some basic concepts and results encountered in the study of vector spaces such as the notions of subspace, independence, basis, dimension, inner-product, norm and orthogonality, and then will focus on the general properties of linear transformations and matrix decompositions (LU, Cholesky, SVD and QR), and finally on various matrix operations and applications. This chapter represents a brief summary of the most important linear algebra concepts and results from the viewpoint of their applicability in the fields of signal processing, communications and networking. In writing this chapter, we made use of several excellent references available in the linear algebra literature [1–8]. For a more detailed description and additional topics, the readers are directed to these excellent references.

3.1   Vector Spaces

A set $\boldsymbol{\mathcal{F}}$ equipped with two operations $+, \cdot : \boldsymbol{\mathcal{F}} \times \boldsymbol{\mathcal{F}} \to \boldsymbol{\mathcal{F}}$ is said to be a field if it satisfies the following conditions:

1. If $s, t \in \boldsymbol{\mathcal{F}}$, then $s + t \in \boldsymbol{\mathcal{F}}$ and $st \in \boldsymbol{\mathcal{F}}$.

2. If $s \in \boldsymbol{\mathcal{F}}$, then there exists an additive inverse $-s \in \boldsymbol{\mathcal{F}}$ such that $s + (-s) = 0$; and for any $s \in \boldsymbol{\mathcal{F}} - 0$, there is a multiplicative inverse $s^{-1} \in \boldsymbol{\mathcal{F}}$ such that $ss^{-1} = 1$.

3. $0, 1 \in \boldsymbol{\mathcal{F}}$.

The operations of addition and multiplication are also assumed to be commutative and associative:

- $\forall r, s, t \in \boldsymbol{\mathcal{F}}$: $s + t = t + s$ and $r + (s + t) = (r + s) + t$.

- $\forall r, s, t \in \boldsymbol{\mathcal{F}}$: $st = ts$ and $r(st) = (rs)t$.

Moreover, multiplication is distributive over addition:

- $\forall r, s, t \in \boldsymbol{\mathcal{F}}$: $r(s + t) = rs + rt$.

Consider now that $\mathbf{F}$ is a field. A vector space $\mathcal{V}$ over
called vectors, that are equipped with the following
and $\cdot : \mathbf{F} \times \mathcal{V} \to \mathcal{V}$ that satisfy the following proper

1. $\forall \mathbf{v}, \mathbf{w} \in \mathcal{V}$: $\mathbf{v} + \mathbf{w} \in \mathcal{V}$ (vector additio

2. $\forall s \in \mathbf{F}$ and $\forall \mathbf{v} \in \mathcal{V}$: $s\mathbf{v} \in \mathcal{V}$ (scalar multipli

and that satisfy further the following properties:

1. $\forall \mathbf{v}, \mathbf{w} \in \mathcal{V}$: $\mathbf{v} + \mathbf{w} = \mathbf{w} + \mathbf{v}$ (co

2. $\forall \mathbf{v}, \mathbf{w}, \mathbf{u} \in \mathcal{V}$: $\mathbf{u} + (\mathbf{v} + \mathbf{w}) = (\mathbf{u} + \mathbf{v}) + \mathbf{w}$ (a

3. There exists a unique vector $\mathbf{0} \in \mathcal{V}$ such that

4. $\forall \mathbf{v} \in \mathcal{V}$, there exists $(-\mathbf{v}) \in \mathcal{V}$ such that $\mathbf{v} + ($
   inverse).

5. $\forall r, s \in \mathbf{F}$, $\forall \mathbf{v} \in \mathcal{V}$: $(rs)\mathbf{v} = r(s\mathbf{v})$ and $(r + s)\mathbf{v}$

6. $\forall r \in \mathbf{F}$, $\forall \mathbf{v}, \mathbf{w} \in \mathcal{V}$: $r(\mathbf{v} + \mathbf{w}) = r\mathbf{v} + r\mathbf{w}$ (dist

7. $\forall \mathbf{v} \in \mathcal{V}$: $1\mathbf{v} = \mathbf{v}$.

3.1.1   Subspaces and Direct Sums

Subspaces

A subspace of a vector space $\mathcal{V}$ defined over the fiel
itself a vector space over the same field. The followi
and sufficient condition for a subset to be a vector s

Theorem 3.1.1. *A nonempty subset* $\mathcal{S}$ *of a vector*
only if (iff)

$$r\mathbf{s}_1 + t\mathbf{s}_2 \in \mathcal{S}, \qquad \forall \mathbf{s}_1, \mathbf{s}_2 \in \mathcal{S},$$

A justification of this result is deferred to Exercise

Direct Sums and Complements

Let $\mathcal{S}_1, \mathcal{S}_2, ..., \mathcal{S}_k$ be subspaces of the vector space $\mathcal{V}$
sets $\mathcal{S}_i$ and $\mathcal{S}_j$, we understand the set of all vectors $\mathbf{v}$
as $\mathbf{v} = \mathbf{v}_i + \mathbf{v}_j$, where $\mathbf{v}_i \in \mathcal{S}_i$ and $\mathbf{v}_j \in \mathcal{S}_j$. The inter
$\mathcal{S}_j$ is the set of all vectors that are common to both
of sum and intersection can be extended to an arbit
that $\mathcal{V}$ can be represented as the sum of subspaces

$$\mathcal{V} = \sum_{i=1}^{k} \mathcal{S}_i,$$

and $\mathcal{S}_i \cap \mathcal{S}_j = \{\mathbf{0}\}$, (i.e., subspaces $\mathcal{S}_i$ and $\mathcal{S}_j$ are disjoint) for any $i \neq j$, then $\mathcal{V}$ is referred to as the *direct sum* of $\mathcal{S}_i$'s and it is represented as

$$\mathcal{V} = \mathcal{S}_1 \oplus \mathcal{S}_2 \oplus ... \oplus \mathcal{S}_k. \tag{3.3}$$

An alternative way to interpret condition (3.3) is to say that for any $\mathbf{v} \in \mathcal{V}$, there are unique vectors $\mathbf{v}_1 \in \mathcal{S}_1$, $\mathbf{v}_2 \in \mathcal{S}_2$, ..., $\mathbf{v}_k \in \mathcal{S}_k$ such that $\mathbf{v} = \sum_{i=1}^{k} \mathbf{v}_i$ [3, 4, 8]. If $\mathcal{S}_1$ and $\mathcal{S}_2$ are two subspaces of vector space $\mathcal{V}$, and

$$\mathcal{V} = \mathcal{S}_1 \oplus \mathcal{S}_2, \tag{3.4}$$

then $\mathcal{S}_2$ is referred to as the complement of $\mathcal{S}_1$ in $\mathcal{V}$. In general, $\mathcal{S}_1$ and $\mathcal{S}_2$ are called complementary subspaces of $\mathcal{V}$.

3.1.2 Spanning and Linear Independency

Let $\mathcal{V}$ be a vector space and $\mathcal{U}$ be a nonempty subset of $\mathcal{V}$. The subspace spanned by $\mathcal{U}$ is the set of all linear combinations of vectors in $\mathcal{U}$ and is represented in terms of the notation:

$$\langle \mathcal{U} \rangle = \text{span} \, (\mathcal{U}) = \{r_1 u_1 + r_2 u_2 ... + r_n u_n | \forall r_i \in \mathcal{F}, \forall u_i \in \mathcal{U}\}. \tag{3.5}$$

If any vector in $\mathcal{V}$ can be represented as a linear combination of vectors in $\mathcal{U}$, then

$$\mathcal{V} = \text{span} \, (\mathcal{S}) \, . \tag{3.6}$$

A subset of vectors $\mathcal{S} \subset \mathcal{V}$ is called linearly independent if

$$\sum_{i=1}^{n} a_i \mathbf{s}_i = 0 \, , \, \mathbf{s}_i \in \mathcal{S}, a_i \in \mathcal{F} \quad \Rightarrow \quad a_i = 0, i = 1, ..., n \, . \tag{3.7}$$

A set that is not linearly independent is called *linearly dependent*.

3.1.3 Bases

A linearly independent subset of vectors $B = \{\mathbf{b}_1, ..., \mathbf{b}_n\}$ in vector space $\mathcal{V}$ which spans $\mathcal{V}$ is referred to as a *basis* for $\mathcal{V}$. As a corollary, every non-zero vector in $\mathcal{V}$ can be represented uniquely as linear combination of vectors from B. The next two theorems establish important properties for the basis of a vector space [8].

Theorem 3.1.2. *The following statements hold:*

1. *For every vector space $\mathcal{V} \neq \{\mathbf{0}\}$, one can construct a basis.*

2. *Every linearly independent subset in vector space $\mathcal{V}$ can be contained in a basis, i.e., every linearly independent set in $\mathcal{V}$ can be extended to a basis for $\mathcal{V}$.*

3. *Every subset S spanning the vector space $\mathcal{V}$ (*
   basis.

Theorem 3.1.3. *Let $\mathcal{V}$ be a vector space that is spa*
and suppose the vectors $\mathbf{v}_1, ..., \mathbf{v}_n$ are linearly indep

3.1.4   Dimension

A vector space is called finite-dimensional, if it admi
number of vectors. The dimension of a finite-dime
cardinality of any basis for $\mathcal{V}$ (i.e., the number of ele
it is denoted by the notation dim $(\mathcal{V})$. By convention
space has the dimension zero. It can be shown that t
is unique.

3.1.5   Ordered Basis

If $\mathcal{V}$ is a vector space of dimension n, then an ordere
the ordered n-tuple $\mathbf{B} = (\mathbf{v}_1, ..., \mathbf{v}_n)$ of vectors that
independent. Every vector $\mathbf{v}$ in $\mathcal{V}$ can be represe
$(\mathbf{v}_1, ..., \mathbf{v}_n)$ as follows:

$$\mathbf{v} = \sum_{i=1}^{n} c_i \mathbf{v}_i .$$

The scalars c_i in representation (3.8) are referred to
with respect to the basis $(\mathbf{v}_1, ..., \mathbf{v}_n)$. An importan
for $\mathcal{V} = \mathbb{R}^n$ is the *standard basis* represented by the

$$\mathbf{e}_1 = (1, 0, 0, \ldots, 0)^T$$
$$\mathbf{e}_2 = (0, 1, 0, \ldots, 0)^T$$
$$\vdots$$
$$\mathbf{e}_n = (0, 0, 0, \ldots, 1)^T .$$

If B is an ordered basis $(\mathbf{v}_1, ..., \mathbf{v}_n)$, the notation:

$$[\mathbf{v}]_B = (c_1, c_2, \ldots, c_n)^T$$

will be used to denote the vector of coordinates of v

3.1.6   Norms

Let $\mathcal{V}$ be a vector space. A norm is a real-valued
associates to every vector $\mathbf{v} \in \mathcal{V}$ a positive real num
that satisfies the following properties:

- $\|\mathbf{v}\| \geq 0 \;\forall \mathbf{v} \in \mathcal{V}$ (with equality iff $\mathbf{v} = \mathbf{0}$) (non-negativity).

- $\|a\mathbf{v}\| = |a|\,\|\mathbf{v}\|, \;\; \forall \mathbf{v} \in \mathcal{V}, \forall a \in \mathcal{F}$ (scaling).

- $\|\mathbf{v} + \mathbf{w}\| \leq \|\mathbf{v}\| + \|\mathbf{w}\| \;\forall \mathbf{v}, \mathbf{w} \in \mathcal{V}$ (triangle inequality).

A vector space equipped with a norm is called a *normed vector space*. In general, there are many different norms that can be defined on a vector space. The following equations define some of the most used vector norms in $\mathbb{R}^n$. For the vector $\mathbf{v} = (v_1, v_2, ..., v_n) \in \mathbb{R}^n$, the following norms can be considered.

- $\|\mathbf{v}\|_1 = \sum_{i=1}^{n} |v_i|$ (l_1 norm).

- $\|\mathbf{v}\|_p = (\sum_{i=1}^{n} |v_i|^p)^{1/p}, \; 1 < p < \infty$ (l_p norm).

- $\|\mathbf{v}\|_\infty = \max_{1 \leq i \leq n} |v_i|$ (l_∞ norm).

3.1.7 Inner-Products

An inner-product on a vector space $\mathcal{V}$, defined over a field of real or complex numbers $\mathcal{F}$, is a function that assigns a scalar $\langle \mathbf{v}|\mathbf{w}\rangle \in \mathcal{F}$ to each pair of vectors $\mathbf{v}, \mathbf{w} \in \mathcal{V}$, such that:

1. $\langle \mathbf{v}|\mathbf{v}\rangle \geq 0 \;\forall \mathbf{v} \in \mathcal{V}$ (equality holds iff $\mathbf{v} = \mathbf{0}$),

2. $\langle \mathbf{v}|\mathbf{w}\rangle = \langle \mathbf{w}|\mathbf{v}\rangle^* \;\forall \mathbf{v}, \mathbf{w} \in \mathcal{V}$ (the superscript $(\cdot)^*$ denotes the complex conjugate of $\langle \mathbf{v}|\mathbf{w}\rangle$) (conjugate symmetry).

3. $\langle \mathbf{u} + \mathbf{v}|\mathbf{w}\rangle = \langle \mathbf{u}|\mathbf{w}\rangle + \langle \mathbf{v}|\mathbf{w}\rangle, \;\forall \mathbf{u}, \mathbf{v}, \mathbf{w} \in \mathcal{V}$ (linearity).

4. $\langle c\mathbf{v}|\mathbf{w}\rangle = c\,\langle \mathbf{v}|\mathbf{w}\rangle \;\;\; \forall \mathbf{u} \in V, \forall c \in \mathcal{F}$ (scaling).

Inner-Product Vector Space

A vector space $\mathcal{V}$ equipped with an inner-product is referred to as an *inner-product vector space*. As an example, for the vector space of n-dimensional complex-valued vectors $\mathbb{C}^n$, the standard inner-product is defined as:

$$\langle \mathbf{v}|\mathbf{w}\rangle = \langle (v_1, v_2, ..., v_n)|(w_1, w_2, ..., w_n)\rangle$$
$$= \mathbf{w}^\dagger \mathbf{v}$$
$$= \sum_{i=1}^{n} v_i w_i^* \,. \tag{3.10}$$

The superscript $(\cdot)^\dagger$ stands for Hermitian conjugation. For the vector space of all continuous, complex-valued functions defined on the same interval $[a, b]$, an inner-product may be defined as follows:

$$\langle f|g\rangle = \int_a^b f(x)g^*(x)dx\,, \tag{3.11}$$

where the superscript $()^*$ denotes the complex conjugation operation.

3.1.8   Induced Norms

Assuming $\langle \cdot | \cdot \rangle$ an inner product defined on the vec
can be defined in terms of the inner-product as follo

$$\|\mathbf{v}\| = \langle \mathbf{v} | \mathbf{v} \rangle^{1/2} \qquad \forall \mathbf{v} \in$$

The induced norm satisfies all the required prope
before. In addition, it also satisfies the *Cauchy-Sch*

$$|\langle \mathbf{v} | \mathbf{w} \rangle| \leq \|\mathbf{v}\| \, \|\mathbf{w}\| \quad \forall \mathbf{v}, \mathbf{w}$$

The proof of this result is left to Exercise 3.23.3.

3.1.9   Orthogonality

Let $\mathcal{V}$ be a vector space and $\mathbf{v}, \mathbf{w} \in \mathcal{V}$. If $\langle \mathbf{v} | \mathbf{w} \rangle = 0$,
to be orthogonal or perpendicular ($\mathbf{v} \perp \mathbf{w}$). A set
$\mathcal{V}$, $\{\mathbf{v}_1, \mathbf{v}_2, ..., \mathbf{v}_n\}$ is called an *orthogonal set*, if

$$\langle \mathbf{v}_i | \mathbf{v}_j \rangle = 0 \quad \forall i, j = 1, ..., n,$$

If in addition to (3.14), the unit-norm condition $\|\mathbf{v}_i$
then the set of vectors $\{\mathbf{v}_1, \mathbf{v}_2, ..., \mathbf{v}_n\}$ is called an
every $\mathbf{v}_i \in \mathcal{V} - \{\mathbf{0}\}$, we can obtain a unit vector $\mathbf{u}$
(operation referred to as *normalization*):

$$\mathbf{u}_i = \frac{1}{\|\mathbf{v}_i\|} \mathbf{v}_i \,.$$

The following general result [8], whose proof is defer

Theorem 3.1.4. *Let $\mathcal{V}$ be an inner-product vector*
non-zero vectors in $\mathcal{V}$ is linearly independent.

   If $B = \{\mathbf{b}_1, \mathbf{b}_2, ..., \mathbf{b}_n\}$ is a set of orthogonal vec
tor space $\mathcal{V}$ of dimension n, then it is referred to
Normalizing $\mathbf{b}_i$'s to unit norm, an orthonormal basi

Orthogonal Subspaces and Complements

Let $\mathcal{W}_1$ and $\mathcal{W}_2$ be two subspaces of the inner-prod\

$$\langle \mathbf{w}_1 | \mathbf{w}_2 \rangle = 0 \quad \forall \mathbf{w}_1 \in \mathcal{W}_1 \,, \, \forall \mathbf{w}_2$$

then $\mathcal{W}_1$ and $\mathcal{W}_2$ are said to be *orthogonal subspace*

$$\mathcal{W}_1^\perp = \{\mathbf{v} \in \mathcal{V} \,|\, \langle \mathbf{v} | \mathbf{w}_1 \rangle = 0 \quad , \quad \forall \mathbf{w}$$

is called the *orthogonal complement* of $\mathcal{W}_1$, and it is also a subspace of $\mathcal{V}$.

If $\mathcal{W}_1 \perp \mathcal{W}_2$ and $\mathcal{V} = \mathcal{W}_1 \oplus \mathcal{W}_2$, then $\mathcal{V}$ is called the *orthogonal direct sum* of the subspaces $\mathcal{W}_1$ and $\mathcal{W}_2$. In particular, it follows that $\mathcal{W}_2 = \mathcal{W}_1^{\perp}$ and

$$\mathcal{V} = \mathcal{W}_1 \oplus \mathcal{W}_1^{\perp}, \tag{3.18}$$

and every vector $\mathbf{v} \in \mathcal{V}$ can be decomposed as $\mathbf{v} = \mathbf{w}_1 + \mathbf{w}_2$, such that $\mathbf{w}_1 \in \mathcal{W}_1$ and $\mathbf{w}_2 \in \mathcal{W}_1^{\perp}$, and

$$\dim(\mathcal{V}) = \dim(\mathcal{W}_1) + \dim(\mathcal{W}_1^{\perp}). \tag{3.19}$$

3.1.10 Gram-Schmidt Orthogonalization

Theorem 3.1.5. *Suppose $\{\mathbf{v}_1, ..., \mathbf{v}_n\}$ is a set of linearly independent vectors in a vector space $\mathcal{V}$ of dimension n. An orthogonal set of vectors $\{\mathbf{w}_1, ..., \mathbf{w}_n\}$ can be constructed such that the set of vectors $\{\mathbf{w}_1, ..., \mathbf{w}_k\}$ represents an orthogonal basis for the subspace spanned by vectors $\{\mathbf{v}_1, ..., \mathbf{v}_k\}$, for all $k = 1, ..., n$.*

The process of constructing the orthogonal set of vectors, $\{\mathbf{w}_1, ..., \mathbf{w}_n\}$, is referred to as *Gram-Schmidt orthogonalization* and is described next. Consider $\mathbf{w}_1 = \mathbf{v}_1$, and then recursively define:

$$\mathbf{w}_{k+1} = \mathbf{v}_{k+1} - \sum_{i=1}^{k} \frac{\langle \mathbf{v}_{k+1} | \mathbf{w}_i \rangle}{\|\mathbf{w}_i\|^2} \mathbf{w}_i, \quad k = 1, ..., n-1. \tag{3.20}$$

At each step $\mathbf{w}_{k+1}$ is built by subtracting from $\mathbf{v}_{k+1}$ those components of $\mathbf{v}_{k+1}$ which lie in the subspace spanned by the vectors $\mathbf{v}_1, ..., \mathbf{v}_k$. The orthogonality of the set $\{\mathbf{w}_1, ..., \mathbf{w}_n\}$ can be checked by showing

$$\langle \mathbf{w}_{k+1} | \mathbf{w}_i \rangle = 0, \quad i = 1, ..., k.$$

Also, it can be checked that the set of vectors $\{\mathbf{w}_1, ..., \mathbf{w}_k\}$ represents an orthogonal basis for the subspace spanned by vectors $\mathbf{v}_1, ..., \mathbf{v}_k$, for all $k = 1, ..., n-1$. The justification of these results is deferred to Exercise 3.23.5.

An important corollary of the Gram-Schmidt orthogonalization approach is that any finite-dimensional vector space admits an orthonormal basis.

3.2 Linear Transformations

If $\mathcal{V}$ and $\mathcal{W}$ are two vector spaces defined over the same field $\mathcal{F}$, then a function (mapping) T from domain $\mathcal{V}$ into co-domain $\mathcal{W}$, $T : \mathcal{V} \to \mathcal{W}$, which satisfies the following linearity condition:

$$T(c_1\mathbf{v}_1 + c_2\mathbf{v}_2) = c_1 T(\mathbf{v}_1) + c_2 T(\mathbf{v}_2) \quad \forall \mathbf{v}_1, \mathbf{v}_2 \in \mathcal{V}, \forall c_1, c_2 \in \mathcal{F}, \tag{3.21}$$

is called a *linear transformation* from $\mathcal{V}$ to $\mathcal{W}$. Note that (3.21) implies $T(\mathbf{0}) = 0$. If there exists a vector $\mathbf{v} \in \mathcal{V} - \{\mathbf{0}\}$ such that $T(\mathbf{v}) = \mathbf{0}$, then T is termed a *singular* transformation. In the contrary case, T is termed a nonsingular transformation.

3.2.1 Range and Nullspace of a Linea

Let T be a linear transformation from $\mathcal{V}$ to $\mathcal{W}$. The
by $\mathcal{R}(T)$ or $\mathcal{I}m(T)$, and is defined as the set:

$$\mathcal{R}(T) = \{\mathbf{w} \in \mathcal{W} | T(\mathbf{v}) = \mathbf{w} \text{ for so}$$
$$= \{T(\mathbf{v}) | \mathbf{v} \in \mathcal{V}\} \ .$$

The *nullspace* (*kernel*) of T is denoted in terms of the
and it is defined as the set:

$$\mathcal{N}\text{ull}\,(T) = \{\mathbf{v} \in \mathcal{V} | T(\mathbf{v}) =$$

For finite dimensional vector spaces $\mathcal{V}$ and $\mathcal{W}$, dim
are called the *rank* and *nullity* of T, respectively. It

$$\dim\,(\mathcal{R}(T)) = \dim\left(\mathcal{N}\text{ull}\,(T)^{\perp}\right)$$
$$\dim\,(\mathcal{V}) = \text{rank}\,(T) + \dim\,\mathcal{N}$$

3.2.2 Composition and Invertibility

Let T_1 and T_2 be two linear transformations from ve
and from $\mathcal{W}$ to vector space $\mathcal{Z}$, respectively. Then t
as follows:
$$T_2 T_1(\mathbf{v}) = T_2(T_1(\mathbf{v})), \quad \forall \mathbf{v}$$

and it represents a linear transformation from $\mathcal{V}$ to
Assume $T_1 : \mathcal{V} \to \mathcal{W}$ is a linear transformation. I
mation T_1^{-1} from $\mathcal{W}$ to $\mathcal{V}$ ($T_1^{-1} : \mathcal{W} \to \mathcal{V}$) such that
$T_1 T_1^{-1}$ are both identity, then T_1 and T_1^{-1} are cal
and they are referred to as the *inverse* of each oth
transformation T_1 is unique, and in general the not
represent the inverse of T_1.

We close this section by introducing the concepts
formations. Let $T : \mathcal{V} \to \mathcal{W}$ denote an arbitrary line
said to be *onto* (or *surjective*) if and only if $\mathcal{R}(T) =$
(or *injective*) if and only if $\mathcal{N}\text{ull}\,(T) = \{\mathbf{0}\}$. A linear
onto and one-to-one is called *bijective* or *invertible*.

3.2.3 Matrix Representation of Linea

Let $\mathbf{A}$ be an $m \times n$ matrix over the field $\mathcal{F}$ and let
the transformation $T_A : \mathcal{V} \to \mathcal{W}$ defined via

$$T_A(\mathbf{v}) = \mathbf{A}\mathbf{v} \ , \quad \forall \mathbf{v} \in \mathcal{V}$$

is a linear transformation, and matrix $\mathbf{A}$ is referred to as the *matrix representation* of transformation T_A. The rank of matrix $\mathbf{A}$ is by definition the dimension of range space of T_A. In addition, the column (row) rank of matrix $\mathbf{A}$ is defined to be the largest number of independent columns (rows) of $\mathbf{A}$. The following result merges together the concepts of row rank, column rank, and rank of matrix.

Theorem 3.2.1. *Let* $\mathbf{A}$ *be an* $m \times n$ *matrix. Then the* row rank $(\mathbf{A})$, *i.e., the dimension of the subspace spanned by the row vectors of* $\mathbf{A}$, *is equal to* column rank $(\mathbf{A})$, *i.e., the dimension of the subspace spanned by the column vectors of A, and equal to the rank of matrix* $\mathbf{A}$:

$$\text{row rank}\,(\mathbf{A}) = \text{column rank}\,(\mathbf{A}) = \text{rank}\,(\mathbf{A}) \ . \tag{3.27}$$

The proof of this result is delegated to Exercise 3.23.6. The next result establishes a unique matrix representation for a linear transformation in terms of the basis vectors of its domain and co-domain [3].

Theorem 3.2.2. *Let* T *be a linear transformation from the vector space* $\mathcal{V}$ *of dimension n to the vector space* $\mathcal{W}$ *of dimension m. If* $B_1 = (\mathbf{b}_1, ..., \mathbf{b}_n)$ *is an ordered basis of* $\mathcal{V}$ *and* B_2 *is an ordered basis of* $\mathcal{W}$, *then there is a matrix representation for the linear transformation* T *with respect to* B_1 *and* B_2 *such that*

$$[T(\mathbf{v})]_{B_2} = \mathbf{A}\,[\mathbf{v}]_{B_1}\ , \quad \forall \mathbf{v} \in \mathcal{V}, \tag{3.28}$$

where the ith *column of* $\mathbf{A}$ *is given by:*

$$\mathbf{A}^{(i)} = \mathbf{A}\,[\mathbf{b}_i]_{B_1} = [T(\mathbf{b}_i)]_{B_2}\ . \tag{3.29}$$

The notation $[\cdot]_{B_1}$ means that $[\mathbf{b}_i]_{B_1} = e_i$, where e_i stands for the ith column of the unit matrix. In other words, the ith column of matrix $\mathbf{A}$ is given by the representation coordinates of $T(\mathbf{b}_i)$ with respect to basis B_2. A quick justification of this theorem can be obtained by observing the fact that any vector $\mathbf{v} \in \mathcal{V}$ can be represented as a linear combination of basis vectors $\mathbf{b}_1, \ldots, \mathbf{b}_n$:

$$\mathbf{v} = \sum_{i=1}^{n} v_i \mathbf{b}_i \ . \tag{3.30}$$

Applying the transformation T to both sides of (3.30) and taking into account the linearity of T, (3.30) leads to the sought result $T(\mathbf{v}) = \sum_{i=1}^{n} v_i T(\mathbf{b}_i)$, which justifies (3.28) and (3.29).

3.2.4 Projection Operators

A linear transformation $T : \mathcal{V} \to \mathcal{W}$ is called *idempotent* or *projection* if it leaves its image unchanged:

$$T^2 = T \cdot T = T\ . \tag{3.31}$$

The following general result holds.

Theorem 3.2.3. *Let $T : \mathcal{V} \to \mathcal{V}$ be a projection oper* *the range and the nullspace of T are disjoint subspa*

$$\mathcal{R}(T) \cap \mathcal{N}\text{ull}\,(T) = \{\mathbf{0}\}$$

The proof of this result is immediate by recognizi implies $\mathbf{v} = \mathbf{0}$, i.e., only $\mathbf{0}$ can belong to both the ra

Orthogonal Projections

Let $T : \mathcal{V} \to \mathcal{V}$ be a projection over an inner-produc

$$\mathcal{R}(T) \perp \mathcal{N}\text{ull}\,(T) \ ,$$

then T is called an *orthogonal projection*.

Theorem 3.2.4. *If $\mathbf{P}$ is an idempotent Hermitian*

$$\mathbf{P}^2 = \mathbf{P} = \mathbf{P}^\dagger \ ,$$

then $\mathbf{P}$ defines an orthogonal projection operator.

Notice that if $\mathbf{P}$ represents the matrix representatior then

$$\mathbf{P}\mathbf{v} \in \mathcal{R}(T) \ ,$$

$$(\mathbf{I} - \mathbf{P})\,\mathbf{v} \in \mathcal{N}\text{ull}\,(T) \ .$$

It can be shown that the range space and the nulls Exercise 3.23.9). This result follows by taking into $\mathbf{P}) = \mathbf{0}$, and $\mathbf{I} = \mathbf{P} + (\mathbf{I} - \mathbf{P})$. Furthermore, $\mathbf{I} - \mathbf{P}$ is its range is the nullspace of $\mathbf{P}$.

3.2.5 Linear Functionals and Dual Sp

Let $\mathcal{V}$ be a vector space over field $\mathcal{F}$. Then a linea $f : \mathcal{V} \to \mathcal{F}$, is referred to as a *linear functional*. functionals defined on vector space $\mathcal{V}$ is a vector spa and is represented in terms of the notation $\mathcal{V}^*$. The procedure for constructing a basis for the dual space

Theorem 3.2.5. *Let $\mathcal{V}$ be a vector space of dimens*

$$B = \{\mathbf{b}_1, ..., \mathbf{b}_n\} \ ,$$

stand for a basis of $\mathcal{V}$. For each $i = 1, ..., n$, a lin *defined such that*

$$b_i^*(\mathbf{b}_j) = \delta_{ij} = \begin{cases} 1 & \text{if } i = \\ 0 & \text{otherwi} \end{cases}$$

Then the set

$$B^* = \{b_1^*, ..., b_n^*\}$$

forms a unique basis (termed the dual basis of B) for the dual space $\mathcal{V}^*$.

A corollary of this result is that for a finite dimensional, $\mathcal{V}$, the following equality holds:

$$\dim(\mathcal{V}^*) = \dim(\mathcal{V}).$$

Notice that the linear independence of dual basis functionals b_i^*, $i = 1, \ldots, n$, follows from the fact that an identity of the form:

$$\sum_{i=1}^{n} c_i b_i^* = 0, \tag{3.37}$$

where c_i's are some scalars, implies always that $c_i = 0$, $i = 1, \ldots, n$. Indeed, applying the linear functional $\sum_{i=1}^{n} c_i b_i^*$ on the basis vector $\mathbf{b}_j$, it follows that: $\sum_{i=1}^{n} c_i b_i^*(\mathbf{b}_j) = \sum_{i=1}^{n} c_i \delta_{i,j} = c_j$, which in accordance to (3.37) implies that $c_j = 0$, $j = 1, \ldots, n$. Notice furthermore that:

$$\sum_{i=1}^{n} f(\mathbf{b}_i) b_i^*(\mathbf{b}_j) = \sum_{i=1}^{n} f(\mathbf{b}_i) \delta_{i,j} = f(\mathbf{b}_j). \tag{3.38}$$

Equation (3.38) proves that an arbitrary functional f lies in the span of B^*. This fact, combined with the independence of functionals b_i^*, $i = 1, \ldots, n$, ensures that B^* is a basis for the dual space $\mathcal{V}^*$.

3.2.6 Adjoint of a Linear Transformation

The following result, called the *Riesz representation theorem*, plays an important role in defining the concept of adjoint of a linear transformation.

Theorem 3.2.6. *(Riesz representation theorem [6]) If $\mathcal{V}$ is an inner-product vector space of dimension n, and f is a linear functional on $\mathcal{V}$ ($f \in \mathcal{V}^*$), then there is a unique $\mathbf{v}_R \in \mathcal{V}$ such that*

$$f(\mathbf{v}) = \langle \mathbf{v} | \mathbf{v}_R \rangle, \quad \forall \mathbf{v} \in \mathcal{V}. \tag{3.39}$$

$\mathbf{v}_R$ *is called the Riesz vector for f.*

A justification for this result can be obtained by considering the linear functional $f_{\mathbf{v}_R} : \mathcal{V} \to \mathbf{F}$,

$$f_{\mathbf{v}_R}(\mathbf{v}) = \langle \mathbf{v} | \mathbf{v}_R \rangle. \tag{3.40}$$

Then $f_{\mathbf{v}_R}$ belongs to $\mathcal{V}^*$. Defining

$$\mathbf{v}_R = \sum_{i=1}^{n} (f(\mathbf{v}_i))^* \mathbf{v}_i, \tag{3.41}$$

in which $\mathbf{v}_i$'s $(i = 1, ..., n)$ form an orthonormal ba
$f_{\mathbf{v}_R}(\mathbf{v}_j) = f(\mathbf{v}_j)$, $\forall \mathbf{v}_j$, and thus $f = f_{\mathbf{v}_R}$.
The following result introduces the concept of adjoi

Theorem 3.2.7. *Let $T : \mathcal{V} \to \mathcal{W}$ be a linear transf*
finite dimensional inner-product vector spaces over $\mathcal{J}$
transformation $T^* : \mathcal{W} \to \mathcal{V}$ such that

$$\langle T(\mathbf{v})|\mathbf{w}\rangle = \langle \mathbf{v}|T^*(\mathbf{w})\rangle \ , \quad \forall \mathbf{v} \in \mathcal{V}$$

T^* is called the adjoint of T. The existence of a
defining the linear functional $f : \mathcal{V} \to \mathcal{F}$ through

$$f(\mathbf{v}) = \langle T(\mathbf{v})|\mathbf{w}\rangle \ , \ \forall \mathbf{v} \in$$

for a fixed vector $\mathbf{w} \in \mathcal{W}$. According to the Riesz re
a unique vector $\mathbf{w}_R \in \mathcal{W}$ which satisfies:

$$\langle T(\mathbf{v})|\mathbf{w}\rangle = \langle \mathbf{v}|\mathbf{w}_R\rangle \ , \quad \forall \mathbf{v}$$

Now define $T^*(\mathbf{w}) = \mathbf{w}_R$. Therefore, it follows that

$$\langle T(\mathbf{v})|\mathbf{w}\rangle = \langle \mathbf{v}|T^*(\mathbf{w})\rangle$$

The linearity of T^* can be justified by exploiting the
and T [6].

Next, some properties of the adjoint transform wi
T_2 be linear transformations from $\mathcal{V}$ to $\mathcal{W}$, and $s \in$
hold:

- $(T_1 + T_2)^* = T_1^* + T_2^*$.

- $(sT)^* = \bar{s}T^*$ (the bar stands for conjugation o

- $T^{**} = T$. From the adjoint's definition, i
  $\langle \mathbf{v}|T(\mathbf{w})\rangle$.

- If $\mathcal{W} = \mathcal{V}$, then $(T_1 T_2)^* = T_2^* T_1^*$.

- For invertible T, $(T^{-1})^* = (T^*)^{-1}$.

As a corollary, assuming that matrix $\mathbf{A}$ denotes tl
linear transformation on $\mathcal{V} = \mathbb{C}^n$, using the stanc
immediately:

$$\langle \mathbf{A}\mathbf{v}|\mathbf{w}\rangle = \mathbf{w}^\dagger \mathbf{A}\mathbf{v}, \quad \forall \mathbf{v}, \mathbf{w}$$

On the other hand, from the definition of adjoint it

$$\langle \mathbf{A}\mathbf{v}|\mathbf{w}\rangle = \langle \mathbf{v}|\mathbf{A}^*\mathbf{w}\rangle = (\mathbf{A}^*\mathbf{w})^\dagger \mathbf{v} =$$

Because these equalities hold for all $\mathbf{v}$, $\mathbf{w} \in \mathcal{V}$, it
$\mathbf{A}^* = \mathbf{A}^\dagger$. As a special case, if matrix $\mathbf{A}$ is defined
it follows that $\mathbf{A}^* = \mathbf{A}^T$.

3.2.7 Four Fundamental Subspaces

Corresponding to every linear transformation $T : \mathcal{V} \to \mathcal{W}$, there exist four fundamental subspaces, the range and the nullspace of T, and the range and the nullspace of $T^* : \mathcal{W} \to \mathcal{V}$. Recall that $\mathcal{R}(T)$ and $\mathcal{N}\text{ull}(T^*)$ are subspaces of $\mathcal{W}$ and likewise, $\mathcal{N}\text{ull}(T)$ and $\mathcal{R}(T^*)$ are subspaces of $\mathcal{V}$. The relations between these four fundamental subspaces are established by the following theorem [3, 6, 8].

Theorem 3.2.8. *Let $T : \mathcal{V} \to \mathcal{W}$ be a linear-transformation, where $\mathcal{V}$ and $\mathcal{W}$ are inner-product finite dimensional vector spaces. Then the range of T is the orthogonal complement of the nullspace of T^*, and the nullspace of T is the orthogonal complement of the range of T^*.*

As a corollary, the following decompositions hold:

$$\begin{aligned} \mathcal{V} &= \mathcal{R}(T^*) \oplus \mathcal{N}\text{ull}(T) \\ &= \mathcal{N}\text{ull}(T)^\perp \oplus \mathcal{N}\text{ull}(T) \\ &= \mathcal{R}(T^*) \oplus \mathcal{R}(T^*)^\perp , \end{aligned} \tag{3.46}$$

and

$$\begin{aligned} \mathcal{W} &= \mathcal{R}(T) \oplus \mathcal{N}\text{ull}(T^*) \\ &= \mathcal{R}(T) \oplus \mathcal{R}(T)^\perp . \end{aligned} \tag{3.47}$$

The justification of these decompositions is left to Exercise 3.23.11. Furthermore, the following corollary is obtained: if $\mathbf{A}$ stands for the matrix representation of a linear transformation from $\mathcal{V} = \mathbb{R}^n$ to $\mathcal{W} = \mathbb{R}^m$, then

$$\mathcal{V} = \mathcal{R}\left(\mathbf{A}^T\right) \oplus \mathcal{N}\text{ull}(\mathbf{A}) , \tag{3.48}$$

$$\mathcal{W} = \mathcal{R}(\mathbf{A}) \oplus \mathcal{N}\text{ull}\left(\mathbf{A}^T\right) . \tag{3.49}$$

3.3 Operator Norms and Matrix Norms

Let $T : \mathcal{V} \to \mathcal{W}$ be a linear transformation, where $\mathcal{V}$ and $\mathcal{W}$ are both normed vector spaces. Then T is said to be *bounded* if

$$\sup_{\mathbf{v} \in \mathcal{V} - \{\mathbf{0}\}} \frac{\|T(\mathbf{v})\|}{\|\mathbf{v}\|} < \infty , \tag{3.50}$$

where the norms $\|T(\mathbf{v})\|$ and $\|\mathbf{v}\|$ are computed in vector spaces $\mathcal{W}$ and $\mathcal{V}$, respectively. The value of supremum in (3.50) is referred to as the *operator norm* of T and is represented by the notation $\|T\|$. It follows immediately that

$$\|T\| = \sup_{\mathbf{v} \in \mathcal{V} - \{\mathbf{0}\}} \frac{\|T(\mathbf{v})\|}{\|\mathbf{v}\|} = \sup_{\mathbf{v} \in \mathcal{V} , \|\mathbf{v}\|=1} \|T(\mathbf{v})\| . \tag{3.51}$$

Consider now the linear transformation $T : \mathbb{R}^n \to \mathbb{R}$

$$T(\mathbf{x}) = \mathbf{A}\mathbf{x},$$

where $\mathbf{A}$ is an $m \times n$ matrix. From (3.51), it turns T is equal to the matrix norm of $\mathbf{A}$, and it can be d

- l_1-norm:

$$\|\mathbf{A}\|_1 = \max_{\|\mathbf{x}\|_1 = 1} \|\mathbf{A}\mathbf{x}\|_1 = \mathrm{ma}_j$$

- l_∞-norm:

$$\|\mathbf{A}\|_\infty = \max_{\|\mathbf{x}\|_\infty = 1} \|\mathbf{A}\mathbf{x}\|_\infty = \mathrm{m}$$

- l_2-norm, also called the spectral norm of matr

$$\|\mathbf{A}\|_2 = \max_{\|\mathbf{x}\|_2 = 1} \|\mathbf{A}\mathbf{x}\|_2 = \mathrm{ma}_{\mathbf{x}^\dagger \mathbf{x} =}$$

where the l_2-norm of the vector $\mathbf{x}$ is given by:

Using the Lagrange multiplier technique and the eigenvalues of a matrix, it can be shown further tha

$$\|\mathbf{A}\|_2 = \sqrt{\rho(\mathbf{A}^\dagger \mathbf{A})},$$

where notation $\rho(\mathbf{A}^\dagger \mathbf{A})$ denotes the spectral radius value of matrix $\mathbf{A}^\dagger \mathbf{A}$ [5, 9]. An alternative immedia on the concept of Rayleigh quotient to be introduce
Another common norm for a matrix $\mathbf{A} = [a_{ij}]$ Euclidean norm, which is defined by

$$\|\mathbf{A}\|_F = \left(\sum_{i=1}^{m} \sum_{j=1}^{n} |a_{ij}|^2\right)^{1/2} = \sqrt{\mathrm{t}}$$

where $\mathrm{tr}\left(\mathbf{A}^\dagger \mathbf{A}\right)$ stands for the trace of a matrix $\mathbf{A}$ elements of $\mathbf{A}^\dagger \mathbf{A}$. Exercise 3.23.15 illustrates variou were defined in this section. References [5, 9] prese properties of operator (matrix) norms.

3.4 Systems of Linear Equations

Consider a system of m linear equations with n unknowns $x_1, x_2, \ldots, x_n$ of the form:

$$a_{11}x_1 + a_{12}x_2 + \ldots + a_{1n}x_n = b_1$$
$$a_{21}x_1 + a_{22}x_2 + \ldots + a_{2n}x_n = b_2$$
$$\vdots \tag{3.57}$$
$$a_{m1}x_1 + a_{m2}x_2 + \ldots + a_{mn}x_n = b_m \,.$$

The coefficients a_{ij}'s and b_i's, $i = 1, \ldots, m$, $j = 1, \ldots, n$, are assumed real-valued and known. Based on the relationship between the number of equations m and the number of unknowns n, a system of linear equations may be classified into one of these categories:

- when $m < n$, the system will be called *underdetermined*.

- when $m = n$, the system will be called *determined*.

- when $m > n$, the system will be called *overdetermined*.

It is quite common for underdetermined systems of equations to present an infinite number of solutions, while for overdetermined systems to lack any exact solution. In general, the system of linear equations just introduced in (3.57) is expressed in the compact matrix form:

$$\mathbf{Ax} = \mathbf{b} \,, \tag{3.58}$$

where

$$\mathbf{A} = \begin{bmatrix} a_{11} & a_{12} & \cdots & a_{1n} \\ a_{21} & a_{22} & \cdots & a_{2n} \\ \vdots & \vdots & \ddots & \vdots \\ a_{m1} & a_{m2} & \cdots & a_{mn} \end{bmatrix}, \quad \mathbf{x} = \begin{bmatrix} x_1 \\ x_2 \\ \vdots \\ x_n \end{bmatrix}, \quad \mathbf{b} = \begin{bmatrix} b_1 \\ b_2 \\ \vdots \\ b_n \end{bmatrix} \,.$$

In general, $\mathbf{A}$ is referred to as the *coefficient matrix*, $\mathbf{x}$ stands for the unknown vector, and $\mathbf{b}$ denotes the vector of constant terms. When $\mathbf{b} = \mathbf{0}$, the system of equations is called *homogeneous*, and it admits at least one trivial solution, the *all-zero* solution. When $\mathbf{b} \neq \mathbf{0}$, the system is called *inhomogeneous*, and its solution is in general expressed as the sum between a particular solution and the general solution of the homogeneous system of equations.

Matrix $\mathbf{A}$ might stand for the matrix representation of a linear transformation $T : \mathbb{R}^n \to \mathbb{R}^m$ defined via

$$T(\mathbf{x}) = \mathbf{Ax} \,.$$

Notice that in this case the range of T is clearly the space spanned by the column vectors of $\mathbf{A}$, i.e., the column space of matrix $\mathbf{A}$. However, the nullspace of T is just the set of solutions of the homogeneous system of equations $\mathbf{Ax} = \mathbf{0}$.

In general, a system of linear equations might ε
infinitely many solutions, or no solution. In the lit
posed a large number of approaches to calculate the
equations. Among the most encountered approache
able elimination (or substitution) method, reduction
Cramer's rule, various factorization approaches (suc
Cholesky factorization), and iterative approaches (
these approaches). In what follows, we will review
concepts and factorizations pertaining to matrices.

3.5 Determinant, Adjoint, and
trix

Let $\mathbf{A} = [a_{ij}]$ be an $n \times n$ matrix with the entry i
denoted by a_{ij}. The *determinant* of $\mathbf{A}$ denoted by d
associated to matrix $\mathbf{A}$, and that can be inductively

$$\det(\mathbf{A}) = \begin{cases} a_{11} & \text{if } n = \\ \sum_{j=1}^{n} a_{ij} A_{ij} (= \sum_{i=1}^{n} a_{ij} A \end{cases}$$

where $A_{ij} = (-1)^{i+j} \det(\mathbf{M}_{ij})$ stands for the *co-fac*
$(n-1) \times (n-1)$ matrix, referred to as the *minor* o
i^{th} row and the jth column of $\mathbf{A}$. Determinants ar
solutions of linear systems of equations via Cramer
values of matrices and evaluating volumes of parall
The adjoint of matrix $\mathbf{A}$ is defined as:

$$\text{adj}(\mathbf{A}) = \begin{bmatrix} A_{11} & A_{21} & \dots & A_n \\ A_{12} & A_{22} & \dots & A_n \\ \vdots & \vdots & \ddots & \vdots \\ A_{1n} & A_{2n} & \dots & A_n \end{bmatrix}$$

Notice that the inverse of matrix $\mathbf{A}$, $\mathbf{A}^{-1}$, can be ca

$$\mathbf{A}^{-1} = \frac{1}{\det(\mathbf{A})} \text{adj}(\mathbf{A})$$

If $det(\mathbf{A}) = 0$, matrix $\mathbf{A}$ is called *singular* or *non*
properties of the determinant are listed next:

- $\det(\mathbf{A}) = \det(\mathbf{A}^T)$, for every square matrix $\mathbf{A}$

- $\det(\mathbf{AB}) = \det(\mathbf{A}) \det(\mathbf{B})$, for square matrice
  this property $(\mathbf{B} = \mathbf{A}^{-1})$ implies that $\det(\mathbf{A}^-$

- The determinant of a lower triangular matrix (a matrix with all the entries above the main diagonal equal to zero), is the product of its diagonal elements. In other words, if $\mathbf{A} = [a_{ij}]$ is lower triangular ($a_{ij} = 0$ for $i < j$), then $\det(\mathbf{A}) = \prod_i a_{ii}$.

- The determinant of a diagonal matrix is equal to the product of its diagonal entries.

- The determinant of a matrix is equal to the product of matrix eigenvalues.

- If $\mathbf{A}$ and $\mathbf{B}$ are matrices of dimensions $m \times n$ and $n \times m$, respectively, then $\det(\mathbf{I}_m + \mathbf{AB}) = \det(\mathbf{I}_n + \mathbf{BA})$. This identity is referred to as Sylvester's determinant theorem, and its proof is delegated to Exercise 3.23.17.

The reader is directed to [5, 7, 8] for additional elementary properties of determinants.

3.6 Cramer's Rule

Cramer's rule determines a solution for the linear system of equations:

$$\mathbf{Ax} = \mathbf{b},$$

where $\mathbf{A}$ is an $n \times n$ nonsingular matrix and $\mathbf{b} \in \mathbb{R}^n$, by expressing the ith entry of solution vector $\mathbf{x}$ as follows:

$$x_i = \frac{\det(\mathbf{A}_i)}{\det(\mathbf{A})}. \tag{3.62}$$

In Equation (3.62), matrix $\mathbf{A}_i$ is obtained by replacing the ith column of matrix $\mathbf{A}$ by vector $\mathbf{b}$.

3.7 Unitary and Orthogonal Operators and Matrices

Let $\mathcal{V}$ be a finite dimensional, inner-product vector space and let $T : \mathcal{V} \to \mathcal{V}$ be a linear transformation. If

$$\langle T(\mathbf{v})|T(\mathbf{w})\rangle = \langle \mathbf{v}|\mathbf{w}\rangle \qquad \forall \mathbf{v}, \mathbf{w} \in \mathcal{V}, \tag{3.63}$$

then T is called a *unitary (orthogonal) operator* in the complex-(real)-valued field. From (3.63), it turns out that a unitary (orthogonal) operator preserves the inner-product value. The adjoint of unitary transformation T satisfies the property:

$$T^* = T^{-1}. \tag{3.64}$$

Stated differently, the inverse of a unitary transform

If the unitary (orthogonal) transform $T : \mathcal{V} \to \mathcal{V}$
form

$$T(\mathbf{v}) = \mathbf{A}\mathbf{v}, \qquad \forall \mathbf{v} \in \mathcal{V}$$

then matrix $\mathbf{A}$ will be called unitary (orthogonal)
implies that

$$\langle \mathbf{A}\mathbf{v} | \mathbf{A}\mathbf{w} \rangle = \langle \mathbf{v} | \mathbf{w} \rangle, \qquad \forall \mathbf{v}, \mathbf{w}$$

which leads further to:

$$\langle \mathbf{A}^* \mathbf{A}\mathbf{v} | \mathbf{w} \rangle = \langle \mathbf{v} | \mathbf{w} \rangle, \qquad \forall \mathbf{v}, \mathbf{w}$$

and thus to

$$\mathbf{A}^* \mathbf{A} = \mathbf{I},$$

where $\mathbf{A}^*$ is the matrix adjoint of $\mathbf{A}$ and it is equal

3.8   LU Decomposition

LU decomposition (or LU factorization) is a popular
of linear equations $\mathbf{A}\mathbf{x} = \mathbf{b}$, with n equations and n u
two triangular-form systems of linear equations. Th
of equations are then solved by forward or backwa
The LU decomposition states that any square $n \times$
into the product of two triangular matrices:

$$\mathbf{A} = \mathbf{L}\mathbf{U},$$

such that $\mathbf{L} = [l_{ij}]$ is a unit lower triangular matr
and $l_{ij} = 0$ for $i < j$), and $\mathbf{U}$ is an upper triang
original system of equations $\mathbf{A}\mathbf{x} = \mathbf{b}$ in terms of th
be expressed as:

$$\mathbf{L}\mathbf{U}\mathbf{x} = \mathbf{b}.$$

Introducing the notation:

$$\mathbf{y} = \mathbf{U}\mathbf{x},$$

(3.69) simplifies to:

$$\mathbf{L}\mathbf{y} = \mathbf{b}.$$

Notice that solving the original system of equations
gular systems of equations (3.71) followed by (3.70)
lower triangular system of linear equations (3.71) ca
substitution approach, while the upper triangular s
be solved via the backward substitution approach.

As a corollary, the LU factorization represents also an efficient way to compute the determinant of matrix $\mathbf{A}$. Since $\det(\mathbf{A}) = \det(\mathbf{L})\det(\mathbf{U})$, and determinant of a triangular matrix is just the product of its diagonal entries, it follows that $\det(\mathbf{L}) = 1$ and $\det(\mathbf{A}) = \det(\mathbf{U}) = \prod_{i=1}^{n} u_{ii}$, where u_{ii} is the entry of $\mathbf{U}$ located on the ith row and ith column.

Next, we will focus on how to factorize the matrix $\mathbf{A}$ into the product $\mathbf{LU}$. The basic idea to obtain the $\mathbf{LU}$ factorization is to perform elementary row operations to convert $\mathbf{A}$ into an upper triangular matrix. In general, there are three basic row operations that can be performed on the rows of a matrix: scaling a row by a constant, subtracting (adding) a scaled version of a row from (to) another row, and swapping of two rows. These elementary row operations can be implemented through left multiplications of the original matrix with lower triangular matrices. The operation of transforming matrix $\mathbf{A}$ into an upper triangular matrix $\mathbf{U}$ is done on a column-by-column basis starting with the first column and zeroing by means of elementary row operations all the matrix entries located on that column and under the main diagonal. Therefore, the ith step in obtaining the desired upper triangular matrix consists of zeroing out the entries on the ith column of matrix $\mathbf{A}$ that are located under its main diagonal by adding possibly a scaled version of the ith row to the $(i+1)$th row up to the nth row. The elementary row operations required in each step can be implemented (represented) by a left multiplication of $\mathbf{A}$ with a unit lower triangular matrix; i.e., at each step we left-multiply $\mathbf{A}$ by a unit lower triangular matrix that implements the row operations that force the under diagonal entries on ith column of $\mathbf{A}$ to be zero. Considering all these steps together, a number of $(n-1)$ unit lower triangular matrices (let's call them $\mathbf{L}_i^{-1}$, $i = 1, \ldots, n-1$) must be left-multiplied with $\mathbf{A}$ to convert $\mathbf{A}$ to an upper triangular matrix $\mathbf{U}$. Therefore, we will end up with this equation:

$$\mathbf{L}_{n-1}^{-1}\mathbf{L}_{n-2}^{-1}\ldots\mathbf{L}_{1}^{-1}\mathbf{A} = \mathbf{U}, \tag{3.72}$$

where $\mathbf{L}_i^{-1}$, $i = 1, \ldots, n-1$, denote the lower triangular matrix representation of the elementary row operations performed in the ith step on the ith column under-diagonal entries. From (3.72), it follows that

$$\mathbf{A} = \mathbf{L}_1\mathbf{L}_2\ldots\mathbf{L}_{n-1}\mathbf{U}, \tag{3.73}$$

where $\mathbf{L}_i$ $(i = 1, \ldots, n-1)$ denotes the inverse of $\mathbf{L}_i^{-1}$, and it represents also a lower triangular matrix. Because the product of a number of lower triangular matrices is also a triangular matrix, it follows that $\mathbf{L} = \mathbf{L}_1\mathbf{L}_2\ldots\mathbf{L}_{n-1}$ is a unit lower triangular matrix. Thus, the required factorization $\mathbf{A} = \mathbf{LU}$ is obtained.

3.9 LDL and Cholesky Decomposition

Suppose the LU factorization for a Hermitian matrix $\mathbf{A} = \mathbf{LU}$ (i.e., $\mathbf{A}^\dagger = \mathbf{A}$). Notice that matrix $\mathbf{U}$ can be factored further as:

$$\mathbf{U} = \mathbf{D}\mathbf{U}_1, \tag{3.74}$$

where $\mathbf{D}$ is diagonal and $\mathbf{U}_1$ is a unit upper triang
(3.74), it follows that:

$$\mathbf{A} = \mathbf{LDU}_1.$$

One can show that such a factorization will be un
unit lower triangular, diagonal, and unit upper tri
furthermore that in the factorization

$$\mathbf{LU} = \mathbf{LDU}_1,$$

the diagonal elements of $\mathbf{D}$ coincide with the diago
Hermitian symmetric, it follows that:

$$\mathbf{A} = \mathbf{A}^\dagger \Rightarrow \mathbf{LDU}_1 = (\mathbf{LDU}_1)^\dagger =$$

Because of the uniqueness of factorization, it turns

$$\mathbf{U}_1 = \mathbf{L}^\dagger.$$

Thus, we obtain the LDL factorization:

$$\mathbf{A} = \mathbf{LDL}^\dagger.$$

LDL decomposition is usually applied to real-valued
If matrix $\mathbf{A}$ is positive definite (i.e., $\mathbf{v}^\dagger \mathbf{A} \mathbf{v} > 0$
entries of matrix $\mathbf{D}$ are positive. Therefore, one can

$$\mathbf{D} = (\mathbf{D}^{1/2})(\mathbf{D}^{1/2}) = (\mathbf{D}^{1/2})(\mathbf{I}$$

where the square-root matrix $D^{1/2}$ is a diagonal mat
taking the square roots of the diagonal entries of $\mathbf{L}$
leads to

$$\mathbf{A} = (\mathbf{LD}^{1/2})(\mathbf{LD}^{1/2})^\dagger = \tilde{\mathbf{L}}$$

where $\tilde{\mathbf{L}} = \mathbf{LD}^{1/2}$ is a lower triangular matrix wit
The factorization (3.81) of positive definite matrix $\mathbf{A}$
decomposition, and it plays an important role in sig
cations applications. In applications, involving the
matrix of a stochastic process or designing beamfor
with improved signal-to-noise power ratios), Cholesk
an important role [10–12]. In linear algebra, Choles
putationally efficient solutions for systems of linear e
where matrix $\mathbf{A}$ is positive definite. As a final re
(3.81) can also be alternatively expressed as $\mathbf{A} = $
gular matrix. This alternative factorization can be

$$\mathbf{R} = \tilde{\mathbf{L}}^\dagger.$$

3.10 QR Decomposition

Let $\mathbf{A}$ be an $m \times n$ matrix of rank n. The QR decomposition consists in the factorization of matrix $\mathbf{A}$ into a product $\mathbf{QR}$, in which $\mathbf{Q}$ is an $m \times n$ matrix consisting of orthonormal column vectors and $\mathbf{R}$ is a non-singular upper triangular $n \times n$ matrix. Let $\mathbf{a}_1, \mathbf{a}_2, ..., \mathbf{a}_n$ be the columns of $\mathbf{A}$ such that the column space of $\mathbf{A}$ coincides with span $(\mathbf{a}_1, \mathbf{a}_2, ..., \mathbf{a}_n)$. Consider also the matrix:

$$\mathbf{Q} = [\mathbf{q}_1, \mathbf{q}_2, ..., \mathbf{q}_n], \tag{3.82}$$

where $\{\mathbf{q}_1, \mathbf{q}_2, ..., \mathbf{q}_n\}$ represents an orthonormal basis for the range space of $\mathbf{A}$. The vectors $\{\mathbf{q}_1, \mathbf{q}_2, ..., \mathbf{q}_n\}$ can be obtained from the column vectors $\{\mathbf{a}_1, \mathbf{a}_2, ..., \mathbf{a}_n\}$ of $\mathbf{A}$ by applying the Gram-Schmidt orthogonalization procedure [8]. If $\mathbf{p}_i$ is the projection vector of $\mathbf{a}_{i+1}$ onto the span $(\mathbf{q}_1, \mathbf{q}_2, ..., \mathbf{q}_i)$, $i = 1, ..., n-1$, then we can define the upper triangular matrix $\mathbf{R}$ as follows

$$\mathbf{R} = \begin{bmatrix} r_{11} & r_{12} & \cdots & r_{1n} \\ 0 & r_{22} & \cdots & r_{2n} \\ \vdots & \cdots & \ddots & \vdots \\ 0 & \cdots & 0 & r_{nn} \end{bmatrix}, \tag{3.83}$$

where

$$r_{11} = \|\mathbf{a}_1\|, \tag{3.84}$$
$$r_{kk} = \|\mathbf{a}_k - \mathbf{p}_{k-1}\|, \qquad k = 1, ..., n, \tag{3.85}$$

and

$$r_{ik} = \mathbf{q}_i^T \mathbf{a}_k, \qquad i = 1, ..., k-1, \; k = 2, ..., n. \tag{3.86}$$

It can be easily shown that the ith column of the product $\mathbf{QR}$ will be

$$\mathbf{Qr}_i = r_{1i}\mathbf{q}_1 + r_{2i}\mathbf{q}_2 + \cdots + r_{ii}\mathbf{q}_i = \mathbf{a}_i, \qquad i = 1, ..., n. \tag{3.87}$$

Thus, the desired QR factorization is obtained:

$$\mathbf{QR} = \mathbf{A}. \tag{3.88}$$

QR factorization is sometimes referred to as the Gram-Schmidt factorization.

3.11 Householder and Givens Transformations

A Householder transform represents a linear transformation that reflects any vector into its mirror image located on the other side of a plane that contains the origin. The Householder transformation assumes the matrix representation:

$$\mathbf{H_v} = \mathbf{I} - 2\frac{\mathbf{vv}^T}{\|\mathbf{v}\|_2^2}, \tag{3.89}$$

where $\mathbf{v}$ represents a fixed vector orthogonal to the pl
will be performed. Thus, an arbitrary point (vect
Householder transform into the mirror image $\mathbf{H_v}(\mathbf{x}$

$$\mathbf{H_v}(\mathbf{x}) = \mathbf{H_v}\mathbf{x} = \mathbf{x} - 2\frac{\langle \mathbf{x}|\mathbf{v}}{\langle \mathbf{v}|\mathbf{v}}$$

In general, $\mathbf{v}$ is *normalized* to unit norm, and theref

$$\mathbf{H_u} = \mathbf{I} - 2\mathbf{u}\mathbf{u}^T,$$

where $\mathbf{u} = \mathbf{v}/\|\mathbf{v}\|_2$. A matrix of this form is called an
and clearly it is both symmetric and orthogonal, si
$2\mathbf{u}\mathbf{u}^T) = \mathbf{I}$, and $\mathbf{H_u} = \mathbf{H_u}^T = \mathbf{H_u}^{-1}$. Also, a Housel
(involution) transformation: $\mathbf{H}^2 = \mathbf{I}$.

Now suppose that we want to solve the linear s
This system of equations can be easily solved using t
$\mathbf{A}$. One of the most important applications of Hou
that it can be used to obtain the desired QR factor
orthogonal transformations as opposed to the Gram-
proach. This is achieved by employing a set of ortho
that are properly designed to introduce zeros on a
the entries located under the main diagonal of $\mathbf{A}$. T
column and ends when the last column is reached.
holder transform can null the entries of a vector, su
we want to find a Householder matrix $\mathbf{H_u}$ such that

$$\mathbf{H_u}(\mathbf{x}) = (a, 0, 0, \ldots, 0)^T =$$

where $\mathbf{e}_1$ represents the first column of identity mat
of $\mathbf{H}$, it follows that $\|\mathbf{H_u}\mathbf{x}\|_2 = \|\mathbf{x}\|_2$, which leads to

$$\|\mathbf{x}\|_2 = \|a\mathbf{e}_1\|_2 = |a| .$$

Therefore, we select:

$$a = \|\mathbf{x}\|_2 .$$

On the other hand, because $\mathbf{H_u}$ is involutory:

$$\mathbf{H_u}(\mathbf{H_u}(\mathbf{x})) = \mathbf{x} = a(\mathbf{e}_1 - 2$$

where $\mathbf{u}_1 = \mathbf{u}^T\mathbf{e}_1$. Therefore, equating all the entrie

$$x_1 = a(1 - 2u_1^2),$$
$$x_2 = -2au_1u_2,$$
$$\vdots$$
$$x_n = -2au_1u_n,$$

where x_i's and u_i's stand for the ith entries of the vectors $\mathbf{x}$ and $\mathbf{u}$, respectively. By solving the system of equations (3.93) in terms of unknowns u_i's, it follows that

$$\mathbf{u} = -\frac{1}{2au_1}[-2aku_1^2, x_2, \ldots, x_n]^T. \tag{3.94}$$

Thus, the Householder matrix ($\mathbf{H_u}$) defined as:

$$\mathbf{H_u} = \mathbf{I} - 2\mathbf{u}\mathbf{u}^T,$$

exhibits the desired property:

$$\mathbf{H_u}(\mathbf{x}) = a\mathbf{e}_1.$$

Notice also that since the Householder matrix is determined just by the vector $\mathbf{u}$, we just need to store this vector rather than all the n^2 elements of whole matrix $\mathbf{H_u}$. In general, one can introduce zeros only in the last $n - i$ entries of a vector $\mathbf{x}$ by employing an appropriate Householder transform. To check how this can be done, first we split the vector $\mathbf{x}$ into two subvectors:

$$\mathbf{x}_1 = [x_1, x_2, \ldots, x_{i-1}]^T, \quad \mathbf{x}_2 = [x_i, x_{i+1}, \ldots, x_n]^T. \tag{3.95}$$

Let $\mathbf{H}_{\mathbf{u},i}$ be the Householder matrix that satisfies:

$$\mathbf{H}_{\mathbf{u},i}(\mathbf{x}_2) = \|\mathbf{x}_2\|_2 \, \mathbf{e}_1. \tag{3.96}$$

Notice now that the transform $\mathbf{H_u}$ defined below via:

$$\mathbf{H_u} = \begin{bmatrix} \mathbf{I}_{(i-1)\times(i-1)} & 0 \\ 0 & \mathbf{H}_{\mathbf{u},i} \end{bmatrix}, \tag{3.97}$$

is a Householder (orthogonal) transform and satisfies the expected requirement:

$$\mathbf{H_u}\mathbf{x} = [x_1, x_2, \ldots, x_{i-1}, \sqrt{\sum_{k=i}^{n} x_k^2}, 0, \ldots 0]^T. \tag{3.98}$$

Thus, by left-multiplying an arbitrary matrix with orthogonal Householder transforms we can introduce zeros on a column-by-column basis in all the matrix entries that are located under the main diagonal. Therefore, by using orthogonal Householder matrices, the original matrix can be transformed into an upper triangular matrix, and the desired QR factorization is obtained. The procedure of zeroing out the entries of a column vector via Householder transforms will be referred to as Householder reduction. Notice also that beside Householder transforms, one can exploit plane rotation matrices as orthogonal transformations to introduce zeros into a vector or matrix. The plane rotation matrices will be introduced in the next section.

3.11.1 Orthogonal Reduction

The plane rotation matrices, also referred to as Give orthogonal and anti-symmetric matrices of the form

$$
\mathbf{P}_{ij} = \begin{array}{c} 1 \\ \vdots \\ i^{th} \rightarrow \\ \vdots \\ j^{th} \rightarrow \\ \vdots \\ n \end{array}
\left(
\begin{array}{cccccc}
1 & & & & & \\
 & \ddots & & & & \\
 & & 1 & & & \\
 & & & a & & \\
 & & & & 1 & \\
 & & & & & \ddots \\
 & & & -b & & & 1
\end{array}
\right)
\overset{i^{th}}{}
$$

where $a^2 + b^2 = 1$. Matrix $\mathbf{P}_{ij}$ is called a plane r $a = \cos(\theta)$ and $b = \sin(\theta)$, then $\mathbf{P}_{ij}\mathbf{x}$ represents a angle θ in plane (i, j).

Similar to a Householder transform, a Givens r zero out entries of a vector. Notice that by applying a to a non-zero vector $\mathbf{x} \in \mathbb{R}^n$, only the ith and j^{th} en

$$
\mathbf{P}_{ij}\mathbf{x} = \begin{array}{c} 1 \\ \vdots \\ i \rightarrow \\ \vdots \\ j \rightarrow \\ \vdots \\ n \end{array}
\left(
\begin{array}{c}
x_1 \\
\vdots \\
ax_i + bx_j \\
\vdots \\
-bx_i + ax_{} \\
\vdots \\
x_n
\end{array}
\right)
$$

By carefully choosing the values of a and b, we can se of $\mathbf{x}$, in particular, the jth entry of $\mathbf{x}$ can be nulle entries of $\mathbf{x}$ except x_i . To this end, notice that the

$$
a = \frac{x_i}{\sqrt{x_i^2 + x_j^2}},
$$

$$
b = \frac{x_j}{\sqrt{x_i^2 + x_j^2}},
$$

enforce the jth entry of vector $\mathbf{P}_{ij}\mathbf{x}$ to be equal to zero. Therefore, by applying several Givens rotations $\mathbf{P}_{ij}$'s on a vector $\mathbf{x}$, we can selectively zero out entries of $\mathbf{x}$. Therefore, Givens transformations represent an alternative approach with respect to Householder reflectors for zeroing entries of a vector, and for performing the QR factorization of matrix.

3.12 Best Approximations and Orthogonal Projections

The following result plays an important role in many signal processing applications involving estimation and optimization tasks such as solving an overdetermined system of equations in the least-squares (LS) sense [10, 11].

Theorem 3.12.1. *(Orthogonal Projection Theorem [6]) Let $\mathcal{V}$ be an inner-product vector space and let $\mathcal{S}$ be a subspace of it. For every vector $\mathbf{v} \in \mathcal{V}$ there may exist a best approximation $\mathbf{s} \in \mathcal{S}$ such that*

$$\|\mathbf{v} - \mathbf{s}\| \leq \|\mathbf{v} - \mathbf{s}'\| , \quad \forall \mathbf{s}' \in \mathcal{S}. \tag{3.100}$$

This condition holds if and only if

$$(\mathbf{v} - \mathbf{s}) \perp \mathbf{s}', \quad \forall \mathbf{s}' \in \mathcal{S}. \tag{3.101}$$

Moreover, if this best approximation exists, then it is unique. Now let an orthogonal basis for $\mathcal{S}$ be denoted by $\{\mathbf{s}_1, \mathbf{s}_2, \ldots, \mathbf{s}_m\}$. It can be shown that the best approximation of $\mathbf{v}$ by vectors in $\mathcal{S}$ can be computed as

$$\mathbf{s} = \sum_{k=1}^{m} \frac{\langle \mathbf{v} | \mathbf{s}_k \rangle}{\|\mathbf{s}_k\|^2} \mathbf{s}_k. \tag{3.102}$$

In other words, the best approximation of $\mathbf{v}$ by vectors in $\mathcal{S}$ is given by the *orthogonal projection* $\mathbf{s}$ of vector $\mathbf{v}$ onto the subspace $\mathcal{S}$. Notice further that in (3.102), the kth term in the sum is the projection of $\mathbf{v}$ on the direction of vector $\mathbf{s}_k$. Consider now that for any vector $\mathbf{v} \in \mathcal{V}$, there exists an orthogonal projection onto subspace $\mathcal{S}$. Thus, a mapping $P : \mathcal{V} \to \mathcal{S}$ can be defined that associates to each $\mathbf{v} \in \mathcal{V}$, its orthogonal projection onto $\mathcal{S}$. Mapping P will be referred to as the orthogonal projection of $\mathcal{V}$ onto subspace $\mathcal{S}$, and some of its properties will be considered in Exercise 3.23.18.

3.13 Least Squares Approximations

Suppose $T : \mathcal{V} \to \mathcal{W}$ is a linear transformation such that for some $\mathbf{v} \in \mathcal{V}$ and $\mathbf{w} \in \mathcal{W}$, the equation

$$T(\mathbf{v}) = \mathbf{w} \tag{3.103}$$

does not present an exact solution; i.e., $\mathbf{w}$ does
T. In practice, such equations are often encounte
(3.103) in an approximative manner must be determi
approaches adopted for solving (3.103) is to find a v
the least-squares (LS) criterion:

$$e = \|T(\mathbf{v}) - \mathbf{w}\|^2,$$

is minimized. This problem reduces to finding the v

$$T(\mathbf{v}) = \mathbf{w}',$$

and $\|\mathbf{w}' - \mathbf{w}\|$ is minimized. The corresponding solu
least squares solution. The following result characte
(see e.g., [6, 8]).

Theorem 3.13.1. *Vector* $\mathbf{v} \in \mathcal{V}$ *is the minimizer o*

$$T^*T(\mathbf{v}) = T^*(\mathbf{w}).$$

If T^*T *is invertible, then the LS solution is unique*

$$\mathbf{v} = (T^*T)^{-1}T^*(\mathbf{w}),$$

and

$$\mathbf{w}' = T(\mathbf{v}) = T(T^*T)^{-1}T^*$$

A justification of this result is delegated to Exercise
Now consider the inconsistent overdetermined m
equations $\mathbf{A}\mathbf{v} = \mathbf{w}$. If $\mathbf{w}$ does not lie in the colur
exact solution. Using the LS method, a unique vec
to $\mathbf{w}$ can be found; or equivalently we can choose $\hat{\mathbf{v}}$
error $\|\mathbf{A}\hat{\mathbf{v}} - \mathbf{w}\|$. Finding the least squares solution

$$\hat{\mathbf{w}} = \mathbf{A}\hat{\mathbf{v}},$$

that is closest to $\mathbf{w}$ than any other vector in the colu
to the new system of equations:

$$\mathbf{A}\mathbf{v} = \hat{\mathbf{w}}$$

is referred to as a least-squares solution for $\mathbf{A}\mathbf{v} = \mathbf{w}$
squares solution to $\mathbf{A}\mathbf{v} = \mathbf{w}$ is any vector $\mathbf{v}$ that min
ing result characterizes the LS solution of minimun
system of linear equations [3, 6, 8].

Theorem 3.13.2. *If* $\mathbf{A}$ *is an* $m \times n$ *matrix* $(m \geq n$
among all the least squares solutions of the system o
a unique LS solution of minimum norm given by:

$$\mathbf{v} = (\mathbf{A}^*\mathbf{A})^{-1}\mathbf{A}^*\mathbf{w}.$$

The matrix $(\mathbf{A}^*\mathbf{A})^{-1}\mathbf{A}^*$ is called the Moore-Penrose pseudoinverse of $\mathbf{A}$. The approximation $\hat{\mathbf{w}}$ of $\mathbf{w}$ by the vectors in $\mathcal{R}(\mathbf{A})$ is given by:

$$\hat{\mathbf{w}} = \mathbf{A}\mathbf{v} = \mathbf{A}(\mathbf{A}^*\mathbf{A})^{-1}\mathbf{A}^*\mathbf{w}, \tag{3.111}$$

and it is obtained by projecting $\mathbf{w}$ onto $\mathcal{R}(\mathbf{A})$. As a corollary, the error term $\mathbf{w} - \hat{\mathbf{w}}$ is orthogonal to $\mathcal{R}(\mathbf{A})$. Reference [3] presents several extensions of these results and a characterization of all the LS solutions.

3.14 Angles Between Subspaces

3.14.1 Principal Angles Between Subspaces

Recall that given two non-zero vectors $\mathbf{v}, \mathbf{w} \in \mathbb{R}^n$, the angle between them is computed as

$$\theta = \cos^{-1} \frac{\langle \mathbf{v}|\mathbf{w}\rangle}{\|\mathbf{v}\| \cdot \|\mathbf{w}\|} \; , \; (0 \leq \theta \leq \pi/2). \tag{3.112}$$

Now suppose that we are given two subspaces represented in terms of the range spaces of two matrices $\mathbf{A}$ and $\mathbf{B}$. A set of angles between them can be defined recursively. These angles will be referred to as *principal (canonical) angles*. Matrices $\mathbf{A}$ and $\mathbf{B}$ are assumed real-valued with the same number n of rows and $\dim \mathcal{R}(\mathbf{A}) = p$ and $\dim \mathcal{R}(\mathbf{B}) = q$ ($q \leq p$). The principal angles $\theta_1, \theta_2, \ldots, \theta_q \in [0, \pi/2]$ between the two column spaces $\mathcal{R}(\mathbf{A})$ and $\mathcal{R}(\mathbf{B})$ are recursively defined via:

$$\cos(\theta_k) = \max_{\mathbf{v} \in \mathcal{R}(\mathbf{A})} \max_{\mathbf{w} \in \mathcal{R}(\mathbf{B})} \mathbf{v}^T\mathbf{w}$$
$$= \mathbf{v}_k^T \mathbf{w}_k \; , \quad k = 1, \ldots, q,$$

subject to these conditions: $\|\mathbf{v}\| = \|\mathbf{w}\| = 1$, $\mathbf{v}^T\mathbf{v}_i = 0$, $\mathbf{w}^T\mathbf{w}_i = 0$, $i = 1, \ldots, k-1$. Notice that the notation $\|.\|$ stands for the standard Euclidean norm of a vector, and the resulting vectors $\mathbf{v}_i$'s and $\mathbf{w}_i$'s are called the *principal vectors*. Notice also that during step k, the algorithm determines θ_k, and the associated principal vectors $\mathbf{v}_k$ and $\mathbf{w}_k$ by searching for vectors in subspaces that are orthogonal with respect to $\mathbf{v}_i$ and $\mathbf{w}_i$, $i = 1, \ldots, k-1$, respectively. Therefore, by searching for vectors that present the largest angle between them and that are orthogonal to the already found principal vectors, the complete sets of principal angles and principal vectors are obtained.

The concept of principal angles between two vector spaces has found applications in assessing the degree of correlation between two sets of random variables and has been used in canonical analysis [13, 14]. Applications of principal angles also include random processes [15, 16], stochastic realization of random processes and system identification [17–19], pattern recognition and machine learning (see e.g., [20, 21]), and signal processing and communications (see e.g., the signal estimation and detection applications presented in [22]).

3.15   Eigenvalues and Eigenvec

Let T be a linear operator on a vector space $\mathcal{V}$, $T :$
a scalar λ such that

$$T(\mathbf{v}) = \lambda \mathbf{v} \, ,$$

for a non-zero vector $\mathbf{v} \in \mathcal{V}$. The vector $\mathbf{v}$ is called a
for an $n \times n$ matrix $\mathbf{A}$ that is the matrix representa
tion T, an eigenvalue is a scalar λ that makes the
eigenvector of matrix $\mathbf{A}$ corresponding to eigenvalu
that satisfies the equation $\mathbf{A}\mathbf{v} = \lambda \mathbf{v}$. The polynomi

$$\chi_A(\lambda) = \det(\mathbf{A} - \lambda \mathbf{I}) \, ,$$

is called the *characteristic polynomial* of $\mathbf{A}$ and the

$$\det(\mathbf{A} - \lambda \mathbf{I}) = 0 \, ,$$

is referred to as the *characteristic equation* of $\mathbf{A}$. Th
istic equation provides the set of eigenvalues, and the
the *spectrum* of $\mathbf{A}$ and denoted by the notation by
λ is a root of the characteristic polynomial. The ei
eigenvalue λ of $\mathbf{A}$ is defined as the subspace

$$E_\lambda = \{ \mathbf{v} \in \mathcal{V} | \mathbf{A}\mathbf{v} = \lambda \mathbf{v} \}$$

If $\mathbf{A}$ has distinct eigenvalues, the eigenspaces are all o
general results, whose proofs are deferred to Exercis

Theorem 3.15.1. *Let $\mathbf{A}$ be an $n \times n$ matrix with a
eigenvectors of $\mathbf{A}$ form a linearly independent set.*

Theorem 3.15.2. *The eigenvectors of a Hermitian
vectors corresponding to distinct eigenvalues are ort*

The concepts of eigenvalue and eigenvector play
high-resolution spectral estimation techniques, desi
localization and tracking in antenna array process
digital filters, system identification, designing efficie
antenna (multi–input multi–output (MIMO)) based
etc.

3.15.1   Diagonalization

A square matrix $\mathbf{A}$ is *diagonalizable* if it can be fact

$$\mathbf{A} = \mathbf{S} \boldsymbol{\Lambda} \mathbf{S}^{-1} \, ,$$

for some non-singular matrix $\mathbf{S}$ and diagonal matrix $\boldsymbol{\Lambda}$. Notice that $\boldsymbol{\Lambda}$ presents the same eigenvalues as $\mathbf{A}$. In general, two matrices $\mathbf{A}$ and $\boldsymbol{\Lambda}$ that are related by means of an equation of the form (3.117) are referred to as *similar matrices*, and they present the same set of eigenvalues. The following result provides an alternative characterization for diagonalizable matrices.

Theorem 3.15.3. *If $\mathbf{A}$ is an $n \times n$ matrix, then it is diagonalizable if it admits n linearly independent eigenvectors or n distinct eigenvalues.*

A justification of this result is left to Exercise 3.23.23.

Using the characteristic function, it can be checked that the eigenvalues of a diagonal matrix coincides with its diagonal entries. Also, it is not difficult to show that the ith column of $\mathbf{S}$ is actually the eigenvector of $\mathbf{A}$ corresponding to the ith diagonal entry (eigenvalue) of $\boldsymbol{\Lambda}$ ($\mathbf{A}$), for any $i = 1, \ldots, n$. The following result shows a simple way to calculate the higher order powers of a matrix in terms of the higher-order powers of its eigenvalues.

Theorem 3.15.4. *Let $\mathbf{A}$ be an $n \times n$ matrix with eigenvalues $\lambda_1, \ldots, \lambda_n$. Then the eigenvalues of $\mathbf{A}^k$ ($k \in \mathbb{Z}$) are $\lambda_1^k, \ldots, \lambda_n^k$. Moreover, it follows that*

$$\mathbf{A}^k = (\mathbf{S}\boldsymbol{\Lambda}\mathbf{S}^{-1})^k$$
$$= \mathbf{S}\boldsymbol{\Lambda}^k\mathbf{S}^{-1}. \tag{3.118}$$

3.16 Schur Factorization and Spectral Theorem

Theorem 3.16.1. *Let $\mathbf{A}$ be an $n \times n$ matrix. There exists a unitary matrix $\mathbf{Q}$ such that the multiplication*

$$\mathbf{T} = \mathbf{Q}^{\dagger}\mathbf{A}\mathbf{Q} \tag{3.119}$$

is upper triangular, or equivalently $\mathbf{A}$ can be factorized as

$$\mathbf{A} = \mathbf{Q}\mathbf{T}\mathbf{Q}^{\dagger}, \tag{3.120}$$

where $\mathbf{T}$ is an upper triangular matrix.

The decomposition (3.120) is referred to as Schur's factorization. Notice that the diagonal elements of $\mathbf{T}$ consist of the eigenvalues of $\mathbf{A}$. The proof of this theorem can be done using the induction principle with respect to n. Alternative justifications for Schur factorization can be found in [3, 5, 24].

Theorem 3.16.2. *(Spectral Theorem) For any Hermitian matrix $\mathbf{A}$, there is a unitary diagonalizing matrix $\mathbf{Q}$.*

Stated differently, if $\mathbf{A}$ is a Hermitian matrix, the matrix $\mathbf{T}$ in Schur's factorization must be diagonal. Assume there exists a unitary matrix $\mathbf{Q}$ such that (3.119) holds, where $\mathbf{T}$ is an upper triangular matrix. However, $\mathbf{A}^{\dagger} = \mathbf{A}$, because $\mathbf{A}$ is Hermitian

symmetric. Hence, $\mathbf{T}^\dagger = \mathbf{Q}^\dagger \mathbf{A}^\dagger \mathbf{Q} = \mathbf{Q}^\dagger \mathbf{A} \mathbf{Q} = \mathbf{T}$
symmetric. Being also an upper triangular matri
an immediate corollary, if $\mathbf{A}$ is a real-valued symm
diagonalized via an orthogonal matrix $\mathbf{Q}$, i.e., the
Since $\mathbf{Q}$ is diagonalizing $\mathbf{A}$ to $\mathbf{T}$, the diagonal eleme
$\mathbf{A}$, and the columns of $\mathbf{Q}$ consist of eigenvectors of A
the spectral decomposition of a symmetric matrix.

3.17 Singular Value Decompos

Let $\mathbf{A}$ be an $m \times n$ real-valued matrix. Then the
(SVD) of matrix $\mathbf{A}$ is given by the factorization [24

$$\mathbf{A} = \mathbf{U}\boldsymbol{\Sigma}\mathbf{V}^T,$$

where $\boldsymbol{\Sigma}$ is an $m \times n$ diagonal matrix, and $\mathbf{U}$ and $\mathbf{V}$
of dimensions $m \times m$ and $n \times n$, respectively. In ge

$$\boldsymbol{\Sigma} = \begin{bmatrix} \boldsymbol{\Sigma}' & \mathbf{0}_{r\times(n-r)} \\ \mathbf{0}_{(m-r)\times r} & \mathbf{0}_{(m-r)\times(n-} \end{bmatrix}$$

where

$$\boldsymbol{\Sigma}' = \begin{bmatrix} \sigma_1 & 0 & \cdots & 0 \\ 0 & \sigma_2 & \ddots & \vdots \\ \vdots & \ddots & \ddots & 0 \\ 0 & \cdots & 0 & \sigma_r \end{bmatrix}_{r\times r} \qquad \sigma_i > 0$$

and r stands for the rank of matrix $\mathbf{A}$. The non-zer
of $\boldsymbol{\Sigma}$, σ_i, $i = 1, \ldots r$, are non-negative and are refe
of $\mathbf{A}$. Notice that using (3.121), it follows that $\mathbf{A}A$
$\mathbf{V}\boldsymbol{\Sigma}^T\boldsymbol{\Sigma}\mathbf{V}^T$. Because matrices $\mathbf{U}$ and $\mathbf{V}$ are orthog
diagonal, it follows that the singular values of matri
roots of non-zero eigenvalues of $\mathbf{A}^\dagger\mathbf{A}$ and $\mathbf{A}\mathbf{A}^\dagger$. Fu
and $\mathbf{V}$ are the eigenvectors of $\mathbf{A}\mathbf{A}^T$ and $\mathbf{A}^T\mathbf{A}$, respec
complex-valued matrices. If $\mathbf{A}$ is a complex-valued
the factorization: $\mathbf{A} = \mathbf{U}\boldsymbol{\Sigma}\mathbf{V}^\dagger$, where $\mathbf{U}$ and $\mathbf{V}$ a
a diagonal matrix with nonnegative entries. Althou
$\sigma_1, \ldots, \sigma_r$ are unique, $\mathbf{U}$ and $\mathbf{V}$ are not.

The SVD (3.121) can be re-expressed in these tw

$$\mathbf{A}\mathbf{V} = \mathbf{U}\boldsymbol{\Sigma},$$

and

$$\mathbf{A}^T\mathbf{U} = \mathbf{V}\boldsymbol{\Sigma}^T.$$

Equating the ith columns in the left-hand-side (LHS) and right-hand side (RHS) of (3.124) and (3.125), it follows that:

$$\mathbf{A}\mathbf{v}_i = \sigma_i \mathbf{u}_i \qquad i = 1, 2, \ldots r, \qquad (3.126)$$

and

$$\mathbf{A}^T \mathbf{u}_i = \begin{cases} \sigma_i \mathbf{v}_i & 1 \leq i \leq r \\ \mathbf{0} & r+1 \leq i \leq m \end{cases}, \qquad (3.127)$$

where vectors $\mathbf{u}_i$ and $\mathbf{v}_i$ stand for the ith column of matrices $\mathbf{U}$ and $\mathbf{V}$, respectively. Vectors $\mathbf{v}_i$'s and $\mathbf{u}_i$'s are called right singular vectors and left singular vectors of $\mathbf{A}$, respectively. It follows also that $\{\mathbf{v}_1, \mathbf{v}_2, \ldots, \mathbf{v}_r\}$ and $\{\mathbf{u}_1, \mathbf{u}_2, \ldots, \mathbf{u}_r\}$ represent orthogonal bases for $\mathcal{R}(A^T)$ and $\mathcal{R}(A)$, respectively. Likewise, $\{\mathbf{v}_{r+1}, \mathbf{v}_{r+2}, \ldots, \mathbf{v}_n\}$ and $\{\mathbf{u}_{r+1}, \mathbf{u}_{r+2}, \ldots, \mathbf{u}_r\}$ are orthogonal bases for $\mathcal{N}\text{ull}(A)$ and $\mathcal{N}\text{ull}(A^T)$, respectively. Thus, the SVD provides bases for all the four fundamental subspaces of a matrix: $\mathcal{R}(A)$, $\mathcal{N}\text{ull}(A)$, $\mathcal{R}(A)^\perp$, and $\mathcal{N}\text{ull}(A)^\perp$. Notice also that the rank r of $\mathbf{A}$ is obtained immediately from its SVD, and it is equal to the number of non-zero entries located on the main diagonal of $\boldsymbol{\Sigma}$. SVD enables also the construction of the orthogonal projections on the four fundamental subspaces of a matrix. E.g., if matrices $\mathbf{U}_{1:r}$ and $\mathbf{V}_{1:r}$ stand for the sub-matrices formed by the columns 1 through r of matrices $\mathbf{U}$ and $\mathbf{V}$, respectively, then $\mathbf{U}_{1:r}\mathbf{U}_{1:r}^T$ and $\mathbf{V}_{1:r}\mathbf{V}_{1:r}^T$ represent the orthogonal projection matrices on $\mathcal{R}(A)$ and $\mathcal{N}\text{ull}(A)^\perp$, respectively [3, 24].

Reference [24] presents a number of applications of SVD in calculating row-rank matrix approximations, principal angles between subspaces, generalized inverses, solving systems of equations in the standard or total least-squares sense, orthogonal projections, etc. During the past decades, SVD proved to be a powerful tool in solving numerous problems in statistics, signal processing, wireless communications, information theory, pattern recognition, machine learning and other fields. Also, SVD played an important role in enabling the calculation of the capacity of a MIMO channel by reducing the MIMO channel to a number of independent single–input single–output (SISO) channels [25], developing efficient high-resolution spectral estimation techniques [12], and designing optimal channel estimation, signaling and equalization techniques.

3.18 Rayleigh Quotient

Suppose $\mathbf{A}$ is an $n \times n$ Hermitian matrix, and denote by $\lambda_1 \geq \lambda_2 \geq \cdots \geq \lambda_n$ and $\mathbf{u}_1, \mathbf{u}_2, \ldots, \mathbf{u}_n$ its eigenvalues and corresponding orthogonal eigenvectors, respectively. Notice also that the Spectral Theorem 3.16.2 guarantees the existence of a factorization of the form $\mathbf{A} = \mathbf{U}\boldsymbol{\Lambda}\mathbf{U}^H$, with $\mathbf{U}$ unitary and $\boldsymbol{\Lambda}$ diagonal matrix containing on its main diagonal the eigenvalues of $\mathbf{A}$. One can observe that orthonormal vectors $\mathbf{u}_1, \mathbf{u}_2, \ldots, \mathbf{u}_n$, representing the columns of $\mathbf{U}$, stand also for the eigenvectors of $\mathbf{A}$ corresponding to eigenvalues $\lambda_1, \ldots, \lambda_n$. If $\mathbf{x}$ is a non-zero

vector in $\mathbb{C}^n$, then the Rayleigh quotient of $\mathbf{x}$ with
is defined as:

$$R(\mathbf{x}) = \frac{\langle \mathbf{A}\mathbf{x}|\mathbf{x}\rangle}{\langle \mathbf{x}|\mathbf{x}\rangle} = \frac{\mathbf{x}^\dagger \mathbf{A}}{\mathbf{x}^\dagger \mathbf{x}}$$

It can be shown that the minimum and the max
quotient are λ_n and λ_1, respectively, i.e.,

$$\lambda_n \leq R(\mathbf{x}) \leq \lambda_1 .$$

A quick justification of (3.129) can be obtained by
$\mathbf{x} \in \mathbb{C}^n$ as a linear combination of the orthogonal eig
(or in terms of the singular vectors of $\mathbf{A}$ obtained v

$$\mathbf{x} = \sum_{i=1}^{n} a_i \mathbf{u}_i .$$

Plugging (3.130) into (3.128), it follows that:

$$R(\mathbf{x}) = \frac{\sum_{i=1}^{n} \lambda_i |a_i|^2}{\sum_{i=1}^{n} |a_i|^2}$$

$$= \frac{\sum_{i=1}^{n} \lambda_i |a_i|^2}{\|\mathbf{a}\|_2^2}$$

where $\mathbf{a} = [a_1, a_2, \ldots, a_n]^T$. Because of the ordering
that the numerator in (3.131) can be lower- and up

$$\lambda_n \|\mathbf{a}\|_2^2 \leq \sum_{i=1}^{n} \lambda_i |a_i|^2 \leq \lambda_1$$

from where (3.129) follows immediately. Rayleigh c
signal processing and wireless communication applic
of filters and equalization schemes with improved p

3.19 Application of SVD and
Principal Component An

Suppose we have an $m \times n$ ($m \geq n$) matrix $\mathbf{X}$ of ran
are possibly correlated (i.e., are not orthogonal). T
ysis (PCA) is a method that transforms a set of co
column vectors of matrix $\mathbf{X}$) into a possibly smalle
$\{\mathbf{y}_1, \ldots \mathbf{y}_r\}$, called *principal components vectors*, th

norm (variance, power or variability) and span the same vector space as the column space of matrix $\mathbf{X}$. Herein application, the "uncorrelatedness" condition is achieved by imposing that the vectors $\mathbf{y}_i$'s, $i = 1, 2, \ldots r$, are orthogonal.

Consider now the SVD (or spectral) decomposition of correlation matrix $\mathbf{C} = \mathbf{X}^T\mathbf{X}/(n-1)$ under the form $\mathbf{C} = \mathbf{U}\boldsymbol{\Lambda}\mathbf{U}^T$, where $\boldsymbol{\Lambda}$ is a diagonal matrix with the diagonal entries $\lambda_1 \geq \lambda_2 \geq \ldots \geq \lambda_r > 0$, and $\mathbf{U} = [\mathbf{u}_1, \mathbf{u}_2, \ldots, \mathbf{u}_n]$ is an orthogonal matrix. The facts that $\mathbf{y}_i$'s must lie in the range space of $\mathbf{X}$ and must be uncorrelated imply that at most r principal component vectors might be defined. Next we will determine the first principal component $\mathbf{y}_1$ as the vector of largest variance or norm that lies in the range space of matrix $\mathbf{X}$. This reduces to finding the vector $\mathbf{y}_1 = \mathbf{X}\mathbf{v}_1$, where $\mathbf{v}_1$ is an arbitrary vector of unit norm, such that

$$Var(\mathbf{y}_1) = \frac{\mathbf{y}_1^T\mathbf{y}_1}{n-1} = \frac{(\mathbf{X}\mathbf{v}_1)^T\mathbf{X}\mathbf{v}_1}{n-1} = \mathbf{v}_1^T\mathbf{C}\mathbf{v}_1, \tag{3.132}$$

is maximized. The Rayleigh quotient implies immediately that $\mathbf{v}_1$ should coincide with the eigenvector $(\mathbf{u}_1)$ corresponding to the largest eigenvalue (λ_1) of correlation matrix $\mathbf{C}$. Now, using the Induction Principle, we will construct the remaining principal directions. Assuming that the first $k-1$ principal component vectors $\mathbf{y}_1, \ldots, \mathbf{y}_{k-1}$ are given by the eigenvectors $\mathbf{u}_1, \ldots, \mathbf{u}_{k-1}$ corresponding to the first $k-1$ largest eigenvalues, respectively, we want to prove that the kth principal component vector is given by the eigenvector $\mathbf{u}_k$ corresponding to the kth eigenvalue λ_k. Consider that $\mathbf{y}_k = \mathbf{X}\mathbf{v}_k$, where $\mathbf{v}_k$ is an arbitrary vector of unit norm. Because $\mathbf{y}_k$ should be orthogonal to $\mathbf{y}_1, \ldots, \mathbf{y}_{k-1}$, it follows that

$$\mathbf{y}_k^T\mathbf{y}_l = 0 \,, \forall l = 1, \ldots, k-1. \tag{3.133}$$

Equation (3.133) leads further to

$$(\mathbf{v}_k^T\mathbf{X}^T)(\mathbf{X}\mathbf{u}_l) = (n-1)\mathbf{v}_k^T\mathbf{C}\mathbf{u}_l = (n-1)\lambda_l\mathbf{v}_k^T\mathbf{u}_l = 0, \;\; \forall l = 1, \ldots, k-1 \tag{3.134}$$

Thus, it follows that necessarily $\mathbf{v}_k$ should lie in the range space of orthogonal eigenvectors $\mathbf{u}_k, \ldots, \mathbf{u}_n$. Consequently, there exist the scalars α_l, $l = k, \ldots, n$, such that:

$$\mathbf{v}_k = \sum_{l=k}^{n} \alpha_l\mathbf{u}_l \,, \tag{3.135}$$

and $\sum_{l=k}^{n} \alpha_l^2 = 1$ is imposed to enforce the unit norm for $\mathbf{v}_k$. Notice that the maximization of the kth principal component norm or variance is equivalent to the maximization of

$$\mathbf{y}_k^T\mathbf{y}_k = (\sum_{j=k}^{n} \alpha_j\mathbf{u}_j)^T\mathbf{X}^T\mathbf{X}(\sum_{l=k}^{n} \alpha_l\mathbf{u}_l) = \sum_{l=k}^{n} \alpha_l^2\lambda_l^2 \,, \tag{3.136}$$

with respect to scalars α_l, $l = k, \ldots, n$, and assuming the constraint $\sum_{l=k}^{n} \alpha_l^2 = 1$. Obviously, the maximum of (3.136) is achieved when $\alpha_l = 0$ for $l > k$, and $|\alpha_k| = 1$.

This concludes the proof. In other words, the eige
correlation matrix $\mathbf{C}$ yield the principal component
applications where the PCA method is employed, P(
such as Karhunen-Loeve transformation, SVD or F
details on PCA, the reference [25] presents a compr
its applications.

3.20 Special Matrices

In this section several special matrices such as bl
dermonde, stochastic, generalized inverse and posi
introduced.

3.20.1 Block Matrices

A matrix $\mathbf{A}$ is referred to as a *block matrix* if it c
sub-matrices $\mathbf{A}_{i,j}$ of appropriate sizes as follows:

$$\mathbf{A} = \begin{bmatrix} \mathbf{A}_{1,1} & \mathbf{A}_{1,2} & \cdots & \mathbf{A}_{1,r} \\ \mathbf{A}_{2,1} & \mathbf{A}_{2,2} & \cdots & \mathbf{A}_{2,r} \\ \vdots & \vdots & \ddots & \vdots \\ \mathbf{A}_{m,1} & \mathbf{A}_{m,2} & \cdots & \mathbf{A}_{m,} \end{bmatrix}$$

A block matrix with non-zero entries on its main diag
is called a *block diagonal matrix*, and assumes the e

$$\mathbf{A} = \begin{bmatrix} \mathbf{A}_1 & \cdots & 0 \\ \vdots & \ddots & \vdots \\ 0 & \cdots & \mathbf{A}_n \end{bmatrix}.$$

Some properties pertaining to the multiplication of
of appropriate sizes are listed below.

- If $\mathbf{A} = \begin{bmatrix} \mathbf{A}_1 \\ \mathbf{A}_2 \end{bmatrix}$, then $\mathbf{AB} = \begin{bmatrix} \mathbf{A}_1\mathbf{B} \\ \mathbf{A}_2\mathbf{B} \end{bmatrix}$.

- If $\mathbf{B} = \begin{bmatrix} \mathbf{B}_1 & \mathbf{B}_2 \end{bmatrix}$, then $\mathbf{A} \begin{bmatrix} \mathbf{B}_1 & \mathbf{B}_2 \end{bmatrix} = \begin{bmatrix} \mathbf{AB}_1 & \mathbf{AI} \end{bmatrix}$

- If $\mathbf{A} = \begin{bmatrix} \mathbf{A}_1 \\ \mathbf{A}_2 \end{bmatrix}$, $\mathbf{B} = \begin{bmatrix} \mathbf{B}_1 & \mathbf{B}_2 \end{bmatrix}$, then $\mathbf{AB} = \begin{bmatrix} \mathbf{A} \\ \mathbf{A} \end{bmatrix}$
.

- If $\mathbf{A} = \begin{bmatrix} \mathbf{A}_{1,1} & \mathbf{A}_{1,2} \\ \mathbf{A}_{2,1} & \mathbf{A}_{2,2} \end{bmatrix}$, $\mathbf{B} = \begin{bmatrix} \mathbf{B}_{1,1} & \mathbf{B}_{1,2} \\ \mathbf{B}_{2,1} & \mathbf{B}_{2,2} \end{bmatrix}$, then

$$\mathbf{AB} = \begin{bmatrix} \mathbf{A}_{1,1}\mathbf{B}_{1,1} + \mathbf{A}_{1,2}\mathbf{B}_{2,1} & \mathbf{A}_{1,1}\mathbf{B}_{1,2} + \mathbf{A}_{1,2}\mathbf{I} \\ \mathbf{A}_{2,1}\mathbf{B}_{1,1} + \mathbf{A}_{2,2}\mathbf{B}_{2,1} & \mathbf{A}_{2,1}\mathbf{B}_{1,2} + \mathbf{A}_{2,2}\mathbf{I} \end{bmatrix}$$

The proof of the last property is delegated to Exercise 3.23.25. Reference [7] presents additional properties of block matrices.

3.20.2 Circulant Matrices

An $n \times n$ matrix $\mathbf{A}$ is called *circulant* if each row of $\mathbf{A}$ is obtained by right-shifting the previous row by one column, and the element in the last column being shifted to the first column. In other words, a circulant matrix can be defined in terms of elements located on its first row, as depicted below:

$$\mathbf{A} = circ(a_1, ..., a_n) = \begin{bmatrix} a_1 & a_2 & ... & a_{n-1} & a_n \\ a_n & a_1 & ... & a_{n-2} & a_{n-1} \\ \vdots & \vdots & & \vdots & \vdots \\ a_2 & a_3 & ... & a_n & a_1 \end{bmatrix}. \tag{3.138}$$

The circularity condition for matrix $A = [a_{k,l}]$ can be described alternatively through the condition:

$$a_{k,l} = \begin{cases} a_{k-1,l-1} & k = 2, ..., n \,, \; l = 2, ..., n \\ a_{k-1,n} & k = 2, ..., n, \; l = 1 \end{cases}.$$

Consider now two circulant matrices $\mathbf{A}$ and $\mathbf{B}$ of the same size and arbitrary scalars α and β. Then the following statements hold:

- $\mathbf{A}^T$ is circulant.

- $\alpha \mathbf{A} + \beta \mathbf{B}$ is circulant.

- $\mathbf{A}^k$ is circulant, for any positive integer k.

- If $\mathbf{A}$ is nonsingular, then $\mathbf{A}^{-1}$ is circulant.

- $\mathbf{AB}$ is circulant.

The proofs of the last two properties are delegated to Exercise 3.23.27.

An important feature of circulant matrices is that they can be diagonalized using Discrete Fourier Transform (DFT). In fact, for an arbitrary $n \times n$ circulant matrix $\mathbf{A}$ the following spectral decomposition holds:

$$\mathbf{A} = \mathbf{F}_n^* \mathbf{\Lambda}_n \mathbf{F}_n, \tag{3.139}$$

where $\mathbf{F}_n$ is the DFT matrix with entries

$$[\mathbf{F}_n]_{k,l} = \frac{1}{\sqrt{n}} e^{-2\pi \sqrt{-1} kl/n}, \quad 0 \le k, l \le n - 1,$$

and $\mathbf{\Lambda}_n$ is a diagonal matrix with its diagonal entr.
matrix $\mathbf{A}$. The diagonal entries of $\mathbf{\Lambda}_n$, i.e., the eige.
can be calculated in $O(n \log n)$ flops (floating point
fast Fourier transform (FFT) of the first column of
denoting by $\mathbf{1}$ the vector of all ones and $\mathbf{e_1} = (1, 0, \ldots$
hold:

$$\mathbf{\Lambda}_n \mathbf{F}_n = \mathbf{F}_n \mathbf{A},$$
$$\mathbf{\Lambda}_n \mathbf{F}_n \mathbf{e}_1 = \mathbf{F}_n \mathbf{A} \mathbf{e}_1,$$
$$\mathbf{\Lambda}_n \mathbf{1} = \mathbf{F}_n \mathbf{A} \mathbf{e_1},$$

where we took into consideration in (3.140) the fact
k entries in the LHS and RHS of (3.140), it follows
Λ_n are given by

$$\lambda_k = \frac{1}{\sqrt{n}} \sum_{j=0}^{n-1} a_j e^{-2\pi ijk/n}, \quad k = 0, \ldots$$

Exercise 3.23.28 describes how the eigenvectors of
ficiently determined. Circular matrices are importa
cations because they arise in the matrix represen
between two discrete-time sequences (see e.g., [26]).
product of two DFTs is equal to the circular conv
whose DFTs are multiplied, circular matrices play a
ing the linear convolution between two sequences [
nice introduction to circular matrices. Additional
can be found in [7].

3.20.3   Toeplitz Matrices

An $n \times n$ matrix $\mathbf{A} = [a_{k,l}]$ whose entries satisfy
called a Toeplitz matrix. In other words, the entry
perfectly described by the difference between its rov
this fact, the following well-known characterization
in the literature: all matrix entries located on diag
diagonal are equal. As a corollary, all Toeplitz mat
by its first column and first row vectors. Notice al
a special class of Toeplitz matrices. Toeplitz matr
sentation of discrete-time linear convolutions betwe
correlation (auto-covariance) matrices of stationary
teresting and quite useful property of Toeplitz mat
systems of equations of the form $\mathbf{Ax} = \mathbf{b}$ can be so
using the Levinson-Durbin algorithm. This result
signing parametric autoregressive-moving average ($\mathbf{A}$

density estimators (solving the Yule-Walker system of equations) [12], calculation of Wiener (minimum mean-square error (MMSE)) equalizers and predictors. For additional information, reference [27] represents an excellent elementary introduction to Toeplitz and circulant matrices.

3.20.4   Hankel Matrices

A matrix $\mathbf{A} = [a_{k,l}]$ whose entries satisfy the condition: $a_{k,l} = a_{k+l}$ is referred to as a *Hankel* matrix. Hankel matrices enjoy the property that the entries located on anti-diagonals are equal. Therefore, the entries of a Hankel matrix are perfectly described by its first row and last column. Hankel matrices found applicability in the design of state-space based system identification and channel estimation algorithms [28]. Hankel matrices enjoy similar properties to Toeplitz matrices in terms of computing their inverses. This can be understood by noticing the result: if $\mathbf{J}$ stands for the *exchange* matrix, i.e., the matrix with all the entries on the main counter-diagonal equal to 1 and with zeros elsewhere, then $\mathbf{JA}$ is Toeplitz for a square Hankel matrix $\mathbf{A}$ [7].

3.20.5   Vandermonde Matrices

A *Vandermonde matrix* $\mathbf{A}$ admits the following structure:

$$\mathbf{A} = \begin{bmatrix} 1 & a_1 & a_1^2 & \dots & a_1^{m-1} \\ 1 & a_2 & a_2^2 & \dots & a_2^{m-1} \\ \vdots & \vdots & \vdots & & \vdots \\ 1 & a_m & a_m^2 & \dots & a_m^{m-1} \end{bmatrix} \tag{3.141}$$

$$= \begin{bmatrix} 1 & 1 & \dots & 1 \\ a_1 & a_2 & \dots & a_m \\ a_1^2 & a_2^2 & \dots & a_m^2 \\ \vdots & \vdots & \vdots & \vdots \\ a_1^{m-1} & a_2^{m-1} & \dots & a_m^{m-1} \end{bmatrix}^T \tag{3.142}$$

In other words, in each row in a Vandermonde matrix are terms of a geometric progression. A square Vandermonde matrix $\mathbf{A}$ enjoys the property that its determinant can be expressed in closed-form expression:

$$\det(\mathbf{A}) = \prod_{1 \leq i < j \leq m} (a_j - a_i). \tag{3.143}$$

Vandermonde matrices arise in applications involving evaluations of polynomials at a specific set of values, polynomial least-squares fitting and polynomial interpolation designs.

3.20.6   Normal Matrices

A complex square matrix $\mathbf{A}$ is said to be a *normal*

$$\mathbf{A}^\dagger \mathbf{A} = \mathbf{A}\mathbf{A}^\dagger ,$$

where we recall that superscript $\cdot^\dagger$ stands for the co
tian transposition) operation. For real-valued matr
the condition:

$$\mathbf{A}^T \mathbf{A} = \mathbf{A}\mathbf{A}^T .$$

One of the most important features of normal mat
unitarily diagonalizable, i.e., there exists a unitary
$\mathrm{diag}(\lambda_1, \ldots, \lambda_n)$ [5]. In particular, notice that a
symmetric) matrix is a normal matrix.

3.20.7   Stochastic Matrices

An $n \times n$ matrix $\mathbf{A} = [a_{k,l}]$ is called *stochastic (pro*
if all its entries are non-negative ($a_{k,l} \geq 0$) and th
to one ($\sum_{l=1}^{n} a_{k,l} = 1$.) A stochastic matrix can b
probability matrix in a Markov chain, and its use
probability, statistics, computer science, optimizati
A matrix with non-negative entries and with each
also referred to as a stochastic matrix. A doubly sto
non-negative entries, and in which each row and ea

3.20.8   Positive and Negative Definite

Assume $\mathbf{x}$ is an $n \times 1$ vector in $\mathbb{R}^n$ and $n \times n$ matrix
form $f(\mathbf{x}) = \mathbf{x}^T \mathbf{A}\mathbf{x}$ is said to be positive (negative)
for any non-zero vector $\mathbf{x}$. By definition, the real sy

- Positive Definite iff $\mathbf{x}^T \mathbf{A}\mathbf{x} > 0 , \quad \forall \mathbf{x} \in \mathbb{R}^n -$

- Positive Semi-Definite iff $\mathbf{x}^T \mathbf{A}\mathbf{x} \geq 0 , \quad \forall \mathbf{x} \in$

- Negative Definite iff $\mathbf{x}^T \mathbf{A}\mathbf{x} < 0 , \quad \forall \mathbf{x} \in \mathbb{R}^n -$

- Negative Semi-Definite iff $\mathbf{x}^T \mathbf{A}\mathbf{x} \leq 0 , \quad \forall \mathbf{x} \in$

The following result provides an alternative way to a
definite.

Theorem 3.20.1. *A real symmetric matrix $\mathbf{A}$ is po*
of its eigenvalues are positive.

The proof of this result is left to Exercise 3.23.30. An alternative method to check for the positive definiteness of a matrix is to verify that the determinants of all principal submatrices are all positive (see e.g., [8]). Additional information about the properties and applications of positive definite matrices can be found in the book [29].

3.20.9 Matrix Condition Number

Consider the system of equations $\mathbf{Ax} = \mathbf{b}$. Matrix $\mathbf{A}$ is said to be *ill-conditioned* if a relatively small change in $\mathbf{b}$ ($\Delta\mathbf{b}$) causes relatively large changes in the solution vector $\mathbf{x}$ ($\Delta\mathbf{x}$). In contrast, matrix $\mathbf{A}$ is called a *well-conditioned* matrix, if a relatively small change in $\mathbf{b}$ causes relatively small changes in solution $\mathbf{x}$. Assume $\mathbf{A}$ is a square nonsingular matrix and

$$\mathbf{Ax'} = \mathbf{A(x + \Delta x)} = \mathbf{b} + \Delta\mathbf{b}, \tag{3.144}$$

where $\mathbf{x'}$ represents the solution error caused by an error $\Delta\mathbf{b}$ in $\mathbf{b}$. Considering $\mathbf{Ax} = \mathbf{b}$, it follows that $\mathbf{A}\Delta\mathbf{x} = \Delta\mathbf{b}$, and hence:

$$\Delta\mathbf{x} = \mathbf{A}^{-1}\Delta\mathbf{b}. \tag{3.145}$$

Therefore,

$$\|\Delta\mathbf{x}\| \leq \|\mathbf{A}^{-1}\| \|\Delta\mathbf{b}\|. \tag{3.146}$$

On the other hand,

$$\|\Delta\mathbf{b}\| = \|\mathbf{A}\Delta\mathbf{x}\| \leq \|\mathbf{A}\| \|\Delta\mathbf{x}\|. \tag{3.147}$$

Therefore,

$$\frac{\|\Delta\mathbf{b}\|}{\|\mathbf{A}\|} \leq \|\Delta\mathbf{x}\| \leq \|\mathbf{A}^{-1}\| \|\Delta\mathbf{b}\|. \tag{3.148}$$

Considering $\mathbf{x} = \mathbf{A}^{-1}\mathbf{b}$, and following the same reasoning it is concluded that:

$$\frac{\|\mathbf{b}\|}{\|\mathbf{A}\|} \leq \|\mathbf{x}\| \leq \|\mathbf{A}^{-1}\| \|\mathbf{b}\|. \tag{3.149}$$

From (3.148) and (3.149), it follows that:

$$\frac{1}{\|\mathbf{A}\| \|\mathbf{A}^{-1}\|} \frac{\|\Delta\mathbf{b}\|}{\|\mathbf{b}\|} \leq \frac{\|\Delta\mathbf{x}\|}{\|\mathbf{x}\|} \leq \|\mathbf{A}\| \|\mathbf{A}^{-1}\| \frac{\|\Delta\mathbf{b}\|}{\|\mathbf{b}\|}. \tag{3.150}$$

The number $\|\mathbf{A}\| \|\mathbf{A}^{-1}\|$ in (3.150) is referred to as the *condition number* of matrix $\mathbf{A}$ and is denoted by $\text{cond}(\mathbf{A})$. Therefore, (3.150) can be expressed as:

$$\frac{1}{\text{cond}(\mathbf{A})} \left(\frac{\|\Delta\mathbf{b}\|}{\|\mathbf{b}\|} \right) \leq \frac{\|\Delta\mathbf{x}\|}{\|\mathbf{x}\|} \leq \text{cond}(\mathbf{A}) \left(\frac{\|\Delta\mathbf{b}\|}{\|\mathbf{b}\|} \right). \tag{3.151}$$

This equation relates the relative solution error ($\|\Delta\mathbf{x}\|/\|\mathbf{x}\|$) to the relative error in ($\|\Delta\mathbf{b}\|/\|\mathbf{b}\|$). If $\text{cond}(\mathbf{A})$ is close to 1, then the relative errors in $\mathbf{x}$ and $\mathbf{b}$ are close. However, if $\text{cond}(\mathbf{A})$ is a large number, then the relative errors in $\mathbf{x}$ can be significantly larger than the relative errors in $\mathbf{b}$.

3.20.10   Sherman-Morrison-Woodbur

Consider that the system of linear equations $\mathbf{A}\mathbf{x} =$
$\mathbf{x}_0$. Consider also the slightly changed version of t.
the form: $\tilde{\mathbf{A}}\mathbf{x} = \mathbf{b}$, and whose solution we want to de
that $\tilde{\mathbf{A}}$ represents a perturbed version of $\mathbf{A}$. The p
address is finding a computationally efficient solutic
equations assuming knowledge of $\mathbf{x}_0$ and that $\tilde{\mathbf{A}}$ repr
of $\mathbf{A}$, i.e.,

$$\tilde{\mathbf{A}} = \mathbf{A} - \mathbf{u}\mathbf{v}^T,$$

in which $\mathbf{u}$ and $\mathbf{v}$ are two arbitrary vectors. Becau
can be expressed in terms of the original matrix $\mathbf{A}$

$$(\tilde{\mathbf{A}})^{-1} = \mathbf{A}^{-1} + \frac{\left(\mathbf{A}^{-1}\mathbf{u}\right)\left(\mathbf{v}^T\right.}{1 - \mathbf{v}^T\mathbf{A}^-}$$

Identity (3.153) is referred to as the *Sherman-Morris*
(3.153) with $\mathbf{b}$, the solution of the perturbed system
as:

$$\tilde{\mathbf{x}} = \mathbf{x}_0 + \Delta\mathbf{x} = \mathbf{x}_0 + \frac{\left(\mathbf{A}^{-1}\mathbf{u}\right)}{1 - \mathbf{v}^T}$$

where we took into account $\mathbf{A}^{-1}\mathbf{b} = \mathbf{x}_0$. Sherman
tended to situations when matrix $\mathbf{A}$ is subject to hi
of the form:

$$\tilde{\mathbf{A}} = \mathbf{A} - \mathbf{U}\mathbf{V}^T,$$

where $\mathbf{U}$ and $\mathbf{V}$ are $n \times m$ matrices and $n \geq m$. I
perturbed matrix $\tilde{\mathbf{A}}$ is calculated by means of a g
Morrison formulation, referred to as the *Woodbury*

$$(\tilde{\mathbf{A}})^{-1} = \mathbf{A}^{-1} + \mathbf{A}^{-1}\mathbf{U}\left(\mathbf{I}_m - \mathbf{V}^T\mathbf{A}^-\right.$$

Equation (3.155) is often referred to as the Sherm
tity. Right-multiplying (3.155) with vector $\mathbf{b}$, it tur:
perturbed system can be expressed in terms of $\mathbf{x}_0$ a

$$\mathbf{x}' = \mathbf{x}_0 + \Delta\mathbf{x}$$

$$= \mathbf{x}_0 + \mathbf{A}^{-1}\mathbf{U}\left(\mathbf{I}_m - \mathbf{V}^T\mathbf{A}^{-1}\mathbf{U}\right.$$

The Sherman-Morrison-Woodbury identities (3.153)
role in the development of recursive (iterative or o
techniques. The RLS algorithm and Kalman filter a
this regard [10, 11, 30, 31].

3.20.11 Schur Complement

Many practical applications require the calculation of the inverse of a block-partitioned matrix. Assume that the matrix $\mathbf{E}$ is block-partitioned as:

$$\mathbf{E} = \begin{bmatrix} \mathbf{A} \ \mathbf{B} \\ \mathbf{C} \ \mathbf{D} \end{bmatrix}, \tag{3.158}$$

where submatrix $\mathbf{A}$ is invertible. The expression $\mathbf{F} = \mathbf{D} - \mathbf{C}\mathbf{A}^{-1}\mathbf{B}$ is called the *Schur complement* of submatrix $\mathbf{A}$ in $\mathbf{E}$ and represents an important ingredient in computing the inverse of $\mathbf{E}$ in terms of the inverse of $\mathbf{A}$. It turns out that the inverse of the partitioned matrix $\mathbf{E}$ in (3.158) can be expressed in terms of the inverses of submatrix $\mathbf{A}$ and its Schur complement $\mathbf{F}$ as follows:

$$\mathbf{E}^{-1} = \begin{bmatrix} \mathbf{A}^{-1} + \mathbf{A}^{-1}\mathbf{B}\mathbf{F}^{-1}\mathbf{C}\mathbf{A}^{-1} & -\mathbf{A}^{-1}\mathbf{B}\mathbf{F}^{-1} \\ -\mathbf{F}^{-1}\mathbf{C}\mathbf{A}^{-1} & \mathbf{F}^{-1} \end{bmatrix}. \tag{3.159}$$

Furthermore, the determinant of matrix $\mathbf{E}$ can be calculated using determinants of submatrix $\mathbf{A}$ and its Schur complement:

$$|\mathbf{E}| = |\mathbf{A}|\,|\mathbf{F}|. \tag{3.160}$$

The proof of (3.160) is deferred to Exercise 3.23.32. Additional properties and applications of Schur complement can be found in [7, 32].

3.20.12 Generalized Inverses

Often it is necessary to extend the concept of matrix invertibility to a wider class of rank-deficient matrices by means of concepts such as *generalized inverse* or *pseudoinverse* of a matrix. One of the major applications of the concept of generalized inverse is in solving systems of equations of the form: $\mathbf{A}\mathbf{x} = \mathbf{b}$, where matrix $\mathbf{A}$ is neither square nor nonsingular. Such a system of equations may not admit any solution or may admit an infinite number of solutions.

Basic Definition of Generalized Inverse

The generalized inverse of a matrix is not unique. Any matrix that satisfies the following equation:

$$\mathbf{A}\mathbf{A}^{-}\mathbf{A} = \mathbf{A}, \tag{3.161}$$

is referred to as a generalized inverse of matrix $\mathbf{A}$ and is denoted by $\mathbf{A}^{-}$. In general, the generalized inverse of an $m \times n$ matrix is an $n \times m$ matrix. Some of the main properties of generalized inverses are enumerated below.

1. If $\mathbf{A}$ is a square and nonsingular matrix, then $\mathbf{A}^{-} = \mathbf{A}^{-1}$, i.e., the generalized matrix coincides with the regular inverse.

2. Transposition: $\left(\mathbf{A}^T\right)^- = \left(\mathbf{A}^-\right)^T$.

3. Assume that matrix $\mathbf{A}$ is of size $m \times n$, then

   (a) $\mathrm{rank}\left(\mathbf{A}\mathbf{A}^-\right) = \mathrm{rank}\left(\mathbf{A}\right)$.

   (b) $\mathrm{rank}\left(\mathbf{I} - \mathbf{A}^-\mathbf{A}\right) = n - \mathrm{rank}\left(A\right)$.

4. Consider $m \times n$ matrices $\mathbf{A}$ and $\mathbf{B}$ and their n
   and $\mathbf{B}^-$, respectively.

   (a) $\mathbf{A}\left(\mathbf{I} + \mathbf{A}\right)^- = \left(\mathbf{I} + \mathbf{A}^-\right)^-$.

   (b) $\left(\mathbf{A}^- + \mathbf{B}^-\right)^- = \mathbf{A}\left(\mathbf{A} + \mathbf{B}\right)^- \mathbf{B}$.

   (c) $\mathbf{A} - \mathbf{A}\left(\mathbf{A} + \mathbf{B}\right)^- \mathbf{A} = \mathbf{B} - \mathbf{B}\left(\mathbf{A} + \mathbf{B}\right)^- \mathbf{B}$

   (d) $\left(\mathbf{I} + \mathbf{A}\mathbf{B}\right)^- = \mathbf{I} - \mathbf{A}\left(\mathbf{I} + \mathbf{B}\mathbf{A}\right)^- \mathbf{B}$.

5. Generalized inverse of block matrices. Conside

$$\mathbf{E} = \begin{bmatrix} \mathbf{A}\,\mathbf{B} \\ \mathbf{C}\,\mathbf{D} \end{bmatrix},$$

where $\mathbf{A}$, $\mathbf{B}$, $\mathbf{C}$, and $\mathbf{D}$ are submatrices of $\mathbf{E}$ an
for their generalized inverses, respectively.
calculation of generalized inverse of $\mathbf{E}$:

$$\mathbf{E}^- = \begin{bmatrix} \mathbf{A}^- + \mathbf{A}^-\mathbf{B}\mathbf{F}^-\mathbf{C}\mathbf{A}^- \\ -\mathbf{F}^-\mathbf{C}\mathbf{A}^- \end{bmatrix}$$

where $\mathbf{F} = \mathbf{D} - \mathbf{C}\mathbf{A}^-\mathbf{B}$.

An excellent reference dealing with the properties a
inverses is [33]. More recently, [7] provides a goo
exhibited by generalized (weak) inverses.

Generalized Inverses to Solve Consistent Sys

Assume $\mathbf{A}\mathbf{x} = \mathbf{b}$ is a consistent under-determined
system of equations admits more than one solution
of matrix $\mathbf{A}$ is assumed to satisfy (3.161). Right-
follows that:

$$\mathbf{A}\mathbf{A}^-\mathbf{A}\mathbf{x} = \mathbf{A}\mathbf{x}.$$

Since $\mathbf{A}\mathbf{x} = \mathbf{b}$, (3.162) simplifies further to:

$$\mathbf{A}(\mathbf{A}^-\mathbf{b}) = \mathbf{b}.$$

Therefore, $\mathbf{x} = \mathbf{A}^-\mathbf{b}$ can be interpreted as a solution of the original system of equations $\mathbf{Ax} = \mathbf{b}$. Notice that in general any vector of the form: $\mathbf{x} = \mathbf{A}^-\mathbf{b} + (\mathbf{I} - \mathbf{A}^-\mathbf{A})\mathbf{y}$, where $\mathbf{y}$ is any $n \times 1$ vector, is a solution of the original system of equations (and in fact all the solutions are of this type [3]). Indeed, one can check directly by performing the required calculations that

$$\begin{aligned}
\mathbf{Ax} &= \mathbf{A}\left(\mathbf{A}^-\mathbf{b} + (\mathbf{I} - \mathbf{A}^-\mathbf{A})\mathbf{y}\right) \\
&= \mathbf{AA}^-\mathbf{b} + (\mathbf{A} - \mathbf{AA}^-\mathbf{A})\mathbf{y} \\
&= \mathbf{b} - \mathbf{0} \\
&= \mathbf{b}.
\end{aligned}$$

The reverse implication can be proved by using the fact that any solution of inhomogeneous system of equations $\mathbf{Ax} = \mathbf{b}$ can be expressed as the sum between a particular solution $(\mathbf{A}^-\mathbf{b})$ of the original system $\mathbf{Ax} = \mathbf{b}$ and the general solution of the homogeneous system of equations $\mathbf{Ax} = \mathbf{0}$, which coincides with the null space of $\mathbf{A}$. It is not difficult to check that $\mathbf{I} - \mathbf{A}^-\mathbf{A}$ is the projector onto the null space of $\mathbf{A}$ [3].

Moore-Penrose Inverse

As mentioned before, the definition of a generalized inverse does not yield a unique generalized inverse, unless it is restricted to satisfy some additional requirements. By imposing several additional constraints to the definition (3.161), a unique generalized inverse can be obtained for any arbitrary matrix $\mathbf{A}$. The resulting unique generalized inverse is referred to as the *Moore-Penrose inverse* of $\mathbf{A}$ and is represented in terms of the notation: $\mathbf{A}^+$. The additional constraints satisfied by the Moore-Penrose inverse are listed below:

$$\mathbf{A}^+\mathbf{AA}^+ = \mathbf{A}^+, \tag{3.164}$$

$$\left(\mathbf{A}^+\mathbf{A}\right)^T = \mathbf{A}^+\mathbf{A}, \tag{3.165}$$

$$\left(\mathbf{AA}^+\right)^T = \mathbf{AA}^+. \tag{3.166}$$

The Moore-Penrose inverse $\mathbf{A}^+$ exists for any arbitrary matrix $\mathbf{A}$ and is unique. There are several methods to calculate the Moore-Penrose inverse of a matrix. One of the most common methods is based on the singular values decomposition of the matrix, explained in Section 3.17. Let the SVD of matrix $\mathbf{A}$ be given by $\mathbf{A} = \mathbf{U\Sigma V}^T$, where $\mathbf{\Sigma}$ is the same as in (3.122), and matrices $\mathbf{U}$, $\mathbf{V}$ are orthogonal. The pseudoinverse of $\mathbf{\Sigma}$, $\mathbf{\Sigma}^+$, is an $n \times m$ diagonal matrix with $1/\sigma_i$, $i = 1, \ldots, r$, on its main diagonal, and 0 elsewhere. The Moore-Penrose inverse of $\mathbf{A}$ is given by (see e.g., [3, 7, 24]):

$$\mathbf{A}^+ = \mathbf{V\Sigma}^+\mathbf{U}^\dagger. \tag{3.167}$$

3.21   Matrix Operations

Several useful matrix operations such as Kronecker
dot product, direct sum, and matrix differentiation
tion.

3.21.1   Kronecker Product

The *Kronecker product* of $m \times n$ matrix $\mathbf{A} = [a_{ij}]$
represented in terms of notation $\otimes$, and it is given
$\mathbf{C}$:

$$\mathbf{C} = \mathbf{A} \otimes \mathbf{B} = \begin{bmatrix} a_{11}\mathbf{B} & a_{12}\mathbf{B} & \dots \\ a_{21}\mathbf{B} & a_{22}\mathbf{B} & \dots \\ \vdots & \vdots & \vdots \\ a_{m1}\mathbf{B} & a_{m2}\mathbf{B} & \dots \end{bmatrix}$$

This Kronecker product is also referred to as *direct*
    Suppose that $\mathbf{A}$, $\mathbf{B}$, $\mathbf{C}$, and $\mathbf{D}$ are matrices wit
also that $\mathbf{a}$ and $\mathbf{b}$ are vectors of appropriate sizes,
following properties hold for the Kronecker produ
properties):

- $(\mathbf{A} + \mathbf{B}) \otimes \mathbf{C} = \mathbf{A} \otimes \mathbf{C} + \mathbf{B} \otimes \mathbf{C}$.

- $\mathbf{A} \otimes (\mathbf{B} + \mathbf{C}) = \mathbf{A} \otimes \mathbf{B} + \mathbf{A} \otimes \mathbf{C}$.

- $(\mathbf{A} \otimes \mathbf{B}) \otimes \mathbf{C} = \mathbf{A} \otimes (\mathbf{B} \otimes \mathbf{C})$.

- $(\mathbf{A} \otimes \mathbf{B})(\mathbf{C} \otimes \mathbf{D}) = (\mathbf{AC}) \otimes (\mathbf{BD})$.

- $(\mathbf{A} \otimes \mathbf{B})^T = \mathbf{A}^T \otimes \mathbf{B}^T$.

- $\mathbf{a} \otimes \mathbf{b}^T = \mathbf{b}^T \otimes \mathbf{a} = \mathbf{ab}^T$.

- $(\mathbf{A} \otimes \mathbf{B})^{-1} = \mathbf{A}^{-1} \otimes \mathbf{B}^{-1}$.

- $\operatorname{rank}(\mathbf{A} \otimes \mathbf{B}) = \operatorname{rank}(\mathbf{A}) \operatorname{rank}(\mathbf{B})$.

3.21.2   Hadamard Product

The *Hadamard product*, also called the *element-wise*
in terms of notation $\odot$, and it is defined for two m
$\mathbf{B} = [b_{k,l}]$ as follows:

$$\mathbf{A} \odot \mathbf{B} = \begin{bmatrix} a_{1,1}b_{1,1} & a_{2,1}b_{2,1} & \dots \\ a_{2,1}b_{2,1} & a_{2,2}b_{2,2} & \dots \\ \vdots & \vdots & \dots \\ a_{m,1}b_{m,1} & a_{m,2}b_{m,2} & \dots \end{bmatrix}$$

For arbitrary matrices $\mathbf{A}$, $\mathbf{B}$, and $\mathbf{C}$, assumed of appropriate sizes, the Hadamard product exhibits the following properties:

- $\mathbf{A} \odot \mathbf{B} = \mathbf{B} \odot \mathbf{A}$

- $(\mathbf{A} \odot \mathbf{B}) \odot \mathbf{C} = \mathbf{A} \odot (\mathbf{B} \odot \mathbf{C})$

- $(\mathbf{A} + \mathbf{B}) \odot \mathbf{C} = \mathbf{A} \odot \mathbf{C} + \mathbf{B} \odot \mathbf{C}$

- $(\mathbf{A} \odot \mathbf{B})^T = \mathbf{A}^T \odot \mathbf{B}^T$

- $\mathbf{C} (\mathbf{A} \odot \mathbf{B}) = (\mathbf{C}\mathbf{A}) \odot \mathbf{B} = \mathbf{A} \odot (\mathbf{C}\mathbf{B})$

3.21.3 Dot Product

The *dot product*, also called the *inner-product* of two $m \times n$ matrices $\mathbf{A} = [\mathbf{a}_i]$ and $\mathbf{B} = [\mathbf{b}_i]$ is defined in terms of their column vectors $\mathbf{a}_i$ and $\mathbf{b}_i$, respectively, as follows:

$$\langle \mathbf{A} | \mathbf{B} \rangle = \sum_{i=1}^{m} a_i^T b_i . \tag{3.170}$$

The dot product satisfies the following properties:

- $\langle \mathbf{A} | \mathbf{B} \rangle = \operatorname{tr}\left(\mathbf{A}^T \mathbf{B} \right)$

- $\langle \mathbf{A} | \mathbf{B} \rangle = \left\langle \mathbf{A}^T \middle| \mathbf{B}^T \right\rangle$

3.21.4 Direct Sum

The *direct sum* of k square matrices $\mathbf{A}_1$, $\mathbf{A}_2$, ..., $\mathbf{A}_k$ with sizes $n_1 \times n_1$, $n_2 \times n_2$, ..., $n_k \times n_k$, respectively, is a block diagonal square matrix of size $(n_1 + n_2 + ... + n_k) \times (n_1 + n_2 + ... + n_k)$ defined by

$$\mathbf{A}_1 \oplus \mathbf{A}_2 \oplus ... \oplus \mathbf{A}_k = \operatorname{diag}(\mathbf{A}_1, \mathbf{A}_2, ..., \mathbf{A}_k) = \begin{bmatrix} \mathbf{A}_1 & 0 & ... & 0 \\ 0 & \mathbf{A}_2 & ... & 0 \\ \vdots & \vdots & \ddots & \vdots \\ 0 & 0 & ... & \mathbf{A}_k \end{bmatrix} . \tag{3.171}$$

The following properties of direct sum hold:

- $\operatorname{tr}(\mathbf{A}_1 \oplus \mathbf{A}_2 \oplus ... \oplus \mathbf{A}_k) = \operatorname{tr}(\mathbf{A}_1) + \operatorname{tr}(\mathbf{A}_2) + ... + \operatorname{tr}(\mathbf{A}_k)$

- $|\mathbf{A}_1 \oplus \mathbf{A}_2 \oplus ... \oplus \mathbf{A}_k| = |\mathbf{A}_1| |\mathbf{A}_2| ... |\mathbf{A}_k|$

- $\operatorname{rank}(\mathbf{A}_1 \oplus \mathbf{A}_2 \oplus ... \oplus \mathbf{A}_k) = \operatorname{rank}(\mathbf{A}_1) + \operatorname{rank}(\mathbf{A}_2) + ... + \operatorname{rank}(\mathbf{A}_k)$

3.21.5   Differentiation of Matrix and

This subsection focuses on presenting some basic con
the calculation of derivatives for matrix- and vector
to a scalar, vector, or matrix-valued independent va

Differentiation with Respect to a Scalar

When the differentiation variable is a scalar, the der
as the operand and its elements are simply deriv
operand. Assuming that the $m \times 1$ vector $\mathbf{x}(t) =$
matrix $\mathbf{Y}(t) = [y_{i,j}(t)]$ are functions of the independe
derivatives with respect to t are given by:

$$\frac{\partial \mathbf{x}(t)}{\partial t} = \left(\frac{\partial x_1(t)}{\partial t}, ..., \frac{\partial x_m(t)}{\partial t} \right)^T,$$

$$\frac{\partial \mathbf{Y}(t)}{\partial t} = \begin{bmatrix} \frac{\partial y_{1,1}(t)}{\partial t} & \cdots & \frac{\partial y_{1,n}(t)}{\partial t} \\ \vdots & \ddots & \vdots \\ \frac{\partial y_{m,1}(t)}{\partial t} & \cdots & \frac{\partial y_{m,n}(t)}{\partial t} \end{bmatrix} =$$

Similarly the higher order derivatives of a vector (m
scalar is a vector (matrix) of the same size, and who
order derivatives of the vector (matrix) entries.

Differentiation with Respect to a Vector

The derivative of a scalar function $x(\mathbf{t})$ with resp
$(t_1, ..., t_n)^T$ is also referred to as the *gradient*, and i

$$\frac{\partial x(\mathbf{t})}{\partial \mathbf{t}^T} = \left(\frac{\partial x(\mathbf{t})}{\partial t_1}, ..., \frac{\partial x(\mathbf{t}}{\partial t_n} \right.$$

An alternative notation for the gradient of x is:

$$\nabla x = \frac{\partial x(\mathbf{t})}{\partial \mathbf{t}^T}.$$

The derivative of any operand with respect to a v
element of the operand. Therefore, the derivative
vector is a matrix. Moreover, the derivative of a
respect to a vector is a three-dimensional array.

Assume the $m \times 1$ vector $\mathbf{x}(\mathbf{t}) = (x_1(\mathbf{t}), ..., x_m(\mathbf{t}$
$[y_{i,j}(\mathbf{t})]$. Then the derivatives of $\mathbf{x}(\mathbf{t})$ and $\mathbf{Y}(\mathbf{t})$ wit

$\mathbf{t} = (t_1, ..., t_n)^T$ are given by:

$$\frac{\partial \mathbf{x}(\mathbf{t})}{\partial \mathbf{t}^T} = \begin{bmatrix} \frac{\partial x_1(\mathbf{t})}{\partial t_1} & \frac{\partial x_1(\mathbf{t})}{\partial t_2} & \cdots & \frac{\partial x_1(\mathbf{t})}{\partial t_n} \\ \frac{\partial x_2(\mathbf{t})}{\partial t_1} & \frac{\partial x_2(\mathbf{t})}{\partial t_2} & \cdots & \frac{\partial x_2(\mathbf{t})}{\partial t_n} \\ \vdots & \vdots & \ddots & \vdots \\ \frac{\partial x_m(\mathbf{t})}{\partial t_1} & \frac{\partial x_m(\mathbf{t})}{\partial t_2} & \cdots & \frac{\partial x_m(\mathbf{t})}{\partial t_n} \end{bmatrix}, \tag{3.176}$$

$$\frac{\partial \mathbf{Y}(\mathbf{t})}{\partial \mathbf{t}} = [y_{i,j,k}] \quad, \quad y_{i,j,k} = \frac{\partial y_{i,j}}{\partial t_k}, i = 1, \ldots, p, \ \ j = 1, \ldots, q, \ \ k = 1, \ldots, n \tag{3.177}$$

The $m \times n$ matrix $\frac{\partial \mathbf{x}(\mathbf{t})}{\partial \mathbf{t}^T}$ is referred to as the *Jacobian* of $\mathbf{x}(\mathbf{t})$ and is usually represented in terms of the notation:

$$\mathbf{J_x} = \frac{\partial \mathbf{x}}{\partial \mathbf{t}^T}. \tag{3.178}$$

Higher order derivatives can be calculated by applying first-order derivatives successively several times; e.g., the second-order derivative of a scalar $x(\mathbf{t})$ with respect to the $n \times 1$ vector $\mathbf{t} = (t_1, ..., t_n)^T$ is the first-order derivative of the vector $\frac{\partial x(\mathbf{t})}{\partial \mathbf{t}^T} = \left(\frac{\partial x(\mathbf{t})}{\partial t_1}, ..., \frac{\partial x(\mathbf{t})}{\partial t_n} \right)^T$ with respect to the vector $\mathbf{t}$, and it is equal to the matrix:

$$\frac{\partial^2 x(\mathbf{t})}{\partial \mathbf{t} \, \partial \mathbf{t}^T} = \frac{\partial}{\partial \mathbf{t}} \left(\frac{\partial x(\mathbf{t})}{\partial \mathbf{t}^T} \right)$$

$$= \frac{\partial}{\partial \mathbf{t}} \left(\frac{\partial x(\mathbf{t})}{\partial t_1}, ..., \frac{\partial x(\mathbf{t})}{\partial t_n} \right)$$

$$= \begin{pmatrix} \frac{\partial^2 x}{\partial t_1^2} & \frac{\partial^2 x}{\partial t_2 t_1} & \cdots & \frac{\partial^2 x}{\partial t_n t_1} \\ \frac{\partial^2 x}{\partial t_1 t_2} & \frac{\partial^2 x}{\partial t_2^2} & \cdots & \frac{\partial^2 x}{\partial t_n t_2} \\ \vdots & \vdots & \ddots & \vdots \\ \frac{\partial^2 x}{\partial t_1 t_n} & \frac{\partial^2 x}{\partial t_2 t_n} & \cdots & \frac{\partial^2 x}{\partial t_n^2} \end{pmatrix}$$

$$= \begin{pmatrix} \frac{\partial^2 x}{\partial t_1^2} & \frac{\partial^2 x}{\partial t_1 t_2} & \cdots & \frac{\partial^2 x}{\partial t_1 t_n} \\ \frac{\partial^2 x}{\partial t_2 t_1} & \frac{\partial^2 x}{\partial t_2^2} & \cdots & \frac{\partial^2 x}{\partial t_2 t_n} \\ \vdots & \vdots & \ddots & \vdots \\ \frac{\partial^2 x}{\partial t_n t_1} & \frac{\partial^2 x}{\partial t_n t_2} & \cdots & \frac{\partial^2 x}{\partial t_n^2} \end{pmatrix}. \tag{3.179}$$

The matrix in (3.179) is called the *Hessian* of $x(\mathbf{t})$ and is represented in general in terms of one of these notations:

$$\nabla \nabla x = \nabla^2 x = \frac{\partial^2 x}{\partial \mathbf{t}^2}. \tag{3.180}$$

Derivatives of some commonly used expressions with respect to the $n \times 1$ vector $\mathbf{x}$ are shown in Table 3.1. In this table, $\mathbf{A}$ stands for an $n \times n$ matrix, and $\mathbf{x}$ and $\mathbf{a}$ are

Table 3.1: Derivatives of commonly used expressi

$f(\mathbf{x})$	$\frac{\partial f(\mathbf{x})}{\partial \mathbf{x}}$	$\frac{\partial}{\partial}$
$\mathbf{x}^T\mathbf{a}$	$\mathbf{a}$	ε
$\mathbf{a}^T\mathbf{A}\mathbf{x}$	$\mathbf{A}^T\mathbf{a}$	$\mathbf{a}$
$\mathbf{A}\mathbf{x}$	$\mathbf{A}^T$	
$\mathbf{x}^T\mathbf{x}$	$2\mathbf{x}$	2
$\mathbf{x}^T\mathbf{A}\mathbf{x}$	$\mathbf{A}\mathbf{x}+\mathbf{A}^T\mathbf{x}$	$\mathbf{x}^T\mathbf{A}^T$

Table 3.2: Derivatives of commonly used expressic

$f(\mathbf{X})$	$\frac{\partial f(\mathbf{X})}{\partial \mathbf{X}}$
$\mathbf{a}^T\mathbf{X}\mathbf{b}$	$\mathbf{a}\mathbf{b}^T$
$\mathbf{a}^T\mathbf{X}^T\mathbf{b}$	$\mathbf{b}\mathbf{a}^T$
$\mathrm{tr}\,(\mathbf{A}\mathbf{X})$	$\mathbf{A}^T$
$\mathrm{tr}\left(\mathbf{X}\mathbf{X}^T\right)$	$2\mathbf{X}$
$\mathrm{tr}\,(\mathbf{A}\mathbf{X}\mathbf{B})$	$\mathbf{A}^T\mathbf{B}^T$
$|\mathbf{X}|$	$|\mathbf{X}|\left(\mathbf{X}^{-1}\right)$

$n \times 1$ vectors. Generally, the following conventions

and $\frac{\partial f(\mathbf{x})}{\partial \mathbf{x}^T} = \left(\frac{\partial f(\mathbf{x})}{\partial \mathbf{x}}\right)^T$.

Differentiation with Respect to a Matrix

The derivative of each element of an object with res
of derivatives of that element with respect to the r
derivatives of a scalar, vector, or a matrix with resp
three-dimensional array, and four-dimensional arra
some commonly used expressions with respect to a
in Table 3.2. In this table, $\mathbf{A}$, $\mathbf{B}$, and $\mathbf{X}$ are $n \times$
$n \times 1$ vectors. Finally, we would like to mention that
more comprehensive description of the results pert
derivatives for vector- and matrix-valued functions.

3.22   References and Further S

This chapter presented a short summary of the main
ear algebra that are used most frequently in the field
communications, and networking. For more detailed
of the basic concepts in linear algebra and matrix th

to the excellent references [3, 5, 8, 34]. For numerical stable implementations of various matrix factorizations and linear algebra algorithms, [24] represents still one of the state-of-the-art references in this area. The handbook [7] presents an encyclopedic coverage of matrix theory from the viewpoint of a statistician. Another excellent reference that provides an in-depth treatment of basic properties of matrices is [5]. Finally, we end this chapter with the remark that linear algebra represents a very important tool for designing efficient algorithms for signal processing, communications, and networking applications (see e.g., [10–12, 30, 35]).

Acknowledgment: This work was supported by NSF Grant 0915444.

3.23 Exercises

Exercise 3.23.1. Show that the subset $\mathcal{S} \subset \mathcal{V}$, ($\mathcal{S} \neq \{0\}$), where $\mathcal{V}$ is a vector space over field $\mathcal{F}$, is a subspace if and only if

$$\forall \mathbf{s}_1, \mathbf{s}_2 \in \mathcal{S}, \quad \forall r_1, r_2 \in \mathcal{F}: \quad r_1 \mathbf{s}_1 + r_2 \mathbf{s}_2 \in \mathcal{S}.$$

Exercise 3.23.2. Let $B = \{\mathbf{b}_1, \mathbf{b}_2, \ldots, \mathbf{b}_n\}$ be an ordered basis for vector space $\mathcal{V}$.

- Show that every inner-product on $\mathcal{V}$ can be computed using the values

$$a_{i,j} = \langle \mathbf{b}_j | \mathbf{b}_i \rangle \quad i = 1, \ldots, n, \, j = 1, \ldots, n.$$

Hint: Take $\mathbf{x}, \mathbf{y} \in \mathcal{V}$ and write them as linear combinations of $\mathbf{b}_i$'s ($i = 1, \ldots n$). Form the inner-product $\langle \mathbf{x} | \mathbf{y} \rangle$ and the coordinate matrices of $\mathbf{x}$ and $\mathbf{y}$ in the ordered basis B, $[\mathbf{x}]_B$ and $[\mathbf{y}]_B$. Show that

$$\langle \mathbf{x} | \mathbf{y} \rangle = [\mathbf{y}]_B^\dagger A [\mathbf{x}]_B,$$

where $n \times n$ matrix $\mathbf{A} = [a_{i,j}]$ consists of the inner-products of the vectors in ordered basis B.

- Show that $\mathbf{A}$ is Hermitian positive definite matrix.

Exercise 3.23.3. This problem establishes the Cauchy-Schwarz inequality for the induced norm. Let $\mathbf{v}$ be a non-zero vector and define:

$$\mathbf{v}' = \mathbf{w} - \frac{\langle \mathbf{w} | \mathbf{v} \rangle}{\|\mathbf{v}\|^2} \mathbf{v}.$$

Show that $\langle \mathbf{v}' | \mathbf{v} \rangle = 0$ and from the fact $0 \leq \|\mathbf{v}'\|^2$, conclude that

$$|\langle \mathbf{v} | \mathbf{w} \rangle|^2 \leq \|\mathbf{v}\|^2 \|\mathbf{w}\|^2.$$

Using this result, prove the triangle inequality for the induced norm.

Exercise 3.23.4. Show that every orthogonal subs
of a vector space $\mathcal{V}$ is linearly independent.

Exercise 3.23.5. Prove that in the Gram-Schmid
troduced in Theorem 3.1.5, the following relationsh

$$\langle \mathbf{w}_{k+1} | \mathbf{w}_i \rangle = 0, \quad i = 1, ..$$

Thus, $\{\mathbf{w}_1, \mathbf{w}_2, \ldots, \mathbf{w}_{k+1}\}$ is an orthogonal set. Do
represent a basis for span $(\mathbf{v}_1, \mathbf{v}_2, \ldots, \mathbf{v}_{k+1})$? Justify

Exercise 3.23.6. Using the fact that for an $m \times$
(column) operations do not affect the column (row)

$$\text{row rank}\,(\mathbf{A}) = \text{column rank}$$

Exercise 3.23.7. If $\mathbf{A}$ is an $m \times n$ matrix over field $\mathcal{J}$
linear transformation $T_A : \mathcal{F}^n \to \mathcal{F}^m$ is injective if
is surjective if and only if rank $(\mathbf{A}) = m$.

Exercise 3.23.8. Show that for a projection opera

$$T(\mathbf{v}) = \mathbf{v}, \quad \forall \mathbf{v} \in \mathcal{R}(T)$$

Use this statement to prove Theorem 3.2.3.

Exercise 3.23.9. By computing $\langle \mathbf{Pu} | (\mathbf{I} - \mathbf{P})\mathbf{v} \rangle$ sho
matrix $\mathbf{P}$ represents an orthogonal projection. Use th

Exercise 3.23.10. Let $B = \{\mathbf{v}_1, \mathbf{v}_2, \ldots, \mathbf{v}_n\}$ be a l

- Show that there is a unique linear functional

$$v_i^*(\mathbf{v}_j) = \delta_{ij}.$$

- Show that the set of n distinct linear function
  mentioned above), are linearly independent.

- From the dimension of $\mathcal{V}^*$ what can we conclu

- Let f be a linear functional on $\mathcal{V}$. Write f a
  $(i = 1, 2, \ldots, n)$. Moreover, let $\mathbf{v}$ be a vector
  a linear combination of $\mathbf{v}_i$'s. Write the expr
  unique expression for $\mathbf{v}$ as a linear combinatio

Exercise 3.23.11. Let $T : \mathcal{V} \to \mathcal{W}$ be a linear trans
product finite dimensional vector spaces $\mathcal{V}$ and $\mathcal{W}$.

- Let $\mathbf{w} \in \mathcal{R}(T)$ and $\mathbf{n} \in \mathcal{N}\text{ull}\,(T^*)$. By comput

$$\mathcal{N}\text{ull}\,(T^*) \subset \mathcal{R}(T)^\perp\,.$$

- Now, let $\mathbf{w} \in \mathcal{R}(T)^\perp$. Using the definition of the adjoint of T, show that

$$\mathbf{w} \in \mathcal{N}\text{ull}\,(T^*)\,,$$

and

$$\mathcal{R}(T)^\perp \subset \mathcal{N}\text{ull}\,(T^*)\,.$$

- Using the previous results, show that

$$\mathcal{R}(T)^\perp = \mathcal{N}\text{ull}\,(T^*)\,.$$

- Show that $\mathcal{R}(T^*)^\perp = \mathcal{N}\text{ull}\,(T)$ using a similar argument.

Exercise 3.23.12. Using the properties of the four fundamental subspaces of linear transformation T, show that T is surjective if and only if T^* is injective, and T is injective if and only if T^* is surjective.

Exercise 3.23.13. Show the following equalities:

$$\mathcal{N}\text{ull}\,(T^*T) = \mathcal{N}\text{ull}\,(T)$$
$$\mathcal{N}\text{ull}\,(TT^*) = \mathcal{N}\text{ull}\,(T^*)$$
$$\mathcal{R}(T^*T) = \mathcal{R}(T^*)$$
$$\mathcal{R}(TT^*) = \mathcal{R}(T)\,.$$

Hint: First, show that $\mathcal{N}\text{ull}\,(T) \subset \mathcal{N}\text{ull}\,(T^*T)$. Then using the definition of the adjoint, show that if $T^*T(\mathbf{v}) = 0$, then $T(\mathbf{v}) = 0$; and hence, $\mathcal{N}\text{ull}\,(T^*T) \subset \mathcal{N}\text{ull}\,(T)$.

Exercise 3.23.14. Show that the two definitions of the operator norm are equivalent, i.e.,

$$\|T\| = \sup_{\mathbf{v} \in \mathcal{V} - \{\mathbf{0}\}} \frac{\|T(\mathbf{v})\|}{\|\mathbf{v}\|}$$
$$= \sup_{\mathbf{v} \in \mathcal{V},\, \|\mathbf{v}\|=1} \|T(\mathbf{v})\|\,.$$

Exercise 3.23.15. This problem derives several matrix norms.

- Using the definition of $\|\mathbf{x}\|_1$, show that:

$$\|\mathbf{A}\|_1 = \max_{\|\mathbf{x}\|_1=1} \|\mathbf{A}\mathbf{x}\|_1$$
$$= \max_j \sum_i |a_{ij}|\,.$$

- Using the definition of $\|\mathbf{x}\|_\infty$, show that

$$\|\mathbf{A}\|_\infty = \max_{\|\mathbf{x}\|_\infty = 1} \|\mathbf{Ax}\|_c$$

$$= \max_i \sum_j |a_{ij}|$$

- In order to find $\|\mathbf{A}\|_2$, use Lagrange's multipli

$$f(\mathbf{x}) = \mathbf{x}^\dagger \mathbf{A}^\dagger \mathbf{Ax} - \lambda(\mathbf{x}^\dagger \mathbf{x} -$$

Hint: Take the gradient of $f(\mathbf{x})$ with respect to $[\mathbf{x} \ \lambda]$
then the eigenvalues and eigenvectors of matrix $\mathbf{A}^\dagger$.

- Show that

$$\|\mathbf{A}\|_F^2 = \mathrm{tr}\left(\mathbf{A}^\dagger \mathbf{A}\right) = \sum_{i=1}^m \sum_{j=1}^n$$

Exercise 3.23.16. Let $\mathbf{A}$ be an $m \times n$ matrix and $\mathbf{U}$
Show that

$$\|\mathbf{UA}\|_F = \|\mathbf{A}\|_F \ .$$

Exercise 3.23.17. Let $\mathbf{A}$ and $\mathbf{B}$ be matrices of
respectively. Show that

$$\det(\mathbf{I}_m + \mathbf{AB}) = \det(\mathbf{I}_n +$$

Exercise 3.23.18. Let $\mathcal{S}$ be a subspace of an inne
let $\mathbf{v} \in \mathcal{V}$. Follow the steps below to prove Theorem

- Suppose $\mathbf{s} \in \mathcal{S}$ and $\mathbf{v} - \mathbf{s}$ is orthogonal to all $\mathbf{s}$

$$\langle \mathbf{v} - \mathbf{s} | \mathbf{s}^* \rangle = 0 \quad \forall \mathbf{s}^* \in \mathcal{S}$$

Let $\mathbf{s}' \in \mathcal{S}$ and $\mathbf{s}' \neq \mathbf{s}$. By rewriting the differenc

$$\mathbf{v} - \mathbf{s}' = (\mathbf{v} - \mathbf{s}) + (\mathbf{s} - \mathbf{s}$$

show that

$$\|\mathbf{v} - \mathbf{s}'\|^2 \geq \|\mathbf{v} - \mathbf{s}\|^2 \ .$$

- Conversely, now let $\|\mathbf{v} - \mathbf{s}'\|^2 \geq \|\mathbf{v} - \mathbf{s}\|^2$, $\forall \mathbf{s}'$

$$2\Re(\langle \mathbf{v} - \mathbf{s} | \mathbf{s}^* \rangle) + \|\mathbf{s}^*\|^2 \geq 0 \ , \ \forall$$

Then by a suitable choice for $\mathbf{s}^*$ show that $\langle \mathbf{v} - \mathbf{s} |$
$\mathbf{v} - \mathbf{s}$ is orthogonal to every vector in $\mathcal{S}$.

- Show that the best approximation is unique.

- Let $\mathbf{s} = \sum_{k=1}^{m} \frac{\langle \mathbf{v} | \mathbf{s}_k \rangle}{\|\mathbf{s}_k\|^2} \mathbf{s}_k$. Show that $\mathbf{v} - \mathbf{s}$ is orthogonal to $\mathbf{s}_j$, $j = 1, \ldots, m$, and it is orthogonal to every vector in $\mathcal{S}$. Therefore, $\mathbf{s}$ is the best approximation of $\mathbf{v}$ by vectors in $\mathcal{S}$.

Exercise 3.23.19. Let $\mathcal{S}$ be a finite-dimensional subspace of inner-product vector space $\mathcal{V}$. Suppose $\mathbf{P}$ is an orthogonal projection of $\mathcal{V}$ onto $\mathcal{S}$.

- Prove that the mapping defined by matrix $\mathbf{I} - \mathbf{P}$ which maps any vector $\mathbf{v} \in \mathcal{V}$ onto $\mathbf{v} - P\mathbf{v}$ is actually the orthogonal projection of $\mathcal{V}$ onto $\mathcal{S}^{\perp}$.

- Show that $\mathbf{P}$ is an idempotent linear mapping from $\mathcal{V}$ to $\mathcal{S}$.

- Show that

$$\mathbf{P}\mathbf{s}' = \mathbf{0} \iff \mathbf{s}' \in \mathcal{S}^{\perp}.$$

- Show that

$$\mathcal{V} = \mathcal{S} \oplus \mathcal{S}^{\perp}.$$

Exercise 3.23.20. This problem pertains to the proof of Theorem 3.13.1.

- Using the best approximation theorem and the four fundamental subspaces related to T, show that the minimum error $(T(\mathbf{v}) - \mathbf{w})$ should be in the nullspace of T^*.

- Conclude that

$$T^* T(\mathbf{v}) = T^*(\mathbf{w}).$$

Exercise 3.23.21. Show that if an $n \times n$ matrix $\mathbf{A}$ has distinct eigenvalues, the corresponding eigenvectors should be linearly independent.

Exercise 3.23.22. Let $\mathbf{v}_i$ and $\mathbf{v}_j$ be two eigenvectors of an $n \times n$ Hermitian matrix $\mathbf{A}$, corresponding to different eigenvalues λ_i and λ_j, respectively. Compute $\mathbf{v}_i^{\dagger} \mathbf{v}_j$ and show that the eigenvectors are orthogonal. Moreover, by computing $\mathbf{v}_i^{\dagger} \mathbf{A} \mathbf{v}_i$, show that $\mathbf{A}$'s eigenvalues are real. Use these results to prove the claims of Theorem 3.15.2.

Exercise 3.23.23. This problem is related to diagonalizing a matrix (Theorem 3.15.3). Suppose $\mathbf{v}_1, \mathbf{v}_2, \ldots, \mathbf{v}_n$ are n linearly independent eigenvectors of an $n \times n$ matrix $\mathbf{A}$, corresponding to eigenvalues $\lambda_1, \lambda_2, \ldots, \lambda_n$, respectively.

- Show that

$$\mathbf{A}[\mathbf{v}_1, \mathbf{v}_2, \ldots, \mathbf{v}_n] = [\mathbf{v}_1, \mathbf{v}_2, \ldots, \mathbf{v}_n]$$

Define $\mathbf{S} = [\mathbf{v}_1, \mathbf{v}_2, \ldots, \mathbf{v}_n]$ and $\mathbf{\Lambda} = \text{diag}(\lambda_1, \ldots$
onal matrix. Note that $\mathbf{S}$ is invertible. Conclude th

- Conversely, let $\mathbf{S}$ be an invertible matrix such

$$\mathbf{S}^{-1}\mathbf{A}\mathbf{S} = \mathbf{\Lambda},$$

where $\mathbf{\Lambda}$ is diagonal. Show that the ith column
corresponding to the ith diagonal entry of $\mathbf{\Lambda}$.

Exercise 3.23.24. Using the SVD of matrix $\mathbf{A}$, sh

$$\|\mathbf{A}\|_F = (\sigma_1^2 + \sigma_2^2 + \cdots + \sigma$$

where n is the rank of $\mathbf{A}$ and σ_i $(i = 1, \ldots, n)$ are t

Exercise 3.23.25. Show that the multiplication of b

and $\mathbf{B} = \begin{bmatrix} \mathbf{B}_{11} & \mathbf{B}_{12} \\ \mathbf{B}_{21} & \mathbf{B}_{22} \end{bmatrix}$ can be expressed as:

$$\mathbf{AB} = \begin{bmatrix} \mathbf{A}_{11}\mathbf{B}_{11} + \mathbf{A}_{12}\mathbf{B}_{21} & \mathbf{A}_{1,1}\mathbf{B}_{12} \\ \mathbf{A}_{21}\mathbf{B}_{11} + \mathbf{A}_{22}\mathbf{B}_{21} & \mathbf{A}_{21}\mathbf{B}_{12} \end{bmatrix}$$

Exercise 3.23.26. Let the permutation matrix $\mathbf{\Pi}_r$

$$\mathbf{\Pi}_m = circ(0, 1, 0, ..., 0) = [e_m, e_1, e$$

with

$$e_i = \begin{pmatrix} 1 & \ldots & i-1 & i & i+1 \\ 0 & \ldots & 0 & 1 & 0 \end{pmatrix}$$

Prove that the square matrix $\mathbf{A}$ is circulant if and c

Exercise 3.23.27. Using the result proved in Exer
ing properties.

- If $\mathbf{A}$ and $\mathbf{B}$ are circulant matrices, then $\mathbf{AB}$ is

- If $\mathbf{A}$ is circulant, then $\mathbf{A}^{-1}$ is a circulant matr

Exercise 3.23.28. Show that:

- The eigenvalues of a circulant matrix $(c_o, c_1, \ldots, c_{n-1})$ are

$$\lambda_k = \sum_{i=0}^{n-1} c_i e^{-j2\pi i k/n}.$$

- The corresponding eigenvectors are

$$\mathbf{x}_k = \frac{1}{\sqrt{n}} [1 \quad e^{-j2\pi k/n} \quad e^{-j2\pi(2k)/n} \quad \ldots \quad e^{-j2\pi(n-1)k/n}]^T.$$

Exercise 3.23.29. Using the equation for determinant of a Vandermonde matrix, what conditions must the scalars a_1, a_2, ..., a_n satisfy in order for Vandermonde matrix $\mathbf{A}$ to be invertible?

$$\mathbf{A} = \begin{bmatrix} 1 & a_1 & a_1^2 & \ldots & a_1^{m-1} \\ 1 & a_2 & a_2^2 & \ldots & a_2^{m-1} \\ \vdots & \vdots & \vdots & \ddots & \vdots \\ 1 & a_m & a_m^2 & \ldots & a_m^{m-1} \end{bmatrix}.$$

Exercise 3.23.30. Consider the square matrix $\mathbf{A}$. Prove that:

1. If $\mathbf{A}$ is positive definite, then all its eigenvalues are positive.

2. If all eigenvalues of $\mathbf{A}$ are positive, $\mathbf{A}$ is positive definite.

Exercise 3.23.31. Prove that the following factorization is valid for the block-partitioned matrix $\mathbf{M}$:

$$\mathbf{M} = \begin{bmatrix} \mathbf{A} & \mathbf{B} \\ \mathbf{C} & \mathbf{D} \end{bmatrix} = \begin{bmatrix} \mathbf{A} & \mathbf{0} \\ \mathbf{C} & \mathbf{E} \end{bmatrix} \begin{bmatrix} \mathbf{I} & \mathbf{A}^{-1}\mathbf{B} \\ \mathbf{0} & \mathbf{I} \end{bmatrix},$$

where $\mathbf{E}$ is the Schur complement of submatrix $\mathbf{A}$ in matrix $\mathbf{M}$.

Exercise 3.23.32. Consider the block matrix

$$\mathbf{M} = \begin{bmatrix} \mathbf{A} & \mathbf{B} \\ \mathbf{C} & \mathbf{D} \end{bmatrix}.$$

Prove that the determinant of matrix $\mathbf{M}$ is equal to the product of determinants of $\mathbf{A}$ and its Schur complement.

Exercise 3.23.33. Show that the generalized inverse of the block matrix

$$\mathbf{M} = \begin{bmatrix} \mathbf{A} & \mathbf{B} \\ \mathbf{C} & \mathbf{D} \end{bmatrix}$$

is given by:

$$\mathbf{M}^- = \begin{bmatrix} \mathbf{A}^- + \mathbf{A}^-\mathbf{B}\mathbf{E}^-\mathbf{C}\mathbf{A}^- & -\mathbf{A}^-\mathbf{B}\mathbf{E}^- \\ -\mathbf{E}^-\mathbf{C}\mathbf{A}^- & \mathbf{E}^- \end{bmatrix},$$

where $\mathbf{E} = \mathbf{D} - \mathbf{C}\mathbf{A}^-\mathbf{B}$.

Exercise 3.23.34. Prove the following properties ο
trices $\mathbf{A}$, $\mathbf{B}$, $\mathbf{C}$ and $\mathbf{D}$ are assumed of appropriate s

1. $(\mathbf{A} + \mathbf{B}) \otimes \mathbf{C} = \mathbf{A} \otimes \mathbf{C} + \mathbf{B} \otimes \mathbf{C}$.

2. $(\mathbf{A} \otimes \mathbf{B}) \otimes \mathbf{C} = \mathbf{A} \otimes (\mathbf{B} \otimes \mathbf{C})$.

3. $(\alpha\mathbf{A}) \otimes (\beta\mathbf{B}) = \alpha\beta (\mathbf{A} \otimes \mathbf{B})$.

4. $(\mathbf{A} \otimes \mathbf{B}) (\mathbf{C} \otimes D) = (\mathbf{A}\mathbf{C}) \otimes (\mathbf{B}D)$.

Exercise 3.23.35. Prove the vector and matrix ο
Tables 3.1 and 3.2, respectively.

References

[1] J. E. Gentle, *Matrix Algebra: Theory, Comp*
Statistics. NY: Springer, 2010.

[2] S. Leon, *Linear Algebra with Applications.* 9tl

[3] A. J. Laub, *Matrix Analysis for Scientists and I*
2004.

[4] P. Lax, *Linear Algebra and Its Applications.* 2r
2007.

[5] C. Meyer, *Matrix Analysis and Applied Linear*
SIAM, 2001.

[6] S. Roman, *Advanced Linear Algebra.* 3rd ed.,

[7] G. A. F. Seber, *A Matrix Handbook for Statist*

[8] G. Strang, *Linear Algebra and Its Applications.*
KY: Brooks Cole, 2005.

[9] R. A. Horn and C. R. Johnson, *Matrix Analy*
University Press, 1985.

[10] S. Kay, *Fundamentals of Statistical Signal Proc*
ory, vol. II: Detection Theory. 1st ed., NY: P

[11] A. H. Sayed, *Adaptive Filters.* 1st ed., NY: W

[12] P. Stoica and R. L. Moses, *Spectral Analysis ο*
2005.

[13] J. Dauxois and G. M. Nkiet, "Canonical analysis of two Euclidean subspaces and its applications," *Linear Algebra Appl.*, vol. 264, no. 10, pp. 355–388, October 1997.

[14] R. Gittins, *Canonical Analysis: A Review with Applications in Ecology, Biomathematics*, Berlin: Springer-Verlag, 1985.

[15] T. Kailath, "A view of three decades of linear filtering theory," *IEEE Transactions on Information Theory*, vol. 20, no. 2, pp. 146–181, February 1974.

[16] I. M. Gelfand and A. M. Yaglom, "Calculation of the amount of information about a random function contained in another such function," *American Mathematical Society Translations, series 2*, vol. 12, pp. 199–246, 1959.

[17] H. Akaike, "Stochastic theory of minimal realization," *IEEE Transactions on Automatic Control*, pp. 667–674, 1974.

[18] P. E. Caines, *Linear Stochastic Systems.* NY: Wiley, 1988.

[19] P. V. Oversche and B. de Moor, *Subspace Identification for Linear Systems: Theory, Implementation and Applications.* Boston: Kluwer Academics Publishers, 1996.

[20] T. Kim, O. Arandjelovic, and R. Cipolla, "Boosted manifold principal angles for image set-based recognition," *Pattern Recognition*, vol. 40, issue 9, pp. 2475–2484, September 2007.

[21] L. Wolf and A. Shashua, "Learning over Sets using Kernel Principal Angles," *Journal of Machine Learning Research*, vol. 4, pp. 913–931, 2003.

[22] R. T. Behrens and L. Scharf, "Signal Processing Applications of Oblique Projection Operators," *IEEE Transactions on Signal Processing*, vol. 42, no. 6, pp. 1413–1414, June 1994.

[23] I. E. Telatar, "Capacity of multi-antenna Gaussian channels," *Eur. Trans. Telecommun.*, vol. 10, no. 6, pp. 585–595, November 1999.

[24] G. Golub and C. F. van Loan, *Matrix Computations.* 3rd. ed., Baltimore: John Hopkins University Press, 1996.

[25] I. Jolliffe, *Principal Component Analysis.* Berlin: Springer, 2002.

[26] S. K. Mitra, *Digital Signal Processing.* NY: McGraw-Hill, 2005.

[27] R. M. Gray, *Toeplitz and Circulant Matrices: A Review.* Boston-Delft: Now Publishers Inc., 2005.

[28] M. Verhaegen, "Identification of the Deterministic Part of MIMO State Space Models given in Innovations Form from Input-Output Data," *Automatica*, vol. 30, no. 1, pp. 61–74, January 1994.

92

[29] R. Bhatia, *Positive Definite Matrices.* Princeto
2006.

[30] S. Haykin, *Adaptive Filter Theory.* NY: Pren

[31] T. Soderstrom and P. Stoica, *System Identificat*

[32] F. Zhang, *The Schur Complement and Its Ap*
2010.

[33] C. R. Rao and S. K. Mitra, *Generalized Invers
tions.* NY: Wiley, 1972.

[34] T. K. Moon and W. C. Stirling, *Mathematica
Signal Processing.* Upper Saddle River, NJ: P

[35] S. B. Wicker, *Error Control Systems for Digita
Upper Saddle River, NJ: Prentice-Hall, 1995.

Chapter 4 Elements of Galois Fields

Tolga M. Duman[‡]
[‡]Arizona State University, Tempe, USA

This chapter deals with finite field algebra particularly useful for various channel coding and cryptography techniques. We are accustomed to operations in the field of real numbers R or rational numbers Q with the usual notions of addition, multiplication, subtraction and division. We normally do this without questioning the underlying assumptions or definitions. For the case of finite fields the operations and properties are quite different; hence, we need to go back to the basics and define these algebraic structures from the first principles.

As the material in this chapter is somewhat abstract, it may not be clear to an engineering student at a first glance why they are needed or useful. While our objective is not to provide a detailed discussion of the uses of finite fields, also called Galois fields, we stress that some of the fundamental engineering applications utilize this theory. For instance, many practical channel coding techniques, e.g., Reed-Solomon codes, BCH codes, etc., can only be studied or understood after a good understanding of Galois fields. Another example is cryptography where many techniques rely on finite fields and their properties.

Our coverage of finite fields, despite being mostly self-contained, is admittedly superficial. The reader is referred to other books on modern algebra for more detailed treatments on the subject, e.g. [1, 2]. Regarding the applications of the finite field algebra, the reader may wish to refer to channel coding books such as [3, 4] and resources on cryptography such as [5].

The chapter starts with the general definitions of groups, rings, and fields. Then finite fields are discussed in some detail and some of the properties are summarized. Our focus then shifts to the extension fields of the binary field. Polynomials with binary coefficients are described and then are employed to construct the finite fields with a number of elements equal to a power of two. Finally, a couple of applications of the theory are highlighted to conclude the chapter.

4.1   Groups, Rings, and Fields

4.1.1   Groups

Let us start with the definition of a group.

Definition 4.1.1. A group G is a set of elements cl
"·" which satisfies the following three properties:

- $\forall a, b, c \in G$, $(a \cdot b) \cdot c = a \cdot (b \cdot c)$, i.e., the bina;

- $\exists e \in G$ such that $\forall a \in G$, $a \cdot e = e \cdot a = a$
  element e,

- $\forall a \in G$, $\exists$ an element $a^{-1} \in G$ such that $a \cdot$
  exists an inverse for all the elements of G.

Furthermore, if $\forall a, b \in G$, $a \cdot b = b \cdot a$, then the grou

An immediate consequence of the group definiti
and the inverse of a group element are unique. To
is true: assume that there are two identity elements
a given group element $a \in G$, the following must ho'
inverse of a, we then can write $(a^{-1} \cdot a) \cdot e_1 = (a^{-1} \cdot$
associativity property. This yields $e_1 = e_2$ which is
the second part of the statement is left as an exerci;
Examples of groups include the set of integers Z
is an abelian group), and the set of all non-zero real n
cation (also an abelian group). Nonsingular (square
under matrix multiplication form a non-abelian gr
does not commute. On the other hand, the set of r
plication or the set of non-negative integers under
(in both cases not all elements of the set have an in
Of particular interest are groups with a finite n
examples are given below.

Example 4.1.1. The set of numbers $G = \{0, 1, 2$
arbitrary positive integer) form an abelian group u:
instance, for $m = 4$, the modulo-m addition table is
the identity element is 0, and the inverses of 0, 1
respectively.

Example 4.1.2. The set of positive integers from 1 t
form an abelian group under modulo-p multiplicati
number.

+	0	1	2	3
0	0	1	2	3
1	1	2	3	0
2	2	3	0	1
3	3	0	1	2

Figure 4.1: Modulo-4 addition table with elements $\{0, 1, 2, 3\}$.

To see why this is the case, we first assume that p is a prime number. Associativity and commutativity of modulo-p multiplication are obvious. The identity element is 1. G is closed since there are no $m, n \in G$ such that $m \cdot n = 0$ modulo-p (otherwise p must divide $m \cdot n$, but p and $m \cdot n$ are relatively prime since $m, n < p$ and p is prime). Furthermore, every element of G has an inverse (see Problem 4.6.2). Therefore, G is a finite group under modulo-p multiplication.

The other direction of the proposition is easily seen by noting that if p is not a prime number, then there exist $m, n \in G$ such that $m \cdot n = 0$ modulo-p, i.e., the set is not closed under the modulo-p multiplication operation. $\square$

Modulo-5 multiplication table with elements $\{1, 2, 3, 4\}$ is shown in Figure 4.2. Clearly, the set is closed under this binary operation, "1" is the identity element, and each element has an inverse ("1" is observed in each row exactly once).

.	1	2	3	4
1	1	2	3	4
2	2	4	1	3
3	3	1	4	2
4	4	3	2	1

Figure 4.2: Modulo-5 multiplication table with elements $\{1, 2, 3, 4\}$.

The number of elements in a group can be finit*
we define the order of a group as the cardinality of t
element $g \in G$ is the smallest positive integer n suc*

$$\underbrace{g \cdot g \ldots g}_{n \quad many} = g^n = 1,$$

where "1" is the identity element of the group.

Example 4.1.3. Consider the group formed by $\{1,$
plication. The order of the group is 4. The orders
(for 4), and 4 (for 2 and 3). Notice that the orders
4 (the order of the group). This is not a coincidenc*
$\square$

Example 4.1.4. Consider the group $\{1, 2, \ldots, 10\}$
tion. This group contains one element of order 1 (1
four elements of order 5 (3, 4, 5 and 9) and four elem*
8). Again the possible orders (1, 2, 5 and 10) are int*
group.

Definition 4.1.2. Let H be a subset of G (where
binary operation "$\cdot$"). If H satisfies the group prope*
then it is a subgroup of G.

Clearly, with this definition, H is closed under the b*
the identity element of the group, and for any ele*
contained in H.

Example 4.1.5. Consider $G = \{0, 1, 2, \ldots, 7\}$ und*
the set $H_1 = \{0, 2, 4, 6\}$ forms a subgroup under t
does the set $H_2 = \{0, 4\}$.

Definition 4.1.3. Let H be a subgroup of G with
$a \in G$, the set of elements $a \cdot H = \{a \cdot h : h \in$
$H \cdot a = \{h \cdot a : h \in H\}$ is a right-coset.

For commutative groups, the left- and right-cose*
they are simply referred to as a *coset*. It can be sh*
coset are identical. To see this, consider an element
Enumerate the elements of H as $H = \{h_1, h_2, \ldots,$*
two distinct elements of H, h_i and h_j such that $a \cdot h_*$
of a to both sides, we obtain $a^{-1} \cdot a \cdot h_i = a^{-1} \cdot a \cdot$
of the binary operation, we obtain $h_i = h_j$ which i*
same argument holds for the case of right-cosets, an*
Similarly, it can also be proved that no two el*
of H are identical (see Problem 4.6.4). In other w*

disjoint. Furthermore, every element of the group G is an element of one of the cosets of H (since for any $b \in G$, we can write $b = (b \cdot h^{-1}) \cdot h$ for some $h \in H$). Combining these results, we can say that the order of a subgroup H divides the order of the group, and further its cosets form a partition of the group. Let us illustrate this result via a simple example.

Example 4.1.6. Consider $G = \{0, 1, 2, \ldots, 15\}$ with modulo-16 addition, and its subgroup $H = \{0, 4, 8, 12\}$. There are four different cosets of H (since this is an abelian group, we do not make a distinction between the left- and right-cosets): $\{0, 4, 8, 12\}, \{1, 5, 9, 13\}, \{2, 6, 10, 14\}$, and $\{3, 7, 11, 15\}$. Clearly, the four cosets do not have any common elements and their union is G. □

4.1.2 Rings

We now define a different algebraic structure with two binary operations.

Definition 4.1.4. A set of elements R with two operations "+" and "·" form a ring if

- R is an abelian group under "+",

- "·" is associative,

- "·" distributes over "+", i.e., $\forall a, b, c \in R,\ a \cdot (b + c) = (a \cdot b) + (a \cdot c)$.

The ring is commutative if $\forall a, b \in R,\ a \cdot b = b \cdot a$.

It is clear that while a ring is an abelian group under the first operation "+", it does not necessarily satisfy the properties of a group with respect to the second operation.

An important example of a ring is the set of polynomials. That is, the set of polynomials with real coefficients form a commutative ring with identity under the usual polynomial addition ("+" operation) and polynomial multiplication ("·" operation). The additive identity is 0, and the multiplicative identity is 1 (constants). A more important example for our purposes is the set of polynomials with binary coefficients that will be studied in more detail later in the chapter.

4.1.3 Fields

We are now in a position to define a field.

Definition 4.1.5. A set of elements F with two binary operations "+" (referred to as addition) and "·" (referred to as multiplication) form a field if

- F is an abelian group under "+". The additive identity of the field is denoted by "0", and the additive inverse of a field element a is denoted by $-a$.

- $F \setminus \{0\}$ is an abelian group under "·". The mu
  by "1", and the multiplicative inverse of a fiel
  a^{-1}.

- The binary operation "·" distributes over the

We further define subtraction "−" and division "/"

$$a - b \triangleq a + (-b)$$

and

$$a/b \triangleq a \cdot (b^{-1}), \text{if } b \neq 0$$

Also, sometimes we find it convenient to drop "·" w
of two elements, i.e., we use "ab" to represent "$a \cdot b$"

Common examples of fields include the set of
tion and multiplication, the set of complex number
complex multiplication, and the set of rational num
and multiplication. These fields clearly contain infin
uncountably infinite, the latter countably infinite).
finite fields are more important.

Example 4.1.7. If p is a prime number, the set
form a field under modulo-p addition and modulo
to see this since modulo-p multiplication distributes
we have already established in Section 4.1.1 that th
abelian group under modulo-p addition (regardless
element 0, and that the set $\{1, 2, 3, \ldots, p - 1\}$ is an
multiplication (if and only if p is a prime number).

Finite fields are also called Galois fields denoted
ample 4.1.7, for instance, the notation is $GF(p)$. I
field denoted by $GF(2)$ where the two operations a
multiplication as shown in Figure 4.3. Notice that bi
are the same since the additive inverses of both elem
We can perform the usual arithmetic operations
ficulty in a field. For instance, for any two elements
$-(a \cdot b) = (-a) \cdot b = a \cdot (-b) = -ab$ (see Problem 4.6.
argue that cancellation of common multiplicative te
(as long as the term being cancelled is non-zero). T
then $b = c$ (easily seen by simply multiplying bot
multiplicative inverse of a which exists since a is a n
argue that if two non-zero elements of the field are
be zero (i.e., the additive identity). Furthermore, if
the additive identity, the result is the additive iden

+	0	1	·	0	1
0	0	1	0	0	0
1	1	0	1	0	1

Figure 4.3: Addition and multiplication tables in $GF(2)$.

then $a \cdot 0 = 0$ since

$$
\begin{aligned}
a \cdot 0 &= a \cdot (b + (-b)) \\
&= a \cdot b + a \cdot (-b) \\
&= a \cdot b + (-(a \cdot b)) \\
&= 0.
\end{aligned}
$$

We note that an immediate implication of an earlier example (Example 4.1.7) is that finite fields of a prime order (for any prime number) exist. However, it does not establish existence of fields of non-prime orders, and it does not address the question whether we can define a finite field of an arbitrary order or not. To partially address this question, we present an example of a field construction with four elements.

Example 4.1.8. Consider the elements $F = \{0, 1, 2, 3\}$ with the multiplication and addition table given in Figure 4.4.
One can easily verify that F along with the defined addition and multiplication operations form a field. For instance, it is clear that additive identity is 0, multiplicative identity is 1, the additive inverse of an element is itself, and the multiplicative inverses of the elements 1, 2, and 3 are 1, 3, and 2, respectively. □

The example clarifies that there is a field of order 4 which is not a prime number. Further, it makes the point that to construct a field we may need to define the binary operations in a somewhat peculiar and nontrivial manner. For instance, using the very first approach that comes to mind, i.e., modulo-4 addition and multiplication operations, does not work.

Similar to the definition of a subgroup, we refer to a subset of field elements with the same two operations as a subfield if the subset satisfies the field properties. Clearly, the finite field in Example 4.1.8 ($GF(4)$) contains $GF(2)$ as a subfield.

+	0	1	2	3	.	0
0	0	1	2	3	0	0
1	1	0	3	2	1	0
2	2	3	0	1	2	0
3	3	2	1	0	3	0

Figure 4.4: Addition and multiplication↦

4.2   Galois Fields

We now describe details of finite fields, also referred
their inventor Évariste Galois, leading up to a full ↦
finite fields and their construction in a subsequent
contains at least two elements, "0" (the additive ide↦
tive identity). $GF(2)$ described in the previous sec
field one can define.

Consider a finite field of order q denoted by GF
a prime number. It is easy to prove that when we ↦
"1" repeatedly we will obtain the additive identity o↦
there are only a finite number of field elements). ↦
the field $GF(q)$ as the smallest positive integer p su↦

$$\underbrace{1 + 1 + \cdots + 1}_{p \ many} = \sum_{i=1}^{p} 1 =$$

We note that the characteristic of a finite field i↦
$\sum_{i=1}^{p} 1 = \left(\sum_{i=1}^{m} 1\right)\left(\sum_{i=1}^{n} 1\right) = 0$ for some non-neg↦
leads to a contradiction).

A second observation is that $GF(p)$ is a subf↦
$\{0, 1, 1 + 1, 1 + 1 + 1, \ldots, \sum_{i=1}^{p-1} 1\}$. This can be co↦
subset along with the two binary operations of the
properties. For instance, when two elements $\sum_{i=1}^{m}$↦
obtain $\sum_{i=1}^{k} 1$ with $k = m + n$ modulo-p (utilizing $\sum$
elements are multiplied, we obtain $\sum_{i=1}^{l} 1$ with $l =$↦
that $\sum_{i=1}^{p} 1 = 0$). Further, the additive inverse o↦
in the set. Similarly, there is a multiplicative inver↦

$\sum_{i=1}^{m} 1$ (with $m \neq 0$) contained in the set. To see this, note that since p and m are relatively prime ($1 \leq m \leq p - 1$, and p is prime), there exist two positive integers r, s such that $mr = ps + 1$ (using Euclid's algorithm for finding the greatest common divisor of two integers), i.e., $mr = 1$ modulo-p, therefore $\sum_{i=1}^{m} 1$ has a multiplicative inverse in the set given by $\sum_{i=1}^{r} 1$.

Let us denote the elements of $GF(p)$ (which is a subfield of $GF(q)$) more compactly as $\{\alpha_0, \alpha_1, \alpha_2, \ldots, \alpha_{p-1}\}$ (with the understanding that $\alpha_0 = 0$ and $\alpha_1 = 1$). Consider a non-zero element of $GF(q)$, denoted by β_1. Then $\alpha_i \beta_1$ ($i \in \{0, 1, \ldots, p - 1\}$) are p distinct elements of $GF(q)$. If all the elements of $GF(q)$ are accounted for in this list then $q = p$. If not, consider another element of $GF(q)$, β_2, not listed previously. Then $\alpha_i \beta_1 + \alpha_j \beta_2$ ($0 \leq i, j \leq p - 1$) are p^2 distinct elements in $GF(q)$. If there are no other elements, then $q = p^2$, otherwise consider an additional element β_3, and form $\alpha_i \beta_1 + \alpha_j \beta_2 + \alpha_k \beta_3$ ($0 \leq i, j, k \leq p-1$) to construct p^3 distinct elements of $GF(q)$. We continue this process until there are no other elements left uncovered in $GF(q)$ to argue that the order of the field $GF(q)$ must be a power of its characteristic, i.e., $q = p^m$ for some integer m.

With the above argument we establish the following theorem.

Theorem 4.2.1. *The order of a finite field $GF(q)$ must be a power of a prime number.*

We already know from an earlier example how to obtain $GF(q)$ when q is prime: we simply can use the elements $F = \{0, 1, 2, \ldots, q - 1\}$ with modulo-q addition and multiplication to define the field. However, the construction of $GF(q)$ when $q = p^m$ where p is a prime number and $m > 1$, is not that straightforward, and we need to approach the problem differently. It may be tempting to propose using modulo-q arithmetic, but this clearly does not work, as $q - 1$ is not prime, and we do not have a group structure for the non-zero elements under the multiplication operation.

Several other results regarding the properties of a finite field are considered in the following.

Theorem 4.2.2. *The $(q - 1)$th power of a non-zero element of a finite field is 1, i.e., if $a \in GF(q)$ and $a \neq 0$, we have $a^{q-1} = \prod_{i=1}^{q-1} a = 1$.*

Proof. See Problem 4.6.7. □

Let us define the order of a non-zero element $a \in GF(q)$ as the smallest integer n such that $\prod_{i=1}^{n} a = a^n = 1$.

Theorem 4.2.3. *The order of a non-zero element $a \in GF(q)$ divides $q - 1$.*

Proof. Consider a non-zero element a in the field with order n, i.e., the smallest integer for which $a^n = 1$ is n. Assume that n does not divide $q - 1$ and let the remainder obtained by dividing $q - 1$ by n be r, i.e., $q - 1 = nk + r$ with $0 < r < n$. We can write $a^{q-1} = a^{nk+r}$, which implies that $a^{q-1} = (a^n)^k a^r$ from which we obtain $a^r = 1$ (since $a^{q-1} = 1$ for any non-zero field element). This is a contradiction as $0 < r < n$, hence r must be zero, i.e., n must divide $q - 1$. □

A non-zero element $a \in GF(q)$ is called primitive
contains at least one primitive element as an immec
theorem.

Theorem 4.2.4. *The number of elements of order*
$q - 1$*) is given by*

$$\varphi(n) = n \prod_{p|n} \left(1 - \frac{1}{p}\right)$$

where p runs over all prime numbers (that divide n

Proof. See [4] or any book on modern algebra.

$\varphi(n)$ is called the Euler totient function and is
$\{1, 2, \ldots, n - 1\}$ that are relatively prime to n (incl
   As a corollary to the theorem, we can easily cc
$\varphi(q - 1)$ primitive elements in $GF(q)$.

Example 4.2.1. Consider $GF(16)$. There are $\varphi(1$
primitive elements, $\varphi(5) = 5(1 - \frac{1}{5}) = 4$ elements of o
2 elements of order 2.

   The rest of the chapter is devoted to the Galois
$p = 2$, i.e., to the construction of extension fields o
results directly extend to the extension fields of G
prime number; however, this generalization is omit
we proceed, we need to review some basic concepts a
coefficients.

4.3   Polynomials with Coefficie

We now discuss some basic properties of polynomials
which is of interest. A polynomial over $GF(2)$ of de

$$f(X) = f_0 + f_1 X + f_2 X^2 + \cdots$$

with the understanding that "X" is an indetermin
are taken from $GF(2)$. We define the polynomia
operations as usual. Assume that $f(X)$ and $g(X)$ a
n and m with coefficients $f_0, f_1, \ldots, f_n$ and $g_0, g_1,$
loss of generality assume that $n \geq m$. We define the

$$f(X) + g(X) = (f_0 + g_0) + (f_1 + g_1)X + \cdots + (f_m + g_m)$$

and, the polynomial multiplication by

$$f(X) \cdot g(X) = h_0 + h_1 X + h_2 X^2 + \cdots$$

where $h_i = \sum_{j=0}^{i} f_j g_{i-j}$ (with the understanding that the $g_{i-j} = 0$ if $i - j > m$).
For instance, if $f(X) = 1 + X + X^3$ and $g(X) = X + X^2$, we obtain

$$f(X) + g(X) = 1 + X^2 + X^3$$

and

$$f(X) \cdot g(X) = X + X^3 + X^4 + X^5,$$

where we employed $GF(2)$ addition and multiplication to determine the coefficients of the powers of the indeterminate X in the resulting polynomials.

With the two binary operations (polynomial addition and polynomial multiplication), the set of polynomials over $GF(2)$ form a ring as can easily be verified.

Factorization of a polynomial in $GF(2)$, or any other field, is an important notion. For instance, we can write

$$X^3 + 1 = (X + 1)(X^2 + X + 1),$$

i.e., $X^3 + 1$ has two factors with coefficients again from $GF(2)$. If a polynomial cannot be factorized into polynomials of smaller degrees over $GF(2)$, it is said to be irreducible. For instance, it is not possible to factorize $X^2 + X + 1$ or $X^3 + X^2 + 1$; hence, they are irreducible polynomials. We note, however, that it is still possible to factorize a polynomial in an extension field as will become apparent later in the chapter.

We digress here to point out that we are already familiar with similar statements in the context of a different field, namely, the complex number field. We can think of complex numbers as two-dimensional vectors (a, b) with $a, b \in R$ (the first component being the real part and the second component being the imaginary part), or equivalently by $a + jb$ where $j = \sqrt{-1}$. The addition of two complex numbers is defined as componentwise addition, i.e., $(a, b) + (c, d) = (a + c, b + d)$, and multiplication of two complex numbers is defined through the operation $(a, b) \cdot (c, d) = (a \cdot c - b \cdot d, a \cdot d + b \cdot c)$. If we have a polynomial with coefficients from the real number field, it may be irreducible over the ground field, i.e. it may not have a real root; however, it is certainly reducible over the extension field of complex numbers. Considering the polynomial $X^2 + 1$ with real coefficients, the roots are complex conjugates of each other and are given by $(0, 1)$ and $(0, -1)$. Since there are no real roots, the polynomial is irreducible over R. As another example, the polynomial (with coefficients in R) given by $X^3 - 1$ has one real root, $(1, 0)$ and two complex roots, $(-1/2, \mp\sqrt{3}/2)$. Since it has a real root, it is reducible over R, and in fact we have $X^3 - 1 = (X - 1)(X^2 + X + 1)$. Clearly the factors are irreducible over R. In general, any polynomial with real coefficients can be written as a unique product of degree-1 and irreducible degree-2 polynomials over R.

The following theorem is useful in the construction of $GF(2^m)$.

Theorem 4.3.1. *An irreducible polynomial with binary coefficients of degree-m divides $X^{2^m-1} + 1$.*

Proof. Refer to [1].

Definition 4.3.1. An irreducible polynomial over
primitive if the smallest n such that the polynomial

It is also a well-known result that primitive polync
are exactly $\varphi(2^m - 1)/m$ degree-m primitive polyn
is the Euler totient function as defined earlier. Such
construction of Galois fields. Examples of primitive
10 are provided in Figure 4.5.

degree	primitive po
2	$1 + X +$
3	$1 + X +$
4	$1 + X +$
5	$1 + X^2 -$
6	$1 + X +$
7	$1 + X +$
8	$1 + X^2 + X^3 -$
9	$1 + X^4$
10	$1 + X^3 +$

Figure 4.5: Examples of primitive polynomials wi

4.4   Construction of $GF(2^m)$

Let $p(X)$ be a primitive polynomial of degree m with
that α is a root of that polynomial, i.e., $p(\alpha) = 0$. U
elements

$$\{0, 1, \alpha, \alpha^2, \ldots, \alpha^{2^m - 2}\}$$

are all distinct. To see this, we simply note that α ca
$X^n + 1$ if $n < 2^m - 1$ (otherwise $p(X)$ for which α i
polynomial). In other words, choosing α as a root

degree-m ensures that all the non-zero elements of $GF(2^m)$ can be generated as its powers.

With this approach, multiplication of two field elements is trivial, we simply need to utilize $\alpha^{2^m-1} = 1$. For the addition operation, we note that $p(\alpha) = 0$ and represent α^m as a polynomial in α with a degree less than m. In other words, we can write the elements as vectors of length m where each component denotes the corresponding coefficient (in $GF(2)$) of $1, \alpha, \alpha^2, \ldots, \alpha^{m-1}$, respectively. Notice that with this m-dimensional vector representation with binary coefficients, we can generate exactly 2^m elements (which are all the elements of the field $GF(2^m)$). Hence a suitable addition operation can also be defined for any two field elements by a simple componentwise addition.

Example 4.4.1. Consider $GF(2^3)$ generated by $p(X) = 1 + X + X^3$. If α is a primitive element of the field, the entire set of elements can be written as $F^* = \{0, 1, \alpha, \alpha^2, \ldots, \alpha^6\}$. Taking α as a root of the primitive polynomial $p(X) = 1 + X + X^3$, we have $\alpha^3 = 1 + \alpha$. Using this relation, we can simply write the elements α^4, α^5 and α^6 in terms of 1, α, and α^2 as:

$$\alpha^4 = \alpha \cdot \alpha^3 = \alpha \cdot (1 + \alpha) = \alpha + \alpha^2,$$
$$\alpha^5 = \alpha \cdot \alpha^4 = \alpha \cdot (\alpha + \alpha^2) = \alpha^2 + \alpha^3 = 1 + \alpha + \alpha^2,$$
$$\alpha^6 = \alpha \cdot \alpha^5 = \alpha + \alpha^2 + \alpha^3 = \alpha + \alpha^2 + 1 + \alpha = 1 + \alpha^2.$$

For instance, for the addition of α^5 and α^3, we obtain $\alpha^5 + \alpha^3 = 1 + \alpha + \alpha^2 + 1 + \alpha = \alpha^2$. The complete addition and multiplication tables are shown in Figure 4.6. □

Since we can construct $GF(2^3)$ using another primitive polynomial, it is clear that we can use a different primitive element, and obtain an entirely different addition table. Hence one may ask the question whether a finite field of a particular order is unique or not. This is answered by a fundamental result in finite field algebra which states that all such constructions are isomorphic, i.e., the Galois field of a particular order is unique up to a renaming of its elements [1].

To illustrate this point further, we present the following example.

Example 4.4.2. Consider $GF(2^3)$ generated by $p(X) = 1 + X^2 + X^3$ which is also a primitive polynomial. Taking a primitive element of the field as a root of this polynomial, we can enumerate all the field elements as 0, 1, β, β^2,

$$\beta^3 = 1 + \beta^2,$$
$$\beta^4 = \beta \cdot \beta^3 = \beta \cdot (1 + \beta^2) = \beta + \beta^3 = 1 + \beta + \beta^2,$$
$$\beta^5 = \beta \cdot \beta^4 = \beta \cdot (1 + \beta + \beta^2) = \beta + \beta^2 + \beta^3 = 1 + \beta,$$
$$\beta^6 = \beta \cdot \beta^5 = \beta(1 + \beta) = \beta + \beta^2.$$

Figure 4.7 shows the addition and multiplication tables obtained with this construction. It may not be obvious, but it is easy to check that by renaming the elements of $GF(2^3)$ as

$$0 \longleftrightarrow 0$$
$$1 \longleftrightarrow 1$$
$$\beta \longleftrightarrow \alpha^6$$
$$\beta^2 \longleftrightarrow \alpha^5$$
$$\beta^3 \longleftrightarrow \alpha^4$$
$$\beta^4 \longleftrightarrow \alpha^3$$
$$\beta^5 \longleftrightarrow \alpha^2$$
$$\beta^6 \longleftrightarrow \alpha$$

we obtain the binary operation table given in Figure
structions for this example and for Example 4.4.1 ar
of the elements.

Now that we have constructed the finite field $GF($
can perform usual arithmetic operations using its el
multiplication and addition rules. For instance, the
the process of solving a set of linear equations.

Example 4.4.3. Consider the finite field $GF(2^4)$.
and a root of $1 + X + X^4$. We are interested in sol
Y which satisfy

$$\alpha^3 X + \alpha^2 Y = \alpha^7$$
$$\alpha^7 X + \alpha^{12} Y = \alpha^3.$$

We first note that the elements $\alpha^4, \alpha^5, \ldots, \alpha^{14}$ of G

$$\alpha^4 = \alpha + 1,$$
$$\alpha^5 = \alpha^2 + \alpha,$$
$$\alpha^6 = \alpha^3 + \alpha^2,$$
$$\alpha^7 = \alpha^3 + \alpha + 1,$$
$$\alpha^8 = \alpha^2 + 1,$$
$$\alpha^9 = \alpha^3 + \alpha,$$
$$\alpha^{10} = \alpha^2 + \alpha + 1,$$
$$\alpha^{11} = \alpha^3 + \alpha^2 + \alpha,$$
$$\alpha^{12} = \alpha^3 + \alpha^2 + \alpha + 1,$$
$$\alpha^{13} = \alpha^3 + \alpha^2 + 1,$$
$$\alpha^{14} = \alpha^3 + 1.$$

$\cdot$	0	1	α	α^2	α^3	α^4	α^5	α^6
0	0	0	0	0	0	0	0	0
1	0	1	α	α^2	α^3	α^4	α^5	α^6
α	0	α	α^2	α^3	α^4	α^5	α^6	1
α^2	0	α^2	α^3	α^4	α^5	α^6	1	α
α^3	0	α^3	α^4	α^5	α^6	1	α	α^2
α^4	0	α^4	α^5	α^6	1	α	α^2	α^3
α^5	0	α^5	α^6	1	α	α^2	α^3	α^4
α^6	0	α^6	1	α	α^2	α^3	α^4	α^5

$+$	0	1	α	α^2	α^3	α^4	α^5	α^6
0	0	1	α	α^2	α^3	α^4	α^5	α^6
1	1	0	α^3	α^6	α	α^5	α^4	α^2
α	α	α^3	0	α^4	1	α^2	α^6	α^5
α^2	α^2	α^6	α^4	0	α^5	α	α^3	1
α^3	α^3	α	1	α^5	0	α^6	α^2	α^4
α^4	α^4	α^5	α^2	α	α^6	0	1	α^3
α^5	α^5	α^4	α^6	α^3	α^2	1	0	α
α^6	α^6	α^2	α^5	1	α^4	α^3	α	0

Figure 4.6: Multiplication (top) and addition (bottom) tables for $GF(2^3)$ generated by the primitive polynomial $1 + X + X^3$.

·	0	1	β	β²	β³	β
0	0	0	0	0	0	0
1	0	1	β	β²	β³	β
β	0	β	β²	β³	β⁴	β
β²	0	β²	β³	β⁴	β⁵	β
β³	0	β³	β⁴	β⁵	β⁶	1
β⁴	0	β⁴	β⁵	β⁶	1	β
β⁵	0	β⁵	β⁶	1	β	β
β⁶	0	β⁶	1	β	β²	β

+	0	1	β	β²	β³	β
0	0	1	β	β²	β³	β
1	1	0	β⁵	β³	β²	β
β	β	β⁵	0	β⁶	β⁴	β
β²	β²	β³	β⁶	0	1	β
β³	β³	β²	β⁴	1	0	β
β⁴	β⁴	β⁶	β³	β⁵	β	0
β⁵	β⁵	β	1	β⁴	β⁶	β
β⁶	β⁶	β⁴	β²	β	β⁵	1

Figure 4.7: Multiplication (top) and addition (bottor
by the primitive polynomial $1 + X^2 + X^3$.

Multiplying the first equation by α^4 and adding the two equations, we obtain

$$(\alpha^7 + \alpha^7)X + (\alpha^6 + \alpha^{12})Y = \alpha^{11} + \alpha^3.$$

Since $\alpha^7 + \alpha^7 = 0$, $\alpha^6 + \alpha^{12} = \alpha^3 + \alpha^2 + \alpha^3 + \alpha^2 + \alpha + 1 = \alpha + 1 = \alpha^4$, and $\alpha^{11} + \alpha^3 = \alpha^3 + \alpha^2 + \alpha + \alpha^3 = \alpha^2 + \alpha = \alpha^5$, we obtain $\alpha^4 Y = \alpha^5$, and $Y = \alpha$. Substituting this into the first original equation gives $\alpha^3 X + \alpha^2\alpha = \alpha^7$, which results in $X = \alpha$. □

Although linear equations are solved with ease, the same may not be true for non-linear equations. In general, we would need to substitute all the elements of the Galois field into the set of equations being solved to identify the solutions. For instance, finding the roots of a polynomial in an extension field is not trivial, nor computing logarithms (i.e., performing inverse of exponentiation).

4.4.1   Conjugate Elements

We can define conjugates of an element of finite field akin to the conjugate of a number in the complex number field. We start with a basic result.

Theorem 4.4.1. *If $b \in GF(2^m)$ is a root of a polynomial $p(X)$ over $GF(2)$, then b^{2^k} is also a root of the same polynomial for any non-negative integer k.*

Proof. Let $p(X) = p_0 + p_1 X + \cdots + p_n X^n$, we can easily write

$$\begin{aligned}
(p(X))^2 &= (p_0 + p_1 X + \cdots + p_n X^n)^2 \\
&= p_0^2 + p_1^2 X^2 + \cdots + p_n^2 X^{2n} \\
&= p_0 + p_1 X^2 + \cdots + p_n X^{2n} \\
&= p(X^2).
\end{aligned}$$

The second and third lines follow since the coefficients and the operations are in $GF(2)$. Continuing in this manner, we find that $(p(X))^{2^k} = p(X^{2^k})$. Hence, if $p(b) = 0$, we also have $p(b^{2^k}) = 0$, and the proof is complete. □

As a side note, consider the complex number field. If a polynomial with real coefficients has a particular complex number as one of its roots, then the complex conjugate of the number is also a root – similar to the result of this theorem (which applies in the context of finite fields).

Definition 4.4.1. The elements of the field of the form b^{2^k} are defined as the conjugates of b. Obviously, we have a finite number of conjugates for any element of the Galois field as there are only finitely many distinct elements. The set of all conjugates of b is referred to as the conjugacy class of the element b.

Definition 4.4.2. The smallest degree polynomial $\phi(X)$ over $GF(2)$ such that $\phi(b) = 0$ ($b \in GF(2^m)$) is called the minimal polynomial of b.

Note that the minimal polynomial is unique. To s
two distinct degree-n minimal polynomials $\phi_1(X), \phi$
Then their sum (which is a polynomial of a smaller c
also has the same element as a root, which is a cont

Theorem 4.4.2. *The minimal polynomial of any ele*
over $GF(2)$.

Proof. Assume that it is not irreducible, and write
$\phi_1(b)\phi_2(b) = 0$, hence either $\phi_1(b) = 0$ or $\phi_2(b) = 0$

Theorem 4.4.3. *If $p(b) = 0$ for a binary polynomi*
the minimal polynomial of b divides $p(X)$.

Proof. See Problem 4.6.9.

As a corollary, we can say that if $p(X)$ is an irred
coefficients and $p(b) = 0$ for $b \in GF(2^m)$, then $p(X)$
b.

Theorem 4.4.4. *The minimal polynomials of th*
$X^{2^m} + X$ (also including the minimal polynomial fc

Theorem 4.4.5. *If the distinct conjugates of $b \in C$*
$b_1 = b$), then its minimal polynomial is given by

$$\phi(X) = \prod_{i=1}^{n}(X + b_i).$$

We further note that $n \leq m$, and hence the degree o
is at most m.

Proof. See [1].

Example 4.4.4. Consider $GF(2^4)$. Let $b = \alpha^5$ wh
the field (and a root of the primitive polynomial $1 +$
of b is $b_2 = b^2 = \alpha^{10}$ (note $b^4 = \alpha^{20} = \alpha^5 = b$). Ther
for $b = \alpha^5$ is

$$\phi(X) = (X + \alpha^5)(X + \alpha^{10}) = 1$$

Similarly, for the element $\alpha^6 \in GF(2^4)$, the conjug
the corresponding minimal polynomial is

$$\phi(X) = (X + \alpha^6)(X + \alpha^{12})(X + \alpha$$
$$= 1 + X + X^2 + X^3 + X^4.$$

The details of this computation are left as an exerci

4.4.2   Factorization of the Polynomial $X^n + 1$

An important issue in channel coding, particularly in the design of cyclic codes, is the factorization of the polynomial of the type $X^n + 1$ over $GF(2)$. For instance, the factors of this polynomial can be used as generator polynomials for a cyclic code of length n bits. For the special case where $n = 2^m - 1$, the process of factoring $X^n + 1$ relies on the ideas developed earlier in the section with the observation that we already have a tool to generate minimal polynomials of all the elements of $GF(2^m)$, and the roots of these polynomials define the corresponding conjugacy classes. A strongly related result further states that the factors of $X^{2^m} + X$ are all irreducible polynomials of degrees that divide m (see [1]), and all the elements of $GF(2^m)$ are the roots of the binary polynomial $X^{2^m} + X$.

The factors of $X^{2^m - 1} + 1$ are simply the minimal polynomials of the non-zero elements of $GF(2^m)$. Let us illustrate this computation by a simple example.

Example 4.4.5. Consider $GF(2^3)$ generated by $p(X) = 1 + X + X^3$, and the minimal polynomials of its elements to factorize the binary polynomial $X^7 + 1$. The minimal polynomial for 1 is $X + 1$, for the elements $\alpha, \alpha^2, \alpha^4$, the minimal polynomial is $(X + \alpha)(X + \alpha^2)(X + \alpha^4) = 1 + X + X^3$ (assuming that α is a root of the primitive polynomial $p(X) = 1 + X + X^3$), and for $\alpha^3, \alpha^6, \alpha^5$, the corresponding minimal polynomial is $(X + \alpha^3)(X + \alpha^6)(X + \alpha^5) = 1 + X^2 + X^3$. Hence, we can factorize $X^7 + 1$ as

$$X^7 + 1 = (X + 1)(X^3 + X + 1)(X^3 + X^2 + 1). \quad \Box$$

We give a similar example in the following.

Example 4.4.6. To factorize the binary polynomial $X^{15} + 1$, let us consider the $GF(2^4)$ generated by the primitive polynomial $1 + X + X^4$. Let α be a root of this polynomial. Then the conjugacy classes of the non-zero elements of the field are

$$\{1\},$$
$$\{\alpha, \alpha^2, \alpha^4, \alpha^8\},$$
$$\{\alpha^3, \alpha^6, \alpha^{12}, \alpha^9\},$$
$$\{\alpha^5, \alpha^{10}\},$$
$$\{\alpha^7, \alpha^{14}, \alpha^{13}, \alpha^{11}\},$$

and the corresponding minimal polynomials (after some algebra) are

$$X + 1,$$
$$X^4 + X + 1,$$
$$X^4 + X^3 + X^2 + X + 1,$$
$$X^2 + X + 1,$$

$$X^4 + X^3 + 1.$$

Therefore, we can factorize $X^{15} + 1$ as

$$X^{15}+1 = (X+1)(X^2+X+1)(X^4+X^3+1)(X^4+X$$

For an arbitrary positive integer n, the factorizat
procedure. The main difference is that there is not
and we cannot work with a primitive element of such
classes. Therefore, we identify an extension field G
of $2^m - 1$. Therefore, there exists an element $\beta \in$
can work with the minimal polynomials of the pow
distinct resulting minimal polynomials are factors of
examples to conclude the section.

Example 4.4.7. To factorize $X^5 + 1$, we consider t
divides $2^4 - 1$. Let b be an element of $GF(2^4)$ with
$b = \alpha^3$ where α is a primitive element of $GF(2^4)$. T
with respect to b: $\{1\}$ and $\{b, b^2, b^4, b^3\}$. The corres
are $X + 1$ and $X^4 + X^3 + X^2 + X + 1$, hence

$$X^5 + 1 = (X + 1)(X^4 + X^3 + X^2$$

Example 4.4.8. Let us factorize $X^{13} + 1$. Conside
field of $GF(2)$ with order 13 (such elements are pre
$2^{12} - 1 = 4095$). Then the conjugacy classes are

$$\{1\},$$

and

$$\{b, b^2, b^4, b^8, b^3, b^6, b^{12}, b^{11}, b^9, b^5$$

Therefore, there are two irreducible factors of $X^{13} -$

$$X^{13}+1 = (X+1)(X^{12}+X^{11}+X^{10}+X^9+X^8+X^7+X$$

4.5   Some Notes on Applicatior

We close this chapter by highlighting a couple of a
particular, for channel coding and cryptography. C
information, 0's and 1's, over a noisy channel. In or
bits against channel errors, it is a common practice
to a coded sequence of bits, and transmit the res
ceiver side, occasional channel errors are identified (
the controlled redundancy in the codewords. This
nel coding and is an integral part of any modern c

The second application area, cryptography, refers to the process of communication with an intended receiver (loosely speaking) while making sure that a third party (eavesdropper) cannot decode the message bits even if it has access to the entire transmitted sequence.

Cyclic codes represent a fairly general and widely employed channel coding method. Let us consider a message sequence of length k-bits as a degree $k-1$ polynomial over $GF(2)$, i.e., represent the message bits m_i, $i \in \{0, 1, \ldots, k-1\}$ with the polynomial $m(X) = m_0 + m_1 X + m_2 X^2 + \cdots + m_{k-1}X^{k-1}$. Define a factor of the polynomial $X^n + 1$ as the generator polynomial of a cyclic code denoted by $g(X)$ (of degree $n-k$). We can then obtain a specific channel code by considering the codeword polynomials as the product of the message polynomial with the generator polynomial, i.e.,

$$c(X) = m(X) \cdot g(X).$$

The coefficients of $c(X)$ determine the components of the n-bit long codeword sequence. It should now be clear that the theory of finite fields is needed for a complete study of such coding schemes. For instance, by using the ideas in the previous section, one can determine possible dimensions of all cyclic codes, i.e., figure out for what values of n and k, cyclic codes do or do not exist.

As an example, consider the factorization of the binary polynomial $X^{23} + 1$ using the techniques described in this chapter. There are three irreducible factors: $X + 1$ and two degree-11 polynomials. Therefore, we can define a cyclic code of dimensions $n = 23$ and $k = 12$ using either one of the two degree-11 factors as its generator polynomial. The so-obtained codes are named "Golay codes" and have a beautiful algebraic structure and desirable properties. They have been employed in practical communication systems including those utilized in deep-space missions of the National Aeronautics and Space Administration (NASA).

Reed-Solomon (RS) codes can also be cited as specific examples of cyclic codes. These codes find widespread applications in existing communication systems, for instance, in magnetic recording systems (hard drives), compact discs (CD), digital versatile discs (DVD), blue-ray disks, etc. When there are scratches on a DVD, the RS code ensures that the digital data can be recovered and accessed reliably. To describe the connection to the theory of Galois fields, consider a specific case with 255 bytes (2040 bits) as codewords correcting t many byte errors. This RS code may be obtained by a generator polynomial with coefficients in $GF(2^8)$ with the roots $\alpha, \alpha^2, \ldots, \alpha^{2t}$ where α is a primitive element of $GF(2^8)$ and t is an integer selected based on the desired error correction capability of the code (which comes at the cost of a reduced code rate). We refer the reader to textbooks on channel coding for the details of different algebraic code constructions, analysis and decoding algorithms, e.g. [3, 4].

As a particular instance of a crypto-system that utilizes finite field algebra, let us consider public key cryptography, where all the users advertise their own keys and encryption takes place by obtaining a common key between user pairs with the publicly available information. For instance, the users can pick an element

of a high order finite field α^m (where α is a primi⁝
make m publicly available (while α^m is kept secr⁝
two users can then be easily generated by perforⁿ
For instance, if users A and B advertise m_A and
$(\alpha^{m_B})^{m_A} = (\alpha^{m_A})^{m_B} = \alpha^{m_A m_B}$ as their commo⁝
knowledge of the primitive element of the field, i⁝
eavesdropper to compute the key due to the inef⁝
calculating logarithms in a finite field.

4.6   Exercises

Exercise 4.6.1. Prove that the inverse of a group

Exercise 4.6.2. Consider $G = \{1, 2, 3, \ldots, p - 1\}$
where p is a prime number. Prove that every eleme⁝
$a \in G$, there exists $b \in G$ such that $a \cdot b = 1$ modulᴄ

Exercise 4.6.3. Consider the group $\{1, 2, \ldots, 6\}$
Show the multiplication table. Specify the inverse of
the orders of all the elements.

Exercise 4.6.4. Prove that no two elements of two⁝
H (of a group G) can be identical.

Exercise 4.6.5. For any two elements of a finite fiel⁝
$(-a) \cdot b = a \cdot (-b)$.

Exercise 4.6.6. If a, b, c are field elements, show ⁝
then $b = c$.

Exercise 4.6.7. Assume that a is a non-zero elem⁝
Prove that $a^{q-1} = 1$.

Exercise 4.6.8. Consider $GF(2^3)$ and assume tha⁝
polynomial $1 + X + X^3$. Solve for X, Y, Z if

$$\alpha X + Y + \alpha^3 Z = \alpha^5$$
$$\alpha^3 Y + \alpha Z = \alpha$$
$$\alpha^5 X + \alpha^2 Z = \alpha^3.$$

Exercise 4.6.9. Let $p(X)$ be a polynomial over G⁝
$GF(2^m)$. Prove that if $f(b) = 0$, then the minimal ⁝
$p(X)$.

Exercise 4.6.10. Prove that degree of the minima⁝
$GF(2^m)$ is at most m.

Exercise 4.6.11. Show the details of the minimal polynomial computation in Example 4.4.4.

Exercise 4.6.12. Find the degrees of the minimal polynomials of the Galois field elements of order 15 and 1023 (with respect to $GF(2)$). Specify examples of Galois fields that contain these elements.

Exercise 4.6.13. How many binary irreducible polynomials exist in the factorization of $X^{25} + 1$? Also find the degrees of the different factors.

References

[1] R. J. McEliece, *Finite Fields for Computer Scientists and Engineers.* Boston: Kluwer Academic Publishers, 1987.

[2] R. Lidl and H. Niederreiter, *Introduction to Finite Fields and Their Applications.* Cambridge: Cambridge University Press, 1986.

[3] S. Lin and D. J. Costello, Jr., *Error Control Coding: Fundamentals and Applications*, 2nd ed. Upper Saddle River, NJ: Prentice Hall, 2004.

[4] S. B. Wicker, *Error Control Systems for Digital Communications and Storage.* Upper Saddle River, NJ: Prentice Hall, 1995.

[5] H. C. A. van Tilborg, *An Introduction to Cryptology.* Boston: Kluwer Academic Publishers, 1988.

Chapter 5   Numerical Analysis

Vivek Sarin[‡]

[‡]Texas A&M University, College Station, USA

Computers provide an indispensable tool for computing the solution of a wide array of problems. These problems often involve numerical calculations in which numbers are manipulated on the computer. These computations include basic arithmetic operations on real numbers. An important limitation of computers is the inaccuracy inherent in storing and manipulating real numbers. Each computation can potentially give rise to a numerical error that propagates through the computational process. Understanding the effect of these errors on the computed solution is extremely important. It can lead to better models, algorithms, and software that are more resilient to numerical errors.

5.1   Numerical Approximation

Numerical error is the difference between the true value and the computed value of a quantity. The significance of error depends on the magnitude of the true value. We define *absolute error* and *relative error* to represent this notion formally:

$$\text{Absolute Error} = \hat{a} - a, \tag{5.1a}$$

$$\text{Relative Error} = \frac{\hat{a} - a}{a}, \tag{5.1b}$$

where $\hat{a}$ is an approximation of a. A useful relation between the true and approximate value is given below

$$\hat{a} = a(1 + \text{relative error}). \tag{5.2}$$

The number of digits of accuracy in the approximation $\hat{a}$ is related to the relative error

$$\text{Decimal digits of accuracy} = -\log_{10}|\text{relative error}|. \tag{5.3}$$

Error in the computed value can be attributed to two sources: *data error* and *computational error*. Data error refers to the error in data that exists prior

to computation, and includes errors from measure
is incurred during the computational process. Bc
through the computational steps. These errors can
to overwhelm the computation, leading to erroneou:

There are two main sources of computational err
cation errors. Roundoff errors are due to the represe
a finite number of digits. Truncation errors occur as
of a true value, such as when using a finite number

Example 5.1.1 (Rounding Error). Consider tw
9.99×10^{-4} whose sum is 9.990999. On a compa
representation for numbers, the result of $a + b$ wil
identical to a. Other arithmetic operations may not
no error in dividing a by b: $a/b = 1.0 \times 10^4$. Mult
$9.98001 \times 10^{-3} \approx 9.98 \times 10^{-3}$ has a relative error o

Example 5.1.2 (Truncation Error). Consider t
the infinite series

$$e^x = 1 + x + \frac{x^2}{2!} + \frac{x^3}{3!} +$$

The truncation error is $e^\eta x^{k+1}/k!$, where $0 < \eta < x$,
k terms. A similar situation arises when approximat
using the standard finite difference scheme:

$$f'(x) \approx \frac{f(x+h) - f(x}{h}$$

The truncation error is computed using Taylor's expa
$x < \eta < x + h$.

It is important to understand how error propag
ations. Let us analyze the case of addition of two
approximations $\hat{a}$ and $\hat{b}$, respectively. Let $\hat{a} = a(1$
relative error in a. Similarly, let $\hat{b} = b(1 + \epsilon_b)$. Ther

$$\hat{a} + \hat{b} = (a + b)\left(1 + \frac{a}{a+b}\epsilon_a + \frac{b}{a+b}\epsilon_b\right) =$$

where ϵ_{a+b} denotes the relative error in addition, ar

$$\epsilon_{a+b} = \frac{a}{a+b}\epsilon_a + \frac{b}{a+b}\epsilon$$

To compute the relative error in multiplication, obse

$$\hat{a} \times \hat{b} = a \times b(1 + \epsilon_a)(1 + \epsilon_b) \approx a \times b$$

where the term $\epsilon_a \epsilon_b$ is ignored since it is negligible compared to ϵ_a and ϵ_b. Thus,

$$\epsilon_{a \times b} = \epsilon_a + \epsilon_b. \tag{5.5}$$

The relative error in division is computed similarly:

$$\frac{\hat{a}}{\hat{b}} = \frac{a}{b} \left(\frac{1 + \epsilon_a}{1 + \epsilon_b} \right) \approx \frac{a}{b} (1 + \epsilon_a - \epsilon_b)$$

and

$$\epsilon_{a/b} = \epsilon_a - \epsilon_b. \tag{5.6}$$

Example 5.1.3 (Error Propagation). Consider the problem of computing the square of a using its approximation $\hat{a}$. Since $\hat{a} = a(1 + \epsilon)$, where ϵ is the relative error, we have $\hat{a}^2 = a^2(1 + 2\epsilon + \epsilon^2)$ that can be approximated by $a^2(1 + 2\epsilon)$ since ϵ^2 is negligible compared to ϵ. This calculation shows that the relative error in computing the square is twice the relative error in the data. □

The relative error in multiplication and division operations depends only on the relative error in the operands. On the other hand, the relative error in addition depends on the values of the operands. The relative error can be large when $a + b$ is small compared to either a or b. Such a situation occurs when adding nearly equal numbers with an opposite sign.

Loss of accuracy when subtracting numbers of similar magnitudes is called *catastrophic cancellation*. Consider adding two numbers $a = 80.0499$ and $b = -79.9999$ on a computer with 6-digit representation. Even though the result $a+b = 0.0500$ has no computational error, it has only three significant digits compared to six in the operands. This loss in accuracy is due to cancellation and is due to the large value of $1/(a + b)$ that multiplies the relative error of the operands.

Example 5.1.4 (Catastrophic Cancellation). Let us compute the roots of the quadratic equation

$$x^2 - 800x + 1 = 0$$

on a computer that uses 3-digit arithmetic. Using the well-known formula, we get

$$x_{\pm} = \frac{800 \pm \sqrt{(-800)^2 - 4}}{2}.$$

In 3-digit arithmetic, $800^2 - 4$ evaluates to $640,000$, which results in $\hat{x}_- = 0$ and $\hat{x}_+ = 800$. Compared to the more accurate values $x_- = 0.00125000$ and $x_+ = 799.99875$, we see that the computed value for x_- has large error. □

There are two ways to avoid catastrophic cancellation. One can modify the algorithm to avoid the steps in which cancellation occurs. For example, one can use the following formula to compute the roots of the quadratic equation $ax^2 + bx + c = 0$:

$$x_1 = \frac{-b - \text{sign}(b)\sqrt{b^2 - 4ac}}{2a},$$

The computation for the numerator of x_1 now invc
same sign, which eliminates the possibility of cancel
roots in the preceding example are computed to be
$x_2 = 1/800 = 0.00125$ using 3-digit arithmetic.

One can use Taylor's expansion to eliminate the
error. Cancellation occurs when computing the d
points that are close to each other. Knowledge of
reformulate the problem:

$$y = f(x + \delta) - f(x) = \left[f(x) + \delta f'(x) + \frac{\delta^2}{2!} f''(\cdot \right.$$

in which there is no need to compute the difference

5.2   Sensitivity and Conditioni

The inaccuracy in the computed solution is not alway
Problems that are highly sensitive to perturbations i
with large error in the computed solution. Sensitive
in input data even in the absence of computation
computation is exact. The *sensitivity* of a problem ca
called the *condition number* that is defined as rat:
computed value to the relative error in the input da

$$\text{Condition number} = \frac{\text{relative err}}{\text{relative err}}$$

The condition number is the magnification underge
the solution process.

A problem is said to be *well-conditioned* if a g
data causes a commensurate change in the solution
conditioned if a given relative change in input data c
the solution. The condition number of the problem p
of its conditioning. The condition number of compu

$$\text{Condition number} = \frac{|(f(\hat{x}) - f}{|(\hat{x} - }$$

Substituting $\hat{x} = x + \Delta x$, and taking the limit as Δ
the following expression

$$\text{Condition number} = \left| f'(x) \cdot \right.$$

Example 5.2.1 (Condition Number). Consider the following system of equations

$$wx + 2y = 1$$
$$2x + wy = 0$$

whose solution is given by

$$x(w) = \frac{w}{w^2 - 4}, \qquad y(w) = \frac{-2}{w^2 - 4},$$

where the values of x and y depend on the parameter w. The problem of computing $x(w)$ has the condition number

$$\kappa_x = \left| w \cdot \frac{x'(w)}{x(w)} \right| = \left| w \cdot \frac{(-w^2 - 4)}{(w^2 - 4)^2} \cdot \frac{(w^2 - 4)}{w} \right| = \left| \frac{w^2 + 4}{w^2 - 4} \right|$$

This shows that the problem becomes ill-conditioned when $|w| \approx 2$ because the condition number κ_x becomes large. For values of w close to 2 or -2, the left-hand sides of the linear system become nearly identical, and the system becomes ill-conditioned.

On the other hand, the problem of computing $x(w) + y(w)$ has the condition number

$$\kappa_{x+y} = \left| w \cdot \frac{x'(w) + y'(w)}{x(w) + y(w)} \right|$$

Since $x(w) + y(w) = 1/(w + 2)$,

$$\kappa_{x+y} = \left| w \cdot \frac{(-1)}{(w + 2)^2} (w + 2) \right| = \left| \frac{w}{(w + 2)} \right|$$

This problem is ill-conditioned when $w \approx -2$, but is well-conditioned for other values of w. □

It may appear that it is not possible to get an acceptable solution to an ill-conditioned problem due to the growth of data and computational errors. To get around this issue, we must revisit the notion of an acceptable solution for ill-conditioned problems. We define *backward error* as a hypothetical perturbation in input data that would yield the same solution through exact computation as the solution computed by the algorithm. Suppose an algorithm to compute $y = f(x)$ actually computes an approximation $\hat{y}$ that includes data and computational errors. Let $\hat{x}$ be the point at which the function matches the computed value, i.e., $\hat{y} = f(\hat{x})$. Since the computed value $\hat{y}$ is the exact result for a perturbed input, it is acceptable as a solution as long as the perturbation is small. For ill-conditioned problems, large error in the computed value $\hat{y}$ can be disregarded if the backward error $\hat{x} - x$ is small.

5.3   Computer Arithmetic

To understand the behavior of computational algor
essary to understand how numbers are represented
represented in a positional number system in which
the value it contributes to the number. An integer
of digits $(a_n a_{n-1} \ldots a_1 a_0)_\beta$, where β is the base c
following restrictions: $1 \leq a_n < \beta$ and $0 \leq a_i < \beta$,
computed as follows.

$$I = (a_n a_{n-1} \ldots a_1 a_0)_\beta = \sum_{i=}^{n}$$

We use the decimal system with $\beta = 10$ in our dai
the binary system with $\beta = 2$ to represent intege
represented as a sequence of digits $(a_{-1} a_{-2} \ldots a_{-n})$

$$z = (a_{-1} a_{-2} \ldots a_{-n})_\beta = \sum_{i=1}^{n}$$

An infinite number of digits are needed to represent
Furthermore, real numbers that require a finite nu
may need an infinite number of digits in another s
$(0.3333\ldots)_{10}$ and $(0.2)_{10} = (0.001100110011\ldots)_2$.

A real number can be represented as $x = \pm\beta^e($
d_i are such that $d_1 \neq 0$ and $0 \leq d_i < \beta, i > 1$. The
of x is defined as $fl(x) = \pm\beta^e(0.\delta_1\delta_2\delta_3 \ldots \delta_p)$, where
The value of $fl(x)$ is given by

$$fl(x) = \pm\beta^e \sum_{i=1}^{p} \delta_i \beta^{-i}$$

For a floating-point number, it is sufficient to store th
$(\delta_1\delta_2\delta_3 \ldots \delta_p)$.

The floating-point numbers used on a computer
gers: base or radix β, precision p, and lower and u
and U. These values determine the number of bits
For example, the IEEE standard uses 64 bits to st
double precision. The exponent uses 11 bits and th
bit is reserved for the sign. In this format, $\beta =$
$U = 1023$. This allows us to store numbers in the
roughly equivalent to $(10^{-308}, 10^{308})$. Numbers wit
L cause *underflow* while those with a value greater
There are important limitations to the floating
numbers that can be represented are a subset of

| 0 | 1 | 11 | 12 | | 63 |

Figure 5.1: IEEE double precision floating-point numbers.

of the exponent e, we can represent $(\beta - 1)\beta^{p-1}$ equally spaced numbers with a spacing of β^{e-p}. To convert a real number to its floating-point equivalent, one needs to use *rounding* or *chopping*. Rounding requires that the real number be converted to the nearest floating-point number. Chopping is done by retaining the first p digits in the floating-point representation.

The absolute and relative errors incurred in rounding depend on the base β and the precision p:

$$\text{Absolute error in rounding} = |fl(x) - x| < 0.5\beta^{e-p}, \qquad (5.11a)$$

$$\text{Relative error in rounding} = \left|\frac{fl(x) - x}{x}\right| < 0.5\beta^{1-p}. \qquad (5.11b)$$

It is useful to define the *unit round-off* error which is the largest relative error possible when converting a real number to its floating-point equivalent. This value is also known as the *machine precision* of a floating-point number system, and is given by

$$\epsilon_{\text{mach}} = 0.5\beta^{1-p}. \qquad (5.12)$$

The machine precision characterizes the accuracy of a floating-point number system. A floating-point number $fl(x)$ has the following relation to x:

$$fl(x) = x(1 + \epsilon), \qquad |\epsilon| \le \epsilon_{\text{mach}}. \qquad (5.13)$$

The machine precision also allows us to analyze the effect of rounding error in basic arithmetic operations. One can assume that all basic arithmetic operations obey the following model:

$$fl(x \circ y) = (x \circ y)(1 + \epsilon), \qquad |\epsilon| \le \epsilon_{\text{mach}}, \qquad (5.14)$$

where $\circ$ is one of the basic arithmetic operations $+, -, \times, /$. This indicates that the relative error in basic arithmetic operations does not exceed ϵ_{mach}.

A computer algorithm consists of a sequence of such operations. The analysis of error propagation in a computer algorithm using this model is called *forward error analysis*.

Example 5.3.1 (Rounding Error in Addition). Let us consider the computation $x + y + z$. In floating-point arithmetic, we ha

$$
\begin{aligned}
fl(x + y + z) &= fl(fl(x + y) + z) \\
&= fl((x + y)(1 + \epsilon_1) + z), \\
&= ((x + y)(1 + \epsilon_1) + z)(1 + \\
&= (x + y)(1 + \epsilon_1)(1 + \epsilon_2) + \\
&\leq (x + y + z)(1 + |\epsilon_1|)(1 + | \\
&\approx (x + y + z)(1 + |\epsilon_1| + |\epsilon_2| \\
&= (x + y + z)(1 + |\epsilon|), \qquad |
\end{aligned}
$$

Thus, the relative error in the floating-point additi
exceed $2\epsilon_{\text{mach}}$.

In general, the bound-on error when computing

$$
\left| fl\left(\sum_{i=1}^{n} x_i\right) - \left(\sum_{i=1}^{n} x_i\right) \right| \leq (n-1)\epsilon
$$

Example 5.3.2 (Error Propagation in Floatin
sider the recurrence to compute a sequence of numk

$$
s_k = 3s_{k-1} - \frac{8}{9}s_{k-2}, \qquad k
$$

with initial values $s_1 = 8$ and $s_2 = 8/3$. The solu
$8/3^{k-1}$. We use the following code in MATLAB to

```
s(1) = 8;
s(2) = 8/3;
for k = 3:50
    s(k) = 3*s(k-1) - (8/9)*s(k-2)
end
```

The values of s_k are shown below. The figure s
by constant factor until $k = 20$ and then grow by cc
The relative error in s_k increases from $-4.44 \times 1($
for $k = 20$ at which point it is comparable to the f
6.8831×10^{-9}. To understand the reason for the gro
general solution for the recurrence is given by

$$
s_k = \alpha_1 \left(\frac{1}{3}\right)^{k-1} + \alpha_2 \left(\frac{8}{3}\right)
$$

The constants α_1 and α_2 are computed to satisfy the
to satisfy $s_1 = 8$ and $s_2 = 8/3$, we must have $\alpha_1 =$

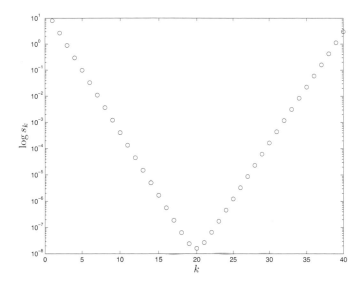

Figure 5.2: Growth of error in floating-point computation.

floating-point representation of the initial value s_2 is not exact, which causes α_2 to be a small non-zero value. The error introduced in α_2 allows the second term to be part of the computed solution. Even though α_1 and α_2 are not computed, the solution implicitly includes the two terms. The second term grows rapidly and dominates the solution for $k > 20$.

5.4   Interpolation

Interpolation is the process of fitting a function to given data such that the function has the same values as the given data. For example, consider the data

$$
\begin{array}{c|ccc}
x & 1 & 2 & 3 \\
y & -3 & 0 & 5
\end{array}
$$

The function $f(x) = x^2 - 4$ interpolates the data because $f(x) = y$ at $x = 1, 2, 3$. This function can be used to compute estimates for the value of y in the interval $[1, 3]$. It can also be used to compute estimates outside this interval.

The interpolating function is called an *interpolant*. The interpolant can be used to estimate the trend shown by the data at points other than those provided. These estimates can be within the data range or outside the range. Estimating values within the data range is called *interpolation*, whereas estimating values outside the data range is called *extrapolation*. When the data represent an underlying unknown function, the interpolant can be used to compute an estimate of the derivative or the integral of the function. The interpolant is sometimes used to

represent a function that is expensive or difficult to
also be used to plot a smooth curve between the da
   Given a set of n data values (x_i, y_i), the interpol
function $f(x)$ such that

$$f(x_i) = y_i, \qquad i = 1, \ldots$$

An infinite number of functions satisfy these condi
function that is a good representation of the data v
It is desirable to use a simple function as an inter
interpolant determines which function should be use

5.4.1   Polynomial Interpolation

The simplest and most common type of interpolat
polate the data. Given n data values (x_i, y_i) suc
the problem is to determine the lowest degree poly
condition

$$p(x_i) = y_i, \qquad i = 1, \ldots$$

Consider the $(n-1)$ degree polynomial

$$p_{n-1}(x) = a_1 + a_2 x + \cdots + a$$

where the coefficients a_i need to be determined to
coefficients satisfy the following system of equations

$$\mathbf{Ax} = \begin{bmatrix} 1 & x_1 & \cdots & x_1^{n-1} \\ 1 & x_2 & \cdots & x_2^{n-1} \\ \vdots & \vdots & \ddots & \vdots \\ 1 & x_n & \cdots & x_n^{n-1} \end{bmatrix} \begin{bmatrix} a_1 \\ a_2 \\ \vdots \\ a_n \end{bmatrix} =$$

The matrix $\mathbf{A}$ is called a Vandermonde matrix.
product of all pairs $(x_i - x_j), i \neq j$ which implies th
x_i's are distinct. This guarantees a solution of the
$(n-1)$ degree polynomial.

Example 5.4.1 (Interpolation). Consider the p
nomial that interpolates the three data points $(-1$
determine the coefficients of the following polynomi

$$p_2(x) = a_1 + a_2 x + a_3 x$$

by solving the system

$$\begin{bmatrix} 1 & -1 & 1 \\ 1 & 0 & 0 \\ 1 & 1 & 1 \end{bmatrix} \begin{bmatrix} a_1 \\ a_2 \\ a_3 \end{bmatrix} = \begin{bmatrix} 1 \\ 1 \\ 3 \end{bmatrix}$$

Gaussian elimination yields the solution $a_1 = 1, a_2 = 1, a_3 = 1$. The interpolating polynomial is

$$p_2(x) = 1 + x + x^2.$$

□

The interpolating polynomial for a given set of data points is unique. Using this property, one can develop efficient means of computing the polynomial. The key idea is to represent the polynomial in a form that is more amenable to computing. Before discussing alternate ways to represent the interpolating polynomial, we need to prove its uniqueness.

Theorem 5.4.1 (Uniqueness of the Interpolating Polynomial). *Given a set of points $x_1 < x_2 < \cdots < x_n$, there exists only one polynomial that interpolates a function at those points.*

Proof Let $p(x)$ and $q(x)$ be two interpolating polynomials of degree at most $(n-1)$ for the same set of points $x_1, x_2, \ldots, x_n$. The difference of these polynomials $r(x) = p(x) - q(x)$ is itself a polynomial of degree no greater than $(n-1)$ with roots at the n interpolating points $x_1, x_2, \ldots, x_n$. This is possible only if $r(x) = 0$, i.e., if $p(x) = q(x)$, which proves that interpolating polynomial is unique. □

The interpolating polynomial can be represented using the *Lagrange* basis functions for $p_{n-1}(x)$, which are defined as

$$L_j(x) = \prod_{i=1, i \neq j}^{n} \frac{(x - x_i)}{(x_j - x_i)}, \qquad j = 1, \ldots, n. \tag{5.19}$$

From the definition, it can be seen that $L_j(x)$ is a polynomial of degree $(n-1)$ such that

$$L_j(x_i) = \begin{cases} 1 & i = j \\ 0 & i \neq j \end{cases}, \qquad i, j = 1 \ldots, n.$$

Due to this property, the interpolating polynomial takes a simple form

$$p_{n-1}(x) = y_1 L_1(x) + y_2 L_2(x) + \cdots + y_n L_n(x). \tag{5.20}$$

Example 5.4.2 (Lagrange Interpolation). Consider the problem of computing the Lagrange interpolating polynomial for the data points $(-1, 1), (0, 1), (1, 3)$ from Example 5.4.1. The Lagrange basis functions are

$$L_1(x) = \frac{(x - 0)(x - 1)}{(-1 - 0)(-1 - 1)} = -\frac{x}{2} + \frac{x^2}{2}$$

$$L_2(x) = \frac{(x + 1)(x - 1)}{(0 + 1)(0 - 1)} = 1 - x^2$$

$$L_3(x) = \frac{(x + 1)(x - 0)}{(1 + 1)(1 - 0)} = \frac{x}{2} + \frac{x^2}{2}$$

and the Lagrange interpolating polynomial is

$$p_2(x) = 1 \cdot \left(-\frac{x}{2} + \frac{x^2}{2} \right) + 1 \cdot \left(1 - x^2 \right)$$

$$= \left(-\frac{x}{2} + \frac{x^2}{2} \right) + \cdot \left(1 - x^2 \right) + \Big($$

Upon simplification we get the same polynomial as

The two methods to compute the interpolating have a different cost associated with computing th them for a given argument. The first one requires matrix $\mathbf{A}$, whereas the second one requires expens the coefficients of the basis functions. A third meth provides a computationally attractive alternative Newton's representation of the interpolating polyno

$$p_{n-1}(x) = c_1 + c_2(x - x_1) + c_3(x - x_1$$
$$+ c_n(x - x_1)(x - x_2)(x - x_3$$

where the coefficients $c_i, i = 1, \ldots, n$ are chosen to er conditions in (5.16). To determine the coefficients, linear system

$$\begin{bmatrix} 1 & 0 & 0 & 0 \cdots & 0 \\ 1 & x_2 - x_1 & 0 & 0 \cdots & 0 \\ 1 & x_3 - x_1 & (x_3 - x_1)(x_3 - x_2) & 0 \cdots & 0 \\ \vdots & \vdots & \vdots & \ddots \vdots & \vdots \\ 1 & x_n - x_1 & (x_n - x_1)(x_n - x_2) & \cdots\cdots & \prod_{i=1}^{n-1}(x_n - \end{bmatrix}$$

The lower triangular structure makes it easy to solv

Example 5.4.3 (Newton Interpolation). Consi the Newton interpolating polynomial for the data p Example 5.4.1. The linear system to be solved is

$$\begin{bmatrix} 1 & 0 & 0 \\ 1 & (0 + 1) & 0 \\ 1 & (1 + 1) & (1 + 1)(1 - 0) \end{bmatrix} \begin{bmatrix} c_1 \\ c_2 \\ c_3 \end{bmatrix}$$

The solution is obtained as $c = \begin{bmatrix} 1 & 0 & 1 \end{bmatrix}^T$, and the i the form

$$p_2(x) = 1 + 0 \cdot (x + 1) + 1 \cdot (x + 1)(x -$$

Simplification yields the same polynomial as in Exa

Newton's representation of the interpolating polynomial has the added advantage of simplicity with which a new point can be added to the data. Adding a new point (x_{n+1}, y_{n+1}) to the data requires computing $p_n(x)$ which has the form

$$p_n(x) = p_{n-1}(x) + c_{n+1} \prod_{i=1}^{n} (x_{n+1} - x_i), \qquad (5.23)$$

where $p_{n-1}(x)$ has already been determined. The unknown coefficient c_{n+1} is computed by evaluating $p_{n-1}(x_{n+1})$ from the relation

$$c_{n+1} = \frac{y_{n+1} - p_{n-1}(x_{n+1})}{\prod_{i=1}^{n} (x_{n+1} - x_i)}. \qquad (5.24)$$

The algorithm to compute Newton's representation of the interpolating polynomial uses this fact to build the polynomial by successively adding data points starting from a single data point.

An alternative method to compute the coefficients of the Newton's polynomial uses the *divided differences* that are defined recursively as follows

$$f[x_1, x_2, \ldots, x_k] = \frac{f[x_2, x_3, \ldots, x_k] - f[x_1, x_2, \ldots, x_{k-1}]}{x_k - x_1}, \qquad (5.25)$$

where $f[x_i] = y_i, i = 1, \ldots, n$. It can be shown that the kth divided difference is identical to the kth coefficient in the Newton's representation, i.e., $c_k = f[x_1, x_2, \ldots, x_k]$. This approach to computing the coefficients is less error prone and is preferred over the others.

Example 5.4.4 (Divided Differences). Consider the use of divided differences in computing the coefficients of the Newton interpolating polynomial for the data points $(-1, 1), (0, 1), (1, 3)$ from Example 5.4.1.

$$f[x_1] = y_1 = 1, \qquad f[x_2] = y_2 = 1, \qquad f[x_3] = y_3 = 3$$
$$f[x_1, x_2] = \frac{f[x_2] - f[x_1]}{x_2 - x_1} = \frac{1-1}{0+1} = 0$$
$$f[x_2, x_3] = \frac{f[x_3] - f[x_2]}{x_3 - x_2} = \frac{3-1}{1-0} = 2$$
$$f[x_1, x_2, x_3] = \frac{f[x_2, x_3] - f[x_1, x_2]}{x_3 - x_1} = \frac{2-0}{1+1} = 1$$

Thus, $c_1 = f[x_1] = 1, c_2 = f[x_1, x_2] = 0, c_3 = f[x_1, x_2, x_3] = 1$. $\qquad \square$

The algorithm to compute the coefficients in Newton's interpolation is given next.

```
c = f
for j = 1 : n
    for i = n : -1 : j
        c_i = (c_i - c_{i-1})/(x_i - x_{i-1})
    end
end
```

Newton's interpolation does not depend on a pa
$x_1, x_2, \ldots, x_n$. The same polynomial is obtained irre
the points are included. In the presence of data ar
ever, different orderings will yield different values c
technique to minimize such errors is to order points .
the mean of the data set.

When interpolating continuous functions with p
assess the quality of the interpolant. The error betw
mial and the function at an arbitrary point other
interpolation is characterized by the following result

Theorem 5.4.2 (Error in Polynomial Interpol
polynomial of degree $(n-1)$ that interpolates a suf
at n points $x_1 < x_2 < \ldots < x_n$. The error at a poin

$$f(x) - p_{n-1}(x) = \frac{f^{(n)}(\xi)}{n!}(x - x_1)(x -$$

where $\xi \in [x_0, x_n] \cup \{x\}$.

Since ξ is an unknown value, the result is not very
a bound for $f^{(n)}(t)$ for $t \in [x_0, x_n] \cup \{x\}$. It does pro
behavior of interpolating polynomials. Increasing t
polynomial by using additional data points does not r
of the interpolation. Although the error is zero at t
be high between two adjacent points.

A lack of uniform convergence of an interpolating
function with increase in degree is seen for equally
equally spaced points to interpolate Runge's function
that the error in the first and last interval increases
polynomial degree even as the error in the middle in
is to use points that are carefully chosen to minimiz
is the *Chebyshev points* that are defined on the inte
$1)\pi/2n), i = 1, \ldots, n$. These points are clustered

interval and result in convergence to the function throughout the interval as the number of points is increased.

5.4.2 Piecewise Polynomial Interpolation

Fitting a large number of data points with a single polynomial of high degree is likely to result in unacceptable errors. Piecewise polynomial interpolation mitigates the difficulties associated with single polynomial interpolants by using several low degree polynomials to interpolate the data over disjoint subintervals. Given a set of data points $(x_i, y_i), i = 1, \ldots, n$, a piecewise linear interpolation scheme uses a different polynomial of degree one in each interval $[x_i, x_{i+1}]$. The interpolating polynomial $p_i(x)$ for the interval $[x_i, x_{i+1}]$ has the form

$$p_i(x) = y_{i+1}\left(\frac{x - x_i}{x_{i+1} - x_i}\right) + y_i\left(\frac{x_{i+1} - x}{x_{i+1} - x_i}\right) \qquad x_i \le x \le x_{i+1} \qquad (5.27)$$

The interpolant is the set of polynomials $p_i(x), i = 1, \ldots, n - 1$.

Piecewise interpolating polynomials suffer from lack of smoothness at the interval end points. Using higher degree polynomials in each interval allows us more freedom in selecting properties of the interpolant. *Splines* are piecewise polynomials of degree k that are continuously differentiable $k - 1$ times at the interval end points. *Cubic splines* are piecewise cubic polynomials that are constructed to have continuous first and second derivatives at internal data points. Each of the $(n - 1)$ intervals $[x_i, x_{i+1}]$ has a cubic polynomial of the form

$$s_i(x) = a_0^{(i)} + a_1^{(i)}x + a_2^{(i)}x^2 + a_3^{(i)}x^3, \qquad x_i \le x \le x_{i+1}, \qquad (5.28)$$

where the coefficients are determined to satisfy the constraints. Matching the data points at the interval ends gives $2(n - 1)$ constraints, and imposing continuity of the first and second derivatives at the internal points gives an additional $2(n - 2)$ constraints. To determine the $4(n - 1)$ coefficients of all the cubic polynomials $s_i(x), i = 1, \ldots, n$ uniquely, one has to assign two more constraints. A common choice of forcing the second derivative to be zero at x_1 and x_n gives *natural cubic splines*.

Natural cubic splines with the above properties are given by

$$s_i(x) = \frac{\sigma_i}{h}(x_{i+1} - x)^3 + \frac{\sigma_{i+1}}{h}(x - x_i)^3 + \left(\frac{y_i}{h} - \sigma_i h\right)(x_{i+1} - x)$$
$$+ \left(\frac{y_{i+1}}{h} - \sigma_{i+1}h\right)(x - x_i), \qquad (5.29)$$

where σ_i denotes the second derivative of the interpolant at the points x_i for $i = 2, \ldots, n - 1$. For simplicity, we discuss only the case with equally spaced points such that the interval size h is fixed. Recall that $\sigma_1 = \sigma_n = 0$. The values of

$\sigma_i, i = 2, \ldots, n-1$ are obtained by solving the follo

$$
\begin{bmatrix}
4 & 1 & & & & \\
1 & 4 & 1 & & & \\
& 1 & 4 & 1 & & \\
& & \ddots & \ddots & \ddots & \\
& & & 1 & 4 & 1 \\
& & & & 1 & 4
\end{bmatrix}
\begin{bmatrix}
\sigma_2 \\
\sigma_3 \\
\sigma_4 \\
\vdots \\
\sigma_{n-3} \\
\sigma_{n-2}
\end{bmatrix}
=
\begin{bmatrix}
g_2 \\
g_3 \\
g_4 \\
\vdots \\
g_{n-3} \\
g_{n-2}
\end{bmatrix}
, \quad g_i = \frac{1}{h}\left(\frac{y}{} \right.
$$

Example 5.4.5 (Natural Spline Interpolation)
terpolating the data given below

$$(-1, 1), (-1/2, 3/4), (0, 1), (1/2,$$

using natural cubic splines. Since there are four i
$s_i(x), i = 1, \ldots, 4$. Each interval has equal width h
elements of the right-hand side vector of the tridiag

$$
g_2 = \frac{1}{1/2}\left(\frac{1 - 3/4}{1/2} - \frac{3/4 - }{1/2} \right.
$$

$$
g_3 = \frac{1}{1/2}\left(\frac{7/4 - 1}{1/2} - \frac{1 - 3/}{1/2} \right.
$$

$$
g_4 = \frac{1}{1/2}\left(\frac{3 - 7/4}{1/2} - \frac{7/4 - }{1/2} \right.
$$

To compute $\sigma_i, i = 2, 3, 4$, we solve the following sys

$$
\begin{bmatrix}
4 & 1 & \\
1 & 4 & 1 \\
& 1 & 4
\end{bmatrix}
\begin{bmatrix}
\sigma_2 \\
\sigma_3 \\
\sigma_4
\end{bmatrix}
=
\begin{bmatrix}
1 \\
1 \\
1
\end{bmatrix}
$$

and obtain $\sigma_2 = 3/14, \sigma_3 = 2/14, \sigma_4 = 3/14$. We a
these values, we obtain the spline interpolant

$$s_1(x) = \frac{0}{7}\left(\frac{-1}{2} - x\right)^3 + \frac{3}{7}(x+1)^3 + \left(2 - \frac{0}{7}\right)\left(\frac{-1}{2} - x\right) + \left(\frac{3}{2} - \frac{3}{7}\right)(x+1)$$

$$= \frac{3}{7}(x+1)^3 + 2\left(-\frac{1}{2} - x\right) + \frac{15}{14}(x+1)$$

$$s_2(x) = \frac{3}{7}(0 - x)^3 + \frac{2}{7}\left(x - \frac{-1}{2}\right)^3 + \left(\frac{3}{2} - \frac{3}{7}\right)(0 - x) + \left(2 - \frac{2}{7}\right)\left(x - \frac{-1}{2}\right)$$

$$= \frac{3}{7} - x^3 + \frac{2}{7}\left(x + \frac{1}{2}\right)^3 + -\frac{15}{14}x + \frac{12}{7}\left(x + \frac{1}{2}\right)$$

$$s_3(x) = \frac{2}{7}\left(\frac{1}{2} - x\right)^3 + \frac{3}{7}(x - 0)^3 + \left(2 - \frac{2}{7}\right)\left(\frac{1}{2} - x\right) + \left(\frac{7}{2} - \frac{3}{7}\right)(x - 0)$$

$$= \frac{2}{7}\left(\frac{1}{2} - x\right)^3 + \frac{3}{7}x^3 + \frac{12}{7}\left(\frac{1}{2} - x\right) + \frac{43}{14}x$$

$$s_4(x) = \frac{3}{7}(1 - x)^3 + \frac{0}{7}\left(x - \frac{1}{2}\right)^3 + \left(\frac{7}{2} - \frac{3}{7}\right)(1 - x) + \left(2 - \frac{0}{7}\right)\left(x - \frac{1}{2}\right)$$

$$= \frac{3}{7}(1 - x)^3 + \frac{43}{14}(1 - x) + 2\left(x - \frac{1}{2}\right)$$

□

5.5 Nonlinear Equations

Mathematical modeling of physical phenomena gives rise to equations that can be linear or nonlinear. These equations may involve a single variable or many variables. An example of a one-dimensional nonlinear equation is $p(x) = 0$, where $p(x)$ is a polynomial of degree greater than one. Computing the roots of a polynomial is equivalent to solving the nonlinear equation $p(x) = 0$.

5.5.1 Interval Bisection

One of the simplest methods to compute a root of a function $f(x)$ is the *interval bisection method*. The method requires an interval $[a, b]$ that encloses a root of $f(x)$. The method successively halves the interval size while ensuring that the root lies within the interval at all times. This approach relies on the *intermediate value theorem* which states that given a function $f(x)$ that is continuous on $[a, b]$ and a real number g that lies between $f(a)$ and $f(b)$, there exists a point $x \in [a, b]$ where $g = f(x)$. A corollary to this states that if $\text{sign}(f(a)) \neq \text{sign}(f(b))$, then there exists $x \in [a, b]$ such that $f(x) = 0$. Using this idea, the method computes the mid-point of the interval $[a, b]$ and determines which half contains the root. The half containing the root is taken to be the interval for the next iteration.

```
while (b − a) > ε
    m = a + (b − a)/2
    if sign(f(a)) = sign(f(m)) then
        a = m
    else
        b = m
    end
end
α = a + (b − a)/2
```

The interval size reduces by a factor of two at ea
The algorithm terminates when the interval size is s
provided by the user. Upon termination of the whi
the mid-point of the last interval as an approximat
the approximate root $\hat{\alpha}$ is bounded as shown below

$$|\hat{\alpha} - \alpha| \leq \frac{\epsilon}{2},$$

where α is the root. Choosing a small value of ϵ
computed root is acceptable. There is, however, a
that prevents ϵ to be arbitrarily small. The interva
and b are within ϵ_{mach} of each other.

Example 5.5.1 (Interval Bisection). Consider t
of the polynomial $p_2(x) = x^2 - \sqrt{3}x$ that lies in the i
bisection method. The root is $\alpha = \sqrt{3}$ which equals
10 digits. The method is called with an initial inter
sequence of iterations given in Table 5.1 shows the
the approximation, and the interval size $(b - a)$.

The number of iterations needed to reduce
$\lceil \log_2(2/10^{-6}) \rceil = 21$.

Each iteration requires a single function evaluati
computational cost of the algorithm. To estimate th
compute the number of iterations required by the
the size of the interval is reduced to $(b - a)/2^k$. The
to converge to the root is given by

$$k = \left\lceil \log_2\left(\frac{b - a}{\epsilon}\right) \right\rceil$$

An interesting feature of this algorithm is that the
pendent of the nature of the function. This makes

Table 5.1: Iterations given for interval bisection method for finding the root of polynomial $p_2(x) = x^2 - \sqrt{3}x$ that lies in the interval $[1, 3]$.

a	$f(a)$	b	$f(b)$	$\hat{\alpha} = a + (b-a)/2$	$\hat{\alpha} - \alpha$	$(b-a)$
1.00000000	-7.32 $\times 10^{-1}$	2.00000000	5.36 $\times 10^{-1}$	1.50000000	2.32 $\times 10^{-1}$	1.00 $\times 10^{0}$
1.50000000	-3.48 $\times 10^{-1}$	2.00000000	5.36 $\times 10^{-1}$	1.75000000	-1.79 $\times 10^{-2}$	5.00 $\times 10^{-1}$
1.50000000	-3.48 $\times 10^{-1}$	1.75000000	3.14 $\times 10^{-2}$	1.62500000	1.07 $\times 10^{-1}$	2.50 $\times 10^{-1}$
1.62500000	-1.74 $\times 10^{-1}$	1.75000000	3.14 $\times 10^{-2}$	1.68750000	4.46 $\times 10^{-2}$	1.25 $\times 10^{-1}$
1.68750000	-7.52 $\times 10^{-2}$	1.75000000	3.14 $\times 10^{-2}$	1.71875000	1.33 $\times 10^{-2}$	6.25 $\times 10^{-2}$
1.71875000	-2.29 $\times 10^{-2}$	1.75000000	3.14 $\times 10^{-2}$	1.73437500	-2.32 $\times 10^{-3}$	3.13 $\times 10^{-2}$
1.71875000	-2.29 $\times 10^{-2}$	1.73437500	4.03 $\times 10^{-3}$	1.72656250	5.49 $\times 10^{-3}$	1.56 $\times 10^{-2}$
1.72656250	-9.48 $\times 10^{-3}$	1.73437500	4.03 $\times 10^{-3}$	1.73046875	1.58 $\times 10^{-3}$	7.81 $\times 10^{-3}$
1.73046875	-2.74 $\times 10^{-3}$	1.73437500	4.03 $\times 10^{-3}$	1.73242188	-3.71 $\times 10^{-4}$	3.91 $\times 10^{-3}$
1.73046875	-2.74 $\times 10^{-3}$	1.73242188	6.43 $\times 10^{-4}$	1.73144531	6.05 $\times 10^{-4}$	1.95 $\times 10^{-3}$
1.73144531	-1.05 $\times 10^{-3}$	1.73242188	6.43 $\times 10^{-4}$	1.73193359	1.17 $\times 10^{-4}$	9.77 $\times 10^{-4}$
1.73193359	-2.03 $\times 10^{-4}$	1.73242188	6.43 $\times 10^{-4}$	1.73217773	-1.27 $\times 10^{-4}$	4.88 $\times 10^{-4}$
1.73193359	-2.03 $\times 10^{-4}$	1.73217773	2.20 $\times 10^{-4}$	1.73205566	-4.86 $\times 10^{-6}$	2.44 $\times 10^{-4}$
1.73193359	-2.03 $\times 10^{-4}$	1.73205566	8.41 $\times 10^{-6}$	1.73199463	5.62 $\times 10^{-5}$	1.22 $\times 10^{-4}$
1.73199463	-9.73 $\times 10^{-5}$	1.73205566	8.41 $\times 10^{-6}$	1.73202515	2.57 $\times 10^{-5}$	6.10 $\times 10^{-5}$
1.73202515	-4.44 $\times 10^{-5}$	1.73205566	8.41 $\times 10^{-6}$	1.73204041	1.04 $\times 10^{-5}$	3.05 $\times 10^{-5}$
1.73204041	-1.80 $\times 10^{-5}$	1.73205566	8.41 $\times 10^{-6}$	1.73204803	2.77 $\times 10^{-6}$	1.53 $\times 10^{-5}$
1.73204803	-4.80 $\times 10^{-6}$	1.73205566	8.41 $\times 10^{-6}$	1.73205185	-1.04 $\times 10^{-6}$	7.63 $\times 10^{-6}$
1.73204803	-4.80 $\times 10^{-6}$	1.73205185	1.80 $\times 10^{-6}$	1.73204994	8.66 $\times 10^{-7}$	3.81 $\times 10^{-6}$
1.73204994	-1.50 $\times 10^{-6}$	1.73205185	1.80 $\times 10^{-6}$	1.73205090	-8.81 $\times 10^{-8}$	1.91 $\times 10^{-6}$
1.73204994	-1.50 $\times 10^{-6}$	1.73205090	1.53 $\times 10^{-7}$	1.73205042	3.89 $\times 10^{-7}$	9.54 $\times 10^{-7}$

convergence is assured regardless of the properties of the function. However, the rate of convergence is considered slow in comparison to other methods.

5.5.2 Newton's Method

The bisection method does not make use of the function values or other characteritics. By using properties of the function, one achieves much faster convergence by comparison. Suppose x_0 is an initial estimate of the root α. The next estimate x_1 is the point of intersection of the tangent to $f(x)$ at the point $(x_0, f(x_0))$ with the x-axis. Since the slope at x_0 is given by $f'(x_0) = f(x_0)/(x_0 - x_1)$, x_1 can be computed as follows

$$x_1 = x_0 - \frac{f(x_0)}{f'(x_0)}.$$

The idea is to use the gradient or slope of $f(x)$ at x_0 to determine a point that would be closer to the root. If $f(x)$ is a linear function, then x_1 is the root, and convergence is achieved in one step. In general, however, $f(x)$ is not linear and the process needs to be repeated until convergence is achieved.

Newton's method for finding roots of a function is
ing from an initial guess x_0, it computes a sequence

$$x_{k+1} = x_k - \frac{f(x_k)}{f'(x_k)}$$

The iterations are terminated when an acceptable
puted. This is ensured when either the magnitude
relative change in the approximate solution $|f(x_k)/f$
One can also force both criteria to be satisfied simu

Select initial guess x_0
$k = 0$
while not converged
      $x_{k+1} = x_k - f(x_k)/f'(x_k)$
      $k = k + 1$
end

Example 5.5.2 (Newton's Method). Consider t
ple 5.5.1 of finding the root of the polynomial $p_2(x$
interval $[1, 3]$ using Newton's method. The method
$x_0 = 3$. The iterations are terminated when either
falls below the specified tolerance $\epsilon = 10^{-6}$. The s
Table 5.2 shows the estimate x_k at the kth iteration
shows the number of digits of accuracy computed as

Table 5.2: Iterations given for Exa

k	x_k	$f(x_k)$	$x_k - \alpha$
1	2.10874113	7.94 $\times 10^{-1}$	-3.77 $\times 10^-$
2	1.78914174	1.02 $\times 10^{-1}$	-5.71 $\times 10^-$
3	1.73381623	3.06 $\times 10^{-3}$	-1.77 $\times 10^-$
4	1.73205260	3.11 $\times 10^{-6}$	-1.80 $\times 10^-$
5	1.73205081	3.22 $\times 10^{-12}$	-1.86 $\times 10^-$

Starting from a different initial guess $x_0 = 1$, we f
more iterations to converge. A small value of $f'(x_0$
root. As tabulated in Table 5.3, subsequent steps c
as in the previous case.

Table 5.3: Iterations given for Example 5.5.2 with a different initial guess $x_0 = 1$.

k	x_k	$f(x_k)$	$x_k - \alpha$	Digits of accuracy
1	3.73205081	7.46×10^0	-2.00×10^0	0.0
2	2.42988133	1.70×10^0	-6.98×10^{-1}	0.2
3	1.88774528	2.94×10^{-1}	-1.56×10^{-1}	0.8
4	1.74391353	2.07×10^{-2}	-1.19×10^{-2}	1.9
5	1.73213096	1.39×10^{-4}	-8.01×10^{-5}	4.1
6	1.73205081	6.42×10^{-9}	-3.71×10^{-9}	8.4
7	1.73205081	4.44×10^{-16}	-2.22×10^{-16}	15.7

Newton's method converges at a much higher rate compared to the bisection method. To characterize the rate of convergence of iterative methods for solving nonlinear equations, we first define the *order of convergence* of a method. A method is order p if

$$\lim_{k \to \infty} \frac{|e_{k+1}|}{|e_k|^p} = c, \tag{5.33}$$

where e_k is the error at the kth step and c is a constant. The digits of accuracy increase by a factor of p at each iteration. Example 5.5.2 shows that the digits of accuracy double at each iteration once the iterates are close enough to the root. This indicates that Newton's method is a second order method.

To show this formally, we define a function

$$\psi(x) = x - \frac{f(x)}{f'(x)},$$

where $f(x)$ is the function with root α. Each step of Newton's method can be written as $x_{k+1} = \psi(x_k)$. At the root α, we have $\alpha = \psi(\alpha)$. Under the assumption that $f'(\alpha) \neq 0$, we obtain

$$x_{k+1} - \alpha = \psi(x_k) - \psi(\alpha).$$

The Taylor series expansion of $\psi(x_k)$ is

$$\psi(x_k) = \psi(\alpha) + e_k \psi'(\alpha) + \frac{e_k^2}{2!} \psi''(\xi_k), \qquad \xi_k \in [x_k, \alpha].$$

It is easy to verify that $\psi'(\alpha) = 0$ since $f(\alpha) = 0$. Thus,

$$e_{k+1} = \frac{1}{2} e_k^2 \psi''(\xi_k). \tag{5.34}$$

The value of $\psi''(\xi_k)$ approaches the constant $\psi''(\alpha)$ as the iterates converge to the root. This proves that Newton's method is a second order method.

The convergence of Newton's method is guarant⟨
is close to the root. Starting guesses farther away
convergence considerably. Newton's method displa⟨
method when the iterates are far from the root. A b⟨
divergence of the iterations. A small value of $f'(x)$ ⟨
very far from the root leading to divergence or to c⟨
the example presented earlier, an initial guess $x_0 =$
the method since $f'(x_0) = 0$.

5.5.3   Secant Method

Newton's method requires computing the function
eration. The derivative may be expensive to com⟨
Replacing the derivative by an approximation com
ates gives the *secant method*. Specifically, the sec⟨
Newton's method with the approximation

$$f'(x_k) \approx \frac{f(x_k) - f(x_{k-}}{x_k - x_{k-1}}$$

The next iterate x_{k+1} is the point of interse⟨
through the points $(x_{k-1}, f(x_{k-1}))$ and $(x_k, f(x_k))$
ton's method, the secant method requires two initia⟨
secant idea, one obtains a method that is less exp⟨
but retains a high order of convergence. The orde⟨
method is $p = 1.618$, which is also known as the gol⟨
method has a slower rate of convergence compare⟨
be seen that two steps of secant are better than c
Assuming that the evaluation of the function and
one Newton's step turns out to be as expensive as tv
that the secant method is a very good alternative t⟨

Select initial guesses x_0 and x_1
$k = 0$
while not converged
    $x_{k+1} = x_k - f(x_k)(x_k - x_{k-1})/(f(x_k) - f($
    $k = k + 1$
end

Example 5.5.3 (Secant Method). Consider th⟨
ple 5.5.1 of finding the root of the polynomial p_2

the interval $[1, 3]$ using the secant method. The method is called with the initial guesses $x_0 = 1, x_1 = 3$. The iterations are terminated when either $|f(x_k)|$ or $|f(x_k)(x_k - x_{k-1})/(f(x_k) - f(x_{k-1}))|/|x_k|$ falls below the specified tolerance $\epsilon = 10^{-6}$. The sequence of iterations given in Table 5.4 shows the estimate x_k at the kth iteration. The last column shows the number of digits of accuracy computed as $-\log_{10}|x_k - \alpha|$.

Table 5.4: Iterations given for Example 5.5.1 with secant method.

k	x_k	$f(x_k)$	$x_k - \alpha$	Digits of accuracy
1	1.32278096	-5.41×10^{-1}	4.09×10^{-1}	0.4
2	1.53174690	-3.07×10^{-1}	2.00×10^{-1}	0.7
3	1.80508424	1.32×10^{-1}	-7.30×10^{-2}	1.1
4	1.72293499	-1.57×10^{-2}	9.12×10^{-3}	2.0
5	1.73168011	-6.42×10^{-4}	3.71×10^{-4}	3.4
6	1.73205277	3.40×10^{-6}	-1.96×10^{-6}	5.7
7	1.73205081	-7.27×10^{-10}	4.20×10^{-10}	9.4

The number of digits of accuracy increase by factor lower than two but higher than one. Two iterations taken together increase the digits of accuracy by a factor close to three. □

The initial guess plays an important role in the convergence of the secant method. A bad choice of x_0 and x_1 can cause the method to diverge or to breakdown. In the previous example, selecting the initial guesses to be $\delta + \sqrt{3}/2$ and $\delta - \sqrt{3}/2$ for any value of δ other than $\sqrt{3}/2$ causes the secant to be parallel to the x-axis, and the denominator in the formula for x_{k+1} to be zero. Guarding against such breakdowns and divergent behavior is necessary to ensure convergence.

5.5.4 Muller's Method

The secant method is an interpolatory method since it interpolates the function at two points x_{k-1} and x_k with a polynomial of degree one and uses its root as the next iterate x_{k+1}. *Muller's method* is an extension of this strategy to three points. A quadratic polynomial is computed to interpolate the function at the three most recent iterates x_{k-2}, x_{k-1}, x_k. One of the roots of the polynomial that is closest to x_k is taken as the next iterate x_{k+1}. Unlike other methods, Muller's method can produce complex iterates from real starting values, which allows it to discover complex roots of polynomials with real coefficients.

5.5.5 Linear Fractional Interpolation

The behavior of the function is critical to the success of a method. Functions with horizontal asymptotes yield horizontal tangents and secants which causes

the breakdown of the secant and Newton's method
difficult to locate roots of functions with vertical asy
be circumvented by using rational functions to inte
these rational functions are used as approximation
fractional interpolation uses the following function t

$$\phi(x) = \frac{x - a}{bx - c},$$

which has a root at a, a vertical asymptote at $x = c$
$y = 1/b$. Interpolating $f(x)$ at the three points $(x_k$
and $(x_k, f(x_k))$ using $\phi(x)$ yields a linear system th
which becomes the next iterate x_{k+1}

$$\begin{bmatrix} 1 & x_{k-2}f(x_{k-2}) & -f(x_{k-2}) \\ 1 & x_{k-1}f(x_{k-1}) & -f(x_{k-1}) \\ 1 & x_k f(x_k) & -f(x_k) \end{bmatrix} \begin{bmatrix} a \\ b \\ c \end{bmatrix}$$

5.5.6 Zeros of Polynomials

The methods described earlier are used to compute
function. For a polynomial of degree n, it is often
n zeros. These zeros may be complex even if the
are real. Newton's method with complex iterates ca
Muller's method can be also used to find a zero or a p
Once a zero has been computed, *deflation* is requir
zeros. To find the second zero, we consider the defla
where α is the zero that has already been computed
polynomial of one lower degree. Deflation is needed
One should also go back and refine the zeros using
to eliminate the undesirable effects of rounding err
polynomials.

5.6   Eigenvalues and Singular V

An $n \times n$ matrix $\mathbf{A}$ has n eigenvalues $\lambda_i, i = 1, \ldots$
characteristic polynomial of the matrix

$$p(z) = \det(z\mathbf{I} - \mathbf{A})$$

Similarly, an arbitrary polynomial of degree n is asso
matrix whose eigenvalues are identical to the roots o
that there is no closed-form formula to express the r
higher than four in terms of its coefficients. The a
roots of polynomials leads to an important conclu

algorithm that can compute the eigenvalues of an arbitrary matrix of size larger than four. The methods to compute eigenvalues are iterative, with termination depending on achieving an acceptable accuracy.

5.6.1 Power Iterations

One of the simplest methods to compute an eigenvalue of an $n \times n$ matrix $\mathbf{A}$ is an iterative method in which an arbitrary starting vector $\mathbf{q}_0$ is multiplied repeatedly by the matrix. Each multiplication results in scaling the components along the eigenvectors by corresponding eigenvalues. Suppose $\mathbf{A}$ has the eigenvalues $\lambda_i, i = 1, \ldots, n$ such that

$$|\lambda_1| > |\lambda_2| > |\lambda_3| > \cdots > |\lambda_n|$$

and corresponding eigenvectors $\mathbf{v}_i, i = 1, \ldots, n$. Let $\mathbf{q}_0 = \sum_{i=1}^{n} \alpha_i \mathbf{v}_i$. Then

$$\mathbf{A}^k \mathbf{q}_0 = \sum_{i=1}^{n} \lambda_i^k \alpha_i \mathbf{v}_i = \lambda_1^k \left(\alpha_1 \mathbf{v}_1 + \sum_{i=2}^{n} \left(\frac{\lambda_i}{\lambda_1} \right)^k \alpha_i \mathbf{v}_i \right).$$

Since $|\lambda_i/\lambda_1| < 1, i = 2, \ldots, n$, the eigenvector $\mathbf{v}_1$ acquires a larger presence in $\mathbf{A}^k \mathbf{q}_0$ relative to other eigenvectors as k is increased. For a sufficiently large k, the components along the other eigenvectors vanish, and $\mathbf{v}_1$ can be recovered by normalization $\mathbf{v}_1 = \mathbf{A}^k \mathbf{q}_0 / \|\mathbf{A}^k \mathbf{q}_0\|$. The largest eigenvalue is obtained from $\mathbf{v}_1$ by computing $\lambda_1 = \mathbf{v}_1^T \mathbf{A} \mathbf{v}_1$.

The *power method* to compute λ_1 and $\mathbf{v}_1$ is a slight variation of this approach in which the iteration vector is normalized at each step. This limits the growth of the magnitude of the elements of the vector and eliminates instability due to numerical errors. The rate of convergence of $\lambda_1^{(k)}$ to λ_1 is determined by the following relation

$$|\lambda_1^{(k)} - \lambda_1| = c \left| \frac{\lambda_2}{\lambda_1} \right|^k, \tag{5.37}$$

where c is a constant. The error decreases by a constant factor of $|\lambda_2/\lambda_1|$ at each iteration. The power method is a first order method with a linear convergence rate.

Select arbitrary vector $\mathbf{q}_0$
for $k = 1, 2, \ldots$
   $\mathbf{z}_k = \mathbf{A} \mathbf{q}_{k-1}$
   $\mathbf{q}_k = \mathbf{z}_k / \|\mathbf{z}_k\|_2$
   $\lambda_1^{(k)} = \mathbf{q}_k^T \mathbf{A} \mathbf{q}_k$
end

In practice, one can use $\mathbf{A} - \sigma\mathbf{I}$ instead of $\mathbf{A}$, w
chosen to improve the convergence of the method.
the error to reduce by $|\lambda_2 - \sigma|/|\lambda_1 - \sigma|$ at each iterat
largest eigenvalue of the shifted matrix. The conver
$\sigma = (\lambda_2 + \lambda_n)/2$. The shift must be added back to t
an estimate of λ_1.

Example 5.6.1 (Power Method). Let us use the
largest eigenvalue and the corresponding eigenvecto

$$\mathbf{A} = \begin{bmatrix} 1 & 2 & 3 \\ 0 & 4 & 5 \\ 0 & 0 & 6 \end{bmatrix}$$

For this matrix, $\lambda_1 = 6$ and $\mathbf{v}_1 = [0.510842 \quad 0.798$
vector is $\mathbf{q}_0 = [1 \quad 1 \quad 1]$. The components of $\mathbf{q}_k$, the
the error in the eigenvalue are tabulated in Table 5.

Table 5.5: The components of $\mathbf{q}_k$, the eigenvalue est
eigenvalue for Example 5.6.1.

k	q_1	q_2	q_3	$\lambda_1^{(k)}$
0	1.0000	1.0000	1.0000	
1	0.4851	0.7276	0.4851	6.941
2	0.4877	0.7664	0.4180	6.597
3	0.4959	0.7809	0.3799	6.373
4	0.5015	0.7878	0.3575	6.237
5	0.5049	0.7917	0.3439	6.153
6	0.5070	0.7941	0.3353	6.100
7	0.5084	0.7955	0.3298	6.065
8	0.5092	0.7964	0.3262	6.043
9	0.5098	0.7970	0.3239	6.028
10	0.5101	0.7974	0.3223	6.019

One drawback of the power method is its inability
than λ_1. To compute the smallest eigenvalue λ_n, on
to the matrix $\mathbf{A}^{-1}$ instead of $\mathbf{A}$. Since the eigenval
eigenvalues of $\mathbf{A}$, they satisfy

$$|\lambda_n^{-1}| > |\lambda_{n-1}^{-1}| > |\lambda_{n-2}^{-1}| > \cdots$$

The *inverse iteration* method applies the power me
λ_n^{-1} and the associated eigenvector $\mathbf{v}_n$. The algorith

Select arbitrary vector $\mathbf{q}_0$
for $k = 1, 2, \ldots$
    Solve $\mathbf{A}\mathbf{z}_k = \mathbf{q}_{k-1}$ for $\mathbf{z}_k$
    $\mathbf{q}_k = \mathbf{z}_k / \|\mathbf{z}_k\|_2$
    $\lambda_1^{(k)} = \mathbf{q}_k^T \mathbf{A} \mathbf{q}_k$
end

$\mathbf{A}^{-1}$. Instead of computing $\mathbf{z}_k = \mathbf{A}^{-1}\mathbf{q}_{k-1}$, we solve the system $\mathbf{A}\mathbf{z}_k = \mathbf{q}_{k-1}$ to obtain $\mathbf{z}_k$.

Shifts can be used in the inverse iteration method to accelerate the rate of convergence. Shifts can also be used to find an arbitrary eigenvalue λ_k if a good estimate to that eigenvalue is available. The rate of convergence is given by the ratio $|\lambda_{k'} - \sigma| / |\lambda_k - \sigma|$, where λ_k is the eigenvalue closest to σ and $\lambda_{k'}$ is the next closest eigenvalue. Convergence is rapid when σ is a good estimate of λ_k. The potential benefits of shifting are far greater for the inverse iterations compared to the power method.

Example 5.6.2 (Inverse Iteration). Let us use the inverse iteration to compute the smallest eigenvalue and the corresponding eigenvector for the matrix considered Example 5.6.1. For the matrix, $\lambda_3 = 1$ and $\mathbf{v}_3 = [1 \ 0 \ 0]$. The starting vector is $\mathbf{q}_0 = [1 \ 1 \ 1]$. The components of $\mathbf{q}_k$, the eigenvalue estimate $\lambda_3^{(k)}$, and the error in the eigenvalue are given in Table 5.6.

Table 5.6: The components of $\mathbf{q}_k$, the eigenvalue estimate $\lambda_3^{(k)}$, and the error in the eigenvalue for Example 5.6.2.

k	q_1	q_2	q_3	$\lambda_3^{(k)}$	$\lambda_3 - \lambda_3^{(k)}$
0	1.0000	1.0000	1.0000		
1	0.9245	0.0925	0.3698	3.0769	-2.08×10^0
2	0.9954	-0.0633	0.0724	1.1054	-1.05×10^{-1}
3	0.9995	-0.0303	0.0118	0.9766	2.34×10^{-2}
4	0.9999	-0.0099	0.0019	0.9863	1.37×10^{-2}
5	1.0000	-0.0029	0.0003	0.9953	4.74×10^{-3}
6	1.0000	-0.0008	0.0001	0.9986	1.40×10^{-3}
7	1.0000	-0.0002	0.0000	0.9996	3.86×10^{-4}
8	1.0000	-0.0001	0.0000	0.9999	1.03×10^{-4}
9	1.0000	-0.0000	0.0000	1.0000	2.66×10^{-5}
10	1.0000	-0.0000	0.0000	1.0000	6.82×10^{-6}

When using a shift $\sigma = 0.9$, convergence is rapid, a

Table 5.7: The components of $\mathbf{q}_k$, the eigenvalue est
eigenvalue for Example 5.6.2 with a shift $\sigma = 0.9$.

k	q_1	q_2	q_3	$\lambda_3^{(k)}$
0	1.0000	1.0000	1.0000	
1	0.9988	0.0016	0.0491	1.162
2	1.0000	-0.0015	0.0010	0.999
3	1.0000	-0.0001	0.0000	0.999
4	1.0000	-0.0000	0.0000	1.000

   The methods discussed so far compute a single e
The method of *orthogonal iterations* is a modificat
can be used to compute several eigenvalues. Cor
vectors $\mathbf{q}_i, i = 1 \ldots, p$ that are simultaneously use
power method. It is convenient to define the $n \times$
$[\mathbf{q}_1, \mathbf{q}_2, \ldots, \mathbf{q}_p]$ whose columns are the starting vec
repeatedly would cause each column to converge to v
the columns of the matrix obtained after each multip
algorithm. The first vector converges to $\mathbf{v}_1$ as in tl
vector, which is forced to be orthogonal to the first
$\mathbf{v}_2$. Orthonormalization at each step ensures that
converges to $\mathbf{v}_i$, $i = 1, \ldots, p$.

Select an $n \times p$ orthogonal matrix $\tilde{\mathbf{Q}}_0$
for $k = 1, 2, \ldots$
   $\tilde{\mathbf{Z}}_k = \mathbf{A}\tilde{\mathbf{Q}}_{k-1}$
   Compute QR factorization: $\tilde{\mathbf{Q}}_k\tilde{\mathbf{R}}_k = \tilde{\mathbf{Z}}_k$
end

   To obtain the largest p eigenvalues, we need to c
is an upper triangular matrix. As $\tilde{\mathbf{Q}}_k$ converges t
converges to an upper triangular form. The eigenva
this matrix. When performing computation in real
to a quasi-upper triangular form. Such a matrix has
eigenvalues must be computed. The rate of conve

$$\mathbf{A}_0 = \mathbf{A}$$
$$\mathbf{V}_0 = \mathbf{I}$$
for $k = 1, 2, \ldots$
      Compute QR factorization: $\mathbf{Q}_k \mathbf{R}_k = \mathbf{A}_{k-1}$
      $\mathbf{A}_k = \mathbf{R}_k \mathbf{Q}_k$
      $\mathbf{V}_k = \mathbf{V}_{k-1} \mathbf{Q}_k$
end

---

given by

$$|\lambda_i^{(k)} - \lambda_i| = c_i \left| \frac{\lambda_{i+1}}{\lambda_i} \right|^k, i = 1, \ldots, p, \tag{5.38}$$

where $|\lambda_{i+1}| < |\lambda_i|$. Extending this idea further, one can start the iterations with an $n \times n$ orthonormal matrix $\tilde{\mathbf{Q}}_0$. The $n \times n$ identity matrix $\mathbf{I}$ can be used instead of the starting matrix $\tilde{\mathbf{Q}}_0$.

5.6.2 QR Algorithm

The algorithms discussed so far compute the eigenvectors of $\mathbf{A}$. To compute eigenvalues, we have to construct $\mathbf{A}_k$, which converges to an upper triangular form that contains the eigenvalues. The QR iterations use a different strategy to compute $\mathbf{A}_k$ without computing $\tilde{\mathbf{Q}}_k$ explicitly. The method uses a sequence of similarity transformations that convert $\mathbf{A}$ to an upper triangular form. A *similarity transformation* of $\mathbf{A}$ is a transformation of the form $\mathbf{X}^{-1} \mathbf{A} \mathbf{X}$ by an arbitrary nonsingular matrix $\mathbf{X}$. For matrices with a full set of eigenvectors, we obtain a nonsingular $\mathbf{X}$ whose columns are the eigenvectors, and that transforms $\mathbf{A}$ into a diagonal matrix with eigenvalues on the diagonal. For other choices of $\mathbf{X}$, the eigenvalues of a matrix remain unchanged after a similarity transform.

Every $n \times n$ matrix can be transformed to a triangular form, known as the *Schur form*, by a similarity transformation using unitary matrices. The QR iterations use this fact to convert $\mathbf{A}$ to an upper triangular form through the use of a sequence of similarity transformations with unitary matrices.

Upon convergence, $\mathbf{A}_k$ is upper triangular with eigenvalues on the diagonal and $\mathbf{V}_k$ is the matrix of eigenvectors. To save computation and storage, one can ignore the steps where $\mathbf{V}_k$ is updated. Since $\mathbf{R}_k = \mathbf{Q}_k^T \mathbf{A}_{k-1}$, we can determine the relation between $\mathbf{A}_k$ and $\mathbf{A}$

$$\mathbf{A}_k = \mathbf{Q}_k^T \mathbf{A}_{k-1} \mathbf{Q}_k = \mathbf{Q}_k^T \mathbf{Q}_{k-1}^T \mathbf{A}_{k-2} \mathbf{Q}_{k-1} \mathbf{Q}_k = \tilde{\mathbf{Q}}_k^T \mathbf{A} \tilde{\mathbf{Q}}_k,$$

where $\tilde{\mathbf{Q}}_k = \mathbf{Q}_1 \mathbf{Q}_2 \cdots \mathbf{Q}_k$. The QR iteration is equivalent to the orthogonal iteration if $\tilde{\mathbf{Q}}_k$ equals the orthonormal matrix $\tilde{\mathbf{Q}}_k$ constructed in the orthogonal

iteration. Recall that the orthogonal iteration meth■
orthonormal basis for $\mathbf{A}^k$. It suffices to show that
$\mathbf{A}^k = \tilde{\mathbf{Q}}_k \tilde{\mathbf{R}}_k$, where $\tilde{\mathbf{R}}_k = \mathbf{R}_k \mathbf{R}_{k-1} \cdots \mathbf{R}_1$. This c◂
argument. The first iteration of the algorithm pr■
i.e., $\mathbf{A} = \mathbf{Q}_1 \mathbf{R}_1$. Assume that the QR factorization■
$\tilde{\mathbf{Q}}_{k-1} \tilde{\mathbf{R}}_{k-1}$. The algorithm provides the relation ◖
which is used to show that

$$\mathbf{A}^k = \tilde{\mathbf{Q}}_{k-1} \mathbf{Q}_k \mathbf{R}_k \mathbf{R}_{k-1} \cdots \mathbf{R}_2 \mathbf{R}$$
$$= \tilde{\mathbf{Q}}_{k-1} \mathbf{R}_{k-1} \mathbf{Q}_{k-1} \mathbf{R}_{k-1} \cdots$$
$$\vdots$$
$$= \tilde{\mathbf{Q}}_{k-1} \mathbf{R}_{k-1} \mathbf{R}_{k-1} \cdots \mathbf{R}_2 \mathbf{R}$$
$$= \tilde{\mathbf{Q}}_{k-1} \tilde{\mathbf{R}}_{k-1} \mathbf{A}.$$

This establishes the equivalence between the QR
iterations.

Example 5.6.3 (QR Iteration). Let us use the
eigenvalues of the following matrix

$$\mathbf{A} = \begin{bmatrix} 1.962729 & 0.310213 & -0.9087\text■ \\ 0.310213 & 3.675702 & -0.4397\text■ \\ -0.908741 & -0.439720 & 2.33341\text■ \\ -0.128730 & -0.346540 & -0.53598\text■ \end{bmatrix}$$

The matrix has the eigenvalues $\lambda_1 = 4, \lambda_2 = 3, \lambda_3 =$
first four iterations are shown below. After eight it■
been computed to some accuracy.

$$\mathbf{A}_1 = \begin{bmatrix} 2.861664 & 0.643809 & -0.6913\text■ \\ 0.643809 & 3.573659 & 0.08786\text■ \\ -0.691310 & 0.087864 & 2.02831\text■ \\ -0.097405 & -0.132538 & -0.5962\text■ \end{bmatrix}$$

$$\mathbf{A}_2 = \begin{bmatrix} 3.401118 & 0.618391 & -0.3981\text■ \\ 0.618391 & 3.345577 & 0.25128\text■ \\ -0.398176 & 0.251286 & 2.07195\text■ \\ -0.040439 & -0.032919 & -0.4106\text■ \end{bmatrix}$$

$$\mathbf{A}_3 = \begin{bmatrix} 3.675257 & 0.506006 & -0.2223\text■ \\ 0.506006 & 3.205732 & 0.24529\text■ \\ -0.222346 & 0.245298 & 2.07058\text■ \\ -0.012452 & -0.006459 & -0.2208\text■ \end{bmatrix}$$

$$\mathbf{A}_4 = \begin{bmatrix} 3.819343 & 0.396899 & -0.120441 & -0.003411 \\ 0.396899 & 3.125506 & 0.195951 & -0.001099 \\ -0.120441 & 0.195951 & 2.043014 & -0.110946 \\ -0.003411 & -0.001099 & -0.110946 & 1.012138 \end{bmatrix},$$

$$\mathbf{A}_8 = \begin{bmatrix} 3.981359 & 0.135401 & -0.008327 & -0.000015 \\ 0.135401 & 3.016400 & 0.046565 & 0.000003 \\ -0.008327 & 0.046565 & 2.002194 & -0.006840 \\ -0.000015 & 0.000003 & -0.006840 & 1.000047 \end{bmatrix}.$$

□

The equivalence of QR iterations and orthogonal iterations gives us an elegant algorithm to compute the eigenvalues directly. The QR iterations can also be interpreted as orthogonal iterations applied to $\mathbf{A}^{-1}$ with a different interpretation of the triangular matrix $\mathbf{R}_k$. This suggests that the shifting strategy described for inverse iterations should work for the QR iteration as well. The QR iteration can be combined with a shift strategy to improve the convergence of eigenvalues. The shifted QR iteration is a slight modification of the precedent method.

$\mathbf{A}_0 = \mathbf{A}$
$\mathbf{V}_0 = \mathbf{I}$
for $k = 1, 2, \ldots$
    Choose shift σ_k
    Compute QR factorization: $\mathbf{Q}_k \mathbf{R}_k = \mathbf{A}_{k-1} - \sigma_k \mathbf{I}$
    $\mathbf{A}_k = \mathbf{R}_k \mathbf{Q}_k + \sigma_k \mathbf{I}$
    $\mathbf{V}_k = \mathbf{V}_{k-1} \mathbf{Q}_k$
end

The eigenvector matrix $\mathbf{V}_k$ plays no role in convergence and can be ignored to save computation and storage. A good choice for the shift is the last diagonal element of the matrix $\mathbf{A}_{k-1}$, $a_{n,n}^{(k-1)}$. The shifted matrix $\mathbf{A}_{k-1} - \sigma_k \mathbf{I}$ is nonsingular as long as $a_{n,n}$ is not an eigenvalue of the matrix. To ensure that the algorithm continues without encountering a singular shifted matrix, it is necessary to identify eigenvalues that have been computed to an acceptable accuracy. Off-diagonal elements in the last row with magnitudes below a threshold $\epsilon \|\mathbf{A}\|$ are forcibly set to zero. The last row and the last column are no longer considered in subsequent iterations. The algorithm continues processing the leading submatrix of dimension $(n - 1)$. The idea is applied each time an eigenvalue is discovered to reduce the size of the submatrix. The process of reducing the dimension of the matrix upon discovery of eigenvalues is called *deflation*.

Example 5.6.4 (QR Iteration with Shifts). T‹
QR Iteration with and without shifts, we compute
used in Example 5.6.3. We use the eigenvalue of ￼
right corner of $\mathbf{A}_{k-1}$ that is closest to $a_{n,n}$ as the sh￼
an eigenvalue has converged, we employ deflation t￼
row and column from subsequent computation. Shi￼
active rows.

The first eigenvalue to be discovered is λ_3, whicl￼
The second eigenvalue to be discovered is λ_4, whicl￼
ations. The remaining eigenvalues are discovered ￼
iterations. The accuracy of the eigenvalues is also v￼

$$\mathbf{A}_1 = \begin{bmatrix} 3.157629 & 0.694244 & -0.784\zeta \\ 0.694244 & 3.357662 & 0.3026\zeta \\ -0.784425 & 0.302620 & 1.5650\zeta \\ -0.057380 & -0.050894 & -0.340\zeta \end{bmatrix}$$

$$\mathbf{A}_2 = \begin{bmatrix} 3.632023 & 0.654695 & -0.564\blacksquare \\ 0.654695 & 2.784889 & 0.7244 \\ -0.564138 & 0.724419 & 1.5840\zeta \\ -0.005255 & -0.001851 & -0.044\zeta \end{bmatrix}$$

$$\mathbf{A}_4 = \begin{bmatrix} 3.985335 & 0.120263 & -0.009\zeta \\ 0.120263 & 3.013443 & 0.0485\zeta \\ -0.009534 & 0.048536 & 1.0012\zeta \\ -0.000011 & -0.000000 & -0.000\blacksquare \end{bmatrix}$$

$$\mathbf{A}_6 = \begin{bmatrix} 4.000000 & -0.000000 & -0.000\zeta \\ 0.000000 & 3.000000 & 0.0000\zeta \\ -0.000000 & 0.000001 & 1.0000\zeta \\ -0.000011 & -0.000000 & -0.000\blacksquare \end{bmatrix}$$

Each iteration of the algorithm requires a QR fac￼
and $\mathbf{R}_k$. This is followed by a step to compute the￼
For an $n \times n$ matrix the QR factorization can be c￼
using Householder transformations. It involves mult￼
by a sequence of $(n-1)$ orthonormal matrices from t￼
upper triangular matrix $\mathbf{R}_k$. The same transforma￼
the right to compute the product $\mathbf{R}_k \mathbf{Q}_k$. Since eacl￼
update of the identity matrix, it can be applied to th￼
The algorithm requires $O(kn^3)$ operations, where ￼
needed for convergence.

To reduce the cost of each iteration, we can prepr￼
it into an upper Hessenberg form. Matrices of this

triangular part below the first subdiagonal. To ensure that the eigenvalues of the matrix remain unchanged, this preprocessing step must be a similarity transformation. We choose a Householder transformation that converts the vector $\mathbf{v}$ defined by the elements in the first column below the diagonal, i.e., $\mathbf{v} = a_{2:n,1}$, to one with a single non-zero element in the first location, i.e., $[\|\mathbf{v}\|_2 \ 0 \ \cdots \ 0]^T$. Specifically, the Householder transformation matrix for $\mathbf{v}$ is defined as

$$\mathbf{H} = \mathbf{I} - \frac{2}{\mathbf{u}^T\mathbf{u}}\mathbf{u}\mathbf{u}^T, \qquad \mathbf{u} = \mathbf{v} + \text{sign}(v_1)\|\mathbf{v}\|_2\mathbf{e}_1 \tag{5.39}$$

where v_1 is the first element of $\mathbf{v}$ and $\mathbf{e}_1$ is the first column of identity matrix. Using $\text{sign}(v_1)$ eliminates the possibility of cancellation error in computing $\mathbf{u}$.

Example 5.6.5 (Householder Transformation). To illustrate the construction and use of Householder transformations, let us consider the vector $\mathbf{v} = [-5 \ 2 \ 4 \ 2]^T$. Then

$$\mathbf{u} = \begin{bmatrix} -5 \\ 2 \\ 4 \\ 2 \end{bmatrix} - 7\begin{bmatrix} 1 \\ 0 \\ 0 \\ 0 \end{bmatrix} = \begin{bmatrix} -12 \\ 2 \\ 4 \\ 2 \end{bmatrix}$$

The Householder matrix $\mathbf{H}$ is not computed explicitly. Multiplication of $\mathbf{H}$ with $\mathbf{v}$ is computed as shown below

$$\mathbf{Hv} = \left(\mathbf{I} - \frac{2}{\mathbf{u}^T\mathbf{u}}\mathbf{u}\mathbf{u}^T\right)\mathbf{v} = \mathbf{v} - \frac{2(\mathbf{u}^T\mathbf{v})}{\mathbf{u}^T\mathbf{u}}\mathbf{u} = \begin{bmatrix} -5 \\ 2 \\ 4 \\ 2 \end{bmatrix} - \frac{2(84)}{168}\begin{bmatrix} -12 \\ 2 \\ 4 \\ 2 \end{bmatrix} = \begin{bmatrix} 7 \\ 0 \\ 0 \\ 0 \end{bmatrix}.$$

Note that $\|\mathbf{v}\|_2 = 7$. Since $\mathbf{H}$ is an orthogonal matrix, the norm of $\mathbf{v}$ is unchanged after the transformation. All the elements of the transformed vector $\mathbf{Hv}$ except for the first one are converted to zero. $\square$

The Householder transformation is applied to the matrix from the left, resulting in zeros in the first column below the first subdiagonal. The same transformation must be applied from the right to ensure that the eigenvalues of the modified matrix are unchanged. Since the transformation from the right does not involve the first column, the zeros introduced by the first transformation are retained.

The non-zero structure of a 5×5 matrix is shown below. The original entries $\times$ that are modified by the left transform are marked $+$, and the ones modified by the right transform are marked $\star$.

$$\begin{bmatrix} \times & \times & \times & \times & \times \\ \times & \times & \times & \times & \times \\ \times & \times & \times & \times & \times \\ \times & \times & \times & \times & \times \\ \times & \times & \times & \times & \times \end{bmatrix} \rightarrow \begin{bmatrix} \times & \times & \times & \times & \times \\ + & + & + & + & + \\ & + & + & + & + \\ & + & + & + & + \\ & + & + & + & + \end{bmatrix} \rightarrow \begin{bmatrix} \times & \star & \star & \star & \star \\ + & \star & \star & \star & \star \\ & \star & \star & \star & \star \\ & \star & \star & \star & \star \\ & \star & \star & \star & \star \end{bmatrix}$$

This process is repeated for the second column to▪ subdiagonal. By applying $(n-1)$ transforms of this▪ to an upper Hessenberg form with zeros below the fi▪ structure of a 5×5 matrix after each step is shown

$$
\begin{bmatrix}
\times & \times & \times & \times & \times \\
\times & \times & \times & \times & \times \\
\times & \times & \times & \times & \times \\
\times & \times & \times & \times & \times \\
\times & \times & \times & \times & \times
\end{bmatrix}
\rightarrow
\begin{bmatrix}
\times & \times & \times & \times & \times \\
\times & \times & \times & \times & \times \\
 & \times & \times & \times & \times \\
 & \times & \times & \times & \times \\
 & \times & \times & \times & \times
\end{bmatrix}
\rightarrow
$$

$$
\rightarrow
\begin{bmatrix}
\times & \times & \times & \times & \times \\
\times & \times & \times & \times & \times \\
 & \times & \times & \times & \times \\
 & & \times & \times & \times \\
 & & & \times & \times
\end{bmatrix}
$$

The main reason to convert the matrix $\mathbf{A}$ to an▪ that it allows computing the QR factorization in C▪ retains the Hessenberg structure at each iteration.▪ the QR factorization cost is available at each iter▪ eigenvalue computation is reduced by a factor of n.

The Hessenberg structure of $\mathbf{A}_{k-1}$ has to be expl▪ factorization in the QR iteraion. To convert an u▪ upper triangular matrix via orthogonal transforms,▪ Givens rotation matrix for a vector $\mathbf{v} = [x \; y]^T$ is de▪

$$
G_{\mathbf{v}} = \frac{1}{\|x^2 + y^2\|_2}
\begin{bmatrix}
x & y \\
-y & x
\end{bmatrix}
$$

This matrix transforms $\mathbf{v}$ to $[\|x^2 + y^2\|_2 \; 0]^T$. A C▪ structed for the two non-zero elements in the first c▪ from the left converts the subdiagonal element in ▪ also modifies all the elements in the first two rows.▪ to introduce a zero in the subdiagonal position in t▪ effects elements in the second and third rows, the▪ first column is retained. The process is repeated (▪ to the upper triangular matrix $\mathbf{R}_k$. To compute $\mathbf{A}$▪ to be applied to $\mathbf{R}_k$ from the right in the same or▪ elements in the first two columns, and reintroduces▪ below the diagonal. Subsequent transformation app▪ Hessenberg structure. Elements below the first subc▪ The changes to the structure of the matrix due ▪

shown below. Elements modified by a transformation are marked $+$.

$$
\begin{bmatrix}
\times & \times & \times & \times & \times \\
\times & \times & \times & \times & \times \\
 & \times & \times & \times & \times \\
 & & \times & \times & \times \\
 & & & \times & \times
\end{bmatrix}
\rightarrow
\begin{bmatrix}
+ & + & + & + & + \\
 & + & + & + & + \\
 & \times & \times & \times & \times \\
 & & \times & \times & \times \\
 & & & \times & \times
\end{bmatrix}
\rightarrow
\begin{bmatrix}
\times & \times & \times & \times & \times \\
 & + & + & + & + \\
 & + & + & + & + \\
 & & \times & \times & \times \\
 & & & \times & \times
\end{bmatrix}
$$

$$
\rightarrow
\begin{bmatrix}
\times & \times & \times & \times & \times \\
 & \times & \times & \times & \times \\
 & & + & + & + \\
 & & + & + & + \\
 & & & \times & \times
\end{bmatrix}
\rightarrow
\begin{bmatrix}
\times & \times & \times & \times & \times \\
 & \times & \times & \times & \times \\
 & & \times & \times & \times \\
 & & & + & + \\
 & & & & +
\end{bmatrix}
$$

The changes to the structure of the matrix due to the right transformations are shown below.

$$
\begin{bmatrix}
\times & \times & \times & \times & \times \\
 & \times & \times & \times & \times \\
 & & \times & \times & \times \\
 & & & \times & \times \\
 & & & & \times
\end{bmatrix}
\rightarrow
\begin{bmatrix}
+ & + & \times & \times & \times \\
+ & + & \times & \times & \times \\
 & & \times & \times & \times \\
 & & & \times & \times \\
 & & & & \times
\end{bmatrix}
\rightarrow
\begin{bmatrix}
\times & + & + & \times & \times \\
\times & + & + & \times & \times \\
 & + & + & \times & \times \\
 & & & \times & \times \\
 & & & & \times
\end{bmatrix}
$$

$$
\rightarrow
\begin{bmatrix}
\times & \times & + & + & \times \\
\times & \times & + & + & \times \\
 & \times & + & + & \times \\
 & & + & + & \times \\
 & & & & \times
\end{bmatrix}
\rightarrow
\begin{bmatrix}
\times & \times & \times & + & + \\
\times & \times & \times & + & + \\
 & \times & \times & + & + \\
 & & \times & + & + \\
 & & & + & +
\end{bmatrix}
$$

The cost of applying a Givens transformation is $O(n)$ operations. The overall cost of the QR factorization for a Hessenberg matrix is $O(n^2)$, which is also the cost of each iteration.

A practical algorithm that incorporates these techniques to improve the efficiency of the QR iterations is given below.

Selecting the last diagonal element as a shift may not always be useful. A more robust strategy is the *Wilkinson shift*, which uses the eigenvalue of the 2×2 matrix in the lower right corner of $\mathbf{A}_{k-1}$ that is closest to $a_{n,n}$.

5.6.3 Computing Singular Values

The SVD of an $m \times n$ real matrix $\mathbf{A}$ has the form

$$
\mathbf{A} = \mathbf{U} \mathbf{\Sigma} \mathbf{V}^T \tag{5.41}
$$

where $\mathbf{U}$ is an $m \times m$ orthogonal matrix, $\mathbf{V}$ is an $n \times n$ orthogonal matrix and $\mathbf{\Sigma}$ is an $m \times n$ diagonal matrix with non-negative diagonal entries $\sigma_i, i = 1, \ldots, \min(m, n)$

Convert $\mathbf{A}_0$ to Hessenberg form: $\mathbf{A}_0 = \mathbf{U}^T \mathbf{A} \mathbf{U}$
$\mathbf{V}_0 = \mathbf{U}$
for $k = 1, 2, \ldots$
  Choose a shift σ_k
  Compute QR factorization: $\mathbf{Q}_k \mathbf{R}_k = \mathbf{A}_{k-}$
  $\mathbf{A}_k = \mathbf{R}_k \mathbf{Q}_k + \sigma_k \mathbf{I}$
  $\mathbf{V}_k = \mathbf{V}_{k-1} \mathbf{Q}_k$
  Set subdiagonal elements satisfying $|\mathbf{A}_{i,i-}$
end

known as the *singular values* of $\mathbf{A}$. The columns
vectors of $\mathbf{A}$ and the columns of $\mathbf{v}_i$ of $\mathbf{V}$ are the *rig*

Singular values and vectors are closely related
vectors of a matrix. The eigenvalues of $\mathbf{A}\mathbf{A}^T$ and
of the singular values of $\mathbf{A}$. The columns of $\mathbf{U}$ are
the columns of $\mathbf{V}$ are the eigenvectors of $\mathbf{A}^T\mathbf{A}$. O
to compute eigenvalues of $\mathbf{A}\mathbf{A}^T$ or $\mathbf{A}^T\mathbf{A}$ and obt
approach suffers from numerical instabilities since t
and $\mathbf{A}^T\mathbf{A}$ introduces error and degrades the conditi

Robust algorithms to compute SVD use a variar
preprocessing step, the matrix is reduced to a bidia
transforms. A Householder transform is applied fro
in the first column below the diagonal to zero. A di
is applied from the right to introduce zeros in the fi
perdiagonal. This does not change the zero elements
step introduces zeros below the diagonal in the sec
in the second row to the right of the superdiagonal.
unchanged. The process is continued until the matri
form with one superdiagonal.

Next, the bidiagonal matrix is transformed to a d
method in which orthogonal transforms are applie
of the matrix. Each iteration applies a QR step to
constructing this product. A shifting strategy is u
convergence. The left and right transforms are app
bidiagonal structure of $\mathbf{A}_k$ throughout the process
magnitude falls below the threshold are set to zero.

5.7 Further Reading

There is a large amount of literature on numerical analysis. The material presented in this chapter has been around for several decades. The algorithms and analysis have evolved over many years to reach a point of maturity. Most textbooks on numerical analysis will provide a detailed discussion on the topics covered here. This includes discussion on numerical approximation, error propagation, finite-precision arithmetic, interpolation, and root finding. A limited selection of useful graduate level references is [1–8].

The algorithms presented in this chapter to compute roots of nonlinear functions are limited to one-dimensional problems. Comprehensive references for algorithms to solve systems of nonlinear equations are [9, 10]. The discussion on eigenvalue computations did not include the important case of sparse matrices. Methods to compute a few eigenvalues of large sparse matrices differ substantially from the algorithms for dense matrices. On the topic of eigenvalues and singular values, the classic reference is [11]. Other references include [12–14].

5.8 Exercises

Exercise 5.8.1. Compute the absolute error and the relative error in approximating a by $\hat{a}$ for the following values of a by $\hat{a}$. In each case, determine the number of decimal digits of accuracy

1. $a = 1.23456 \times 10^3$, $\hat{a} = 1234$

2. $a = \sin(\pi/4)$, $\hat{a} = 0.707$

3. $a = \pi$, $\hat{a} = 22/7$

4. $a = 1/11$, $\hat{a} = 1.11$

5. $a = 1000/1001$, $\hat{a} = 1$

6. $a = e$, $\hat{a} = 2.72$

Exercise 5.8.2. Find the relative error in computing the value of a using 3-digit arithmetic with rounding.

1. $1/3$

2. $\sum_{i=0}^{10} x^i$, where $x = 0.8$

3. $1/(1 - x)$, where $x = 0.8$

4. $\sum_{i=0}^{10} x^i$, where $x = -0.8$

5. $1/(1 - x)$, where $x = -0.8$

Exercise 5.8.3. Determine the relative error in co▮
the approximation

$$e^x = 1 + x + \frac{x^2}{2!} + \frac{x^3}{3!} + \cdots$$

for four different values of k: $k = 1, 4, 8, 16$.

Exercise 5.8.4. Find the relative error in computi▮
approximation

$$e^x = 1 + x + \frac{x^2}{2!} + \frac{x^3}{3!} + \cdots$$

for four different values of k: $k = 1, 4, 8, 16$. Give ar▮

Exercise 5.8.5. Compute an approximation to the▮
$\tan\theta$ using the centered difference formula

$$f'(x) \approx \frac{f(x+h) - f(x-}{2h}$$

at three points: $x = 0, \pi/4, 7\pi/16$. Use $h = \pi/32$.
with the accurate value obtained by the formula $f'($▮

Exercise 5.8.6. Consider the recurrence that defin▮

$$y_k = \frac{7}{3}y_{k-1} - \frac{2}{3}y_{k-2}, \qquad y_0 = 1$$

The solution is $y_k = 1/3^k$. The value for y_{50} comput▮
is -0.03137225. Explain.

Exercise 5.8.7. For what range of values of x wil▮
given below have large relative error? Assume $0 \le$ ▮

1. $f(x) = 10 - \sqrt{(100 - x)}$

2. $f(x) = 1/(100 - x) - 1/(100 + x)$

In each case, suggest an alternative scheme to comp▮

Exercise 5.8.8. Find the condition number of the
$\tan\theta$ when

1. $\theta = \pi/4$

2. $\theta = \pi/2$

Exercise 5.8.9. Compute the value of ϵ_{mach} for a ▮
$31, L = -128$, and $U = 127$. Estimate the number ▮
available on this system.

Exercise 5.8.10. On a computer using a decimal floating-point system with $\epsilon_{\text{mach}} = 10^{-8}$ and an exponent range of $[-16, 15]$, what is the result of the following computations?

1. $10^4 + 10^{-5}$

2. $1 - 10^{-4}$

3. $10^{10}/10^9$

4. $10^{-10}/10^9$

5. $10^{15} \times 10^{-16}$

6. $10^{15} \times 10^{16}$

Exercise 5.8.11. Which of the following two formulas to compute the midpoint of an interval $[a, b]$ is preferable in floating-point arithmetic?

$$m = \frac{a + b}{2}, \qquad m = a + \frac{b - a}{2}$$

Is it possible to compute m that lies outside the interval with either one of these formulas?

Exercise 5.8.12. The standard deviation of a set of real numbers can be computed in two mathematically equivalent ways

$$\sigma = \left[\frac{1}{n-1} \sum_{i=1}^{n} (x_i - \bar{x})^2 \right]^{1/2}, \qquad \sigma = \left[\frac{1}{n-1} \sum_{i=1}^{n} x_i^2 - \bar{x}^2 \right]^{1/2}$$

where $\bar{x} = \sum_{i=1}^{n} x_i$ is the mean. Which one of these formulas is preferable in floating-point arithmetic?

Exercise 5.8.13. Determine the polynomial of degree two that interpolates the data $(-1, 2), (0, -1), (1, 1)$ using

1. Vandermonde matrix approach

2. Lagrange approach

3. Newton's approach

Show that the three methods give the same polynomial.

Exercise 5.8.14. Compute the interpolating polynomial in Newton's form for the data

x	1	2	3	4	5
y	36	100	144	144	100

incrementally as data points are added one at a time.

Exercise 5.8.15. Compute the natural spline interp

x	1	2	3	4	
y	36	100	144	144	1

Exercise 5.8.16. The error function is defined as

$$\mathrm{erf}(x) = \frac{2}{\sqrt{\pi}} \int_0^x e^{-t^2} d$$

Construct an interpolating polynomial of degree two
tion using its value at three points: erf$(1/4) = 0.276$
erf$(3/4) = 0.711156$. Use the interpolating polynom

1. erf$(5/8)$

2. erf$(7/8)$

Determine the relative error using the exact values ro
erf$(5/8) = 0.623241$ and erf$(7/8) = 0.784075$.

Exercise 5.8.17. What is the form of Newton's it
the polynomial $f(x) = ax + b$, where a and b are kn

Exercise 5.8.18. The square root of a positive real
finding the roots of the polynomial $f(x) = x^2 - a$. V
for computing the square root of number?

Exercise 5.8.19. Implement the bisection method
functions

1. $x^3 - x$

2. $x^3 - 4x^2 + 5x - 2$

3. $e^{-x} - x$

4. $x \cos(x)$

What were the starting intervals and stopping crite
was the rate of convergence of the iterations?

Exercise 5.8.20. Implement the Newton's method
and use it to find a root of the functions given in E
initial guesses that resulted in convergence? What
in each case? What was the rate of convergence of t

Exercise 5.8.21. Implement the secant method for
use it to find a root of the functions given in Exercise
guesses that resulted in convergence? What was the
case? What was the rate of convergence of the itera

Exercise 5.8.22. Suppose we use the power method to compute the largest eigenvalue of the matrix

$$A = \begin{bmatrix} 1 & 2 \\ 0 & 3 \end{bmatrix}$$

Determine the eigenvalue and eigenvector that are obtained when the starting vector is $\mathbf{v}_0 = [1 \ 1]^T$. What is the outcome if we use $\mathbf{v}_0 = [1 \ 0]^T$?

Exercise 5.8.23. Implement the power method to compute the largest eigenvalue and the corresponding eigenvector of the matrix

$$A = \begin{bmatrix} 1 & 7 & 2 \\ 5 & 0 & 1 \\ 8 & 2 & 2 \end{bmatrix}$$

Use the starting vector $\mathbf{v}_0 = [0 \ 0 \ 1]^T$.

Exercise 5.8.24. Implement the inverse iteration to compute the smallest eigenvalue and the corresponding eigenvector of the matrix

$$A = \begin{bmatrix} 8 & 2 & 1 \\ 1 & 5 & 1 \\ 5 & 2 & 3 \end{bmatrix}$$

Use the starting vector $\mathbf{v}_0 = [0 \ 0 \ 1]^T$.

Exercise 5.8.25. Implement the QR iteration with shifts and use it to compute the eigenvalues of the matrix in Exercise 5.8.24. Use the lower rightmost entry of the matrix as the shift. Implement deflation to eliminate discovered eigenvalues and the corresponding rows and columns from subsequent processing.

References

[1] G. Dahlquist and A. Bjorck, *Numerical Methods*. Prentice Hall, Englewood Cliffs, NJ, 1974.

[2] J. Stoer and R. Bulirsch, *Introduction to Numerical Analysis*. Springer-Verlag, New York, 2002.

[3] G. Stewart, *Afternotes on Numerical Analysis*. Society for Industrial and Applied Mathematics, Philadelphia, 1996.

[4] M. Heath, *Scientific Computing*, 2nd ed. McGraw-Hill, New York, 1997.

[5] E. Isaacson and H. Keller, *Analysis of Numerical Methods*. Dover Publishers, New York, 1994.

158

[6] W. Cheney and D. Kincaid, *Numerical I*
Brooks/Cole Pub Co., Pacific Grove, CA, 200

[7] R. Kress, *Numerical Analysis.* Springer-Verlag

[8] J. Ortega, *Numerical Analysis: A Second Cou*
Applied Mathematics, Philadelphia, 1990.

[9] J. Dennis and R. Schnabel, *Numerical Method*
tion and Nonlinear Equations. Society for Indu
ics, Philadelphia, 1996.

[10] J. Ortega and W. Rheinboldt, *Iterative Solut*
Several Variables. Society for Industrial and A
phia, 2000.

[11] J. Wilkinson, *The Algebraic Eigenvalue Prob*
Oxford, 1965.

[12] G. Golub and C. Van Loan, *Matrix computc*
Univ. Press, Baltimore, 1996.

[13] A. Gourlay and G. Watson, *Computational M*
lems. Wiley, London, 1973.

[14] B. Parlett, *The Symmetric Eigenvalue Proble*
Applied Mathematics, Philadelphia, 1998.

Chapter 6   Combinatorics

Walter D. Wallis[‡]
[‡]Southern Illinois University, Carbondale, USA

Broadly speaking, combinatorics is the branch of mathematics that deals with different ways of selecting objects from a set or arranging objects. It tries to answer two major kinds of questions, namely the *existence* question (does there exist a selection or arrangement of objects with a particular set of properties?) and the *enumerative* question (how many ways can a selection or arrangement be chosen with a particular set of properties?). But you may be surprised by the depth of problems that arise in combinatorics.

The main point to remember is that it really doesn't matter what sort of objects are being discussed. We shall often assume that we are talking about sets of numbers, and sometimes use their arithmetical or algebraic properties, but these methods are used to prove results that apply to all sorts of objects.

6.1   Two Principles of Enumeration

In this section we introduce two easy rules for enumerating the numbers of elements of certain sets or certain types of arrangements.

The Addition Principle states that if we are building an arrangement of objects starting with a set of mutually exclusive beginning states, then the number of arrangements is the sum of the number of arrangements starting from each beginning state. This rather complicated-sounding sentence is far more trivial than it might appear. For instance, suppose that a young woman wants to go on a job interview, and has only one clean dress skirt and one clean pair of dress slacks. With the skirt, she can create 3 presentable outfits; with the pants, five. It follows that she has $3 + 5 = 8$ ways to dress appropriately.

The Multiplication Principle states that if we are building an arrangement of objects in stages, and the number of choices at each stage does not depend on the choice made at any earlier stage, then the number of arrangements is the product of the number of choices at each stage. Again, this principle is simpler than it appears; if a restaurant, for example, offers two kinds of salad, nine entrées, and seven desserts, the number of meals consisting of one salad and one entrée and one dessert is $2 \times 9 \times 7 = 126$.

Example 6.1.1. How many subsets of a set of n d

We line up the elements of the set in some orde
a mark if the element is in our subset, and no mark
make n decisions as we go through the list of eleme
alternatives, mark or no mark. Since no decision wi
see that there are 2^n subsets of a set of n elements.

Example 6.1.2. Suppose we have five tunes stored
player, and we wish to listen to each of them one a
are there to arrange the playlist?

We have five tunes, so there are five possibilitie
have chosen this tune, then *regardless of which tun*
are four tunes left (because we don't wish to repeat
second tune, there are three possibilities left, and so
choose a playlist in $5! = 5 \times 4 \times 3 \times 2 \times 1$ ways, by
In more general terms, we find that there are $n!$ wa
so we can arrange 7 books on a shelf in 7! ways, pla
ways, and so forth.

Example 6.1.3. A large corporation gives employe
or three letters and two digits. How many codes a
of the 26 uppercase letters and any of the digits fro
digits?

The multiplication principle tells us that there a
two-letter codes, and $26^3 \cdot 10^2$ for the three-letter
tells us that there are

$$26^2 \cdot 10^2 + 26^3 \cdot 10^2 = 67,600 + 1,757,$$

codes.

6.2 Permutations and Combina

In practice, two particular applications of the Multi
times. Because of this, they have their own notation
or ordered subsets, and *combinations*, or unordered

Theorem 6.2.1. *The number of ways to arrange k
n objects is $n!/(n-k)!$, commonly denoted $P(n,k)$*

Proof.   There are n possible choices for the first c
and so forth, until we find that there are $n - k +$
object. The Multiplication Principle tells us that the
arrangements altogether. When we divide $n!$ by $(n-$

Example 6.2.1. How many ways are there for an awards committee to award three distinct scholarships among 12 students entered?

The formula gives us at once $12!/9! = 1,320$.

We next consider what happens if we do not wish to order the k chosen objects. For example, if we have three (identical) scholarships and 12 students, how many ways can we award the scholarships? The answer will be much smaller than the $1,320$ that we got for the previous example, because we do not distinguish among scholarship winners. Thus, if students A, B, and C are selected to win a scholarship, that one case corresponds to six cases in the last example; for we might have A first, then B and C last, or A, then C, then B, and so on. Because there are three winners, there are $3!$ ways to order them. We can employ the Multiplication Principle by asking how many ways there are to order the three students. Let the number of ways to choose 3 of 12 students be denoted $\binom{12}{3}$ or $C(12,3)$. Both notations are commonly used; we say "12 choose 3." Then there are $C(12,3)$ ways to choose the students, and $3!$ ways to order the chosen students; this says that $C(12,3) \times 3! = P(12,3)$. It follows that $C(12,3)$ is just $P(12,3)/3! = 220$. This example also may be generalized. □

Theorem 6.2.2. *The number of ways to choose k objects from a set of n distinct objects is $\binom{n}{k} = n!/(k!(n-k)!)$, also commonly denoted $C(n,k)$.*

Proof. By Theorem 6.2.1, there are $P(n,k)$ ways to choose k elements in a particular order, and there are $k!$ ways of ordering the given k elements. It follows that $\binom{n}{k} \times k! = P(n,k)$, and dividing by $k!$ completes the proof.

Clearly $(x+y)^n = (y+x)^n$, so $\binom{n}{k} = \binom{n}{n-k}$.

Example 6.2.2. A school decides to offer a new bioinformatics program, and decides to appoint a committee consisting of two mathematicians, three biologists, and three computer scientists to plan the implementation. If there are six mathematicians, ten biologists, and five computer scientists available to serve on this committee, in how many ways can the committee be formed?

There are six mathematicians who could serve and we must choose two of them; this may be done in $C(6,2) = 15$ ways. We select the biologists in one of $C(10,3) = 120$ ways. The computer scientists are chosen in one of $C(5,3) = 10$ ways. By the Multiplication Principle, we have a total of $15 \times 120 \times 10 = 18,000$ possible committees.

The symbol $C(n,k)$ is called a *binomial coefficient*, because of the following result (The *Binomial Theorem*): □

Theorem 6.2.3.

$$(x+y)^n = \sum_{k=0}^{n} \binom{n}{k} x^k y^{n-k}$$

Proof. Write the product as $(x_1 + y_1)(x_2 + y_2) \ldots (x_n + y_n)$. Then each term of the product will be of the form $a_1 a_2 \ldots a_{n-1} a_n$, where each a_i is either an x or a

y. To find a term that will contribute to the coeffic
of the coefficients a_i to be x and the other $n - k$ of t
$\binom{n}{k}$ ways to pick k of the as to be x. Removing the
collect all such terms together to get $\binom{n}{k}x^k y^{n-k}$.
from 0 to n.

Corollary 6.2.1.

$$\sum_{k=0}^{n} \binom{n}{k} = 2^n$$

Proof. Set $x = y = 1$ in Theorem 6.2.3.

Corollary 6.2.2.

$$\sum_{k=0}^{n} (-1)^k \binom{n}{k} = 0$$

Proof. Set $x = 1$ and $y = -1$ in Theorem 6.2.3.

There are many other identities and formulas re
An important one is *Pascal's Identity*:

Theorem 6.2.4.

$$\binom{n-1}{k-1} + \binom{n}{k-1} = ($$

Proof. Consider a set of n distinct objects, one c
label of some kind from the others. We count the num
objects. Clearly, there are $\binom{n}{k}$ ways to do this. Nov
subsets of size k according to whether or not they co
how many of each kind are there? The sets conta
may be formed by taking the object and choosing
of the $n - 1$; so there are $\binom{n-1}{k-1}$ subsets of size k t
element. Similarly, if we do not include the disting
ways of choosing k objects from the $n - 1$ not-disti
subset either does or does not contain the distinguish

Binomial coefficients are usually presented in a t
Triangle (although it certainly predates Pascal; see
Chinese, Indian, and European sources). In the fig
and column k is $\binom{n}{k}$.

$n\backslash k$	0	1	2	3	4	5	6
0	1						
1	1	1					
2	1	2	1				
3	1	3	3	1			
4	1	4	6	4	1		
5	1	5	10	10	5	1	
6	1	6	15	20	15	6	1

6.3   The Principle of Inclusion and Exclusion

The addition principle could be restated as follows: if X and Y are two disjoint sets then

$$|X \cup Y| = |X| + |Y|.$$

But suppose X and Y are *any* two sets, not necessarily disjoint, and you need to list all members of $X \cup Y$. If you list all members of X, then list all the members of Y, you certainly cover all of $X \cup Y$, but the members of $X \cap Y$ are listed twice. To fix this, you could count all members of both lists, then subtract the number of duplicates:

$$|X \cup Y| = |X| + |Y| - |X \cap Y|. \tag{6.1}$$

A Venn diagram illustrating this equation is shown in Figure 6.1. To enumerate $X \cup Y$, first count all elements in the shaded areas and then subtract the number of elements in the heavily shaded area.

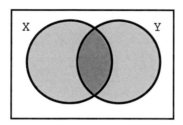

Figure 6.1: Enumerating $X \cup Y$ using (6.1).

If X is the collection of all the objects of type I in some universal set and Y consists of all objects of type II, then (6.1) expresses the way to count the objects that are either of type I or type II:

(i) count the objects of type I;

(ii) count the objects of type II;

(iii) count the objects that are of both types;

(iv) add answers (i) and (ii) and then subtract answer (iii).

Another rule, sometimes called the *rule of sum*, says that the number of objects of type I equals the number that are both of type I and of type II, plus the number that are of type I but not of type II; in terms of the sets X and Y, this is

$$|X| = |X \cap Y| + |X \backslash Y|. \tag{6.2}$$

For example, suppose a professor teaches a cou
operating systems (OS). There are 26 students who
take the C^{++} course but not the other. Obviously
C^{++}. This illustrates Equation (6.2): if X is the C
classlist, then

$$46 = |X| = |X \cap Y| + |X\text{\textbackslash}$$
$$= \quad 26 \quad + \quad 2($$

The two rules (6.1) and (6.2) can be combined t

$$|X \cup Y| = |Y| + |X\text{\textbackslash}Y$$

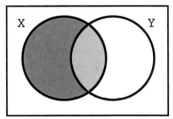

Figure 6.2: Illustration of (6.2) (left) a

Figure 6.2 illustrates the last two rules. In the le
and $X \cap Y$ light, and (6.2) tells us to enumerate T
the right-hand diagram, Y is dark and $X\text{\textbackslash}Y$ light,
contents of the shaded set $X \cup Y$ by adding those t

We now extend the above work to three sets. In
a union, $Y \cup Z$. Then

$$|X \cup Y \cup Z| = |X| + |Y \cup Z| - |X$$

The term $|Y \cup Z|$ can be replaced by

$$|Y| + |Z| - |Y \cap Z|;$$

as a consequence of the distributive laws for interse
$(Y \cup Z) = (X \cap Y) \cup (X \cap Z)$, and the last term ca

$$|X \cap Y| + |X \cap Z| - |X \cap Y$$

So the whole expression is

$$|X \cup Y \cup Z| = |X| + |Y| + |Z|$$
$$- |X \cap Y| - |Y \cap X$$
$$+ |X \cap Y \cap Z|.$$

This formula can be generalized. The generalization is called the *Principle of Inclusion and Exclusion*:

Theorem 6.3.1. *The number of elements in $X_1 \cup X_2 \cup \cdots \cup X_n$ is*

$$
\begin{aligned}
&|X_1 \cup X_2 \cup \cdots \cup X_n| \\
&= \sum_{i_1=1}^{n} |X_{i_1}| - \sum_{i_1=1}^{n} \sum_{i_2=i_1+1}^{n} |X_{i_1} \cap X_{i_2}| + \cdots \\
&\quad + (-1)^{k-1} \sum_{i_1=1}^{n} \sum_{i_2=i_1+1}^{n} \cdots \sum_{i_k=i_{k-1}+1}^{n} |X_{i_1} \cap X_{i_2} \cap \cdots \cap X_{i_k}| \\
&\quad + \cdots + (-1)^{n-1} |X_{i_1} \cap X_{i_2} \cap \cdots \cap X_{i_n}|.
\end{aligned}
\tag{6.4}
$$

Proof. We can think of the process as:
1. Count the elements of all the sets (with repetitions), $\sum |X_{i_1}|$.
2. Subtract all elements that were counted at least twice (that is, count intersections of two sets), $\sum \sum |X_{i_1} \cap X_{i_2}|$.
3. Add in all elements that were counted at least three times (that is, count intersections of three sets), $\sum \sum \sum |X_{i_1} \cap X_{i_2} \cap X_{i_3}|$.
   $\cdots$

Now suppose an object belongs to precisely r of the sets, say sets $X_{j_1}, X_{j_2}, \ldots,$ X_{j_r}. The element is counted r times in calculating $\sum |X_{i_1}|$, and contributes r to that part of the sum (6.4). It appears $\binom{r}{2}$ times in subsets of the form $X_{i_1} \cap X_{i_2}$, since it appears in precisely those for which $\{i_l, i_2\}$ is a 2-set of $\{j_l, j_2, \ldots, j_r\}$; so it contributes $-\binom{r}{2}$ to that part of (6.4). Continuing, we obtain total contribution

$$
\sum_{k=1}^{r} (-1)^{k-1} \binom{r}{k},
\tag{6.5}
$$

which equals 1 by Corollary 6.2.2. So (6.4) counts 1 for every member of $X_1 \cup X_2 \cup \cdots \cup X_n$; therefore, its total is $|X_1 \cup X_2 \cup \cdots \cup X_n|$.

Suppose all the sets X_i are subsets of some set S (that is, S is a universal set for the discussion). Using the general form of de Morgan's Laws, we get the following alternative form of (6.4):

$$
\begin{aligned}
&|\overline{X_1} \cap \overline{X_2} \cap \cdots \cap \overline{X_n}| \\
&= |S| - |X_1 \cup X_2 \cup \cdots \cup X_n| \\
&= |S| - \sum_{i_1=1}^{n} |X_{i_1}| + \sum_{i_1=1}^{n} \sum_{i_2=i_1+1}^{n} |X_{i_1} \cap X_{i_2}| + \cdots \\
&\quad + (-1)^{k} \sum_{i_1=1}^{n} \sum_{i_2=i_1+1}^{n} \cdots \sum_{i_k=i_{k-1}+1}^{n} |X_{i_1} \cap X_{i_2} \cap \cdots \cap X_{i_k}| \\
&\quad + \cdots + (-1)^{n} |X_{i_1} \cap X_{i_2} \cap \cdots \cap X_{i_n}|.
\end{aligned}
\tag{6.6}
$$

Say we want to know how many of the positiv
perfect squares, perfect cubes, or perfect higher po
find out how many of the positive integers less than
than the first. Trivially, 1 is a perfect k-th power f
our calculations, and add 1 to the answer. In this
to seventh powers: $2^8 = 256 > 200$, so every eighth
than 200 whenever $x \geq 2$. For $i = 2, 3, \ldots, 6$ let us
powers of integers that lie between 2 and 199 (inc
number of integers that belong to at least one of th

$$
\begin{aligned}
&|P_2| + |P_3| + |P_4| + |P_5| + |P_6| + |P_7| \\
&- (|P_2 \cap P_3| + |P_2 \cap P_4| + \ldots + |P_5 \cap P_7|) \\
&+ (|P_2 \cap P_3 \cap P_4| + |P_2 \cap P_3 \cap P_5| + \ldots + \\
&- (|P_2 \cap P_3 \cap P_4 \cap P_5| + \ldots + |P_4 \cap P_5 \cap \\
&+ (|P_2 \cap P_3 \cap P_4 \cap P_5 \cap P_6| + \ldots + |P_3 \cap \\
&- |P_2 \cap P_3 \cap P_4 \cap P_5 \cap P_6 \cap P_7|.
\end{aligned}
$$

All of these sets and their sizes are easy to calcula
199 is 14^2, so $P_2 = \{2, 3, 4, \ldots, 14\}$ and $|P_2| = 13$,
$|P_4| = 3$, $|P_5| = |P_6| = |P_7| = 1$. If $k > 7$ then P_k is
where l is the least common multiple of i and j, an
three or more, so $P_2 \cap P_3 = P_2 \cap P_6 = P_3 \cap P_6 = P_2 \cap$
$P_2 \cap P_4 = P_4$, and all the others are empty. Therefo

$$
\begin{aligned}
&|P_2| + |P_3| + |P_4| + |P_5| + |P_6| + \\
&- |P_2 \cap P_3| - |P_2 \cap P_4| - |P_2 \cap P_6 \\
&+ |P_2 \cap P_3 \cap P_4| + |P_2 \cap P_3 \cap P_6|
\end{aligned}
$$

which equals $13 + 4 + 3 + 1 + 1 + 1 - 1 - 3 - 1 - 1$
the original question is 20 (adding 1 for the integer

In some applications of Inclusion/Exclusion, all
as are all the intersections of two sets (like $X_{i_1} \cap X_{i_2}$
sets, and so on. If the intersection of any k of the s
then (6.4) becomes

$$
ns_1 - \binom{n}{2} s_2 + \binom{n}{3} s_3 - \cdots + \Big(
$$

This is called the *uniform form* of the Principle of I

Example 6.3.1. A bank routing number is a string o
of this example, let us assume that any digit can be
there are 10^9 possible routing numbers (the numbe
symbols). How many routing numbers contain all th

To calculate how many routing numbers contain all the odd digits, we find out how many *do not* contain all the odd digits, then subtract that number from 10^9. Write R_x for the set of all routing numbers that *do not* contain x. Then we want to calculate $|R_1 \cup R_3 \cup R_5 \cup R_7 \cup R_9|$. For any x, the number of routing numbers not containing x equals the number of sequences of length 9 on 9 symbols, so $|R_x| = 9^9$; similarly, if $x \neq y$ then $|R_x \cap R_y| = 8^9$; if x, y, z are all different, then $|R_x \cap R_y \cap R_z| = 7^9$, and so on. So, from (6.7), the sum is

$$5 \times 9^9 - \binom{5}{2} \times 8^9 + \binom{5}{3} \times 7^9 - \binom{5}{4} \times 6^9 + 5^9$$

and the answer we require is

$$10^9 - 5 \times 9^9 + 10 \times 8^9 - 10 \times 7^9 + 5 \times 6^9 - 5^9.$$

$\square$

6.4 Generating Functions

Some kinds of combinatorial calculations are modeled by the behavior of infinite power series or even finite power series (i.e., polynomials). For instance, we considered in Section 6.2 how to determine the number of ways of selecting k items from a set of n; for simplicity, we will consider $n = 4$, $k = 2$. Suppose we wish to multiply $(1 + x_1)(1 + x_2)(1 + x_3)(1 + x_4)$. In the product, we ask how many terms of the form $x_i x_j$ there will be; clearly, $\binom{4}{2} = 6$ of them, since that is how many ways we can choose two of the factors to contribute an x_i. Now, if we set $x_1 = x_2 = x_3 = x_4 = x$, the product is the polynomial $(1 + x)^4$. The coefficient of x^2 in this product is $\binom{4}{2} = 6$. (The reader will recall that this subscripting technique is how we proved Theorem 6.2.3, the Binomial Theorem.) In the same way, some numbers of combinatorial interest (such as a sequence of integers) have properties that become clear or easier to prove when they are used as coefficients of a polynomial or an infinite series. These considerations motivate us to define the *generating function* of a sequence $a_0, a_1, \ldots$ to be the formal power series $\sum_{i \geq 0} a_i x^i$.

The expression "formal power series" is understood to mean that we do not evaluate the power series at any point, or concern ourselves with the question of convergence necessarily. Indeed, some uses of generating functions do not require infinite series at all; in this case, we shall assume that $a_n = 0$ for all n greater than some specific integer N.

Example 6.4.1. Suppose we wish to buy a dozen donuts from a shop with varieties v_1, v_2, v_3, and v_4. We want between 3 and 5 donuts of variety v_1 (because the shop has only 5 available, and at least three people will want one); similarly, we require $1 \leq |v_2| \leq 10$, $4 \leq |v_3| \leq 6$, and $0 \leq |v_4| \leq 7$.

In Section 6.3, we introduced the Principle of I
could be used to eliminate the cases that violate the
case the process would be enormously difficult. Inst

$$(x^3 + x^4 + x^5) \times (x + x^2 + x^3 + x^4 + x^5 + x^6$$
$$(x^4 + x^5 + x^6) \times (1 + x + x^2 + x^3 + x^4 + x^5 +$$

and ask where terms involving x^{12} originate. If we
nomials and do not collect like powers of x, a given
the x^4 in the first factor, the x^2 in the second facto
1 in the final factor. This would correspond to a de
type v_1, 2 of type v_2, 6 of type v_3, and none of type
of making up a dozen donuts satisfying the given c
exactly one term x^{12} of the product. Thus we see th
x^n in the product above determines the number of
donuts satisfying the constraints. The product is

$$x^8 + 4x^9 + 10x^{10} + 18x^{11} + 27x^{12} + 36x^{13} +$$
$$62x^{16} + 68x^{17} + 70x^{18} + 68x^{19} + 62x^{20} + 54$$
$$27x^{24} + 18x^{25} + 10x^{26} + 4x^{27} + x^{28}$$

Of course, multiplying out the product (6.8) is
merating possible solutions by hand; but the same
no further work required, if we are faced with the sa
10. And in some problems, algebraic identities can i
of calculating the product.

Example 6.4.2. Assume that we again want a doze
but now the shop has an unlimited supply (i.e., at
lower bound constraints still apply.

Rather than limiting each factor to x^{12}, we allo
infinite series. We have

$$(x^3 + x^4 + \dots)(x + x^2 + \dots)(x^4 + x^5 + \dots$$

and since each factor is a geometric series we can us
sum of such a series to see that it is equivalent to th

$$\frac{x^8}{(1-x)^4} = x^8 \frac{1}{1 - 4x + 6x^2 - 4}$$

Now, polynomial long division (done with the co
will allow us to compute a Maclaurin polynomial, an
the solution to our problem. We get

$$x^8 + 4x^9 + 10x^{10} + 20x^{11} + 35x^{12} + 56x^{13} + 84x^{14} + 1$$

These figures conform well to our intuition that the numbers should be larger, as there are fewer constraints (so there are more ways to make up a box of donuts).

Just as we can calculate the values for the donut problem above, we can use infinite series in a formal way to describe any infinite sequence of integers. Consider the most basic example of this sort, the sequence $\alpha_n = 1$. It will be useful to recall $f_\alpha(x) = \sum_{i=1}^\infty 1x^i = x/(1-x)$. In calculus, we are concerned with the values of x for which the series converges; in combinatorics, we ignore that question.

Another sequence of combinatorial import has a generating function that may be simply computed. Suppose $\beta_n = n$, $n \geq 0$. Then $f_\beta(x) = \sum_1^\infty a_i x^i = x/(1-x)^2$. This formula is easily derived by differentiating the ordinary infinite geometric series, and multiplying by x. Note that since $a_0 = 0$ we begin the summation at $i = 1$; if we begin at $i = 0$, we get the same series. Differentiating again, with another multiplication by x, gives us another formula: $\sum_1^\infty i^2 x^i = (1+x)/(1-x)^3$. With these formulas, we are able to find the generating function for any integer sequence produced by a quadratic polynomial. $\qquad \square$

Example 6.4.3. Find the generating function for $a_n = \binom{n}{2}$.

We may write $a_n = \frac{1}{2}n^2 - \frac{1}{2}n$, so that the generating function will be $f(x) = \frac{1}{2}\left(\sum_2^\infty n^2 - \sum_1^\infty n\right)$. We sum from 2 because $\binom{1}{2} = 0$; we could as easily sum from 0 or 1. Using the values of these sums given above, we see $f(x) = x^2/(1-x)^3$.

There are a number of similar results. For example, consider the triangular numbers $T_n = 1 + 2 + \cdots + n$, $n \geq 1$. The well-known formula is just $T_n = \binom{n+1}{2}$. We may re-index the series given in the foregoing example by letting $j = i+1$. So:

$$f_T(x) = \sum_1^\infty T_i x^i = \sum_2^\infty T_{j-1} x^{j-1} = \sum_2^\infty \binom{j}{2} x^{j-1}$$

Multiplying the right-hand side by x must yield, by the previous example, $x^2/(1-x)^3$; so the desired result is $f_T(x) = \sum_1^\infty T_i x^i = x/(1-x)^3$.

Suppose the sequence α_n is related to β_n by $\beta_n = \sum_{i=1}^n \alpha_i$. We also notice $f_\alpha(x) = (1-x)f_\beta(x)$. Upon noting that $T_n = \sum_{i=1}^n \beta_i$, we look at the generating functions; $f_T(x) = (1-x)f_\beta(x)$. This relationship between the generating functions holds in general: $\qquad \square$

Theorem 6.4.1. *Given sequences a_n and b_n with $b_n = \sum_{i \leq n} a_i$, their generating functions are related by $f_a(x) = (1-x)f_b(x)$.*

Other series manipulations are convenient from time to time; for example:

Theorem 6.4.2. *Given $f_a(x) = \sum_{i=i}^\infty a_i x^i$, we have $\sum_{i=k}^\infty a_i x^i = f_a(x) - a_0 - a_1 x - \cdots - a_{k-1} x^{k-1}$.*

Theorem 6.4.3. *For a_n and $f_a(x)$ as above, we have $x f_a'(x) = \sum_{i \leq \infty} i a_i x^i$.*

Proof. Differentiate both sides of $f_a(x) = \sum a_i x^i$ and multiply by x.

Theorem 6.4.4. *If the sequences a_n and b_n have*
respectively, then the sequence $c_1a_n + c_2b_n$ has gene

There are other identities involving generating fu

Example 6.4.4. Let $a_n = 3^n - 2^n$, for $n \geq 0$. Then
this sum.

This may be summed using the geometric series.

$$\sum_{i=0}^{\infty}(3^i - 2^i)x^i = \sum_{i=0}^{\infty}(3x)^i - \sum_{i=0}^{\infty}(2x)^i = \frac{1}{1-3x} -$$

so $f_a(x) = \frac{x}{(1-3x)(1-2x)}$. More generally, it is tru
defined as a linear combination of nth powers, th
rational. Furthermore, its denominator will have ro
the bases of the nth powers. In the last example w
3^n and 2^n, and the denominator of the generating fu
We will explore other properties of the denominator

6.5 Recurrence Relations

Frequently in studying a sequence of numbers, we d
not have a simple, "closed-form" expression for eac
lationship by which a given term can be derived fro
before. This is called a *recurrence relation*.

A recurrence relation is said to be *linear* if the v
sum, each summand of which is a multiple of the firs
or a constant or function of n; so, for example, a_n
$a_n = (a_{n-1})^2 - 2n$ is nonlinear.

Similarly, we say that the recurrence relation is
a_n is given as a finite sum, each summand of which
terms. So, $a_n = a_{n-1} + 2n + 1$ is *nonhomogeneous*
is homogeneous (but nonlinear). The coefficients i
$a_n = na_{n-1}$ is both linear and homogeneous.

The *degree* of a recurrence is the largest i such
upon a_{n-i}. We give $a_n = na_{n-3} - 1$ as an example
recurrence relation of degree 3. The degree is impo
degree, the more initial values we need to define a s

If we wish to write down the first several values
a_{n-2}, we must know two initial values, say $a_0 = 1$
may compute $a_2 = a_1^2 - a_0 = 3$, $a_3 = a_2^2 - a_1 = 7$, a
are called the *initial conditions*. The reader who has
equations will notice the similarity of the termino

sequence of integers may satisfy many recurrence relations, including linear and nonlinear, homogeneous and nonhomogeneous, of many different degrees.

It will occasionally be useful to prove that two sequences that appear to be the same really are the same. One way to do this is to show that the two sequences satisfy the same recurrence relation with the same initial conditions. This is a short form of what more formally could be a proof by induction of the equality of the two sequences.

Example 6.5.1. Consider the formula $a_n = a_{n-1} + 2a_{n-2}$, for $a \geq 2$, $a_0 = 0$, and $a_1 = 3$. Clearly, we can compute any a_n with this formula by repeatedly applying the equation to the initial conditions. So $a_2 = a_1 + 2a_0 = 3$, and $a_3 = a_2 + 2a_1 = 9$. Can we find a closed-form formula for a_n?

The generating function is one of several ways to solve this problem. Consider $f_a(x) = \sum_{i \geq 0} a_i x^i$. We can use the recurrence to evaluate this function. Multiply both sides of the equation by x^n and sum over n to obtain:

$$\sum_{n=2}^{\infty} a_n x^n = \sum_{n=2}^{\infty} a_{n-1} x^n + 2 \sum_{n=2}^{\infty} a_{n-2} x^n.$$

Each summation may be interpreted in terms of $f_a(x)$, using Theorem 6.4.2. We get $f_a(x) - a_0 - a_1 x = x(f_a(x) - a_0) + 2x^2 f_a(x)$. Now we can solve for $f_a(x)$. We get $f_a(x)(1 - x - 2x^2) = a_0 + a_1 x - a_0 x$ or $f(x) = 3x/(1 - x - 2x^2)$. We simplify using partial fractions.

$$\frac{3x}{(1 - 2x)(1 + x)} = \frac{A}{1 - 2x} + \frac{B}{1 + x} \Rightarrow A = 1, \quad B = -1$$

This yields $f_a(x) = \frac{1}{1-2x} - \frac{1}{1+x}$; we can use the geometric series formula here. We know $\frac{1}{1-2x} = \sum_{i \geq 0} 2^i x^i$ and $\frac{1}{1+x} = \sum_{i \geq 0} (-1)^i x^i$, so $f_a(x) = \sum_{i \geq 0} (2^i - (-1)^i) x^i = \sum_{i \geq 0} a_i x^i$. It follows that $a_n = 2^n - (-1)^n$.

We could compute this formula in another way. Suppose we were to assume that the sequence a_n had a representation as a linear combination of exponential functions. Then $a_n = r^n$ would give us, via the recurrence, an equation $r^{n+1} = r^n + 2r^{n-1}$. Dividing by r^{n-1} yields a quadratic equation, $r^2 - r - 2 = 0$. This is called the *characteristic equation* of the recurrence. Finding the roots gives us $r = 2$ or $r = -1$, the bases of the exponential functions. We are now able to surmise that $a_n = c_1 2^n + c_2 (-1)^n$. Using the given values for a_0 and a_1 we are able to solve for c_1 and c_2 to get the same formula as above.

That method works for distinct roots. If there are repeated roots, we must work a little harder. □

Example 6.5.2. Let us consider $a_{n+1} = 2a_n - a_{n-1}$, with $a_0 = 1$, $a_1 = 2$.

It is easy to find a closed expression for a_n using a guess and a proof by induction, but we will try generating functions. As with our first example, we multiply by x^{n+1} and sum over n to arrive at:

$$f_a(x) - a_0 - a_1 x = 2x f_a(x) - 2a_0 x - x^2 f_a(x$$

Partial fractions will not assist us in this case. Inste
an earlier example, the sequence β_n, which had as i
$x(x-1)^{-2}$. We write this sequence as $\beta_n = (n+0)($
of the type $a_n = (c_1 n + c_2)1^n$ will work. A little wo

Theorem 6.5.1. *If the characteristic equation has*
corresponds to a term of the form $(c_{k-1}n^{k-1} + \cdots +$
a_n.

Example 6.5.3. Solve the recurrence $a_{n+1} = 4a_n -$

The characteristic equation is $x^2 - 4x + 4 = 0$, so t
for some c_1 and c_2. Using the given values of a_0
$2 = 2c_1 + 2c_2$. Solving for c_1 gives $a_n = (3n-2)2^n$.
As a concrete example, we introduce the read
involved with a problem that is over 800 years old,
The *Fibonacci numbers* $f_1, f_2, f_3, \ldots$ are defined
n is any integer greater than 2, $f_n = f_{n-1} + f_{n-2}$.
solution to a problem posed by Leonardo of Pisa, or I
means *son of Bonacci*) in 1202:

> A newly born pair of rabbits of opposite sexes
> at the beginning of the year. Beginning with
> female gives birth to a pair of rabbits of opp
> Each new pair also gives birth to a pair of r
> each month, beginning with their second mont

The number of pairs of rabbits in the enclosure at th
An English translation of Fibonacci's work *Liber A*
those interested.
Some interesting properties of the Fibonacci nu
gruence modulo a positive integer. We say a is *con*
"$a \equiv b(\bmod n)$," if and only if a and b leave the s
n. In other words n is a divisor of $a - b$, or in symb

Example 6.5.4. Prove by induction that the Fibor
only if n is divisible by 3.

Assume n is at least 4. $f_n = f_{n-1} + f_{n-2} = (f_{n-2}$
so $f_n \equiv f_{n-3}(\bmod 2)$.
We first prove that, for $k > 0$, f_{3k} is even. Call
$P(1)$ is true because $f_3 = 2$. Now suppose k is any

true: $f_{3k} \equiv 0(\bmod 2)$. Then (putting $n = 3k + 3$) $f_{3(k+1)} \equiv f_{3k}(\bmod 2) \equiv 0(\bmod 2)$ by the induction hypothesis. So $P(k + 1)$ is true; the result follows by induction. To prove that, for $k > 0$, f_{3k-1} is odd — call this proposition $Q(k)$ — we note that $Q(1)$ is true because $f_1 = 1$ is odd, and if $Q(k)$ is true, then f_{3k-1} is odd, and $f_{3(k+1)-1} \equiv f_{3k-2}(\bmod 2) \equiv 1(\bmod 2)$. We have $Q(k + 1)$ and again the result follows by induction. The proof in the case $k \equiv 1(\bmod 3)$ is similar. $\square$

This sequence is one of the most-studied sequences of integers in all of mathematics. Generating functions make it possible for us to understand many properties of the sequence. Suppose $f(x)$ is the generating function for the Fibonacci numbers. We multiply both sides of the recurrence by x^i and sum.

$$\sum_{i=2}^{\infty} f_{i+1} x^i = \sum_{i=2}^{\infty} f_i x^i + \sum_{i=2}^{\infty} f_{i-1} x^i$$

Careful use of the formula for $f(x)$ will help us to represent this equation in more understandable terms. So, according to Theorem 6.4.2, the first term in the equation is $\sum_{i=2}^{\infty} f_{i+1} x^i = f(x) - f_1 - f_2 x$. We substitute similarly for the other two terms to get

$$f(x) - f_1 - f_2 x = x(f(x) - f_1) + x^2 f(x)$$

which we may solve for $f(x)$, to get

$$-1 - x + x = f(x)(x^2 + x - 1)$$

or $f(x) = -1/(x^2 + x - 1) = 1/(1 - x - x^2)$.

It may be shown that polynomial long division, as in an earlier example, produces the Fibonacci numbers. But there is a more useful and fascinating consequence of our calculation of $f(x)$. Recall that for the earlier case of the sequence $a_n = 3^n - 2^n$, the generating function was also a rational function, and that the bases of the exponents were reciprocals of the roots of the denominator. Alternatively, we may use partial fractions or the characteristic equation.

First, we assume that the Fibonacci numbers may be represented by an expression of the form $f_n = r^n$, $r \geq 1$. Then the recurrence gives us $r^{n+1} = r^n + r^{n-1}$, and dividing by r^{n-1} gives us a quadratic $r^2 - r - 1$ with two real roots, traditionally denoted $\phi_1 = (1 + \sqrt{5})/2$ and $\phi_2 = (1 - \sqrt{5})/2$. To find f_n we simply assume $f_n = c_1 \phi_1^n + c_2 \phi_2^n$ and use the initial values to solve for c_1 and c_2.

$$f_1 = 1 = c_1 \frac{1 + \sqrt{5}}{2} + c_2 \frac{1 - \sqrt{5}}{2}, \qquad f_2 = 1 = c_1 \frac{6 + 2\sqrt{5}}{4} + c_2 \frac{6 - 2\sqrt{5}}{4}$$

This gives us $c_1 = 1/\sqrt{5}$ and $c_2 = -1/\sqrt{5}$. This formula for the Fibonacci numbers is sometimes called *Binet's formula*, although Binet was not its first discoverer. Different starting values yield a different sequence, but it isn't hard to see that the values $f_1 = 0$, $f_2 = 1$ will give the same numbers with different

subscripts, and likewise $f_1 = 1$, $f_2 = 2$. So, the "firs"
values that gives us a different sequence is $f_1 = 1$
also been studied extensively; it is more commonly
numbers, after the mathematician Edouard Lucas.

The Fibonacci numbers represent the number o
an ordered sum of 1s and 2s. More precisely, f_n is
$n - 1$ as such a sum, for $n \geq 2$. The proof of this
how the sum begins. There are f_n ways to write n
beginning of such sequences to create some of the wa
Then there are f_{n-1} ways to write $n - 2$; we may pla
of these sequences to obtain ways of writing n. It f
Now, $f_1 = 1$ (which counts the empty sum only)
write 1 as a sum of a single 1 and no 2s. Thus we
initial conditions as the Fibonacci sequence; it follo
that the sequences are the same. The Fibonacci nu
as well; see Problem 5.7 for an example.

6.6   Graphs

A great deal has been done in the special case whe
elements. In that case the members of the unive
graphically, as points in a diagram, and the set {
drawn joining x and y. Provided the universal set
is called a *graph*. The sets are called *edges* (or *li*
universal set are *vertices* (or *points*). The univers
graph. For any graph G, we write $V(G)$ and $E(G)$
edges of G.

The edge $\{x, y\}$ is simply written xy, when no
are called its *endpoints*. When x and y are endpoi
are *adjacent* and write $x \sim y$ for short; the vertice
neighbors. The set of all neighbors of x is its (open)
y are *not* adjacent we write $x \nsim y$.

Two vertices either constitute a set or not, so a
edges with the same pair of vertices. However, ther
two edges joining the same vertices might make sense
talk about *networks* or *multigraphs* in which there ca
same pair of vertices; those edges are called *multipl*

Another generalization is to allow *loops*, edges
very good term for a graph-type structure in which
usually call one of these a "looped graph" or "loope
loops" although strictly speaking it is not a graph
when no confusion arises, the word "graph" can be u

Any binary relation can be represented by a diag

on the set S, the elements of S are shown as vertices, and if $x\rho y$ is true, then an edge is shown from x to y, with its direction indicated by an arrow. Provided the set S is finite, all information about any binary relation on S can be shown in this way. Such a diagram is called a *directed graph* or *digraph*, and the edge together with its arrow is called an *arc*. If ρ is symmetric, the arrows may be dropped and the result is a graph (possibly with loops).

Several families of graphs have been studied. Given a set S of v vertices, the graph formed by joining each pair of vertices in S is called the *complete graph* on S and denoted K_S. K_v denotes any complete graph with v vertices. As you would expect, we often call K_3 a *triangle*. The *complete bipartite graph* on V_1 and V_2 has two disjoint sets of vertices, V_1 and V_2; two vertices are adjacent if and only if they lie in different sets. We write $K_{m,n}$ to mean a complete bipartite graph with m vertices in one set and n in the other. $K_{1,n}$ in particular is called an *n-star*. Figure 6.3 shows copies of K_6 and $K_{3,4}$.

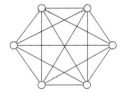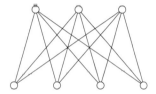

Figure 6.3: K_6 and $K_{3,4}$.

Suppose H is a graph all of whose vertices and edges are vertices and edges of some graph G — that is, $V(H) \subseteq V(G)$ and $E(H) \subseteq E(G)$. Then H is a *subgraph* of G; we write $H \le G$. Every graph G has itself as a subgraph; if H is a subgraph of G but $H \ne G$, H is a *proper* subgraph of G, and we write $H < G$. In particular, if S is some set of vertices of G, then $\langle S \rangle_G$ is the subgraph consisting of all edges of G with both endpoints in S. If G is a complete graph whose vertex-set contains S, then the subscript "G" is dropped, and $\langle S \rangle$ is the complete subgraph based on S. Any subgraph of a complete bipartite graph is itself called *bipartite*.

Instead of saying Figure 6.3 shows two graphs, we could say it is a single graph that consists of two separate subgraphs, with no edges joining one part to the other. We call such a graph *disconnected*; a graph that is all in one piece is called *connected*. The different connected parts of a disconnected graph are called its *components*.

The graph G is trivially a subgraph of the complete graph $K_{V(G)}$. The set of all edges of $K_{V(G)}$ that are *not* edges of G will form a graph with $V(G)$ as its vertex set; this new graph is called the *complement* of G, and written $\overline{G}$. More generally, if G is a subgraph of H, then the graph formed by deleting all edges of G from H is called the *complement of G in H*, denoted $H - G$. The complement $\overline{K}_S$ of the complete graph K_S on vertex set S is called a *null graph*; we also write $\overline{K}_v$ as a general notation for a null graph with v vertices. Figure 6.4 shows a graph and its complement.

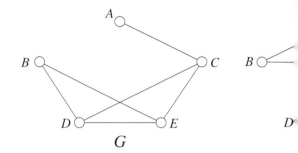

We define the *degree* or *valency* $d(x)$ of the verte
that have x as an endpoint. If $d(x) = 0$, then x is
graph is called *regular* if all its vertices have the
degree is r, it is called *r-regular*. In particular, a 3-r

Theorem 6.6.1. *In any graph or multigraph, the su
equals twice the number of edges (and consequently
even integer).*

Proof. Suppose the graph or multigraph has e edg
..., y_e. Consider a list in which each edge appears
For example, if y_1 has endpoints x_4 and x_7, you
$y_1{:}x_7$. Vertex x will appear in precisely $d(x)$ entries,
equals the sum of the degrees of the vertices. On the
twice, so the total number of entries equals twice th

Corollary 6.6.1. In any graph or multigraph, the n
is even. In particular, a regular graph of odd degree l

Suppose you encounter two graphs with exactly t
vertex-sets. For example, look at Figure 6.4 again; t
sets $\{B, D, E\}$ and $\{C, D, E\}$ are both triangles. The
them until you need to consider the meaning of the
graphs G and H are *isomorphic* if there is a one-to
their vertex-sets $V(G)$ and $V(H)$ such that two ver
only if the corresponding vertices of H are adjacent.
graphs with v vertices are isomorphic.

Not all graphs with the same number of vertices
the graph G of Figure 6.4 and its complement are
five vertices.

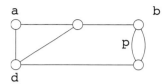

Figure 6.5: An example of road systems.

6.7 Paths and Cycles in Graphs

Much of the terminology of graph theory comes from the application of networks to modeling road systems. For example, the multigraph shown in Figure 6.5 could represent five camps denoted a, b, c, d, e and the walking tracks between them; there are two tracks joining c to e, and they are labeled p and q, and other tracks are denoted by their endpoints. (These representations are ordered.) A sequence of vertices and edges of the form $(\ldots, x, xy, y, yz, z, \ldots)$ will be called a *walk*, because on the diagram it represents a trail that one could walk among the camps.

Consider the following walks:

$$W_1 = (a, ab, b, bd, d, da, a, ad, d, de, e);$$
$$W_2 = (a, ad, d, de, e, p, c, cb, b);$$
$$W_3 = (c, p, e, p, c, cb, b, bd, d, de, e, q, c).$$

Each edge of W_1 is uniquely determined by its endpoints — it does not contain p or q — so we could simply write it as (a, b, d, a, d, e), omitting the edges. This will always be true of walks in graphs. The walk W_2 has no repeated edges. A walk with this property is called *simple*. W_3 returns to its starting point; a walk that does this is called *closed*. We shall make the following formal definitions:

A *walk of order n* (or of *length $n-1$*) in a network is a finite sequence

$$(x_1, y_1, x_1, y_2, x_2, y_3, \ldots, y_{n-1}, x_n)$$

of vertices $x_1, x_2, \ldots$ and edges $y_1, y_2, \ldots$ where the endpoints of y_i are x_i and x_{i+1}. x_1 and x_n are the *start* and *finish*, respectively, of the walk, and we refer to a *walk from x_1 to x_n*. The following are special types of walk:

simple walk: a walk in which no edge is repeated;

path: a simple walk in which no vertex is repeated;

circuit or *closed walk*: a simple walk in which the start and finish are the same;

cycle: a simple closed walk in which $x_1, x_2, \ldots, x_{n-1}$ are all different. No vertex is repeated, except that the start and finish are equal.

A cycle might also be called a *simple circuit*. Man
phrase "closed path," although strictly speaking th
and finish make up a repetition.

Given a path in a graph or multigraph G, you
up of the vertices and edges in the path. This subg
we think of it as a graph by itself, an n-vertex path
with n vertices $x_1, x_2, \ldots, x_n$ and $n-1$ edges x_1x_2, x
n-vertex *cycle* or n-cycle C_n is a graph with n verti
$x_1x_2, x_2x_3, \ldots, x_{n-1}x_n, x_nx_1$, and we can think abo
graphs. A 3-cycle is often called a *triangle*.

Suppose a walk W from x_1 to x_n, where $x_1 \neq x_n$
than once; say x_i is the first occurrence of the poi
walk $x_1, x_2, \ldots, x_i, x_{j+1}, \ldots, x_n$ is part of W — we
contains fewer repeats. If we continue this process,
repeats, a path. That is, any x_1-x_n walk contains a

Theorem 6.7.1. *If a graph contains a walk from*
a path from x_1 to x_n.

In the same way, one could argue that any clos
there is one exception to be made: it is possible to
walk back by retracing your steps. The graph of s
cycle, unless the x_1-x_n walk contained one. But any
We say two vertices are *connected* if there is a w
view of Theorem 6.7.1, we could replace "walk" by "
definition of connectivity: a graph is connected if and
are connected. We can now define a *component* as t
a maximal set of vertices with every pair connecte
them.

It is reasonable to say that graph theory was fou
cian and physicist Leonhard Euler. In 1735, Euler
Academy in 1735 on the problem of the Königsberg
Königsberg was set on both sides of the River Preg
called "The Kneiphof," and the river branched to
four main land masses — let's call them the four *pa*
A, B, C, D — connected to each other by seven bri
is shown on the left in Figure 6.6.

The problem was to find a walk through the city
once and only once, and visit all four parts. The
different pieces of land was to cross the bridges. In n
ogy that would not have existed without Euler's wo
could be represented by a multigraph, with the par
the bridges represented by edges, as shown in the r
A solution to the problem would be a walk in the m

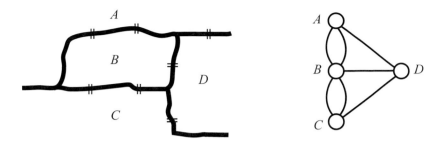

Figure 6.6: The original Königsberg bridges.

edge; even better would be a closed simple walk, so that a person could start and finish from home.

Euler proved that no such walk is possible. But he did much more. He essentially invented graph theory and showed how it could be used to represent any set of islands and bridges, or indeed any set of towns and roads joining them, and concocted an algorithm for traversability problems.

In graph terminology, we define an *Euler walk* to be a simple walk that contains every edge of the network, and an *Euler circuit* to be a closed Euler walk. A network that has an Euler circuit is called *eulerian*.

Suppose G has an Euler walk, and suppose X is a vertex of G. (So there are $d(X)$ bridges leading to the area X in the corresponding map.) Every time the walk passes through X, you traverse two of these bridges. By the end of the walk, this accounts for an even number of edges. The only way X can have odd degree is if it is either the start or the finish of the walk. A walk can have at most one start and one finish, so a network cannot have an Euler walk if it has more than two odd vertices. Similarly, an eulerian graph must have all its vertices even.

Euler showed [6] that these conditions are sufficient, if we add the obvious proviso that the graph must be connected:

Theorem 6.7.2. *If a connected network has no odd vertices, then it has an Euler circuit. If a connected network has two odd vertices, then it has an Euler walk whose start and finish are those odd vertices.*

In 1857 Thomas Kirkman posed the following problem in a paper that he submitted to the Royal Society: given a graph, does there exist a cycle passing through every vertex? Two years later Sir William Rowan Hamilton, who was Astronomer Royal of Ireland, invented a puzzle called the *Icosian game*. The board is a regular graph with 20 vertices and 30 edges. (The graph is a representation of a dodecahedron, with the vertices representing the corners.) The object of the game was to visit every vertex using each edge at most once. This was essentially an example of Kirkman's problem, because the path generated in visiting the vertices would be the sort of cycle Kirkman was discussing, but historically a cycle that

contains every vertex is called a *Hamilton cycle*. (Ki
in the history of block designs.) The name "Hamil
of a path that contains every vertex but is not close

A graph is called *hamiltonian* if it has a Hamilton
is whether or not a given graph is hamiltonian. Som
graph larger than K_2 has a Hamilton cycle, and in
vertices gives rise to one of them.

The following necessary condition was found by

Theorem 6.7.3. *Suppose G is a graph with v vert
of nonadjacent vertices x and y,*

$$deg(x) + deg(y) \geq v$$

then G is hamiltonian.

Proof.   Suppose the theorem is not true. We
nonadjacent vertices satisfy the given degree condi
nonadjacent vertices then the graph formed by add
will be hamiltonian (if not, then join pq and use the
would say G is *maximal* for the condition.

As $v \geq 3$, K_v is hamiltonian, so G is not comple
y, that are not adjacent. Then G is not hamiltonia
must lie in every Hamilton cycle of G^*. Trace such
and delete the first edge; the result is a Hamilton pa
If $x \sim x_i$ for some i with $2 < i < p$, then $y \not\sim x_{i-1}$,

$$(xx_i x_{i+1} \ldots y x_{i-1} x_{i-2} \ldots$$

would be a Hamilton cycle in G, which is impossible

Say the neighbors of x are $x_2, x_{j_2}, x_{j_3}, \ldots, x_{j_d}$, wh
$d = deg(x)$. Then y is *not* adjacent to any of x_{j_2-1}
there are d vertices not adjacent to y, and therefore d
Therefore
$$deg(x) + deg(y)d + deg(y) \leq d + v$$

a contradiction.

This gives a sufficient condition, but it is not
graphs do not satisfy the theorem. In fact, we k
whether a graph is hamiltonian. At first sight, th
cycle problems look similar, but there is a simple
is eulerian, and a straightforward algorithm to con
exists.

Often there are costs associated with the edges, s
In those cases, we define the *cost* of a Hamilton cycl
the edges. These ideas were first studied in the 193(

For example, say a traveling salesman needs to stop at several cities (ending back at home); finding the shortest/cheapest route that can be used is called the *Traveling Salesman Problem*.

The obvious approach to the problem is to list all possible Hamilton cycles, work out the cost of each, and search for the cheapest. However, this can be an extremely long computation. In the case of a complete graph, the time required for a complete search grows exponentially with the number of cities. So several algorithms have been constructed that give an approximate solution in a shorter time. We shall look briefly at two of these, the *sorted edges algorithm* and the *nearest-neighbor algorithm*.

To explain the ideas behind the algorithms, suppose an explorer wants to visit all the villages in a newly discovered area, and wants to do it as efficiently as possible. The "costs" are the distances between settlements; the distance from x_1 to x_2 will be denoted $c(x_1, x_2)$.

The sorted edges algorithm models the way the explorer might proceed if she has a map of the whole area, possibly produced by an aerial survey, with distances marked. She might hope to find an efficient route by taking the shortest feasible connections.

Sorted edges algorithm: Initially sort the edges from shortest (cheapest) to longest and list them in order. Select the shortest edge. At each stage delete from the list the edge you just used, and any edges that would form a circuit or if added to those already chosen, or would be the third edge chosen that was adjacent to some vertex (we'll call this a *threeway*), then select the shortest remaining edge. In this case the result does not depend on a starting point.

In the nearest neighbor algorithm, the explorer starts at one of the villages. She asks the locals, "What is the nearest village to here?" If possible, she goes to that village next.

Nearest neighbor algorithm: First choose any vertex x_1 as a starting point. Find the vertex x_2 for which $c(x_1, x_2)$ is smallest. Continue as follows: at each stage go to the nearest neighbor, except that you never close off a cycle (until the end) or backtrack. In other words, after x_i is chosen ($i < n$), x_{i+1} will be the vertex *not* in $\{x_1, x_2, \ldots, x_i\}$ for which $c(x_i, x_{i+1})$ is minimum. Finally, x_n is joined to x_1. Notice that the result of this algorithm depends on the starting point.

In either example, it may be that two candidate edges have the same cost, and either may be chosen. However, if the algorithms are to be implemented on a computer, some sort of choice must be made. Let us assume we always take the edge that occurs earlier in alphabetical order (if edges bc and bd have the same length, choose bc).

We illustrate these two algorithms using Figure 6.7. In the nearest neighbor algorithm starting at A, the first stop is D, because $c(AD) < c(AB) < c(AC)$ (that is, $35 < 55 < 75$). Next is B ($c(DB) < c(DC)$, AD is not allowed). Finally we go to C. The cycle is $ADBCA$, or equivalently $ACBDA$, cost 200. Nearest neighbor from B yields $BDACB$ (the same cycle), from C, $CDBAC$ (cost 205),

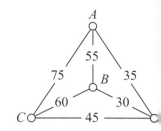

Figure 6.7: A copy of K_4, wit

and from D, $DBACD$ (same cycle as from C).

In sorted edges we first make the list

edge	BD	AD	CD	AB	
cost	30	35	45	55	

Then we proceed: select BD; AD is usable, so select
edges at D); AB is not acceptable (triangle ABD)
The cycle is $ACBDA$, cost 200.

Note that the cheapest route, (A, B, C, D), cost
method. So these algorithms do not always achieve
ever, the results are usually reasonably good and the
is linear function of the number of vertices.

It should be noted that these approximate algo
for the case of a graph that is complete or nearly s
assumption. In cases where there are relatively few
often feasible; and in those cases, it is possible that
the nearest-neighbor algorithm might yield no solut

6.8   Trees

A *tree* is a connected graph that contains no cycle
amples of trees. A tree is a minimal connected g
the resulting graph is not connected. A vertex is c
disconnects the graph (or, in the case of a disconr
component in which it lies); a connected graph is a t
of degree greater than 1 is a cutpoint.

Theorem 6.8.1. *A connected graph G with v vert
has exactly $v - 1$ edges.*

This theorem is easily proven by induction.

It follows that any tree has at least two vertices of degree 1. For example, suppose a tree has v vertices, of which u have degree 1. Every other vertex has degree at least 2, so the sum of the degrees is at least $u + 2(v - u) = 2v - u$. There are $v - 1$ edges, so the sum of the degrees is $2v - 2$. So $2v - u \leq 2v - 2$, and $u \geq 2$.

Suppose you have to connect a number of computers in an office building. For speed and reliability, the management wishes to have a hard-wired connection between every pair of computers. The connection need not be direct; it is possible to relay a message through an intermediate computer. To model this, the computers are represented as vertices and the possible direct connections are represented as edges. The set of edges that are chosen must form a connected subgraph. Moreover this subgraph must contain all the vertices — this is called a *spanning* subgraph.

There will be a positive cost associated with making a direct link between a pair of computers. So, instead of making direct connections between *every* pair of computers for which it is feasible, management might plan to avoid any cycles. The resulting graph would still need to span the system, so they want to find a connected subgraph that contains all the vertices, a *spanning tree* in the original graph.

Every connected graph contains a spanning tree. To see this, consider any connected graph. If there is no cycle, it is a tree. So suppose the graph contains a cycle, such as $abcdea$. The pairs of vertices ab, bc, cd, de and ea are all joined by edges. If you need a path from a to e (for example, to transmit a message between computers), you could use the single edge, or the longer path $a - b - c - d - e$. So remove one edge from the cycle. The resulting spanning subgraph is still connected. If it still contains a cycle, delete an edge from it. Continue in this way: graphs are finite, so the process must stop eventually, and you have a spanning tree.

Spanning trees are important for a number of reasons. A spanning tree is a subgraph with a relatively small number of edges — and therefore easy to examine — but it contains a significant amount of information about the original graph. They have proven very useful in designing routing algorithms and in a number of areas of network design.

In the problem of connecting computers in an office building, consider the graph whose vertices are the computers and whose edges represent hard-wired direct links. As we said, it is cheapest if the connections form a tree. But which tree should be chosen?

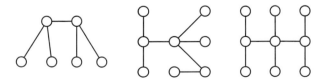

Figure 6.8: Some examples of trees.

We shall assume that the costs are additive. S⟨
as weights on the edges, and the cost of installing t⟨
equal the sum of the costs of the installed edges. W
of minimal cost, or *minimal spanning tree.*

A finite graph can contain only finitely many sp⟨
could list all spanning trees and their weights, and fi⟨
choosing one with minimum weight. But this proce⟨
however, since $\tau(G)$ can be very large. So efficient a⟨
spanning tree are useful.

We shall outline an algorithm due to Prim [18] f⟨
tree in a connected graph G with v vertices. It finds
To start, choose a vertex x. Trivially the cheapest t⟨
only tree with vertex-set $\{x\}$ — is the K_1 with vert⟨
the cheapest tree with two vertices, one of which i⟨
tree that can be formed by adding one edge to T_1.
this way: if you have constructed T_i, then T_{i+1} is
formed by adding one edge to it. The process can a⟨
$i < v$ — the graph is connected, so it must always b⟨
that is joined to one of the vertices of T_i, so there
for the choice of T_{i+1}, and finally T_v is the desired ⟨

It may be shown that the spanning tree constr⟨
minimal.

The new edge used to construct T_{i+1} may no⟨
algorithm, and indeed T_v may not be uniquely def⟨
after all, there may be more than one minimal spa⟨
decide on a "tiebreaking" strategy before you begin⟨
take the edge that comes earlier in alphabetical ord⟨

Example 6.8.1. Apply Prim's algorithm to the gr⟨

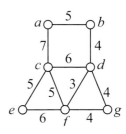

(1) We start at a. The two possible edges are ab an⟨
$y_1 = ab$.
(2) The only possible choices for the next edge are ⟨
bd, is chosen. T_3 is the path abd.
(3) There are now four possible edges: ac (cost 7), c⟨
choice is df.

(4) Now there is a choice. The possible edges are ac (7), cd (6), cf (5), dg (4), ef (6) and fg (4). Either dg or fg could be used. Using alphabetical ordering, we choose dg.

(5) We have ac (7), cd (6), cf (5) and ef (6) (fg is not allowed). The choice is cf.

(6) Finally, from ce (5) and ef (6) we choose ce.

The sequence of trees is

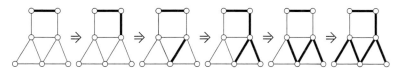

□

6.9 Encoding and Decoding

When we send messages by computer, the message must be converted to a string of instances of two states, "power on" and "power off;" we shall write 1 for "power on" and 0 for "power off." A short message might look like

100110110011000011011100010110011011100010110011011001001011000.

Such a sequence of 0's and 1's is called a *binary string*. The conversion process is called *encoding*, and the corresponding process of going back from the string to the original message is called *decoding*. (The words "encryption" and "decryption" are also used, but these are more appropriate when the main concern is the transmission of secret messages.)

There are a number of ways to represent a message as a binary string. We could represent a as 000001, b as 000010, ..., A as 011011, and so on. These 64 ($= 2^6$) strings will enable us to represent the 26 capitals and 26 lowercase numbers, together with 12 punctuation marks. This type of representation is widely used in computers — the ASCII (American Standard Code for Information Interchange) code, which was developed for computers, uses seven binary symbols for each letter, giving a wide range of punctuation and special symbols. *Extended ASCII* uses eight binary symbols for each.

Sometimes there will be errors in a message. In electronic transmissions they are often due to *noise*. The medium through which the message is transmitted is called a *channel*, and some channels are noisier than others.

To see a way around the noise problem, consider what happens in everyday English, when there is a misprint. If you are reading a book and you come across the word "bive," you are fairly sure this is not a real word. Assuming one letter is wrong, there are many possibilities: *dive, five, live, bile, bite* are only a few. If this were all the information you had, you could not tell which was intended. But the context — the other words nearby — can often help. For example, if the

sentence is "...the next bive years ..." then the au
it says "...a review of a bive performance ..." ther
phrase "...teeth used in the bive ..." suggests "bit
example ..." probably should be "give." The conte

However, if the encoding results in a binary strin
context. You do not even have a context until after
around this, we add some more symbols. One very
message up into substrings or *words*, say of length
add another symbol (a *check digit*), 0 or 1, chosen s
even number of 1's. For example, suppose you want

$$10011011001100001101$$

First, break it up into the form

$$1001 \quad 1011 \quad 0011 \quad 0000 \quad 1$$

and then insert check digits:

$$10010 \quad 10111 \quad 00110 \quad 00000$$

This method is useful when there is a way of checking
if the message was received as

$$10110 \quad 10111 \quad 00110 \quad 00000$$

you could say, "I got the second through fifth wor
first?" But this is not useful for electronic communi
of symbols are sent every second.

Another problem is that the method is useless
in a word. The same occurs in everyday English: i
there are many more possibilities: "bone," "bane,"
no solution to this problem. There are encoding r
correct!) many errors, but it is always possible that

We shall examine one example of an encoding
strings, called *Venn diagram coding* for reasons that
you are happy to assume that no more than one e
transmitted symbols: the cases where more errors
that you can recover the message. Proceed as follc
up into substrings, or *message words*, of four symbo
word by a *codeword* as follows. If the message is
D represents a 1 or a 0), define three more binary
follows so that each of the sums $A + B + C + E$, $A +$
is even.

Example 6.9.1. How do you encode 110011010110

First break the message up as

$$1100 \quad 1101 \quad 0110.$$

If $ABCD$ is 1100, then $A + B + C + E = 1 + 1 + 0 + E$; for this to be even, $E = 0$. Similarly $A + C + D + F = 1 + 0 + 0 + F$, $F = 1$, and $B + C + D + G = 1 + 0 + 0 + G$, $G = 1$. The first codeword is 1100011. The other words are 1101000 and 0110010.

The easiest way to represent this process is as follows. Draw a diagram (essentially a three-set *Venn diagram*) of three overlapping circles, and write the numbers A, B, C and D in the positions shown in the left-hand part of Figure 6.9. Then the other numbers are chosen so that the sum in each circle is even. These are E, F and G, as we defined them, and shown in the right-hand part.

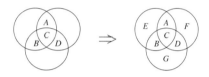

Figure 6.9: Venn diagram encoding.

After the codeword is sent, the receiver, (*decoder*) must reconstruct the original message. Suppose a string of seven binary symbols is received. The decoder assumes that at most one error has been made in the encoding. If more than one error occurs in seven symbols, the wrong message will be found. We have to assume that this will occur very infrequently. (If errors are more common, other methods should be used.) □

Example 6.9.2. Suppose you receive 1111010. How is it decoded?

$A + B + C + E = 3$, so one of these symbols must be wrong. (Since there is assumed to be at most one error, there cannot be three wrong symbols.)

$A + C + D + F = 3$, so one of these symbols must be wrong.

$B + C + D + G = 2$, so these four symbols are all correct.

The first and third conditions tell us that exactly one of A and E is incorrect; the second and third conditions tell us that exactly one of A and F is incorrect. So A is wrong; it should have been 0, not 1, the correct codeword is 0110010 and the original message was 0110.

This decoding process is very simple when the Venn diagram is used. The decoder fills in the diagram and puts the number of 1's next to each circle, as shown on the left-hand side of Figure 6.10. The two top circles are odd; the wrong symbol must be the one in the unique cell common to both of those circles. This process will always work: provided there is at most one error, the Venn diagram method will yield the correct codeword. Moreover, every binary string of length 7 can be decoded by this method.

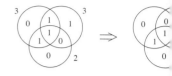

Figure 6.10: Venn diagram de

The Venn diagram code is an elementary exampl
a large literature on Hamming codes and other error
study belongs to linear algebra, rather than combin
should consult texts such as [11].

In addition to the problems of encoding messa
errors, there is another important aspect: secrecy
electronically, others may be able to access your m
study of secret codes, or cryptography. While secre
thousands of years, modern improvements in messag
topic far more important. The area of cryptograp
issues of computational complexity as well as linear
reference.

6.10   Latin Squares

"There are thirty-six military officers, six each from s
ficers from any one regiment include one of each of six
colonel, a major, a captain, a lieutenant, and a s
parade in a square formation, so that every row ar
member of each regiment and one officer of each ran

This problem fascinated Euler in the eighteenth
solve it. In fact, it was not until 1900 that Tarry
search, that no solution is possible.

Suppose we are given an arrangement of the of
every column contains one representative of each re
as a 6×6 array; if the officer in row i, column j be
array has (i, j) entry k. For example, one possible ar
from regiments 1, 2, 3, 4, 5, 6 could be represented

Euler discussed these arrays at length in [7], and
definition:

A *Latin square* of *side* (or *order*) n is an $n \times$
of n symbols (treatments), with the property that
contains every symbol exactly once. In other words

1	3	6	2	4	5
5	1	3	6	2	4
4	5	1	3	6	2
3	6	2	4	5	1
6	2	4	5	1	3
2	4	5	1	3	6

Figure 6.11: An array for the thirty-six officers.

is a permutation of S.

The arithmetical properties of the symbols in a Latin square are not relevant to the definition, so their nature is often immaterial; unless otherwise specified, we assume a Latin square of order n to be based on $\{1, 2, \ldots, n\}$. (Other symbol sets will be used when appropriate.)

Figure 6.12 shows some small Latin squares.

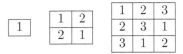

Figure 6.12: Some small Latin squares.

Theorem 6.10.1. *There exists a Latin square of every positive integer order.*

Euler's problem of the 36 officers cannot be solved. However, if we replace 36 by 9, and have three regiments a, b, c with three ranks α, β, γ represented in each, there is a solution. An example is

$a\alpha$	$b\beta$	$c\gamma$
$b\gamma$	$c\alpha$	$a\beta$
$c\beta$	$a\gamma$	$b\alpha$

This array was constructed by superposing two 3×3 Latin squares, namely

a	b	c
b	c	a
c	a	b

and

α	β	γ
γ	α	β
β	γ	α

.

The two squares have the following property: the positions occupied by a fixed symbol in the first square form a transversal in the second square. (For example, the positions containing a in rows 1, 2 and 3 contain α, β and γ, respectively, in the second.) We say the second Latin square is *orthogonal* to the first.

In general, suppose A and B are Latin squares (S_A and S_B, respectively. B is *orthogonal* to A (writt(the set of n positions in A occupied by x contain ev

It is clear that each member of S_B will occur prec by a fixed element in A.

Equivalently, suppose we construct from A anc ordered pairs, where (x, y) occurs in position (i, j) if ((i, j) of A and y occurs in position (i, j) of B, then possible ordered pair with first element in S_A and s(the new array.

Obviously orthogonality is symmetric: B is ort. thogonal to B.

The array formed by superposing two orthogon called a *graecolatin square*. (In fact, the name "Lat: habit of using the Roman (that is, Latin) alphabet and when representing a pair of orthogonal squares for the second square, just as we did.) But it is mor of orthogonal Latin squares," and to use the symb have exhibited a pair of orthogonal Latin squares of is no pair of orthogonal Latin squares of order 6. Or there exist a pair of orthogonal Latin squares? Car squares that are *mutually orthogonal* — each one is

Before we discuss this problem, we define a *star* Latin square whose symbols are the first n positive i

$$\boxed{1\ 2\ 3\ \ldots\ n}.$$

If A is any Latin square, we can convert it to stand symbols. If A and B are orthogonal, we can assume we rename the symbols in each separately.

Theorem 6.10.2 (Maximum Number of MO *orthogonal Latin squares of side n, $n > 1$, then $k <$*

Proof.   Suppose $A_1, A_2, \ldots, A_k$ are Latin squares (is orthogonal to each other one. Without loss of ger been standardized, so that each has first row (6.10) in A_i. No a_i can equal 1 (since the first columns ca the a_i must be different (if $a_i = a_j$, then the n cel contain a repetition in A_j; both the $(1, a_i)$ and $(2,$ a_i). So $\{a_1, a_2, \ldots, a_k\}$ contains k distinct elements

Let us write $N(n)$ for the number of squares mutually orthogonal Latin squares of side n. In this

$$N(n) \leq n - 1 \text{ if } n > 1.$$

For example, $N(4) \leq 3$. In fact, $N(4) = 3$; one set of three mutually orthogonal Latin squares of side 4 is shown in Figure 6.13.

1	2	3	4
2	1	4	3
3	4	1	2
4	3	2	1

1	2	3	4
3	4	1	2
4	3	2	1
2	1	4	3

1	2	3	4
4	3	2	1
2	1	4	3
3	4	1	2

Figure 6.13: Three orthogonal Latin squares of side 4.

On the other hand, the theorem tells us only that $N(6) \leq 5$, but we know from Tarry's result that $N(6) = 1$.

Theorem 6.10.3. $N(n) = n - 1$ *whenever n is a prime power.*

We know that there cannot be a pair of orthogonal Latin squares of order 6, and order 2 is also impossible. Are there any further values n for which $N(n) = 1$? Euler thought so; he conjectured that $N(n) = 1$ whenever $n \equiv 2$ modulo 4. However, the opposite is true:

Theorem 6.10.4. $N(n) \geq 2$ *for every integer $n > 6$.*

The proof can be found in most books on design theory (such as [26]). The history of this problem and its solution can be found in [8].

Further results are known. For example, many authors have contributed to the proof of the following theorem.

Theorem 6.10.5 (Three MOLS). *There exist three mutually orthogonal Latin squares of every side except 2, 3, 6, and possibly 10.*

In general, for any given constant k there exists a lower bound n_k such that $N(n) \geq k$ whenever $n > n_k$, but the only known value is $n_2 = 6$; we do not know $N(10)$, so n_3 may be either 10 or 6. For the best known lower bounds on n_k, as well as a table of the largest known numbers of Latin squares of orders less than 10,000, see [5].

6.11 Balanced Incomplete Block Designs

Suppose you want to compare six varieties of corn, varieties x_1, x_2, x_3, x_4, x_5 and x_6, in order to decide which variety yields the most grain in your area. You could find a large field, divide it into six smaller areas (*plots*), and plant one variety in each plot. Eventually you could weigh the different yields and make your comparisons.

However, suppose one part of the field is more fer
planted there would seem better than the other variet
the quality of the soil, not the corn. The way arour
into four midsized areas — *blocks* — so that the fe
uniform. You could then plant six plots, one for ea«
total of 24.

But now there is another problem. Not all seec
all plants grow equally well, even in identical enviro
uniform, and do not develop uniformly. If the plo
variation could affect the yields enough to outweigh
to study. Maybe 24 plots are too many, the individu
The solution is to make the plots larger; not eve
block. For example, you might restrict each block
block contains x_1, x_2 and x_3, it makes sense to com
in that block, and in every block where both appea.

For conciseness, let us call the blocks $B_1, B_2, .$
with the set of varieties, so $B_1 = \{x_1, x_2, x_3\}$. One p
is

$$\begin{aligned}
B_1 &= \{x_1, x_2, x_3\} \\
B_2 &= \{x_4, x_5, x_6\} \\
B_3 &= \{x_1, x_2, x_5\} \\
B_4 &= \{x_3, x_4, x_6\}.
\end{aligned}$$

We could compare the yield for varieties x_1 and x_2 tv
Such an arrangement of varieties into blocks is «
because they were first studied as tools for conduc
agricultural experiment we have outlined, or *desigr*
case when all the blocks are the same size (like our ex
is used. When the blocks are proper subsets of the
incomplete.

Formally, we define a *block design with parameter.*
b subsets of size k (called *blocks*) from a set V of
treatments). We say the design is *based on V*. A b
variety occurs in the same number of blocks. This
when the data is analyzed, because we have the same
each variety. In this case the number of blocks in whi
the *replication number* or *frequency* of the design, a
The word *incidence* is used to describe the rel.
varieties in a design. We say block B is incident with
incident with block B, to mean that t is a member «
we shall primarily be involved with incidence and o
collection of subsets of the varieties. It makes no d:
are labels representing types of experimental mater

integers, or whatever. We are purely interested in the combinatorics of the situation. For this reason, the objects we are discussing are often called *combinatorial designs*.

The four design parameters we have defined are not independent:

Theorem 6.11.1. *In a regular block design,*

$$bk = vr. \tag{6.11}$$

If x and y are any two different varieties in a block design, we refer to the number of blocks that contain both x and y as the *covalency* of x and y, and write it as λ_{xy}. Many important properties of block designs are concerned with this covalency function. The one that has most frequently been studied is the property of *balance*: a *balanced incomplete block design*, or BIBD, is a regular incomplete design in which λ_{xy} is a constant, independent of the choice of x and y; this constant covalency, denoted λ, is called the *index* of the design. We often refer to a balanced incomplete block design by using the five parameters (v, b, r, k, λ), and call it a (v, b, r, k, λ)-design or (v, b, r, k, λ)-BIBD.

Balanced incomplete block designs were defined (essentially as a puzzle) by Woolhouse [30], in the annual *Lady's and Gentleman's Diary*, a "collection of mathematical puzzles and aenigmas" that he edited. The Reverend T. P. Kirkman [12–15] studied the case of block size 3; in particular he introduced his famous "schoolgirl problem" [14]:

> A schoolmistress has 15 girl pupils and she wishes to take them on a daily walk. The girls are to walk in five rows of three girls each. It is required that no two girls should walk in the same row more than once per week. Can this be done?

The solution is a balanced incomplete block design with $v = 15, k = 3, \lambda = 1$ with some additional properties.

Yates [32] first studied balanced incomplete block designs from a statistical viewpoint.

There is a further relation between the parameters of a balanced incomplete block design:

Theorem 6.11.2. *In a (v, b, r, k, λ)-BIBD,*

$$r(k - 1) = \lambda(v - 1). \tag{6.12}$$

Relations (6.11) and (6.12) allow us to find all parameters of a BIBD when only three are given. For example, if we know v, k and λ, we can calculate r from (6.12) and then deduce b using (6.11). As an example, if $v = 13$, $k = 3$ and $\lambda = 1$, we have $r = \lambda(v - 1)/(k - 1) = 12/2 = 6$ and $b = vr/k = 13 \times 6/3 = 26$, and a $(13, 26, 6, 3, 1)$-design exists. But not all sets $\{v, k, \lambda\}$ give a design. First, all the parameters must be integers; for example, if $v = 11$, $k = 3$ and $\lambda = 1$, we would

get $r = 5$ and $b = 55/3$; you can't have a fractio
is impossible. And even when the parameters are a
corresponding design; for example, there is no (22, 2
6.11.4), even though both equations are satisfied. I
design corresponding to a given set of parameters is

There is no requirement that all the blocks in a
two blocks have the same set of elements, we say t
design that has no repeated block is called *simple*.

If a design has b blocks $B_1, B_2, \ldots, B_b$ and v var
$v \times b$ matrix A with (i, j) entry a_{ij} as follows:

$$a_{ij} = \begin{cases} 1 & \text{if } t_i \in B_j; \\ 0 & \text{otherwise} \end{cases}$$

This matrix A is called the *incidence matrix of the*
that each block corresponds to a column of the incic
corresponds to a row.

Theorem 6.11.3. *In any balanced incomplete bloc*

Proof.   Suppose A is the incidence matrix c
$\det(AA^T) = (r - 1)^{v-1}[r + \lambda(v - 1)]$. From (6.12
$r(k - 1)$, so

$$\det(AA^T) = (r - 1)^{v-1}[r + r(k - 1)]$$

On the other hand, (6.12) can be rewritten as

$$r = \frac{v - 1}{k - 1}\lambda$$

by incompleteness, $k < v$, so $r > \lambda$, and $(r - \lambda)$ is r
is non-zero, so the $v \times v$ matrix AA^T has rank v. T

$$v = \text{rank}(AA^T) \leq \text{rank}(A) \leq \text{n}$$

because A is $v \times b$. Therefore $v \leq b$.

This relation is called Fisher's inequality. Desigi
where the relation is an equality) are called *symmetr*
also satisfies $r = k$, so only three parameters need
phrase is "(v, k, λ)-design." Another abbreviation
Equation (6.12) takes on the simpler form

$$\lambda(v - 1) = k(k - 1).$$

The following important theorem is called the *Br*
3, 21].

Theorem 6.11.4. *If there exists a symmetric balanced incomplete block design with parameters (v, k, λ), then:*

(i) if v is even, $k - \lambda$ must be a perfect square;

(ii) if v is odd, there must exist integers x, y and z, not all zero, such that

$$x^2 = (k - \lambda)y^2 + (-1)^{(v-1)/2}\lambda z^2. \tag{6.14}$$

6.12 Conclusion

We have introduced three main branches of combinatorics: enumeration, graph theory, and design theory. We have borrowed from the introductory text [27] on a number of occasions.

Enumerative problems occur in many places as well as counting. For further treatment of this topic, the reader is referred to [19] and [23], and for work on generating functions, [29].

Graph theory is perhaps the most extensive area of combinatorics, and has widespread applications in the description of computer and social networks, as well and many other areas of science and engineering. Some introductory texts are [25] and [28]. For further material on trees, see trees such as [31].

Combinatorial designs were first studied in order to design and analyze experiments, but have applications in geometry and other areas. Two books on this topic are [16] and [26].

The applications of combinatorics are numerous. We have looked at some counting problems and the Traveling Salesman Problem (see [10]). A number of other applications will be found in the texts cited above.

6.13 Exercises

Exercise 6.13.1. A computer store has four refurbished Gateway computer systems, three refurbished Dell systems, and seven refurbished Acer systems. In how many ways can we purchase a system from this store?

Exercise 6.13.2. Suppose a set X has $2n + 1$ elements. How many subsets of X have $n + 1$ or fewer elements?

Exercise 6.13.3. A furniture store has four kinds of dining tables, six different sets of dining room chairs, and two kinds of buffets. In how many ways can we purchase a dining room set consisting of a table, a set of chairs, and a buffet?

Exercise 6.13.4. A state has a license plate that consists of two letters, three digits, and one further letter, in that order. How many license plates are possible?

Exercise 6.13.5. A graph has 48 edges. There are no isolated vertices. What are the minimum and maximum numbers of vertices the graph can have?

Exercise 6.13.6. A program for generating random
lowercase letter, one uppercase letter, two digits (0 th
may be uppercase or lowercase. How many possibl
produce?

Exercise 6.13.7. A store receives a set of 24 print
owner wishes to display one print in each of four wir
this be done?

Exercise 6.13.8. At one time, area codes were
second digit 0 or 1 and the first digit non-zero. How
under this system? How many are possible if one ca
second and third position (so as not to conflict with

Exercise 6.13.9. How many of the first two mil
exactly one 3, one 5 and one 7?

Exercise 6.13.10. A number is called a *palindro*
forward as backward.

(i) Find the number of palindromes on seven digi

(ii) Find the number of seven-digit palindromes
more than twice.

Exercise 6.13.11. A student decides that time is a
one social networking site. Of four sites recomm
acquaintances on the first, five on the second, six
fourth. Once joined to any site, the student has for
to befriend or not to befriend that acquaintance.
are there of the process of choosing a site and befrie
acquaintance on the site?

Exercise 6.13.12. You need to choose a s
$\{a, b, c, d, e, f, g, h\}$ that contains either a or b.

(i) In how many ways can this be done?

(ii) In how many ways can it be done if you may

Exercise 6.13.13. Find the number of positive inte
and are relatively prime to 21.

Exercise 6.13.14. Find the number of integers betv
are not divisible by 4, 5 or 6.

Exercise 6.13.15. How many elements are there i
$|B| = 34$, and:

(i) A and B are disjoint?

(ii) $|A \cup B| = 47$?

(iii) $A \subseteq B$?

Exercise 6.13.16. Find the generating function for the sequences:

(i) $a_n = n^3$.

(ii) $b_n = 3n^2 - 2n - 1$.

(iii) $c_n = \binom{n}{3}$.

(iv) $d_n = 3^n - \frac{3}{2}2^n$.

Exercise 6.13.17. A bakery sells plain, chocolate chip, and banana-walnut muffins; on a particular day, the bakery has 3 of each kind.

(i) How many ways are there to purchase a selection of n muffins?

(ii) How many ways are there if we wish at least one of each kind?

(iii) How many ways are there if we wish at least 2 banana-walnut muffins, but have no restrictions on the numbers of the other kinds of muffins?

(iv) How many ways are there to purchase n muffins if the bakery has an unlimited supply of each kind? Find a recurrence relation for this sequence.

Exercise 6.13.18. Find the solution for the recurrence with given initial values:

(i) $a_{n+1} = a_n + 12a_{n-1}$, $a_0 = 1$, $a_1 = 4$.

(ii) $a_{n+1} = 4a_n - 4a_{n-1}$, $a_0 = -2$, $a_1 = -2$.

(iii) $a_{n+1} = a_n + 12a_{n-1}$, $a_0 = 1$, $a_1 = 3$.

(iv) $a_{n+1} = 4a_n - 4a_{n-1}$, $a_0 = 1$, $a_1 = 2$.

(v) $a_{n+1} = a_n + 12a_{n-1}$, $a_0 = 0$, $a_1 = 1$.

(vi) $a_{n+1} = 4a_n - 4a_{n-1}$, $a_0 = 0$, $a_1 = 1$.

Exercise 6.13.19. The n-*wheel* W_n has $n+1$ vertices $\{x_0, x_1, \ldots, x_n\}$; x_0 is joined to every other vertex and the other edges are

$$x_1x_2, x_2x_3, \ldots, x_{n-1}x_n, x_nx_1.$$

How many edges does W_n have? What are the degrees of its vertices?

Exercise 6.13.20. A graph has 19 vertices and each vertex has degree at least 4. What are the minimum and maximum numbers of edges the graph can have?

Exercise 6.13.21. Prove that no graph has all its

Exercise 6.13.22. Eight people attend a meeting.
ple shake hands. In total there are 18 handshakes.

(i) Show that there is at least one person who sh

(ii) Is it possible there were exactly three people wl
of times?

Exercise 6.13.23. Five married couples attend a dir
some of the people shake hands; no one shakes han
the end of the party, the hostess asks each person (
"With how many people did you shake hands?" It
shook hands with the same number.

(i) With how many people did the hostess shake

(ii) With how many people did her husband shake

(iii) Generalize this to the case of n couples.

Exercise 6.13.24. Find all paths from x to f in tl

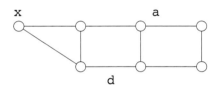

Exercise 6.13.25. The graph G contains two pat
namely $(x, a, \ldots, y)$ and $(x, b, \ldots, y)$. Assuming vert
that G contains a cycle that passes through x. Shov
b are distinct" is necessary.

Exercise 6.13.26. Are the following networks euleri
If not, what is the eulerization number?

(i)   (ii)

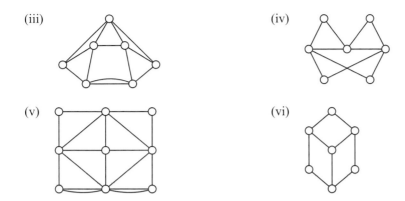

Exercise 6.13.27. Find a connected cubic graph with ten vertices, that is not Hamiltonian.

Exercise 6.13.28. The complete graph K_5 has vertices a, b, c, d, e. In each part we list the set of costs associated with the edges. Find the costs of the routes generated by the nearest neighbor algorithm starting at a, by the nearest neighbor algorithm starting at c, and by the sorted edges algorithm.

(i) $ab = 44$ $ac = 39$ $ad = 30$ $ae = 51$ $bc = 49$
         $bd = 46$ $be = 42$ $cd = 37$ $ce = 38$ $de = 44.$

(ii) $ab = 23$ $ac = 30$ $ad = 27$ $ae = 28$ $bc = 34$
         $bd = 29$ $be = 41$ $cd = 32$ $ce = 42$ $de = 37.$

(iii) $ab = 8$ $ac = 5$ $ad = 7$ $ae = 11$ $bc = 9$
         $bd = 7$ $be = 8$ $cd = 11$ $ce = 9$ $de = 11.$

(iv) $ab = 34$ $ac = 37$ $ad = 32$ $ae = 45$ $bc = 29$
         $bd = 34$ $be = 29$ $cd = 42$ $ce = 43$ $de = 39.$

Exercise 6.13.29. Find minimal spanning trees in the following graphs.

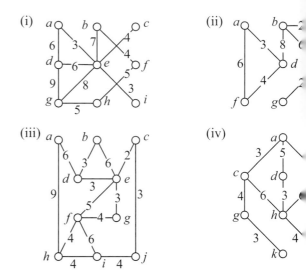

Exercise 6.13.30. What is the distance between t.

(i) 101010111 and 101100110

(ii) 10101010 and 01011101

(iii) 000010110 and 001010110

(iv) 0101101011 and 0011000110

Exercise 6.13.31. Encode the following messages

(i) 1011   (ii) 001▌

(iii) 1101   (iv) 101C

Exercise 6.13.32. Decode the following messages ·

(i) 1101011   (ii) 001▌

(iii) 1001101   (iv) 1101

(v) 1011000   (vi) 101C

Exercise 6.13.33. We say that two Latin squares
obtained from the other by reordering the rows, re·
muting the symbols. A *reduced Latin square* of side
and first column are $(1, 2, \ldots, n)$ in that order.

(i) Prove that every Latin square is equivalent to

(ii) Say there are r_n different reduced Latin squares of side n. Prove that the number of Latin squares of side n is

$$n!(n-1)!r_n.$$

(iii) Find r_n for $n = 3$ and 4.

(iv) Exhibit all reduced Latin squares of order 4.

Exercise 6.13.34. A Latin square is called *self-orthogonal* if it is orthogonal to its own transpose.

(i) Prove that the diagonal elements of a self-orthogonal Latin square of side n must be $1, 2, \ldots, n$ in some order.

(ii) Prove that there is no self-orthogonal Latin square of side 3.

(iii) Find self-orthogonal Latin squares of sides 4 and 5.

Exercise 6.13.35. You need to construct a block design with five blocks of size 2, based on variety-set $\{1, 2, 3\}$, in which every possible pair of varieties occurs at least once. Prove that there are exactly two non-isomorphic ways to do this. Give examples.

Exercise 6.13.36. Suppose v is any positive integer greater than 2. Prove that a $(v, v, v-1, v-1, v-2)$-design exists.

Exercise 6.13.37. In each row of the following table, fill in the blanks so that the parameters are possible parameters for a balanced incomplete block design, or else show that this is impossible.

v	b	r	k	λ
7	14		3	
		6	4	2
		13	6	1
17		8	5	
	30		7	3
33	44		6	

Exercise 6.13.38. Prove that, in any balanced incomplete block design, $\lambda < r$.

Exercise 6.13.39. Show that, in a symmetric balanced incomplete block design, λ cannot be odd when v is even.

Exercise 6.13.40. Show that there do not exist symmetric balanced incomplete block designs with the following parameters: $(46, 10, 2)$, $(52, 18, 6)$.

References

[1] R. H. Bruck and H. J. Ryser, "The non-existe planes," *Canad. J. Math.* **1** (1949), 88–93.

[2] D. M. Burton, *The History of Mathematics* McGraw-Hill, Boston, 2003.

[3] S. Chowla and H. J. Ryser, "Combinatorial p (1950), 93–99.

[4] R. Cooke, *The History of Mathematics: A Brie* & Sons, Hoboken, NJ, 2005.

[5] C. J. Colbourn and J. H. Dinitz, *Handbook of C* Chapman & Hall/CRC, Boca Raton, 2007.

[6] L. Euler, "Solutio Problematis ad geometriam si *Sci. Imp. Petropolitanae* **8** (1736), 128–140.

[7] L. Euler, "Recherches sur une nouvelle espece *hand. Zeeuwsch Gen. Wet. Vlissingen* **9** (1782)

[8] M. Gardner, "Euler's spoilers," *Martin Gardne sions from Scientific American*, Allen & Unwin

[9] S. Giberson and T. J. Osler, "Extending Theor *College Math. J.* **35** (2004), 222.

[10] G. Gutin and A. P. Punnen, *The Traveling Sal tions*, Springer, 2007.

[11] D. C. Hankerson, G. Hoffman, D. A. Leonard, C. A. Rodger, J. R. Wall, *Coding Theory and* 2nd ed., Chapman & Hall/CRC Press, 2000.

[12] T. P. Kirkman, "On a problem in combinations," *J.* **2** (1847), 191–204.

[13] T. P. Kirkman, "Note on an unanswered pri *Dublin Math. J.* **5** (1850), 255–262.

[14] T. P. Kirkman, "Query VI," *Lady's and Gentle*

[15] T. P. Kirkman, "Solution to Query VI," *Lady's c* 48.

[16] C. C. Lindner, C. A. Rodger, *Design Theory*, 2r Press, 2001.

[17] O. Ore, "Note on Hamilton circuits," *Amer. Math. Monthly* **67** (1960), 55.

[18] R. C. Prim, "Shortest connection networks and some generalizations," *Bell Syst. Tech. J.* **36** (1957), 1389–1401.

[19] J. Riordan, *Combinatorial Identities*, Wiley, New York, 1968.

[20] S. Roman, *Introduction to Coding and Information Theory*, Springer-Verlag, 1996.

[21] M. P. Schutzenberger, "A non-existence theorem for an infinite family of symmetrical block designs," *Ann. Eugenics* **14** (1949), 286–287.

[22] L. E. Sigler, *Fibonacci's Liber Abaci: Leonardo Pisano's Book of Calculation*, Springer-Verlag, New York, 2002.

[23] R. P. Stanley, *Enumerative Combinatorics* (2 vol.), Cambridge U. P., 1997, 1999.

[24] G. Tarry, "Le problème des 36 officiers," *Comptes Rend. Assoc. Fr.* **1** (1900), 122–123; **2** (1901), 170–203.

[25] W. D. Wallis, *A Beginner's Guide to Graph Theory*, 2nd ed., Birkhäuser, Boston, 2007.

[26] W. D. Wallis, *Introduction to Combinatorial Designs*, 2nd ed., Chapman & Hall/CRC, Boca Raton, 2007.

[27] W. D. Wallis and J. C. George, *Introduction to Combinatorics*, Chapman & Hall/CRC, Boca Raton, 2010.

[28] D. B. West, *Introduction to Graph Theory*, 3rd ed., Prentice Hall, 2007.

[29] H. Wilf, *Generating Functionology*, 3rd ed., A. K. Peters, Wellesley, 2006.

[30] W. S. B. Woolhouse, "Prize question 1733," *Lady's and Gentleman's Diary* (1844), 84.

[31] B. Y. Wu and K.-M. Chao, *Spanning Trees and Optimization Problems*, Chapman & Hall/CRC, Boca Raton, 2004.

[32] F. Yates, "Incomplete randomized blocks," *Ann. Eugenics* **7** (1936), 121–140.

Chapter 7 Probability, Random Variables, and Stochastic Processes

Dinesh Rajan[‡]
[‡]Southern Methodist University, Dallas, USA

7.1 Introduction to Probability

Probability theory essentially provides a framework and tools to quantify and predict the chance of occurrence of an event in the presence of uncertainties. Probability theory also provides a logical way to make decisions in situations where the outcomes are uncertain. Probability theory has widespread applications in a plethora of different fields such as financial modeling, weather prediction, and engineering. The literature on probability theory is rich and extensive. A partial list of excellent references includes [1–5]. The goal of this chapter is to focus on the basic results and illustrate the theory with several numerical examples. The proofs of the major results are not provided and relegated to the references.

While there are many different philosophical approaches to define and derive probability theory, Kolmogorov's axiomatic approach is the most widely used. This axiomatic approach begins by defining a small number of precise axioms or postulates and then deriving the rest of the theory from these postulates.

Before formally defining Kolmogorov's axioms, we first specify the basic framework to understand and study probability theory. Probability is essentially defined in the context of a repeatable random experiment. An experiment consists of a procedure for conducting the experiment and a set of outcomes/observations of the experiment. A model is assigned to the experiment which affects the occurrence of the various outcomes. A sample space, S, is a collection of finest grain, mutually exclusive and collectively exhaustive set of all possible outcomes. Each element ω of the sample space S represents a particular outcome of the experiment. An event E is a collection of outcomes.

Example 7.1.1. A fair coin is tossed three
$S = \{HHH, HHT, HTH, HTT, THH, THT, T\}$
$\{HTT, THT, TTH\}$ is the set of all outcomes with
coin flips.

Example 7.1.2. The angle that the needle makes
observed. The sample space $S = \{\theta : 0 \leq \theta < 2\pi\}$.

Events E_j and E_k are said to be mutually exclu
are no outcomes that are common to both events, i.
A collection of events $\boldsymbol{F}$ defined over a sample sp

- $\boldsymbol{F}$ includes both the impossible event ϕ and th

- For every set $A \subset \boldsymbol{F}$, it implies that $A^c \subset \boldsymbol{F}$.

- $\boldsymbol{F}$ is closed under countable set operations o
  $A \cap B \subset \boldsymbol{F}$ and $A \cup B \subset \boldsymbol{F}$, $\forall A, B \subset \boldsymbol{F}$.

Given a sigma Field $\boldsymbol{F}$, a probability measure F
event $A \subset \boldsymbol{F}$ to a real number $\Pr(A)$ called the pr
the following three axioms:

1. $\Pr(A) \geq 0$.

2. $\Pr(S) = 1$.

3. For a countable collection of mutually
   $\Pr(A_1 \cup A_2 \cup A_3 \cup \ldots) = \Pr(A_1) + \Pr(A_2) +$

A probability space consists of the triplet $(S, \boldsymbol{F}, P)$.

Example 7.1.3. A fair coin is flipped 1 time. In
sigma field $\boldsymbol{F}$ consists of the sets, $\{H\}$, $\{T\}$, $\{\phi\}$,
maps these sets to the probabilities as follows: $\Pr($
and $\Pr(S) = 1$.

The following simple and intuitive properties of t
be readily derived from these axioms:

- The probability of the null set equals 0, i.e., P

- The probability of any event A is no greater t

- The sum of the probability of an event and the
  equals 1, i.e., $\Pr(A^c) = 1 - \Pr(A)$.

- If $A \subset B$ then $\Pr(A) \leq \Pr(B)$.

- The probability of the union of events A and B can be expressed in terms of the probability of events A, B and their intersection $A \cap B$, i.e.,

$$\Pr(A \cup B) = \Pr(A) + \Pr(B) - \Pr(A \cap B). \qquad (7.1)$$

To prove (7.1), we can express $A \cup B$ in terms of three mutually exclusive sets $A_1 = A \cap B$, $A_2 = A - B$ and $A_3 = B - A$. Hence, $\Pr(A \cup B) = \Pr(A_1) + \Pr(A_2) + \Pr(A_3)$. Then by applying Axiom 3, we obtain $\Pr(A) = \Pr(A_1) + \Pr(A_2)$ and $\Pr(B) = \Pr(A_1) + \Pr(A_3)$. Property (7.1) readily follows. The other properties stated above can be similarly proved.

The conditional probability $\Pr(A|B)$ for events A and B is defined as

$$\Pr(A|B) = \frac{\Pr(A \cap B)}{\Pr(B)}, \qquad (7.2)$$

if $\Pr(B) > 0$. This conditional probability represents the probability of occurrence of event A given the knowledge that event B has already occurred.

If events $A_1, A_2, \ldots A_n$ form a set of mutually exclusive events ($A_i \cap A_j = \phi, \forall i, j$) that partition the sample space ($A_1 \cup A_2 \cup \ldots A_n = S$) then

$$\Pr(A_j|B) = \frac{\Pr(B|A_j)\Pr(A_j)}{\sum_{i=1}^{n} \Pr(B|A_i)\Pr(A_i)}. \qquad (7.3)$$

Conditional probabilities are useful to infer the probability of events that may not be directly measurable.

Example 7.1.4. A card is selected at random from a standard deck of cards. Let event A_1 represent the event of picking a diamond and let event B represent the event of picking a card with the number 7. Then the probability of the various events are $\Pr(A_1) = 1/4$ and $\Pr(B) = 1/13$. Further, $\Pr(A_1|B) = \frac{\Pr(A_1 \cap B)}{\Pr(B)} = \frac{1/52}{1/13} = 1/4$. Also, $\Pr(B|A_1) = \frac{\Pr(A_1 \cap B)}{\Pr(A_1)} = 1/13$.

Let events A_2, A_3 and A_4 represent the event of picking, respectively, a heart, spade and clubs. Clearly, events $A_i, i = 1, 2, 3$, and 4 are mutually exclusive and partition the sample space. Now, we evaluate $\Pr(A_1|B)$ using Bayes results (7.3) as

$$\Pr(A_1|B) = \frac{\Pr(B|A_1)\Pr(A_1)}{\sum_{i=1}^{4} \Pr(B|A_i)\Pr(A_i)} = \frac{(1/13)(1/4)}{4(1/13)(1/4)} = 1/4 \qquad (7.4)$$

which is the same value as calculated directly.

$\square$

Example 7.1.5. Consider the transmission of a equiprobable binary bit sequence over a binary symmetric channel (BSC) with crossover probability α, i.e., a bit gets flipped by the channel with probability α. For simplicity, we consider the transmission of a single bit and let event A_0 denote the event that a bit 0 was sent and event A_1 denote the event that a bit 1 was sent. Similarly, let B_0 and B_1

denote, respectively, the event that bit 0 and bit 1
conditional probability that a bit 0 was sent given th
calculated as

$$\Pr\left(A_0|B_0\right) = \frac{\Pr\left(B_0|A_0\right)\Pr\left(A_0\right)}{\Pr\left(B_0|A_0\right)\Pr\left(A_0\right)+\Pr\left(B_0|A_1\right)\Pr\left(A_0\right)}$$

Events A and B are independent events if

$$\Pr\left(A\cap B\right) = \Pr\left(A\right)\Pr\left(\right)$$

Equivalently, the events are independent if $\Pr\left(A|\right.$
$\Pr\left(B\right)$. Intuitively, if events A and B are indepen
nonoccurrence of event A does not provide any add
occurrence or nonoccurrence of event B.

Multiple events $E_1, E_2, \ldots E_n$ are jointly indep
collection of events, the probability of their intersect
individual probabilities. It should be noted that pa
does not imply joint independence as the following

Example 7.1.6. A fair coin is flipped $n-1$ times, w
$1, 2, \ldots, n-1$ represents the event of receiving a Hea
represent the event that there are even number of He
we can evaluate the probability of the various ev
$1, 2, \ldots, n$. It is also clear that $\Pr\left(E_i \cap E_j\right) = 1/$
the events are pairwise independent. It can also b
these events are independent for $k < n$. However,

independent, since $\Pr\left(E_1 \cap E_2 \cap \ldots \cap E_n\right) = (1/2)^n$

7.2   Random Variables

A random variable, $\mathcal{X}(\omega)$, is a mapping that assigns
in the set of outcomes of the random experiment
such that all outcomes that are mapped to the valu
probability 0. Further, for all values x, the set $\{\mathcal{X}$
Random variables are typically used to quantify and s
associated with a random experiment.

A complex random variable is defined as $\mathcal{Z} = \mathcal{X}$
valued random variables. For simplicity, most of the
focus on real valued random variables.

The cumulative distribution function (CDF) or probability distribution function, $F_{\mathcal{X}}$, of random variable $\mathcal{X}$ is defined as

$$F_{\mathcal{X}}(x) = \Pr(\mathcal{X} \leq x) \tag{7.7}$$

The following properties of the CDF immediately follow:

- The CDF is a number between 0 and 1, i.e., $0 \leq F_{\mathcal{X}}(x) \leq 1$.

- The CDF of a random variable evaluated at infinity and negative infinity equals, 1 and 0, respectively, i.e., $F_{\mathcal{X}}(\infty) = 1$ and $F_{\mathcal{X}}(-\infty) = 0$.

- The CDF $F_{\mathcal{X}}(x)$ is a nondecreasing function of x.

- The probability that the random variable takes values between x_1 and x_2 is given by the difference in the CDF at those values, i.e., $\Pr(x_1 < \mathcal{X} \leq x_2) = F_{\mathcal{X}}(x_2) - F_{\mathcal{X}}(x_1)$, if $x_1 < x_2$.

- The CDF is right continuous, i.e., $\lim_{\epsilon \to 0} F_{\mathcal{X}}(x + \epsilon) = F_{\mathcal{X}}(x)$, when $\epsilon > 0$.

A random variable is completely defined by its CDF in the sense that any property of the random variable can be calculated from the CDF. A random variable is typically categorized as being a discrete random variable, continuous random variable, or mixed random variable.

7.2.1 Discrete Random Variables

Random variable $\mathcal{X}$ is said to be a discrete random variable if the CDF is constant except at a countable set of points. For a discrete random variable, the probability mass function (PMF), $P_{\mathcal{X}}(x)$, is equal to the probability that random variable $\mathcal{X}$ takes on value x. Thus $P_{\mathcal{X}}(x) = F_{\mathcal{X}}(x) - F_{\mathcal{X}}(x^-)$. Clearly, since the PMF represents a probability value, $P_{\mathcal{X}}(x) \geq 0$ and $\sum_x P_{\mathcal{X}}(x) = 1$. Similar to the CDF, the PMF also completely determines the properties of a discrete random variable.

Example 7.2.1. Let random variable $\mathcal{X}$ be defined as the number of Heads that appear in 3 flips of a biased coin with probability of Head in each flip equal to 0.3. Figure 7.1 shows a plot of the PMF and corresponding CDF for random variable $\mathcal{X}$. □

Certain random variables appear in many different contexts and consequently they have been assigned special names. Moreover, their properties have been thoroughly studied and documented. We now highlight a few of the common discrete random variables, their distributions, and a typical scenario where they are applicable.

- A *Bernoulli* random variable takes values 0 and 1 with probabilities α and $1 - \alpha$, respectively. A Bernoulli random variable is commonly used to model scenarios in which there are only two possible outcomes such as in a coin toss or in a pass or fail testing.

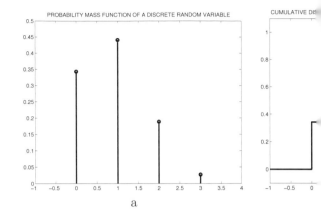

Figure 7.1: Illustration of the PMF and CDF of a si
in Example 7.2.1.

- A binomial random variable $\mathcal{X}$ takes values
  could represent the number of Heads in N i
  the probability of receiving a Head in each flip
  PMF of the binomial random variable is given

$$P_{\mathcal{X}}(x) = \binom{N}{x} p^x (1-p)^{(N-x}$$

Example 7.2.2. In error control coding, a ra
sists of transmitting n identical copies of each
mitted over a binary symmetric channel (BSC
In this case, random variable $\mathcal{X}$ that represen
received correctly has the binomial distributio

- A geometric random variable has a PMF of th

$$P_{\mathcal{X}}(x) = (1-p)^x p, x = 0$$

Example 7.2.3. Consider a packet commu
packet is retransmitted by the transmitter unti
is received. In this case, if the probability of
in each attempt equals p and each transmissi
the others, then random variable $\mathcal{X}$ which repr
until successful packet reception has a geomet
□

- A discrete uniform random variable $\mathcal{X}$ has PN

$$P_{\mathcal{X}}(x) = \frac{1}{b-a+1}, x = a,$$

where without loss of generality $b \geq a$.

- A Pascal random variable has PMF

$$P_{\mathcal{X}}(x) = \binom{N}{x}(1-p)^x p, x = L, L+1, L+2, \ldots \qquad (7.11)$$

Consider a sequence of independent Bernoulli trials in which the probability of success in each trial equals p. The experiment is repeated until exactly L successes. The random variable $\mathcal{X}$ that represents the number of trials has a Pascal distribution.

- A Poisson random variable has PMF of the form

$$P_{\mathcal{X}}(x) = e^{-a}\frac{a^x}{x!}, x = 0, 1, \ldots \qquad (7.12)$$

The Poisson random variable is obtained as the limit of the binomial random variable in the limit that $n \to \infty$ and $p \to 0$ but the product np is a constant. The Poisson random variable represents the number of occurrences of an event in a given time period. For instance, the number of radioactive particles emitted in a given period of time by a radioactive source is modeled as a Poisson random variable. Similarly, in queueing theory a common model for packet arrivals is a Poisson process, in which the number of packet arrivals per unit time is given by (7.12).

7.2.2 Continuous Random Variables

Random variable $\mathcal{X}$ is said to be a continuous random variable if the CDF of $\mathcal{X}$ is continuous. The probability density function (PDF) $f_{\mathcal{X}}(x)$ of random variable $\mathcal{X}$ is defined as

$$F_{\mathcal{X}}(x) = \int_{-\infty}^{x} f_{\mathcal{X}}(u)du \qquad (7.13)$$

Note that unlike the PMF, the PDF may take values greater than 1. The PDF is only proportional to the probability of an event. The interpretation of the PDF is that the probability of $\mathcal{X}$ taking values between x and $x + \delta_x$ approximately equals $f_{\mathcal{X}}(x)\delta_x$, for small positive values of δ_x. Similar to the CDF, the PDF is also a complete description of the random variable. The PDF of $\mathcal{X}$ satisfies the following properties:

- Since the CDF is a nondecreasing function, the PDF is non-negative, i.e., $f_{\mathcal{X}}(x) \geq 0$.

- The integral of the PDF over a certain interval represents the probability of the random variable taking values in that interval, i.e., $\int_a^b f_{\mathcal{X}}(x)dx = \Pr(a < \mathcal{X} \leq b)$.

- Extending the above property, the integral of
  equals 1, i.e., $\displaystyle\int_{-\infty}^{\infty} f_{\mathcal{X}}(x)dx = 1.$

Example 7.2.4. Suppose a random point on the n
the values 0 and 3. Let random variable $\mathcal{X}$ represen
Then the PDF of $\mathcal{X}$ is given by $f_{\mathcal{X}}(x) = \frac{1}{3}$, $0 < x <$
given by

$$F_{\mathcal{X}}(x) = \begin{cases} 0 & x \leq 0 \\ \frac{x}{3} & 0 < x < \\ 1 & x \geq 3 \end{cases}$$

A plot of this PDF and CDF are given in Figure 7.2

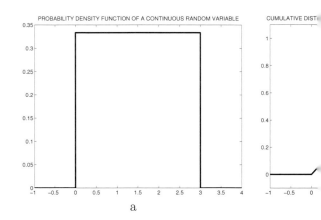

Figure 7.2: Illustration of the PMF and CDF of a si
in Example 7.2.4.

Similar to the discrete case, several commonly
variables have been studied including the following:

- The PDF of a uniform random variable is give

$$f_{\mathcal{X}}(x) = \frac{1}{b - a}, x \in ($$

- The PDF of a Gaussian random variable (also
  variable) is given by

$$f_{\mathcal{X}}(x) = \frac{1}{\sqrt{2\pi\sigma^2}}e^{-\frac{(x-}{2}}$$

The special case of a Gaussian random variable with 0 mean and unit variance is called a standard normal random variable. As will become clear in our study of the Central Limit Theorem, the distribution of any sum of independent random variables asymptotically approaches that of a Gaussian. Consequently, many noise and other realistic scenarios are modeled as Gaussian. The CDF of a Gaussian random variable is unfortunately not known in closed form. However, the CDF of a standard normal random variable has been computed numerically for various values and provided in the form of tables in several books. This CDF, denoted by Φ is defined as

$$\Phi(x) = \int_{-\infty}^{x} \frac{1}{\sqrt{2\pi}} e^{-\frac{u^2}{2}} du \tag{7.17}$$

The CDF of any other Gaussian random variable $\mathcal{X}$ with mean $\mu_{\mathcal{X}}$ and variance $\sigma_{\mathcal{X}}^2$ can be evaluated using the CDF tables of a standard normal random variable as follows

$$\Pr\left(\mathcal{X} \leq x\right) = \int_{-\infty}^{x} \frac{1}{\sqrt{2\pi\sigma_{\mathcal{X}}^2}} e^{-\frac{(u-\mu_{\mathcal{X}})^2}{2\sigma_{\mathcal{X}}^2}} du = \Phi\left(\frac{x - \mu_{\mathcal{X}}}{\sigma_{\mathcal{X}}}\right). \tag{7.18}$$

Note that several authors use other variants of Φ to numerically calculate the CDF and tail probabilities of a Gaussian random variable. For instance, the error function $erf(x)$ is defined as

$$erf(x) = \frac{2}{\sqrt{\pi}} \int_{0}^{x} e^{-t^2} dt \tag{7.19}$$

This error function $erf(x)$ and the Φ function can be expressed in terms of each other as $\Phi(x) = 0.5 + erf(x/\sqrt{2})$ and $erf(x) = \Phi(x\sqrt{2}) - 0.5$ for positive values of x.

- An exponential random variable has a PDF given by

$$f_{\mathcal{X}}(x) = ae^{-ax}, x > 0 \tag{7.20}$$

The exponential distribution is frequently used to model the interarrival time between packets in the queueing theory (see Chapter 17). The exponential distribution has the special memoryless property as demonstrated by the following example.

Example 7.2.5. Let the lifetime of a fluorescent bulb be modeled as an exponential random variable $\mathcal{X}$ with a mean of 10 years. Then the probability that the lifetime $\mathcal{X}$ exceed 15 years is given by

$$\Pr\left(\mathcal{X} > 15\right) = \int_{x=15}^{\infty} 1/10 e^{-x/10} dx = e^{-15/10} \tag{7.21}$$

Now suppose the bulb has already been workir
conditional probability that the lifetime $\mathcal{X}$ exe

$$\Pr\left(\mathcal{X} > 15/\mathcal{X} > 6\right) = \frac{e^{-15/}}{e^{-6/}}$$

which is the same as the probability that the lif
The exponential random variable is the only co
has this memoryless property.

7.3   Joint Random Variables

Recall that a random variable is a mapping from th
iment to real numbers. Clearly, for a given set of
could be numerous mappings representing different
sets of real numbers. To understand the relationship
variables, it is not sufficient to study their propertie
joint study of these random variables is required.

The joint CDF $F_{\mathcal{X},\mathcal{Y}}(x, y)$ of two random variab

$$F_{\mathcal{X},\mathcal{Y}}(x, y) = \Pr\left(\mathcal{X} \leq x, \mathcal{Y}\right.$$

Similar to the case of a single random variable, the j
the properties of the random variables. From the joi
RVs $\mathcal{X}$ and $\mathcal{Y}$ can be obtained as $F_{\mathcal{X}}(x) = F_{\mathcal{X},\mathcal{Y}}(x,$
The joint CDF satisfies the following properties:

- $0 \leq F_{\mathcal{X},\mathcal{Y}}(x, y) \leq 1$.

- $F_{\mathcal{X},\mathcal{Y}}(-\infty, -\infty) = 0$ and $F_{\mathcal{X},\mathcal{Y}}(\infty, \infty) = 1$.

- $\Pr\left(a < \mathcal{X} \leq b, c < \mathcal{Y} \leq d\right) = F_{\mathcal{X},\mathcal{Y}}(a, c) + F_{\mathcal{X},\mathcal{Y}}$

- $F_{\mathcal{X},\mathcal{Y}}(x, y) = \lim_{\epsilon \to 0, \epsilon > 0} F_{\mathcal{X},\mathcal{Y}}(x+\epsilon, y)$ and $F_{\mathcal{X},}$
  ϵ).

The joint PDF of RVs $\mathcal{X}$ and $\mathcal{Y}$ is given by any

$$F_{\mathcal{X},\mathcal{Y}}(x, y) = \int_{-\infty}^{x} \int_{-\infty}^{y} f_{\mathcal{X},\mathcal{Y}}(u$$

Example 7.3.1. Consider random variables $\mathcal{X}$ and

$$f_{\mathcal{X},\mathcal{Y}}(x, y) = \begin{cases} a(x + y + xy) & 0 \leq x \\ 0 & \text{else} \end{cases}$$

The value of constant $a = 1/4$ can be computed using the property that the integral of the PDF over the entire interval equals 1. In this case, the CDF can be computed as

$$F_{\mathcal{X},\mathcal{Y}}(x,y) = \begin{cases} 1 & x > 1, y > 2 \\ a(x^2 y/2 + y^2 x/2 + x^2 y^2/4) & 0 \le x \le 1, 0 \le y \le 2 \\ 0 & \text{else} \end{cases} \tag{7.26}$$

The probability of various events can be computed either from the PDF or from the CDF. For example, let event $A = \{0 \le \mathcal{X} \le 1/2, 1 \le \mathcal{Y} \le 2\}$. The probability of A can be calculated using the PDF as

$$\Pr(A) = \int_0^{1/2} \int_1^2 a(x + y + xy) \, dy \, dx \tag{7.27}$$

$$= a \int_0^{1/2} (x + 3/2 + 3/2x) \, dx \tag{7.28}$$

$$= 17/64 \tag{7.29}$$

The same probability can also be calculated using the joint CDF as

$$\Pr(A) = F_{\mathcal{X},\mathcal{Y}}(1/2,2) + F_{\mathcal{X},\mathcal{Y}}(0,1) - F_{\mathcal{X},\mathcal{Y}}(0,2) - F_{\mathcal{X},\mathcal{Y}}(1/2,1) \tag{7.30}$$

$$= 3/8 + 0 - 0 - 7/64 \tag{7.31}$$

$$= 17/64 \tag{7.32}$$

The marginal PDFs of $\mathcal{X}$ and $\mathcal{Y}$ can now be computed as

$$f_{\mathcal{X}}(x) = \int_{y=0}^2 a(x + y + xy) \, dy = a(4x + 2), \tag{7.33}$$

and

$$f_{\mathcal{Y}}(y) = \int_{x=0}^1 a(x + y + xy) \, dx = a(3y + 1)/2. \tag{7.34}$$

□

The conditional PDF $f_{\mathcal{X}|\mathcal{Y}}(x|y)$ is defined as

$$f_{\mathcal{X}|\mathcal{Y}}(x|y) = \frac{f_{\mathcal{X},\mathcal{Y}}(x,y)}{f_{\mathcal{Y}}(y)} \tag{7.35}$$

when $f_{\mathcal{Y}}(y) > 0$. For instance, in Exercise 7.3.1, the conditional PDF $f_{\mathcal{X}|\mathcal{Y}}(x|y) = \frac{2(x+y+xy)}{3y+1}$, $0 < x < 1$ and conditional PDF $f_{\mathcal{Y}|\mathcal{X}}(y|x) = \frac{(x+y+xy)}{4x+2}$, $0 < y < 2$.

Continuous random variables $\mathcal{X}$ and $\mathcal{Y}$ are said to be independent if and only if

$$f_{\mathcal{X},\mathcal{Y}}(x,y) = f_{\mathcal{X}}(x) f_{\mathcal{Y}}(y), \quad \forall x, y. \tag{7.36}$$

Example 7.3.2. Let the joint PDF of random
by $f_{\mathcal{X},\mathcal{Y}}(x,y) = xy$ for $0 \leq x \leq 1, 0 \leq y \leq 2$. The
PDFs are given by $f_{\mathcal{X}}(x) = \int_{y=0}^{2} xy\,dy = 2x, 0 \leq x$
$y/2, 0 \leq y \leq 2$. Clearly, $f_{\mathcal{X},\mathcal{Y}}(x,y) = f_{\mathcal{X}}(x)f_{\mathcal{Y}}(y$
variables $\mathcal{X}$ and $\mathcal{Y}$ are independent.

Similarly, discrete random variables $\mathcal{X}$ and $\mathcal{Y}$ are
only if

$$P_{\mathcal{X},\mathcal{Y}}(x,y) = P_{\mathcal{X}}(x)P_{\mathcal{Y}}(y),$$

It is important to note that for independence of ran
be satisfied for all values of the random variables $\mathcal{X}$

Example 7.3.3. Let the joint PMF of random var

$$P_{\mathcal{X},\mathcal{Y}}(x,y) = \begin{cases} 1/6 & \mathcal{X} = 1, \\ 1/4 & \mathcal{X} = 1, \\ 1/12 & \mathcal{X} = 1, \\ 1/6 & \mathcal{X} = 2, \\ 1/12 & \mathcal{X} = 1, \\ 1/4 & \mathcal{X} = 1, \end{cases}$$

The marginal PMFs of $\mathcal{X}$ and $\mathcal{Y}$ can be computed to
form densities over their respective alphabets. In thi
that the events $\mathcal{X} = 1$ and $\mathcal{Y} = 1$ are independent
and $\mathcal{Y} = 2$ are not independent. Thus, the random
independent.

Example 7.3.4. Consider a network in which pack
node to the destination node using a routing protoc
bility α of packet loss at each node due to buffer ove
order to increase the overall chances of success, the
copies of each packet over different mutually exclus
have a_1, a_2 and a_3 hops between the source and des
that the probability of success in each hop is indep
this case the overall probability that at least one c

correctly at the destination node can be calculated a

7.3.1   Expected Values, Characteristic

As noted before, the PDF, CDF and PMF are all co
dom variable and can be used to evaluate any pro

However, for many complex scenarios, computing the exact distribution can be challenging. In contrast, there are several statistical values that are computationally simple, but provide only partial information about the random variable. In this section, we highlight some of the frequently utilized statistical measures.

The expected value, $E\{\mathcal{X}\}$, of random variable $\mathcal{X}$ is defined as

$$E\{\mathcal{X}\} = \mu_{\mathcal{X}} = \int x f_{\mathcal{X}}(x)dx \qquad (7.39)$$

In general the expected value of any function $g(\mathcal{X})$ of a random variable $\mathcal{X}$ is given by

$$E\{g(\mathcal{X})\} = \int g(x) f_{\mathcal{X}}(x)dx \qquad (7.40)$$

The term, $E\{\mathcal{X}^k\}$ is known as the kth moment of $\mathcal{X}$. The variance $\sigma_{\mathcal{X}}^2$ of $\mathcal{X}$ is related to the second moment and is given by

$$\sigma_{\mathcal{X}}^2 = E\{(\mathcal{X} - \mu_{\mathcal{X}})^2\} = E\{\mathcal{X}^2\} - \mu_{\mathcal{X}}^2 \qquad (7.41)$$

As another variation, the kth central moment of random variable is defined as $E\{(\mathcal{X} - \mu_{\mathcal{X}})^k\}$.

The covariance between random variables $\mathcal{X}$ and $\mathcal{Y}$ is defined as

$$Cov(\mathcal{X}, \mathcal{Y}) = E\{(\mathcal{X} - \mu_{\mathcal{X}})(\mathcal{Y} - \mu_{\mathcal{Y}})\} \qquad (7.42)$$

The correlation coefficient $\rho_{\mathcal{X},\mathcal{Y}}$ is defined as

$$\rho_{\mathcal{X},\mathcal{Y}} = \frac{E\{(\mathcal{X} - \mu_{\mathcal{X}})(\mathcal{Y} - \mu_{\mathcal{Y}})\}}{\sigma_{\mathcal{X}}\sigma_{\mathcal{Y}}} \qquad (7.43)$$

Example 7.3.5. Jointly Gaussian Vector The joint PDF of the Gaussian vector $[\mathcal{X}_1, \mathcal{X}_2, \ldots, \mathcal{X}_N]$ is given by

$$f_{\mathcal{X}_1, \mathcal{X}_2, \ldots, \mathcal{X}_n}(x_1, x_2, \ldots x_n) = \frac{1}{(2\pi)^{N/2} |R_{\mathcal{X}}|^{1/2}} e^{-\frac{1}{2}(\mathbf{x} - \mu_{\mathbf{x}})^T R_{\mathcal{X}}^{-1}(\mathbf{x} - \mu_{\mathbf{x}})} \qquad (7.44)$$

where $\mathbf{x} = [x_1, x_2, \ldots x_n]^T$ and $\mu_{\mathbf{x}} = [\mu_{\mathcal{X}_1}\ \mu_{\mathcal{X}_1}\ \ldots \mu_{\mathcal{X}_N}]^T$ is the mean of the different random variables and $R_{\mathcal{X}}$ is the covariance matrix with i^{th} row and j^{th} column element given by $Cov(\mathcal{X}_i, \mathcal{X}_j)$. Gaussian random vectors are frequently used in several signal processing applications. For instance, when estimating a vector parameter in the presence of additive noise. The reasons for the popularity of these Gaussian vector models are: i) by central limit theorem, the noise density is well approximated as a Gaussian, ii) several closed form analytical results can be derived using the Gaussian model, and iii) the results derived using a Gaussian approximation serves as a bound for the true performance.

The marginal density of a jointly Gaussian vector are a Gaussian random variable. However, marginal densities being Gaussian does not necessarily imply that the joint density is also Gaussian.

$\square$

The random variables $\mathcal{X}$ and $\mathcal{Y}$ are said to be ι
and $\mathcal{Y}$ are independent, then

$$E\left\{(\mathcal{X} - \mu_\mathcal{X})(\mathcal{Y} - \mu_\mathcal{Y})\right\} = E\left\{\mathcal{X} - \mu_\mathcal{X}\right\}$$

which implies that the random variables are also un
lated random variables are not always independent
ing example.

Example 7.3.6. Let $\mathcal{X}_1$ be uniformly distributed
$\mathcal{X}_2 = cos(\mathcal{X}_1)$ and $\mathcal{X}_3 = sin(\mathcal{X}_1)$. Then it is cle
$E\left\{\mathcal{X}_2\mathcal{X}_3\right\} = 0$. Consequently, $\rho_{\mathcal{X}_2,\mathcal{X}_3} = 0$. However,
dependent random variables since $\mathcal{X}_2^2 + \mathcal{X}_3^2 = 1$ and
of $\mathcal{X}_3$ is known except for its sign.

In the special case that $\mathcal{X}_2$ and $\mathcal{X}_3$ are jointly (
correlated they are also independent. This result
crosscorrelation values of 0 in the autocorrelation
joint PDF of $\mathcal{X}_2$ and $\mathcal{X}_3$. Matrix R becomes a diag
the joint PDF then simply becomes the product of
The characteristic function $\phi_\mathcal{X}(\omega)$ of $\mathcal{X}$ is define

$$\phi_\mathcal{X}(\omega) = E\left\{e^{j\omega\mathcal{X}}\right\} = \int f_\mathcal{X}(x$$

The characteristic function and the PDF form a un
istic function also completely defines the random va
The characteristic function can be used to easily
random variable. Using the Taylors series expansio
characteristic function as

$$\begin{aligned}
\phi_\mathcal{X}(\omega) &= E\left\{e^{j\omega\mathcal{X}}\right\} \\
&= E\left\{1 + j\omega x + \frac{(j\omega x)^2}{2!}\right. \\
&= 1 + j\omega E\left\{\mathcal{X}\right\} + \frac{(j\omega)^2}{2!}
\end{aligned}$$

Now to compute the kth moment $E\left\{\mathcal{X}^k\right\}$, we can di
respect to ω and then evaluate the result at $\omega = 0$.

Example 7.3.7. Let $\mathcal{X}$ be an exponential random v
characteristic function of this random variable $\mathcal{X}$ is

$$\begin{aligned}
\phi_\mathcal{X}(\omega) = E\left\{e^{j\omega x}\right\} &= \int f_\mathcal{X}(x \\
&= \int_0^\infty \lambda e \\
&= \frac{\lambda}{\lambda - j\omega}
\end{aligned}$$

The mean of $\mathcal{X}$ can be calculated as

$$\mu_{\mathcal{X}} = \frac{1}{j}\frac{d}{d\omega}\phi_{\mathcal{X}}(\omega)|_{\omega=0} = \frac{1}{j}\frac{\lambda j}{(\lambda - j\omega)^2}|_{\omega=0} = \frac{1}{\lambda} \tag{7.51}$$

The second order moment can be evaluated as

$$E\left\{\mathcal{X}^2\right\} = \frac{1}{j^2}\frac{d^2}{d\omega^2}\phi_{\mathcal{X}}(\omega)|_{\omega=0} = \frac{1}{j^2}\frac{-2\lambda}{(\lambda - j\omega)^3}|_{\omega=0} = \frac{2}{\lambda^2} \tag{7.52}$$

Consequently, the variance can be calculated as

$$\sigma_{\mathcal{X}}^2 = E\left\{\mathcal{X}^2\right\} - \mu_{\mathcal{X}}^2 = \frac{1}{\lambda^2} \tag{7.53}$$

$\square$

The second characteristic function $\Psi_{\mathcal{X}}(\omega)$ is defined as the natural logarithm of the function $\phi_{\mathcal{X}}(\omega)$. The cumulants λ_n are

$$\lambda_n = \frac{d\Psi(s)}{ds^n}|s = 0 \tag{7.54}$$

The various cumulants are related to the moments as follows:

$$\lambda_1 = E\left\{\mathcal{X}\right\} = \mu_{\mathcal{X}} \tag{7.55}$$
$$\lambda_2 = E\left\{(\mathcal{X} - E\left\{\mathcal{X}\right\})^2\right\} = \sigma_{\mathcal{X}}^2 \tag{7.56}$$
$$\lambda_3 = E\left\{(\mathcal{X} - E\left\{\mathcal{X}\right\})^3\right\} \tag{7.57}$$
$$\lambda_4 = E\left\{\mathcal{X}^4\right\} - 4E\left\{\mathcal{X}^3\right\}E\left\{\mathcal{X}\right\}$$
$$-3(E\left\{\mathcal{X}^2\right\})^2 + 12E\left\{\mathcal{X}^2\right\}(E\left\{\mathcal{X}\right\})^2 - 6(E\left\{\mathcal{X}\right\})^4 \tag{7.58}$$

The cumulants of order higher than 3 are not the same as the central moment. Of special interest is the fourth-order cumulant, which is also referred to as kurtosis. The kurtosis is typically used as a measure of the deviation from Gaussianity of a random variable. The kurtosis of a Gaussian random variable equals 0. Further, for a distribution with a heavy tail and a peak at zero, the kurtosis is positive and for distribution with a fast decaying tail the kurtosis is negative.

Example 7.3.8. Consider the uniform random variable $\mathcal{X}$ with support over $(0,1)$. The first four moments of $\mathcal{X}$ can be calculated as

$$E\left\{\mathcal{X}\right\} = 0, \ E\left\{\mathcal{X}^2\right\} = 1/12, \ E\left\{\mathcal{X}^3\right\} = 0, \ \text{and} \ E\left\{\mathcal{X}^4\right\} = 1/80$$

The kurtosis of $\mathcal{X}$ can be now calculated using (7.58) as $\lambda_4 = -2/15$. $\square$

7.3.2　Inequalities

For many complex systems, analyzing the exact pe
and even analytically intractable. In many such si
to bound the system performance. Such bounding
in communications to derive closed form approxim
various systems. The closed form approximations
that changes in a certain parameter has on perfor
also used in learning theory to study the problem of
given training data. A detailed discussion on using i
theory is given in Chapter 9. In this section, we
frequently used inequalities.

For a non-negative random variable $\mathcal{X}$ with me
inequality provides a bound on the tail probability

$$\Pr\left(\mathcal{X} \geq a\right) \leq \frac{E\left\{\mathcal{X}\right\}}{a}.$$

Clearly, this inequality only provides meaningful res
in that case, the bound is quite weak since it only
to evaluate a bound on the tail probability. A tight
ity is given by the Chebychev inequality which use
variance $\sigma_{\mathcal{X}}^2$ as

$$\Pr\left(\left|\mathcal{X} - \mu_{\mathcal{X}}\right| \geq a\right) \leq \frac{\sigma_{\cdot}^{\cdot}}{a^{\cdot}}$$

The Chebychev inequality is obtained by applying t
$\mu_{\mathcal{X}})^2$.

Jensen's inequality states that for a convex funct
the function of a random variable is no smaller than
value of that random variable, i.e.,

$$g(E\left\{\mathcal{X}\right\}) \leq E\left\{g(\mathcal{X})\right\}$$

The Union bound simply states that the probability
is lesser than the sum of the probabilities of the ind

$$\Pr\left(\cup_i A_i\right) \leq \sum_i \Pr\left(A_i\right.$$

The Schwarz inequality is given by

$$\left|E\left\{\mathcal{X}\mathcal{Y}\right\}\right| \leq \sqrt{E\left\{\mathcal{X}^2\right\} E\left\{\right.}$$

The Chernoff bound is given by

$$
\begin{aligned}
\Pr\left(\mathcal{X} > a\right) &= \Pr\left(e^{t\mathcal{X}} > e^{ta}\right) \forall t > 0 \\
&\leq \frac{E\left\{e^{t\mathcal{X}}\right\}}{e^{ta}} \text{(using Markov inequality)} \\
\implies \Pr\left(\mathcal{X} > a\right) &\leq \min_{t>0} \frac{E\left\{e^{t\mathcal{X}}\right\}}{e^{ta}}
\end{aligned}
\tag{7.64}
$$

Recognize that the tail probability fall of exponentially with the Chernoff bound but only in polynomial terms with the Chebychev inequality. On the other hand, while the Chebychev inequality is general enough to apply to the sum of dependent random variables, the Chernoff bound is only valid for the sum of independent random variables. Hoeffdings inequality generalizes the Chernoff bound to the case of arbitrary bounded random variables [7].

7.3.3 Functions of Multiple Random Variables

Let random variable $\mathcal{Z} = g(\mathcal{X}_1, \mathcal{X}_2, \ldots, \mathcal{X}_n)$. Then, the distribution of $\mathcal{Z}$ can be computed from the joint distribution of the random variables $\mathcal{X}_i, i = 1, 2, \ldots n$. Depending on whether the random variables are continuous or discrete, several cases arise.

- Random variables $\mathcal{X}_1, \mathcal{X}_2, \ldots, \mathcal{X}_n$ and $\mathcal{Z}$ are continuous. In this case, the CDF of $\mathcal{Z}$ can be computed as

$$
F_{\mathcal{Z}}(z) = \int \cdots \int_{(x_1, \ldots x_n):g(x_1, \ldots, x_n) \leq z} f_{\mathcal{X}_1, \mathcal{X}_2, \ldots, \mathcal{X}_n}(x_1, x_2, \ldots, x_n) dx_1 \ dx_2 \ \ldots dx_n,
\tag{7.65}
$$

where the integral is over the $n-$tuples of $(\mathcal{X}_1, \mathcal{X}_2, \ldots, \mathcal{X}_n)$ that are mapped by the function g to a value lesser than z. The PDF of $\mathcal{Z}$ can be computed by differentiating the CDF obtained in (7.65). While this same principle can be used in case the random variables are not continuous, it is also possible to directly compute the PMF of random variable $\mathcal{Z}$ if it's discrete as discussed in the next 2 cases.

- Random variables $\mathcal{X}_1, \mathcal{X}_2, \ldots, \mathcal{X}_n$ and $\mathcal{Z}$ are discrete. In this case the PMF of $\mathcal{Z}$ can be computed as

$$
P_{\mathcal{Z}}(z) = \sum_{(x_1, \ldots x_n):g(x_1, \ldots, x_n) = z} P_{\mathcal{X}_1, \mathcal{X}_2, \ldots, \mathcal{X}_n},
\tag{7.66}
$$

where the summation is over the $n-$tuples of $(\mathcal{X}_1, \mathcal{X}_2, \ldots, \mathcal{X}_n)$ that are mapped by the function g to the value z.

- Random variables $X_1, X_2, \ldots, X_n$ are continu
  discrete. In this case the PMF of Z can be co

$$P_Z(z) = \int \cdots \int_{(x_1,\ldots x_n):g(x_1,\ldots,x_n)=z}$$

where the integral is over the $n-$tuples of $(X_1$
by the function g to the value z.

Example 7.3.9. Let the joint PDF of random vari

$$f_{X_1,X_2}(x_1, x_2) = 1, \quad 1 \le x_1 \le 2,$$

Now let $Z = X_1 X_2$. The CDF of Z can be calculate

- For the values $1 \le Z \le 2$

$$
\begin{aligned}
F_Z(z) &= \int_1^z \int_1^{z/x_1} 1 \\
&= \int_1^z \left(\frac{z}{x_1} - \right. \\
&= z\log(z) - z
\end{aligned}
$$

- For the values $2 < Z \le 4$

$$
\begin{aligned}
F_Z(z) &= \int_1^{z/2} \int_1^2 1 dx_2 \, dx_1 + \\
&= z/2 - 1 + \int_{z/2}^2 \left(\frac{z}{x_1} - \right. \\
&= z/2 - 1 + z(\log(2) - \text{lo} \\
&= z(1 + \log(2) - \log(z/2)
\end{aligned}
$$

The corresponding PDF of Z is given by

$$f_Z(z) = \begin{cases} \log(z) & 1 \\ 2\log(2) - \log(z) & 2 \end{cases}$$

Now suppose $Y = \lfloor 2X_1 + 2X_2 \rfloor$. Then the PMF of Y

$$P_Y(y) = \begin{cases} \int_1^{1.5} \int_1^{2.5-x_1} dx_2 dx_1 = 1/8 \\ \int_1^{1.5} \int_{2.5-x_1}^{3-x_1} dx_2 dx_1 + \int_{1.5}^2 \int_1^{3-x_1} dx_2 \\ \int_1^{1.5} \int_{3-x_1}^2 dx_2 dx_1 + \int_{1.5}^2 \int_{3-x_1}^{3.5-x_1} dx_2 \\ \int_{1.5}^2 \int_{3.5-x_1}^2 dx_2 dx_1 = 1/8 \end{cases}$$

In many applications it turns out that we are interested in computing the distribution of the sum of multiple independent random variables. In this case, the PDF of $\mathcal{Z}$ can be computed simply as the convolution of the marginal PDF's of random variables $\mathcal{X}_1, \mathcal{X}_2, \ldots, \mathcal{X}_n$. For simplicity, we prove this result for the case that $\mathcal{Z} = \mathcal{X}_1 + \mathcal{X}_2$. The CDF of $\mathcal{Z}$ can be calculated as,

$$F_{\mathcal{Z}}(z) \;=\; \int_{x_1+x_2\leq z} f_{\mathcal{X}_1,\mathcal{X}_2}(x_1,x_2)dx_1\,dx_2, \tag{7.78}$$

$$=\; \int_{-\infty}^{\infty}\int_{-\infty}^{z-x_2} f_{\mathcal{X}_1}(x_1)f_{\mathcal{X}_2}(x_2)dx_1\,dx_2 \tag{7.79}$$

(due to independence of $\mathcal{X}_1$ and $\mathcal{X}_2$)

The PDF of $\mathcal{Z}$ can now be computed by taking the derivative of the CDF as

$$f_{\mathcal{Z}}(z) = \int_{-\infty}^{\infty} f_{\mathcal{X}_1}(x_1)f_{\mathcal{X}_2}(z-x_1)dx_1, \tag{7.80}$$

which can be recognized as the convolution of the marginal PDFs of $\mathcal{X}_1$ and $\mathcal{X}_2$. In the case of discrete random variables, the convolution integral is given by

$$P_{\mathcal{Z}}(k) = \sum_k P_{\mathcal{X}_1}(k)f_{\mathcal{X}_2}(z-k). \tag{7.81}$$

The PDF of $\mathcal{Z}$ can also be computed using the properties of the characteristic functions as,

$$\phi_{\mathcal{Z}}(w) \;=\; E\left\{e^{jw(\mathcal{X}_1+\mathcal{X}_2)}\right\} \tag{7.82}$$

$$=\; E\left\{e^{jw\mathcal{X}_1}\right\}E\left\{e^{jw\mathcal{X}_2}\right\} \tag{7.83}$$

$$=\; \phi_{\mathcal{X}_1}(w)\phi_{\mathcal{X}_2}(w) \tag{7.84}$$

This product relationship is similar to the analysis of linear time invariant systems in which the output of the system equals the convolution of the input and impulse response. Or equivalently, the Fourier transform of the output equals the product of the Fourier transform of the input and the frequency response of the system.

Example 7.3.10. Let $\mathcal{X}_1 \approx$ Uniform (a,b) and $\mathcal{X}_2 \approx$ Uniform (c,d) be independent uniformly distributed continuous random variables. For simplicity, assume that $d-c = b-a$. Let $\mathcal{Y} = \mathcal{X}_1 + \mathcal{X}_2$. The characteristic function of $\mathcal{X}_1$ equals

$$\phi_{\mathcal{X}_1}(w) = \frac{e^{jwb} - e^{jwa}}{jw(b-a)} \tag{7.85}$$

Using (7.84), the characteristic function of $\mathcal{Y}$ can be evaluated as,

$$\phi_{\mathcal{Y}}(w) \;=\; \frac{e^{jwb} - e^{jwa}}{jw(b-a)}\,\frac{e^{jwd} - e^{jwc}}{jw(d-c)} \tag{7.86}$$

By taking the inverse Fourier transform of (7.86) it c
equals

$$
f_{\mathcal{Y}}(y) = \begin{cases} \frac{4(y-a-c)}{(b+d-a-c)(b+d-a-c)} = \frac{y-a-c}{(d-c)(b-a)} & (a+ \\ \frac{4(b+d-y)}{(b+d-a-c)(b+d-a-c)} = \frac{b+d-y}{(d-c)(b-a)} & (b+ \end{cases}
$$

The PDF of $\mathcal{Y}$ can also be derived by convolvir
shown below. For $(a+c) \le y < (a+b+c+d)/2$

$$
\begin{aligned}
f_{\mathcal{Y}}(y) &= \int_{a+c}^{y} \frac{1}{b-a} \frac{1}{d-} \\
&= \frac{y-a-c}{(b-a)(d-c)}
\end{aligned}
$$

For $(a+b+c+d)/2 < y \le (b+d)$, the PDF is give

$$
\begin{aligned}
f_{\mathcal{Y}}(y) &= \int_{y}^{b+d} \frac{1}{b-a} \frac{1}{d-} \\
&= \frac{b+d-y}{(b-a)(d-c)}
\end{aligned}
$$

It can be recognized that this PDF corresponds to a

Example 7.3.11. Let $\mathcal{X}_1$ and $\mathcal{X}_2$ be independent G
means $\mu_{\mathcal{X}_1}$ and $\mu_{\mathcal{X}_2}$, respectively. Let their variances
The distribution of $\mathcal{Y} = a\mathcal{X}_1 + b\mathcal{X}_2$ can be comp
functions as follows. The characteristic function of

$$
\phi_{\mathcal{X}_1}(\omega) = e^{j\mu_{\mathcal{X}}\omega - \sigma_{\mathcal{X}}^2 \omega^2/}
$$

Now, the mean and variance of $a\mathcal{X}_1$ are $a\mu_{\mathcal{X}_1}$ and $a^2\sigma$
the characteristic function of $\mathcal{Y}$ can be evaluated as

$$
\begin{aligned}
\phi_{\mathcal{Y}}(\omega) &= e^{j\omega a\mu_{\mathcal{X}_1} - a^2\sigma_{\mathcal{X}_1}^2 \omega^2/2} e^{j\omega b\mu} \\
&= e^{j\omega(a\mu_{\mathcal{X}_1} + b\mu_{\mathcal{X}_2}) - \omega^2(a^2\sigma_{\mathcal{X}}^2}
\end{aligned}
$$

From (7.94), it can be clearly seen that $\mathcal{Y}$ is also a C
mean $a\mu_{\mathcal{X}_1} + b\mu_{\mathcal{X}_2}$ and variance $a^2\sigma_{\mathcal{X}_1}^2 + b^2\sigma_{\mathcal{X}_2}^2$. Th
Gaussian variable results in another Gaussian varial

Example 7.3.12. Let $\mathcal{X}_i$ be a sequence of i.i.d.
with PDF $f_{\mathcal{X}}(x) = \alpha e^{-\alpha x}$. Let $\mathcal{Y}_n = \sum_{i=1}^{n} \mathcal{X}_n$ repr

variables $\mathcal{X}_i$. It can be shown using the principles of mathematical induction that the PDF and CDF of $\mathcal{Y}_n$ are given by

$$f_{\mathcal{Y}_n}(x) \quad = \quad \alpha e^{-\alpha x} \frac{(\alpha x)^{n-1}}{(n-1)!} \tag{7.95}$$

$$F_{\mathcal{Y}_n}(x) \quad = \quad 1 - e^{-\alpha x}\left(1 + \frac{\alpha x}{1!} + \ldots + \frac{(\alpha x)^{n-1}}{(n-1)!}\right) \tag{7.96}$$

Now, define a new family of discrete random variables $\mathcal{N}(t)$ as follows

$$\mathcal{N}(t) = \arg\max_k \mathcal{Y}_k \le t \tag{7.97}$$

In other words, $\mathcal{N}(t)$ takes on the value n iff $\mathcal{Y}_n \le t$ and $\mathcal{Y}_{n+1} > t$. The probability of this event can be calculated as

$$\Pr\left(\mathcal{N}(t) = n\right) = F_{\mathcal{Y}_n}(t) - F_{\mathcal{Y}_{n+1}}(t) = e^{-\alpha t}\frac{(\alpha t)^n}{n!} \tag{7.98}$$

This demonstrates the relationship between the discrete Poisson random variable and the continuous exponential random variables. □

7.3.4 Convergence of Random Variables

Before formally introducing the law of large numbers, in this section, we introduce the various types of convergences of random variables.

First, consider a sequence $a_1, a_2, \ldots a_n$ of real numbers. This sequence is said to converge to a real number a if for any given value of $\epsilon > 0$, there exists an integer N such that $|a_n - a| < \epsilon$, $\forall n \ge N$.

As noted before a random variable is a mapping from the set of outcomes to real numbers. Consider a sequence of random variables $\mathcal{X}_1, \mathcal{X}_2, \ldots$.

Convergence everywhere: If for every value of $w \in S$, the sequence $\mathcal{X}_1(w), \mathcal{X}_2(w), \ldots$ converges to a value that could depend on w then the sequence is said to converge everywhere.

Convergence almost everywhere: If the set of outcomes w for which the sequence $\mathcal{X}_1(w), \mathcal{X}_2(w), \ldots$ converges has a probability equal to 1, then the sequence is said to converge almost everywhere. In other words,

$$\Pr\left(\mathcal{X}_n \to \mathcal{X}\right) = 1, \text{ as } n \to \infty. \tag{7.99}$$

Convergence in probability or stochastic convergence: For a given value of $\epsilon > 0$, consider the sequence of real numbers given by $\Pr\left(|\mathcal{X}_n - \mathcal{X}| > \epsilon\right)$. If this sequence of real numbers converges to 0 for all values of $\epsilon > 0$, then the sequence $\mathcal{X}_1, \mathcal{X}_2, \ldots$ is said to converge in probability to the random variable $\mathcal{X}$.

It should be noted that the expression "converges almost everywhere" also implies convergence in probability while the reverse is not always true. Further, several other forms of convergences are defined in the literature [8].

7.3.5 Law of Large Numbers (LLN) a
orem (CLT)

The LLN allows the approximation of the average of
dom variables by a single number. The LLN essenti
of the behavior of populations becomes more predict
increases. In its simple form, the weak law of large
follows:

Theorem 7.3.1 (Weak Law of Large Number
$\ldots + \mathcal{X}_n$), where the random variables $\mathcal{X}_i$ are all in
means $\mu_{\mathcal{X}}$ and a finite variance no greater than $\sigma_{\mathcal{X}}^2$.

$$\Pr\left(|\mathcal{M}_n - \mu_{\mathcal{X}}| \geq \delta\right) \leq \frac{c}{n}$$

In the limit of large n, the RHS of (7.100) equals 0.
to $\mu_{\mathcal{X}}$ in probability.

In simple terms, the WLLN states that $\mathcal{M}_n$ ca
should be noted that the finite variance requirement
needed; it only makes the proof easier.

Theorem 7.3.2 (Strong Law of Large Numbe
$\ldots + \mathcal{X}_n$), where the random variables $\mathcal{X}_i$ are all in
means $\mu_{\mathcal{X}}$. Then, $\mathcal{M}_n$ converges to $\mu_{\mathcal{X}}$ almost ever

$$\Pr\left(\mathcal{M}_n \to \mu_{\mathcal{X}}\right) = 1 \quad as \; n \text{ -}$$

Theorem 7.3.3 (Central Limit Theorem). Let
where the random variables $\mathcal{X}_i$ are all independent
and a finite variance $\sigma_{\mathcal{X}}^2$. Then,

$$\lim_{n\to\infty} \Pr\left(\frac{\mathcal{M}_n - \mu_{\mathcal{X}}}{\sqrt{n^{-1}\sigma_{\mathcal{X}}^2}} \leq c\right) =$$

The CLT allows the sum or mean of a finite set of
to be approximated by a Gaussian distribution and
better for large population sizes.

The LLN can be used to determine the probabil
able and has independent probabilities in each tri
variable $\mathcal{X}_i = 1$ if event A occurs in the ith trial
$\Pr(A) = E\{\mathcal{X}_i\}$ The sample mean $\hat{\mu}_{\mathcal{X}} = \frac{1}{n}\sum_{k=1}^{n}\mathcal{X}_k$ is
ity $\Pr(A)$. By the LLN, as the sample size gets large
the true probability. Other applications of the laws c
theory are given in Chapter 8.

The Monte Carlo simulation method is based on this principle and is commonly used to evaluate the performance of a variety of systems. For instance, to detect the probability of bit error of a communication system, a large number of bits are transmitted and the average number of errors are calculated. In information theory, the asymptotic equipartition property (AEP) [9] is a consequence of the LLN.

7.4 Random Processes

A random process $\mathcal{X}(t, \omega)$ is a mapping from each outcome ω to functions of time. A random process can also be considered as an indexed collection of random variables. The index set can be continuous or discrete. The values taken by the process can also belong to a discrete set or continuous set. Thus, there are 4 possible types of random processes:

- Continuous time, continuous valued processes such as seismic measurements and Brownian motion.

- Continuous time, discrete valued processes such as population size and stock prices.

- Discrete time, continuous valued processes such as sampled speech and temperature signals.

- Discrete time, discrete valued processes such as digitized speech and video signals.

The collection of temporal functions over all values of ω is sometimes referred to as the ensemble of function. For simplicity of notation, we do not explicitly show the dependence of the random process $\mathcal{X}(t)$ on the outcome ω. To completely define the random process, the joint distribution of the set of random variables $\mathcal{X}(t_1), \mathcal{X}(t_2), \ldots, \mathcal{X}(t_K)$ needs to be specified for all values of K and $t_1, t_2, \ldots, t_K$. Clearly, this is a daunting task and practically impossible except in some very narrow and specialized cases such as a process defined by an i.i.d. sequence of random variables. Consequently, most analysis of random processes is restricted to understanding some statistical properties.

The mean function $\mu_{\mathcal{X}(t)}$ of a random process is defined as

$$\mu_{\mathcal{X}(t)} = \int x f_{\mathcal{X}(t)}(x) dx \qquad (7.103)$$

The autocorrelation $R_{\mathcal{X}}(t, \tau)$ of the random process is defined as

$$R_{\mathcal{X}}(t, \tau) = E\left\{ \mathcal{X}(t)\mathcal{X}^*(t + \tau) \right\} \qquad (7.104)$$

The autocovariance of the process is defined as

$$C_{\mathcal{X}}(t,\tau) = E\left\{(\mathcal{X}(t) - \mu_{\mathcal{X}(t)})(\mathcal{X}(t+\tau) - \mu_{\mathcal{X}(t+\tau)})^*\right.$$

The crosscorrelation $R_{\mathcal{X},\mathcal{Y}}(t,\tau)$ of two random p₁

as

$$R_{\mathcal{X},\mathcal{Y}}(t,\tau) = E\left\{\mathcal{X}(t)\mathcal{Y}^*(t\right.$$

Properties of autocorrelation: The autocorrelatio
function. Further, it can be shown that given any p₀
we can compute a random process $\mathcal{X}(t)$ for which t₁
tion. It can also be easily shown that $R_{\mathcal{X}}(t+\tau, -\tau)$

7.4.1   Stationary Process

Random process $\mathcal{X}(t)$ is said to be strictly s₁
of $(\mathcal{X}(t_1), \mathcal{X}(t_2), \ldots, \mathcal{X}(t_k))$ is the same as t₁
$T), \mathcal{X}(t_2 + T), \ldots, \mathcal{X}(t_k + T))$ for all values of
random process $\mathcal{X}_t$ and $\mathcal{Y}_t$ are jointly statio₁
$(\mathcal{X}(t_1), \mathcal{X}(t_2), \ldots, \mathcal{X}(t_k), \mathcal{Y}(t_{k+1}), \mathcal{Y}(t_{k+2}), \ldots, \mathcal{Y}(t_K$
of $(\mathcal{X}(t_1 + T), \mathcal{X}(t_2 + T), \ldots, \mathcal{X}(t_k + T), \mathcal{Y}(t_{k+1} + T$
for all values of $k, K, t_1, t_2, \ldots, t_K$ and T.

Random process $\mathcal{X}(t)$ is said to be wide-sense st₁
its mean is independent of time $(\mu_{\mathcal{X}(t)} = \mu_{\mathcal{X}})$ and its
on time difference, i.e., $R_{\mathcal{X}}(t,\tau) = R_{\mathcal{X}}(\tau)$. Two ra₁
jointly WSS if both $\mathcal{X}(t)$ and $\mathcal{Y}(t)$ are WSS and th
depends only on time difference.

Random process $\mathcal{X}(t)$ is said to be cyclosta₁
$(\mathcal{X}(t_1), \mathcal{X}(t_2), \ldots, \mathcal{X}(t_k))$ is the same as the joint
$T), \ldots, \mathcal{X}(t_k + T))$ for all values of $k, t_1, t_2, \ldots, t_k$ an₁
the time period of the random process.

The change in the process $\mathcal{X}(t_2) - \mathcal{X}_{t_1}, \mathcal{X}(t_3) - \mathcal{X}_0$
successive sampling instants, is called the increment
which the increments are all independent is said to l
process. Similarly, a process with stationary increm
increments process. The stationary properties of di₁
are similarly defined.

A random process is said to be ergodicity if the
temporal sequence of the process are the same as the
functions.

Random Walk: The random walk process $\mathcal{D}_n$ is de
of independent random variables $\mathcal{L}_i$ that takes value
and $1 - \alpha$, respectively. Thus, $\mathcal{D}_n = \sum_{i=1}^{n} L_i$. In other v

process either increases by 1 or decreases by 1. The random walk process is clearly memoryless in the sense that given the current state of the process, the future state is independent of the past. It can also be easily seen that the increments of the process are independent and stationary.

ARMA Process: Consider a random process $\mathcal{Y}_n$ which is obtained by passing the White noise process $\mathcal{X}_n$ through a linear filter $H(f)$. If the transfer function of the filter is of the form

$$H(f) = \frac{\sum_{k=0}^{B} b_k e^{-j2\pi fk}}{1 - \sum_{k=0}^{A} a_k e^{-j2\pi fk}} \tag{7.107}$$

then the evolution of the process can be represented as

$$\mathcal{Y}_n = \sum_{k=0}^{A} a_k \mathcal{Y}_{n-k} + \sum_{k=0}^{B} b_k \mathcal{X}_{n-k} \tag{7.108}$$

Such a process is referred to as an ARMA process and is frequently used in several signal processing applications and for time-series analysis.

Wiener random process and Brownian motion: A Wiener random process is constructed as the limit of a random walk process. Specifically, consider the symmetric random walk process which has equal probability to increase or decrease by 1. Now let the time increments be denoted by δ and the increase in process at each time instant be set to $\sqrt{\alpha\delta}$. In the limit of $\delta \to 0$, the continuous time random process $\mathcal{X}(t)$ becomes a process with zero mean and a variance that increases linearly with time. Further, since each $\mathcal{X}(t)$ is now the sum of an infinite number of independent random variables, it has a Gaussian distribution. Such a process is referred to as the Wiener random process and is commonly used to model Brownian motion. Clearly, by construction, it can be seen that the Wiener process has independent and stationary increments. The autocorrelation of this process can be shown to equal $R_\mathcal{X}(t, \tau) = \alpha \min\{t, t + \tau\}$.

7.4.2 Random Process as the Input to a Linear System

When a random process $\mathcal{X}(t)$ is the input to a LTI system then the output $\mathcal{Y}(t)$ is also a random process. Further, if the input is WSS then the output is also WSS as will be clear from (7.111) and (7.116). The mean and autocorrelation of the output can be derived in terms of the mean and autocorrelation of the input to the system and the impulse response of the LTI system. Let $h(t)$ and $H(f)$ denote, respectively, the impulse response and frequency response of the LTI system.

Now, the mean of the output process $\mathcal{Y}(t)$ can be derived as

$$E\{\mathcal{Y}(t)\} = E\left\{\int h(t-u)\mathcal{X}(u)du\right\} \tag{7.109}$$

$$= \int h(t-u)E\{\mathcal{X}(u)\}\,du \tag{7.110}$$

$$= \mu_\mathcal{X} H(0) \tag{7.111}$$

Thus, the mean of the output process $\mathcal{Y}(t)$ is indepe
lation of the output is given by,

$$
\begin{aligned}
R_{\mathcal{Y}}(t,\tau) &= E\left\{\mathcal{Y}_t \mathcal{Y}^*_{t+\tau}\right\} \\
&= E\left\{\int_{-\infty}^{\infty} h(u)\mathcal{X}(t-u)du \int_{-\infty}^{\infty} \right. \\
&= \int_{-\infty}^{\infty}\int_{-\infty}^{\infty} h(u)h^*(v)E\left\{\mathcal{X}(t-\imath\right.
\end{aligned}
$$

Since, $\mathcal{X}(t)$ is WSS, (7.114) can be rewritten as

$$
\begin{aligned}
R_{\mathcal{Y}}(t,\tau) &= \int_{-\infty}^{\infty} h(u) \int_{-\infty}^{\infty} h^*(v) R_{\mathcal{X}} \\
&= R_{\mathcal{X}}(\tau) * h^*(\tau) * h(-\tau)
\end{aligned}
$$

Thus, the autocorrelation of the output process
difference τ.

The cross-correlation is given by

$$
\begin{aligned}
R_{\mathcal{X},\mathcal{Y}}(t,\tau) &= E\left\{\int_{-\infty}^{\infty} \mathcal{X}(t)h^*(u)\mathcal{X}^* \right. \\
&= \int_{-\infty}^{\infty} h^*(u)E\left\{\mathcal{X}(t)\mathcal{X}^* \right. \\
&= \int_{-\infty}^{\infty} h^*(u) R_{\mathcal{X}}(\tau-u)
\end{aligned}
$$

If random processes $\mathcal{X}(t)$ and $\mathcal{Y}(t)$ are jointly wide

$$
R_{\mathcal{X},\mathcal{Y}}(\tau) = R_{\mathcal{Y},\mathcal{X}}(-\tau).
$$

The power spectral density (PSD), denoted $W_{\mathcal{X}}$
defined as the Fourier transform of the autocorrelat

$$
W_{\mathcal{X}}(f) = \int_{-\infty}^{\infty} R_{\mathcal{X}}(t)e^{-j2\pi}
$$

Similarly, the cross-power spectral density, denoted
dom processes $\mathcal{X}(t)$ and $\mathcal{Y}(t)$ is defined as,

$$
W_{\mathcal{X},\mathcal{Y}}(f) = \int_{-\infty}^{\infty} R_{\mathcal{X},\mathcal{Y}}(t)e^{-j}
$$

For this linear time invariant system with input
can be computed using the frequency transfer functi
of the input as,

$$
W_{\mathcal{Y}}(f) = W_{\mathcal{X}}(f)\,|H(f)|
$$

The cross-power spectral density is related to the PSD of $\mathcal{X}(t)$ and $\mathcal{Y}(t)$ as

$$
\begin{aligned}
W_{\mathcal{X},\mathcal{Y}}(f) &= H(f)W_{\mathcal{X}}(f) & (7.125) \\
W_{\mathcal{Y}}(f) &= H^*(f)W_{\mathcal{X},\mathcal{Y}}(f) & (7.126)
\end{aligned}
$$

Example 7.4.1. Consider a WSS random process $\mathcal{X}(t) = A cos(2\pi f_0 t + \theta)$, where A is a constant and θ is uniformly distributed between $(0, 2\pi)$. It can be easily shown that the autocorrelation $R_{\mathcal{X}}(\tau) = \frac{A^2}{2} cos(2\pi f + 0\tau)$. Let this process be input to differentiator system with frequency response $H(f) = j2\pi f$. The output $\mathcal{Y}(t)$ has spectral density given by

$$
S_{\mathcal{Y}}(f) = 4\pi^2 f^2 S_{\mathcal{X}}(f) = A2\pi^2 f_0^2 \left(\delta(f - f_0) + \delta(f + f_0) \right). \tag{7.127}
$$

$\square$

Discrete Time Processes and Systems

For a discrete time random process, $R_{\mathcal{X}}(m, n) = E\{\mathcal{X}_n \mathcal{X}_{n+m}\}$. If the process is wide sense stationary, then $R_{\mathcal{X}}(m, n) = R_{\mathcal{X}}(n)$. Since $R_{\mathcal{X}}(n)$ is a discrete function, the PSD is a periodic function. Thus, it is sufficient to consider one period of the PSD. The relationship between the autocorrelation and PSD is now given by,

$$
W_{\mathcal{X}}(f) = \sum_k R_{\mathcal{X}}(k)e^{-je\pi kf} \tag{7.128}
$$

$$
R_{\mathcal{X}}(k) = \int_{-1/2}^{1/2} W_{\mathcal{X}}(f)e^{j2\pi fk}df \tag{7.129}
$$

Example 7.4.2. Let the input to a simple averaging filter $h(n)$ be a WSS process $\mathcal{X}(n)$ with autocorrelation

$$
R_{\mathcal{X}}(n) = \delta(n) \tag{7.130}
$$

Let $h(n) = 1/(2M + 1), n = -M, \ldots - 1, 0, 1, \ldots M$ and $h(n) = 0$ for all other values of n. The autocorrelation of the output process $\mathcal{Y}(n)$ can be calculated as

$$
\begin{aligned}
R_{\mathcal{Y}}(n) &= \sum_{i=-M}^{M} \sum_{j=-M}^{M} h_i h_j R_{\mathcal{X}}(n + i - j) & (7.131) \\
&= \begin{cases} \frac{1}{(2M+1)^2}(2M + 1 - |n|) & n = -(2M + 1), \ldots, -1, 0, 1, \ldots (2M + 1) \\ 0 & \text{otherwise} \end{cases}
\end{aligned}
$$

$$(7.132)$$

$\square$

7.5   Markov Process

A random process with discrete states is said to
memoryless, i.e., it satisfies the following property

$$f_{\mathcal{X}_t | \mathcal{X}_{\tau < 0}} = f_{\mathcal{X}_t | \mathcal{X}_\tau}$$

Without loss of generality, the states of the r
by $\{0, 1, \ldots\}$.

7.5.1   Markov Chains

A discrete time Markov process is also referred to as

$$\Pr\left(\mathcal{X}_{n+1} | \mathcal{X}_n, \mathcal{X}_{n-1}, \ldots\right) = \Pr\left(\mathcal{A}\right.$$

A Markov chain is completely defined by its one step
denoted by p_{ij} and the initial starting condition. Fo
probabilities are arranged in a matrix form represent
row and j^{th} column element such that $p_{ij} = \Pr\left(\mathcal{X}\right.$
transition probabilities can be derived from p_{ij} usi
equations as follows:

$$p_{ij}^{(n)} = \sum_k p_{ik}^{(m)} p_{kj}^{(n-m)}, 0 \leq r$$

To intuitively understand (7.135), we consider the tr
in exactly n steps. After $m < n$ steps, let the Marko
all the paths that start at state i and is in state k a
of all such paths is given by $p_{ik}^{(m)}$. Similarly, the pro
to state j in $n - m$ steps equals $p_{kj}^{(n-m)}$. Thus the
start at state i and end at state j and passing throu
by $p_{ik}^{(m)} p_{kj}^{(n-m)}$. Thus, the total probability equals th
intermediate states k.

     A Markov chain is sometimes represented pictoria
a group of states with arcs representing the transitio
In a discrete time MC, there is a transition at every
time MC, the transition between states occur at ran

     The states of a Markov chain are classified as foll
if there is a non-zero probability of reaching state j
state i. States i and j are said to communicate if c
from state j and vice versa. A communicating class
can communicate with each other. Clearly, all the st
partitioned into disjoint communicating classes. An
it contains only one class, i.e., all states communica

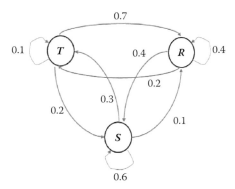

Figure 7.3: Illustration of the transitions between various states of a discrete Markov chain.

The first passage time or hitting time $\mathcal{T}_{ij}$ is the random time it takes to first reach state j starting from state i. Let $f_{jj}^{(n)} = \Pr\left(\mathcal{T}_{jj} = n\right)$ denote the probability that the MC comes back to state j in exactly n-steps. Let $f_{jj} = \sum_{n=1}^{\infty} f_{jj}^{(n)}$.

State j is transient if $f_{jj} < 1$. State j is recurrent if $f_{jj} = 1$. The recurrent states are positive recurrent if $E\left\{\mathcal{T}_j j\right\} < \infty$ and null recurrent if $E\left\{\mathcal{T}_{jj}\right\} = \infty$. The positive recurrent states are said to be *periodic* if revisits must occur in multiples of a certain period d. The positive recurrent states are said to be *a periodic* if revisits have period 1.

If the Markov chain is irreducible, and all its states are aperiodic and positive recurrent, then the Markov chain is said to be ergodic. In this case, as time continues the chain approaches a steady state or equilibrium condition. The probability distribution of the states in equilibrium are unique and can be computed from the transition probability matrix of the MC. Let π_i denote the probability of being in state i and let $\mathbf{\Pi} = [\pi_1 \ \pi_2 \ \ldots]$. These probabilities π_i also represent a time average, i.e., the fraction of time spent in state i. The steady-state probabilities are not dependent on the initial state probabilities and can be calculated as the solution to the balance equations which is given in matrix form as,

$$\mathbf{\Pi} = \mathbf{\Pi}\mathbf{P} \tag{7.136}$$

with the additional constraint that the sum of the pr
equal 1, i.e.,

$$\sum_i \pi_i = 1$$

Example 7.5.1. Consider a node in an ad-hoc ne
possible states during each time slot: transmittin
data (R), or in sleep mode (S). The transition proba
are given in Figure 7.3. Compute the steady-stat
states. If the energy consumed in the T, R, and S s
30 mJ, and 1 mJ, compute the average energy cons
   The transition probability matrix is given by

$$\mathbf{P} = \begin{bmatrix} 0.1 & 0.7 & 0.2 \\ 0.2 & 0.4 & 0.4 \\ 0.3 & 0.1 & 0.6 \end{bmatrix}$$

The steady-state probability distribution for the trar
can be computed respectively as $2/9, 1/3$, and $4/9$.
tion equals 32.67 mJ.

   Time reversibility. Let $\mathcal{Y}_n = \mathcal{X}_{-n}$ represent the M
of the Markov chain $\mathcal{X}_n$. Clearly, the time reverse
the original MC. Also, the time spent in each state
original MC or the time-reversed MC, so the stea $($
identical. Further, the transition probabilities of the
to those of the original MC as

$$\Pr\left(\mathcal{Y}_{n-1} = i | \mathcal{Y}_n = j\right) = \frac{\pi}{\pi}$$

7.5.2   Continuous Time Markov Chai

As noted before, a continuous time Markov chain ϵ
characterized by arcs representing state transitions.
states can happen at any time. The state transiti
continuous time Markov chains can also be define
discrete time Markov chain. However, the transition
on time as follows

$$p_{ij}(s,t) = \Pr\left(\mathcal{X}(s+t) = j | \mathcal{X}\right.$$

For simplicity, it is sometimes assumed that $p_{ij}(s,$
time s and only depends on time difference t. Eve
complicated since the transition probabilities must b
   An alternate approach to the analysis of such s
tinuous time MC at the time instants when a state

$\mathcal{Y}_n = \mathcal{X}(t_n)$ is an embedded discrete time Markov chain where the transition times are $t_1, t_2, \ldots$. Further, for a given value of $\mathcal{Y}_n$, the time intervals $\tau_n = t_{n+1} - t_n$ are independent exponential random variables with mean a_i that depends only on the current state.

The infinitesimal rate a_{ij} is defined as

$$a_{ij} = \lim_{t \to 0} \frac{p_{ij}(t)}{t} = a_i \Pr\left(\mathcal{Y}_{n+1} = j \mid \mathcal{Y}_n = i\right) \tag{7.141}$$

The time spent by the MC in state i before a transition to state j is exponentially distributed with parameter a_{ij}. Clearly, $a_i = \sum_j a_{ij} \forall i$. These rates are arranged in matrix form for convenience and denoted by matrix A with the ith row and j^{th} column entry being a_{ij}.

The steady-state distribution for the continuous time MC can be computed using the same principle as for its discrete time counterpart. Recall that π_i denotes the probability of being in state i. Now the probability of transition from state i to state j is given by a_{ij}/a_i. Thus,

$$\pi_j = \sum_{i \neq j} \pi_i \frac{a_{ij}}{a_j} \tag{7.142}$$

$$\Longrightarrow \pi_j a_j = \sum_{i \neq j} \pi_i a_{ij} \tag{7.143}$$

$$\Longrightarrow \sum_{i \neq j} \pi_i a_{ij} - \pi_j a_j = 0 \tag{7.144}$$

with the additional constraint that $\sum_i \pi_i = 1$. Grouping (7.144) for all i together into matrix form results in the following balance equation

$$\Pi Q = 0 \tag{7.145}$$

where

$$\mathbf{Q} = \begin{bmatrix} -a_0 & a_{01} & a_{02} & \cdots \\ a_{10} & -a_1 & a_{12} & \cdots \\ a_{20} & a_{21} & -a_2 & \cdots \end{bmatrix} \tag{7.146}$$

Example 7.5.2. Each node in an ad-hoc network is either in transmit or receive mode. The time spent in the transmit mode is exponentially distributed with parameter α and the time spent in the receive mode is exponentially distributed with parameter β. Assume that all transmit and receive durations are independent. In this case, the $\mathbf{Q}$ matrix is given by

$$\mathbf{Q} = \begin{bmatrix} -\alpha & \alpha \\ \beta & -\beta \end{bmatrix} \tag{7.147}$$

The steady-state probabilities can easily be computed as $\frac{\beta}{\alpha+\beta}$ and $\frac{\alpha}{\alpha+\beta}$.   □

7.5.3   Hidden Markov Model

In many applications the output variables being c
the Markov property. However, there is an underlyi
variable which exhibits the Markov property. Furthe
there is a probability distribution that governs the ou
Such a model is referred to as a HMM and such m
modeling and coding applications as illustrated by
An excellent introduction to HMMs is given in [10].

Let random variables $\mathcal{X}_n$, for $n = 1, 2, \ldots$ represe
system. Let the observed random variables be repre
The hidden Markov model is characterized by two co
A and **B**. Matrix **A** with elements $a_{ij} = \Pr(\mathcal{X}_{n+}$
state transition probabilities for the underlying stat
elements $b_{ik} = \Pr(\mathcal{Y}_n = k | \mathcal{X}_n = i)$ determines the c
output variable for each input variable. In some ap
tion, $\pi_i = \Pr(\mathcal{X}_1 = i)$, of the states of the system is
of the hidden Markov model is depicted in Figure 7

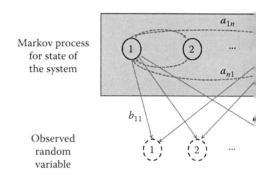

Figure 7.4: Structure of a general Hidde

There are essentially three main problems in HM

1. *Enumeration/Evaluation Problem:* Given a ce
   HMM and a sequence of observations, $y_1, y_2, \ldots$
   ity $\Pr(y_1, y_2, \ldots y_N | \mathbf{A}, \mathbf{B}, \mathbf{\Pi})$ that the model c

2. *State Estimation:* Given a certain model (
   quence $y_1, y_2, \ldots y_N$, find the best sequences $\hat{x}$
   that could result in the observation.

3. *Model Parameter Optimization:* Given a certain sequence of observations $y_1, y_2, \ldots y_N$, find the model probabilities $(\mathbf{A}, \mathbf{B}, \mathbf{\Pi})$ that best fit the given data. This problem is typically solved during the model training phase and is critical in the practical use of HMMs.

Solution to the Enumeration problem

The problem of enumeration can be evaluated using elementary tools of conditional probability. Formally, given a sequence of observations $y_1, y_2, \ldots, y_N$, this sequence can be obtained from a given state sequence as

$$\Pr(y_1, y_2, \ldots, y_N | s_1, s_2, \ldots, s_N) = \Pi_{k=1}^{N} \Pr(y_k | s_k), \tag{7.148}$$

assuming the observations are independent. The probability of this state sequence is given by

$$\Pr(s_1, s_2, \ldots s_N) = \Pr(\mathcal{X}_1 = s_1) \Pi_{k=2}^{N} \Pr(\mathcal{X}_k = s_k | \mathcal{X}_{k-1} = s_{k-1}). \tag{7.149}$$

The desired probability of the observations given the model can then be calculated as

$$\Pr(y_1, y_2, \ldots, y_N | A, B, \pi) = \sum_{s_1, s_2, \ldots, s_N} \Pr(\mathcal{X}_1 = s_1) \Pr(y_1 | s_1)$$
$$\Pi_{k=2}^{N} \Pr(\mathcal{X}_k = s_k | \mathcal{X}_{k-1} = s_{k-1}) \Pr(y_k | s_k) \tag{7.150}$$

The difficulty with this direct evaluation (7.150) of $\Pr(y_1, y_2, \ldots, y_N | A, B, \pi)$ is that the number of computations required is of the order of NM^N, where M is the number of possible states of the underlying Markov chain. Clearly, this poses a significant computational challenge even for reasonable lengths N of the observed sequence. In practice, a computationally efficient algorithms known as the *Forward-Backward* procedure is used to reduce the computational complexity. It should be mentioned that this *Forward-Backward* procedure can be derived as a special case of a message passing algorithm in graphs as described in Chapter 13.

The main steps in the *Forward* algorithm are as follows. Define $\alpha_i(m)$ as the probability of the partial observation sequence $y_1, y_2, \ldots, y_i$ up to time i and having state $x_i = m$. Now these probabilities $\alpha_i(m)$ can be calculated using the following recursive approach:

1. The Initialization Step: First set

$$\alpha_1(m) = \pi_m \Pr(y_1 | x_1 = m), \forall 1 \leq m \leq M \tag{7.151}$$

2. The Induction Step:

$$\alpha_{i+1}(k) = \left[\sum_{m=1}^{M} \alpha_i(m) \Pr(x_{i+1} = k | x_i = m) \right] \Pr(y_{i+1} | x_{i+1} = k),$$
$$\forall k = 1, \ldots M, \text{and } i = 1, \ldots N - 1. \tag{7.152}$$

3. The Termination Step:

$$\Pr\left(y_1, y_2, \ldots, y_N | A, B, \pi\right) =$$

The *Backward* algorithm proceeds in a si *ward* algorithm. Define $\beta_i(m)$ as the joint pr quence $y_{i+1}, y_{i+2}, \ldots, y_N$ given that the state of is $x_i = m$. Now β_i is calculated using the following

1. The Initialization Step: First set $\beta_N(m) = 1, \forall$

2. The Induction Step:

$$\beta_i(k) = \sum_{m=1}^{M} \Pr\left(x_{i+1} = m | x_i = k\right) \Pr$$

$$\forall k = 1, \ldots M, \text{and } i = N - 1, \ldots$$

3. The Termination Step:

$$\Pr\left(y_1, y_2, \ldots, y_N | A, B, \pi\right) = \sum_{m=1}^{M} \beta_1(m)\pi$$

Solution to the State Estimation Problem

Unlike the enumeration problem, the optimal solution lem depends on the definition of optimality. One p to compute the state at each step that maximizes t at that step. Clearly this approach could result in infeasible depending on the state transition probabi of optimality that is commonly used is to find the st probability. The well-known Viterbi algorithm provi method to compute this optimal state sequence. In rithm is also shown to be a special case of a generaliz on graphs.

Similar to the forward algorithm, in this case, we of the partial state sequence and observations up to ti as

$$\delta_i(m) = \max_{x_1, x_2, \ldots, x_{i-1}} \Pr\left(x_1, x_2, \ldots, x_{i-1}, x_i =\right.$$

Using the principle of induction, it can be easily ties $\delta_i(m)$ can be computed as

$$\delta_{i+1}(m) = \left[\max_k \delta_i(k) \Pr\left(x_{i+1} = m | x_i = k\right)\right] \Pr$$

It turns out that to find the optimal state sequence, we also need to keep track of the maximizing argument at each step in (7.157). This maximizing argument is thus stored in the $\psi_i(m)$ variables as outlined in the formal Viterbi procedure:

1. The initialization step

$$\delta_1(m) = \pi_m \Pr(y_1 | x_1 = m), \quad \psi_1(m) = 0, \forall 1 \leq m \leq M \quad (7.158)$$

2. The recursive step

$$\delta_{i+1}(m) = \max_{1 \leq k \leq M} [\delta_i(k) \Pr(x_{i+1} = m | x_i = k)] \Pr(y_{i+1} | x_{i+1} = m) \quad (7.159)$$

$$\psi_{i+1}(m) = \arg\max_{1 \leq k \leq M} [\delta_i(k) \Pr(x_{i+1} = m | x_i = k)] \quad (7.160)$$

3. The termination step

$$P^* = \max_{1 \leq m \leq M} \delta_N(m) \quad (7.161)$$

$$\hat{x}_N = \arg\max_{1 \leq m \leq M} \delta_N(m) \quad (7.162)$$

4. Backtracking step

$$\hat{x}_n = \psi_{n+1}(\hat{x}_{n+1}), \quad n = N-1, N-2, \ldots, 1. \quad (7.163)$$

Solution to the Model Parameter Estimation Problem

This problem is the most challenging of all problems in HMMs and there exists no analytical solutions to find the optimal model parameters given a finite observation sequence. However, the model parameters can be selected as a local optimum that maximizes the probability of the observation sequence for that model, using the iterative Baum-Welch algorithm.

Define $\epsilon_i(m, n)$ as the conditional probability of being in state $x_i = m$ and $x_{i+1} = n$ given the observation. This conditional probability can be expressed in terms of the $\alpha_i(m)$ and $\beta_i(n)$ variables as,

$$\epsilon_i(m, n) = \frac{\alpha_i(m) \beta_{i+1}(n) \Pr(x_{i+1} = n | x_i = m) \Pr(y_{i+1} | x_{i+1} = n)}{\sum_{l=1}^{M} \sum_{l=1}^{M} \alpha_i(k) \beta_{i+1}(l) \Pr(x_{i+1} = l | x_i = k) \Pr(y_{i+1} | x_{i+1} = l)} \quad (7.164)$$

The iterative procedure is as follows:

1. Begin with an initial estimate for the parameters of the HMM.

2. Calculate a new estimate for the parameters a

$$\pi_i \;\;=\;\; \sum_{j=}^{N}$$

$$a_{ij} = \Pr\left(x_{n+1} = j | x_n = i\right) \;\;=\;\; \frac{}{\sum}$$

$$b_{ik} \;\;=\;\; \frac{su}{\sum}$$

3. Using the new calculated parameters recalcul
   repeat step 2 above, i.e., evaluate (7.165)-(7.1

This iterative process is repeated until convergence; a
the global optimum but a local optimum. It should
rithm can be interpreted as being similar to the Ex
algorithm [11, 12].

7.6   Summary and Further Rea

In this chapter, we provided a brief introduction t
vided several simple examples to understand the ma
the basis of several chapters in this text. For instanc
processes are frequently used in Chapter 17 on que
models are considered as a special case of a generali
lation in Chapter 13. The ideas on conditional prob
extensively used in Detection (Chapter 11) and Est
As noted before, several excellent textbooks and pa
this chapter in great detail [1–5].

Acknowledgment: This work is supported in part
0546519.

7.7   Exercises

Exercise 7.7.1. A box contains 100 identical bal
100. Let there be 10 balls picked from the box w
the probability that the largest numbered ball sele
probability that the smallest numbered ball selected

Exercise 7.7.2. A biased coin has a probability of head equal to p. This biased coin is flipped 5 times. The outcome of each flip is independent of the outcome of the other flips. Let discrete random variable $\mathcal{X}$ denote the number of heads in the first 3 flips and discrete random variable $\mathcal{Y}$ denote the number of tails in the last 3 flips. Find the joint PMF $P_{\mathcal{X},\mathcal{Y}}(x,y)$. Are $\mathcal{X}$ and $\mathcal{Y}$ independent?

Exercise 7.7.3. There are two different types of radioactive elements available to a scientist. Type 1 emits radioactive particles with a Poisson distribution and mean of α. Type 2 emits radioactive particles with a Poisson distribution and a mean of β. Let $\alpha > \beta$. The scientist does not know apriori which of the two materials he is given, i.e., both materials have a probability of 0.5 of being selected. The scientist can only observe the number of emissions from the source in a given period of time. How should the scientist decide whether the material is type 1 or type 2? In other words, what is the decision rule? What is the probability of making an error using this decision rule?

Exercise 7.7.4. We have a set of M coins, numbered $1, 2, \ldots, M$. The probability of getting a head when coin k is tossed equals $1/k$. We select one of the coins at random and toss it N times. The sequence of N observations is then given to you. What is your decision rule to determine which of the coins was selected based on your observations?

Exercise 7.7.5. Let $\mathcal{X}(t)$ be a random process defined by $\mathcal{X}(t) = \sum_{i=1}^{2} N_i sin(2\pi f_0 t + \theta_i)$, where f_0 is a known frequency and all N_i and θ_j are independent random variables. The characteristic function for N_i is $\Phi_{N_i}(\omega) = e^{(\lambda_i[e^{j\omega} - 1])}$, where λ_i is a positive constant. Random variable θ_i is uniformly distributed on $[-\pi, +\pi]$. Find the mean and covariance of the process $\mathcal{X}(t)$. Is the process $\mathcal{X}(t)$ wide sense stationary?

Exercise 7.7.6. Consider the motion of a *Queen* piece on a 3×3 chessboard in which the piece moves its position at every integer times starting from $t > 0$. For simplicity assume that the squares are numbered sequentially from 1 through 9, with the top left square being 1 and the bottom right square being 9. Let $\mathcal{X}_t$ denote which vertex the queen is at time t. Assume $\mathcal{X}_t, t > 0$ is a discrete time Markov process, such that given $\mathcal{X}_t$, the next state $\mathcal{X}_{t+1}$ is equally likely to be any one of the legal squares to which the queen may move. (a) Sketch the one step transition probability diagram for $\mathcal{X}_t$. (b) Compute the steady-state probabilities.

Exercise 7.7.7. Let $\mathcal{X}$ and $\mathcal{Y}$ be continuous random variables with a joint PDF given by

$$f_{\mathcal{X},\mathcal{Y}}(x,y) = \begin{cases} cxy & 0 < x < 1, 0 < y < 1 \\ 0 & \text{otherwise} \end{cases} \tag{7.168}$$

Let $\mathcal{A} = \min\{\mathcal{X}, \mathcal{Y}\}$ and $\mathcal{B} = \max\{\mathcal{X}, \mathcal{Y}\}$. Compute the joint PDF, $f_{\mathcal{A},\mathcal{B}}(a,b)$ and marginal PDFs $f_{\mathcal{A}}(a)$ and $f_{\mathcal{B}}(b)$.

Exercise 7.7.8. The value of a particular stock ec
uary of a particular year. Every day the price of th
probability 0.1 and decreases by 1% with probabili
0.7 the value of the stock remains the same for tha'
represent the value of the stock after 1 year of trad'
that the price of the stock exceeds $105 after 1 year

Exercise 7.7.9. Let continuous random variables .
random variables with uniform distribution over th
random variable $\mathcal{N}$ be independent of $\mathcal{X}_i$ and have !

$$P_{\mathcal{N}}(n) = \begin{cases} 1/2 & n = 1, 2 \\ 0 & \text{otherwi} \end{cases}$$

Let random variable $M = \sum_{i=1}^{n} \mathcal{X}_j$. Find the PDF of .

Exercise 7.7.10. Let $\mathcal{Y}(t) = \mathcal{X}(t) - a_1\mathcal{X}(t - d_1) -$
are real constants. Find the autocorrelation and spe

Exercise 7.7.11. Consider a random walk over the
probability of transitioning from state i to state $i+$
of transitioning from state i to state $i-1$ equals $1 - q$
of transitioning to state 1 equals 1. The probability
to state M equals q. Find the long-term probability

Exercise 7.7.12. Compute the autocorrelation of a
form
$$Y_n = X_n + \beta X_{n-1}$$
where $\{X_n\}$ are i.i.d. with mean μ and variance σ^2.

Exercise 7.7.13. Consider a packet retransmissic
packet is repeatedly sent by the transmitter until th
correct versions of the packet. Assume that each tra
probability of error of $1 - q$ and is independent of t.
is the expected number of packet transmissions?

Exercise 7.7.14. In source coding, a prefix code or ε
which no codeword is a prefix of any other codeword.
code: Symbols $\{a, b, c, d\}$ are mapped to codeword:
Consider an i.i.d. sequence of symbols with probabili
$0.5, \Pr(b) = 0.3, \Pr(c) = \Pr(d) = 0.1$. Find the
representation of the code. Let $\mathcal{B}_n$ represent the sec
result from the encoding. Find the PMF of $\mathcal{B}_n$.

Exercise 7.7.15. In error control coding, one metric used to measure the performance of a block code is the number of bit errors that it can correct. For instance, in a 2-error correcting code, any bit error combination up to 2 errors can be *corrected* by the decoder at the receiver. Consider a 5-error correcting code with a resulting codeword of length 100. This length 100 code is transmitted over a binary symmetric channel with crossover probability $p = 0.05$. Compute the exact probability that all the errors made by the channel can be corrected by the code. Now, estimate this probability using the Central Limit Theorem.

References

[1] W. Feller, *An Introduction to Probability Theory and Its Applications*. New York, Wiley, 1968.

[2] S. Kay, *Intuitive Probability and Random Processes Using MATLAB*, 8th ed. Springer, New York, 2005.

[3] A. Leon-Garcia, *Probability, Statistics and Random Processes for Electrical Engineering*, 3rd ed. Prentice Hall, Upper Saddle River, NJ, 2008.

[4] S. Ross, *A First Course in Probability*. Prentice Hall, Upper Saddle River, NJ, 2009.

[5] R. Yates and D. J. Goodman, *Probability and Stochastic Processes: A Friendly Introduction for Electrical and Computer Engineers*, 2nd ed. Hoboken, NJ: John Wiley and Sons Inc., 2004.

[6] S. Lin and D. J. Costello, *Error Control Coding: Fundamentals and Applications*, 2nd ed., Prentice Hall, Upper Saddle River, NJ, 2004.

[7] W. Hoeffding, "Probability inequalities for sums of bounded random variables," *Journal of the American Statistical Association*, vol. 58, no. 301, pp. 13–30, March 1963.

[8] A. Papoulis and U. S. Pillai, *Probability, Random Variables and Stochastic Processes*. McGraw-Hill, New York, NY, 2002.

[9] T. M. Cover and J. A. Thomas, *Elements of Information Theory*, 2nd ed. Wiley, New York, 2006.

[10] L. Rabiner, "A tutorial on hidden Markov models and selected applications in speech recognition," *Proceedings of the IEEE*, vol. 77, no. 2, pp. 257–286, February 1989.

[11] A. P. Dempster, N. M. Laird, and D. B. Rubin, "Maximum likelihood from incomplete data via the EM algorithm," *Journal of the Royal Statistical Society*, vol. 39, pp. 1–38, 1977.

244

[12] M. Gupta and Y. Chen, "Theory and use of EN
Trends in Signal Processing, vol. 4, pp. 223–29

Chapter 8 Random Matrix Theory

Romain Couillet[‡] and Merouane Debbah[‡]

[‡]L'École Supérieure D'Electricité (SUPELEC), France

Random matrix theory deals with the study of *matrix-valued random variables*. It is conventionally considered that random matrix theory dates back to the work of Wishart in 1928 [1] on the properties of matrices of the type $\mathbf{X}\mathbf{X}^{\dagger}$ with $\mathbf{X} \in \mathbb{C}^{N \times n}$ a random matrix with independent Gaussian entries with zero mean and equal variance. Wishart and his followers were primarily interested in the joint distribution of the entries of such matrices and then on their eigenvalues distribution. It then dawned to mathematicians that, as the matrix dimensions N and n grow large with ratio converging to a positive value, its eigenvalue distribution converges weakly and almost surely to some deterministic distribution, which is somewhat similar to a law of large numbers for random matrices. This triggered a growing interest in particular among the signal processing community, as it is usually difficult to deal efficiently with large dimensional data because of the so-called curse of dimensionality. Other fields of research have been interested in large dimensional random matrices, among which the field of wireless communications, as the eigenvalue distribution of some random matrices is often a sufficient statistics for the performance evaluation of multidimensional wireless communication systems.

In the following, we introduce the main notions, results and details of classical as well as recent techniques to deal with large random matrices.

8.1 Probability Notations

In this chapter, an *event* will be the element ω of some set Ω. Based on Ω, we will consider the probability space $(\Omega, \mathcal{F}, P)$, with $\mathcal{F}$ some σ-field on Ω and P a probability measure on $\mathcal{F}$. If $\mathcal{X}$ is a random variable on Ω, we will denote

$$\mu_{\mathcal{X}}(A) \triangleq \Pr\left(\{\omega, \mathcal{X}(\omega) \in A\}\right)$$

the probability distribution of $\mathcal{X}$.

When $\mu_{\mathcal{X}}$ has a PDF, it will be denoted $P_{\mathcal{X}}$, i.e
Lebesgue measure and for all measurable f,

$$\int f(x)P_{\mathcal{X}}(x)\,dx \triangleq \int f(x)\mu_{\ldots}$$

To differentiate between multidimensional rando
variables, we may denote $p_{\mathcal{X}}(x) \triangleq P_{\mathcal{X}}(x)$, in lower
The CDF of a real random variable will often be de
$x \in \mathbb{R}$,

$$F(x) \triangleq p_{\mathcal{X}}((-\infty, x])$$

denotes the CDF of $\mathcal{X}$.

We further denote, for $\mathcal{X}$, $\mathcal{Y}$ two random variable
that $P_{\mathcal{Y}}(y) > 0$,

$$P_{\mathcal{X}|\mathcal{Y}}(x, y) \triangleq \frac{P_{\mathcal{X},\mathcal{Y}}(x, y)}{P_{\mathcal{Y}}(y)}$$

the conditional probability density of $\mathcal{X}$ given $\mathcal{Y}$.

8.2   Spectral Distribution of Ra

We start this section with a formal definition of a
duction of necessary notations.

Definition 8.2.1. An $N \times n$ matrix $\mathbf{X}$ is said to
matrix-valued random variable on some probability
some measurable space $(\mathcal{R}, \mathcal{G})$, where $\mathcal{F}$ is a σ-field
P and $\mathcal{G}$ is a σ-field on $\mathcal{R}$. As per conventional n
realization of the variable $\mathbf{X}$ at point $\omega \in \Omega$.

We shall in particular often consider the margina
tion of the eigenvalues of random Hermitian matrice
the distribution function (d.f.) of the *real* eigenvalue
We now discuss the properties of the so-called
known results on unitarily invariant random matrice
to the characterization, e.g., of Neyman-Pearson test
[2], [3].

8.2.1   Wishart Matrices

We start with the definition of a Wishart matrix.

Definition 8.2.2. The $N \times N$ random matrix $\mathbf{XX}^{\dagger}$
Wishart matrix with n degrees of freedom and covar

of the $N \times n$ matrix $\mathbf{X}$ are zero mean independent (real or complex) Gaussian vectors with covariance matrix $\mathbf{R}$. This is denoted

$$\mathbf{X}\mathbf{X}^\dagger \sim \mathcal{W}_N(n, \mathbf{R}).$$

Defining the *Gram matrix* associated to any matrix $\mathbf{X}$ as being the matrix $\mathbf{X}\mathbf{X}^\dagger$, $\mathbf{X}\mathbf{X}^\dagger \sim \mathcal{W}_N(n, \mathbf{R})$ is by definition the Gram matrix of a matrix with Gaussian i.i.d. columns with zero mean and variance $\mathbf{R}$. When $\mathbf{R} = \mathbf{I}_N$, it is usual to refer to $\mathbf{X}$ as a *standard Gaussian matrix*.

One interest of Wishart matrices in signal processing applications lies in the following remark.

Remark 8.2.1. Let $\mathbf{x}_1, \ldots, \mathbf{x}_n \in \mathbb{C}^N$ be n independent samples of the random process $\mathbf{x}_1 \simeq \mathcal{CN}(0, \mathbf{R})$. Then, denoting $\mathbf{X} = [\mathbf{x}_1, \ldots, \mathbf{x}_n]$,

$$\sum_{i=1}^n \mathbf{x}_i \mathbf{x}_i^\dagger = \mathbf{X}\mathbf{X}^\dagger.$$

For this reason, the random matrix $\mathbf{R}_n = \frac{1}{n}\mathbf{X}\mathbf{X}^\dagger$ is often referred to as an *(empirical) sample covariance matrix* associated to the random process $\mathbf{x}_1$. This is to be contrasted with the *population covariance matrix* $E\left\{\mathbf{x}_1\mathbf{x}_1^\dagger\right\} = \mathbf{R}$. Of particular importance is the case when $\mathbf{R} = \mathbf{I}_N$. In this situation, $\mathbf{X}\mathbf{X}^\dagger$, sometimes referred to as a *zero (or null) Wishart matrix*, is proportional to the sample covariance matrix of a white Gaussian process. The zero (or null) terminology is due to the signal processing problem of hypothesis testing, in which one has to decide whether the observed $\mathbf{X}$ emerges from a white noise process or from an information plus noise process.

Wishart provides us with the joint probability density function of the entries of Wishart matrices, as follows:

Theorem 8.2.1 ([1]). *The PDF of the complex Wishart matrix* $\mathbf{X}\mathbf{X}^\dagger \simeq \mathcal{W}_N(n, \mathbf{R})$, $\mathbf{X} \in \mathbb{C}^{N \times n}$, *for* $n \geq N$ *is*

$$P_{\mathbf{X}\mathbf{X}^\dagger}(\mathbf{B}) = \frac{\pi^{N(N-1)/2}}{\det(\mathbf{R}^n) \prod_{i=1}^N (n-i)!} e^{-\operatorname{tr}(\mathbf{R}^{-1}\mathbf{B})} \det\left(\mathbf{B}^{n-N}\right). \qquad (8.1)$$

Note in particular that for $N = 1$, this is a conventional chi-square distribution with n degrees of freedom.

For null Wishart matrices, notice that $P_{\mathbf{X}\mathbf{X}^\dagger}(\mathbf{B}) = P_{\mathbf{X}\mathbf{X}^\dagger}(\mathbf{U}\mathbf{B}\mathbf{U}^\dagger)$, for any unitary $N \times N$ matrix $\mathbf{U}$.[1] Otherwise stated, the eigenvectors of the random variable $\mathbf{X}\mathbf{X}^\dagger$ are uniformly distributed over the space $\mathcal{U}(N)$ of unitary $N \times N$ matrices. As such, the eigenvectors do not carry relevant information, and $P_{\mathbf{X}\mathbf{X}^\dagger}(\mathbf{B})$ is only

---

[1]We remind that a unitary matrix $\mathbf{U} \in \mathbb{C}^{N \times N}$ is such that $\mathbf{U}\mathbf{U}^\dagger = \mathbf{U}^\dagger\mathbf{U} = \mathbf{I}_N$.

a function of the eigenvalues of **B**. This property
derivation of further properties of Wishart matrices

The joint PDF of the eigenvalues of zero Wishar
taneously in 1939 by different authors [4–7]. The tw
in the following,

Theorem 8.2.2. *Let the entries of* $\mathbf{X} \in \mathbb{C}^{N \times n}$,
identically distributed (i.i.d.) Gaussian with zero r
joint PDF $P_{(\lambda_i)}$ *of the ordered eigenvalues* $\lambda_1 \geq$.
matrix $\mathbf{XX}^\dagger$, *is given by*

$$P_{(\lambda_i)}(\lambda_1, \ldots, \lambda_N) = e^{-\sum_{i=1}^{N} \lambda_i} \prod_{i=1}^{N} \frac{}{(n-i)}$$

where, for a Hermitian non-negative $N \times N$ *matrix*
monde determinant of its eigenvalues $\lambda_1, \ldots, \lambda_N$,

$$\Delta(\mathbf{\Lambda}) \triangleq \prod_{1 \leq i < j \leq N} (\lambda_j - \lambda$$

The marginal PDF p_λ ($\triangleq P_\lambda$) *of the unordered*

$$p_\lambda(\lambda) = \frac{1}{M} \sum_{k=0}^{N-1} \frac{k!}{(k+n-N)!} [L_k^{n-N}$$

where $L_n^k(\lambda)$ *are the Laguerre polynomials defined a*

$$L_n^k(\lambda) = \frac{e^\lambda}{k! \lambda^n} \frac{d^k}{d\lambda^k} (e^{-\lambda} \lambda^{n-}$$

The generalized case of (non-zero) central Wish
since it requires advanced tools of multivariate anal
Harish-Chandra integral [8]. We will mention the re
is at the core of the results in signal sensing present

Theorem 8.2.3. *For nonsingular* $N \times N$ *positive*
and **B** *of respective eigenvalues* $a_1, \ldots, a_N$ *and* $b_1,$.
$a_i \neq a_j$ *and* $b_i \neq b_j$, *we have*

$$\int_{\mathbf{U} \in \mathcal{U}(N)} e^{\kappa \operatorname{tr}(\mathbf{AUBU}^\dagger)} d\mathbf{U} = \left(\prod_{i=1}^{N-1} i! \right) \kappa^{\frac{1}{2}N(N-}$$

where, for any bivariate function f, $\{f(i,j)\}_{1 \leq i,j \leq N}$
(i,j) *entry* $f(i,j)$, *and* $\mathcal{U}(N)$ *is the space of* $N \times N$

---

[2]Throughout this work, we will respect the convention tha
matrix) is non-negative if $x \geq 0$, while x is positive if $x > 0$.

This result enables the calculation of the marginal joint-eigenvalue distribution of (non-zero) central Wishart matrices [9], given as follows:

Theorem 8.2.4. *Let the columns of* $\mathbf{X} \in \mathbb{C}^{N \times n}$ *be independent and identically distributed (i.i.d.) zero mean Gaussian with positive definite covariance* $\mathbf{R}$*. The joint PDF* $P_{(\lambda_i)}$ *of the ordered positive eigenvalues* $\lambda_1 \geq \ldots \geq \lambda_N$ *of the central Wishart matrix* $\mathbf{X}\mathbf{X}^\dagger$*, reads*

$$P_{(\lambda_i)}(\lambda_1, \ldots, \lambda_N) = \frac{\det\left(\{e^{-r_j^{-1}\lambda_i}\}_{1 \leq i,j \leq N}\right)}{\Delta(\mathbf{R}^{-1})} \Delta(\mathbf{\Lambda}) \prod_{j=1}^{N} \frac{\lambda_j^{n-N}}{r_j^n (n-j)!}$$

where $r_1 \geq \ldots \geq r_N$ *denote the ordered eigenvalues of* $\mathbf{R}$ *and* $\mathbf{\Lambda} =$ *diag* $(\lambda_1, \ldots, \lambda_N)$.

This is obtained from the joint distribution of Wishart matrices $\mathbf{X}\mathbf{X}^\dagger$ which, up to a variables change, leads to the joint distribution of the couples $(\mathbf{U}, \mathbf{L})$ of unitary matrices and diagonal eigenvalue matrices such that $\mathbf{X}\mathbf{X}^\dagger = \mathbf{U}\mathbf{L}\mathbf{U}^\dagger$. In performing this variable change, the Jacobian $\Delta(\mathbf{L})^2$ arises. Integrating over $\mathbf{U}$ to obtain the marginal distribution of $\mathbf{L}$, we recognize the Harish-Chandra equality which finally leads to the result.

These are the tools we need for the study of Wishart matrices. As it appears, the above properties hold due to the rotational invariance of Gaussian matrices. For more involved random matrix models, e.g., when the entries of the random matrices under study are no longer Gaussian, the study of the eigenvalue distribution is much more involved, if not unfeasible.

However, it turns out that, as the matrix dimensions grow large, nice properties arise that can be studied much more efficiently than when the matrix sizes are kept fixed. A short introduction to these large matrix considerations is described hereafter.

8.2.2 Limiting Spectral Distribution

Consider an $N \times N$ (non-necessarily random) Hermitian matrix $\mathbf{X}_N$. Define its *empirical spectral distribution (e.s.d.)* $F^{\mathbf{X}_N}$ to be the d.f. of the eigenvalues of $\mathbf{X}_N$, i.e., for $x \in \mathbb{R}$,

$$F^{\mathbf{X}_N}(x) = \frac{1}{N} \sum_{j=1}^{N} 1_{\lambda_j \leq x}(x),$$

where $\lambda_1, \ldots, \lambda_N$ are the eigenvalues of $\mathbf{X}_N$.[3]

The relevant aspect of large $N \times N$ Hermitian matrices $\mathbf{X}_N$ is that their (random) e.s.d. $F^{\mathbf{X}_N}$ often converges, with $N \to \infty$, towards a nonrandom distribution

---

[3]The Hermitian property is fundamental to ensure that all eigenvalues of $\mathbf{X}_N$ belong to the real line. However, the extension of the empirical spectral distribution (e.s.d.) to non-Hermitian matrices is sometimes required; for a definition, see (1.2.2) of [10].

F. This function F, if it exists, will be called the *lim*
of $\mathbf{X}_N$. Weak convergence [11] of $F^{\mathbf{X}_N}$ to F, i.e., fo
$F^{\mathbf{X}_N}(x) - F(x) \to 0$, is often sufficient to obtain rel

$$F^{\mathbf{X}_N} \Rightarrow F.$$

In most cases though, the weak convergence of $F^{\mathbf{X}_N}$
of matrices $\mathbf{X}_N = \mathbf{X}_N(\omega)$ of measure one. This will
$F^{\mathbf{X}_N} \Rightarrow F$ *almost surely*.

The Marčenko-Pastur Law

In signal processing, one is often interested in samp
more general matrices such as independent and iden
trices with left and right correlation, or i.i.d. matric
One of the best known results with a large range of
ing is the convergence of the empirical spectral dist
matrix of a random matrix with i.i.d. entries of zero
(not necessarily a Wishart matrix). This result is due
so that the limiting e.s.d. of the Gram matrix is cal
The result unfolds as follows.

Theorem 8.2.5. *Consider a matrix* $\mathbf{X} \in \mathbb{C}^{N \times n}$ *with*

that $\mathcal{X}_{ij}^{(N)}$ *has zero-mean and variance 1. As* $n, N -$
e.s.d. of $\mathbf{R}_n = \mathbf{X}\mathbf{X}^\dagger$ *converges almost surely to a n*
f_c *given by*

$$f_c(x) = (1 - c^{-1})^+ \delta(x) + \frac{1}{2\pi c x} \sqrt{(x}$$

where $a = (1 - \sqrt{c})^2$, $b = (1 + \sqrt{c})^2$ *and* $\delta(x) = 1$
$\delta(x) = 0$ *otherwise).*

The d.f. F_c is named the Marčenko-Pastur law
is depicted in Figure 8.1 for different values of th
particular that, when c tends to be small and app
Pastur law reduces to a single mass in 1, as the law
probability theory requires.

Several approaches can be used to derive the Ma
the original technique proposed by Marčenko and Pas
tool, the *Stieltjes transform*, which will be constantl
following we present the Stieltjes transform, along v
before we introduce several applications based on th

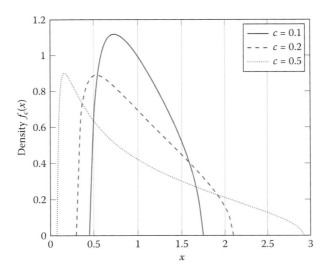

Figure 8.1: Marčenko-Pastur law for different limit ratios $c = \lim N/n$.

The Stieltjes Transform and Associated Lemmas

Definition 8.2.3. Let F be a real-valued bounded measurable function over $\mathbb{R}$. Then the Stieltjes transform $m_F(z)$,[4] for $z \in \text{Supp}\,(F)^c$, the complex space complementary to the support of F,[5] is defined as

$$m_F(z) \triangleq \int_{-\infty}^{\infty} \frac{1}{\lambda - z} dF(\lambda). \tag{8.3}$$

For all F that admit a Stieltjes transform, the inverse transformation exists and is given by [10, Theorem B.8]:

Theorem 8.2.6. *If x is a continuity points of F, then*

$$F(x) = \frac{1}{\pi} \lim_{y \to 0^+} \int_{-\infty}^{x} \Im\,[m_F(x + iy)] \, dx. \tag{8.4}$$

In practice here, F will be a distribution function. Therefore, there exists an intimate link between distribution functions and their Stieltjes transforms. More precisely, if F_1 and F_2 are two distribution functions (therefore right-continuous by definition [16, see §14]) that have the same Stieltjes transform, then F_1 and F_2 coincide everywhere and the converse is true. As a consequence, m_F uniquely

---

[4]We borrow here the notation m due to a large number of contributions from Bai, Silverstein et al. [14, 15]. In other works, the notation s or S for the Stieltjes transform is used.

[5]We recall that the support $\text{Supp}\,(F)$ of a real function F is the set $\{x \in \mathbb{R}, |F(x)| > 0\}$.

determines F and vice versa. It will turn out that, wh
functions of the empirical eigenvalues of large rande
task, the approach via Stieltjes transforms greatly si
intuition behind the Stieltjes transform approach fo
following remark: for an Hermitian matrix $\mathbf{X} \in \mathbb{C}^N$

$$
m_{F^{\mathbf{X}}}(z) = \int \frac{1}{\lambda - z} dF^{\mathbf{X}}(\lambda
$$

$$
= \frac{1}{N} \operatorname{tr} (\mathbf{\Lambda} - z\mathbf{I}_N
$$

$$
= \frac{1}{N} \operatorname{tr} (\mathbf{X} - z\mathbf{I}_N
$$

in which we denoted $\mathbf{\Lambda}$ the diagonal matrix of eige
the Stieltjes transform of $F^{\mathbf{X}}$ therefore boils dowr
$(\mathbf{X} - z\mathbf{I}_N)^{-1}$, and more specifically with the sum
matrix inversion lemmas and several fundamental ma
simple to derive limits of traces $\frac{1}{N} \operatorname{tr} (\mathbf{X} - z\mathbf{I}_N)^{-1}$,
Stieltjes transform of the weak limit of $F^{\mathbf{X}}$. For
denote $m_{\mathbf{X}} \triangleq m_{F^{\mathbf{X}}}$ the Stieltjes transform of the e
$\mathbf{X}$, and call $m_{\mathbf{X}}$ the *Stieltjes transform of* $\mathbf{X}$.

An identity of particular interest is the relation b
of $\mathbf{X}\mathbf{X}^{\dagger}$ and $\mathbf{X}^{\dagger}\mathbf{X}$, for $\mathbf{X} \in \mathbb{C}^{N \times n}$. Note that both
actually non-negative definite, so that the Stieltjes tra

Lemma 8.2.1. For $z \in \mathbb{C} \setminus \mathbb{R}^+$, we have

$$
\frac{n}{N} m_{F^{\mathbf{X}\mathbf{X}^{\dagger}}}(z) = m_{F^{\mathbf{X}\mathbf{X}^{\dagger}}}(z) + \frac{N}{}
$$

On the wireless communication side, it turns ou
is directly connected to the expression of the mutua
called *Shannon transform*, initially coined by Tulir

Definition 8.2.4. Let F be a probability distril
Shannon-transform $\mathcal{V}_F$ of F is defined, for $x \in \mathbb{R}^+$,

$$
\mathcal{V}_F(x) \triangleq \int_0^{\infty} \log(1 + x\lambda) d
$$

The Shannon-transform of F is related to its St
the expression

$$
\mathcal{V}_F(x) = \int_{\frac{1}{x}}^{\infty} \left(\frac{1}{t} - m_F(-t) \right)
$$

This last relation is fundamental to derive a lir
distribution (l.s.d.) of a random matrix and the m
dimensional channel, whose model is based on this r

We complete this section by the introduction of fundamental lemmas, required to derive the l.s.d. of random matrix models with independent entries, among which the Marčenko-Pastur law, and that will be necessary to the derivation of deterministic equivalents. These are recalled briefly below.

The first lemma is called the *trace lemma*, introduced in [15] (and extended in [18] under the form of a central limit theorem), which we formulate in the following theorem.

Theorem 8.2.7. *Let* $\mathbf{A}_1, \mathbf{A}_2, \ldots, \mathbf{A}_N \in \mathbb{C}^{N \times N}$, *be a series of matrices with uniformly bounded spectral norm. Let* $\mathbf{x}_1, \mathbf{x}_2, \ldots$ *be random vectors of i.i.d. entries such that* $\mathbf{x}_N \in \mathbb{C}^N$ *has zero mean, variance* $1/N$ *and finite eighth order moment, independent of* $\mathbf{A}_N$. *Then*

$$\mathbf{x}_N^\dagger \mathbf{A}_N \mathbf{x}_N - \frac{1}{N} \operatorname{tr}(\mathbf{A}_N) \xrightarrow{\text{a.s.}} 0, \tag{8.7}$$

as $N \to \infty$.

Several alternative versions of this result exist in the literature, which can be adapted to different application needs, see e.g., [12, 14].

The second important ingredient is the rank-1 perturbation lemma, given below [14, Lemma 2.6]:

Theorem 8.2.8. *(i) Let* $z \in \mathbb{C} \setminus \mathbb{R}$, $\mathbf{A} \in \mathbb{C}^{N \times N}$, $\mathbf{B} \in \mathbb{C}^{N \times N}$ *with* $\mathbf{B}$ *Hermitian, and* $\mathbf{v} \in \mathbb{C}^N$. *Then*

$$\left| \frac{1}{N} \operatorname{tr}\left(\mathbf{A} \left((\mathbf{B} - z\mathbf{I}_N)^{-1} - (\mathbf{B} + \mathbf{v}\mathbf{v}^\dagger - z\mathbf{I}_N)^{-1} \right) \right) \right| \leq \frac{\|\mathbf{A}\|}{N |\Im(z)|},$$

with $\|\mathbf{A}\|$ *the spectral norm of* $\mathbf{A}$.

(ii) Moreover, if $\mathbf{B}$ *is non-negative definite, for* $z \in \mathbb{R}^-$,

$$\left| \frac{1}{N} \operatorname{tr}\left(\mathbf{A} \left((\mathbf{B} - z\mathbf{I}_N)^{-1} - (\mathbf{B} + \mathbf{v}\mathbf{v}^\dagger - z\mathbf{I}_N)^{-1} \right) \right) \right| \leq \frac{\|\mathbf{A}\|}{N |z|}.$$

Generalizations of the above result can be found e.g., in [12].

Based on the above ingredients and classical results from probability theory, it is possible to prove the almost sure weak convergence of the e.s.d. of $\mathbf{X}\mathbf{X}^\dagger$, where $\mathbf{X} \in \mathbb{C}^{N \times n}$ has i.i.d. entries of zero mean and variance $1/n$, the Marčenko-Pastur law, as well as the convergence of the e.s.d. of more involved random matrix models based on matrices with independent entries. In particular, we will be interested in Section 8.5.2 in limiting results on the e.s.d. of sample covariance matrices.

Limiting Spectrum of Sample Covariance Matrices

The limiting spectral distribution of the sample covariance matrix unfolds from the following result, originally provided by Bai and Silverstein in [14], and further extended in e.g., [10],

Theorem 8.2.9. *Consider the matrix* $\mathbf{B}_N = \mathbf{A}_N$
$\mathbf{X}_N = \left(\frac{1}{\sqrt{n}} \mathcal{X}_{ij}^N\right) \in \mathbb{C}^{N \times n}$ *with entries* $\mathcal{X}_{ij}^N$ *independ*
and finite order $2+\varepsilon$ *moment for some* $\varepsilon > 0$ *(ε is in*
$F^{\mathbf{T}_N}$ *of* $\mathbf{T}_N = \mathrm{diag}\left(t_1^N, \ldots, t_N^N\right) \in \mathbb{R}^{N \times N}$ *converge*
F^T, $\mathbf{A}_N$ *is* $n \times n$ *Hermitian whose e.s.d. converge*
F^A, N/n *tends to* c, *with* $0 < c < \infty$ *as* n, N *grow*
of $\mathbf{B}_N$ *converges weakly and almost surely to* F^B *s*
satisfies

$$m_{F^B}(z) = m_{F^A}\left(z - c \int \frac{t}{1 + tm_{F^B}}\right.$$

The solution of the implicit equation (8.8) in the dum
on the set $\{z \in \mathbb{C}^+, m_{F^B}(z) \in \mathbb{C}^+\}$. *Moreover, if the*
entries, then the result holds without requiring that c

In the following, using the tools from the previou
the proof of Theorem 8.2.9.

Proof. The fundamental idea to infer the final formu
guess the form it should take. For this, write

$$m_{F^{\mathbf{B}_N}}(z) \triangleq \frac{1}{n} \mathrm{tr}\left(\mathbf{A}_N + \mathbf{X}_N^\dagger \mathbf{T}_N \mathbf{X}\right.$$

and take $\mathbf{D}_N \in \mathbb{C}^{N \times N}$ to be some deterministic ma

$$m_{F^{\mathbf{B}_N}}(z) - m_N(z) \xrightarrow{\text{a.s.}}$$

with

$$m_N(z) \triangleq \frac{1}{n} \mathrm{tr}\left(\mathbf{A}_N + \mathbf{D}_N - z\right)$$

as $N, n \to \infty$ with $N/n \to c$. We then have, fror
$\mathbf{A}^{-1}(\mathbf{B} - \mathbf{A})\mathbf{B}^{-1}$,

$$m_{F^{\mathbf{B}_N}}(z) - m_N(z) = \frac{1}{n} \mathrm{tr}\left((\mathbf{B}_N - z\mathbf{I}_N)^{-1}(\mathbf{D}_N - \mathbf{X}\right.$$

Taking $\mathbf{D}_N = a_N \mathbf{I}_N$, and writing

$$\mathbf{X}_N^\dagger \mathbf{T}_N \mathbf{X}_N = \sum_{k=1}^N t_k^N \mathbf{x}_k \mathbf{x}$$

with $\mathbf{x}_k$ the kth column of $\mathbf{X}_N^\dagger$, we further have

$$m_{F^{\mathbf{B}_N}}(z) - m_N(z) = \frac{a_N}{n} \mathrm{tr}\left((\mathbf{B}_N - z\mathbf{I}_N)^{-1}(\mathbf{A}_N\right.$$

$$- \frac{1}{n} \sum_{k=1}^N t_k^N \mathbf{x}_k^\dagger (\mathbf{A}_N + \mathbf{D}_N - z)$$

Using the matrix inversion identity

$$(\mathbf{A} + \mathbf{v}\mathbf{v}^\dagger - z\mathbf{I}_N)^{-1}\mathbf{v} = \frac{1}{1 + \mathbf{v}^\dagger(\mathbf{A} - z\mathbf{I}_N)^{-1}\mathbf{v}}(\mathbf{A} - z\mathbf{I}_N)^{-1}\mathbf{v},$$

each term in the sum of the right-hand side can further be expressed as

$$t_k^N \mathbf{x}_k^\dagger (\mathbf{A}_N + \mathbf{D}_N - z\mathbf{I}_N)^{-1}(\mathbf{B}_N - z\mathbf{I}_N)^{-1}\mathbf{x}_k$$
$$= \frac{t_k^N \mathbf{x}_k^\dagger (\mathbf{A}_N + \mathbf{D}_N - z\mathbf{I}_N)^{-1}(\mathbf{B}_{(k)} - z\mathbf{I}_N)^{-1}\mathbf{x}_k}{1 + t_k^N \mathbf{x}_k^\dagger (\mathbf{B}_{(k)} - z\mathbf{I}_N)^{-1}\mathbf{x}_k}$$

where $\mathbf{B}_{(k)} = \mathbf{B}_N - t_k^N \mathbf{x}_k \mathbf{x}_k^\dagger$ and where now $\mathbf{x}_k$ and $(\mathbf{A}_N + \mathbf{D}_N - z\mathbf{I}_N)^{-1}(\mathbf{B}_{(k)} - z\mathbf{I}_N)^{-1}$ are independent. But then, using the trace lemma, Theorem 8.2.7, we have that

$$\mathbf{x}_k^\dagger (\mathbf{A}_N + \mathbf{D}_N - z\mathbf{I}_N)^{-1}(\mathbf{B}_{(k)} - z\mathbf{I}_N)^{-1}\mathbf{x}_k$$
$$- \frac{1}{n} \operatorname{tr}\left((\mathbf{A}_N + \mathbf{D}_N - z\mathbf{I}_N)^{-1}(\mathbf{B}_{(k)} - z\mathbf{I}_N)^{-1}\right) \xrightarrow{\text{a.s.}} 0.$$

Replacing the quadratic form by the trace in the Stieltjes transform difference, we then have for all large N,

$$m_{F^{\mathbf{B}_N}}(z) - m_N(z) \simeq \frac{a_N}{n} \operatorname{tr}\left((\mathbf{B}_N - z\mathbf{I}_N)^{-1}(\mathbf{A}_N + \mathbf{D}_N - z\mathbf{I}_N)^{-1}\right)$$
$$- \frac{1}{n} \sum_{k=1}^N \frac{t_k^N \frac{1}{n} \operatorname{tr}\left((\mathbf{A}_N + \mathbf{D}_N - z\mathbf{I}_N)^{-1}\left(\mathbf{B}_{(k)} - z\mathbf{I}_N\right)^{-1}\right)}{1 + t_k^N \frac{1}{n} \operatorname{tr}\left(\left(\mathbf{B}_{(k)} - z\mathbf{I}_N\right)^{-1}\right)}.$$

But then, from the rank-1 perturbation lemma, Theorem 8.2.8, this is further approximated, for all large N by

$$m_{F^{\mathbf{B}_N}}(z) - m_N(z) \simeq \frac{a_N}{n} \operatorname{tr}\left((\mathbf{B}_N - z\mathbf{I}_N)^{-1}(\mathbf{A}_N + \mathbf{D}_N - z\mathbf{I}_N)^{-1}\right)$$
$$- \frac{1}{n} \sum_{k=1}^N \frac{t_k^N \frac{1}{n} \operatorname{tr}\left((\mathbf{A}_N + \mathbf{D}_N - z\mathbf{I}_N)^{-1}\left(\mathbf{B}_N - z\mathbf{I}_N\right)^{-1}\right)}{1 + t_k^N \frac{1}{n} \operatorname{tr}\left(\left(\mathbf{B}_N - z\mathbf{I}_N\right)^{-1}\right)}$$

where we recognize in the right-hand side the Stieltjes transform $m_{F^{\mathbf{B}_N}}(z) = \frac{1}{n} \operatorname{tr}\left((\mathbf{B}_N - z\mathbf{I}_N)^{-1}\right)$. Taking

$$a_N = \frac{1}{n} \sum_{k=1}^N t_k^N \frac{1}{1 + t_k^N m_{F^{\mathbf{B}_N}}(z)} \simeq c \int \frac{t}{1 + t m_{F^{\mathbf{B}_N}}(z)} dF^T(t),$$

it is clear that the difference $m_{F^{\mathbf{B}_N}}(z) - m_N(z)$ becomes increasingly small for large N and therefore $m_{F^{\mathbf{B}_N}}(z)$ is asymptotically close to

$$\frac{1}{n} \operatorname{tr}\left(\mathbf{A}_N + c \int \frac{t dF^T(t)}{1 + t m_{F^{\mathbf{B}_N}}(z)}\mathbf{I}_N - z\mathbf{I}_N\right)^{-1}$$

which is exactly

$$m_{F^{\mathbf{A}_N}}\left(z - c\int \frac{t\,dF^T(t)}{1 + tm_{F^{\mathbf{B}_N}}}\right.$$

Hence the result.

The sample covariance matrix model correspond $\mathbf{A}_N = 0$. In that case, (8.8) becomes

$$m_{\underline{F}}(z) = -\left(z - c\int \frac{t}{1 + tm_{\underline{F}}(z)}\right.$$

where we denoted $\underline{F} \triangleq F^B$ in this special case. This used to differentiate the l.s.d. F of the matrix $\mathbf{T}_N^{\frac{1}{2}}$ of the reversed Gram matrix $\mathbf{X}_N^{\dagger}\mathbf{T}_N\mathbf{X}_N$. Remark i the Stieltjes transform $m_{\underline{F}}$ of the l.s.d. $\underline{F}$ of $\mathbf{X}_N^{\dagger}\mathbf{T}$ transform m_F of the l.s.d. F of $\mathbf{T}_N^{\frac{1}{2}}\mathbf{X}_N\mathbf{X}_N^{\dagger}\mathbf{T}_N^{\frac{1}{2}}$ thro

$$m_{\underline{F}}(z) = cm_F(z) + (c - \mathbf{1}$$

and then we also have access to a characterizatio asymptotic eigenvalue distribution of the sample co the denormalized columns $\sqrt{n}\mathbf{x}_1, \ldots, \sqrt{n}\mathbf{x}_n$ of $\sqrt{n}$. pendent vectors with zero mean and covariance m illustrative simulation example is given in Figure 8. three distinct eigenvalues.

Secondly, in addition to the uniqueness of th $\{z \in \mathbb{C}^+, m_{\underline{F}}(z) \in \mathbb{C}^+\}$ solution of (8.9), an inverse form can be written in closed-form, i.e., we can $\{\underline{m} \in \mathbb{C}^+, z_{\underline{F}}(\underline{m}) \in \mathbb{C}^+\}$, such that

$$z_{\underline{F}}(\underline{m}) = -\frac{1}{\underline{m}} + c\int \frac{t}{1 + t\underline{m}}d$$

This will turn out to be extremely useful to cha More on this topic is discussed in Section 8.3.

8.3   Spectral Analysis

In this section, we summarize some important result zation of the support of the eigenvalues of a sample position of the individual eigenvalues of a sample c

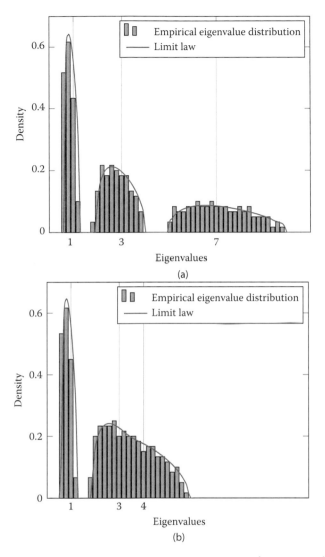

Figure 8.2: Histogram of the eigenvalues of $\mathbf{B}_N = \mathbf{T}_N^{\frac{1}{2}} \mathbf{X}_N \mathbf{X}_N^{\dagger} \mathbf{T}_N^{\frac{1}{2}}$, $N = 300$, $n = 3000$, with $\mathbf{T}_N$ diagonal composed of three evenly weighted masses in (a) 1, 3, and 7, (b) 1, 3, and 4.

(i) is obviously a must-have from a pure mathematical viewpoint, but is also fundamental to the study of estimators based on large dimensional random matrices. We will provide in Section 8.4 and in Section 8.5.2 estimators of functionals of the eigenvalues of a population covariance matrix based on the observation of a sample covariance matrix. We will in particular investigate large dimensional sample covariance matrix models with population covariance matrix composed of a few

eigenvalues with large multiplicities. The validity of
tantly on the fact that the support of the l.s.d. of
is formed of disjoint so-called *clusters*; each cluster
eigenvalues of the population covariance matrix. Cl
port is therefore paramount to the study of the estin
(ii) is even more important for the estimators des
position of the individual eigenvalues allows one to
second point is also fundamental to the derivation
large dimensional matrix analysis, that will be intro
we will show in particular is that, under mild assun
model, all eigenvalues are asymptotically contained
Also, when the limiting support is divided into dis
sample eigenvalues in each cluster corresponds exac
population eigenvalue attached to this cluster. For
mental as the observation of a sample eigenvalue
support of the pure noise hypothesis (called hypothe
is present in the observed data.

    We start with the point (ii).

8.3.1   Exact Eigenvalue Separation

The results of interest here are due to Bai and Silve
the following theorems.

Theorem 8.3.1 ([15]). *Let* $\mathbf{X}_N = \left(\frac{1}{\sqrt{n}}\mathcal{X}_{ij}^N\right) \in \mathbb{C}$
that $\mathcal{X}_{ij}^N$ *has zero mean, variance 1 and finite four*
$\mathbb{C}^{N \times N}$ *be nonrandom, whose e.s.d.* $F^{\mathbf{T}_N}$ *converge*
8.2.9, the e.s.d. of $\mathbf{B}_N = \mathbf{T}_N^{\frac{1}{2}}\mathbf{X}_N\mathbf{X}_N^{\dagger}\mathbf{T}_N^{\frac{1}{2}} \in \mathbb{C}^{N \times N}$
surely towards some distribution function F, *as* N
$c_N = N/n \to c,\ 0 < c < \infty.$ *Similarly, the e.s.d.*
converges towards $\underline{F}$ *given by*

$$\underline{F}(x) = cF(x) + (1 - c)\,1_{[0,\infty}$$

Denote $\underline{F}_N$ *the distribution of Stieltjes transform* m
of the following equation in m

$$m = -\left(z - \frac{N}{n}\int \frac{\tau}{1 + \tau m}dF^{\mathbf{T}_N}\right)$$

and define F_N *the d.f. such that*

$$\underline{F}_N(x) = \frac{N}{n}F_N(x) + \left(1 - \frac{N}{n}\right)$$

Let $N_0 \in \mathbb{N}$, and choose an interval $[a, b]$, $a > 0$, outside the union of the supports of F and F_N for all $N \geq N_0$. For $\omega \in \Omega$, the random space generating the series $\mathbf{X}_1, \mathbf{X}_2, \ldots$, denote $\mathcal{L}_N(\omega)$ the set of eigenvalues of $\mathbf{B}_N(\omega)$. Then,

$$P(\omega, \mathcal{L}_N(\omega) \cap [a, b] \neq \emptyset, \text{ i.o.}) = 0.$$

This means concretely that, given a segment $[a, b]$ outside the union of the supports of F and $F_{N_0}, F_{N_0+1}, \ldots$, for all series $\mathbf{B}_1(\omega), \mathbf{B}_2(\omega), \ldots$, with ω in some set of probability one, there exists $M(\omega)$ such that, for all $N \geq M(\omega)$, there will be no eigenvalue of $\mathbf{B}_N(\omega)$ in $[a, b]$.

As an immediate corollary of Theorem 8.2.5 and Theorem 8.3.1, we have the following results on the extreme eigenvalues of $\mathbf{B}_N$, with $\mathbf{T}_N = \mathbf{I}_N$.

Corollary 8.3.2. Let $\mathbf{B}_N \in \mathbb{C}^{N \times N}$ be defined as $\mathbf{B}_N = \mathbf{X}_N \mathbf{X}_N^\dagger$, with $\mathbf{X}_N \in \mathbb{C}^{N \times n}$ with i.i.d. entries of zero mean, variance $1/n$ and finite fourth order moment. Then, denoting $\lambda_{\min}^N$ and $\lambda_{\max}^N$ the smallest and largest eigenvalues of $\mathbf{B}_N$, respectively, we have

$$\lambda_{\min}^N \xrightarrow{\text{a.s.}} (1 - \sqrt{c})^2$$
$$\lambda_{\max}^N \xrightarrow{\text{a.s.}} (1 + \sqrt{c})^2$$

as $N, n \to \infty$ with $N/n \to c$.

This result further extends to the case when $\mathbf{B}_N = \mathbf{X}_N \mathbf{T}_N \mathbf{X}_N^\dagger$, with $\mathbf{T}_N$ diagonal with ones on the diagonal but for a few entries different from one. This model, often referred to as *spiked model* lets some eigenvalues escape the limiting support of $\mathbf{B}_N$ (which is still the support of the Marčenko-Pastur law). Note that this is not inconsistent with Theorem 8.3.1 since here, for all finite N_0, the distribution functions $F_{N_0}, F_{N_0+1}, \ldots$ may all have a non-zero mass outside the support of the Marčenko-Pastur law. The segments $[a, b]$ where no eigenvalues are found asymptotically must be away from these potential masses. The theorem, due to Baik, is given precisely as follows

Theorem 8.3.3 ([19]). Let $\bar{\mathbf{B}}_N = \bar{\mathbf{T}}_N^{\frac{1}{2}} \mathbf{X}_N \mathbf{X}_N^\dagger \bar{\mathbf{T}}_N^{\frac{1}{2}}$, where $\mathbf{X}_N \in \mathbb{C}^{N \times n}$ has i.i.d. entries of zero mean and variance $1/n$, and $\bar{\mathbf{T}}_N \in \mathbb{R}^{N \times N}$ is diagonal given by

$$\bar{\mathbf{T}}_N = \text{diag}\left(\underbrace{\alpha_1, \ldots, \alpha_1}_{k_1}, \ldots, \underbrace{\alpha_M, \ldots, \alpha_M}_{k_M}, \underbrace{1, \ldots, 1}_{N - \sum_{i=1}^M k_i} \right)$$

with $\alpha_1 > \ldots > \alpha_M > 0$ for some positive integer M. We denote here $c = \lim_N N/n$. Call $M_0 = \#\{j | \alpha_j > 1 + \sqrt{c}\}$. For $c < 1$, take also M_1 to be such that $M - M_1 = \#\{j | \alpha_j < 1 - \sqrt{c}\}$. Denote additionally $\lambda_1, \ldots, \lambda_N$ the eigenvalues of $\bar{\mathbf{B}}_N$, ordered as $\lambda_1 \geq \ldots \geq \lambda_N$. We then have

- *for $1 \leq j \leq M_0$, $1 \leq i \leq k_j$,*

$$\lambda_{k_1+\ldots+k_{j-1}+i} \xrightarrow{\text{a.s.}} \alpha_j +$$

- *for the other eigenvalues, we must discriminat*

  - *if $c < 1$,*
    * *for $M_1 + 1 \leq j \leq M$, $1 \leq i \leq k_j$,*

    $$\lambda_{N-k_j-\ldots-k_M+i} \xrightarrow{\text{a.s.}}$$

    * *for the indexes of eigenvalues of $\bar{\mathbf{T}}_N$*

    $$\lambda_{k_1+\ldots+k_{M_0}+1} \xrightarrow{\text{a.}}$$

    $$\lambda_{N-k_{M_1+1}-\ldots-k_M} \xrightarrow{\text{a.}}$$

  - *if $c > 1$,*

    $$\lambda_n \xrightarrow{\text{a.s.}}$$

    $$\lambda_{n+1} = \ldots = \lambda_N = 0,$$

  - *if $c = 1$,*

    $$\lambda_{\min(n,N)} \xrightarrow{\text{a.s.}}$$

The important part of this result is that all α_j duces an eigenvalue of $\mathbf{B}_N$ outside the support of asymptotically at the position $\alpha_j + \frac{c\alpha_j}{\alpha_j - 1}$.

Now Theorem 8.3.1 and Theorem 8.3.3 ensure t value of $\mathbf{B}_N$ is found outside the support of F_{N_0}, *h* do not say where the eigenvalues of $\mathbf{B}_N$ are appro swer to this question is provided by Bai and Silverst separation properties of the l.s.d. of such matrices I

Theorem 8.3.4 ([20]). *Assume $\mathbf{B}_N$ is as in Theore definite and $F^{\mathbf{T}_N}$ converging weakly to the distributi converging to c. Consider also $0 < a < b < \infty$ s support of F, the l.s.d. of $\mathbf{B}_N$. Denote additionally of $\mathbf{B}_N$ and $\mathbf{T}_N$ in decreasing order, respectively. Th*

1. *If $c(1 - H(0)) > 1$, then the smallest eigenva positive and $\lambda_N \to x_0$ almost surely, as $N \to c$*

2. If $c(1 - H(0)) \leq 1$, or $c(1 - H(0)) > 1$ but $[a, b]$ is not contained in $[0, x_0]$, then
$$\Pr\left(\omega, \lambda_{i_N} > b, \lambda_{i_N + 1} < a\right) = 1,$$
for all N large, where i_N is the unique integer such that
$$\tau_{i_N} > -1/m_F(b),$$
$$\tau_{i_N + 1} < -1/m_F(a).$$

Theorem 8.3.4 states in particular that, when the limiting spectrum can be divided in disjoint clusters, then the index of the sample eigenvalue "for which a jump from one cluster (right to b) to a subsequent cluster (left to a) arises" corresponds exactly to the index of the population eigenvalue where a jump arises in the population eigenvalue spectrum (from $-1/m_F(b)$ to $-1/m_F(a)$). Therefore, the sample eigenvalues distribute as one would expect between the consecutive clusters. This result will be used in Section 8.4 and Section 8.5.2 to find which sample eigenvalues are present in which cluster. This is necessary because we will perform complex integration on contours surrounding specific clusters and that residue calculus will demand that we know exactly what eigenvalues are found inside these contours.

Nonetheless, this still does not exactly answer the question of the exact characterization of the limiting support, which we treat in the following.

8.3.2 Support of l.s.d.

Remember from the inverse Stieltjes transform formula (8.4) that it is possible to determine the support of the l.s.d. F of a random matrix once we know its limiting Stieltjes transform $m_F(z)$ for all $z \in \mathbb{C}^+$. Thanks to Theorem 8.2.9, we know in particular that we can determine the support of the l.s.d. of a sample covariance matrix. Nonetheless, (8.4) features a limit for the imaginary part y of the argument $z = x + iy$ of $m_F(z)$ going to zero, which has not been characterized to this point (even its existence everywhere is not ensured). Choi and Silverstein proved in [21] that this limit does exist for the case of sample covariance matrices and goes even further in characterizing exactly what this limit is. This uses the important Stieltjes transform composition inverse formula (8.11) and is summarized as follows.

Theorem 8.3.5 ([21]). *Denote S_X^c the complementary of S_X, the support of some d.f. X. Let $\mathbf{B}_N = \mathbf{X}_N^\dagger \mathbf{T}_N \mathbf{X}_N \in \mathbb{C}^{n \times n}$ have l.s.d. $\underline{F}$, where $\mathbf{X}_N \in \mathbb{C}^{N \times n}$ has i.i.d. entries of zero mean and variance $1/n$, $\mathbf{T}_N$ has l.s.d. H and $N/n \to c$. Let $B = \{\underline{m} \mid \underline{m} \neq 0, -1/\underline{m} \in S_H^c\}$ and $x_{\underline{F}}$ be the function defined on B by*

$$x_{\underline{F}}(\underline{m}) = -\frac{1}{\underline{m}} + c \int \frac{t}{1 + t\underline{m}} dH(t). \tag{8.12}$$

For $x_0 \in \mathbb{R}^*$, we can then determine the limit of $m_{\underline{F}}(z)$ as $z \to x_0$, $z \in \mathbb{C}^+$, along the following rules,

(R.I) If $x_0 \in S_{\underline{F}}^c$, then the equation $x_0 = x_{\underline{F}}(\underline{m})$
a unique real solution $m_0 \in B$ such that $x'_{\underline{F}}$
of $m_{\underline{F}}(z)$ when $z \to x_0$, $z \in \mathbb{C}^+$. Conve
$x'_{\underline{F}}(m_0) > 0$, $x_0 = x_{\underline{F}}(m_0) \in S_{\underline{F}}^c$.

(R.II) If $x_0 \in S_{\underline{F}}$, then the equation $x_0 = x_{\underline{F}}(\underline{m})$ i
unique complex solution $m_0 \in B$ with a po
is the limit of $m_{\underline{F}}(z)$ when $z \to x_0$, $z \in \mathbb{C}^+$

From Theorem 8.3.5.(R.I), it is possible to deter
It indeed suffices to draw $x_{\underline{F}}(\underline{m})$ for $-1/\underline{m} \in \mathbb{R} \setminus S$
on an interval I, $x_{\underline{F}}(I)$ is outside $S_{\underline{F}}$. The support
(modulo the mass in 0), is then defined exactly by

$$S_{\underline{F}} = \mathbb{R} \setminus \bigcup_{\substack{a,b \in \mathbb{R} \\ a < b}} \left\{ x_{\underline{F}}((a,b)) \mid \forall \underline{m} \in (a, \right.$$

This is depicted in Figure 8.3 in the case when H
weighted masses t_1, t_2, t_3 in $\{1,3,5\}$ or $\{1,3,10\}$ a
the case where $t_3 = 10$, F is divided into three clu
divided into only two clusters, which is due to the fa
the interval $(-1/3, -1/5)$.

From Figure 8.3 and Theorem 8.3.5, we now ob
$2K_F$ roots with K_F the number of clusters in F. D
$\underline{m}_2^- < \underline{m}_2^+ < \ldots \leq \underline{m}_{K_F}^- < \underline{m}_{K_F}^+$. Each pair $(\underline{m}_j^-, \underline{m}$
is the jth cluster in F. We therefore have a way to
asymptotic spectrum through the function $x'_{\underline{F}}$. Thi
result.

Theorem 8.3.6 ([22, 23]). *Let* $\mathbf{B}_N \in \mathbb{C}^{N \times N}$ *be*
Then the support S_F *of the l.s.d.* F *of* $\mathbf{B}_N$ *is define*

$$S_F = \bigcup_{j=1}^{K_F} [x_j^-, x_j^+],$$

where $x_1^-, x_1^+, \ldots, x_{K_F}^-, x_{K_F}^+$ *are defined as*

$$x_j^- = -\frac{1}{\underline{m}_j^-} + \sum_{r=1}^{K} c_r \frac{t_r}{1 + t_r}$$

$$x_j^+ = -\frac{1}{\underline{m}_j^-} + \sum_{r=1}^{K} c_r \frac{t_r}{1 + t_r}$$

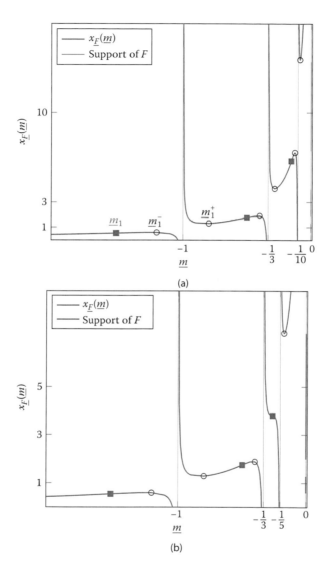

Figure 8.3: $x_{\underline{F}}(\underline{m})$ for $\mathbf{T}_N$ diagonal composed of three evenly weighted masses in 1, 3, and 10 (a) and 1, 3, and 5 (b), $c = 1/10$ in both cases. Local extrema are marked in circles, inflexion points are marked in squares. The support of F can be read on the right vertical axes.

with $\underline{m}_1^- < \underline{m}_1^+ \leq \underline{m}_2^- < \underline{m}_2^+ \leq \ldots \leq \underline{m}_{K_F}^- < \underline{m}_{K_F}^+$ *t*
multiplicity) real roots of the equation in $\underline{m}$,

$$\sum_{r=1}^{K} c_r \frac{t_r^2 \underline{m}^2}{(1 + t_r \underline{m}^2)^2} = 1$$

Notice further from Figure 8.3 that, while $x'_F(\underline{m})$
intervals $(-1/t_{k-1}, -1/t_k)$, it always has a unique
proved in [23] by observing that $x''_F(\underline{m}) = 0$ is equiv

$$\sum_{r=1}^{K} c_r \frac{t_r^3 \underline{m}^3}{(1 + t_r \underline{m})^3} - 1 =$$

the left-hand side of which has always a positive der
in the neighborhood of t_r; hence the existence of a u
interval $(-1/t_{k-1}, -1/t_k)$, for $1 \leq k \leq K$, with
x_F increases on an interval $(-1/t_{k-1}, -1/t_k)$, it mu
point of positive derivative (from the concavity chang
Therefore, to verify that cluster k_F is disjoint from
(when they exist), it suffices to verify that the $(k-1)$
of $x''_F(\underline{m})$ are such that $x'_F(\underline{m}_{k-1}) > 0$ and $x'_F(\underline{m}_k)$
following result states for the case of a sample covari
covariance matrix has few eigenvalues, each with a

Theorem 8.3.7 ([23, 24]). *Let* $\mathbf{B}_N$ *be defined as*
diag $(\tau_1, \ldots, \tau_N) \in \mathbb{R}^{N \times N}$, *diagonal containing* K
$\ldots < t_K$, *for some fixed* K. *Denote* N_k *the multiplici*
counted with multiplicity (assuming ordering of the τ
$\tau_{N_1} = t_1, \ldots, \tau_{N-N_K+1} = \ldots = \tau_N = t_K$). *Assum*
$N_r/n \to c_r > 0$, *and* $N/n \to c$, *with* $0 < c < \infty$. *The*
the eigenvalue t_k *in the l.s.d.* F *of* $\mathbf{B}_N$ *is distinct f*
$(k+1)_F$ *(when they exist), associated to* t_{k-1} *and*
only if

$$\sum_{r=1}^{K} c_r \frac{t_r^2 \underline{m}_k^2}{(1 + t_r \underline{m}_k^2)^2} <$$

$$\sum_{r=1}^{K} c_r \frac{t_r^2 \underline{m}_{k+1}^2}{\left(1 + t_r \underline{m}_{k+1}^2\right)^2} <$$

where $\underline{m}_1, \ldots, \underline{m}_K$ *are such that* $\underline{m}_{K+1} = 0$ *and* $\underline{m}_1$
solutions of the equation in $\underline{m}$,

$$\sum_{r=1}^{K} c_r \frac{t_r^3 \underline{m}^3}{(1 + t_r \underline{m})^3} = 1.$$

For $k = 1$, *this condition ensures* $1_F = 2_F - 1$; *for* $k = K$, *this ensures* $K_F = (K-1)_{F+1}$; *and for* $1 < k < K$, *this ensures* $(k-1)_{F+1} = k_F = (k+1)_{F-1}$.

This result is again fundamental in the sense that the separability of subsequent clusters in the support of the l.s.d. of $\mathbf{B}_N$ will play a fundamental role in the validity of statistical inference methods. In the subsequent section, we introduce the key ideas that allow statistical inference for sample covariance matrices.

8.4 Statistical Inference

Statistical inference allows for the estimation of deterministic parameters present in a stochastic model based on observations of random realizations of the model. In the context of sample covariance matrices, statistical inference methods consist in providing estimates of functionals of the eigenvalue distribution of the population covariance matrix $\mathbf{T}_N \in \mathbb{C}^{N \times N}$ based on the observation $\mathbf{Y}_N = \mathbf{T}_N^{\frac{1}{2}} \mathbf{X}_N$ with $\mathbf{X}_N \in \mathbb{C}^{N \times n}$ a random matrix of independent and identically distributed entries. Different methods exist that allow for statistical inference that mostly rely on the study of the l.s.d. of the sample covariance matrix $\mathbf{B}_N = \frac{1}{n} \mathbf{Y}_N \mathbf{Y}_N^\dagger$. One of these methods relates to free probability theory [25], and more specifically to free deconvolution approaches, see e.g., [26], [27]. The idea behind free deconvolution is based on the fact that the moments of the l.s.d. of some random matrix models can be written as a polynomial function of the moments of the l.s.d. of another (random) matrix in the model, under some proper conditions. Typically, the moments of the l.s.d. of $\mathbf{T}_N$ can be written as a polynomial of the moments of the (almost sure) l.s.d. of $\mathbf{B}_N$, if $\mathbf{X}_N$ has Gaussian entries and the e.s.d. of $\mathbf{T}_N$ has uniformly bounded support. Therefore, to put it simply, one can obtain all moments of $\mathbf{T}_N$ based on a sufficiently large observation of $\mathbf{B}_N$; this allows one to recover the l.s.d. of $\mathbf{T}_N$ (since Carleman condition is satisfied) and therefore any functional of the l.s.d. However natural, this method has some major drawbacks. From a practical point of view, a reliable estimation of moments of high order requires extremely large dimensional matrix observations. This is due to the fact that the estimate of the moment of order k of the l.s.d. is based on polynomial expressions of the estimates of moments of lower orders. A small error in the estimate in a low order moment therefore propagates as a large error for higher moments; it is therefore compelling to obtain accurate first order estimates, hence large dimensional observations.

We will not further investigate the moment-based approach above, which we discuss in more detail with a proper introduction to free probability theory in [28]. Instead, we introduce the methods based on the Stieltjes transform and those rely strongly on the results described in the previous section. We will introduce this method for the sample covariance matrix model discussed so far, because it will be instrumental to understanding the power estimator introduced in Section 8.5.2. Similar results have been provided for other models of interest to telecommunications, as for instance the so-called information-plus-

The central idea is based on a trivial applicatior
gration formula [30]. Consider f some complex hol
H a distribution function and denote G the functior

$$G(f) = \int f(z) dH(z).$$

From the Cauchy integration formula, we have, f
path γ enclosing the support of H and with winding

$$G(f) = \frac{1}{2\pi i} \int \oint_\gamma \frac{f(\omega)}{z - \omega} d\omega$$

$$= \frac{1}{2\pi i} \oint_\gamma \int \frac{f(\omega)}{z - \omega} dH$$

$$= \frac{1}{2\pi i} \oint_\gamma f(\omega) m_H(\omega)$$

the integral inversion being valid since $f(\omega)/(z - \omega$
that the sign inversion due to the negative contour
the sign reversal of $(\omega - z)$ in the denominator.

If dH is a sum of finite or countable masses and
$f(\lambda_k)$, with λ_k the value of the kth mass with we
oriented contour γ_k enclosing λ_k and excluding λ_j,

$$l_k f(\lambda_k) = \frac{1}{2\pi i} \oint_{\gamma_k} f(\omega) m_H($$

This last expression is particularly convenient whe
through an expression of its Stieltjes transform.

Now, in terms of random matrices, for the sam
$\mathbf{T}_N^{\frac{1}{2}} \mathbf{X}_N \mathbf{X}_N^\dagger \mathbf{T}_N^{\frac{1}{2}}$, we already noticed that the l.s.d.
l.s.d. $\underline{F}$ of $\underline{\mathbf{B}}_N = \mathbf{X}_N^\dagger \mathbf{T}_N \mathbf{X}_N)$ can be rewritten und
further be rewritten

$$\frac{c}{m_{\underline{F}}(z)} m_H \left(-\frac{1}{m_{\underline{F}}(z)} \right) = -z m_{\underline{F}}($$

where H is the l.s.d. of $\mathbf{T}_N$. Note that it is allowed
for $z \in \mathbb{C}^+$ since $-1/m_{\underline{F}}(z) \in \mathbb{C}^+$.

As a consequence, if one only has access to $F^{\mathbf{B}_l}$
then the only link from the observation to H is
$F^{\underline{\mathbf{B}}_N} \Rightarrow \underline{F}$ almost surely and (ii) the fact that $\underline{F}$ and
Evaluating a functional f of the eigenvalue λ_k of
(8.15). The relations (8.15) and (8.16) are the ess
derivation of a consistent estimator for $f(\lambda_k)$.

We now concentrate specifically on the sample covariance matrix $\mathbf{B}_N = \mathbf{T}_N^{\frac{1}{2}} \mathbf{X}_N \mathbf{X}_N^{\dagger} \mathbf{T}_N$ defined as in Theorem 8.3.1 with $\mathbf{T}_N$ composed of K distinct eigenvalues $t_1, \ldots, t_K$ of multiplicities $N_1, \ldots, N_K$, respectively. We further denote $c_k \triangleq \lim_n N_k/n$ and will discuss the question of estimating t_k itself. What follows summarizes the original ideas of Mestre in [22] and [24]. We have from (8.15) that, for any continuous f and for any *negatively oriented* contour $\mathcal{C}_k$ that encloses t_k and t_k only, $f(t_k)$ can be written under the form

$$\frac{N_k}{N} f(t_k) = \frac{1}{2\pi i} \oint_{\mathcal{C}_k} f(\omega) m_H(\omega) d\omega$$

$$= \frac{1}{2\pi i} \oint_{\mathcal{C}_k} \frac{1}{N} \sum_{r=1}^{K} N_r \frac{f(\omega)}{t_r - \omega} d\omega$$

with H the limit $F^{\mathbf{T}_N} \Rightarrow H$. This provides a link between $f(t_k)$ for all continuous f and the Stieltjes transform $m_H(z)$.

Letting $f(x) = x$ and taking the limit $N \to \infty$, $N_k/N \to c_k/c$, with $c \triangleq c_1 + \ldots + c_K$ the limit of N/n, we have

$$\frac{c_k}{c} t_k = \frac{1}{2\pi i} \oint_{\mathcal{C}_k} \omega m_H(\omega) d\omega. \tag{8.17}$$

We now want to express m_H as a function of m_F, the Stieltjes transform of the l.s.d. F of $\mathbf{B}_N$. For this, we have the two relations (8.10), i.e.,

$$m_{\underline{F}}(z) = c m_F(z) + (c-1)\frac{1}{z}$$

and (8.16) with $F^T = H$, i.e.,

$$\frac{c}{m_{\underline{F}}(z)} m_H \left(-\frac{1}{m_{\underline{F}}(z)} \right) = -z m_{\underline{F}}(z) + (c-1).$$

Together, those two equations give the simpler expression

$$m_H \left(-\frac{1}{m_{\underline{F}}(z)} \right) = -z m_{\underline{F}}(z) m_F(z).$$

Applying the variable change $\omega = -1/m_{\underline{F}}(z)$ in (8.17), we obtain

$$\frac{c_k}{c} t_k = \frac{1}{2\pi i} \oint_{\mathcal{C}_{\underline{F},k}} z \frac{m_{\underline{F}}(z) m_{\underline{F}}'(z)}{c} + \frac{1-c}{c} \frac{m_{\underline{F}}(z)'}{m_{\underline{F}}^2(z)} dz$$

$$= \frac{1}{c} \frac{1}{2\pi i} \oint_{\mathcal{C}_{\underline{F},k}} z \frac{m_{\underline{F}}'(z)}{m_{\underline{F}}(z)} dz, \tag{8.18}$$

where $\mathcal{C}_{\underline{F},k}$ is the preimage of $\mathcal{C}_k$ by $-1/m_{\underline{F}}$. The second equality (8.18) comes from the fact that the second term in the previous relation is the derivative of

$(c-1)/(cm_{\underline{F}}(z))$, which therefore integrates to 0 o
real or complex integration rules [30]. Obviously,
$-1/m_{\underline{F}}(z) \in \mathbb{C}^+$ (the same being true if $\mathbb{C}^+$ is rep
continuous and of non-zero imaginary part wheneve
careful about the exact choice of $\mathcal{C}_{\underline{F},k}$.

We make the important assumption that the inc
conditions of Theorem 8.3.7. That is, the cluster k_F
from $(k-1)_F$ and $(k+1)_F$ (whenever they exist).
two real values such that

$$x^+_{(k-1)_F} < x^{(l)}_F < x^-_{k_F} < x^+_{k_F} < x^{(r}_F$$

with $\{x_1^-, x_1^+, \ldots, x_{K_F}^-, x_{K_F}^+\}$ the support bounda
rem 8.3.6. Now remember Theorem 8.3.5 and Figu
viously, $m_{\underline{F}}(z)$ has a limit $m^{(l)} \in \mathbb{R}$ as $z \to x^{(l)}_F$, $z \in$
$z \to x^{(r)}_F$, $z \in \mathbb{C}^+$, those two limits verifying

$$t_{k-1} < x^{(l)} < t_k < x^{(r)} < t$$

with $x^{(l)} \triangleq -1/m^{(l)}$ and $x^{(r)} \triangleq -1/m^{(r)}$.

This is the most important outcome of the integ
$\mathcal{C}_{\underline{F},k}$ to be *any* continuous contour surrounding clus
the real axis in only two points, namely $x^{(l)}_F$ and x
and $-1/m_{\underline{F}}(\mathbb{C}^-) \subset \mathbb{C}^-$, $\mathcal{C}_k$ does not cross the real
is obviously continuously differentiable there; now $\mathcal{C}$
and $x^{(r)}$, and is in fact continuous there. Because of
is (at least) continuous and piecewise continuously d
t_k. This is what is required to ensure the validity of

The difficult part of the proof is completed. The r
We start by considering the following expression,

$$\hat{t}_k \triangleq \frac{1}{2\pi i} \frac{n}{N_k} \oint_{\mathcal{C}_{\underline{F},k}} z \frac{m'_{F\underline{\mathbf{B}}_N}(z)}{m_{F\underline{\mathbf{B}}_N}(z)} d$$

$$= \frac{1}{2\pi i} \frac{n}{N_k} \oint_{\mathcal{C}_{\underline{F},k}} z \frac{\frac{1}{n}\sum_{i=1}^n \overline{(\lambda_i}}{\frac{1}{n}\sum_{i=1}^n \overline{\lambda}}$$

where we remind that $\underline{\mathbf{B}}_N \triangleq \mathbf{X}_N^\dagger \mathbf{T}_N \mathbf{X}_N$ and where,
$\ldots = \lambda_n = 0$.

The value $\hat{t}_k$ can be viewed as the empirical cou
from Theorem 8.2.9 that $m_{F\underline{\mathbf{B}}_N}(z) \xrightarrow{\text{a.s.}} m_{\underline{F}}(z)$ and
difficult to verify, from the fact that $m_{\underline{F}}$ is holomorph
holds for the successive derivatives.

At this point, we need the two fundamental results that are Theorem 8.3.1 and Theorem 8.3.4. We know that, for all matrices $\mathbf{B}_N$ in a set of probability one, all the eigenvalues of $\mathbf{B}_N$ are contained in the support of F for all large N, and that the eigenvalues of $\mathbf{B}_N$ contained in cluster k_F are exactly $\{\lambda_i, i \in \mathcal{N}_k\}$ for these large N, with $\mathcal{N}_k = \{\sum_{j=1}^{k-1} N_j + 1, \ldots, \sum_{j=1}^{k} N_j\}$. Take such a $\mathbf{B}_N$. For all large N, $m_{\mathbf{B}_N}(z)$ is uniformly bounded over N and $z \in \mathcal{C}_{F,k}$, since $\mathcal{C}_{F,k}$ is away from the support of F. The integrand on the right-hand side of (8.20) is then uniformly bounded for all large N and for all $z \in \mathcal{C}_{F,k}$. By the dominated convergence theorem, Theorem 16.4 in [16], we then have that $\hat{t}_k - t_k \xrightarrow{\text{a.s.}} 0$.

It then remains to evaluate $\hat{t}_k$ explicitly. This is performed by residue calculus [30], i.e., by determining the poles in the expanded expression of $\hat{t}_k$ (when developing $m_{F^{\mathbf{B}_N}}(z)$ in its full expression). Those poles are found to be $\lambda_1, \ldots, \lambda_N$ (indeed, the integrand of (8.20) behaves like $O(1/(\lambda_i - z))$ for $z \simeq \lambda_i$) and $\mu_1, \ldots, \mu_N$, the N real roots of the equation in μ, $m_{F^{\mathbf{B}_N}}(\mu) = 0$ (indeed, the denominator of the integrand cancels for $z = \mu_i$ while the numerator is non-zero). Since $\mathcal{C}_{F,k}$ encloses only those values λ_i such that $i \in \mathcal{N}_k$, the other poles are discarded. Noticing now that $m_{F^{\mathbf{B}_N}}(\mu) \to \pm\infty$ as $\mu \to \lambda_i$, we deduce that $\mu_1 < \lambda_1 < \mu_2 < \ldots < \mu_N < \lambda_N$, and therefore we have that $\mu_i, i \in \mathcal{N}_k$ are all in $\mathcal{C}_{F,k}$ but maybe for $\mu_j, j = \min \mathcal{N}_k$. It can in fact be shown that μ_j is also in $\mathcal{C}_{F,k}$. To notice this last remaining fact, observe simply that

$$\frac{1}{2\pi i} \oint_{\mathcal{C}_k} \frac{1}{\omega} d\omega = 0.$$

since 0 is not contained in the contour $\mathcal{C}_k$. Applying the variable change $\omega = -1/m_{\underline{F}}(z)$ as previously, this gives

$$\oint_{\mathcal{C}_{F,k}} \frac{m'_{\underline{F}}(z)}{m_{\underline{F}}^2(z)} dz = 0. \tag{8.21}$$

From the same reasoning as above, with the dominated convergence theorem argument, we have that for sufficiently large N and almost surely,

$$\left| \oint_{\mathcal{C}_{F,k}} \frac{m'_{F^{\mathbf{B}_N}}(z)}{m_{F^{\mathbf{B}_N}}^2(z)} dz \right| < \frac{1}{2}. \tag{8.22}$$

At this point, we need to proceed to residue calculus in order to compute the integral on the left-hand side of (8.22). We will in fact prove that the value of this integral is an integer, hence necessarily equal to zero from the inequality (8.22). Notice indeed that the poles of (8.21) are the λ_i and the μ_i that lie inside the integration contour $\mathcal{C}_{F,k}$, all of order one with residues equal to -1 and 1, respectively. Therefore, (8.21) equals the number of such λ_i minus the number of such μ_i (remember that the integration contour is negatively oriented, so we need to reverse the signs). We, however, already know that this difference, for large N, equals either 0 or 1, since only the position of the leftmost μ_i is unknown yet. But since the integral is asymptotically less than $1/2$, this implies that it is

identically zero, and therefore the leftmost μ_i (inde$\;$
the integration contour.

From this point on, we can evaluate (8.20), whi$\;$
we know exactly which eigenvalues of $\mathbf{B}_N$ are conta$\;$
all large N) within the integration contour. This c$\;$
the steps of which are detailed below. Denoting

$$f(z) = z \frac{m'_{F\mathbf{B}_N}(z)}{m_{F\mathbf{B}_N}(z)},$$

we find that λ_i (inside $\mathcal{C}_{\underline{F},k}$) is a pole of order 1 wit$\;$

$$\lim_{z \to \lambda_i} (z - \lambda_i) f(z) = -\lambda$$

which is straightforwardly obtained from the fact tha$\;$
μ_i (inside $\mathcal{C}_{\underline{F},k}$) is a pole of order 1 with residue

$$\lim_{z \to \mu_i} (z - \mu_i) f(z) = \mu_i$$

Since the integration contour is chosen to be n
kept in mind that the signs of the residues need be $\;$

Noticing finally that $\mu_1, \ldots, \mu_N$ are also the eigen$\;$
with $\boldsymbol{\lambda} \triangleq (\lambda_1, \ldots, \lambda_N)^T$, from a lemma provided i$\;$
finally have the following statistical inference result f$\;$

Theorem 8.4.1 ([24]). *Let* $\mathbf{B}_N = \mathbf{T}_N^{\frac{1}{2}} \mathbf{X}_N \mathbf{X}_N^{\dagger} \mathbf{T}_N^{\frac{1}{2}}$ $\;$
orem 8.3.7, i.e., $\mathbf{T}_N$ *has* K *distinct eigenvalues* t_1
$N_1, \ldots, N_K$, *respectively, for all* r, $N_r/n \to c_r$, $0 <$
conditions (8.13) *are satisfied. Further denote* $\lambda_1 \leq$
$\mathbf{B}_N$ *and* $\boldsymbol{\lambda} = (\lambda_1, \ldots, \lambda_N)^T$. *Let* $k \in \{1, \ldots, K\}$, *an$\;$*

$$\hat{t}_k = \frac{n}{N_k} \sum_{m \in \mathcal{N}_k} (\lambda_m - \mu_m$$

with $\mathcal{N}_k = \left\{ \sum_{j=1}^{k-1} N_j + 1, \ldots, \sum_{j=1}^{k} N_j \right\}$ *and* μ_1
eigenvalues of the matrix $\operatorname{diag}(\boldsymbol{\lambda}) - \frac{1}{n} \sqrt{\boldsymbol{\lambda}} \sqrt{\boldsymbol{\lambda}}^T$.
Then, if condition (8.13) *is fulfilled, we have*

$$\hat{t}_k - t_k \to 0$$

almost surely as $N, n \to \infty$, $N/n \to c$, $0 < c < \infty$.

Similarly, for the quadratic form, the following h$\;$

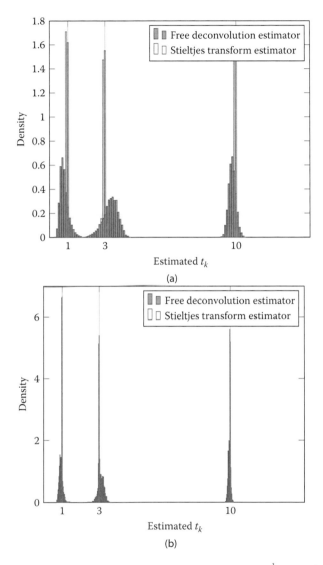

Figure 8.4: Estimation of t_1, t_2, t_3 in the model $\mathbf{B}_N = \mathbf{T}_N^{\frac{1}{2}} \mathbf{X}_N \mathbf{X}_N^\dagger \mathbf{T}_N^{\frac{1}{2}}$ based on the first three empirical moments of $\mathbf{B}_N$ and Newton-Girard inversion, see [32], for $N_1/N = N_2/N = N_3/N = 1/3$, $N/n = 1/10$, for 100,000 simulation runs; (a) $N = 30$, $n = 90$; (b) $N = 90$, $n = 270$. Comparison is made against the Stieltjes transform estimator of Theorem 8.4.1.

Theorem 8.4.2 ([24]). *Let* $\mathbf{B}_N$ *be defined as in Th*
$\sum_{k=1}^{N} \lambda_k \mathbf{b}_k \mathbf{b}_k^\dagger$, $\mathbf{b}_k^\dagger \mathbf{b}_i = \delta_k^i$, *the spectral decomposit*
$\mathbf{T}_N = \sum_{k=1}^{K} t_k \mathbf{U}_k \mathbf{U}_k^\dagger$, $\mathbf{U}_k^\dagger \mathbf{U}_k = \mathbf{I}_{n_k}$, *with* $\mathbf{U}_k \in \mathbb{C}^N$
to t_k. *For given vectors* $\mathbf{x}, \mathbf{y} \in \mathbb{C}^N$, *denote*

$$u(k; \mathbf{x}, \mathbf{y}) \triangleq \mathbf{x}^\dagger \mathbf{U}_k \mathbf{U}_k^\dagger \mathbf{y}$$

Then we have

$$\hat{u}(k; \mathbf{x}, \mathbf{y}) - u(k; \mathbf{x}, \mathbf{y}) \xrightarrow{\text{a.s}}$$

as $N, n \to \infty$ *with ratio* $c_N = N/n \to c$, *where*

$$\hat{u}(k; \mathbf{x}, \mathbf{y}) \triangleq \sum_{i=1}^{N} \theta_k(i) \mathbf{x}^\dagger \mathbf{b}_k$$

and $\theta_k(i)$ *is defined by*

$$\theta_i(k) = \left\{ \begin{array}{ll} -\phi_k(i) & , \ i \notin \\ 1 + \psi_k(i) & , \ i \in \end{array} \right.$$

with

$$\phi_k(i) = \sum_{r \in \mathcal{N}_k} \left(\frac{\lambda_r}{\lambda_i - \lambda_r} - \frac{}{\lambda_i -} \right.$$

$$\psi_k(i) = \sum_{r \notin \mathcal{N}_k} \left(\frac{\lambda_r}{\lambda_i - \lambda_r} - \frac{}{\lambda_i -} \right.$$

and $\mathcal{N}_k$, $\mu_1, \ldots, \mu_N$ *defined as in Theorem 8.4.1.*

The estimator proposed in Theorem 8.4.1 is ext
much more flexible and precise than free deconvoluti
parison is proposed in Figure 8.4 for the same scena
free deconvolution (also called moment-based) met
techniques proposed in e.g., [26, 32]. Nonetheless,
cluster separability condition, necessary to the valic
approach, is mandatory and sometimes a rather stro
number of observations must be rather large compa
in order to be able to resolve close values of t_k.

8.5   Applications

In this section, we apply the random matrix met
problems of multidimensional binary hypothesis test
More details on these applications as well as a more
notably in the field of wireless communications, are

8.5.1 Binary Hypothesis Testing

We first consider the problem of detecting the presence of a signal source impaired by white Gaussian noise. The question is therefore to decide whether only noise is being sensed or if some data plus noise are sensed.

Precisely, we consider a signal source or transmitter of dimension K and a sink or receiver composed of N sensors. The linear filter between the transmitter and the receiver is modelled by the matrix $\mathbf{H} \in \mathbb{C}^{N \times K}$, with $(i, j)^{th}$ entries h_{ij}. If at time l the transmitter emits data, those are denoted by the K-dimensional vector $\mathbf{x}^{(l)} = \left(x_1^{(l)}, \ldots, x_K^{(l)} \right)^T \in \mathbb{C}^K$. The additive white Gaussian noise at the receiver is modelled, at time l, by the vector $\sigma \mathbf{w}^{(l)} = \sigma \left(w_1^{(l)}, \ldots, w_N^{(l)} \right)^T \in \mathbb{C}^N$, where σ^2 denotes the variance of the noise vector entries. Without generality restriction, we consider in the following zero mean and unit variance of the entries of both $\mathbf{w}^{(l)}$ and $\mathbf{x}^{(l)}$, i.e., $E \left\{ \left| w_i^{(l)} \right|^2 \right\} = 1$, $E \left\{ \left| x_i^{(l)} \right|^2 \right\} = 1$ for all i. We then denote $\mathbf{y}^{(l)} = \left(y_1^{(l)}, \ldots, y_N^{(l)} \right)^T$ the N-dimensional data received at time l. Assuming the filter is static during at least M sampling periods, we finally denote $\mathbf{Y} = \left[\mathbf{y}^{(1)}, \ldots, \mathbf{y}^{(M)} \right] \in \mathbb{C}^{N \times M}$ the matrix of the concatenated received vectors.

Depending on whether the transmitter emits data, we consider the following hypotheses

- $\mathcal{H}_0$. Only background noise is received.

- $\mathcal{H}_1$. Data plus background noise are received.

Therefore, under condition $\mathcal{H}_0$, we have the model

$$\mathbf{Y} = \sigma \mathbf{W}$$

with $\mathbf{W} = \left[\mathbf{w}^{(1)}, \ldots, \mathbf{w}^{(M)} \right] \in \mathbb{C}^{N \times M}$ and under condition $\mathcal{H}_1$

$$\mathbf{Y} = \left(\mathbf{H} \quad \sigma \mathbf{I}_N \right) \begin{pmatrix} \mathbf{X} \\ \mathbf{W} \end{pmatrix} \tag{8.24}$$

with $\mathbf{X} = \left[\mathbf{x}^{(1)}, \ldots, \mathbf{x}^{(M)} \right] \in \mathbb{C}^{N \times M}$.

Under this hypothesis, we further denote $\boldsymbol{\Sigma}$ the covariance matrix of $\mathbf{y}^{(1)}$,

$$\boldsymbol{\Sigma} = E \left\{ \mathbf{y}^{(1)} \left(\mathbf{y}^{(1)} \right)^\dagger \right\} = \mathbf{H}\mathbf{H}^\dagger + \sigma^2 \mathbf{I}_N = \mathbf{U}\mathbf{G}\mathbf{U}^\dagger$$

where $\mathbf{G} = \mathrm{diag} \left(\nu_1 + \sigma^2, \ldots, \nu_N + \sigma^2 \right) \in \mathbb{R}^{N \times N}$, with $\{\nu_1, \ldots, \nu_N\}$ the eigenvalues of $\mathbf{H}\mathbf{H}^\dagger$ and $\mathbf{U} \in \mathbb{C}^{N \times N}$ a certain unitary matrix.

The receiver is entitled to decide whether data were transmitted or not. It is a common assumption to be in the scenario where σ^2 is known in advance, although it is uncommon to know the transfer matrix $\mathbf{H}$. This is true in particular of

the wireless signal sensing scenario where $\mathbf{H}$ is the
between two antenna arrays. We consider specificall
ability distribution of $\mathbf{H}$ is unitarily invariant, whic
wireless communications with channel models that
directions of energy propagation. This is in particu
presents rotational invariance properties. For simpli
i.i.d. Gaussian with zero mean and $E\left\{|h_{ij}|^2\right\} = 1$,
go well beyond the Gaussian case.

For simplicity, we consider in the following $K = 1$
exists for $K \geq 1$ [2]. The Neyman-Pearson criterio
whether data were transmitted is based on the ratic

$$C\left(\mathbf{Y}\right) = \frac{\mathrm{Pr}_{\mathcal{H}_1|\mathbf{Y}}\left(\mathbf{Y}\right)}{\mathrm{Pr}_{\mathcal{H}_0|\mathbf{Y}}\left(\mathbf{Y}\right)},$$

where $\mathrm{Pr}_{\mathcal{H}_i|\mathbf{Y}}\left(\mathbf{Y}\right)$ is the probability of the event $\mathcal{H}_i$
$\mathbf{Y}$. For a given receive space-time matrix $\mathbf{Y}$, if $C(\mathbf{Y}$
an informative signal was transmitted, while if $C(\mathbf{Y})$
background noise was captured. To ensure a low pro
positive), i.e., the probability to declare a pure noise
signal, a certain threshold ξ is generally set such that
declares data were transmitted, while when $C(\mathbf{Y})$
no data were sent. The question of what ratio ξ sh
maximally acceptable false alarm rate will not be t
provide an explicit expression of (8.25) for the afo
compare its performance to that achieved by the c
results provided in this section are taken from [2].

Applying Bayes' rule, (8.25) becomes

$$C\left(\mathbf{Y}\right) = \frac{\mathrm{Pr}_{\mathcal{H}_0} \cdot \mathrm{Pr}_{\mathcal{H}_1|\mathbf{Y}}\left(\mathbf{Y}\right.}{\mathrm{Pr}_{\mathcal{H}_1} \cdot \mathrm{Pr}_{\mathcal{H}_0|\mathbf{Y}}\left(\mathbf{Y}\right.}$$

with $\mathrm{Pr}_{\mathcal{H}_i}$ the a priori probability for hypothesis $\mathcal{H}_i$
side information allows the receiver to consider tha
than $\mathcal{H}_0$, and therefore set $\mathrm{Pr}_{\mathcal{H}_0} = \mathrm{Pr}_{\mathcal{H}_1} = \frac{1}{2}$, so tha

$$C\left(\mathbf{Y}\right) = \frac{\mathrm{Pr}_{\mathcal{H}_1|\mathbf{Y}}\left(\mathbf{Y}\right)}{\mathrm{Pr}_{\mathcal{H}_0|\mathbf{Y}}\left(\mathbf{Y}\right)},$$

reduces to a maximum likelihood ratio.

Likelihood under $\mathcal{H}_0$. In this first scenario, the r
and independent. The probability density of $\mathbf{Y}$, whic
tor with NM entries, is then an NM multivariate u

with covariance matrix $\sigma^2 \mathbf{I}_{NM}$,

$$\Pr_{\mathbf{Y}|\mathcal{H}_0} (\mathbf{Y}) = \frac{1}{(\pi\sigma^2)^{NM}} e^{-\frac{1}{\sigma^2} \operatorname{tr}(\mathbf{Y}\mathbf{Y}^\dagger)}. \tag{8.27}$$

Denoting $\boldsymbol{\lambda} = (\lambda_1, \dots, \lambda_N)^T$ the eigenvalues of $\mathbf{Y}\mathbf{Y}^\dagger$, (8.27) only depends on $\sum_{i=1}^N \lambda_i$, as follows

$$\Pr_{\mathbf{Y}|\mathcal{H}_0} (\mathbf{Y}) = \frac{1}{(\pi\sigma^2)^{NM}} e^{-\frac{1}{\sigma^2} \sum_{i=1}^N \lambda_i}.$$

Likelihood under $\mathcal{H}_1$.   Under the data plus noise hypothesis $\mathcal{H}_1$, the problem is more involved. The entries of the channel matrix $\mathbf{H}$ are modeled as jointly uncorrelated Gaussian, with $E\left\{|h_{ij}|^2\right\} = 1/K$. Therefore, since here $K = 1$, $\mathbf{H} \in \mathbb{C}^{N \times 1}$ and $\boldsymbol{\Sigma} = \mathbf{H}\mathbf{H}^\dagger + \sigma^2 \mathbf{I}_N$ has $N - 1$ eigenvalues $g_2 = \cdots = g_N$ equal to σ^2 and another distinct eigenvalue $g_1 = \nu_1 + \sigma^2 = (\sum_{i=1}^N |h_{i1}|^2) + \sigma^2$. The density of $g_1 - \sigma^2$ is a complex chi-square distribution of N degrees of freedom (denoted χ_N^2), which up to a scaling factor 2 is equivalent to a real χ_{2N}^2 distribution. Hence, the eigenvalue distribution of $\boldsymbol{\Sigma}$, defined on $\mathbb{R}^{+N} = [0, \infty)^N$, reads

$$\Pr_{\mathbf{G}} (\mathbf{G}) = \frac{1}{N} \left((g_1 - \sigma^2)_+^{N-1} \right) \frac{e^{-(g_1 - \sigma^2)}}{(N-1)!} \prod_{i=2}^N \delta(g_i - \sigma^2).$$

From the model $\mathcal{H}_1$, $\mathbf{Y}$ is distributed as correlated Gaussian, as follows

$$\Pr_{\mathbf{Y}|\boldsymbol{\Sigma}, I_1} (\mathbf{Y}, \boldsymbol{\Sigma}) = \frac{1}{\pi^{MN} \det{(\mathbf{G})}^M} e^{-\operatorname{tr}(\mathbf{Y}\mathbf{Y}^\dagger \mathbf{U}\mathbf{G}^{-1}\mathbf{U}^\dagger)},$$

where I_k denotes the prior information at the receiver "$\mathcal{H}_1$ and $K = k$."

Since $\mathbf{H}$ is unknown, we need to integrate out all possible linear filters for the transmission model under $\mathcal{H}_1$ over the probability space of $N \times K$ matrices with Gaussian i.i.d. distribution. From the invariance of Gaussian i.i.d. random matrices by left and right products with unitary matrices, this is equivalent to integrating out all possible covariance matrices $\boldsymbol{\Sigma}$ over the space of such non-negative definite Hermitian matrices, as follows

$$\Pr_{\mathbf{Y}|\mathcal{H}_1} (\mathbf{Y}) = \int_{\boldsymbol{\Sigma}} \Pr_{\mathbf{Y}|\boldsymbol{\Sigma}, \mathcal{H}_1} (\mathbf{Y}, \boldsymbol{\Sigma}) \Pr_{\boldsymbol{\Sigma}} (\boldsymbol{\Sigma}) \, d\boldsymbol{\Sigma}.$$

Eventually, after complete integration calculus given in the proof below, the Neyman-Pearson decision ratio (8.25) for the single-input multiple-output channel takes an explicit expression, given by the following theorem.

Theorem 8.5.1. *The Neyman-Pearson test ratio C*
reads

$$C_{\mathbf{Y}}(\mathbf{Y}) = \frac{1}{N} \sum_{l=1}^{N} \frac{\sigma^{2(N+M-1)} e^{\sigma^2 + \frac{\lambda_l}{\sigma^2}}}{\prod_{\substack{i=1 \\ i \neq l}}^{N} (\lambda_l - \lambda_i)} J$$

with $\lambda_1, \ldots, \lambda_N$ *the eigenvalues of* $\mathbf{YY}^\dagger$ *and where*

$$J_k(x, y) \triangleq \int_x^{+\infty} t^k e^{-t - \frac{y}{t}}$$

The proof of Theorem 8.5.1 is provided below. An
(8.28), note that the Neyman-Pearson test does only
$\mathbf{YY}^\dagger$. This suggests that the eigenvectors of $\mathbf{YY}^\dagger$ d
regarding the presence of data. The essential reaso
$\mathcal{H}_1$, the eigenvectors of $\mathbf{Y}$ are isotropically distribut
complex sphere due to the Gaussian assumptions
realization of the eigenvectors of $\mathbf{Y}$ does indeed not
to the hypothesis test. The Gaussian assumption fo:
entropy principle is in fact essential here. Note howe
to a function of the sum $\sum_i \lambda_i$ of the eigenvalues, a:
detector.

On the practical side, note that the integral J_k
form expression, but for $x = 0$, [33, see e.g., pp. 56]
for practical purposes, since $J_k(x, y)$ must either b
tabulated. It is also difficult to get any insight o
detector for different values of σ^2, N and K. We
proof of Theorem 8.5.1, in which classical multidime
are introduced. In particular, the tools introduced i
be key ingredients of the derivation.

Proof. We start by noticing that $\mathbf{H}$ is Gaussian and t
of its entries is invariant by left and right unitary pre
distribution of the matrix $\mathbf{\Sigma} = \mathbf{HH}^\dagger + \sigma^2 \mathbf{I}$ is unitari
write

$$\Pr_{\mathbf{Y}|I_1}(\mathbf{Y}) = \int_{\mathbf{\Sigma}} \Pr_{\mathbf{Y}|\mathbf{\Sigma}, \mathcal{H}_1}(\mathbf{Y}, \mathbf{\Sigma}) \Pr_{\mathbf{\Sigma}}(\mathbf{\Sigma}) d$$

$$= \int_{\mathcal{U}(N) \times (\mathbb{R}^+)^N} \Pr_{\mathbf{Y}|\mathbf{\Sigma}, \mathcal{H}_1}(\mathbf{Y}, \mathbf{\Sigma})$$

$$= \int_{\mathcal{U}(N) \times \mathbb{R}^+} \Pr_{\mathbf{Y}|\mathbf{\Sigma}, \mathcal{H}_1}(\mathbf{Y}, \mathbf{\Sigma}) \Pr_{g_1}$$

with $\mathcal{U}(N)$ the space of $N \times N$ unitary matrices and $\boldsymbol{\Sigma} = \mathbf{U}\mathbf{G}\mathbf{U}^\dagger$.

The latter can further be equated to

$$\Pr_{\mathbf{Y}|I_1} (\mathbf{Y}) = \int\limits_{\mathcal{U}(N)\times\mathbb{R}^+} \frac{e^{-\operatorname{tr}(\mathbf{Y}\mathbf{Y}^\dagger \mathbf{U}\mathbf{G}^{-1}\mathbf{U}^\dagger)}}{\pi^{NM}\det(\mathbf{G})^M} \left((g_1 - \sigma^2)_+^{N-1}\right) \frac{e^{-(g_1-\sigma^2)}}{N!} d\mathbf{U}d(g_1)$$

with $(x)_+ \triangleq \max(x,0)$ here.

To go further, we use the Harish–Chandra identity provided in Theorem 8.2.3. Denoting $\Delta(\mathbf{Z})$ the Vandermonde determinant of matrix $\mathbf{Z} \in \mathbb{C}^{N\times N}$ with eigenvalues $z_1 \leq \ldots \leq z_N$

$$\Delta(\mathbf{Z}) \triangleq \prod_{i>j} (z_i - z_j), \tag{8.29}$$

the likelihood $\Pr_{\mathbf{Y}|I_1} (\mathbf{Y})$ reads

$$\Pr_{\mathbf{Y}|I_1} (\mathbf{Y}) = \left(\lim_{g_2,\ldots,g_N \to \sigma^2} \frac{e^{\sigma^2}(-1)^{\frac{N(N-1)}{2}} \prod_{j=1}^{N-1} j!}{\pi^{MN}\sigma^{2M(N-1)}N!} \right) \int\limits_{\sigma^2}^{+\infty} \frac{1}{g_1^M} (g_1 - \sigma^2)^{N-1}$$

$$e^{-g_1} \frac{\det\left(e^{-\frac{\lambda_i}{g_j}}\right)}{\Delta\left(\mathbf{Y}\mathbf{Y}^\dagger\right)\Delta\left(\mathbf{G}^{-1}\right)} d(g_1) \tag{8.30}$$

in which we remind that $\lambda_1,\ldots,\lambda_N$ are the eigenvalues of $\mathbf{Y}\mathbf{Y}^\dagger$. Note the trick of replacing the known values of $g_2,\ldots,g_N$ by limits of scalars converging to these known values, which allows us to use correctly the Harish–Chandra formula. The remainder of the proof consists of deriving the explicit limits, which in particular relies on the following result [34, Lemma 6].

Theorem 8.5.2. *Let $f_1,\ldots,f_N$ be a family of infinitely differentiable functions and let $x_1,\ldots,x_N \in \mathbb{R}$. Denote*

$$R(x_1,\ldots,x_N) \triangleq \frac{\det\left(\{f_i(x_j)\}_{i,j}\right)}{\prod_{i>j} (x_i - x_j)}.$$

Then, for $p \leq N$ and for $x_0 \in \mathbb{R}$,

$$\lim_{x_1,\ldots,x_p \to x_0} R(x_1,\ldots,x_N) = \frac{\det\left(f_i(x_0), f_i'(x_0),\ldots,f_i^{(p-1)}(x_0), f_i(x_{p+1}),\ldots,f_i(x_N)\right)}{\prod_{p<j<i} (x_i - x_j) \prod_{i=p+1}^{N} (x_i - x_0)^p \prod_{j=1}^{p-1} j!}.$$

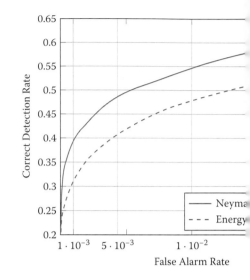

Figure 8.5: ROC curve for single-source detection
SNR $= -3$ dB, FAR range of practical interest.

The integral in the numerator is then extended o
der of the matrix having a Vandermonde determina
integral form of the g_1 parameter. Hence the result.

The scenario where $K \geq 1$ unfolds similarly. Th
in [2]. The receiver operating characteristics (ROC)
test against that of the energy detector is provide
$M = 8$ and $\sigma^2 = 3$ dBm. We observed a significa
of detection rate incurred by the Neyman-Pearson
energy detector.

This completes this section on hypothesis testing
go beyond the hypothesis test and move to the ques
a slightly more complex data plus noise model than

8.5.2   Parameter Estimation

We consider a similar scenario as in the previous sect
use different transmit powers $P_1, \ldots, P_K$, which th
from successive observations.

Consider K data sources which are transmitting
mitter $k \in \{1, \ldots, K\}$ has power P_k and has space di
of n_k antennas. We denote $n \triangleq \sum_{k=1}^{K} n_k$ the tota
sions. Consider also a sink or receiver with space d

$\mathbf{H}_k \in \mathbb{C}^{N \times n_k}$ the multidimensional filter matrix between transmitter k and the receiver. We assume that the entries of $\sqrt{N}\mathbf{H}_k$ are independent and identically distributed with zero mean, unit variance and finite fourth order moment. At time instant m, transmitter k emits the signal $\mathbf{x}_k^{(m)} \in \mathbb{C}^{n_k}$, with entries assumed to be independent, independent along m, k, identically distributed along m, and have all zero mean, unit variance and finite fourth order moment (the $\mathbf{x}_k^{(m)}$ need not be identically distributed along k). Assume further that at time instant m the receive signal is impaired by additive white Gaussian noise with entries of zero mean and variance σ^2, denoted $\sigma\mathbf{w}^{(m)} \in \mathbb{C}^N$. At time m, the receiver therefore senses the signal $\mathbf{y}^{(m)} \in \mathbb{C}^N$ defined as

$$\mathbf{y}^{(m)} = \sum_{k=1}^{K} \sqrt{P_k}\mathbf{H}_k\mathbf{x}_k^{(m)} + \sigma\mathbf{w}^{(m)}.$$

Assuming the filter coefficients are constant over at least M consecutive sampling periods, by concatenating M successive signal realizations into $\mathbf{Y} = \left[\mathbf{y}^{(1)}, \ldots, \mathbf{y}^{(M)}\right] \in \mathbb{C}^{N \times M}$, we have

$$\mathbf{Y} = \sum_{k=1}^{K} \sqrt{P_k}\mathbf{H}_k\mathbf{X}_k + \sigma\mathbf{W},$$

where $\mathbf{X}_k = [\mathbf{x}_k^{(1)}, \ldots, \mathbf{x}_k^{(M)}] \in \mathbb{C}^{n_k \times M}$, for every k, and $\mathbf{W} = \left[\mathbf{w}^{(1)}, \ldots, \mathbf{w}^{(M)}\right] \in \mathbb{C}^{N \times M}$. This can be further rewritten as

$$\mathbf{Y} = \mathbf{HP}^{\frac{1}{2}}\mathbf{X} + \sigma\mathbf{W}, \tag{8.31}$$

where $\mathbf{P} \in \mathbb{R}^{n \times n}$ is diagonal with first n_1 entries P_1, subsequent n_2 entries P_2, etc. and last n_K entries P_K, $\mathbf{H} = [\mathbf{H}_1, \ldots, \mathbf{H}_K] \in \mathbb{C}^{N \times n}$ and $\mathbf{X} = \left[\mathbf{X}_1^T, \ldots, \mathbf{X}_K^T\right]^T \in \mathbb{C}^{n \times M}$. By convention, we assume $P_1 \leq \ldots \leq P_K$.

Our objective is to infer the values of the powers $P_1, \ldots, P_K$ from the realization of a single random matrix $\mathbf{Y}$. This is successively performed from different approaches in the following sections. We first consider the conventional approach that assumes n small, N much larger than n, and M much larger than N. This will lead to a simple although largely biased estimation algorithm. This algorithm will be improved using Stieltjes transform approaches in the same spirit as in Section 8.4.

Conventional Approach

The first approach assumes numerous sensors in order to have much diversity in the observation vectors, as well as an even larger number of observations, in order to create an averaging effect on the incoming random data. In this situation, let us rewrite (8.31) under the form

$$\mathbf{Y} = \left(\mathbf{HP}^{\frac{1}{2}} \quad \sigma \mathbf{I}_N \right) \binom{\mathbf{X}}{\mathbf{W}}$$

We shall denote $\lambda_1 \leq \ldots \leq \lambda_N$ the ordered eigen
eigenvalues of which are almost surely different).

Appending $\mathbf{Y} \in \mathbb{C}^{N \times M}$ into the larger matrix $\underline{\mathbf{Y}}$

$$\underline{\mathbf{Y}} = \begin{pmatrix} \mathbf{HP}^{\frac{1}{2}} & \sigma \mathbf{I}_N \\ 0 & 0 \end{pmatrix} \binom{\mathbf{X}}{\mathbf{W}}$$

we recognize that, conditioned on $\mathbf{H}$, $\frac{1}{M} \underline{\mathbf{Y}} \underline{\mathbf{Y}}^\dagger$ is a
which the *population covariance matrix* is

$$\mathbf{T} \triangleq \begin{pmatrix} \mathbf{HPH}^\dagger + \sigma^2 \mathbf{I}_N & (\\ 0 & (\end{pmatrix}$$

and the random matrix

$$\binom{\mathbf{X}}{\mathbf{W}}$$

has independent (non-necessarily identically distrib
and unit variance. The population covariance matr
also form a matrix unitarily equivalent to a sample co
almost sure limit spectral distribution as N grows la
n. Extending Theorem 8.2.9 and Theorem 8.3.3 to c
(once for the population covariance matrix $\mathbf{T}$ and
have that, as $M, N, n \to \infty$ with $M/N \to \infty$ and l
the largest n eigenvalues of $\frac{1}{M} \mathbf{Y} \mathbf{Y}^\dagger$ is asymptoticall
mass $\sigma^2 + P_1$ of weight $\lim n_1/n$, a mass $\sigma^2 + P_2$ o
mass $\sigma^2 + P_K$ of weight $\lim n_K/n$. As for the distr
eigenvalues of $\frac{1}{M} \mathbf{Y} \mathbf{Y}^\dagger$, it converges to a single mass
If σ^2 is a priori known, a rather trivial estimator

$$\frac{1}{n_k} \sum_{i \in \mathcal{N}_k} \left(\lambda_i - \sigma^2 \right),$$

where $\mathcal{N}_k = \left\{ \sum_{j=1}^{k-1} n_j + 1, \ldots, \sum_{j=1}^{k} n_j \right\}$ and we r
the ordered eigenvalues of $\frac{1}{M} \mathbf{Y} \mathbf{Y}^\dagger$.

This means in practice that P_K is asymptotica
averaged value of the n_K largest eigenvalues of $\frac{1}{M} \mathbf{Y}$
by the averaged value of the n_{K-1} eigenvalues befor
that σ^2 is perfectly known at the receiver. If it were
value of the $N - n$ smallest eigenvalues of $\frac{1}{M} \mathbf{Y} \mathbf{Y}^\dagger$ is

This therefore leads to the second estimator $\hat{P}_k^\infty$ for P_k, that will constitute our reference estimator,

$$\hat{P}_k^\infty = \frac{1}{n_k} \sum_{i \in \mathcal{N}_k} \left(\lambda_i - \hat{\sigma}^2 \right),$$

where

$$\hat{\sigma}^2 = \frac{1}{N - n} \sum_{i=1}^{N-n} \lambda_i.$$

Incidentally, although not derived on purpose, the refined (n, N, M)-consistent estimator of Section 8.5.2 will appear not to depend on a prior knowledge of σ^2. Note that the estimation of P_k only relies on n_k contiguous eigenvalues of $\frac{1}{M}\mathbf{Y}\mathbf{Y}^\dagger$, which suggests that the other eigenvalues are asymptotically uncorrelated from these. It will turn out that the improved (n, N, M)-consistent estimator does take into account all eigenvalues for each k, in a certain manner.

The Stieltjes Transform Method

The Stieltjes transform approach relies heavily on the techniques from Mestre, established in [24] and introduced in Section 8.4. We provide hereafter only the main steps of the method. The details can be found in [23].

Limiting spectrum of $\mathbf{B}_N$. In this section, we prove the following result.

Theorem 8.5.3. *Let $\mathbf{B}_N = \frac{1}{M}\mathbf{Y}\mathbf{Y}^\dagger$, with $\mathbf{Y}$ defined as in (8.31). Then, for M, N, n growing large with limit ratios $M/N \to c$, $N/n_k \to c_k$, $0 < c, c_1, \ldots, c_K < \infty$, the empirical spectral distribution $F^{\mathbf{B}_N}$ of $\mathbf{B}_N$ converges almost surely to the distribution function F, whose Stieltjes transform $m_F(z)$ satisfies, for $z \in \mathbb{C}^+$,*

$$m_F(z) = cm_{\underline{F}}(z) + (c - 1)\frac{1}{z}, \tag{8.33}$$

where $m_{\underline{F}}(z)$ is the unique solution with positive imaginary part of the implicit equation in $m_{\underline{F}}$,

$$\frac{1}{m_{\underline{F}}} = -\sigma^2 + \frac{1}{f} - \sum_{k=1}^{K} \frac{1}{c_k} \frac{P_k}{1 + P_k f} \tag{8.34}$$

in which we denoted f the value

$$f = (1 - c)\, m_{\underline{F}} - czm_{\underline{F}}^2.$$

Proof. First remember that the matrix $\mathbf{Y}$ in (8.31) can be extended into the larger sample covariance matrix $\underline{\mathbf{Y}} \in \mathbb{C}^{(N+n) \times M}$

$$\underline{\mathbf{Y}} = \begin{pmatrix} \mathbf{H}\mathbf{P}^{\frac{1}{2}} & \sigma\mathbf{I}_N \\ 0 & 0 \end{pmatrix} \begin{pmatrix} \mathbf{X} \\ \mathbf{W} \end{pmatrix}.$$

From Theorem 8.2.9, since $\mathbf{H}$ has independent ϵ
moments, we have that the e.s.d. of $\mathbf{HPH}^\dagger$ converge
a limit distribution G as $N, n_1, \ldots, n_K \to \infty$ with
the Stieltjes transform $m_G(z)$ of G is the unique sol
part of the equation in m_G,

$$z = -\frac{1}{m_G} + \sum_{k=1}^{K} \frac{1}{c_k} \frac{P_k}{1 + P_k}$$

The almost sure convergence of the e.s.d. of $\mathbf{H}$
convergence of the e.s.d. of the matrix $\left(\frac{\mathbf{HPH}^\dagger + \sigma^2 \mathbf{I}_N}{0} \right.$
$z \in \mathbb{C}^+$ is the Stieltjes transform of the l.s.d. of $\mathbf{HPH}$
adding n zero eigenvalues, we finally have that the e
almost surely to a distribution H whose Stieltjes tra

$$m_H(z) = \frac{c_0}{1 + c_0} m_G \left(z - \sigma^2 \right) -$$

for $z \in \mathbb{C}^+$, where we denoted by c_0 the limit
$\left(c_1^{-1} + \ldots + c_K^{-1} \right)^{-1}$.

As a consequence, the sample covariance matrix
variance matrix which is not deterministic but whe
limit H for increasing dimensions. Since $\mathbf{X}$ and $\mathbf{W}$
order moment, we can again apply Theorem 8.2.9 a
$\underline{\mathbf{B}}_N \triangleq \frac{1}{M} \underline{\mathbf{Y}}^\dagger \underline{\mathbf{Y}}$ converges almost surely to the limit
$m_{\underline{F}}(z)$ is the unique solution in $\mathbb{C}^+$ of the equation

$$z = -\frac{1}{m_{\underline{F}}} + \frac{1}{c} \left(1 + \frac{1}{c_0} \right) \int \frac{t}{1 + t}$$

$$= -\frac{1}{m_{\underline{F}}} + \frac{1 + \frac{1}{c_0}}{c m_{\underline{F}}} \left(1 - \frac{1}{m_{\underline{F}}} m_H \right)$$

for all $z \in \mathbb{C}^+$.

For $z \in \mathbb{C}^+$, $m_{\underline{F}}(z) \in \mathbb{C}^+$. Therefore $-1/m_{\underline{F}}(z)$
(8.36) at $-1/m_{\underline{F}}(z)$. Combining (8.36) and (8.37),

$$z = -\frac{1}{c} \frac{1}{m_{\underline{F}}(z)^2} m_G \left(-\frac{1}{m_{\underline{F}}(z)} - \sigma^2 \right) +$$

where, according to (8.35), $m_G(-1/m_{\underline{F}}(z) - \sigma^2)$ sat

$$\frac{1}{m_{\underline{F}}(z)} = -\sigma^2 + \frac{1}{m_G \left(-\frac{1}{m_{\underline{F}}(z)} - \sigma^2 \right)} - \sum_{k=1}^{K} \frac{1}{c_k} \frac{1}{1}$$

Together with (8.38), this is exactly (8.34), with $f(z) = m_G \left(-\frac{1}{m_{\underline{F}}(z)} - \sigma^2 \right) = (1 - c) \, m_{\underline{F}}(z) - czm_{\underline{F}}(z)^2$.

Since the eigenvalues of the matrices $\mathbf{B}_N$ and $\underline{\mathbf{B}}_N$ only differ by $M - N$ zeros, we also have that the Stieltjes transform $m_F(z)$ of the l.s.d. of $\mathbf{B}_N$ satisfies

$$m_F(z) = cm_{\underline{F}}(z) + (c - 1) \frac{1}{z}. \tag{8.40}$$

This completes the proof of Theorem 8.5.3.   □

For further usage, notice here that (8.40) provides a simplified expression for $m_G \left(-1/m_{\underline{F}}(z) - \sigma^2 \right)$. Indeed we have,

$$m_G \left(-1/m_{\underline{F}}(z) - \sigma^2 \right) = -zm_F(z)m_{\underline{F}}(z). \tag{8.41}$$

Therefore, the support of the (almost sure) l.s.d. F of $\mathbf{B}_N$ can be evaluated as follows: for any $z \in \mathbb{C}^+$, $m_F(z)$ is given by (8.33), in which $m_{\underline{F}}(z)$ is solution of (8.34); the inverse Stieltjes transform formula (8.4) allows then to evaluate F from $m_F(z)$, for values of z spanning over the set $\{z = x + iy, \, x > 0\}$ and y small.

Multisource power inference. In the following, we finally prove the main result of this section, which provides the G-estimator $\hat{P}_1, \ldots, \hat{P}_K$ of the transmit powers $P_1, \ldots, P_K$.

Theorem 8.5.4. *Let* $\mathbf{B}_N \in \mathbb{C}^{N \times N}$ *be defined as* $\mathbf{B}_N = \frac{1}{M} \mathbf{Y} \mathbf{Y}^\dagger$ *with* $\mathbf{Y}$ *defined as in (8.31), and* $\boldsymbol{\lambda} = (\lambda_1, \ldots, \lambda_N)$, $\lambda_1 \leq \ldots \leq \lambda_N$, *be the vector of the ordered eigenvalues of* $\mathbf{B}_N$. *Further assume that the limiting ratios* $c_0, c_1, \ldots, c_K, c$ *and* $\mathbf{P}$ *are such that the cluster mapped to* P_k *in* $\mathbf{B}_N$ *does not map another* P_i, $i \neq k$. *Then, as* N, n, M *grow large, we have*

$$\hat{P}_k - P_k \xrightarrow{\text{a.s.}} 0,$$

where the estimate $\hat{P}_k$ *is given by*

- *if* $M \neq N$,

$$\hat{P}_k = \frac{NM}{n_k (M - N)} \sum_{i \in \mathcal{N}_k} (\eta_i - \mu_i),$$

- *if* $M = N$,

$$\hat{P}_k = \frac{N}{n_k (N - n)} \sum_{i \in \mathcal{N}_k} \left(\sum_{j=1}^{N} \frac{\eta_i}{(\lambda_j - \eta_i)^2} \right)^{-1},$$

in which $\mathcal{N}_k = \left\{ \sum_{i=1}^{k-1} n_i + 1, \ldots, \sum_{i=1}^{k} n_i \right\}$, $\eta_1 \leq \ldots \leq \eta_N$ *are the ordered eigenvalues of the matrix* $\operatorname{diag}(\boldsymbol{\lambda}) - \frac{1}{N} \sqrt{\boldsymbol{\lambda}} \sqrt{\boldsymbol{\lambda}}^\dagger$ *and* $\mu_1 \leq \ldots \leq \mu_N$ *are the ordered eigenvalues of the matrix* $\operatorname{diag}(\boldsymbol{\lambda}) - \frac{1}{M} \sqrt{\boldsymbol{\lambda}} \sqrt{\boldsymbol{\lambda}}^\dagger$.

Remark 8.5.1. We immediately notice that, if N
with l the largest integer such that $N - \sum_{i=l}^{K} n_i <$
clusters may be empty. The case $N \leq n$ turns out t
clusters always merge and no consistent estimate of

Proof. The approach pursued to prove Theorem 8.5
nal idea of [22], which was detailed for the case of
Section 8.4. From Cauchy's integration formula,

$$P_k = c_k \frac{1}{2\pi i} \oint_{\mathcal{C}_k} \frac{1}{c_k} \frac{\omega}{P_k - \omega} d\omega = c_k \frac{1}{2\pi i} \oint_{\mathcal{C}_k}$$

for any negatively oriented contour $\mathcal{C}_k \subset \mathbb{C}$, such
surface described by the contour, while for every $i \neq$
The strategy is very similar to that used for the sam
Section 8.4. It comes as follows: we first propose a c
$\mathcal{C}_k$ which is parametrized by a functional of the Sti
l.s.d. of $\mathbf{B}_N$. We proceed to a variable change in (8.4
of $m_F(z)$. We then evaluate the complex integral
limiting $m_F(z)$ in (8.42) by its empirical counterpart
This new integral, whose value we name $\hat{P}_k$, is show
P_k in the large N limit. It then suffices to evaluate
residue calculus.

Similar to Section 8.4, it turns out that the clust
of $\mathbf{B}_N$ can be mapped to one or many power values I
that the clusters are disjoint so that no holomorph
prove the following (given in detail in [23] and [28
outside the support of F, on either side of cluster
is uniquely mapped to P_k) such that $m_{\underline{F}}(z)$ has l
$m_{\underline{F},k_G}^{(r)} \triangleq m_{\underline{F}}^{\circ}(x_{k_F}^{(r)})$, as $z \to x_{k_F}^{(l)}$ and $z \to x_{k_F}^{(r)}$, respe
extension of $m_{\underline{F}}$ in the points $x_{k_F}^{(l)} \in \mathbb{R}$ and $x_{k_F}^{(r)} \in$
$m_{\underline{F},k_G}^{(r)}$ are on either side of cluster k_G (i.e., the clus
P_k) in the support of $-1/H$, and therefore $-1/m_{\underline{F}}^{(l)}$
are on either side of cluster k_G in the support of G.

Consider any continuously differentiable complex
and $x_{k_F}^{(r)}$, and interior points of positive imaginary
$\mathcal{C}_{F,k}$ as the union of $\Gamma_{F,k}$ oriented from $x_{k_F}^{(l)}$ to $x_k^{(}$
$\Gamma_{F,k}^{*}$ oriented backwards from $x_{k_F}^{(r)}$ to $x_{k_F}^{(l)}$. The cont
and piecewise continuously differentiable. Also, the
completely inside $\mathcal{C}_{F,k}$, while the support of the neig
$\mathcal{C}_{F,k}$. The support of cluster k_G in H is then inside

the support of cluster k_G in G is inside $\mathcal{C}_{G,k} \triangleq -1/m_{\underline{F}}(\mathcal{C}_{F,k}) - \sigma^2$. Since $m_{\underline{F}}$ is continuously differentiable on $\mathbb{C} \setminus \mathbb{R}$ (it is in fact holomorphic [21]) and has limits in $x_{k_F}^{(l)}$ and $x_{k_F}^{(r)}$, $\mathcal{C}_{G,k}$ is also continuous and piecewise continuously differentiable. Going one last step in this process, we finally have that P_k is inside the contour $\mathcal{C}_k \triangleq -1/m_G(\mathcal{C}_{G,k})$, while P_i, for all $i \neq k$, is outside $\mathcal{C}_k$. Since m_G is also holomorphic on $\mathbb{C} \setminus \mathbb{R}$ and has limits in $-1/m_{\underline{F}}^{\circ}(x_{k_F}^{(l)}) - \sigma^2$ and $-1/m_{\underline{F}}^{\circ}(x_{k_F}^{(r)}) - \sigma^2$, $\mathcal{C}_k$ is a continuous and piecewise continuously differentiable complex path, which is sufficient to perform complex integration [30].

Recall now that P_k was defined as

$$P_k = c_k \frac{1}{2\pi i} \oint_{\mathcal{C}_k} \sum_{r=1}^{K} \frac{1}{c_r} \frac{\omega}{P_r - \omega} d\omega.$$

With the variable change $\omega = -1/m_G(t)$, this becomes

$$P_k = \frac{c_k}{2\pi i} \oint_{\mathcal{C}_{G,k}} \sum_{r=1}^{K} \frac{1}{c_r} \frac{-1}{1 + P_r m_G(t)} \frac{m'_G(t)}{m_G(t)^2} dt$$

$$= \frac{c_k}{2\pi i} \oint_{\mathcal{C}_{G,k}} \left(m_G(t) \left[-\frac{1}{m_G(t)} + \sum_{r=1}^{K} \frac{1}{c_r} \frac{P_r}{1 + P_r m_G(t)} \right] + \frac{c_0 - 1}{c_0} \right) \frac{m'_G(t)}{m_G(t)^2} dt.$$

From Equation (8.35), this simplifies into

$$P_k = \frac{c_k}{c_0} \frac{1}{2\pi i} \oint_{\mathcal{C}_{G,k}} (c_0 t m_G(t) + c_0 - 1) \frac{m'_G(t)}{m_G(t)^2} dt. \qquad (8.43)$$

Using (8.38) and proceeding with the further change of variable $t = -1/m_{\underline{F}}(z) - \sigma^2$, (8.43) becomes

$$P_k = \frac{c_k}{2\pi i} \oint_{\mathcal{C}_{F,k}} (1 + \sigma^2 m_{\underline{F}}(z)) \left[-\frac{1}{z m_{\underline{F}}(z)} - \frac{m'_{\underline{F}}(z)}{m_{\underline{F}}(z)^2} - \frac{m'_{\underline{F}}(z)}{m_F(z) m_{\underline{F}}(z)} \right] dz. \qquad (8.44)$$

This whole process of variable changes allows us to describe P_k as a function of $m_F(z)$, the Stieltjes transform of the almost sure limiting spectral distribution of $\mathbf{B}_N$, as $N \to \infty$. It then remains to exhibit a relation between P_k and the empirical spectral distribution of $\mathbf{B}_N$ for finite N. This is what the subsequent section is dedicated to.

Let us now define $\hat{m}_F(z)$ and $\hat{m}_{\underline{F}}(z)$ as the Stieltjes transforms of the empirical eigenvalue distributions of $\mathbf{B}_N$ and $\underline{\mathbf{B}}_N$, respectively, i.e.,

$$\hat{m}_F(z) = \frac{1}{N} \sum_{i=1}^{N} \frac{1}{\lambda_i - z} \qquad (8.45)$$

and

$$\hat{m}_{\underline{F}}(z) = \frac{N}{M}\hat{m}_F(z) - \frac{M -}{M}$$

Instead of going further with (8.44), define $\hat{P}_k$, t
P_k, as

$$\hat{P}_k = \frac{n}{n_k}\frac{1}{2\pi i}\oint\limits_{\mathcal{C}_{F,k}} \frac{N}{n}\left(1 + \sigma^2\hat{m}_{\underline{F}}(z)\right)\left[-\frac{1}{z\hat{m}_{\underline{F}}(z)} - \imath\right.$$

The integrand can then be expanded into nine ter
can easily be performed. Denote first $\eta_1,\ldots,\eta_N$ th
and $\mu_1,\ldots,\mu_N$ the N real roots of $\hat{m}_{\underline{F}}(z) = 0$. We
poles for the nine aforementioned terms: (i) the set
the set $\{\eta_1,\ldots,\eta_N\}\cap[x_{k_F}^{(l)},x_{k_F}^{(r)}]$, and (iii) the set $\{$
$M \neq N$, the full calculus leads to

$$\hat{P}_k = \frac{NM}{n_k(M-N)}\left[\sum_{\substack{1\leq i\leq N \\ x_{k_F}^{(l)}\leq\eta_i\leq x_{k_F}^{(r)}}}\eta_i - \sum_{\substack{1\leq i\leq N \\ x_{k_F}^{(l)}\leq\mu_i\leq x_{k_F}^{(r)}}}\mu_i\right.$$

$$+ \frac{N}{n_k}\left[\sum_{\substack{1\leq i\leq N \\ x_{k_F}^{(l)}\leq\eta_i\leq x_{k_F}^{(r)}}}\sigma^2 - \sum_{\substack{1\leq i\leq N \\ x_{k_F}^{(l)}\leq\lambda_i\leq x_{k_F}^{(r)}}}\sigma^2\right] + \frac{N}{n_k}\left[\right.$$

Now, we know from Theorem 8.5.3 that $\hat{m}_F(z)$
$m_{\underline{F}}(z)$ as $N \to \infty$. Observing that the integrand ir
on the compact $\mathcal{C}_{F,k}$, the dominated convergence th
ensures $\hat{P}_k \xrightarrow{\text{a.s.}} P_k$.

To go further, we now need to determine which
$\mu_1,\ldots,\mu_N$ lie inside $\mathcal{C}_{F,k}$. It can be proved, by ϵ
Theorem 8.3.4 to the current model, that there wil
$\mathbf{B}_N$) outside the support of F, and the number of
is exactly n_k. Since $\mathcal{C}_{F,k}$ encloses cluster k_F and is
$\{\lambda_1,\ldots,\lambda_N\}\cap\left[x_{k_F}^{(l)},x_{k_F}^{(r)}\right] = \{\lambda_i, i \in \mathcal{N}_k\}$ almost sur
any $i \in \{1,\ldots,N\}$, it is easy to see from (8.45) th
and $\hat{m}_F(z) \to -\infty$ when $z \downarrow \lambda_i$. Therefore $\hat{m}_F(z) =$
each interval $(\lambda_{i-1}, \lambda_i)$, with $\lambda_0 = 0$, hence $\mu_1 < \lambda_1$
implies that, if k_0 is the index such that $\mathcal{C}_{F,k}$ contair

then $\mathcal{C}_{F,k}$ also contains $\{\mu_{k_0+1}, \ldots, \mu_{k_0+(n_k-1)}\}$. The same result holds for $\eta_{k_0+1}, \ldots, \eta_{k_0+(n_k-1)}$. When the indexes exist, due to cluster separability, η_{k_0-1} and μ_{k_0-1} belong, for N large, to cluster $k_F - 1$. We are then left with determining whether μ_{k_0} and η_{k_0} are asymptotically found inside $\mathcal{C}_{F,k}$.

For this, we use the same approach as in [22] by noticing that, since 0 is not included in $\mathcal{C}_k$, one has

$$\frac{1}{2\pi i} \oint_{\mathcal{C}_k} \frac{1}{\omega} d\omega = 0.$$

Performing the same changes of variables as previously, we have

$$\oint_{\mathcal{C}_{F,k}} \frac{-m_{\underline{F}}(z)m_F(z) - zm'_{\underline{F}}(z)m_F(z) - zm_{\underline{F}}(z)m'_F(z)}{z^2 m_{\underline{F}}(z)^2 m_F(z)^2} dz = 0. \tag{8.48}$$

For N large, the dominated convergence theorem ensures again that the left-hand side of the (8.48) is close to

$$\oint_{\mathcal{C}_{F,k}} \frac{-\hat{m}_{\underline{F}}(z)(z) - z\hat{m}'_{\underline{F}}(z)(z) - z\hat{m}_{\underline{F}}(z)'(z)}{z^2 \hat{m}_{\underline{F}}(z)^2 (z)^2} dz. \tag{8.49}$$

Residue calculus of (8.49) then leads to

$$\sum_{\substack{1 \le i \le N \\ \lambda_i \in [x_{k_F}^{(l)}, x_{k_F}^{(r)}]}} 2 - \sum_{\substack{1 \le i \le N \\ \eta_i \in [x_{k_F}^{(l)}, x_{k_F}^{(r)}]}} 1 - \sum_{\substack{1 \le i \le N \\ \mu_i \in [x_{k_F}^{(l)}, x_{k_F}^{(r)}]}} 1 \xrightarrow{\text{a.s.}} 0. \tag{8.50}$$

Since the cardinalities of $\left\{i, \eta_i \in \left[x_{k_F}^{(l)}, x_{k_F}^{(r)}\right]\right\}$ and $\left\{i, \mu_i \in \left[x_{k_F}^{(l)}, x_{k_F}^{(r)}\right]\right\}$ are at most n_k, (8.50) is satisfied only if both cardinalities equal n_k in the limit. As a consequence, $\mu_{k_0} \in \left[x_{k_F}^{(l)}, x_{k_F}^{(r)}\right]$ and $\eta_{k_0} \in \left[x_{k_F}^{(l)}, x_{k_F}^{(r)}\right]$. For N large, $N \ne M$, this allows us to simplify (8.47) into

$$\hat{P}_k = \frac{NM}{n_k(M-N)} \sum_{\substack{1 \le i \le N \\ \lambda_i \in \mathcal{N}_k}} (\eta_i - \mu_i) \tag{8.51}$$

with probability one. The same reasoning holds for $M = N$. This is our final relation. It now remains to show that the η_i and the μ_i are the eigenvalues of $\mathrm{diag}\,(\boldsymbol{\lambda}) - \frac{1}{N}\sqrt{\boldsymbol{\lambda}}\sqrt{\boldsymbol{\lambda}}^T$ and $\mathrm{diag}\,(\boldsymbol{\lambda}) - \frac{1}{M}\sqrt{\boldsymbol{\lambda}}\sqrt{\boldsymbol{\lambda}}^T$, respectively. But this is merely a consequence of [23, Lemma 1].

This concludes the proof of Theorem 8.5.4.

$\square$

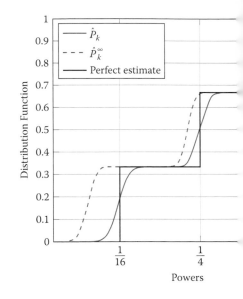

Figure 8.6: Distribution function of the estimators
$P_1 = 1/16$, $P_2 = 1/4$, $P_3 = 1$, $n_1 = n_2 = n_3 = $
sensors, $M = 128$ samples, and SNR $= 20$ dB. Op
dashed lines.

We now evaluate the performance difference be
Stieltjes transform inference methods, for $K = 3$ s
$N = 24$ sensors, $M = 128$ samples, and $n_1 = n_2 = $
Figure 8.6 which compares the distribution functio
by both methods. As anticipated, we observe a sig
reduction for the Stieltjes transform method.

8.6   Conclusion

Random matrix theory for signal processing is a
whose interest is mainly motivated by the increase o
plexity of today's systems. While the first years o
mainly focusing on Gaussian and invariant matrix d
of research were mainly targeting large dimensional
tries. This provided interesting results in particula
sample covariance matrices, which led to new results
dimensional systems. These results are often surpr
they perform well against exact maximum likelihoo
of not too large dimensions. Much more is however

viewpoint relative in particular to second order statistics, see e.g., [35, 36], in order to evaluate theoretically the performance of these methods as well as a generalization to more intricate random matrix structures, such as Vandermonde matrices for array processing, see e.g., [37], or unitary random matrices, see e.g., [38]. A more exhaustive account of random matrix methods as well as more details on the methods presented here can be found in [10, 17, 28].

8.7 Exercises

Exercise 8.7.1 (Sampling and Signal Energy). Based on Theorem 8.2.9, prove the Marčenko-Pastur law, Theorem 8.2.5.

Hint 8.7.1. Observe that the fixed-point equation in m_{FB} reduces now to a second order polynomial from which $m_{FB}(z)$ takes an explicit form. The inverse Stieltjes transform formula 8.4 gives the expression of F^B.

Exercise 8.7.2. Let $\mathbf{X}_N \in \mathbb{C}^{N \times n}$ be a random matrix with i.i.d. Gaussian entries of zero mean and variance $1/n$. For $\mathbf{R}_N \in \mathbb{C}^{N \times N}$ and $\mathbf{T}_N \in \mathbb{C}^{n \times n}$ deterministic and of uniformly bounded spectral norm such that $F^{\mathbf{R}_N} \Rightarrow F^R$ and $F^{\mathbf{T}_N} \Rightarrow F^T$, as $N, n \to \infty$, determine an expression of the Stieltjes transform of the limiting eigenvalue distribution of $\mathbf{B}_N = \mathbf{R}_N^{\frac{1}{2}} \mathbf{X}_N \mathbf{T}_N \mathbf{X}_N^\dagger \mathbf{R}_N^{\frac{1}{2}}$ as $N/n \to c$.

Hint 8.7.2. Follow the proof of Theorem 8.2.9 by looking for a deterministic equivalent of $\frac{1}{N} \operatorname{tr} \left(\mathbf{A} \left(\mathbf{B}_N - z \mathbf{I}_N \right)^{-1} \right)$ for some deterministic $\mathbf{A}$, taken to be successively $\mathbf{R}_N$ and $\mathbf{I}_N$. A good choice of the matrix $\mathbf{D}_N$ is $\mathbf{D}_N = a_N \mathbf{R}_N$.

Exercise 8.7.3. Based on the definition of the Shannon-transform and on the G-estimator for the Stieltjes transform, determine a G-estimator for

$$
\mathcal{V}_{\mathbf{T}_N}(x) = \frac{1}{N} \log \det \left(x \mathbf{T}_N + \mathbf{I}_N \right)
$$

based on the observations

$$
\mathbf{y}_k = \mathbf{T}_N^{\frac{1}{2}} \mathbf{x}_k
$$

with $\mathbf{x}_k \in \mathbb{C}^N$ with i.i.d. entries of zero mean and variance $1/N$, independent across k, for $k \in \{1, \ldots, n\}$.

Hint 8.7.3. Write the expression of $\mathcal{V}_{\mathbf{T}_N}(x)$ as a function of the Stieltjes transform of $\mathbf{T}_N$ and operate a variable change in the resulting integral using Theorem 8.2.9.

Exercise 8.7.4. From the result of Theorem 8.3.3, propose a hypothesis test for the presence of a signal transmitted by a signal source and observed by a large array of sensors, assuming that the additive noise variance is either perfectly known or not.

290

Hint 8.7.4. Observe that the ratio of the extreme €
hypotheses is asymptotically independent of the noi

Exercise 8.7.5. For $\mathbf{W} \in \mathbb{C}^{N \times n}$, $n < N$, the n c
invariant unitary matrix, $\mathbf{w}$ a column vector of $\mathbf{W}$,
matrix with bounded spectral norm, function of all €
$N, n \to \infty$ with $n/N \to c < 1$,

$$\mathbf{w}^{\dagger} \mathbf{B}_N \mathbf{w} - \frac{1}{N - n} \operatorname{tr} \left(() \, \mathbf{I}_N - \mathbf{W} \mathbf{W} \right.$$

Hint 8.7.5. write $\mathbf{w}$ as the normalized projectio.
the subspace orthogonal to the space spanned by t
$\mathbf{w} = \mathbf{\Pi} \mathbf{x}$, with $\mathbf{\Pi} = \mathbf{I}_N - \mathbf{W} \mathbf{W}^{\dagger} + \mathbf{w} \mathbf{w}^{\dagger}$.

References

[1] J. Wishart, "The generalized product moment ◄
    normal multivariate population," *Biometrika*, v◄
    1928.

[2] R. Couillet and M. Debbah, "A Bayesian fram◄
    source signal detection," *IEEE Transactions* ◄
    no. 10, pp. 5186–5195, Oct. 2010.

[3] P. Bianchi, J. Najim, M. Maida, and M. Debbah.
    based Hypothesis Tests for Collaborative Sensi
    Workshop on Statistical Signal Processing, SSF

[4] R. A. Fisher, "The sampling distribution of s
    non-linear equations," *The Annals of Eugenics,*

[5] M. A. Girshick, "On the sampling theory of root
    The Annals of Math. Statistics, vol. 10, pp. 203

[6] P. L. Hsu, "On the distribution of roots of cert
    The Annals of Eugenics, vol. 9, pp. 250–258, 1∫

[7] S. Roy, "p-statistics or some generalizations in t
    priate to multi-variate problems," *Sankhya: Th*
    vol. 4, pp. 381–396, 1939.

[8] Harish-Chandra, "Differential operators on a se◄
    ican Journal of Mathematics, vol. 79, pp. 87–12

[9] C. Itzykson and J. B. Zuber, *Quantum Field*
    Dover Publications, 2005, p. 705.

[10] Z. Bai and J. W. Silverstein, *Spectral Analysis of Large Dimensional Random Matrices*, Springer Series in Statistics, 2009.

[11] P. Billingsley, *Convergence of Probability Measures*. Hoboken, NJ: John Wiley & Sons, Inc., 1968.

[12] S. Wagner, R. Couillet, M. Debbah, and D. T. M. Slock, "Large System Analysis of Linear Precoding in MISO Broadcast Channels with Limited Feedback," 2010. [Online]. Available: http://arxiv.org/abs/0906.3682

[13] V. A. Marčenko and L. A. Pastur, "Distributions of eigenvalues for some sets of random matrices," *Math USSR-Sbornik*, vol. 1, no. 4, pp. 457–483, Apr. 1967.

[14] J. W. Silverstein and Z. D. Bai, "On the empirical distribution of eigenvalues of a class of large dimensional random matrices," *Journal of Multivariate Analysis*, vol. 54, no. 2, pp. 175–192, 1995.

[15] Z. D. Bai and J. W. Silverstein, "No Eigenvalues Outside the Support of the Limiting Spectral Distribution of Large Dimensional Sample Covariance Matrices," *Annals of Probability*, vol. 26, no. 1, pp. 316–345, Jan. 1998.

[16] P. Billingsley, *Probability and Measure*, 3rd ed. Hoboken, NJ: John Wiley & Sons, Inc., 1995.

[17] A. M. Tulino and S. Verdú, "Random matrix theory and wireless communications," *Foundations and Trends in Communications and Information Theory*, vol. 1, no. 1, 2004.

[18] D. N. C. Tse and O. Zeitouni, "Linear multiuser receivers in random environments," *IEEE Transactions on Information Theory*, vol. 46, no. 1, pp. 171–188, 2000.

[19] J. Baik and J. W. Silverstein, "Eigenvalues of large sample covariance matrices of spiked population models," *Journal of Multivariate Analysis*, vol. 97, no. 6, pp. 1382–1408, 2006.

[20] Z. D. Bai and J. W. Silverstein, "Exact Separation of Eigenvalues of Large Dimensional Sample Covariance Matrices," *The Annals of Probability*, vol. 27, no. 3, pp. 1536–1555, 1999.

[21] J. W. Silverstein and S. Choi, "Analysis of the limiting spectral distribution of large dimensional random matrices," *Journal of Multivariate Analysis*, vol. 54, no. 2, pp. 295–309, 1995.

[22] X. Mestre, "On the asymptotic behavior of the sample estimates of eigenvalues and eigenvectors of covariance matrices," *IEEE Transactions on Signal Processing*, vol. 56, no. 11, pp. 5353–5368, Nov. 2008.

292

[23] R. Couillet, J. W. Silverstein, and M. Debbah
Estimation of Multiple Sources," 2011. [Online
org/abs/1001.3934

[24] X. Mestre, "Improved estimation of eigenvalu
their associated subspaces using their sample e
on Information Theory, vol. 54, no. 11, pp. 511

[25] F. Hiai and D. Petz, The semicircle law, free r
Mathematical Surveys and Monographs No. 77.
can Mathematical Society, 2006.

[26] Ø. Ryan and M. Debbah, "Free deconvolution
tions," in (ISIT'07), Nice, France, June 2007, p

[27] N. R. Rao and A. Edelman, "The polynomial r
Foundations of Computational Mathematics, vc
2008.

[28] R. Couillet and M. Debbah, Random Matrix M
cations, 1st ed. New York, NY: Cambridge Ur

[29] P. Vallet, P. Loubaton, and X. Mestre, "Impro
methods with large arrays: The deterministic s
2009, pp. 2137–2140.

[30] W. Rudin, Real and Complex Analysis, 3rd ed.
Mathematics, May 1986.

[31] D. Gregoratti and X. Mestre, "Random DS/CI
ward relay channel," IEEE Transactions on Wi
no. 2, pp. 1017–1027, 2009.

[32] R. Couillet and M. Debbah, "Free deconvoluti
detection," in (PIMRC'08), Cannes, France, 20

[33] I. S. Gradshteyn and I. M. Ryzhik, "Table of In
Academic Press, 6th edition, 2000.

[34] S. H. Simon, A. L. Moustakas, and L. Marinelli
pansions: Moment generating function and othe
related channels," IEEE Transactions on Inforr
pp. 5336–5351, 2006.

[35] Z. D. Bai and J. W. Silverstein, "CLT of linear
mensional sample covariance matrices," Annals
pp. 553–605, 2004.

[36] J. Yao, R. Couillet, J. Najim, E. Mouline, and M. Debbah, "CLT for eigen-inference methods in cognitive radios," in *(ICASSP'11)*, Prague, Czech Republic, 2011, pp. 2980–2983.

[37] Ø. Ryan and M. Debbah, "Asymptotic behavior of random Vandermonde matrices with entries on the unit circle," *IEEE Transactions on Information Theory*, vol. 55, no. 7, pp. 3115–3148, July 2009.

[38] R. Couillet, J. Hoydis, and M. Debbah, "Deterministic equivalents for the analysis of unitary precoded systems," *IEEE Transactions on Information Theory*, 2011.

Chapter 9   Large Deviations

Hongbin Li[‡]
[‡]Stevens Institute of Technology, USA

9.1   Introduction

The theory of large deviations is concerned about the probabilities of rare events. Consider, for example, tossing a fair die n times. There are 6 possible outcomes per toss and a total of 6^n possible outcomes. What is the probability of the average of the throws being close to $\frac{1+2+3+4+5+6}{6} = 3.5$? This is a small deviation event for large n, since by the law of large numbers, the probability is close to 1 and, in average, each face of the die appears about $n/6$ times. What is the probability of the average of the throws being about 4, or the probability of getting each of faces 1 to 5 with about 1 percent of the throws and face 6 with about 95 percent of the throws? Both are rare or large deviation events with vanishing probabilities as n increases. Although the probabilities of such rare events can be computed precisely, given knowledge of the probability distribution, such exact calculations are usually complex and render little insights into the problem. It is often of interest to examine how fast the probability of a rare event decreases with increasing n. Large deviation techniques offer simple and insightful ways to compute the decaying rate of rare event probabilities, and have found numerous applications in communication and computer networks, sensing systems, computational biology, statistical mechanics, and risk analysis.

This chapter provides an introduction to some basic large deviation concepts and techniques including concentration inequalities, rate function, Cramér's theorem, type analysis, and Sanov's theorem. For applications, we consider hypothesis testing and present large deviation results pertaining to the error exponents of several standard hypothesis testing methods.

9.2   Concentration Inequalities

Concentration inequalities are bounds on the probability of a random variable X concentrating on a region of its domain, e.g., the tail region of the distribution as encountered in large deviation analysis. In the following, we discuss some fundamental concentration inequalities. Throughout, we use $\Pr(\cdot)$ to denote the prob-

ability of a random event and $E\{\cdot\}$ the statistical
Unless explicitly specified, the probability distributi
calculations is usually clear from context and thus c

We first consider Markov's inequality, which g
probability of a non-negative random variable no le
constant.

Theorem 9.2.1 (Markov). *For any non-negative*

$$\Pr(X \geq a) \leq \frac{\overline{X}}{a}$$

where $\overline{X} \triangleq E\{X\}$.

Proof: Consider an indicator function

$$\mathbf{1}_S(x) = \begin{cases} 1, & \text{if } x \in S \\ 0 & \text{if } x \notin S \end{cases}$$

where S denotes a set of real numbers. It is easy to

$$\Pr(X \geq a) = E\left\{\mathbf{1}_{[a,\infty)}(X)\right\}$$

where the last inequality is due to the linear bou
$\mathbf{1}_{[a,\infty)}(x) \leq x/a$ (see Figure 9.1).
Markov's inequality implies that if $g(\cdot)$ is a mono
negative function, then for any random variable X

$$\Pr(X \geq a) = \Pr(g(X) \geq g(a)) \leq$$

An application of (9.1) with $g(x) = x^2$ leads to *Che*

$$\Pr(|X - \overline{X}| \geq a) = \Pr(|X - \overline{X}|^2 \;$$

where $\sigma_X^2 \triangleq E\left\{|X - \overline{X}|^2\right\}$ denotes the variance.
obtained by using $g(x) = x^p$ for $p > 0$:

$$\Pr(|X - \overline{X}| \geq a) \leq \frac{E\{|X -}{a^p}$$

Chebyshev's inequality gives a loose bound for
bound can be obtained by taking $g(x)$ to be an expor
exponent. This leads to *Chernoff's inequality*. Fig
quadratic, and exponential bounds on the indicator
Markov, Chebyshev, and Chernoff's inequalities.

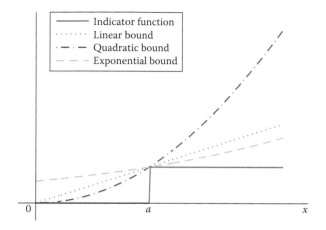

Figure 9.1: Bounds of the indicator function $\mathbf{1}_{[a,\infty)}(x)$.

Theorem 9.2.2 (Chernoff). *For any random variable X and real-valued number a,*

$$\Pr(X \geq a) \leq \inf_{\theta > 0} e^{-\theta a} E\left\{e^{\theta X}\right\} = e^{-\theta^* a} E\left\{e^{\theta^* X}\right\} \tag{9.2}$$

where θ^* is the non-negative solution to

$$E\left\{X e^{\theta X}\right\} = a E\left\{e^{\theta X}\right\}. \tag{9.3}$$

Proof: The inequality in (9.2) follows by using $g(x) = e^{\theta x}$ in (9.1) and noting that this holds for any $\theta > 0$. The tightest bound is obtained by solving

$$0 = \frac{d}{d\theta} E\left\{e^{\theta(X-a)}\right\} = E\left\{(X-a)e^{\theta(X-a)}\right\}$$
$$= e^{-\theta a}\left[E\left\{X e^{\theta X}\right\} - a E\left\{e^{\theta X}\right\}\right]$$

or equivalently (9.3). ∎

Example 9.2.1. Consider an exponential random variable X with PDF

$$f(x) = \lambda e^{-\lambda x}, \quad x \geq 0 \text{ and } \lambda > 0. \tag{9.4}$$

We compare several bounds for the upper tail probability $\Pr(X \geq a)$ for $a > 0$, which is

$$\Pr(X \geq a) = \int_a^\infty \lambda e^{-\lambda x} dx = e^{-\lambda a}. \tag{9.5}$$

A bound given by the Markov inequality is

$$\Pr(X \geq a) \leq \frac{\overline{X}}{a} = \frac{1}{\lambda a}$$

where $\overline{X} = 1/\lambda$ for the exponential distribution. (
used to provide a bound. For simplicity, we assume

$$\Pr(X \geq a) = \Pr(|X - \overline{X}| \geq a - \overline{X})$$

$$= \frac{1}{(\lambda a - 1)^2}, \quad \text{for } a \geq$$

where $\sigma_X^2 = 1/\lambda^2$ for the exponential X. Finally, to
we note

$$E\left\{Xe^{\theta X}\right\} = \frac{\lambda}{(\lambda - \theta)^2}$$

$$E\left\{e^{\theta X}\right\} = \frac{\lambda}{\lambda - \theta}.$$

Substituting these results into (9.3) and solving the e
solution $\theta^* = \lambda - \frac{1}{a}$, for $a \geq 1/\lambda$. Hence, the Cherno

$$\Pr(X \geq a) \leq e^{-\theta^* a} E\left\{e^{\theta^* X}\right\} = \lambda a e^{1-\lambda}$$

It is noted that just like the exact tail probabilit
decreases exponentially as a increases. In contrast
bounds decrease only polynomially and are loose for

9.3   Rate Function

The exponent of the Chernoff bound is related to a so-
an important role in large deviation analysis. Spec
generating function (CGF) of random variable X

$$\Lambda(\theta) \triangleq \log M(\theta) \triangleq \log E\left\{e\right.$$

Chernoff's inequality (9.2) can be expressed as

$$\Pr(X \geq a) \leq \inf_{\theta > 0} e^{-\theta a} M(\theta) = e^{-\text{su}}$$

The *rate function* $I(x)$ is defined as [1, p.26]

$$I(x) \triangleq -\log\left[\inf_{\theta \in \mathbb{R}} e^{-\theta x} M(\theta)\right] = \sup_{\theta \in \mathbb{I}}$$

which is the *Legendre transform* of $\Lambda(\theta)$. Geomet.
gap between the linear function $x\theta$ and the CGF Λ
When multiple observations of random variable X a

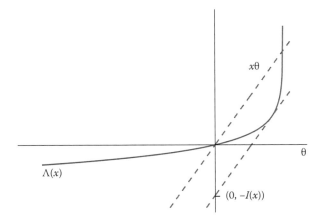

Figure 9.2: Geometric relation between the CGF and the rate function.

characterizes the logarithmic rate of the decaying probability of the empirical mean deviating from the statistical mean (more on this in Section 9.4). We examine here some important properties of the cumulant generating function (CGF) $\Lambda(\theta)$ and rate function $I(x)$.

Theorem 9.3.1. *The CGF* $\Lambda(\theta)$ *and rate function* $I(x)$ *of a random variable* X *have the following properties:*

(a) *Both* $\Lambda(\theta)$ *and* $I(x)$ *are convex functions.*

(b) $I(x) \geq 0, \forall x$ *and if* $\overline{X}$ *is finite,* $I(\overline{X}) = 0$.

(c) $I(x)$ *is a nondecreasing function for* $x > \overline{X}$, *and a nonincreasing function for* $x < \overline{X}$.

(d) $\Lambda(\theta)$ *is differentiable in the interior of its domain with*

$$\Lambda'(\theta) = \frac{E\left\{X e^{\theta X}\right\}}{M(\theta)}. \tag{9.9}$$

Furthermore,

$$I(x) = \theta^* x - \Lambda(\theta^*) \tag{9.10}$$

where θ^* *is such that* $\Lambda'(\theta^*) = x$.

Proof: (a) The convexity of $\Lambda(\theta)$ follows from Hölder's inequality:

$$\Lambda\left(\rho\theta_1 + (1-\rho)\theta_2\right) = \log E \left\{ \left(e^{\theta_1 X}\right)^\rho \left(e^{\theta_2 X}\right)^{1-\rho}\right\}$$
$$\leq \log([E(e^{\theta_1 X})]^\rho [E(e^{\theta_2 X})]^{1-\rho})$$
$$= \rho\Lambda\left(\theta_1\right) + (1-\rho)\Lambda\left(\theta_2\right)$$

where $0 \leq \rho \leq 1$. Note that $\theta x - \Lambda(\theta)$ in (9.8) is
x. The convexity of $I(x)$ follows immediately since
convex function is also convex [2, Section 3.2.3].

(b) The non-negativity of $I(x)$ follows from the
and, therefore, $I(x) \geq 0 - \Lambda(0) = 0$. Now, by Jensen

$$\Lambda(\theta) = \log E\left\{e^{\theta X}\right\} \geq E\left\{\log e^{\theta X}\right\} =$$

which implies that $I(\overline{X}) = 0$.

(c) For $x \geq \overline{X}$, we have

$$I(x) = \sup_{\theta \geq 0}\left[\theta x - \Lambda(\theta)\right]$$

This is because for $\theta < 0$, $\theta x - \Lambda(\theta) \leq \theta\overline{X} - \Lambda(\theta) \leq$
also implies that $I(x)$ is a nondecreasing function
negative θ, the function $\theta x - \Lambda(\theta)$ is nondecreasing in
supremum. Similarly, for $x \leq \overline{X}$, we have

$$I(x) = \sup_{\theta \leq 0}\left[\theta x - \Lambda(\theta)\right]$$

which implies that $I(x)$ is a nonincreasing function

(d) Equation (9.9) is obtained by interchanging t
integration (which can be justified by the dominated
while, (9.10) follows by applying the second equality

Example 9.3.1. (a) Consider the *exponential rando*
The CGF is [see (9.7)]

$$\Lambda(\theta) = \log\frac{\lambda}{\lambda - \theta}$$

which is differentiable. Hence, θ^* is obtained by solv

$$x = \Lambda'(\theta) = \frac{1}{\lambda - \theta}$$

yielding $\theta^* = \lambda - \frac{1}{x}$. It follows that

$$I(x) = \lambda x - 1 - \log(\lambda x)$$

(b) The CGF of a *Gaussian random variable* X with

$$\Lambda(\theta) = \theta m + \frac{1}{2}\theta^2\sigma^2 = \frac{1}{2}(\theta\sigma + m)$$

A calculation similar to the previous example yields

$$I(x) = \frac{1}{2}\left(\frac{x - m}{\sigma}\right)^2.$$

(c) Let X be a *Bernoulli random variable* with $\Pr X = 1 = p$ and $\Pr X = 0 = 1-p$. Its CGF and rate function are given by

$$\Lambda(\theta) = \log(1 - p + pe^\theta).$$

$$I(x) = x \log \frac{x}{p} + (1 - x) \log \frac{1 - x}{1 - p}. \tag{9.11}$$

□

9.4 Cramér's Theorem

Cramér's theorem is concerned about the large deviations of the empirical mean of a sequence of independent and identically distributed (i.i.d.) random variables. Specifically, consider the empirical mean [1, p.26]

$$\hat{S}_n = \frac{1}{n} \sum_{i=1}^{n} X_i$$

for i.i.d. random variables X_n. Suppose $\overline{X} \triangleq E\{X_i\}$ exists and is finite, then by the weak law of large number, $\hat{S}_n \xrightarrow{\text{prob.}} \overline{X}$ as $n \to \infty$. This implies that the probability of $\hat{S}_n$ deviating from $\overline{X}$ diminishes with increasing n. Cramér's theorem characterizes the logarithmic rate of this convergence.

Theorem 9.4.1 (Cramér). *Let $X_i \in \mathbb{R}, i = 1, \ldots, n, \ldots$, be i.i.d. random variables with rate function $I(x)$. For any nonempty closed set $\mathcal{F} \subset \mathbb{R}$,*

$$\limsup_{n \to \infty} \frac{1}{n} \log \Pr(\hat{S}_n \in \mathcal{F}) \leq - \inf_{x \in \mathcal{F}} I(x). \tag{9.12}$$

For any open set $\mathcal{G} \subset \mathbb{R}$,

$$\liminf_{n \to \infty} \frac{1}{n} \log \Pr(\hat{S}_n \in \mathcal{G}) \geq - \inf_{x \in \mathcal{G}} I(x). \tag{9.13}$$

Proof: For simplicity, we assume $\overline{X}$ is finite, although Cramér's theorem holds also for infinite $\overline{X}$. First consider the upper bound (9.12). Let $I_{\mathcal{F}} = \inf_{x \in \mathcal{F}} I(x)$. Clearly, (9.12) holds if $I_{\mathcal{F}} = 0$. Assume $I_{\mathcal{F}} > 0$ (note that $I_{\mathcal{F}}$ is non-negative by Theorem 9.3.1). For any x and $\theta \geq 0$,

$$\Pr(\hat{S}_n \geq x) = E\left\{\mathbf{1}_{[x,\infty)}\left(\hat{S}_n\right)\right\} \leq E\left\{e^{\theta \sum_i (X_i - x)}\right\} = e^{-n\theta x} \prod_i E\left\{e^{\theta X_i}\right\}$$

$$= e^{-n[\theta x - \Lambda(\theta)]}$$

where we used Chernoff's inequality (9.2) and the fact that X_i are i.i.d. Therefore, for $x > \overline{X}$,

$$\Pr(\hat{S}_n \geq x) \leq e^{-nI(x)}. \tag{9.14}$$

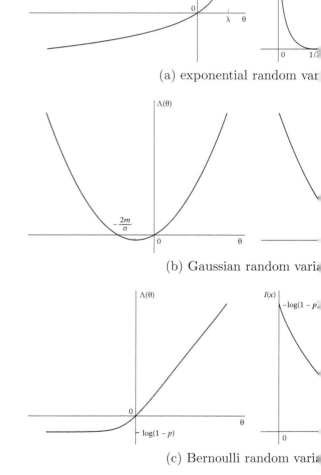

(a) exponential random var

(b) Gaussian random varia

(c) Bernoulli random varia

Figure 9.3: Cumulant generating function (CGF)

Similarly, for $x < \overline{X}$,

$$\Pr(\hat{S}_n \leq x) \leq e^{-nI(x)}. \tag{9.15}$$

Note that $I(\overline{X}) = 0$ ($\overline{X}$ is finite) and since $I_F > 0$, $\overline{X}$ must be within the open set F^c. Let (x_-, x_+) be the union of all open intervals (a, b) that contains $\overline{X}$. It is clear that $\mathcal{F} \subseteq (-\infty, x_-] \cup [x_+, \infty)$. Furthermore, either x_- or x_+ must be finite since $\mathcal{F}$ is not empty. If x_- is finite, then $x_- \in \mathcal{F}$ and, in turn, $I(x_-) \geq I_F$. Likewise, if x_+ is finite, $x_+ \in \mathcal{F}$ and $I(x_+) \geq I_F$. Using (9.14) with $x = x_+$ and (9.15) with $x = x_-$, we have

$$\Pr(\hat{S}_n \in \mathcal{F}) \leq \Pr(\hat{S}_n \in (-\infty, x_-]) + \Pr(\hat{S}_n \in [x_+, \infty)) \leq 2e^{-nI_F}$$

from which (9.12) follows immediately.

The lower bound (9.13) can be proved by using the following result [1, p. 31]:

$$\liminf_{n \to \infty} \frac{1}{n} \Pr(\hat{S}_n \in (-\delta, \delta)) \geq \inf_{\theta \in \mathbb{R}} \Lambda(\theta) = -I(0), \quad \text{for any } \delta > 0. \tag{9.16}$$

Specifically, consider the transformation $Y = X - a$. It is easy to verify that $\Lambda_Y(\theta) = \Lambda(\theta) - \theta a$ and $I_Y(x) = I(x + a)$. It follows from (9.16) that for any x and any $\delta > 0$,

$$\liminf_{n \to \infty} \frac{1}{n} \log \Pr(\hat{S}_n \in (x - \delta, x + \delta)) \geq -I(x).$$

For any $x \in \mathcal{G}$ and sufficiently small $\delta > 0$, $(x - \delta, x + \delta) \subset \mathcal{G}$. Hence, (9.13) follows. ∎

Corollary 9.4.1. For any $a \in \mathbb{R}$,

$$\lim_{n \to \infty} \frac{1}{n} \log \Pr(\hat{S}_n \in [a, \infty)) = -\inf_{x \geq a} I(x). \tag{9.17}$$

Proof: The proof goes by applying Cramér's theorem on the closed set $\mathcal{F} \triangleq [a, \infty)$ and the open set $\mathcal{G} \triangleq (a, \infty)$, and observing that the upper and lower bounds are identical: $\inf_{x \in \mathcal{F}} I(x) = \inf_{x \in \mathcal{G}} I(x) = I(a)$. ∎

Remarks:

1. It is clear from Chernoff's inequality that the upper bound (9.12) holds even for finite n:

$$\Pr(\hat{S}_n \geq a) \leq e^{-n \inf_{x \geq a} I(x)}.$$

If $a \geq \overline{X}$, then the above bound reduces to

$$\Pr(\hat{S}_n \geq a) \leq e^{-nI(a)}$$

since $I(x)$ is nondecreasing for $a \geq \overline{X}$ by Theorem 9.3.1.

2. Let Γ denote an arbitrary set contained in $\mathbb{R}$,
   closure of Γ. If

$$\inf_{x \in \Gamma^o} I(x) = \inf_{x \in \overline{\Gamma}} I(x)$$

   then

$$\lim_{n \to \infty} \frac{1}{n} \log \Pr(\hat{S}_n \in \Gamma)$$

A set Γ satisfying (9.18) is referred to as an
Cramér's theorem has a precise limit (9.19) on
lary 9.4.1 is such a case.

9.5   Method of Types

So far, we have considered the large deviation behav
The method of types is a powerful approach for analy
of a sequence. Let $\mathbf{X} = (X_1, \ldots, X_n)$ be a seque
following a probability distribution $P \in \mathcal{P}(\mathcal{A})$, whe
all probability distributions on an alphabet $\mathcal{A} \triangleq \{$
next section, it is assumed that the alphabet $\mathcal{A}$ is fi

Definition 9.5.1. ([1, p.12] and [3, p.348]) The *type*
$\mathcal{P}(\mathcal{A})$ of a sequence $\mathbf{x} = (x_1, \ldots, x_n) \in \mathcal{A}^n$ is a len
empirical probability distribution of sequence $\mathbf{x}$, wh

$$T_{\mathbf{x}}(a_i) \triangleq \frac{1}{n} \sum_{j=1}^{n} \mathbf{1}_{a_i}(x_j), \quad i = 1$$

which is the relative proportion of occurrence of syn

Let $\mathcal{T}_n$ denote the set of all possible types of seq
number of distinct sequences grows exponentially w
types is significantly less. This is illustrated by a
next example, followed by an upper bound for a mo:

Example 9.5.1. For binary alphabet $\mathcal{A} = \{0, 1\}$, w

$$\mathcal{T}_n = \left\{ (T_{\mathbf{x}}(0), T_{\mathbf{x}}(1)) : \left(\frac{0}{n}, \frac{n}{n} \right), \left(\frac{1}{n}, \frac{n-}{n} \right) \right.$$

Theorem 9.5.1 (Cardinality of set of types). *F*
size,

$$|\mathcal{T}_n| \le (n+1)^{|\mathcal{A}|}.$$

Proof: The proof goes by observing that $T_{\mathbf{x}}(a_i)$, the i-th element of the length-$|\mathcal{A}|$ vector $T_{\mathbf{x}}$, belongs to $\{\frac{0}{n}, \frac{1}{n}, \ldots, \frac{n}{n}\}$, which has a cardinality of $n+1$. ∎

A bound tighter than (9.20) is $|\mathcal{T}_n| \leq (n+1)^{|\mathcal{A}|-1}$, which is due to the fact that there are only $|\mathcal{A}| - 1$ independent elements in $T_{\mathbf{x}}$ which is a PMF that should sum to one. However, the modification is insignificant. Most importantly, there is at most a polynomial number of types of length n. Hence, some types contain exponentially many sequences.

Definition 9.5.2. For any $Q \in \mathcal{T}_n$, the collection of all sequences of length n and type Q is called the *type class* of Q, denoted by $\mathcal{T}_n(Q) \triangleq \{\mathbf{x} \in \mathcal{A}^n : T_{\mathbf{x}} = Q\}$.

Example 9.5.2. For the binary example in Example 9.5.1, let $\mathbf{x} = 11001$. The type of $\mathbf{x}$ is $T_{\mathbf{x}} = (T_{\mathbf{x}}(0), T_{\mathbf{x}}(1)) = (2/5, 3/5)$. There are a total of $|\mathcal{T}_n(T_{\mathbf{x}})| = \binom{5}{3} = 10$ sequences of the same type: 11100, 11010, 10110, 01110, 11001, 10101, 01101, 10011, 01011, 00111. More generally, the exact size of a type class $\mathcal{T}_n(T_{\mathbf{x}})$ on an arbitrary $\mathcal{A}$ is the number of distinct ways of forming a length-n sequence from $\mathcal{A}$, under the constraints that a_1 is used $nT_{\mathbf{x}}(a_1)$ times, a_2 used $nT_{\mathbf{x}}(a_2)$ times, and so on. The solution to this combinatorial problem is

$$|\mathcal{T}_n(T_{\mathbf{x}})| = \frac{n!}{(nT_{\mathbf{x}}(a_1))! \ldots (nT_{\mathbf{x}}(a_{|\mathcal{A}|}))!}. \tag{9.21}$$

However, the combinatorial number is hard to analyze. A simple exponential bound can be obtained through an information theoretic approach as discussed next. □

Definition 9.5.3. The *entropy* of a probability distribution $P \in \mathcal{P}(\mathcal{A})$ is

$$H(P) \triangleq -\sum_{i=1}^{|\mathcal{A}|} P(a_i) \log P(a_i). \tag{9.22}$$

The *relative entropy* or *Kullback-Leibler distance* of a probability distribution P relative to another probability distribution Q is

$$D(P\|Q) \triangleq \sum_{i=1}^{|\mathcal{A}|} P(a_i) \log \frac{P(a_i)}{Q(a_i)}. \tag{9.23}$$

In the above definition, we use the convention that $0 \log 0 \triangleq 0$ and $0 \log \frac{0}{0} \triangleq 0$. It is easy to show that for any P and Q, we have $H(P) \geq 0$, $D(P\|Q) \geq 0$, and $D(P\|Q) = 0$ if and only if $P = Q$. However, the Kullback-Leibler distance is not symmetric, $D(P\|Q) \neq D(Q\|P)$, and does not satisfy the triangle inequality. Therefore, it is not a "true" distance between two distributions.

Theorem 9.5.2 (Sequence probability). *Let $\mathbf{X} = (X_1, \ldots, X_n)$ be drawn i.i.d. according to distribution $P \in \mathcal{P}(\mathcal{A})$, the probability of $\mathbf{X} = \mathbf{x}$ depends only on the type of the sequence $\mathbf{x}$:* [1, p.14] *and* [3, p.349]

$$\Pr(\mathbf{X} = \mathbf{x}) = e^{-n[H(T_{\mathbf{x}}) + D(T_{\mathbf{x}}\|P)]}. \tag{9.24}$$

If $P \in \mathcal{T}_n$ (i.e., P is also a type) and $T_{\mathbf{x}} = P$, then

$$\Pr(\mathbf{X} = \mathbf{x}) = e^{-nH(P)}$$

Proof: Equation (9.24) follows by

$$\Pr(\mathbf{X} = \mathbf{x}) = \prod_{i=1}^{n} \Pr(X_i = x_i) = \prod_{i=1}^{|\mathcal{A}|} P$$

$$= \prod_{i=1}^{|\mathcal{A}|} e^{nT_{\mathbf{x}}(a_i) \log P(a_i)}$$

$$= \prod_{i=1}^{|\mathcal{A}|} e^{-n\left[T_{\mathbf{x}}(a_i) \log \frac{T_{\mathbf{x}}(a_i)}{P(a_i)} -\right.}$$

$$= e^{-n[H(T_{\mathbf{x}})+D(T_{\mathbf{x}}\|P)]}.$$

The special case of $T_{\mathbf{x}} = P$ is proved by using the fa

We now discuss an upper and a lower bound on
are easier to use than the exact combinatorial expre

Theorem 9.5.3 (Size of a type class). *For any*
p.350]

$$(n+1)^{-|\mathcal{A}|} e^{nH(Q)} \le |\mathcal{T}_n(Q)| \le$$

Proof: Since any type class has probability at mo
type class are equally likely,

$$1 \ge \Pr(T_{\mathbf{x}} = Q) = |\mathcal{T}_n(Q)| \Pr$$

where $\mathbf{x} \in \mathcal{T}_n(Q)$. Thus, $|\mathcal{T}_n(Q)| \le \frac{1}{\Pr(\mathbf{X}=\mathbf{x})}$. From
of the probability distribution of the sequence. Th
holds for any distribution. Let us consider the case
drawn i.i.d. from distribution Q. By (9.25), $\Pr(\mathbf{X} =$
bound in (9.26) follows immediately.

For the lower bound, we first note that the typ
probability among all type classes under the distribu

$$\Pr_Q(T_{\mathbf{x}} = Q) \ge \Pr_Q(T_{\mathbf{x}} = Q'),$$

where the dependence on Q is shown explicitly (as a
This is because

$$\frac{\Pr_Q(T_{\mathbf{x}} = Q)}{\Pr_Q(T_{\mathbf{x}} = Q')} = \frac{|\mathcal{T}_n(Q)| \prod_{i=1}^{|\mathcal{A}|} Q(a_i}{|\mathcal{T}_n(Q')| \prod_{i=1}^{|\mathcal{A}|} Q(a_i}$$

$$= \prod_{i=1}^{|\mathcal{A}|} \frac{(nQ'(a_i))!}{(nQ(a_i))!} Q(a_i}$$

where the second equality uses (9.21). The last expression has a factor of the form $\frac{k!}{l!}$, which can be bounded as follows (see Problem 9.9.4) [3, p.352]

$$\frac{k!}{l!} \geq l^{k-l}, \quad k, l \in \mathbb{Z}_+.$$

Applying the bound, we have

$$\frac{\Pr_Q(T_{\mathbf{X}} = Q)}{\Pr_Q(T_{\mathbf{X}} = Q')} \geq \prod_{i=1}^{|\mathcal{A}|} [nQ(a_i)]^{nQ'(a_i) - nQ(a_i)} Q(a_i)^{nQ(a_i) - nQ'(a_i)}$$

$$= \prod_{i=1}^{|\mathcal{A}|} n^{n[Q'(a_i) - Q(a_i)]}$$

$$= n^{n\left[\sum_{i=1}^{|\mathcal{A}|} Q'(a_i) - \sum_{i=1}^{|\mathcal{A}|} Q(a_i)\right]} = 1$$

which proves (9.27). Therefore,

$$1 = \sum_{Q' \in \mathcal{T}_n} \Pr_Q(T_{\mathbf{X}} = Q') \leq |\mathcal{T}_n| \Pr_Q(T_{\mathbf{X}} = Q) \leq |\mathcal{T}_n| e^{-nH(Q)} |\mathcal{T}_n(Q)|$$

and the lower bound on $\mathcal{T}_n(Q)|$ given in (9.26) follows by applying Theorem 9.5.1. ∎

The lower and upper bounds of Theorem 9.5.3 on the size of an arbitrary type class allows one to determine how likely the type Q of a sequence $\mathbf{x}$ deviates the underlying probability distribution P from which $\mathbf{x}$ is generated.

Theorem 9.5.4 (Large deviation probabilities). *For any $Q \in \mathcal{T}_n$, the probability of type class $\mathcal{T}_n(Q)$ under any probability distribution P is approximately $e^{-nD(Q\|P)}$, where the approximation is to first order in the exponent. The probability can be bounded more exactly as [1, p.15] and [3, p.354]*

$$(n+1)^{-|\mathcal{A}|} e^{-nD(Q\|P)} \leq \Pr(T_{\mathbf{X}} = Q) \leq e^{-nD(Q\|P)}. \tag{9.28}$$

Proof: By Theorem 9.5.2,

$$\Pr(T_{\mathbf{X}} = Q) = |\mathcal{T}_n(Q)| \Pr(\mathbf{X} = \mathbf{x}, \mathbf{x} \in \mathcal{T}_n(Q))$$

$$= |\mathcal{T}_n(Q)| e^{-n[H(Q) + D(Q\|P)]}.$$

Equation (9.28) follows by applying Theorem 9.5.3. ∎

To summarize, the main result from the analysis of types is that there are only a polynomial number of types (Theorem 9.5.1), and an exponential number of sequences of each type (Theorem 9.5.3). Since the probability of each type class $\mathcal{T}_n(Q)$ is an exponential function of the Kullback-Leibler distance between the type Q and the probability distribution P, type classes far from the distribution P are large deviation events with exponentially smaller probability (Theorem 9.5.4).

Example 9.5.3. Consider an i.i.d. sequence of Bern
$1, \ldots, n$, with $\Pr(X = 1) = p$. What is the prob.
$\hat{S}_n = \frac{1}{n} \sum_n X_i$ being near $0 \leq q \leq 1$? If $q = p$,
with increasing n by the law of large number. If q
event and, by Cramér's theorem, the probability is r
function $I(x)$ is given by (9.11). Since the event is als
supposing that this is a valid type (i.e., qn is an inte
the probability is about $e^{-nD((1-q,q)\|(1-p,p))}$. It c
$D((1-q,q)\|(1-p,p))$ and hence, the two approach

9.6 Sanov's Theorem

While we have seen that the method of types can b
the questions addressed by Cramér's theorem, it is a
offers ways for characterizing the large deviation of
just the empirical mean. Sanov's theorem is one su
probability of a set of nontypical types.

Theorem 9.6.1 (Sanov). *Let $X_1, \ldots, X_n$ be i.i.d.*
distribution $P \in \mathcal{P}(\mathcal{A})$. For any subset $\mathcal{Q} \subseteq \mathcal{P}(\mathcal{A})$,

$$\Pr(T_{\mathbf{X}} \in \mathcal{Q}) = \Pr(T_{\mathbf{X}} \in \mathcal{Q} \cap \mathcal{T}_n) \leq (n +$$

where $Q^* = \arg\inf_{Q \in \mathcal{Q}} D(Q\|P)$ is the distribution
relative entropy. If $\mathcal{Q}$ is also the closure of its inter

$$\lim_{n \to \infty} \frac{1}{n} \log \Pr(T_{\mathbf{X}} \in \mathcal{Q}) = -D$$

Proof: The upper bound (9.29) follows by applying
9.5.4:

$$\Pr(T_{\mathbf{X}} \in \mathcal{Q}) = \sum_{Q \in \mathcal{Q} \cap \mathcal{T}_n} \Pr(T_{\mathbf{X}} = Q) \leq \sum_{Q \in}$$

$$\leq |\mathcal{Q} \cap \mathcal{T}_n| e^{-nD(Q^*\|P)} \leq (n -$$

It is noted that Q^* need not belong to $\mathcal{T}_n$. For th
dense in $\mathcal{P}(\mathcal{A})$, $\mathcal{Q} \cap \mathcal{T}_n$ is nonempty for all $n \geq n_0$
sequence of distributions $Q_n \in \mathcal{Q} \cap \mathcal{T}_n$ such that $D(Q$
As such, by the lower bound of Theorem 9.5.4 and f

$$\Pr(T_{\mathbf{X}} \in \mathcal{Q}) = \sum_{Q \in \mathcal{Q} \cap \mathcal{T}_n} \Pr(T_{\mathbf{X}} = Q) \geq$$

$$\geq (n+1)^{-|\mathcal{A}|} e^{-nD(Q_n\|P}$$

Since $\lim_{n\to\infty} \frac{1}{n}\log(n+1)^{-|\mathcal{A}|} = 0$, the normalized logarithmic limit of the above equation is

$$\liminf_{n\to\infty} \frac{1}{n}\log\Pr(T_{\mathbf{X}} \in \mathcal{Q}) \geq \liminf_{n\to\infty} D(Q_n\|P) = -D(Q^*\|P).$$

Finally, (9.30) follows by combining the normalized logarithmic limit of the upper and lower bounds. ∎

For illustration, we consider using Sanov's theorem to find $\Pr(\frac{1}{n}\sum_{i=1}^{n} g_j(X_i) \geq b_j, j = 1,\ldots,k)$. The joint event $\frac{1}{n}\sum_{i=1}^{n} g_j(X_i) \geq b_j, j = 1,\ldots,k$, is equivalent to the event $T_{\mathbf{X}} \in \mathcal{Q} \cap \mathcal{T}_n$, where

$$\mathcal{Q} = \left\{Q \in \mathcal{P}(\mathcal{A}) : \sum_{a\in\mathcal{A}} g_j(a)Q(a) \geq b_j, j = 1,\ldots,k\right\}. \tag{9.31}$$

This is because $b_j \leq \frac{1}{n}\sum_{i=1}^{n} g_j(x_i) = \sum_{a\in\mathcal{A}} g_j(a)T_{\mathbf{x}}(a)$, which implies $T_{\mathbf{x}} \in \mathcal{Q}\cap\mathcal{T}_n$. To find the closest distribution Q^* in $\mathcal{Q}$ to P, we minimize $D(Q\|P)$ subject to the constraints in (9.31). Using Lagrange multipliers, we minimize

$$J(Q) = \sum_{x\in\mathcal{A}} Q(x)\log\frac{Q(x)}{P(x)} + \sum_j \lambda_j \sum_{x\in\mathcal{A}} Q(x)g_j(x) + \nu \sum_{x\in\mathcal{A}} Q(x)$$

which leads to the following solution

$$Q^*(x) = \frac{P(x)e^{\sum_j \lambda_j g_j(x)}}{\sum_{a\in\mathcal{A}} P(a)e^{\sum_j \lambda_j g_j(x)}}, \quad x \in \mathcal{A} \tag{9.32}$$

where the constants λ_j should be chosen to satisfy the constraints. A more specific example is considered next.

Example 9.6.1. We want to estimate the probability of observing more than 700 heads in a series of $n = 1000$ tosses of a fair coin. This problem is similar to Example 9.5.3. Here, we use Sanov's theorem, which says that [3, p.365]

$$\Pr(T_{\mathbf{X}} \in \mathcal{Q}) \approx e^{-nD(Q^*\|P)}$$

where Q^* minimizes $D(Q\|P)$ over the set $\mathcal{Q} = \left\{Q \in \mathcal{P}(\{0,1\}) : \sum_{i=0}^{1} iQ(i) \geq 0.7\right\}$. From (9.32), Q^* is given by (note that $P(0) = P(1) = \frac{1}{2}$)

$$Q^*(x) = \frac{e^{\lambda x}}{1 + e^{\lambda x}}, \quad x = 0, 1 \tag{9.33}$$

where λ is chosen to satisfy the constraint $\sum_{i=0}^{1} iQ^*(i) = Q^*(1) = \frac{e^\lambda}{1+e^\lambda} = 0.7$, which yields $e^\lambda = \frac{7}{3}$. Substituting this back to (9.33), we find $Q^*(0) = 0.3$ and $Q^*(1) = 0.7$. Therefore, we have

$$\Pr(T_{\mathbf{X}} \in \mathcal{Q}) \approx e^{-nD((0.3,0.7)\|(0.5,0.5))}$$

which is the same as what was found in Example 9.5.3. □

9.7   Hypothesis Testing

We consider applications of large deviation analysis
is frequently encountered in various sensing, commu
systems. In the following, we first briefly introduce t
and then move on to several large deviation results
are available on hypothesis testing (e.g., [4] and refe

Let $Y_1, \ldots, Y_n$ be a sequence of random variables
either the probability distribution P_0 (hypothesis $H_{\bullet}$
hypothesis testing problem is concerned with discrii
potheses H_0 and H_1, given observations $y_1, \ldots, y_n$ o
the objective is to perform inference about the unc
servations. The hypotheses H_0 and H_1 are sometim
alternative hypotheses, respectively. Let $\mathbf{Y} = (Y_1,$
For simplicity, we only consider the case when the obs
an alphabet $\mathcal{A}$ that may be finite or infinite.

Definition 9.7.1. A *decision rule* or a *test* is a m
that H_0 is accepted (H_1 rejected) if $\delta_n(\mathbf{y}) = 0$, while
if $\delta_n(\mathbf{y}) = 1$. Equivalently, a test is a partition o
subsets $\Gamma_1^n = \{\mathbf{y} \in \mathcal{A}^n : \delta_n(\mathbf{y}) = 1\}$ and $\Gamma_0^n = \{\mathbf{y} \in {}$

The performance of a decision rule δ_n is often a
probabilities:

$$\alpha_n \triangleq \mathrm{Pr}_0(\delta_n(\mathbf{Y}) = 1 | H_0 \text{true}) = \mathrm{P}$$

$$\beta_n \triangleq \mathrm{Pr}_1(\delta_n(\mathbf{Y}) = 0 | H_1 \text{true}) = \mathrm{P}$$

where the probability $\mathrm{Pr}_j(\cdot)$ is computed using dist
tentimes, α_n is referred to as the *probability of false*
of missing. In general, it would be desirable to cho
mize both error probabilities, but there is a tradeoff
$\beta_n = 0$ by using the test $\delta_n(\mathbf{y}) \equiv 1$, but at the pr
feasible approach is to minimize β_n subject to a cor
test following this approach is a *likelihood ratio test*
the likelihood functions $L_1(\mathbf{y})$ and $L_0(\mathbf{y})$ with a thre
$L_j(\mathbf{y})$ is either the probability mass function $P_j(\mathbf{y})$ f
probability density function $p_j(\mathbf{y})$ for continuous dis
by the Neyman-Pearson lemma (see [3, p.376-377])

Theorem 9.7.1 (Neyman-Pearson lemma). *L*
log likelihood ratio. For any $n \in \mathbb{Z}_+$, the Neyman-Pe
observation set with decision region

$$\Gamma_1^n = \left\{ \mathbf{y} \in \mathcal{A}^n : \hat{S}_n(\mathbf{y}) \geq \right.$$

where $\hat{S}_n$ is the sample mean of the log likelihood ratio

$$\hat{S}_n = \frac{1}{n} \sum_{i=1}^{n} X_i$$

and γ_n is a threshold. Let α_n and β_n be the false alarm and missing probabilities, respectively, of the Neyman-Pearson test. Let $\bar{\Gamma}_1^n$ represent the decision region of any other partition/test with associated false alarm and missing probabilities $\bar{\alpha}_n$ and $\bar{\beta}_n$, respectively. If $\bar{\alpha}_n \leq \alpha_n$, then $\bar{\beta}_n \geq \beta_n$.

In other words, the Neyman-Pearson test is optimum in the sense there are neither tests with the same false alarm probability and a smaller missing probability nor tests with the same missing probability and a smaller false alarm probability.

It is often of interest to determine the exponential rate of the error probabilities α_n and β_n of the Neyman-Pearson test with a fixed threshold $\gamma \in (E\{X_1\}_0, E\{X_1\}_1)$, where $E\{X_1\}_j$ denotes the statistical mean of the log likelihood ratio X_1 under hypothesis H_j. These can be readily obtained by examining the large deviations of $\hat{S}_n$.

Theorem 9.7.2. *The Neyman-Pearson test with a constant threshold [1, p.92] $\gamma \in (E\{X_1\}_0, E\{X_1\}_1)$ satisfies*

$$\lim_{n\to\infty} \frac{1}{n} \log \alpha_n = -I_0(\gamma) < 0 \tag{9.34}$$

$$\lim_{n\to\infty} \frac{1}{n} \log \beta_n = \gamma - I_0(\gamma) < 0 \tag{9.35}$$

where $I_0(\cdot)$ is the rate function defined in (9.8), i.e., the Legendre transform of the CGF $\Lambda_0(\theta) \triangleq \log E\{e^{\theta X_1}\}_0$.

Proof: Equation (9.34) follows by a direct application of Corollary 9.4.1. Likewise, the exponential rate of β_n is $-I_1(\gamma)$. Since

$$\Lambda_1(\theta) = E_1\{e^{\theta X_1}\} = E_1 \left\{ \left[\frac{L_1(Y_1)}{L_0(Y_1)} \right]^{\theta} \right\}$$

$$= E_0 \left\{ \left[\frac{L_1(Y_1)}{L_0(Y_1)} \right]^{\theta+1} \right\} = E_0 \left\{ e^{(\theta+1)X_1} \right\} = \Lambda_0(\theta+1).$$

It follows from the definition (9.8) that

$$I_1(x) = I_0(x) - x$$

and thus (9.35) is established. ∎

It can be readily verified that $\Lambda_0(0) = \log E_0\{e^0\} = 0$ and $\Lambda_0(1) = \Lambda_1(0) = \log E_1\{e^0\} = 0$. By Jensen's inequality,

$$E_0\{X_1\} = E_0\{\log e^{X_1}\} < \log E_0\{e^{X_1}\} = \Lambda_0(1) = 0$$

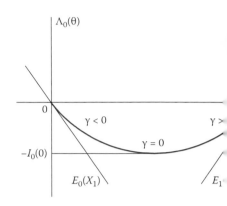

Figure 9.4: Geometric illustration of the CGF $\Lambda_0(\theta$
information $I_0(0)$ (adapted from [1, p.93]).

and, likewise,

$$E_1\{X_1\} = E_1\{-\log e^{-X_1}\} > \log E_1\{e^{\lambda}$$

These inequalities are strict, provided that the distr
and H_1 are distinct, i.e., they differ on a set with
in which case X_1 is not identically a constant. Giv
recalling that $\Lambda_0(\theta)$ is convex (Theorem 9.3.1(a)), w
the relation to the test threshold γ in Figure 9.4.

An immediate result of the previous theorem is
exponential rate of the average or Bayesian error pr

$$P_{e,n} \triangleq \Pr(H_0)\alpha_n + \Pr(H_1$$

where $\Pr H_j$ denotes the prior probability of hypoth

Corollary 9.7.1 (Chernoff's information). For

$$\inf_{\delta} \liminf_{n\to\infty} \frac{1}{n} \log P_{e,n} = -I_0$$

where the outer infimum is over all tests. The bes
Bayesian probability of error $I_0(0)$ is also called C
distributions P_1 and P_0.

Proof: It suffices to consider likelihood ratio test
Neyman-Pearson lemma. Let α_n^* and β_n^* be the error
ratio test with zero threshold. Consider any othe

threshold γ_n and associated error probabilities α_n and β_n. We have

$$\alpha_n \geq \alpha_n^*, \quad \text{if } \gamma_n \leq 0$$
$$\beta_n \geq \beta_n^*, \quad \text{if } \gamma_n \geq 0$$

since the false alarm probability α_n is a nonincreasing function of the threshold γ_n while the missing probability β_n a nondecreasing function of γ_n. Therefore,

$$P_{e,n} \geq \min\{\Pr(H_0), \Pr(H_1)\}\,(\alpha_n + \beta_n)$$
$$\geq \min\{\Pr(H_0), \Pr(H_1)\}\min\{\alpha_n^*, \beta_n^*\}.$$

Hence,

$$\frac{1}{n}\log P_{e,n} \geq \frac{1}{n}\log\left(\min\{\Pr(H_0), \Pr(H_1)\}\right) + \min\left\{\frac{1}{n}\log\alpha_n^*, \frac{1}{n}\log\beta_n^*\right\}.$$

Since $0 < \Pr(H_j) < 1$ is constant with zero exponential rate,

$$\inf_{\delta}\liminf_{n\to\infty}\frac{1}{n}\log P_{e,n} \geq \liminf_{n\to\infty}\min\left\{\frac{1}{n}\log\alpha_n^*, \frac{1}{n}\log\beta_n^*\right\}.$$

It follows from (9.34) and (9.34) that

$$\liminf_{n\to\infty}\frac{1}{n}\log\alpha_n^* = \liminf_{n\to\infty}\frac{1}{n}\log\beta_n^* = -I_0(0).$$

Therefore,

$$\liminf_{n\to\infty}\frac{1}{n}\log P_{e,n} \geq -I_0(0)$$

and the equality is achieved by the zero-threshold likelihood ratio test.   ∎

The likelihood ratio test that minimizes (9.36) is often referred to as the *minimum error probability test* or *Bayesian test*. It is well known that the threshold for the Bayesian test depends on the prior probabilities $\Pr(H_j)$ (e.g., [4, Chap. II.B]), which seems to contradict the zero threshold suggested by Corollary 9.7.1. In fact, the dependence occurs only for finite n. Specifically, it can be shown that the Bayesian threshold $\gamma_n = \log\frac{1}{n}\frac{\Pr(H_0)}{\Pr(H_1)}$, and the effect of the prior probabilities vanishes as $n \to 0$.

Finally, we include Stein's lemma which addresses the best achievable exponential rate for the missing probability β_n when the false alarm probability α_n is bounded away from 1. The proof is omitted and can be found at [1, p.94-95].

Lemma 9.7.1 (Stein's lemma). Let β_n^ϵ be the infimum of β_n among all tests with $\alpha_n < \epsilon$. For any $\epsilon < 1$, [1, p.94]

$$\lim_{n\to\infty}\frac{1}{n}\log\beta_n^\epsilon = E_0\{X_1\}. \tag{9.38}$$

Remark: The results discussed in this section apply to both continuous and discrete distributions. For discrete random variables drawn from a finite alphabet, the results can be equivalently expressed in terms of the relative entropy defined in (9.23).

9.8   Further Readings

There are numerous papers and books on large devia
can be found in [5] along with the author's opinions o
chapter is mainly based on [1, Chaps. 2 and 3] and
considered an excellent book on the theory, but it is
mathematical level. Several efforts have been mad
more accessible to engineering audiences, by highligh
fying or including only sketches of some proofs, pro
graphical illustrations. Cover and Thomas's book [3]
amples and discussions on large deviations. However
an information theoretic point of view and, furtherm
distributions on finite alphabets. Our discussions o
are mainly based on [3].

9.9   Exercises

Exercise 9.9.1. Let X be a Gaussian random vari
variance. The tail probability

$$\Pr(X \geq x) = \frac{1}{\sqrt{2\pi}} \int_x^\infty e^{-\frac{u^2}{2}} \, du$$

does not have a closed form expression. Show that t
given by

$$Q(x) \leq \frac{1}{2} e^{-\frac{x^2}{2}}.$$

Exercise 9.9.2. Consider a Poisson random varia
function

$$P(k) = \frac{\lambda^k e^{-\lambda}}{k!}.$$

Show that the rate function, i.e., the Legendre transfo
generating function $\Lambda(\theta) = \log E\left\{e^{\theta X}\right\}$, is given by

$$I(x) = x \log \frac{x}{\lambda} + \lambda - x.$$

Exercise 9.9.3. Let $X_1, \ldots, X_n$ be i.i.d. Gaussian
mean and unit variance. Find the exponential rate o

$$\Pr\left(\frac{1}{n} \sum_{i=1}^n X_i^2 \geq a^2\right).$$

Exercise 9.9.4. Prove

$$\frac{k!}{l!} \geq l^{k-l}, \quad k, l \in \mathbb{Z}_+.$$

You can consider separately the cases $k \geq l$ and $k < l$.

Exercise 9.9.5. Suppose that we toss a fair die n times. To first order in exponent, what is the probability of the average of the throws being greater than 4, and what is the probability of getting each of faces 1 to 5 with 1 percent of the throws and face 6 with 95 percent of the throws (assume that $0.01n$ is an integer)?

Exercise 9.9.6. Let $X_1, \ldots, X_n$ be a sequence of random variables drawn i.i.d. from probability distributions P. Consider the hypothesis test H_0: $P = P_0 = \left(\frac{1}{4}, \frac{1}{4}, \frac{1}{2}\right)$ vs. hypothesis H_1: $P = P_1 \left(\frac{1}{2}, \frac{1}{4}, \frac{1}{4}\right)$. Find the error exponent of the best achievable probability of missing subject to the probability of false alarm being less than $\frac{1}{2}$.

References

[1] A. Dembo and O. Zeitouni, *Large Deviations Techniques and Applications*, 2nd ed. New York, NY: Springer, 1998.

[2] S. Boyd and L. Vandenberghe, *Convex Optimization*. Cambridge, UK: Cambridge University Press, 2004.

[3] T. M. Cover and J. A. Thomas, *Elements of Information Theory*, 2nd ed. Hoboken, NJ: John Wiley & Sons, Inc., 2006.

[4] H. V. Poor, *An Introduction to Signal Detection and Estimation*, 2nd ed. New York, NY: Springer-Verlag, 1994.

[5] A. Weiss, "An introduction to large deviations for communication networks," *IEEE Journal on Selected Areas in Communications*, vol. 13, no. 6, August 1995.

Chapter 10 Fundamentals of Estimation Theory

Yik-Chung Wu[‡]
[‡]University of Hong Kong

Parameter estimation is prevalent in communications and signal processing applications, e.g., in channel estimation, synchronization, parametric spectral estimation, direction-of-arrival estimation, etc. Estimation theory is extensively used in modern research in various fields related to signal processing. This chapter presents an overview of basic parameter estimation techniques and discusses the relationships among them. Applications in problems involving recent research are provided as examples.

10.1   Introduction

In an estimation problem, we have some observations which depend on the parameters we want to estimate. For example, the observations can be written as

$$\mathbf{x} = \mathbf{f}(\boldsymbol{\theta}) + \mathbf{w}, \tag{10.1}$$

where $\mathbf{x}$ stands for the observations, $\boldsymbol{\theta}$ is the parameter we want to estimate, $\mathbf{f}$ is a known function, and $\mathbf{w}$ denotes the random noise corrupting the observations. All the above quantities can be in vector form. In this chapter, it is assumed that the dimension of $\mathbf{x}$ is $N \times 1$ while that of $\boldsymbol{\theta}$ is $p \times 1$ with $N > p$. In general, $\mathbf{x}$ can depend on $\boldsymbol{\theta}$ in a linear or nonlinear way. It is even possible that some of the elements of $\boldsymbol{\theta}$ are related to $\mathbf{x}$ in a linear way, while others are related to $\mathbf{x}$ in a nonlinear way. The task of parameter estimation is to construct a function (called estimator) such that $\hat{\boldsymbol{\theta}} := \mathbf{g}(\mathbf{x})$ is a good estimate of $\boldsymbol{\theta}$. The first question we encounter is how to define a "good" estimate. One natural criterion is that the average squared difference between $\hat{\theta}_i$ and θ_i should be the smallest, where θ_i is the ith element of $\boldsymbol{\theta}$. Such squared difference is referred to as the MSE:

$$\text{MSE}(\hat{\theta}_i) = E\left\{(\hat{\theta}_i - \theta_i)^2\right\},$$

where expectation is taken with respect to the rand
above expression as

$$\text{MSE}(\hat{\theta}_i) \;=\; E\left\{\left[\left(\hat{\theta}_i - E\left\{\hat{\theta}_i\right\}\right) + \Big(\right.\right.$$

$$\;=\; \underbrace{\left[\hat{\theta}_i - E\left\{\hat{\theta}_i\right\}\right]^2}_{\text{variance}} + \underbrace{\left[E\left\{\hat{\theta}_i\right.\right.}_{\text{(b}}$$

it can be seen that the mean-square error (MSE)
terms. The first term, variance, measures the vari
its average. The second term, squared bias, represen
average estimate and the true value θ_i. Unfortuna
depends on the true value θ_i. Therefore, in the cla
where θ_i is deterministic but unknown, the estimat
general not realizable. In view of this, one "good"
the variance of estimate while constraining the avera
value (i.e., $E\left\{\hat{\theta}_i\right\} = \theta_i$). Such estimator is referred
unbiased estimator (MVUE).

Example 10.1.1. [1] Consider a very simple exam

$$x[n] = A + w[n] \qquad n = 0, 1, ..$$

where $x[n]$ represent the observations, A is the param
$w[n]$ denote the independent and identically distrib
with variance σ^2. A reasonable estimator for A is th

$$\hat{A} = \frac{1}{N} \sum_{n=0}^{N-1} x[n].$$

If we compute the mean of the estimate, we have

$$E\left\{\hat{A}\right\} = \frac{1}{N} \sum_{n=0}^{N-1} E\left\{x[n]\right\} = \frac{1}{N} \sum_{n}^{N}$$

Therefore, the sample mean estimator is unbiased. N
tor

$$\check{A} = \frac{a}{N} \sum_{n=0}^{N-1} x[n],$$

where a is an adjustable constant such that the MSE

shown that $E\left\{\check{A}\right\} = aA$. Furthermore, the variance of $\check{A}$ can be computed as

$$\text{var}\left(\check{A}\right) = a^2 E\left\{\left[\frac{1}{N}\sum_{n=0}^{N-1}x[n] - A\right]^2\right\}$$

$$= a^2 E\left\{\frac{1}{N^2}\sum_{n=0}^{N-1}\sum_{m=0}^{N-1}(A + w[n])(A + w[m]) - \frac{2A}{N}\sum_{n=0}^{N-1}(A + w[n]) + A^2\right\}$$

$$= a^2\left[A^2 + \frac{1}{N}\sigma^2 - 2A^2 + A^2\right]$$

$$= \frac{a^2\sigma^2}{N}.$$

Putting the mean and variance expressions into the bias-variance decomposition of MSE expression (10.2), the MSE can be shown to be

$$\text{MSE}(\check{A}) = \frac{a^2\sigma^2}{N} + (a-1)^2 A^2.$$

Differentiating the MSE with respect to a and setting the result to zero yields

$$a_{opt} = \frac{A^2}{A^2 + \sigma^2/N}.$$

It is obvious that the optimal a depends on the unknown parameter A, and therefore, the optimal estimator might not be realizable. $\square$

10.2 Bound on Minimum Variance – Cramér-Rao Lower Bound

Before we discuss a general method of finding MVUE, we first introduce the Cramér-Rao lower bound (CRLB) in this section. CRLB is a lower bound on the variance of any unbiased estimator. This bound is important, as it provides a benchmark performance for any unbiased estimator. If we have an unbiased estimator that can reach the CRLB, then we know that it is the MVUE, and there is no other unbiased estimator that can perform better (at most another estimator would have equal performance). Even though a given unbiased estimator cannot touch the CRLB, we still have an idea of how far away the performance of this estimator is from the theoretical limit.

10.2.1 Computation of CRLB

The CRLB is derived from the likelihood function, which is the probability density function (PDF) of the observations viewed as a function of the unknown parameter.

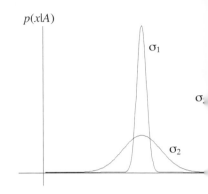

Figure 10.1: Sharpness of likelihood function a

In (10.1), if the noise **w** is Gaussian with covariance
function is

$$p(\mathbf{x}|\boldsymbol{\theta}) = \frac{1}{(2\pi)^{N/2} \det^{1/2}[\mathbf{C}]} \exp\left[-\frac{1}{2}(\mathbf{x} - \mathbf{f}(\boldsymbol{\theta}))\right.$$

In likelihood function, **x** is considered to be fixed, an
the relative probability that $\boldsymbol{\theta}$ being the true par
that $\boldsymbol{\theta}$ generates the given observation **x**). With th
method for estimation is the maximum likelihood (M.
searches for $\boldsymbol{\theta}$ that maximizes $p(\mathbf{x}|\boldsymbol{\theta})$. We will discuss
later in this chapter.

The "sharpness" of the likelihood function conta
timate accuracy any estimator can achieve. For exa
likelihood functions for the simple model $x = A + u$
two different variances. It can be seen that the lik
noise variance has a much sharper peak, and the val
lihood lie within a small range. Thus, the varianc
will be smaller. Mathematically, the "sharpness" is c
derivative of the likelihood function. This is exactly

Theorem 10.2.1. *(Cramér-Rao Lower Bound) [1]*
$p(\mathbf{x}|\boldsymbol{\theta})$ satisfies the regularity condition

$$E\left\{\left[\frac{\partial \ln p(\mathbf{x}|\boldsymbol{\theta})}{\partial \boldsymbol{\theta}}\right]\right\} = 0$$

where the expectation is taken with respect to $p(\mathbf{x}|\boldsymbol{\theta})$
of any unbiased estimator $\hat{\boldsymbol{\theta}}$ satisfies

$$\mathbf{C}_{\hat{\theta}} \geq \mathbf{I}^{-1}(\boldsymbol{\theta})$$

where $\mathbf{A} \geq \mathbf{B}$ means $\mathbf{A} - \mathbf{B}$ is positive semidefinite. Matrix $\mathbf{I}(\boldsymbol{\theta})$ is the Fisher information matrix with the $(i,j)^{th}$ element given by

$$[\mathbf{I}(\boldsymbol{\theta})]_{ij} = -E\left\{\left[\frac{\partial^2 \ln p(\mathbf{x}|\boldsymbol{\theta})}{\partial \theta_i \partial \theta_j}\right]\right\} \tag{10.5}$$

where the expectation is taken with respect to $p(\mathbf{x}|\boldsymbol{\theta})$, and θ_i stands for the ith element of $\boldsymbol{\theta}$.

Equation (10.4) is an inequality on the covariance matrix, which includes as a corollary the lower bound on the variance of individual element-wise estimates: $\text{var}(\theta_i) = [\mathbf{C}_{\hat{\theta}}]_{ii} \geq [\mathbf{I}^{-1}(\boldsymbol{\theta})]_{ii}$. Also notice that the discussion here is about the potential accuracy, and is irrelevant to the specific estimation method being used. Of course, although the potential ultimate accuracy might be high, using a bad estimator might yield poor performance.

Example 10.2.1. [3] In clock synchronization in a wireless sensor node, the observations can be modeled as

$$x_1[n] = \beta_1 t_1[n] + \beta_0 + \beta_1(d + w[n]), \qquad n = 0, ..., N-1, \tag{10.6}$$

$$x_2[n] = t_2[n]/\beta_1 - \beta_0/\beta_1 + (d + z[n]), \qquad n = 0, ..., N-1, \tag{10.7}$$

where $\{x_1[n], x_2[n]\}_{n=0}^{N-1}$ are the observations, $\{t_1[n], t_2[n]\}_{n=0}^{N-1}$ are known quantities and can be considered as the training data, β_1, β_0 and d are the unknowns to be estimated, and $w[n]$ and $z[n]$ are i.i.d. observation noise sequences with zero mean and variance σ^2.

Since the random delays $w[n]$ and $z[n]$ follow i.i.d. Gaussian distributions, the PDF of the observations conditioned on β_1, β_0 and d can be expressed from (10.6)-(10.7) as

$$\ln p(\{x_1[n], x_2[n]\}_{n=0}^{N-1}|\beta_1, \beta_0, d)$$
$$= \ln \frac{N}{2\pi\sigma^2} - \frac{1}{2\sigma^2} \times \sum_{n=0}^{N-1}\left[\left(\frac{x_1[n]}{\beta_1} - t_1[n] - \frac{\beta_0}{\beta_1} - d\right)^2 + \left(x_2[n] - \frac{t_2[n]}{\beta_1} + \frac{\beta_0}{\beta_1} - d\right)^2\right]. \tag{10.8}$$

To derive the CRLB, we first check the regularity condition. Taking β_0 as an example, it can be shown that

$$\frac{\partial \ln p}{\partial \beta_0} = -\frac{1}{\sigma^2}\sum_{n=0}^{N-1}\left[-\frac{1}{\beta_1}\left(\frac{x_1[n] - \beta_0}{\beta_1} - t_1[n] - d\right) + \frac{1}{\beta_1}\left(x_2[n] - d - \frac{t_2[n] - \beta_0}{\beta_1}\right)\right]. \tag{10.9}$$

Putting (10.6) and (10.7) into (10.9), it yields:

$$\frac{\partial \ln p}{\partial \beta_0} = -\frac{1}{\sigma^2}\sum_{i=0}^{N-1}\frac{z[n] - w[n]}{\beta_1}. \tag{10.10}$$

Since $w[n]$ and $z[n]$ have zero means, it is easy t[
(10.10) is zero (notice that taking the expectation
equivalent to taking expectation with respect to $w[n$
satisfies the regularity condition for β_0. The regulari
be checked in a similar way.

From (10.5), the Fisher information matrix is gi

$$
\mathbf{I}(\beta_1, \beta_0, d) = \begin{bmatrix} -E\left\{\frac{\partial^2 \ln p}{\partial \beta_1^2}\right\} & -E\left\{\frac{\partial^2 \ln p}{\partial \beta_1 \partial \beta_0}\right. \\ -E\left\{\frac{\partial^2 \ln p}{\partial \beta_0^2}\right\} & -E\left\{\frac{\partial^2 \ln p}{\partial \beta_0 \partial \beta_1}\right. \\ -E\left\{\frac{\partial^2 \ln p}{\partial d^2}\right\} & -E\left\{\frac{\partial^2 \ln p}{\partial d \partial \beta_1}\right. \end{bmatrix}
$$

and each element in the matrix can be obtained
$-E\left\{\frac{\partial^2 \ln p}{\partial \beta_1 \partial \beta_0}\right\}$, based on (10.9), we obtain

$$
\frac{\partial^2 \ln p}{\partial \beta_1 \partial \beta_0} = -\frac{1}{\sigma^2} \sum_{n=0}^{N-1} \left[\frac{2(x_1[n] + t_2[n] - 2\beta_0)}{\beta_1^3} - \right.
$$

Putting (10.6) and (10.7) into (10.11), we have

$$
\frac{\partial^2 \ln p}{\partial \beta_1 \partial \beta_0} = -\frac{1}{\sigma^2} \sum_{n=0}^{N-1} \left(\frac{t_1[n] + d + 2w[n] - z[n]}{\beta_1^2}\right.
$$

Taking expectation of (10.12) with respect to rando

$$
-E\left\{\frac{\partial^2 \ln p}{\partial \beta_1 \partial \beta_0}\right\} = \frac{1}{\sigma^2 \beta_1^3} \sum_{n=0}^{N-1} [\beta_1(t_1[n] +
$$

where we have used the fact that $w[n]$ and $z[n]$ ha
other elements of $\mathbf{I}(\beta_1, \beta_0, d)$ can be obtained and t
can be expressed as

$$
\mathbf{I}(\beta_1, \beta_0, d) = \frac{1}{\sigma^2} \begin{bmatrix} \mathcal{A} & \mathcal{B} \\ \mathcal{B} & \frac{2N}{\beta_1^2} \\ \mathcal{C} & 0 \end{bmatrix}
$$

where $\mathcal{A} := \beta_1^{-4} \sum_{n=0}^{N-1} [\beta_1^2(t_1[n] + d)^2 + \beta_1^2 \sigma^2 + (t_2[$
$\mathcal{B} := \beta_1^{-3} \sum_{n=0}^{N-1} [\beta_1(t_1[n] + d) + (t_2[n] - \beta_0)]$, and
$\mathcal{C} := \beta_1^{-2} \sum_{n=0}^{N-1} [\beta_1(t_1[n] + d) - (t_2[n] - \beta_0)]$. By in
can be shown that the CRLB for each parameter (

$\mathbf{I}^{-1}(\beta_1, \beta_0, d))$ is given, respectively, by:

$$\mathrm{CRLB}(\beta_1) = \frac{2N\sigma^2}{2N\mathcal{A} - \beta_1^2\mathcal{B}^2 - \mathcal{C}^2} ,$$

$$\mathrm{CRLB}(\beta_0) = \frac{\sigma^2\beta_1^2(2N\mathcal{A} - \mathcal{C}^2)}{2N(2N\mathcal{A} - \beta_1^2\mathcal{B}^2 - \mathcal{C}^2)} ,$$

$$\mathrm{CRLB}(d) = \frac{\sigma^2(2N\mathcal{A} - \beta_1^2\mathcal{B}^2)}{2N(2N\mathcal{A} - \beta_1^2\mathcal{B}^2 - \mathcal{C}^2)} .$$

As can be seen from this example, CRLB in general depends on the true parameters.   □

10.2.2 Finding MVUE Attaining the CRLB

With the lower bound on minimum variance, one may wonder if we can construct an unbiased estimator that can reach the CRLB. The answer is yes. Such an estimator is called *efficient* and might be obtained during the evaluation of the CRLB.

Theorem 10.2.2. *[1] [2] An estimator $\hat{\boldsymbol{\theta}} = \mathbf{g}(\mathbf{x})$ is the MVUE if and only if*

$$\frac{\partial \ln p(\mathbf{x}|\boldsymbol{\theta})}{\partial \boldsymbol{\theta}} = \mathbf{I}(\boldsymbol{\theta})(\mathbf{g}(\mathbf{x}) - \boldsymbol{\theta}) \tag{10.14}$$

for some matrix $\mathbf{I}(\boldsymbol{\theta})$. Furthermore, the covariance matrix of the MVUE is $\mathbf{I}(\boldsymbol{\theta})^{-1}$, which is also the CRLB.

Example 10.2.2. [1] Let us consider a linear model

$$\mathbf{x} = \mathbf{H}\boldsymbol{\theta} + \mathbf{s} + \mathbf{w}$$

where $\mathbf{x}$ is the observation vector, $\mathbf{H}$ represents a known matrix, $\boldsymbol{\theta}$ denotes the parameter vector to be estimated, $\mathbf{s}$ is a known vector, and $\mathbf{w}$ stands for a noise vector with PDF $\mathcal{N}(\mathbf{0}, \mathbf{C})$. Subtracting $\mathbf{s}$ on both sides, and applying the whitening matrix $\mathbf{D}$ with $\mathbf{D}^T\mathbf{D} = \mathbf{C}^{-1}$, we have

$$\mathbf{D}(\mathbf{x} - \mathbf{s}) = \mathbf{D}\mathbf{H}\boldsymbol{\theta} + \mathbf{D}\mathbf{w}$$
$$\mathbf{x}' = \mathbf{H}'\boldsymbol{\theta} + \mathbf{w}'.$$

It can be easily shown that $\mathbf{w}'$ is Gaussian with zero mean and the covariance matrix is an identity matrix. Taking the first order derivative of logarithm of the likelihood function in the form of (10.3), it follows that:

$$\frac{\partial \ln p(\mathbf{x}'|\boldsymbol{\theta})}{\partial \boldsymbol{\theta}} = -\frac{1}{2}\frac{\partial}{\partial \boldsymbol{\theta}}[\mathbf{x}'^T\mathbf{x}' - 2\mathbf{x}'^T\mathbf{H}'\boldsymbol{\theta} + \boldsymbol{\theta}^T\mathbf{H}'\boldsymbol{\theta} + \boldsymbol{\theta}^T\mathbf{H}'^T\mathbf{H}\boldsymbol{\theta}].$$

Since

$$\frac{\partial \mathbf{x}'^T \mathbf{H}' \boldsymbol{\theta}}{\partial \boldsymbol{\theta}} = \mathbf{H}'^T \mathbf{x}',$$

$$\frac{\partial \boldsymbol{\theta}^T \mathbf{H}'^T \mathbf{H}' \boldsymbol{\theta}}{\partial \boldsymbol{\theta}} = 2\mathbf{H}'^T \mathbf{H}'$$

we have

$$\frac{\partial \ln p(\mathbf{x}'|\boldsymbol{\theta})}{\partial \boldsymbol{\theta}} = \mathbf{H}'^T \mathbf{x}' - \mathbf{H}'^T \mathbf{H}' \boldsymbol{\theta}$$

$$= \mathbf{H}'^T \mathbf{H}'[(\mathbf{H}'^T \mathbf{H}')^{-1}]$$

where in the last step, it is assumed that $\mathbf{H}'^T \mathbf{H}'$ is in
equation with (10.14), it is obvious that we obtain

$$\hat{\boldsymbol{\theta}} = (\mathbf{H}'^T \mathbf{H}')^{-1} \mathbf{H}'^T \mathbf{x}' = (\mathbf{H}^T \mathbf{C}^{-1} \mathbf{H})^{-1} \mathbf{H}$$

$$\mathbf{C}_{\hat{\boldsymbol{\theta}}} = \mathbf{I}(\boldsymbol{\theta})^{-1} = (\mathbf{H} \mathbf{C}^{-1} \mathbf{H})^{-1},$$

the MVUE and its corresponding covariance, respec
does not require the computation of second order de
CRLB.

10.3   MVUE Using Rao-Blackw Scheffe Theorem

In the previous section, we introduced a method for
the CRLB. However, even if there is no estimator th
may still exist a MVUE (the MVUE in such case
In this section, we introduce another technique th
from the likelihood function, and this technique req
statistics.

10.3.1   Sufficient Statistics

In a general estimation problem, we have access to a
$x[0]$, $x[1]$, ..., $x[N-1]$. These observations contain inf
we want to estimate. In general, making use of all av
best estimation performance (how to use them is a
wonder: Is there any way to compress the data wh
parameter estimation is not lost? If so, we can cc
into sufficient statistics and use them for parameter
the original data set is a sufficient statistic itself. F

observations, there are many sufficient statistics. The one with the minimum number of elements is called the minimal sufficient statistic.

Formally, the concept of sufficient statistic is defined as follows. Let $\mathbf{x}$ be a vector of observations, and $\boldsymbol{\theta}$ be the parameter of interest, with the likelihood function given by $p(\mathbf{x}|\boldsymbol{\theta})$. A statistic $\mathbf{T}(\mathbf{x})$ is sufficient for $\boldsymbol{\theta}$ if $p(\mathbf{x}|\mathbf{T}(\mathbf{x}), \boldsymbol{\theta})$ is independent of $\boldsymbol{\theta}$. Intuitively, if $\mathbf{T}(\mathbf{x})$ is observed, and if $p(\mathbf{x}|\mathbf{T}(\mathbf{x}), \boldsymbol{\theta})$ still depends on $\boldsymbol{\theta}$, there must be some information contain in $\mathbf{x}$ but not in $\mathbf{T}(\mathbf{x})$, and therefore $\mathbf{T}(\mathbf{x})$ cannot be a sufficient statistic.

Example 10.3.1. [2] Consider a sequence of independent Bernoulli trial $x[0]$, $x[1]$, ..., $x[N-1] \in \{0,1\}$. The probability for $x[n] = 1$ is θ. Then the likelihood function for estimation of θ is

$$p(\mathbf{x}|\theta) = \prod_{n=0}^{N-1} \theta^{x[n]}(1-\theta)^{1-x[n]} = \theta^k(1-\theta)^{N-k} \qquad (10.18)$$

where $k := \sum_{n=0}^{N-1} x[n]$. Now consider

$$p(\mathbf{x}|k,\theta) = \frac{p(\mathbf{x},k|\theta)}{p(k|\theta)}.$$

Since k completely depends on $\mathbf{x}$, $p(\mathbf{x},k|\theta)$ is zero except when k equals the number of ones in $\mathbf{x}$. When it does, $p(\mathbf{x},k|\theta) = p(\mathbf{x}|\theta)$. On the other hand, $p(k|\theta)$ is the distribution of number of ones in N independent Bernoulli trials, and it is given by

$$p(k|\theta) = \binom{N}{k}\theta^k(1-\theta)^{N-k}.$$

Therefore,

$$p(\mathbf{x}|k,\theta) = \frac{\theta^k(1-\theta)^{N-k}}{\binom{N}{k}\theta^k(1-\theta)^{N-k}} = \frac{1}{\binom{N}{k}}.$$

It is obvious that $p(\mathbf{x}|k,\theta)$ is independent of θ, and $k = \sum_{n=0}^{N-1} x[n]$ is a sufficient statistic. That is, for estimation of θ, we only need to store the number of ones in the observations rather than the original sequence of ones and zeros. □

Finding sufficient statistic using the basic definition involves guessing the sufficient statistic and verifying the conditional PDF being independent of the parameter of interest. Both steps can be challenging. Fortunately, we have the following theorem to help us in practice.

Theorem 10.3.1. *(Neyman-Fisher Factorization) [1] [2] Let $\mathbf{x}$ be a vector of observations with the likelihood function $p(\mathbf{x}|\boldsymbol{\theta})$. The statistic $\mathbf{T}(\mathbf{x})$ is sufficient if and only if the likelihood function can be factorized as follows*

$$p(\mathbf{x}|\boldsymbol{\theta}) = g(\mathbf{T}(\mathbf{x}), \boldsymbol{\theta})h(\mathbf{x})$$

where g is a function depending on $\mathbf{x}$ only through $\mathbf{T}(\mathbf{x})$ and h is a function depending on $\mathbf{x}$ only.

Now, let us revisit the previous example of Be
function in (10.18) can be expressed as

$$p(\mathbf{x}|\theta) = \underbrace{\theta^k (1-\theta)^{N-k}}_{g(T(\mathbf{x})=k,\theta)} \cdot$$

The first term g depends on $\mathbf{x}$ through k only, while
fore, from the Neyman-Fisher factorization Theore
Obviously, identifying the sufficient statistic using
Below is another example.

Example 10.3.2. [1] For the mean estimation from
for $n = 0, 1, ..., N-1$, with $w[n]$ being i.i.d. zero me
variance σ^2, the likelihood function is given by

$$p(\mathbf{x}|A) = \frac{1}{(2\pi\sigma^2)^{N/2}} \exp\left[-\frac{1}{2\sigma^2} \sum_{n=0}^{N-1} (x[n]-A)^2\right]$$

$$= \underbrace{\frac{1}{(2\pi\sigma^2)^{N/2}} \exp\left[-\frac{1}{2\sigma^2}\left(NA^2 - 2A\sum_{n=0}^{N-1} x\right)\right.}_{g(T(\mathbf{x})=\sum_{n=0}^{N-1} x[n], A)}$$

Now, g depends on $\mathbf{x}$ through $\sum_{n=0}^{N-1} x[n]$ only, wh
and is independent of A. Clearly, $T(\mathbf{x}) = \sum_{n=0}^{N-1} x$
estimating A. Notice that $T'(\mathbf{x}) = 2\sum_{n=0}^{N-1} x[n]$ is a
In fact, sufficient statistics are unique only to within
□

Another concept that is important to the determin
of *complete* sufficient statistic. In essence, a statist
one function of the statistic that forms an unbiased
sufficient statistic for $\boldsymbol{\theta}$. Furthermore, suppose there
$\mathbf{T}(\mathbf{x})$ that are unbiased estimators of $\boldsymbol{\theta}$. Equivalentl

$$E\left\{\mathbf{g}_1(\mathbf{T}(\mathbf{x})) - \mathbf{g}_2(\mathbf{T}(\mathbf{x}))\right\}$$

for all $\boldsymbol{\theta}$, and the expectation is taken with respect
alently the sufficient statistics). In order to prove
statistic, we must prove that if (10.21) holds, then **g**

Example 10.3.3. [1] In the estimation of A from
with $w[n]$ being i.i.d. Gaussian noise with zero me
shown in the previous example that a sufficient statist
Now suppose there are two unbiased estimators $g_1($

$$E\left\{g_1(T(\mathbf{x})) - g_2(T(\mathbf{x}))\right\} =$$

for all A. Since $\tau := T(\mathbf{x}) \sim \mathcal{N}(NA, N\sigma^2)$, and defining $v(\tau) := g_1(\tau) - g_2(\tau)$, the above equation is equivalent to

$$\int_{-\infty}^{-\infty} v(\tau) \frac{1}{\sqrt{2\pi N\sigma^2}} \exp\left[-\frac{1}{2N\sigma^2}(NA - \tau)^2\right] d\tau = 0.$$

It can be recognized that the above equation is the convolution of $v(\tau)$ and a Gaussian pulse. Since the Gaussian pulse is non-negative, in order to have the convolution equal to zero for all A, $v(\tau)$ must be identically zero. Therefore, $g_1(T(\mathbf{x})) = g_2(T(\mathbf{x}))$, implying the unbiased estimator is unique, and the sufficient statistic $T(\mathbf{x}) = \sum_{n=0}^{N-1} x[n]$ is complete. $\square$

10.3.2   Finding MVUE from Sufficient Statistics

With the background of sufficient statistics, we now present a theorem that helps us to identify the MVUE.

Theorem 10.3.2. *(Rao-Blackwell-Lehmann-Scheffe) [1] [2] Let $\check{\boldsymbol{\theta}}$ be an unbiased estimator of $\boldsymbol{\theta}$, and $\mathbf{T}(\mathbf{x})$ be a sufficient statistic for $\boldsymbol{\theta}$. Then $\hat{\boldsymbol{\theta}} = E\left\{\check{\boldsymbol{\theta}}|\mathbf{T}(\mathbf{x})\right\}$ improves on $\check{\boldsymbol{\theta}}$ in the following ways:*

1. $\hat{\boldsymbol{\theta}}$ is a valid unbiased estimator for $\boldsymbol{\theta}$;

2. $\mathrm{var}\left(\hat{\boldsymbol{\theta}}\right) \leq \mathrm{var}\left(\check{\boldsymbol{\theta}}\right).$

The result of this theorem is important in the sense that it provides a way to improve the variance of any unbiased estimator using sufficient statistics. However, this theorem does not immediately offer the MVUE, since it is not guaranteed that the improved estimator is of minimum variance within the class of unbiased estimators. But if we have the additional knowledge that the sufficient statistic is complete, the estimator $\hat{\boldsymbol{\theta}} = E\left\{\check{\boldsymbol{\theta}}|\mathbf{T}(\mathbf{x})\right\}$ is the only unbiased estimator for $\boldsymbol{\theta}$ while making use of the sufficient statistic. Therefore, it must be the MVUE. In fact, if the estimator we start with $\check{\boldsymbol{\theta}}$ is also an unbiased estimator using complete sufficient statistics, it is automatically the MVUE, since there is only one unbiased estimator if the sufficient statistic is complete. There is even no need to compute the expectation $E\left\{\check{\boldsymbol{\theta}}|\mathbf{T}(\mathbf{x})\right\}$. In summary, the procedure of determining MVUE using Rao-Blackwell-Lehmann-Scheffe theorem is as follows.

1. Find a sufficient statistic for $\boldsymbol{\theta}$, $\mathbf{T}(\mathbf{x})$, using Neyman-Fisher factorization Theorem.

2. Determine if the sufficient statistic is complete.

3. If so, find a function $\hat{\boldsymbol{\theta}} = \mathbf{g}(\mathbf{T}(\mathbf{x}))$ which is an unbiased estimator of $\boldsymbol{\theta}$. The MVUE will be $\hat{\boldsymbol{\theta}}$.

Example 10.3.4. [4] In the clock offset estimation
observation equations can be written as

$$U_n = \delta + \phi + w_n, \qquad n = 1$$
$$V_n = \delta - \phi + z_n, \qquad n = 1$$

where δ symbolizes the fixed portions of the transmis
the variable portions of delays and assume i.i.d. e
means α and β, respectively, and ϕ stands for the
reference time. We want to determine the MVUE o
procedure introduced above.

First, we need to determine a sufficient statistic
by

$$L(\boldsymbol{\theta}) = \underbrace{\alpha^{-N} e^{-\frac{1}{\alpha} \sum_{n=1}^{N} (U_n - \delta - \phi)}}_{:=g_1(\sum_{n=1}^{N} U_n, \delta, \phi, \alpha)} \underbrace{u\left[U_{(}\right.}_{:=g}$$

$$\underbrace{\beta^{-N} e^{-\frac{1}{\beta} \sum_{k=1}^{N} (V_k - \delta + \phi)}}_{:=g_2(\sum_{n=1}^{N} V_n, \delta, \phi, \alpha)} \underbrace{u\left[V_{(1)} - \delta + \phi\right.}_{:=g_4(V_{(1)}, \delta, \phi)}$$

where $u[\cdot]$ denotes the unit step function, and $U_{(1)}$ ar
statistics of U_k and V_k, respectively. In the above exp
dent of the unknown parameter vector $\boldsymbol{\theta}$, whereas g
depending on the data through $\mathbf{T} = \{\sum_{n=1}^{N} U_n, U_{(1)}$
according to Neyman-Fisher factorization Theorem 1
for $\boldsymbol{\theta}$.

Next, we need to prove that $\mathbf{T}$ is complete, a st
$\mathbf{T}$. Unfortunately, since $\sum_{n=1}^{N} U_n$ and $U_{(1)}$, and sin
not independent, the PDF of $\mathbf{T}$ is difficult to be obt
statistic $\mathbf{T}' := \{\sum_{n=1}^{N} (U_n - U_{(1)}), U_{(1)}, \sum_{n=1}^{N} (V_n - V_{(1)})$
is also a sufficient statistic because it is obtained usin
from $\mathbf{T}$. It can be proved that the elements of $\mathbf{T}'$
them obeys the three-parameter Gamma distributio

$$r := \sum_{n=1}^{N} (U_n - U_{(1)}) \sim \Gamma (N - 1$$

$$s := \sum_{n=1}^{N} (V_n - V_{(1)}) \sim \Gamma (N - 1$$

$$U_{(1)} \sim \Gamma (1, \alpha/N$$
$$V_{(1)} \sim \Gamma (1, \beta/N$$

Now, suppose there are two functions $g(\mathbf{T}')$ and $h(\mathbf{T}')$ that are unbiased estimators of $\boldsymbol{\theta}$. Then we have

$$E\left\{g(\mathbf{T}') - h(\mathbf{T}')\right\} = E\left\{\pi(\mathbf{T}')\right\} = 0 \qquad \forall\ \boldsymbol{\theta}$$

where $\pi(\mathbf{T}') := g(\mathbf{T}') - h(\mathbf{T}')$. As a result,

$$\int_{\delta-\phi}^{\infty}\int_{0}^{\infty}\int_{\delta+\phi}^{\infty}\int_{0}^{\infty} \pi\left(r, U_{(1)}, s, V_{(1)}\right) \cdot \frac{(\alpha\beta)^{-(N-1)}}{\{\Gamma(N-1)\}^2}(rs)^{N-2}e^{-\frac{r}{\alpha}-\frac{s}{\beta}}$$
$$\times \frac{N^2}{\alpha\beta}e^{-\frac{N}{\alpha}\{U_{(1)}-\delta-\phi\}-\frac{N}{\beta}\{V_{(1)}-\delta+\phi\}}\ dr\ dU_{(1)}\ ds\ dV_{(1)} = 0$$

for all $\boldsymbol{\theta}$. The above relation can also be expressed as

$$\int_{-\infty}^{\infty}\int_{-\infty}^{\infty}\int_{-\infty}^{\infty}\int_{-\infty}^{\infty} \left[\pi\left(r, U_{(1)}, s, V_{(1)}\right)(rs)^{N-2}u(s)u(r)u(U_{(1)}-\delta-\phi)u(V_{(1)}-\delta+\phi)\right]$$
$$\times e^{-\left\{\frac{r}{\alpha}+\frac{NU_{(1)}}{\alpha}+\frac{s}{\beta}+\frac{NV_{(1)}}{\beta}\right\}}\ dr\ dU_{(1)}\ ds\ dV_{(1)} = 0.$$

The expression on the left above is the four-dimensional Laplace transform of the function within the square brackets. It follows from the uniqueness theorem for the two-sided Laplace transform that $\pi(\mathbf{T}')(rs)^{N-2}u(s)u(r)u(U_{(1)}-\delta-\phi)u(V_{(1)}-\delta+\phi) = 0$ almost everywhere. Since r, s, $u(s)$, $u(r)$, $u(U_{(1)}-\delta-\phi)$, $u(V_{(1)}-\delta+\phi)$ are not identically zero, we can conclude that $\pi(\mathbf{T}') = 0$, resulting in $g(\mathbf{T}') = h(\mathbf{T}')$. This proves that the statistic $\mathbf{T}'$, or equivalently $\mathbf{T}$, is complete for estimating $\boldsymbol{\theta}$.

Finally, we have to find an unbiased estimator for $\boldsymbol{\theta}$ as a function of $\mathbf{T}$. A careful inspection of the sufficient statistics reveals that

$$\hat{\boldsymbol{\theta}} = \frac{1}{2(N-1)}\begin{bmatrix} N\left(U_{(1)} + V_{(1)}\right) - \frac{1}{N}\left(\sum_{n=1}^{N}U_n + \sum_{n=1}^{N}V_n\right) \\ N\left(U_{(1)} - V_{(1)}\right) - \frac{1}{N}\left(\sum_{n=1}^{N}U_n - \sum_{n=1}^{N}V_n\right) \\ 2N\left(\frac{1}{N}\sum_{n=1}^{N}U_n - U_{(1)}\right) \\ 2N\left(\frac{1}{N}\sum_{n=1}^{N}V_n - V_{(1)}\right) \end{bmatrix}$$

is an unbiased estimator of $\boldsymbol{\theta}$, which is also the MVUE according to the Rao-Blackwell-Lehmann-Scheffe Theorem 10.3.2. $\qquad\square$

10.4   Maximum Likelihood Estimation

In general, determining MVUE is difficult due to the fact that we do not know whether the MVUE for a particular problem exists. Even if it does exist, the procedure for determining the MVUE is overwhelmingly challenging for most problems. Therefore, in practice, MVUEs are only reported in isolated cases. In this section, we introduce the maximum likelihood (ML) estimation, a popular technique that usually results in practical estimators.

10.4.1   ML Estimation Principle

As it was introduced previously in Section 10.2 in the
bound, the likelihood function can be interpreted a
rameter $\boldsymbol{\theta}$ with fixed observation $\mathbf{x}$. The maximum l
resumes to finding the parameter vector that maxim

$$\hat{\boldsymbol{\theta}} = \arg\left[\max_{\tilde{\boldsymbol{\theta}}} p(\mathbf{x}|\tilde{\boldsymbol{\theta}})\right],$$

where $\tilde{\boldsymbol{\theta}}$ is the trial value of the unknown paramete
is the parameter that is the most probable for gener

Example 10.4.1.   [1] Consider the received dat
$0, 1, ..., N-1$, with A unknown, and $w[n]$ is i.i.d. G
mean and variance σ^2. We want to derive the M
likelihood function of A is given by (10.19). Takin
function gives

$$\ln p(\mathbf{x}|A) = -\frac{N}{2}\ln 2\pi - N\ln\sigma - \frac{1}{2\sigma^2}$$

Since the value of A that maximizes $p(\mathbf{x}|A)$ is t
$\ln p(\mathbf{x}|A)$, we can proceed to maximize the log-like
stead. Differentiating $\ln p(\mathbf{x}|A)$ with respect to A, w

$$\frac{\partial \ln p(\mathbf{x}|A)}{\partial A} = \frac{1}{\sigma^2}\sum_{n=0}^{N-1}(x[n]$$

Setting (10.22) to zero, the ML estimator of A is giv

$$\hat{A} = \frac{1}{N}\sum_{n=0}^{N-1} x[n],$$

which can be interpreted as the sample mean estima

Notice that the first derivative of the log-likelihoo
example can also be written as

$$\frac{\partial \ln p(\mathbf{x}|A)}{\partial A} = \frac{N}{\sigma^2}\left(\frac{1}{N}\sum_{n=0}^{N-1} x[n\right.$$

Comparing (10.23) to (10.14), it is obvious that the
MVUE. Furthermore, in the general linear model c
the first derivative of the log-likelihood function is
(10.15) to zero and solving for $\boldsymbol{\theta}$, the ML estimator
identical to the MVUE in (10.16). This can be viewe
important result.

Theorem 10.4.1. *(Optimality of the MLE for the Linear Model) [1] [2] For the data* **x** *described by the general linear model*

$$\mathbf{x} = \mathbf{H}\boldsymbol{\theta} + \mathbf{w}$$

where **H** *stands for a known* $N \times p$ *matrix with* $N > p$ *and of full rank,* $\boldsymbol{\theta}$ *represents a* $p \times 1$ *vector of parameters to be estimated, and* **w** *denotes the Gaussian noise vector with PDF* $\mathcal{N}(\mathbf{0}, \mathbf{C})$, *the ML estimate of* $\boldsymbol{\theta}$ *is given by*

$$\hat{\boldsymbol{\theta}} = (\mathbf{H}^T \mathbf{C}^{-1} \mathbf{H})^{-1} \mathbf{H}^T \mathbf{C}^{-1} \mathbf{x}. \qquad (10.24)$$

Furthermore, $\hat{\boldsymbol{\theta}}$ *is also the MVUE and its variance* $(\mathbf{H}^T \mathbf{C}^{-1} \mathbf{H})^{-1}$ *attains the CRLB.*

The above two examples are not pure coincidence. In fact, if an unbiased estimator attains the CRLB (i.e., an efficient estimator exists), the ML procedure will produce it. This can be easily explained by the fact that if an efficient estimator exists, (10.14) holds. On the other hand, ML estimate is obtained by setting the first derivative of the log-likelihood function, i.e., (10.14), to zero. This automatically produces $\hat{\boldsymbol{\theta}} = \mathbf{g}(\mathbf{x})$, which is also the MVUE.

10.4.2 Properties of the ML Estimator

However, being identical to the MVUE in certain situations is not the main reason for the popularity of the ML estimation, as these situations do not frequently occur. On the other hand, the following property itself is strong enough for the ML estimation to be useful in practice.

Theorem 10.4.2. *(Asymptotic Property of the ML estimator) [1] [2] If the likelihood function* $p(\mathbf{x}|\boldsymbol{\theta})$ *of data* **x** *satisfies certain "regularity" conditions (existence of the derivatives of the log-likelihood function, as well as the Fisher information being nonzero), then the ML estimate of* $\boldsymbol{\theta}$ *is asymptotically distributed (for large data records) according to*

$$\hat{\boldsymbol{\theta}} \overset{a}{\sim} \mathcal{N}(\boldsymbol{\theta}, \mathbf{I}^{-1}(\boldsymbol{\theta}))$$

where $\mathbf{I}(\boldsymbol{\theta})$ *is the Fisher information matrix evaluated at the true value of the unknown parameter.*

This theorem states that when the observation data length tends to infinity, the ML estimate is asymptotically unbiased and asymptotically attains the CRLB. The ML estimate is therefore asymptotically efficient.

Example 10.4.2. [1] Consider $x[n] = A + w[n]$ for $n = 0, 1, ..., N - 1$, with $A > 0$ is the parameter of interest, and $w[n]$ is i.i.d. Gaussian noise with unknown variance A. The likelihood function of A is

$$p(\mathbf{x}|A) = \frac{1}{(2\pi A)^{N/2}} \exp\left[-\frac{1}{2A} \sum_{n=0}^{N-1} (x[n] - A)^2 \right].$$

Notice that this problem is not the same as that in t
as both the mean of received data and the varianc
the log-likelihood function with respect to A, it follc

$$\frac{\partial \ln p(\mathbf{x}|A)}{\partial A} = -\frac{N}{2A} + \frac{1}{A}\sum_{n=0}^{N-1}(x[n] - A) + \frac{1}{2A^2}$$

Setting the result to zero produces

$$A^2 + A - \frac{1}{N}\sum_{n=0}^{N-1}x^2[n] =$$

which yields the ML estimate of A:

$$\hat{A} = -\frac{1}{2} + \sqrt{\frac{1}{N}\sum_{n=0}^{N-1}x^2[n]}$$

and the other solution is dropped to guarantee $A >$
Now, we will examine the asymptotic mean and v
To simply the notation, we let $u = \frac{1}{N}\sum_{n=0}^{N-1}x^2[n]$, a

$$\hat{A} = g(u) = -\frac{1}{2} + \sqrt{u+}$$

As $N \to \infty$, $u \to E\{x^2[n]\} = A^2 + A := u_o$. Us
approximation for $g(u)$ around u_o yields

$$\hat{A} = g(u) \approx g(u_o) + \left.\frac{dg(u)}{du}\right|_{u=u_o}(u$$

$$= A + \frac{1/2}{A+1/2}\left[\frac{1}{N}\sum_{n=0}^{N-1}x^2[n] -\right.$$

Therefore, $E\left\{\hat{A}\right\} = A$, so $\hat{A}$ is asymptotically unbia
totic variance is computed as

$$\text{var}\left(\hat{A}\right) = \left(\frac{1/2}{A+1/2}\right)^2 \text{var}\left(\frac{1}{N}\right.$$

$$= \frac{1/4}{N(A+1/2)^2}\text{var}\left(x^2[n\right.$$

Since $x[n]$ is Gaussian distributed with mean A anc
that $\text{var}\left(x^2[n]\right) = 4A^3 + 2A^2$, and therefore

$$\text{var}\left(\hat{A}\right) = \frac{A^2}{N(A+1/2)}$$

To check whether this asymptotic variance equals the CRLB, we differentiate (10.25) with respect to A again, and obtain

$$\frac{\partial^2 \ln p(\mathbf{x}|A)}{\partial A^2} = \frac{N}{2A^2} - \frac{1}{A^2}\sum_{n=0}^{N-1}(x[n]-A) - \frac{N}{A} - \frac{1}{A^3}\sum_{n=0}^{N-1}(x[n]-A)^2 - \frac{1}{A^2}\sum_{n=0}^{N-1}(x[n]-A).$$

The CRLB of A is then given by

$$\left[-E\left\{ \frac{\partial^2 \ln p(\mathbf{x}|A)}{\partial A^2} \right\} \right]^{-1} = -\left(\frac{N}{2A^2} - \frac{N}{A} - \frac{N}{A^2} \right)^{-1}$$

$$= \frac{A^2}{N(A+1/2)}.$$

The asymptotic variance (10.26) coincides with the CRLB. Therefore, the ML estimator for A is asymptotic efficient. $\qquad\square$

Example 10.4.3. [5] In orthogonal frequency division multiple access (OFDMA) uplink, multiple signals from different users arrive at the base station at the same time. However, their carrier frequency offsets (CFO) and channels are different and unknown. Before the data can be detected, we need to estimate all these parameters. In general, the received signal model in such systems can be expressed in the form of matrix equation:

$$\mathbf{x} = \mathbf{Q}(\boldsymbol{\omega})\mathbf{h} + \mathbf{w}, \qquad (10.27)$$

where $\mathbf{x}$ stands for the received data from one OFDM symbol, $\mathbf{h}$ denotes an unknown vector containing the channel coefficients of different users, $\mathbf{Q}(\boldsymbol{\omega})$ is a matrix with known structure but parameterized nonlinearly by the unknown vector $\boldsymbol{\omega} = [\omega_1\ \omega_2\ ...\omega_K]^T$ of frequency offsets, $\mathbf{w}$ represents the i.i.d. complex Gaussian noise with zero mean and variance σ^2. The ML estimate of parameters $\{\mathbf{h}, \boldsymbol{\omega}\}$ is obtained by maximizing the likelihood function

$$p(\mathbf{x}|\tilde{\mathbf{h}}, \tilde{\boldsymbol{\omega}}) = \frac{1}{(\pi\sigma^2)^N} \cdot \exp\left\{ -\frac{1}{\sigma^2}[\mathbf{x} - \mathbf{Q}(\tilde{\boldsymbol{\omega}})\tilde{\mathbf{h}}]^H[\mathbf{x} - \mathbf{Q}(\tilde{\boldsymbol{\omega}})\tilde{\mathbf{h}}] \right\}, \qquad (10.28)$$

where $\tilde{\mathbf{h}}$ and $\tilde{\boldsymbol{\omega}}$ are trial values of $\mathbf{h}$ and $\boldsymbol{\omega}$, respectively. Notice that since $\mathbf{w}$ is complex, the PDF expression (10.28) assumes a slightly different form than that when $\mathbf{w}$ is real-valued in (10.3). Taking the logarithm of (10.28), and ignoring the constant irrelevant terms, the ML estimate can be equivalently found by minimizing

$$\Lambda(\mathbf{x}|\tilde{\mathbf{h}}, \tilde{\boldsymbol{\omega}}) = [\mathbf{x} - \mathbf{Q}(\tilde{\boldsymbol{\omega}})\tilde{\mathbf{h}}]^H[\mathbf{x} - \mathbf{Q}(\tilde{\boldsymbol{\omega}})\tilde{\mathbf{h}}]. \qquad (10.29)$$

Due to the linear dependence of parameter $\mathbf{h}$ in (10.27), the ML estimate for the channel vector $\mathbf{h}$ (when $\tilde{\boldsymbol{\omega}}$ is fixed) is given by (see (10.24))

$$\hat{\mathbf{h}} = (\mathbf{Q}^H(\tilde{\boldsymbol{\omega}})\mathbf{Q}(\tilde{\boldsymbol{\omega}}))^{-1}\mathbf{Q}^H(\tilde{\boldsymbol{\omega}})\mathbf{x}.$$

Plugging $\hat{\mathbf{h}}$ into (10.29), and retaining only those
estimate of $\boldsymbol{\omega}$ can be expressed as

$$\hat{\boldsymbol{\omega}} = \arg\max_{\tilde{\boldsymbol{\omega}}}\Big\{ J(\tilde{\boldsymbol{\omega}}) := \mathbf{x}^H \mathbf{Q}(\tilde{\boldsymbol{\omega}})(\mathbf{Q}^H(\tilde{\boldsymbol{\omega}})\mathbf{Q}(\tilde{\boldsymbol{\omega}}$$

Now the question resumes to how to maximize
One straightforward way is to use a grid search, in w
parameter is divided into many subregions and the p
by one. For example, if the parameter ω is one-dim
we can divide this range into 10,000 equally spaced i
function for 10,000 different values of ω. This appro
presence of a single unknown parameter. However,
eter vector increases, the number of points to be sea
For example, if there are four users in the OFDMA
$\boldsymbol{\omega}$ is four dimensional. If we divide each dimension
number of points to be searched is $(10,000)^4$. This i
if the parameter of interest is not bounded to a finite
is also practically infeasible.

Another possible solution is to use iterative nu
as the Newton-Raphson or steepest descent method
the optimization of a nonlinear function by a serie
with respect to a previously obtained solution. In t
the second order Taylor series expansion is applied t
around a previous estimate $\hat{\boldsymbol{\omega}}^{(i)}$:

$$J(\boldsymbol{\omega}) \approx J(\hat{\boldsymbol{\omega}}^{(i)}) + (\boldsymbol{\omega} - \hat{\boldsymbol{\omega}}^{(i)})\frac{\partial J(\boldsymbol{\omega})}{\partial \boldsymbol{\omega}}\Big|_{\boldsymbol{\omega} = \hat{\boldsymbol{\omega}}^{(i)}} + \frac{1}{2}(\boldsymbol{\omega} - \hat{\boldsymbol{\omega}}^{(i)}$$

where

$$\frac{\partial J(\boldsymbol{\omega})}{\partial \boldsymbol{\omega}} = \left[\frac{\partial J}{\partial \omega_1}, \frac{\partial J}{\partial \omega_2}, \cdots \frac{\partial}{\partial} \right.$$

is the gradient vector, and

$$\frac{\partial^2 J(\boldsymbol{\omega})}{\partial \boldsymbol{\omega}\partial \boldsymbol{\omega}^H} = \left(\begin{array}{ccc} \frac{\partial^2 J}{\partial \omega_1 \partial \omega_1} & \frac{\partial^2 J}{\partial \omega_1 \partial \omega_2} & \cdots \\ \frac{\partial^2 J}{\partial \omega_2 \partial \omega_1} & \frac{\partial^2 J}{\partial \omega_2 \partial \omega_2} & \cdots \\ \vdots & & \\ \frac{\partial^2 J}{\partial \omega_K \partial \omega_1} & \frac{\partial^2 J}{\partial \omega_K \partial \omega_2} & \cdots \end{array} \right.$$

is the Hessian of $J(\boldsymbol{\omega})$. Differentiating (10.31) with re
the result to zero leads to

$$\frac{\partial J(\boldsymbol{\omega})}{\partial \boldsymbol{\omega}}\Big|_{\boldsymbol{\omega} = \hat{\boldsymbol{\omega}}^{(i)}} + \left[\frac{\partial^2 J(\boldsymbol{\omega})}{\partial \boldsymbol{\omega}\partial \boldsymbol{\omega}^H} \right]_{\boldsymbol{\omega} = \hat{\boldsymbol{\omega}}^{(i)}} (\boldsymbol{\omega}$$

Solving the above equation gives the value of $\boldsymbol{\omega}$ that maximizes the second order approximated Taylor series expansion:

$$\hat{\boldsymbol{\omega}}^{(i+1)} = \hat{\boldsymbol{\omega}}^{(i)} - \left[\frac{\partial^2 J(\boldsymbol{\omega})}{\partial \boldsymbol{\omega} \partial \boldsymbol{\omega}^H}\right]^{-1}_{\boldsymbol{\omega}=\hat{\boldsymbol{\omega}}^{(i)}} \frac{\partial J(\boldsymbol{\omega})}{\partial \boldsymbol{\omega}}\bigg|_{\boldsymbol{\omega}=\hat{\boldsymbol{\omega}}^{(i)}}.$$

On the other hand, in the steepest descent method, the update is given by the equation:

$$\hat{\boldsymbol{\omega}}^{(i+1)} = \hat{\boldsymbol{\omega}}^{(i)} - \alpha \frac{\partial J(\boldsymbol{\omega})}{\partial \boldsymbol{\omega}}\bigg|_{\boldsymbol{\omega}=\hat{\boldsymbol{\omega}}^{(i)}},$$

where $\hat{\boldsymbol{\omega}}^{(i)}$ is the estimate of $\boldsymbol{\omega}$ at the ith iteration, the gradient vector $\partial J(\boldsymbol{\omega})/\partial \boldsymbol{\omega}$ points in the direction of the maximum increase of $J(\boldsymbol{\omega})$ at the point $\boldsymbol{\omega}$, and α is a parameter that determines how far away each update moves along the direction opposite to the gradient. Notice that the steepest descent method is derived to minimize a function, but minimizing a function is equivalent to maximizing the negative of the same function. Therefore, we do not explicitly describe the maximization algorithm.

Both methods share some similarities. For example, both methods require the initial estimate near the global optimal solution. Otherwise, in general, in the presence of multi-modal objective functions, only convergence to a local optimum can be guaranteed. Furthermore, both methods require the use of the derivative information. However, the Newton-Raphson method requires the additional computation of the Hessian, which may be difficult or impossible to obtain. Notice that even if the Hessian can be obtained, it may not be invertible. However, the use of the Hessian information (if available) ensures a faster convergence rate for the Newton-Raphson method relative to the steepest descent method.

Newton-Raphson and steepest descent methods may indeed reduce the computational complexity compared to the grid search. However, they are still multidimensional optimization approaches. To further reduce the computational burden brought by the multidimensional searches in the ML estimator, the alternating projection algorithm can be exploited. The alternating projection method reduces a K-dimensional maximization problem (10.30) into a series of one-dimensional maximization problems, by updating one parameter at a time, while keeping the other parameters fixed at the previous estimated values. Let $\hat{\omega}_k^{(i)}$ be the estimate of ω_k at the ith iteration. Further, let

$$\hat{\boldsymbol{\omega}}_{-k}^{(i)} = [\hat{\omega}_1^{(i+1)} \ \dots \ \hat{\omega}_{k-1}^{(i+1)} \ \hat{\omega}_{k+1}^{(i)} \ \dots \ \hat{\omega}_K^{(i)}]^T.$$

Given the initial estimates $\{\hat{\omega}_k^{(0)}\}_{k=0}^K$, the ith ($i \geq 1$) iteration of the alternating projection algorithm for maximizing (10.30) assumes the following form:

$$\hat{\omega}_k^{(i)} = \arg\max_{\tilde{\omega}_k} \left\{ J(\tilde{\omega}_k, \hat{\boldsymbol{\omega}}_{-k}^{(i-1)}) \right\},$$

for $k = 1, ..., K$. Grid search, steepest descent or Ne
used in each of the one-dimensional search. Multiple
until the estimates of ω_k converge to a stable solutior
gorithms, in general, the alternating projection can c
to at least a local optimum. Therefore, a good initia

Figure 10.2(a) shows the MSE performance of tl
ternating projection and grid search in each dimensic
length N. There are two users (K=2), the signal–
10dB, and the number of iterations for alternating
shown in the figure. It can be seen that the perform
proaches the CRLB when N increases, confirming tl
efficient. On the other hand, Figure 10.2(b) shows
ML estimator as a function of SNR. The OFDMA s
and $K = 4$ users. It is obvious that the MSE coincic
to high SNRs. Notice that increasing the number
effect as decreasing the noise variance in observatior
also asymptotically efficient at high SNRs.

The ML estimate of a parameter $\boldsymbol{\theta}$ is obtained
function $p(\mathbf{x}|\boldsymbol{\theta})$. But suppose we are interested not
question is if we can obtain the ML estimate of $\boldsymbol{\alpha}$ fro
a situation occurs when the ML estimate of $\boldsymbol{\theta}$ is rela
direct derivation of ML estimate of $\boldsymbol{\alpha}$ is difficult. T
this question and is another major factor in making

Theorem 10.4.3. *(Invariance property of the ML*
$\mathbf{g}(\boldsymbol{\theta})$, *where* $\boldsymbol{\theta}$ *is a* $p \times 1$ *vector with likelihood fu*
dimensional function of $\boldsymbol{\theta}$. *The ML estimator of* $\boldsymbol{\alpha}$

$$\hat{\boldsymbol{\alpha}} = \mathbf{g}(\hat{\boldsymbol{\theta}})$$

where $\hat{\boldsymbol{\theta}}$ *is the ML estimate of* $\boldsymbol{\theta}$.

Example 10.4.4. In this example, we consider ag
of wireless sensor node discussed in Example 10.2.
(10.6) and (10.7). However, in this example, the ob
are modeled as i.i.d. exponential random variables w
parameter λ. The goal is to estimate β_0, β_1, λ anc
$\{x_1[n], x_2[n]\}_{n=0}^{N-1}$.

To derive the ML estimator, we rewrite the obs
(10.7) as

$$w[n] = \frac{1}{\beta_1} \cdot x_1[n] - t_1[n] - \frac{\beta_0}{\beta_1}$$

$$z[n] = -\frac{1}{\beta_1} \cdot t_2[n] + x_2[n] +$$

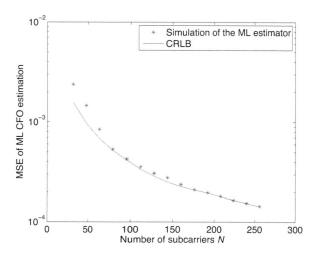

(a) CRLB and MSE of the ML estimator for ω versus N.

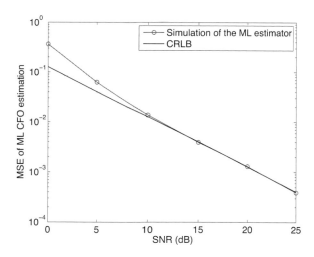

(b) CRLB and MSE of the ML estimator for ω versus SNR.

Figure 10.2: Asymptotic property of the ML estimator.

Since $\{w[n], z[n]\}_{n=0}^{N-1}$ are i.i.d. exponential rand
$p(w[n]) = \lambda \exp(-\lambda w[n])$ and $p(z[n]) = \lambda \exp(-\lambda z[n$
function of $\{x_1[n], x_2[n]\}_{n=0}^{N-1}$ is then given by $\prod_{n=0}^{N-}$
(10.32) and (10.33) into the likelihood function, we

$$p(\{x_1[n], x_2[n]\}_{n=0}^{N-1} | \lambda, \theta_1, \theta_0, d)$$

$$= \lambda^{2N} \exp \left\{ -\lambda \sum_{n=0}^{N-1} [(x_1[n] - t_2[n])\theta_1 \right.$$

$$\cdot \prod_{n=0}^{N-1} I\left[x_1[n]\theta_1 - \theta_0 - d - t_1[n] \geq 0\right]$$

$$\cdot \prod_{n=0}^{N-1} I\left[-t_2[n]\theta_1 + \theta_0 - d + x_2[n] \geq 0\right]$$

where we have used the transformations $\theta_0 := \beta_0/$
the indicator function. Notice that from the invarian
of β_0, β_1 and d is equivalent to that of θ_0, θ_1 and d
invertible one-to-one transformation.

For given θ_1, θ_0 and d, the conditional ML estin
differentiating the logarithm of (10.34) with respect
zero. It follows that

$$\hat{\lambda} = \frac{2N}{\sum_{n=0}^{N-1} [(x_1[n] - t_2[n])\theta_1 + (x_2[n]}$$

Plugging $\hat{\lambda}$ back into (10.34) and discarding some in
the concentrated likelihood function for $\{\theta_0, \theta_1, d\}$ a

$$p(\{x_1[n], x_2[n]\}_{n=1}^{N} | \theta_1, \theta_0, d) \propto \left\{ \sum_{n=0}^{N-1} [(x_1[n] - t_2[n])\right.$$

$$\cdot \prod_{n=0}^{N-1} I\left[x_1[n]\theta_1 - \theta_0 - d\right.$$

$$\cdot \prod_{n=0}^{N-1} I\left[-t_2[n]\theta_1 + \theta_0 - \right.$$

Finally, the ML estimate that maximizes (10.35) is

tion to the following linear programming problem:

$$[\hat{\theta}_0, \hat{\theta}_1, \hat{d}] = \underset{\theta_1,\theta_0,d}{\arg\max} \sum_{n=0}^{N-1} [(t_2[n] - x_1[n])\theta_1 + 2d]$$

$$\text{subject to} \begin{cases} \theta_0 - t_2[0]\theta_1 + x_2[0] - d & \geq & 0 \\ & \vdots & \\ \theta_0 - t_2[N-1]\theta_1 + x_2[N-1] - d & \geq & 0 \\ \theta_0 - x_1[0]\theta_1 + t_1[0] + d & \leq & 0 \\ & \vdots & \\ \theta_0 - x_1[N-1]\theta_1 + t_1[N-1] + d & \leq & 0 \\ d & \geq & 0. \end{cases}$$

Once $\hat{\theta}_0$ and $\hat{\theta}_1$ are obtained, we can obtain $\hat{\beta}_0 = \hat{\theta}_0/\hat{\theta}_1$ and $\hat{\beta}_1 = 1/\hat{\theta}_1$. $\square$

10.5 Least Squares Estimation

The optimal estimators introduced in the previous sections require the knowledge of PDF of the observations. If such information is not available, one reasonable approach is to minimize the sum of squared differences between the received data and the assumed noiseless signal. This approach is called Least Squares (LS) estimation. The advantage is that no probabilistic assumption is needed for the received data. However, in general, no claim of optimality can be made for the LS method.

Example 10.5.1. [6] In digital communications, direct-conversion receivers become more and more popular because of their low cost and power consumption. Nonetheless, due to mismatches in the inphase and quadrature phase (I/Q) branches, it is subjected to the challenging I/Q imbalance problem, which manifests into the mirrored inter-carrier interference in the received signal. In the analog form, a received OFDM signal with I/Q imbalance can be expressed as

$$x(t) = [s(t) \otimes h(t)] \otimes c_1(t) + [s^*(t) \otimes h^*(t)] \otimes c_2(t) + \underbrace{z(t) \otimes c_1(t) + z^*(t) \otimes c_2(t)}_{:=w(t)},$$

where $s(t)$ is the transmitted signal, $h(t)$ is the unknown channel response, $c_1(t)$ and $c_2(t)$ are the unknown I/Q imbalance filters at the receiver, $\otimes$ denotes the convolution operation, and $z(t)$ is white Gaussian noise. After sampling, the digital received signal can be expressed as

$$\mathbf{x} = \mathbf{H}\boldsymbol{\theta} + \mathbf{w}, \tag{10.36}$$

where $\boldsymbol{\theta}$ contains the combined effect of channel and I/Q imbalance filters, $\mathbf{H}$ is a known $N \times p$ matrix with $N > p$, and $\mathbf{w}$ is the effective noise. We want to estimate

$\boldsymbol{\theta}$. Equation (10.36) belongs to the familiar linear
the noise term $\mathbf{w}$ depends on the unknown I/Q in
of the unknown in $\boldsymbol{\theta}$, the variance of $\mathbf{w}$ is unknown
estimator are difficult to obtain. On the other hand
require the statistical information of $\mathbf{x}$, we can estim

$$J(\boldsymbol{\theta}) = (\mathbf{x} - \mathbf{H}\boldsymbol{\theta})^H(\mathbf{x} - \mathbf{H}\boldsymbol{\theta})$$
$$= \mathbf{x}^H\mathbf{x} - 2\mathbf{x}^H\mathbf{H}\boldsymbol{\theta} + \boldsymbol{\theta}^H\mathbf{F}$$

Differentiating $J(\boldsymbol{\theta})$ with respect to $\boldsymbol{\theta}$ yields $\partial J(\boldsymbol{\theta}$
Setting the result to zero, the LS estimate is obtaine

$$\hat{\boldsymbol{\theta}} = (\mathbf{H}^H\mathbf{H})^{-1}\mathbf{H}^H\mathbf{x}.$$

Interestingly, the LS estimate in the linear model has
and ML estimator when the observation noise has a
notice that these estimators are based on different as

10.5.1 Geometrical Interpretation

Now, we examine the geometrical interpretation of
$\mathbf{H}$ in terms of its column vectors, and $\boldsymbol{\theta}$ in terms of i
and $\boldsymbol{\theta} = [\theta_1 \ \theta_2 \ ... \ \theta_p]^T$. Furthermore, let

$$\hat{\mathbf{s}} = \mathbf{H}\boldsymbol{\theta} = \sum_{i=1}^{p} \theta_i \mathbf{h}_i$$

be the estimate of the signal without the observa
function assumes the form:

$$J(\boldsymbol{\theta}) = \left\| \mathbf{x} - \sum_{i=1}^{p} \theta_i \mathbf{h}_i \right\|^2$$

The estimated signal $\hat{\mathbf{s}} = \sum_{i=1}^{p} \theta_i \mathbf{h}_i$ can be interpret
the basis vectors $\{\mathbf{h}_1 \ \mathbf{h}_2 \ ... \ \mathbf{h}_p\}$. Since there are only
dimensional space. However, the received vector $\mathbf{x}$ is
it lies in an N-dimensional space. With these inter
the LS estimation aims to find the best $\boldsymbol{\theta}$ such that
provides the closest approximation to the observation
Figure 10.3 for $p=2$ and $N=3$.

In order to have the closest approximation, $\hat{\mathbf{s}}$ sho
tion of $\mathbf{x}$ onto the space spanned by $\{\mathbf{h}_1 \ \mathbf{h}_2 \ ... \ \mathbf{h}_p\}$
be orthogonal to $\{\mathbf{h}_1 \ \mathbf{h}_2 \ ... \ \mathbf{h}_p\}$. This is the well-k
Therefore, we can write

$$(\mathbf{x} - \mathbf{H}\boldsymbol{\theta})^T\mathbf{h}_i = 0 \qquad\qquad \text{for} \quad i =$$

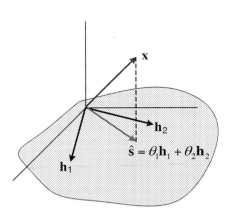

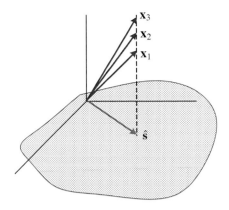

Figure 10.3: LS estimation as orthogonal projection.

Figure 10.4: Possibility of multiple $\mathbf{x}$'s with the same projection makes the projection matrix singular.

Combining all these equations into one matrix equation yields

$$(\mathbf{x} - \mathbf{H}\boldsymbol{\theta})^T \mathbf{H} = \mathbf{0}^T,$$

from which we can express the LS estimate as $\hat{\boldsymbol{\theta}} = (\mathbf{H}^T \mathbf{H})^{-1} \mathbf{H}^T \mathbf{x}$.

With the LS estimate of $\boldsymbol{\theta}$, the estimated signal is

$$\hat{\mathbf{s}} = \mathbf{H}\hat{\boldsymbol{\theta}} = \underbrace{\mathbf{H}(\mathbf{H}^T \mathbf{H})^{-1} \mathbf{H}^T}_{:=\mathbf{P}} \mathbf{x}.$$

$\mathbf{P}$ is an orthogonal projection matrix that projects $\mathbf{x}$ onto the p-dimensional subspace spanned by the columns of $\mathbf{H}$. The orthogonal projection matrix $\mathbf{P}$ presents the following two properties: $\mathbf{P}^T = \mathbf{P}$ and $\mathbf{P}^2 = \mathbf{P}$. In particular, the second property states that if $\mathbf{P}$ is applied to $\mathbf{Px}$, then the result would not change. This is reasonable since $\mathbf{Px}$ is already in the subspace spanned by the columns of $\mathbf{H}$, and projecting $\mathbf{Px}$ one more time has no effect. On the other hand, the error $\mathbf{x} - \hat{\mathbf{s}} = (\mathbf{I} - \mathbf{P})\mathbf{x}$ can be interpreted as projecting $\mathbf{x}$ onto the subspace orthogonal to the signal subspace. The matrix $\mathbf{P}^\perp := \mathbf{I} - \mathbf{P}$ is also an orthogonal projection matrix and satisfies the above two properties. In fact, for any matrix to qualify to be an orthogonal projection matrix, these two properties have to be satisfied. As a final remark, a projection matrix must be singular, otherwise $\mathbf{x}$ can be recovered from $\hat{\mathbf{s}}$. This is impossible since there may be many $\mathbf{x}$'s having the same projection, as shown in Figure 10.4.

10.5.2 Recursive LS Estimation

Previous discussion on LS estimation assumed a batch mode implementation, meaning that all the data is collected before making an inference. However, if

the data arrival is ongoing as time progresses, and
each time instant during the data collection proces
is whether the LS-estimate can be updated each t
answer is yes and such a procedure is called recursiv
we present the derivation of recursive LS estimation

Denote the model for n received data samples
particular, $\mathbf{x}[n]$ and $\mathbf{H}[n]$ can be expressed as

$$\mathbf{x}[n] = [\mathbf{x}[n-1] \ x[n]]^T$$

$$\mathbf{H}[n] = \left[\begin{array}{c} \mathbf{H}[n-1] \\ \mathbf{h}^T[n] \end{array} \right],$$

where $\mathbf{h}^T[n]$ is the additional row of $\mathbf{H}[n]$ due to the
at time n. The LS solution of $\boldsymbol{\theta}$ at time n is therefo

$$\hat{\boldsymbol{\theta}}[n] = (\mathbf{H}^T[n]\mathbf{H}[n])^{-1}\mathbf{H}^T[n]\mathbf{x}[n]$$

$$= \left([\mathbf{H}^T[n-1] \ \mathbf{h}[n]] \left[\begin{array}{c} \mathbf{H}[n-1] \\ \mathbf{h}^T[n] \end{array} \right] \right)^{-1} \cdot [\mathbf{H}^T[n$$

$$= \left(\underbrace{\mathbf{H}^T[n-1]\mathbf{H}[n-1]}_{:=\boldsymbol{\Sigma}^{-1}[n-1]} + \mathbf{h}[n]\mathbf{h}^T[n] \right)^{-1} \cdot \left(\mathbf{H}^T \right.$$

where matrix $\boldsymbol{\Sigma}[n-1]$ defined above can be interpre
at time $n-1$. Since $\mathbf{h}[n]\mathbf{h}^T[n]$ is a rank one matrix
the above equation is called a rank one update. U
(Sherman-Woodbury-Morrison Identity), we have

$$\left(\boldsymbol{\Sigma}^{-1}[n-1] + \mathbf{h}[n]\mathbf{h}^T[n] \right)^{-1} = \boldsymbol{\Sigma}[n-1] - \frac{\boldsymbol{\Sigma}[}{1}$$

$$= (\mathbf{I} - \mathbf{k}[n]\mathbf{h}^T[n]$$

where

$$\mathbf{k}[n] := \frac{\boldsymbol{\Sigma}[n-1]\mathbf{h}[n]}{1 + \mathbf{h}^T[n]\boldsymbol{\Sigma}[n-1]\mathbf{h}}$$

Plugging (10.39) into (10.38) and noticing that $\boldsymbol{\Sigma}[n-$
it follows that

$$\hat{\boldsymbol{\theta}}[n] = \hat{\boldsymbol{\theta}}[n-1] - \mathbf{k}[n]\mathbf{h}^T[n]\hat{\boldsymbol{\theta}}[n-1] + \left(\boldsymbol{\Sigma}[n-1]\mathbf{h}[n] \right.$$

Furthermore, from (10.40), we can obtain $\boldsymbol{\Sigma}[n-1]\mathbf{h}$
$\mathbf{k}[n]$. Therefore, the recursion is

$$\hat{\boldsymbol{\theta}}[n] = \hat{\boldsymbol{\theta}}[n-1] + \mathbf{k}[n]\left(x[n] - \mathbf{h}^T[n\right.$$

The term $(x[n] - \mathbf{h}^T[n]\hat{\boldsymbol{\theta}}[n-1])$ is the error between the newly received data $x[n]$ and the predicted value based on the previous estimate $\hat{\boldsymbol{\theta}}[n-1]$. The error is then multiplied with a gain vector $\mathbf{k}[n]$ and used as a correction to the previous estimate $\hat{\boldsymbol{\theta}}[n-1]$ to obtain the new estimate $\hat{\boldsymbol{\theta}}[n]$.

In summary, the recursive LS involves updating both the estimate $\hat{\boldsymbol{\theta}}[n]$ and its covariance matrix $\boldsymbol{\Sigma}[n-1]$ for every newly collected data $x[n]$:

$$\hat{\boldsymbol{\theta}}[n] = \hat{\boldsymbol{\theta}}[n-1] + \mathbf{k}[n]\Big(x[n] - \mathbf{h}^T[n]\hat{\boldsymbol{\theta}}[n-1]\Big),$$

$$\boldsymbol{\Sigma}[n] = (\mathbf{I} - \mathbf{k}[n]\mathbf{h}^T[n])\boldsymbol{\Sigma}[n-1],$$

where $\mathbf{k}[n]$ is defined in (10.40). There are two ways to start the recursion. Suppose the dimension of the unknown vector $\boldsymbol{\theta}$ is p, and assume $\mathbf{H}[p-1]$ has rank p. The first method is to run the batch mode LS estimator at $n = p-1$, and then employ the recursive LS method for $n \geq p$. The other method is to set $\boldsymbol{\theta}[-1] = \mathbf{0}$, and $\boldsymbol{\Sigma}[-1] = \alpha\mathbf{I}$ with large α. Although this method would bias the solution toward $\mathbf{0}$, a large value of α expresses little confidence in the initialization. The effect of initialization will be negligible when $n \to \infty$.

10.5.3 Weighted LS and Iterative Reweighted LS

Previous discussions on LS estimation focused on minimization of ℓ_2 (squared) norm. In this section, we examine the ℓ_q optimization problem with $q \neq 2$. The method is based on the weighted LS. Weighted LS is similar to basic LS estimation, but the cost function to be minimized involves a weighting matrix

$$J_W(\boldsymbol{\theta}) = (\mathbf{x} - \mathbf{H}\boldsymbol{\theta})^T\mathbf{W}(\mathbf{x} - \mathbf{H}\boldsymbol{\theta}),$$

where the weighting matrix $\mathbf{W}$ is a positive definite matrix, and is used to control the relative importance of different error terms in the estimation. Let $\mathbf{W} = \mathbf{S}^T\mathbf{S}$, where $\mathbf{S}$ is the Cholesky square root of $\mathbf{W}$, then the cost function $J_W(\boldsymbol{\theta})$ can be expressed as follows

$$J_W(\boldsymbol{\theta}) = (\mathbf{x}' - \mathbf{H}'\boldsymbol{\theta})^T(\mathbf{x}' - \mathbf{H}'\boldsymbol{\theta}),$$

with $\mathbf{x}' = \mathbf{S}\mathbf{x}$ and $\mathbf{H}' = \mathbf{S}\mathbf{H}$. Then, the weighted LS estimation is not much different from basic LS estimation, and we can write

$$\hat{\boldsymbol{\theta}} = (\mathbf{H}'^T\mathbf{H}')^{-1}\mathbf{H}'^T\mathbf{x}' = (\mathbf{H}^T\mathbf{S}^T\mathbf{S}\mathbf{H})^{-1}\mathbf{H}^T\mathbf{S}^T\mathbf{S}\mathbf{x}.$$

Now, consider the ℓ_q optimization problem with the cost function

$$J_q(\boldsymbol{\theta}) = \|\mathbf{x} - \mathbf{H}\boldsymbol{\theta}\|_q^q = \sum_{n=0}^{N-1} \big|x[n] - [\mathbf{H}\boldsymbol{\theta}]_n\big|^q,$$

where $[\mathbf{H}\boldsymbol{\theta}]_n$ denotes the nth element of $\mathbf{H}\boldsymbol{\theta}$. Notice that the above cost function can be rewritten into the form of a weighted LS problem [7]

$$\sum_{n=0}^{N-1} w_n\big|x[n] - [\mathbf{H}\boldsymbol{\theta}]_n\big|^2,$$

with $w_n = |x[n] - [\mathbf{H}\boldsymbol{\theta}]_n|^{q-2}$. However, the weighting
want to estimate. One way to tackle this is to set t
estimate of $\hat{\boldsymbol{\theta}}^{(k)}$, and then obtain an updated estima
estimation. In particular, the error at the kth iterat

$$\mathbf{e}^{(k)} := \mathbf{x} - \mathbf{H}\hat{\boldsymbol{\theta}}^{(k)},$$

and the new weighting matrix $\mathbf{S}^{(k+1)}$ is set to

$$\mathbf{S}^{(k+1)} = \text{diag}\left(|e[0]^{(k)}|^{(q-2)/2},\ |e[1]^{(k)}|^{(q-2)/2},\ \right.$$

where $e[i]^{(k)}$ is the ith element of $\mathbf{e}^{(k)}$. Then the sol
LS problem is given by

$$\hat{\boldsymbol{\theta}}^{(k+1)} = (\mathbf{H}^T(\mathbf{S}^{(k+1)})^T\mathbf{S}^{(k+1)}\mathbf{H})^{-1}\mathbf{H}^T(\mathbf{S}^{(k}$$

Equations (10.41)-(10.43) are iterated until the estim
value. The iterative process can be initialized by t
(corresponding to the weighting matrix $\mathbf{S}^{(0)}$ being t

10.5.4   Constrained LS Estimation

Suppose we have a linear model $\mathbf{x} = \mathbf{H}\boldsymbol{\theta} + \mathbf{w}$, an
parameters are related to each other. Such relationsh
as $\mathbf{A}\boldsymbol{\theta} = \mathbf{b}$, where $\mathbf{A}$ is a $r \times p$ matrix and it is as
(meaning the constraints are independent). Then th
$\boldsymbol{\theta}$ under such a constraint. The solution of this pro
technique of Lagrangian multipliers, and the derivat

Under the linear equality constraint, the cost fur
expressed as

$$J_c(\boldsymbol{\theta}) = (\mathbf{x} - \mathbf{H}\boldsymbol{\theta})^T(\mathbf{x} - \mathbf{H}\boldsymbol{\theta}) + \boldsymbol{\gamma}^T(\mathbf{A}\boldsymbol{\ell}$$
$$= \mathbf{x}^T\mathbf{x} - 2\boldsymbol{\theta}^T\mathbf{H}^T\mathbf{x} + \boldsymbol{\theta}^T\mathbf{H}^T\mathbf{H}\boldsymbol{\theta}$$

where $\boldsymbol{\gamma}$ is a $r \times 1$ vector of Lagrangian multipliers.
with respect to $\boldsymbol{\theta}$ produces

$$\frac{\partial J_c(\boldsymbol{\theta})}{\partial\boldsymbol{\theta}} = -2\mathbf{H}^T\mathbf{x} + 2\mathbf{H}^T\mathbf{H}\boldsymbol{\theta}$$

Setting the result to zero, it follows that

$$\hat{\boldsymbol{\theta}}_c = (\mathbf{H}^T\mathbf{H})^{-1}\mathbf{H}^T\mathbf{x} - \frac{1}{2}(\mathbf{H}^T\mathbf{H}$$
$$= \hat{\boldsymbol{\theta}} - \frac{1}{2}(\mathbf{H}^T\mathbf{H})^{-1}\mathbf{A}^T\boldsymbol{\gamma},$$

where $\hat{\boldsymbol{\theta}}$ is the unconstrained LS estimate and $\boldsymbol{\gamma}$ is to be determined by the constraint. Applying the constraint $\mathbf{A}\boldsymbol{\theta} = \mathbf{b}$ to (10.44) gives

$$\mathbf{A}\hat{\boldsymbol{\theta}}_c = \mathbf{A}\hat{\boldsymbol{\theta}} - \mathbf{A}(\mathbf{H}^T\mathbf{H})^{-1}\mathbf{A}^T\frac{\boldsymbol{\gamma}}{2} = \mathbf{b},$$

implying

$$\frac{\boldsymbol{\gamma}}{2} = [\mathbf{A}(\mathbf{H}^T\mathbf{H})^{-1}\mathbf{A}^T]^{-1}(\mathbf{A}\hat{\boldsymbol{\theta}} - \mathbf{b}). \tag{10.45}$$

Finally, substituting (10.45) into (10.44) yields

$$\hat{\boldsymbol{\theta}}_c = \hat{\boldsymbol{\theta}} - (\mathbf{H}^T\mathbf{H})^{-1}\mathbf{A}^T[\mathbf{A}(\mathbf{H}^T\mathbf{H})^{-1}\mathbf{A}^T]^{-1}(\mathbf{A}\hat{\boldsymbol{\theta}} - \mathbf{b})$$

where $\hat{\boldsymbol{\theta}} = (\mathbf{H}^T\mathbf{H})^{-1}\mathbf{H}^T\mathbf{x}$. The constrained LS estimate is a corrected version of the unconstrained LS estimate.

10.6 Regularized LS Estimation

In the linear LS estimation (10.37), if we express $\mathbf{H}$ using SVD $\mathbf{H} = \mathbf{U}\boldsymbol{\Sigma}\mathbf{V}^T$, we can rewrite the LS solution as (assuming the matrices are real)

$$\hat{\boldsymbol{\theta}} = (\mathbf{V}\boldsymbol{\Sigma}^T\mathbf{U}^T\mathbf{U}\boldsymbol{\Sigma}\mathbf{V}^T)^{-1}\mathbf{V}\boldsymbol{\Sigma}^T\mathbf{U}^T\mathbf{x},$$

where the dimensions of $\mathbf{U}$, $\boldsymbol{\Sigma}$, $\mathbf{V}$ are $N \times N$, $N \times p$ and $p \times p$, respectively. Since $\mathbf{U}$ and $\mathbf{V}$ are orthogonal matrices, the above equation can be simplified to

$$\begin{aligned}
\hat{\boldsymbol{\theta}} &= (\mathbf{V}\boldsymbol{\Sigma}^T\boldsymbol{\Sigma}\mathbf{V}^T)^{-1}\mathbf{V}\boldsymbol{\Sigma}^T\mathbf{U}^T\mathbf{x}, \\
&= (\mathbf{V}^T)^{-1}(\boldsymbol{\Sigma}^T\boldsymbol{\Sigma})^{-1}\mathbf{V}^{-1}\mathbf{V}\boldsymbol{\Sigma}^T\mathbf{U}^T\mathbf{x}, \\
&= \mathbf{V}(\boldsymbol{\Sigma}^T\boldsymbol{\Sigma})^{-1}\boldsymbol{\Sigma}^T\mathbf{U}^T\mathbf{x}. \tag{10.46}
\end{aligned}$$

Expressing matrices $\mathbf{V}$, $\mathbf{U}$ and $\boldsymbol{\Sigma}$ in terms of their columns vectors and diagonal entries, respectively: $\mathbf{V} = [\mathbf{v}_1 \ \mathbf{v}_2 \ ... \ \mathbf{v}_p]$, $\mathbf{U} = [\mathbf{u}_1 \ \mathbf{u}_2 \ ... \ \mathbf{u}_N]$ and $\boldsymbol{\Sigma} = \text{diag}(\sigma_1 \ \sigma_2 \ ... \sigma_p)$, (10.46) can be equivalently expressed as

$$\hat{\boldsymbol{\theta}} = \sum_{i=1}^{p} \frac{\mathbf{u}_i^T\mathbf{x}}{\sigma_i}\mathbf{v}_i. \tag{10.47}$$

The LS solution is the sum of a number of orthogonal components. It is obvious that if there are some small σ_i (e.g., if the matrix $\mathbf{H}$ is ill-conditioned), the norm of the solution $\|\hat{\boldsymbol{\theta}}\|^2 = \sum_{i=1}^{p}(\mathbf{u}_i^T\mathbf{x}/\sigma_i)^2$ will be very large, unless the energy of $\mathbf{x}$ in the direction of $\mathbf{u}_i$ is so small that $|\mathbf{u}_i^T\mathbf{x}| < \sigma_i$. Unfortunately, whenever errors are present in $\mathbf{x}$, this requirement is very unlikely to be satisfied. Therefore, the solution will be dominated by a few components with small σ_i and the resultant solution will be useless.

If the matrix $\mathbf{H}$ is just rank-deficient with a well-
and small singular values, the most common appro:
the solution is to replace the small non-zero singula
by excluding them in (10.47):

$$\hat{\boldsymbol{\theta}}_k = \sum_{i=1}^{k} \frac{\mathbf{u}_i^T \mathbf{x}}{\sigma_i} \mathbf{v}_i.$$

This solution is referred to as the truncated SVD so
equivalent to approximating the matrix $\mathbf{H}$ with anot
with a well-defined rank k. Usually, the rank k is ch
of $\mathbf{H}$ $(k < p)$.

10.6.1 ℓ_2 Regularization

However, it is possible that the singular values of $\mathbf{H}$ (
there is no gap in the singular value spectrum, and th
rank for these matrices. Estimation problems involv:
matrices are referred to as ill-posed problems. A pos
problem is to find a balance between the estimatic
solution. In particular, a well-known method is the ℓ
regularization), which aims to minimize [8]

$$\min_{\boldsymbol{\theta}} \left[(\mathbf{x} - \mathbf{H}\boldsymbol{\theta})^T (\mathbf{x} - \mathbf{H}\boldsymbol{\theta}) + \lambda^2 (
$$

where $\boldsymbol{\theta}^T \mathbf{L}^T \mathbf{L} \boldsymbol{\theta}$ controls the properties of the regula
ing different regularization matrices $\mathbf{L}$ (e.g., identity
first derivative for maximum flatness); the variable >
parameter that balances the minimization of the tv
the second term is a penalty term which penalizes
Equation (10.48) is also known as shrinkage because
zero. Effectively, it sacrifices a little bias to reduce
and hence may improve the overall MSE. In order t
we rewrite (10.48) as

$$\min_{\boldsymbol{\theta}} \left\| \begin{bmatrix} \mathbf{x} \\ \mathbf{0} \end{bmatrix} - \begin{bmatrix} \mathbf{H} \\ \lambda \mathbf{L} \end{bmatrix} \boldsymbol{\theta} \right\|^2$$

For a fixed λ, the solution is readily obtained as

$$\hat{\boldsymbol{\theta}}_\lambda = (\mathbf{H}^T \mathbf{H} + \lambda^2 \mathbf{L}^T \mathbf{L})^{-1} \mathbf{H}$$

Example 10.6.1. [9] In decode-and-forward coope
lays send their data simultaneously to the destinatio

of the relays, their signals arrive at the destination at slightly different times. In the training phase, the signal received at the destination is expressed as

$$\mathbf{d} = \underbrace{\begin{bmatrix} \mathbf{A}_{\epsilon_1} \ \mathbf{A}_{\epsilon_2} \ \cdots \ \mathbf{A}_{\epsilon_K} \end{bmatrix}}_{:=\mathbf{A}_{\epsilon}} \underbrace{\begin{bmatrix} h_1 \mathbf{x}_1 \\ h_2 \mathbf{x}_2 \\ \vdots \\ h_K \mathbf{x}_K \end{bmatrix}}_{:=\mathbf{X}} + \mathbf{v} = \mathbf{A}_{\epsilon}\mathbf{X} + \mathbf{v}$$

where ϵ_k denotes the timing delay of the signal from the kth relay, $\mathbf{A}_{\epsilon_k}$ stands for a known matrix parameterized by ϵ_k, h_k is the unknown channel coefficient between the kth relay and the destination, $\mathbf{x}_k$ represents the training data from the kth relay, and $\mathbf{v}$ is the observation noise vector. Assume that $\boldsymbol{\epsilon} := [\epsilon_1, \ \epsilon_2, \ \cdots, \ \epsilon_K]^T$ and $\{h_1, \ h_2, \ ..., \ h_K\}$ have been estimated. Since different signals with different delays are superimposed together at the receiver, we need to resynchronize the received signal. One way to achieve this is to design a resynchronization filter $\mathbf{f}$ using the weighted LS criterion:

$$\min_{\mathbf{f}} \ \left(\mathbf{A}_{\hat{\epsilon}}^H \mathbf{f} - \mathbf{b}\right)^H \mathbf{\Pi}\left(\mathbf{A}_{\hat{\epsilon}}^H \mathbf{f} - \mathbf{b}\right), \tag{10.50}$$

where $\mathbf{b}$ stands for the ideal sampled waveform without intersymbol interference, and $\mathbf{\Pi}$ is the weighting matrix that depends on the estimated channel coefficients.

Unfortunately, due to close values of ϵ_k, the columns of $\mathbf{A}_{\hat{\epsilon}}^H$ are quite similar to each other, and hence, (10.50) is an ill-posed problem. Figure 10.5 shows an example of the singular value distribution of $\mathbf{A}_{\hat{\epsilon}}^H$. It can be seen that there is no significant gap between the singular values, thus unlike a rank deficient problem which can be solved by discarding the zero or close-to-zero singular values, there are no general rules to determine which singular values are to be discarded.

Therefore, the LS estimation approach with ℓ_2 regularization is employed, which is casted as follows:

$$\min_{\mathbf{f}} \left[\left(\mathbf{G}_{\Pi}\mathbf{f} - \mathbf{b}_{\Pi}\right)^H \left(\mathbf{G}_{\Pi}\mathbf{f} - \mathbf{b}_{\Pi}\right) + \lambda^2 \mathbf{f}^H \mathbf{L}^H \mathbf{L}\mathbf{f} \right],$$

where $\mathbf{G}_{\Pi} = \mathbf{\Pi}^{\frac{1}{2}} \mathbf{A}_{\hat{\epsilon}}^H$ and $\mathbf{b}_{\Pi} = \mathbf{\Pi}^{\frac{1}{2}} \mathbf{b}$, with $\mathbf{\Pi}^{\frac{1}{2}}$ representing the Cholesky square root of the weighting matrix $\mathbf{\Pi}$. Therefore, the solution can be expressed as

$$\mathbf{f}_{\lambda} = (\mathbf{G}_{\Pi}^H \mathbf{G}_{\Pi} + \lambda^2 \mathbf{L}^H \mathbf{L})^{-1} \mathbf{G}_{\Pi}^H \mathbf{b}_{\Pi}.$$

Figure 10.6 shows a plot of the norm $\| \mathbf{L}\mathbf{f}_{\lambda} \|^2$ of the regularized solution versus the corresponding residual error norm $\| \mathbf{G}_{\Pi}\mathbf{f}_{\lambda} - \mathbf{b}_{\Pi} \|^2$ for different λ (with $\mathbf{L} = \mathbf{I}$). The curve clearly displays the compromise between minimization of these two quantities. □

Notice that in (10.49), if λ is too large, the solution is over-regularized and the residual error may be overwhelming, while if it is too small, the solution becomes

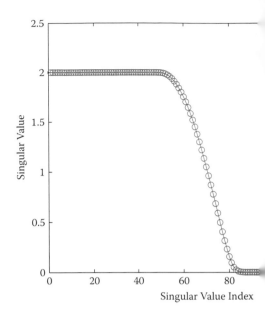

Figure 10.5: An example of the singular values dist
matrix $\mathbf{A}_\epsilon^H$.

under-regularized and the stability of the solution
treme case, when $\lambda = 0$, (10.49) reduces to the ord
performance of the regularized solution $\hat{\boldsymbol{\theta}}_\lambda$ depends
parameter λ. A conceptually simple tool for choosing
is the *L-curve* technique [8], which is a plot of $\parallel$
for different λ (see Figure 10.6). The L-curve cor
curvature point on the curve $\left(\log \parallel \mathbf{x} - \mathbf{H}\hat{\boldsymbol{\theta}}_\lambda \parallel, \log \parallel \right.$
compromise that balances the regularization errors a
regularized solution $\hat{\boldsymbol{\theta}}_\lambda$. Therefore, this point is cho
rameter λ. In order to find the λ that correspon
$\rho = \parallel \mathbf{x} - \mathbf{H}\hat{\boldsymbol{\theta}}_\lambda \parallel$ and $\xi = \parallel \mathbf{L}\hat{\boldsymbol{\theta}}_\lambda \parallel$. Then the curvatu

$$\kappa(\lambda) = 2\frac{\xi\rho}{\xi'}\frac{\lambda^2\xi'\rho + 2\lambda\xi\rho + }{(\lambda^2\xi^2 + \rho^2)^{3/}}$$

where

$$\xi' = \frac{4}{\lambda}\hat{\boldsymbol{\theta}}_\lambda^T(\mathbf{H}^T\mathbf{H} + \lambda^2\mathbf{L}^T\mathbf{L})^{-1}\mathbf{H}^T($$

Note that ρ, ξ and ξ' can be computed for a given λ.
parameter λ with maximum curvature can be locate

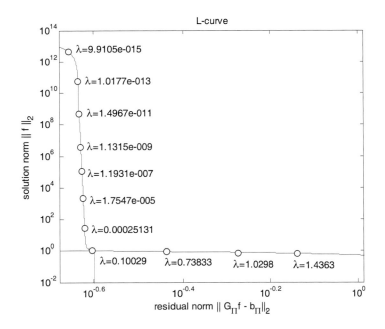

Figure 10.6: An example of L-curve.

10.6.2 LS Estimation with Quadratic Constraint

Another technique for regularization is LS with a quadratic constraint, which can be stated as

$$\min \| \mathbf{x} - \mathbf{H}\boldsymbol{\theta} \|^2 \quad \text{s.t.} \quad \| \mathbf{L}\boldsymbol{\theta} \| \leq \alpha \tag{10.51}$$

where α is a design parameter that bounds the norm of the solution. In the following, the relationship between the ℓ_2 regularization approach and the LS with quadratic constraint is derived. In order to do this, the generalized singular value decomposition (GSVD) of the matrix pair $(\mathbf{H}, \mathbf{L})$ is needed [8]

$$\mathbf{H} = \mathbf{U}\boldsymbol{\Sigma}\mathbf{Y}^{-1}, \qquad \mathbf{L} = \mathbf{V}\mathbf{M}\mathbf{Y}^{-1}$$

where $\boldsymbol{\Sigma} = \mathrm{diag}(\sigma_1, \cdots, \sigma_p)$, $\mathbf{M} = \mathrm{diag}(\mu_1, \cdots, \mu_p)$ are the singular values of $\mathbf{H}$ and $\mathbf{L}$, respectively. Matrices $\mathbf{U}$ and $\mathbf{V}$ are unitary matrices, while $\mathbf{Y}$ is a nonsingular matrix. With the GSVD of $(\mathbf{H}, \mathbf{L})$, the regularized solution $\hat{\boldsymbol{\theta}}_\lambda$ (10.49) is expressed as

$$\hat{\boldsymbol{\theta}}_\lambda = \mathbf{Y}(\boldsymbol{\Sigma}^T\boldsymbol{\Sigma} + \lambda^2\mathbf{M}^T\mathbf{M})^{-1}\boldsymbol{\Sigma}^T\mathbf{U}^T\mathbf{x}. \tag{10.52}$$

For the LS with quadratic constraint problem (10.51), the solution would occur on the boundary $\| \mathbf{L}\boldsymbol{\theta} \| = \alpha$. Substituting (10.52) into the boundary, it follows that

$$\| \mathbf{V}\mathbf{M}\mathbf{Y}^{-1} \cdot \mathbf{Y}(\boldsymbol{\Sigma}^T\boldsymbol{\Sigma} + \lambda^2\mathbf{M}^T\mathbf{M})^{-1}\boldsymbol{\Sigma}^T\mathbf{U}^T\mathbf{x} \| = \alpha.$$

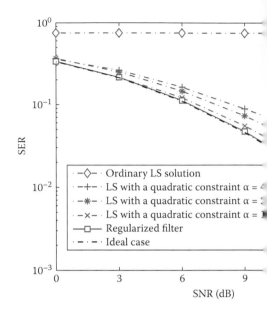

Figure 10.7: Symbol Error Rate (SER) performanc
designed by regularized LS against the performanc
solution and the LS with a quadratic constraint w
modulation.

After some tedious but straightforward manipulatio
and α is given by

$$\sum_{i=1}^{p} \frac{(\sigma_i \mu_i)^2}{(\sigma_i^2 + \lambda^2 \mu_i^2)^2} [\mathbf{U}^T \mathbf{x}]_i^2 =$$

where $[\mathbf{U}^T \mathbf{x}]_i$ is the ith element of $\mathbf{U}^T \mathbf{x}$. From (1
with a quadratic constraint method is a special ca
regularization, when the constraint value of α is o
(10.53). Although LS with a quadratic constraint n
regularization method under some circumstances, g
for choosing optimal α.

We now consider again the previous example
ter design in a cooperative communication system
synchronization filters are designed using the regul
several LS methods, and then in the data transmiss
are applied at the destination before data detection
error rate (SER) performance of different regularize
considered system, there are two relays (i.e., $K = 2$
are uniformly generated in the range $(-0.1, 0.1)$, and

duration. In all simulations, quadrature phase shift keying (QPSK) modulation is used. Alamouti space-time block code is employed at the relays, and each point is obtained by averaging over 10^4 runs. It can be seen that the performance of the regularized filter overlaps with the ideal case while the ordinary LS solution provides disappointing results due to the ill-posed nature of the problem. On the other hand, for the LS with a quadratic constraint, it is obvious that the nonoptimal quadratic constraint parameters α lead to significant performance degradation, especially in the case when $\alpha = 4$.

10.6.3 ℓ_1 Regularization

In previous discussion, we have seen that a ℓ_2 regularization term can be applied to control the norm of the LS solution. In fact, we can use different regularization terms to impose different properties on the LS solution. In general, the regularized estimation problem takes the form

$$\min_{\boldsymbol{\theta}} \left[(\mathbf{x} - \mathbf{H}\boldsymbol{\theta})^T (\mathbf{x} - \mathbf{H}\boldsymbol{\theta}) + \lambda^2 \sum_{i=1}^{p} |\theta_i|^q \right]. \tag{10.54}$$

When $q = 2$, it corresponds to ℓ_2 regularization. On the other hand, when $q = 1$, it is known that some of the θ_i will be driven to zero, leading to a sparse representation. To see the reason, we first note that (10.54) is equivalent to

$$\min_{\boldsymbol{\theta}} \left[(\mathbf{x} - \mathbf{H}\boldsymbol{\theta})^T (\mathbf{x} - \mathbf{H}\boldsymbol{\theta}) \right] \qquad \text{subject to} \qquad \sum_{i=1}^{p} |\theta_i|^q \leq c$$

for an appropriate value of c. Then, the estimation problem with $q = 2$ and $q = 1$ can be represented as in Figure 10.8 for the estimator of two parameters. The ellipses are the contours of the quadratic error surface, the ℓ_2 and ℓ_1 constraints are represented as circle and rotated square, respectively. It can be seen that under the ℓ_1-constraint, the regularized solution has a higher chance occurring at the corners, resulting in a sparse solution, compared to the ℓ_2-constraint. Following this argument, using the ℓ_q-regularization with smaller q (with $0 < q < 1$) will even further enhance the sparsity in the solution. Unfortunately, there is no closed-form solution for the non-ℓ_2 regularized LS problem, and it must be computed numerically.

10.7 Bayesian Estimation

At the beginning of this chapter, it is argued that if the parameter of interest is deterministic, the MMSE estimator is usually not realizable. That is why we search for the minimum variance estimator within the class of unbiased estimator. However, in this section, we will model the parameter of interest as a random

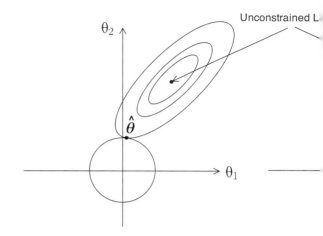

Figure 10.8: Parameter estimation under ℓ_2

variable with a certain prior PDF. This prior PDF
about the parameter before obtaining any observatic
the parameter using not only the observations, whi
the parameter, but also the prior information abou
scenario, the MMSE estimator exists, and in many
turns out to be convenient for implementation.   T
information and observations is the Bayes' rule, and t
is referred to as a Bayesian estimator.   The discus
adopted from [1].

10.7.1   Minimum Mean Square Error

Recall that the classical MSE is defined as

$$\text{MSE}(\hat{\theta}_i) = \int (\hat{\theta}_i - \theta_i)^2 p(\mathbf{x}|\theta$$

where the integration does not involve θ_i since it
However, in the Bayesian approach, since the para
random, the MSE is modified as

$$\text{BMSE}(\hat{\theta}_i) = \int \int (\hat{\theta}_i - \theta_i)^2 p(\mathbf{x},$$

Now, using Bayes' rule, we have

$$p(\mathbf{x}, \theta_i) = p(\theta_i|\mathbf{x})p(\mathbf{x}),$$

and the BMSE can be rewritten as

$$\text{BMSE}(\hat{\theta}_i) = \int \left[\int (\hat{\theta}_i - \theta_i)^2 p(\theta_i|\mathbf{x}) d\theta_i \right] p(\mathbf{x}) d\mathbf{x}. \tag{10.55}$$

Since $p(\mathbf{x}) \geq 0$, in order to minimize the BMSE, the integral inside the square brackets must be minimized for each $\mathbf{x}$. Therefore, the MMSE estimator can be obtained by differentiating the integral with respect to $\hat{\theta}_i$:

$$\frac{\partial}{\partial \hat{\theta}_i} \int (\hat{\theta}_i - \theta_i)^2 p(\theta_i|\mathbf{x}) d\theta_i = \int -2(\hat{\theta}_i - \theta_i) p(\theta_i|\mathbf{x}) d\theta_i$$

$$= -2\hat{\theta}_i \int p(\theta_i|\mathbf{x}) d\theta_i + 2 \int \theta_i p(\theta_i|\mathbf{x}) d\theta_i,$$

and setting the result to zero. It follows that

$$\hat{\theta}_i = \int \theta_i p(\theta_i|\mathbf{x}) d\theta_i = E\{\theta_i|\mathbf{x}\}. \tag{10.56}$$

Thus, the MMSE estimator is the mean of the posterior PDF $p(\theta_i|\mathbf{x})$ of θ_i. Notice that (10.56) can be equivalently written as

$$\hat{\theta}_i = \int \theta_i \left[\int \cdots \int p(\boldsymbol{\theta}|\mathbf{x}) d\theta_1 \ldots \theta_{i-1}\theta_{i+1} \ldots \theta_p \right] d\theta_i = \int \theta_i p(\boldsymbol{\theta}|\mathbf{x}) d\boldsymbol{\theta}.$$

Stacking the result into a vector form, we have

$$\hat{\boldsymbol{\theta}} = \begin{bmatrix} \int \theta_1 p(\boldsymbol{\theta}|\mathbf{x}) d\boldsymbol{\theta} \\ \int \theta_2 p(\boldsymbol{\theta}|\mathbf{x}) d\boldsymbol{\theta} \\ \vdots \\ \int \theta_p p(\boldsymbol{\theta}|\mathbf{x}) d\boldsymbol{\theta} \end{bmatrix}$$

$$= \int \boldsymbol{\theta} p(\boldsymbol{\theta}|\mathbf{x}) d\boldsymbol{\theta} = E\{\boldsymbol{\theta}|\mathbf{x}\}.$$

The vector MMSE estimator minimizes the Bayesian MSE for each component of the estimate $\hat{\boldsymbol{\theta}}$.

Example 10.7.1. [1] Consider the problem of estimating A from $x[n] = A + w[n]$ for $n = 0, 1, ..., N - 1$, and $w[n]$ are i.i.d. obeying $w[n] \sim \mathcal{N}(0, \sigma^2)$. This is basically the same problem as in Example 10.4.1. However, in this example, we also consider A to be a random variable satisfying $A \sim \mathcal{N}(\mu_A, \sigma_A^2)$. We want to obtain the MMSE estimator of A. First, we write down the likelihood function $p(\mathbf{x}|A)$ and the prior PDF $p(A)$. The likelihood function is given by (10.20) and it is again depicted below:

$$p(\mathbf{x}|A) = \frac{1}{(2\pi\sigma^2)^{N/2}} \exp\left[-\frac{1}{2\sigma^2} \sum_{n=0}^{N-1} x^2[n] \right] \exp\left[-\frac{1}{2\sigma^2} \left(NA^2 - 2NA\bar{x} \right) \right] \tag{10.57}$$

where $\bar{x} := \frac{1}{N} \sum_{n=0}^{N-1} x[n]$. On the other hand, since
prior PDF of A is

$$p(A) = \frac{1}{\sqrt{2\pi\sigma_A^2}} \exp\exp\left[-\frac{1}{2\sigma_A^2}(A\right.$$

Since the MMSE estimator is the mean of the poste
it first. For this, we employ the Bayes' rule:

$$p(A|\mathbf{x}) = \frac{p(\mathbf{x}|A)p(A)}{p(\mathbf{x})} = \frac{p(\mathbf{x}|A)}{\int p(\mathbf{x}|A)}$$

Notice that the prior PDF $p(A)$ is the PDF of A
while the posterior PDF $p(A|\mathbf{x})$ is the PDF of A *aft*
ging (10.57) and (10.58) into (10.59), and canceling
numerator and denominator lead to

$$p(A|\mathbf{x}) = \frac{\exp\left[-\frac{1}{2}\left(\frac{1}{\sigma^2}\left(NA^2 - 2NA\bar{x}\right) + \frac{1}{\sigma_A^2}\right.\right.}{\int_{-\infty}^{\infty} \exp\left[-\frac{1}{2}\left(\frac{1}{\sigma^2}\left(NA^2 - 2NA\bar{x}\right) + \frac{1}{\sigma}\right.\right.}$$

It can be shown that the expression within the
written as

$$Q(A) := \frac{1}{\sigma^2}\left(NA^2 - 2NA\bar{x}\right) + \frac{1}{\sigma_A^2}(A-\mu_A)^2 = \left(\frac{N}{\sigma^2} + \right.$$

By completing the square, it follows that

$$Q(A) = \frac{1}{\sigma_{A|x}^2}(A^2 - \mu_{A|x})^2 - \frac{\mu^2}{\sigma^2}$$

where

$$\sigma_{A|x}^2 = \left(\frac{N}{\sigma^2} + \frac{1}{\sigma_A^2}\right)^{-1},$$

$$\mu_{A|x} = \left(\frac{N}{\sigma^2}\bar{x} + \frac{\mu_A}{\sigma_A^2}\right)\sigma_{A|}^2$$

Plugging this result into (10.60), and canceling the
ator and denominator, we obtain

$$p(A|\mathbf{x}) = \frac{\exp\left[-\frac{1}{2\sigma_{A|x}^2}(A - \mu\right.}{\int_{-\infty}^{\infty} \exp\left[-\frac{1}{2\sigma_{A|x}^2}(A - \right.}$$

Notice that in the denominator, A is being integrat
is just a normalization constant independent of A

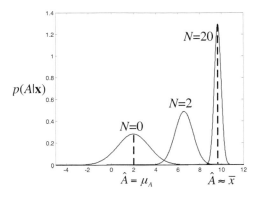

Figure 10.9: Variation of posterior PDF of A as N changes.

quadratic function of A, therefore, the posterior PDF $p(A|\mathbf{x})$ must be a Gaussian PDF. From these two facts, there is no need to explicitly compute the denominator, and we can write

$$p(A|\mathbf{x}) = \frac{1}{\sqrt{2\pi\sigma_{A|x}^2}} \exp\left[-\frac{1}{2\sigma_{A|x}^2}(A - \mu_{A|x})^2\right],$$

since a PDF must be integrated to 1. Finally, the MMSE estimator of A is the mean of $p(A|\mathbf{x})$, which is given in (10.62).

Notice that the posterior mean (10.62) can be equivalently expressed as

$$\hat{A} = \underbrace{\frac{\sigma_A^2}{\sigma_A^2 + \sigma^2/N}}_{:=\alpha} \bar{x} + \underbrace{\frac{\sigma^2/N}{\sigma_A^2 + \sigma^2/N}}_{=1-\alpha} \mu_A, \tag{10.63}$$

where $0 < \alpha < 1$ is a weighting factor that determines how to combine the sample mean estimator and the prior mean. When N is small such that $\sigma_A^2 \ll \sigma^2/N$, $\alpha \approx 0$ and $\hat{A} \approx \mu_A$. On the other hand, when the number of observed data $N \to \infty$, $\alpha \to 1$ and $\hat{A} \to \bar{x}$. For the posterior variance (10.61), it is also a combination of prior variance σ_A^2 and the variance given by the sample mean estimator σ^2/N. When N is small, it depends more heavily on the prior variance. However, as N increases, the posterior variance decreases and will rely more and more on that of the sample mean estimator. The above properties of MMSE estimate of A can be visualized in Figure 10.9, where the posterior PDF is plotted for three different observation lengths (N). □

As shown in the previous example, if the likelihood function is Gaussian, choosing a Gaussian prior ensures that the posterior distribution is also Gaussian. This is due to the property of the multivariate Gaussian distribution.

Theorem 10.7.1. *(Conditional PDF of Multivaria*
a $k \times 1$ vector and $\mathbf{y}$ is a $l \times 1$ vector. If $\mathbf{x}$ and $\mathbf{y}$ a
$[\boldsymbol{\mu}_x^T \ \boldsymbol{\mu}_y^T]^T$ and covariance matrix

$$
\mathbf{C} = \left[\begin{array}{cc} \mathbf{C}_{xx} & \mathbf{C}_{xy} \\ \mathbf{C}_{yx} & \mathbf{C}_{yy} \end{array} \right],
$$

then the conditional PDF $p(\mathbf{y}|\mathbf{x})$ is also Gaussian wit
given by

$$
E\{\mathbf{y}|\mathbf{x}\} = \boldsymbol{\mu}_y + \mathbf{C}_{yx}\mathbf{C}_{xx}^{-1}(\mathbf{x}
$$
$$
\mathbf{C}_{y|x} = \mathbf{C}_{yy} - \mathbf{C}_{yx}\mathbf{C}_{xx}^{-1}\mathbf{C}
$$

Applying this result to the linear model $\mathbf{x} = \mathbf{H}\boldsymbol{\theta}$
Gaussian, we have the following MMSE estimator fc

Theorem 10.7.2. *(MMSE estimator for linear mode*

$$
\mathbf{x} = \mathbf{H}\boldsymbol{\theta} + \mathbf{w}
$$

where $\mathbf{x}$ is an $N \times 1$ vector, $\mathbf{H}$ stands for a known
random vector with PDF $\mathcal{N}(\boldsymbol{\mu}_\theta, \mathbf{C}_\theta)$, and $\mathbf{w}$ denote
PDF $\mathcal{N}(\mathbf{0}, \mathbf{C}_w)$, and is independent of $\boldsymbol{\theta}$. Then the

$$
\hat{\boldsymbol{\theta}} = \boldsymbol{\mu}_\theta + \mathbf{C}_\theta \mathbf{H}^T (\mathbf{H}\mathbf{C}_\theta \mathbf{H}^T + \mathbf{C}_w)^{-}
$$

with the covariance matrix of estimation error $\boldsymbol{\epsilon} = \boldsymbol{\ell}$

$$
\mathbf{C}_\epsilon = \mathbf{C}_\theta - \mathbf{C}_\theta \mathbf{H}^T (\mathbf{H}\mathbf{C}_\theta \mathbf{H}^T + \mathbf{C}_*
$$

Notice that an alternative form of the MMSE est

$$
\hat{\boldsymbol{\theta}} = \boldsymbol{\mu}_\theta + (\mathbf{C}_\theta^{-1} + \mathbf{H}^T \mathbf{C}_w^{-1} \mathbf{H})^{-1} \mathbf{H}^T \mathbf{C}
$$

Now, consider $\mathbf{C}_w = \sigma^2 \mathbf{I}$ and $\boldsymbol{\mu}_\theta = 0$, the MMSE est

$$
\hat{\boldsymbol{\theta}} = (\mathbf{C}_\theta^{-1} + \mathbf{H}^T \mathbf{H}/\sigma^2)^{-1} \mathbf{H}^T
$$
$$
= (\mathbf{H}^T \mathbf{H} + \sigma^2 \mathbf{C}_\theta^{-1})^{-1} \mathbf{H}^T \mathbf{x}
$$

Comparing (10.67) to the ℓ_2-regularized LS solution
the MMSE estimator, σ^2 is in the place of the regular
the MMSE estimator exhibits a build-in regularizati
of the diagonal loading term $\sigma^2 \mathbf{C}_\theta^{-1}$, the matrix $(\mathbf{H}^T\mathbf{I}$
for non-zero σ^2 and finite $\mathbf{C}_\theta$. Interpreted from anoth
can be viewed as adding Gaussian prior on the unk

with ℓ_1 regularization can be derived from the Bayesian framework with the double-exponential prior PDF [10]

$$p(\boldsymbol{\theta}) = \frac{\lambda^2}{2} \exp\left(-\lambda^2 \sum_{i=1}^{p} |\theta_i|\right).$$

Figure 10.10 compares the double-exponential density and the Gaussian density. It is clear that the double-exponential density puts more mass near 0 and in the tail, thus favoring a sparse solution, compared to the Gaussian prior.

As a final remark, in Theorem 10.7.2, if the noise vector $\mathbf{w}$ is not Gaussian but still with the same mean and covariance, the estimator (10.64) is not the MMSE estimator anymore. However, it can be shown to be the linear minimum mean-square error (LMMSE) estimator. That is, it has the MMSE within the class of linear estimator (see Exercise 10.9.12).

10.7.2 General Bayesian Estimator

In the MMSE estimation, the criterion is the mean squared error. However, one can also adopt numerous other criteria, giving rise to estimators with different properties. A general cost function can be expressed as follows

$$E\left\{\mathcal{C}(\theta_i - \hat{\theta}_i)\right\} = \int \left[\int \mathcal{C}(\hat{\theta}_i - \theta_i) p(\theta_i | \mathbf{x}) d\theta_i\right] p(\mathbf{x}) d\mathbf{x}.$$

It is obvious that the general cost function includes the Bayesian MSE (10.55) as a special case when $\mathcal{C}(\hat{\theta}_i - \theta_i) = (\hat{\theta}_i - \theta_i)^2$, which penalizes the errors quadratically. Other commonly used criteria are the absolute error:

$$\mathcal{C}(\hat{\theta}_i - \theta_i) = |\hat{\theta}_i - \theta_i|,$$

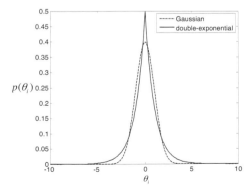

Figure 10.10: Comparison of double-exponential density to Gaussian density.

which penalizes the errors proportionally, and the h

$$
\mathcal{C}(\hat{\theta}_i - \theta_i) = \begin{cases} 0 & \text{if } |\hat{\theta}_i - \theta_i| \\ 1 & \text{if } |\hat{\theta}_i - \theta_i| \end{cases}
$$

where $\delta > 0$ is a threshold. The hit-or-miss criter
error is smaller than δ, and assigns equal penalty fo

Using different $\mathcal{C}(\hat{\theta}_i - \theta_i)$ result in different Baye
if we use the absolute error, the cost function to be

$$
g_1(\hat{\theta}_i) = \int |\hat{\theta}_i - \theta_i| p(\theta_i|\mathbf{x}) d\theta_i
$$

$$
= \int_{-\infty}^{\hat{\theta}_i} (\hat{\theta}_i - \theta_i) p(\theta_i|\mathbf{x}) d\theta_i + \int_{\hat{\theta}_i}^{\infty} (
$$

Differentiating $g_1(\hat{\theta}_i)$ with respect to θ_i and making
be shown that

$$
\frac{\partial g_1(\hat{\theta}_i)}{\partial \theta_i} = \int_{-\infty}^{\hat{\theta}_i} p(\theta_i|\mathbf{x}) d\theta_i - \int_{-\infty}^{\hat{\theta}_i}
$$

Setting the result to zero gives the Bayesian estimat

$$
\int_{-\infty}^{\hat{\theta}_i} p(\theta_i|\mathbf{x}) d\theta_i = \int_{-\infty}^{\hat{\theta}_i} p(\theta_i|
$$

which is the median of the posterior PDF.

On the other hand, if we use the hit-or-miss crit
minimized is

$$
g_2(\hat{\theta}_i) = \int_{-\infty}^{\hat{\theta}_i - \delta} p(\theta_i|\mathbf{x}) d\theta_i + \int_{\hat{\theta}_i + \delta}^{\infty}
$$

$$
= 1 - \int_{\hat{\theta}_i - \delta}^{\hat{\theta}_i + \delta} p(\theta_i|\mathbf{x}) d\theta_i.
$$

Minimizing $g_2(\hat{\theta}_i)$ is equivalent to maximizing $\int_{\hat{\theta}_i - \delta}^{\hat{\theta}_i + }$
of $p(\theta_i|\mathbf{x})$ within a window of length 2δ. For δ arbit
the area is maximized if we put $\hat{\theta}_i$ at the peak of th
is called the "maximum a posteriori (MAP)" estima

To conclude, Bayesian estimation assumes the fo
terior PDF which contains all the information about
information and the observed data), then the estim
the characteristic of the posterior PDF. Figure 10.1
median estimator and MAP estimator for a posterio
be seen that these estimators are looking at differen
rior PDF. Notice that in the case of Gaussian poste
and has only a single peak, all three criteria provide

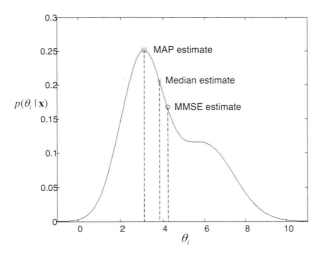

Figure 10.11: Comparison of different Bayesian estimators in a posterior PDF with two peaks.

Example 10.7.2. [1] Consider the problem in the previous Example 10.7.1: estimation of A from $x[n] = A + w[n]$. All the assumptions are the same as those in the previous example, except

1. the variance of A is assumed to be $\sigma_A^2 = \alpha\sigma^2$ with known α,

2. σ^2 is assumed to be unknown with prior PDF obeying the inverted gamma PDF:

$$p(\sigma^2) = \begin{cases} \frac{\lambda\exp(-\lambda/\sigma^2)}{\sigma^4} & \sigma^2 > 0 \\ 0 & \sigma^2 < 0 \end{cases}, \qquad (10.68)$$

where the parameter $\lambda > 0$ is known.

We want to find the MAP of A.

Since σ^2 is also unknown, the likelihood function now depends on two random variables $p(\mathbf{x}|A, \sigma^2)$, and its expression is still given by (10.57). The only difference is that σ^2 is also a parameter to be estimated. Furthermore, the prior of A is still given by (10.58) but we have to express the prior as $p(A|\sigma^2)$ since it is now dependent on σ^2 through $\sigma_A^2 = \alpha\sigma^2$. Applying Bayes' rule yields the posterior PDF:

$$p(A, \sigma^2|\mathbf{x}) = \frac{p(\mathbf{x}|A, \sigma^2)p(A|\sigma^2)p(\sigma^2)}{p(\mathbf{x})}. \qquad (10.69)$$

The MAP estimate corresponds to the values (A, σ^2) that maximize the posterior PDF $p(A, \sigma^2|\mathbf{x})$. Since $p(\mathbf{x})$ is independent of the parameters to be estimated, we only need to maximize the numerator. First, let us look at the maximization with

respect to A. Since $p(\sigma^2)$ does not depend on A, m ⌐
respect to A is equivalent to maximization of $p(\mathbf{x}|A,$
It has been shown in the previous example that for
proportional to a Gaussian distribution. Thus, the M⌐
to the MMSE estimate of A, and the solution is giv⌐
is assumed in this example, plugging this into (10.6⌐

$$\hat{A} = \frac{N\bar{x} + \mu_A/\alpha}{N + 1/\alpha}.$$

As (10.70) does not depend on σ^2, it is the MAP es⌐
We now proceed to find the MAP estimate for σ^2
merator (10.69), and using (10.57), (10.58) and (10.⌐
tions, it can be shown that the numerator of (10.69⌐

$$p(\hat{A}, \sigma^2|\mathbf{x}) \propto \frac{1}{(\sigma^2)^{(N+5)/2}} \exp ⌐$$

where

$$a = \frac{1}{2}\left[\sum_{n=1}^{N} x^2[n] + \frac{\mu_A^2}{\alpha} - \hat{A}^2(N + ⌐\right.$$

is a constant independent of σ^2. Differentiating (1⌐
setting it to zero yields:

$$\hat{\sigma}^2 = \frac{2}{N+5}a$$

$$= \frac{N}{N+5}\left[\frac{1}{N}\sum_{n=1}^{N} x^2[n] - \hat{A}^2\right] + \frac{1}{(N+5)⌐}$$

Notice that as $N \to \infty$, the MAP estimate of A (10⌐
the ML estimator in Example 10.4.1. Similarly, as I⌐

$$\hat{\sigma}^2 \to \frac{1}{N}\sum_{n=1}^{N} x^2[n] - \bar{x}^2 = \frac{1}{N}\sum_{n=1}^{N}(⌐$$

which is also the ML estimate of σ^2 (see Exercise 1⌐

In general, when the number of observations N ⌐
comes the ML estimator. This is because the MAP e⌐
of the posterior PDF:

$$\hat{\boldsymbol{\theta}} = \arg\max_{\tilde{\boldsymbol{\theta}}} \ln p(\tilde{\boldsymbol{\theta}}|\mathbf{x})$$

$$= \arg\max_{\tilde{\boldsymbol{\theta}}}[\ln p(\mathbf{x}|\tilde{\boldsymbol{\theta}}) + \ln ⌐$$

where the second line is due to Bayes' rule $p(\boldsymbol{\theta}|\mathbf{x}) = p(\mathbf{x}|\boldsymbol{\theta})p(\boldsymbol{\theta})/p(\mathbf{x})$ and the fact that $p(\mathbf{x})$ does not affect the maximization. Notice that taking the logarithm does not change the position of maximum in the PDF, and the above equation is indeed the MAP estimator. Now, if $N \to \infty$, the likelihood function $p(\mathbf{x}|\boldsymbol{\theta})$ will be more and more concentrated around the true value $\boldsymbol{\theta}$, and the prior PDF $p(\boldsymbol{\theta})$ will be relatively flat over the nonzero region of $p(\mathbf{x}|\boldsymbol{\theta})$. Therefore, the prior information does not change the maximization, and the MAP estimator is now equivalent to the ML estimator.

10.7.3 Handling Nuisance Parameters

Suppose we want to estimate $\boldsymbol{\theta}$, but the observations also depend on another unknown vector $\boldsymbol{\alpha}$. We are only interested in $\boldsymbol{\theta}$, but not $\boldsymbol{\alpha}$. This situation occurs frequently in practical applications, and the parameter $\boldsymbol{\alpha}$ is called nuisance parameter. In the Bayesian framework, we know that an important step is to compute the posterior PDF $p(\boldsymbol{\theta}|\mathbf{x})$. Now, since the observations depend on both $\boldsymbol{\theta}$ and $\boldsymbol{\alpha}$, the likelihood function is $p(\mathbf{x}|\boldsymbol{\theta}, \boldsymbol{\alpha})$. Denote the prior PDF of $\boldsymbol{\theta}$ and $\boldsymbol{\alpha}$ as $p(\boldsymbol{\theta})$ and $p(\boldsymbol{\alpha})$, respectively, and assume that they are independent. The joint posterior distribution of $\boldsymbol{\theta}$ and $\boldsymbol{\alpha}$ can be obtained from Bayes' rule:

$$p(\boldsymbol{\theta}, \boldsymbol{\alpha}|\mathbf{x}) = \frac{p(\mathbf{x}|\boldsymbol{\theta}, \boldsymbol{\alpha})p(\boldsymbol{\theta})p(\boldsymbol{\alpha})}{p(\mathbf{x})}.$$

Then, the posterior PDF $p(\boldsymbol{\theta}|\mathbf{x})$ can be obtained by integration

$$p(\boldsymbol{\theta}|\mathbf{x}) = \int \frac{p(\mathbf{x}|\boldsymbol{\theta}, \boldsymbol{\alpha})p(\boldsymbol{\theta})p(\boldsymbol{\alpha})}{p(\mathbf{x})} d\boldsymbol{\alpha}, \tag{10.72}$$

from which various Bayesian estimators can be derived.

Example 10.7.3. [11] In this example, we consider the MAP carrier frequency offset (CFO) estimation in single-user OFDM system. The observation model can be written as

$$\mathbf{x} = \underbrace{\boldsymbol{\Gamma}(\omega_o)\mathbf{F}^H\mathbf{D}\mathbf{F}_L}_{:=\mathbf{G}(\omega_o)}\mathbf{h} + \mathbf{w}.$$

In the above equation, $\mathbf{x}$ is the observation vector of length N, $\boldsymbol{\Gamma}(\omega_o) =$ diag $\left(1, ..., e^{j(N-1)\omega_o}\right)$ denotes the matrix parameterized by the unknown CFO ω_o, $\mathbf{F}$ stands for the fast Fourier transform matrix, $\mathbf{F}_L$ is a $N \times L$ matrix containing the first L columns of $\mathbf{F}$, $\mathbf{D}$ is a diagonal matrix containing known training, $\mathbf{h}$ is the $L \times 1$ unknown channel vector which can be considered as nuisance parameter, and $\mathbf{w}$ represents the observation noise with elements obeying i.i.d. Gaussian distribution with zero mean and known variance σ^2. This model is similar to that in Example 10.4.3 of orthogonal frequency division multiple access (OFDMA) CFO and channel estimation. The main difference is the treatment of unknown channel $\mathbf{h}$. In the ML estimation, $\mathbf{h}$ is being jointly estimated. On the other hand, in

Bayesian approach, $\mathbf{h}$ is treated as nuisance parame‐
tion.

The prior PDF of $\mathbf{h}$ is modeled by

$$p(\mathbf{h}) = \frac{1}{\pi^L \det(\mathbf{Q})} \exp(-\mathbf{h}^H \mathbf{($$

where $\mathbf{Q}$ is a known channel covariance matrix, wł
profile information, while the prior PDF for ω_o is

$$p(\omega_o) = \frac{1}{\sqrt{2\pi}\sigma_{\omega_o}} \exp\left(-\frac{\omega}{2\sigma}\right)$$

where $\sigma_{\omega_o}^2$ is the variance of the CFO distribution
prior, since in case there is no prior information, v
distribution becomes uninformative and flat.

Now, we derive the posterior PDF $p(\omega_o|\mathbf{x})$. Frc
$p(\mathbf{x})$ is independent of the parameter of interest, we

$$p(\omega_o|\mathbf{x}) \propto \left(\int p(\mathbf{x}|\omega_o, \mathbf{h}) p(\mathbf{h}) d\mathbf{h}\right.$$

The likelihood function of $\mathbf{x}$ given ω_o and $\mathbf{h}$ is

$$p(\mathbf{x}|\omega_o, \mathbf{h}) = \frac{1}{(\pi\sigma^2)^N} \exp\left\{ -\frac{[\mathbf{x} - \mathbf{G}(\omega_o)\mathbf{h}]^H}{\sigma^2} \right.$$

Using (10.73) and (10.75), we have

$$\int p(\mathbf{x}|\omega_o, \mathbf{h}) p(\mathbf{h}) d\mathbf{h} = \int \frac{1}{(\pi\sigma^2)^N} \exp\left\{ -\frac{[\mathbf{x}}{} \right.$$
$$\times \frac{1}{\pi^L \det(\mathbf{Q})} \exp(-\mathbf{h}^H \mathbf{Q}^{-1}\mathbf{h}) d\mathbf{h}.$$

By combining the terms related to $\mathbf{h}$ into a quadratic
to

$$\int p(\mathbf{x}|\omega_o, \mathbf{h}) p(\mathbf{h}) d\mathbf{h} \propto \frac{1}{(\pi\sigma^2)^N} \exp$$
$$\times \int \exp\left\{ -\frac{[\mathbf{h} - \mathbf{B}(\omega_o)\mathbf{x}]^H \mathbf{A}^{-1}}{\sigma^2} \right.$$

where

$$\mathbf{C}(\omega_o) := \mathbf{I} - \mathbf{G}(\omega_o)[\mathbf{G}(\omega_o)^H \mathbf{G}(\omega_o) + \sigma$$
$$\mathbf{B}(\omega_o) := [\mathbf{G}(\omega_o)^H \mathbf{G}(\omega_o) + \sigma^2 \mathbf{Q}]^{-1} \mathbf{G}(\omega$$
$$\mathbf{A} := \underbrace{\mathbf{G}(\omega_o)^H \mathbf{G}(\omega_o)}_{=\mathbf{F}_L^H \mathbf{D}^H \mathbf{D} \mathbf{F}_L \text{ independent of } \omega_o} + \sigma$$

Note that

$$\int \frac{1}{(\pi\sigma^2)^L \det(\mathbf{A})} \times \exp\left\{-\frac{[\mathbf{h} - \mathbf{B}(\omega_o)\mathbf{x}]^H \mathbf{A}^{-1}[\mathbf{h} - \mathbf{B}(\omega_o)\mathbf{x}]}{\sigma^2}\right\} d\mathbf{h} = 1,$$

and this result leads to

$$\int p(\mathbf{x}|\omega_o, \mathbf{h})p(\mathbf{h})d\mathbf{h} \propto \frac{\det(\mathbf{A})}{(\pi\sigma^2)^{(N-L)}} \exp\left\{-\frac{\mathbf{x}^H \mathbf{C}(\omega_o)\mathbf{x}}{\sigma^2}\right\}. \tag{10.77}$$

Plugging (10.77) into (10.74), and ignoring the terms not related to ω_o, the posterior distribution of ω_o becomes

$$p(\omega_o|\mathbf{x}) \propto \exp\left(-\frac{\omega_o^2}{2\sigma_{\omega_o}^2}\right) \exp\left\{-\frac{\mathbf{x}^H \mathbf{C}(\omega_o)\mathbf{x}}{\sigma^2}\right\}. \tag{10.78}$$

Then the MAP estimate of ω_o is obtained by maximizing (10.78) and is equivalent to

$$\hat{\omega}_o = \arg\max_{\omega_o}\left\{-\frac{\omega_o^2}{2\sigma_{\omega_o}^2} - \frac{\mathbf{x}^H \mathbf{C}(\omega_o)\mathbf{x}}{\sigma^2}\right\}, \tag{10.79}$$

which represents a trade-off between the prior information on CFO and the observation data statistics.

Now, let us look at the special case where we have no prior knowledge on the distribution of CFO and channel. When there is no prior information on CFO, this corresponds to the case $\sigma_{\omega_o} \to \infty$, and the first term inside the bracket of (10.79) goes to zero. Furthermore, the lack of knowledge on channel covariance could be addressed by assigning an uninformative prior to $\mathbf{h}$. Here we assign $\mathbf{Q} = \mathrm{diag}(\delta_1^2, \delta_2^2, \cdots, \delta_{L-1}^2)$ with $\delta_1^2, \delta_2^2, \cdots, \delta_{L-1}^2$ approaching infinity, where δ_i^2 is the variance of the ith tap of the channel. When $\delta_1^2, \delta_2^2, \cdots, \delta_{L-1}^2 \to \infty$, we have $\mathbf{Q}^{-1} \to \mathbf{0}$ and

$$\mathbf{C}(\omega_o) \approx \mathbf{I} - \mathbf{G}(\omega_o)[\mathbf{G}(\omega_o)^H \mathbf{G}(\omega_o)]^{-1}\mathbf{G}(\omega_o)^H.$$

Thus, the estimator without any prior information is obtained as

$$\hat{\omega}_o = \arg\max_{\omega_o}\left\{\mathbf{x}^H \mathbf{G}(\omega_o)[\mathbf{G}(\omega_o)^H \mathbf{G}(\omega_o)]^{-1}\mathbf{G}(\omega_o)^H \mathbf{x}\right\}. \tag{10.80}$$

Comparing (10.80) to (10.30), it is obvious that the MAP estimator reduces to the ML estimator. $\qquad\qquad\square$

In general, the integration required for eliminating the nuisance parameter is not always possible. In this case, numerical methods or Monte Carlo techniques (such as importance sampling, Gibbs sampler, Metropolis-Hastings algorithm) are frequently used. Details on Monte Carlo statistical techniques are discussed in Chapter 12 of this book.

10.8   References and Further R

Many of the theories and discussions in this chapt
results reported in [1] and [2], which represent exc
timation theory from the signal processing perspec
treatment on estimation theory, readers are referre
mation theory [12]. Our brief introduction to ℓ_2 reg
with quadratic constraints is more extensively covere
one of the earliest work is presented in [10], while for
referred to [15]. For Bayesian estimation, the class
recent text [14] is also worth reading. Some of the
taken from the research papers [3–6], [9], [11].

10.9   Exercises

Exercise 10.9.1. Given the observed data $x[n] = $
$n = 0, 1, ..., N - 1$. The parameter A and f_o are ass
noise samples $w[n]$ are i.i.d. Gaussian distributed wi
Derive the CRLB for ϕ. Hint:

$$\frac{1}{N} \sum_{n=0}^{N-1} \cos(4\pi f_o n + 2\phi) \approx$$

would help simplify the expression of the CRLB.

Exercise 10.9.2. Consider the problem of estimati
$Bn + w[n]$ with $n = 0, 1, ..., N - 1$, and $w[n]$ are i.i.
zero mean and variance σ^2. Derive the CRLB for A
the MVUE using the CRLB and Theorem 10.2.2.

Exercise 10.9.3. Let $x[n]$, $n = 0, 1, ..., N - 1$ be
within $[0, \theta]$, where $\theta > 0$ is a parameter to be estima
factorization Theorem 10.3.1, prove that $T = \max x$
Derive the PDF of T; (c) Prove the sufficient statis
MVUE using Theorem 10.3.2.

Exercise 10.9.4. Consider the same setting as in E
estimate for θ.

Exercise 10.9.5. Consider the data model in Exa
the variance σ^2 are unknown. Derive the ML estima

Exercise 10.9.6. Consider the signal model in Exe
that A, f_o, and ϕ are all unknown. Derive the M
parameters. Hint: making use of the transformat
$-A \sin \phi$ would help simplify the problem.

Exercise 10.9.7. For an orthogonal projection matrix $\mathbf{P}$, it is defined such that $\mathbf{P}\mathbf{x}_1 \perp (\mathbf{I} - \mathbf{P})\mathbf{x}_2$ for any vectors $\mathbf{x}_1, \mathbf{x}_2$. Using the above definition, prove that a matrix $\mathbf{P}$ is an orthogonal projection matrix if and only if $\mathbf{P} = \mathbf{P}^T$ and $\mathbf{P}^2 = \mathbf{P}$.

Exercise 10.9.8. In this chapter, we focused on the LS parameter estimation with the number of observations larger than the number of unknown parameters. However, when the number of observations is smaller than the number of parameters in $\mathbf{x} = \mathbf{H}\boldsymbol{\theta}$, there will be many solutions satisfying this equation. We need an extra criterion to select a desirable solution out of many possible solutions. One common way is to select the solution with the minimum norm:

$$\min \boldsymbol{\theta}^T \boldsymbol{\theta} \qquad \text{subject to} \quad \mathbf{x} = \mathbf{H}\boldsymbol{\theta}. \tag{10.82}$$

Find the solution for $\boldsymbol{\theta}$ using the technique of Lagrangian multipliers.

Exercise 10.9.9. Another form of LS estimation with quadratic constraint is

$$\min \| \mathbf{L}\boldsymbol{\theta} \|^2 \quad \text{s.t.} \quad \| \mathbf{x} - \mathbf{H}\boldsymbol{\theta} \| \leq \beta \tag{10.83}$$

where β is a user-defined threshold. In this form, the norm of the solution is minimized subject to the constraint that the estimation error is within a certain limit. By using GSVD, find the relationship between β and the ℓ_2-regularization parameter λ, similar to (10.53).

Exercise 10.9.10. Prove the expressions for MMSE estimator and its covariance matrix in Theorem 10.7.2 by applying the results in Theorem 10.7.1. Furthermore, show that the alternative form (10.66) can be derived from (10.64).

Exercise 10.9.11. Derive the results (10.63) and (10.61) by applying Theorem 10.7.2 of MMSE estimator for linear model.

Exercise 10.9.12. Suppose we have a number of observations $\mathbf{x} = [x[0], x[1], ..., x[N-1]]^T$ which contain information about a parameter θ. Consider the linear estimator

$$\hat{\theta} = \sum_{n=0}^{N-1} a_n x[n] + a_N. \tag{10.84}$$

The coefficient a_n's are designed to minimize the BMSE:

$$\text{BMSE} = E\left\{(\theta - \hat{\theta})^2\right\}, \tag{10.85}$$

where the expectation is taken with respect to θ and $x[n]$. The resultant estimator is called the LMMSE estimator. Prove that the LMMSE estimator is given by

$$\hat{\theta} = E\{\theta\} + \mathbf{c}_{\theta x}^T \mathbf{C}_{xx}^{-1}(\mathbf{x} - E\{\mathbf{x}\}), \tag{10.86}$$

where $\mathbf{C}_{xx} = E\{\mathbf{x}\mathbf{x}^T\}$ and $\mathbf{c}_{\theta x} = E\{\theta \mathbf{x}\}$. Derive the LMMSE estimator for the vector parameter $\boldsymbol{\theta} = [\theta_1, \theta_2, ..., \theta_p]^T$ by stacking a number of LMMSE estimators for scalars into a vector form. Finally, assume that the data obey $\mathbf{x} = \mathbf{H}\boldsymbol{\theta} + \mathbf{w}$, where $\mathbf{w}$ is a noise vector with zero mean and covariance $\mathbf{C}_w$, prove (10.64) is the LMMSE estimator. Notice that we have not assumed any statistical distribution about the noise vector $\mathbf{w}$.

References

[1] S. M. Kay, *Fundamentals of Statistical Signal* NJ: Prentice-Hall, 1993.

[2] L. L. Scharf, *Statistical Signal Processing*, New

[3] M. Leng and Y.-C. Wu, "On clock synchroni. sensor networks under unknown delay," *IEEE* vol. 59, no.1, pp. 182-190, Jan. 2010.

[4] Q. Chaudhari, E. Serpedin and K. Qaraqe, "O estimation of clock offset in a Two-Way Message *Trans. Information Theory*, vol. 56, no. 6, pp. 2

[5] J. Chen, Y.-C. Wu, S. C. Chan and T.-S. Ng, "J and channel estimation for OFDMA uplink using *Trans. on Vehicular Technology*, vol. 57, no. 6,

[6] X. Cai, Y.-C. Wu, H. Lin and K. Yamashita, "I of CFO and I/Q Imbalance in OFDM System *IEEE Trans. on Vehicular Technology*, vol. 6C 2011.

[7] T. K. Moon and W. C. Stirling, *Mathematica Signal Processing*, Upper Saddle River, NJ: Pre

[8] P.C. Hansen, *Rank Deficient and Discrete Ill-F pects of Linear Inversion*, SIAM, 1998.

[9] X. Li, Y.-C. Wu and E. Serpedin, "Timing Sy Forward Cooperative Communication Systems,' *ing*, vol. 57, no. 4, pp. 1444-1455, Apr. 2009.

[10] R. Tibshirani, "Regression Shrinkage and Sele *of the Royal Statistical Society*. Series B (Methc 267-288, 1996.

[11] K. Cai, X. Li, and Y.-C. Wu, "Bayesian CFO Es *in Proc. IEEE Wireless Communications and* 2009.

[12] E. L. Lehmann and G. Casella, *Theory of Point* York: Springer, 1998.

[13] G. E. P. Box and G. C. Tiao, *Bayesian Inference* ken, NJ: Wiley-Interscience, 1992.

[14] C. P. Robert, *The Bayesian Choice: From Decision-Theoretic Foundations to Computational Implementation*, 2nd edition, New York: Springer-Verlag, 2007.

[15] E. J. Candes, M. B. Wakin, and S. P. Boyd, "Enhancing Sparsity by Reweighted ℓ_1-Minimization," *The Journal of Fourier Analysis And Applications*, 14(5):877-905. Special Issue on Sparsity, Dec. 2008.

Chapter 11 Fundamentals of Detection Theory

Venugopal V. Veeravalli[‡]
[‡]University of Illinois at Urbana-Champaign, USA

11.1 Introduction

Detection problems arise in a number of engineering applications such as radar, communications, surveillance, and image analysis. In the basic setting of the problem, the goal is to detect the presence or absence of a signal in noise. This chapter will provide the mathematical and statistical foundations for solving such problems.

11.1.1 Statistical Decision Theory Framework

Detection problems fall under the umbrella of statistical decision theory [1], where the goal is to make a right (optimal) choice from a set of alternatives in a noisy environment. There are five basic ingredients in a typical decision theory problem.

- $\mathcal{S}$: The set of states (of nature). For detection problems, the number of states is finite, i.e., $|\mathcal{S}| = M < \infty$. For binary detection problems, which are prevalent in applications, $\mathcal{S} = \{0, 1\}$. We denote a typical state for detection problems by the variable j, i.e., $j \in \mathcal{S}$.

- $\mathcal{D}$: The set of decisions or actions. This set is the set of decisions about the state. Elements in $\mathcal{D}$ would typically correspond to elements in $\mathcal{S}$. In some applications such as communications with erasure, the set $\mathcal{D}$ could have larger cardinality than the set $\mathcal{S}$. We denote a typical decision by the variable i, i.e., $i \in \mathcal{D}$.

- $C(i, j)$ or $C_{i,j}$: The cost function between decisions and states, $C : \mathcal{D} \times \mathcal{S} \mapsto \mathbb{R}^+$. In order to be able to talk about optimizing the decision, we need to quantify the cost incurred from each decision. The cost function C serves this purpose. An example of cost function, which is relevant in many applications, is the *uniform* cost function for which

$$C_{i,j} = \begin{cases} 0 & \text{if } i = j \\ 1 & \text{if } i \neq j \end{cases} . \tag{11.1}$$

- $\mathcal{Y}$: The set of observations. The decision abou but based on some *random* observation[1] Y tal

- Δ: The set of decision rules or tests. Since t observations, we need to have mappings fro decision set. These are the decision rules, i.e.,

Detection problems are also referred to as *hypothe.* understanding that each element of $\mathcal{S}$ corresponds to of the observations. The hypothesis corresponding t

11.1.2   Probabilistic Structure for Ob

We associate with $\mathcal{Y}$, a sigma algebra $\mathcal{G}$ of subsets abilities. The pair $(\mathcal{Y}, \mathcal{G})$ is the *observation space.* in this chapter, we will almost exclusively have $\mathcal{Y}$ countable set. In the case that $\mathcal{Y} = \mathbb{R}^n$, we take $\mathcal{G}$ to containing all the n-dimensional rectangles in $\mathbb{R}^n$, i.e In the case when $\mathcal{Y} = \{\gamma_1, \gamma_2, \ldots\}$, we take $\mathcal{G}$ to be
For $\mathcal{Y} = \mathbf{R}^n$, we assume that probabilities c dimensional PDF. For $\mathcal{Y} = \{\gamma_1, \gamma_2, \ldots\}$, probabili PMF. We will use the term density for both PDF density function by p, and use a common notation i in [2]:

For $A \in \mathcal{G}$,

$$\mathrm{P}(A) = \int_{y \in A} p(y)\mu(dy) = \begin{cases} \int_{y \in A} p(y)dy & \text{for} \\ \sum_{\gamma_i \in A} p(\gamma_i) & \text{for} \end{cases}$$

Let g be a function on $\mathcal{Y}$. Then the expected value c given by

$$\mathrm{E}[g(Y)] = \int_{\mathcal{Y}} p(y)g(y)\mu(dy) = \begin{cases} \int_{\mathcal{Y}} p(y)g(y)dy \\ \sum_{\mathcal{Y}} p(\gamma_i)g(\gamma_i) \end{cases}$$

11.1.3   Conditional Density and Cond

In order to make a decision about the state j bas need to know how Y depends on j statistically. T conditional density (PDF/PMF) of Y conditioned we denote by $p_j(y)$) is available for each $j \in \mathcal{S}$. In random variable J (see below), $p_j(y)$ is the usual

---

[1] As will be the convention in the rest of the chapter, we der letters and their corresponding realizations by lowercase lette is denoted by y.

Table 11.1: Decision rules and conditional risks for Example 11.1.1.

δ	a	b	c	R_0	R_1
δ_1	0	0	0	0	1
δ_2	0	0	1	0	0.5
δ_3	0	1	0	0	0.5
δ_4	0	1	1	0	0
δ_5	1	0	0	1	1
δ_6	1	0	1	1	0.5
δ_7	1	1	0	1	0.5
δ_8	1	1	1	1	0

but otherwise we can think of the set $\{p_j(y), j \in \mathcal{S}\}$ as simply an indexed set of densities, with p_j being the density for Y that corresponds to the state being j.

The cost associated with a decision rule $\delta \in \Delta$ is a random quantity (because Y is random) given by $C(\delta(Y), j)$. Therefore, to *order* decision rules according to their "merit" we use the quantity

$$R_j(\delta) = E_j\left[C(\delta(Y), j)\right] = \int C(\delta(y), j) p_j(y) \mu(dy).$$

which we call the *conditional risk* associated with δ when the state is j.

The conditional risk function can be used to obtain a (partial) ordering of the decision rules in Δ, in the following sense.

Definition 11.1.1. A decision rule δ is *better* than decision rule δ' if

$$R_j(\delta) \leq R_j(\delta'), \quad \forall j \in \mathcal{S}$$

and

$$R_j(\delta) < R_j(\delta') \text{ for at least one } j \in \mathcal{S}$$

Sometimes it may be possible to find a decision rule $\delta^\star \in \Delta$ which is better than any other $\delta \in \Delta$. In this case, the statistical decision problem is solved. Unfortunately, this usually happens only for trivial cases as in the following example.

Example 11.1.1. Suppose $\mathcal{S} = \mathcal{D} = \{0, 1\}$ with the uniform cost function as in (11.1). Furthermore suppose the observation Y takes values in the set $\mathcal{Y} = \{a, b, c\}$ and the conditional p.m.f.'s of Y are:

$$p_0(a) = 1, p_0(b) = p_0(c) = 0, \qquad p_1(a) = 0, p_1(b) = p_1(c) = 0.5.$$

Then it is easy to see that we have the conditional risks for the eight possible decision rules depicted in Table 11.1. Clearly, δ_4 is the best rule according to Definition 11.1.1, but this happens only because the conditional PMFs p_0 and p_1 have disjoint supports (see Exercise 2). □

Since conditional risks cannot be used directly i
statistical decision making problems except in trivia
approaches for finding optimal decision rules: *Bayes*

11.1.4   Bayesian Approach

Here we assume that we are given an *a priori* probab
states $\mathcal{S}$. The state is then denoted by a random var
(since, for detection problems, the state space is fi
average risk or *Bayes* risk associated with a decision

$$r(\delta) = E\left[R_J(\delta)\right] = \sum_{j \in \mathcal{S}} \pi_j F$$

We can then obtain an ordering on the δ's by using
we choose the decision rule δ_{B} that has minimum B.

$$\delta_{\mathrm{B}} = \arg\ \min_{\delta \in \Delta} r(\delta).$$

The decision rule δ_{B} is called *Bayes* rule.

11.1.5   Minimax Approach

What if we are not given a prior distribution on th
a distribution on $\mathcal{S}$ (for example, a uniform distrib
approach. On the other hand, one may want to guar
mance for all choices of state. In this case, we use a
of the minimax approach is to find the decision rul
case cost:

$$\delta_{\mathrm{m}} = \arg\ \min_{\delta \in \Delta} \max_{j \in \mathcal{S}} R_j(\delta$$

The decision rule δ_{m} is called the *minimax* rule.

In addition to Bayes and minimax approaches
techniques that are specific to special classes of de
example, in binary hypothesis testing, a third approa
approach (see Section 11.4) is often used in practice

11.1.6   Randomized Decision Rules

Even though this might seem counter-intuitive at f
to get a better decision rule by randomly choosing b
decision rules.

Definition 11.1.2. A *randomized* decision rule $\tilde{\delta}$ is

$$\tilde{\delta}(y) = \delta_\ell(y) \text{ with probability } \beta_\ell,$$

for some L and some $\{\beta_\ell\}$, with $\beta_\ell > 0$ and $\sum_\ell \beta_\ell =$

Table 11.2: Decision rules, and Bayes and minimax risks for Example 11.1.2.

δ	a	b	c	R_0	R_1	$0.5(R_0 + R_1)$	$\max(R_0, R_1)$
δ_1	0	0	0	0	1	0.5	1
δ_2	0	0	1	0	0.5	0.25	0.5
δ_3	0	1	0	0.5	0.5	0.5	0.5
δ_4	0	1	1	0.5	0	0.25	0.5
δ_5	1	0	0	0.5	1	0.75	1
δ_6	1	0	1	0.5	0.5	0.5	0.5
δ_7	1	1	0	1	0.5	0.75	1
δ_8	1	1	1	1	0	0.5	1

The set $\tilde{\Delta}$ of randomized decision rules obviously contains the set Δ, and thus optimizing over $\tilde{\Delta}$ will necessarily result in at least as good a decision rule as that obtained by optimizing over Δ.

Theorem 11.1.1. *Randomization does not improve Bayes rules:*

$$\min_{\delta \in \Delta} r(\delta) = \min_{\tilde{\delta} \in \tilde{\Delta}} r(\tilde{\delta}).$$

Proof: Since $\Delta \subset \tilde{\Delta}$, it is clear that the right-hand side (RHS) is less than or equal to the left-hand side (LHS). To prove the reverse inequality, suppose $\tilde{\delta}$ chooses δ_ℓ with probability β_ℓ, $\ell = 1, \ldots, L$. Then

$$r(\tilde{\delta}) = \sum_{\ell=1}^{L} \beta_\ell \, r(\delta_\ell) \geq \sum_{\ell=1}^{L} \beta_\ell \min_{\delta \in \Delta} r(\delta) = \min_{\delta \in \Delta} r(\delta).$$

Taking the minimum over $\tilde{\delta} \in \tilde{\Delta}$ on the LHS gives us the desired inequality. ∎ However, as we see in the following example, randomization could result in a better minimax rule. We will also see later in Section 11.4 that randomization can yield better Neyman-Pearson rules for binary detection problems.

Example 11.1.2. Consider the same setup as in Example 11.1.1 with the following conditional PMF's.

$$p_0(a) = p_0(b) = 0.5, p_0(c) = 0, \qquad p_1(a) = 0, p_1(b) = p_1(c) = 0.5.$$

We can compute the conditional risks for the eight possible decision rules as shown in Table 11.2. Clearly there is no "best" rule based on conditional risks alone in this case. Now consider finding a Bayes rule for priors $\pi_0 = \pi_1 = 0.5$. It is clear from the table that δ_2 and δ_4 are both Bayes rules. Also, δ_2, δ_3, δ_4 and δ_6 are all minimax rules with minimax risk equal to 0.5. Finally, randomizing between δ_2 and δ_4 with equal probability results in a rule with minimax risk equal to 0.25. Thus, we see that randomization can improve minimax rules. □

11.1.7 General Method for Finding E

In the Bayesian framework, we can define the *a poste* state j, given observation y. By Bayes probability la

$$\pi(j|y) = \frac{p_j(y)\pi_j}{p(y)}.$$

We can write the Bayes risk of (11.4) in terms of $\pi($

$$r(\delta) = E[E[C(\delta(Y), J)|Y]] = \int_{y \in \mathcal{Y}} \left[\sum_{j \in \mathcal{S}} \pi(j|y) C \right]$$

Define the *a posteriori* cost of decision $i \in \mathcal{D}$, given

$$C(i|y) \triangleq \sum_{j \in \mathcal{S}} \pi(j|y) C(i, $$

Then it is easy to see that minimizing $r(\delta)$ in (11.$C(\delta(y)|y)$ for each y. Thus

$$\delta_B(y) = \arg\min_{i \in \mathcal{D}} C(i|y)$$

11.2 Bayesian Binary Detectio

We now study the special case of *binary* detection detail. Here $\mathcal{S} = \mathcal{D} = \{0, 1\}$, and hence any determin the observation space into disjoint sets $\mathcal{Y}_0$ and $\mathcal{Y}_1$, co 0 and $\delta(y) = 1$, respectively. The conditional risk written as:

$$R_j(\delta) = C_{0,j} \, P_j(\mathcal{Y}_0) + C_{1,j} \, P_j(\mathcal{Y}_1$$

Assumption 11.2.1. The cost of a correct decisi smaller than that of a wrong decision:

$$C_{0,0} < C_{1,0}, \qquad C_{1,1} < C$$

Using (11.8), we can find a Bayes decision rule fc

$$\delta_B(y) = \arg\min_{i \in \{0,1\}} C(i|y) = \begin{cases} 1 & \text{if } C \\ 0 & \text{if } C \end{cases}$$

Clearly, the Bayes solution need not be unique since whether we assign the decision of "0" or "1" to obs

$C(0|y)$. Using (11.7) and (11.5), we obtain:

$$\delta_B(y) = \begin{cases} 1 & \text{if } \pi(1|y)[C_{0,1} - C_{1,1}] \geq \pi(0|y)[C_{1,0} - C_{0,0}] \\ 0 & \text{otherwise} \end{cases} \tag{11.9}$$

$$= \begin{cases} 1 & \text{if } \dfrac{p_1(y)}{p_0(y)} \geq \dfrac{\pi_0}{\pi_1} \dfrac{C_{1,0} - C_{0,0}}{C_{0,1} - C_{1,1}} \\ 0 & \text{otherwise} \end{cases} \tag{11.10}$$

11.2.1 Likelihood Ratio Test

Definition 11.2.1. The *likelihood ratio* is given by

$$L(y) = \frac{p_1(y)}{p_0(y)}, \quad y \in \mathcal{Y}$$

with the understanding that $\frac{0}{0} = 0$, and $\frac{x}{0} = \infty$, for $x > 0$.

If we further define the threshold τ by:

$$\tau = \frac{\pi_0}{\pi_1} \frac{C_{1,0} - C_{0,0}}{C_{0,1} - C_{1,1}} \tag{11.11}$$

then we can write

$$\delta_B(y) = \begin{cases} 1 & \text{if } L(y) \geq \tau \\ 0 & \text{otherwise} \end{cases}.$$

Thus Bayes rule is a "LRT."

11.2.2 Uniform Costs

For uniform costs (see (11.1)), $C_{0,0} = C_{1,1} = 0$, and $C_{0,1} = C_{1,0} = 1$. Therefore, the threshold for the LRT simplifies to $\tau = \frac{\pi_0}{\pi_1}$ in this case. We can also see from (11.9) that

$$\delta_B(y) = \begin{cases} 1 & \text{if } \pi(1|y) \geq \pi(0|y) \\ 0 & \text{otherwise} \end{cases}.$$

Thus, for uniform costs, Bayes rule is a MAP rule. Furthermore, for uniform costs, the Bayes risk of a decision rule δ is given by

$$r(\delta) = \pi_0 P_0(\mathcal{Y}_1) + \pi_1 P_1(\mathcal{Y}_0).$$

The RHS is the average probability of error, denoted by P_e. Thus, for uniform costs, Bayes rule is also a *minimum probability of error* (MPE) rule.

Finally if we have uniform costs and equal priors (i.e., $\pi_0 = \pi_1 = 0.5$), then

$$\delta_B(y) = \begin{cases} 1 & \text{if } p_1(y) \geq p_0(y) \\ 0 & \text{otherwise} \end{cases}$$

and Bayes rule is a *maximum likelihood* (ML) decision rule.

11.2.3   Examples

Example 11.2.1. *Signal Detection in Gaussian N*
arises in a number of engineering applications, includ
nications, and can be described by the hypotheses t

$$H_0 :\ Y = \mu_0 + Z \ \text{ versus } \ H_1 :\ Y$$

where the constants μ_0 and μ_1 represent deterministi
Gaussian random variable with variance σ^2, denote
loss of generality, we may assume that $\mu_1 > \mu_0$.
   The conditional PDFs are given by:

$$p_j(y) = \frac{1}{\sqrt{2\pi\sigma^2}} \, \exp\left[-\frac{(y-\mu_j)^2}{2\sigma^2}\right]$$

and the likelihood ratio is given by:

$$L(y) = \frac{p_1(y)}{p_0(y)} = \exp\left[\frac{\mu_1 - \mu_0}{\sigma^2}\left(y - \right.\right.$$

It is easy to show that comparing $L(y)$ to τ of (11.
y with τ', where

$$\tau' = \frac{\sigma^2}{\mu_1 - \mu_0}\log\tau + \frac{\mu_1 +}{2}$$

Thus Bayes rule is equivalent to a threshold test on

$$\delta_{\mathrm{B}} = \begin{cases} 1 & \text{if } y \geq \tau' \\ 0 & \text{if } y < \tau' \end{cases}.$$

For uniform costs and equal priors, $\tau = 1$ and $\tau' = $

$$r(\delta_{\mathrm{B}}) = \mathrm{P}_{\mathrm{e}}(\delta_{\mathrm{B}}) = 0.5\mathrm{P}_0(\mathcal{Y}_1) + 0$$

where

$$\mathrm{P}_0(\mathcal{Y}_1) = \mathrm{P}_0\{Y \geq \tau'\} = 1 - \Phi\left(\frac{\tau' - \mu_0}{\sigma}\right) = 1 - \Phi$$

and

$$\mathrm{P}_1(\mathcal{Y}_0) = \mathrm{P}_1\{Y < \tau'\} = \Phi\left(\frac{\tau' - \mu_1}{\sigma}\right) = \Phi\left(\frac{\mu_0}{}\right.$$

where Φ is CDF of a $\mathcal{N}(0,1)$ random variable

$$\Phi(x) = \int_{-\infty}^{x} \frac{1}{\sqrt{2\pi}}\, e^{-t^2/2}$$

and Q is the complement of Φ, i.e., $Q(x) = 1 - \Phi(x) = \Phi(-x)$, for $x \in \mathbb{R}$. Thus

$$r(\delta_B) = P_e(\delta_B) = Q\left(\frac{\mu_1 - \mu_0}{2\sigma}\right).$$

$\square$

Example 11.2.2. *Discrete Observations.* Consider the detection problem of Example 11.1.2 with uniform costs, equal priors and

$$p_0(a) = p_0(b) = 0.5, p_0(c) = 0, \qquad p_1(a) = 0, p_1(b) = p_1(c) = 0.5.$$

The likelihood ratio is given by

$$L(y) = \begin{cases} 0 & \text{if } y = a \\ 1 & \text{if } y = b \\ \infty & \text{if } y = c \end{cases}.$$

With uniform costs and equal priors, the threshold $\tau = 1$. Therefore

$$\delta_B = \begin{cases} 1 & \text{if } L(y) \geq 1 \\ 0 & \text{if } L(y) < 1 \end{cases} = \begin{cases} 1 & \text{if } y = b, c \\ 0 & \text{if } y = a \end{cases}.$$

This rule is nothing but δ_4 of Example 11.1.2. Note that if we had chosen 0 when $L(y) = \tau$, then we would have obtained δ_2, which is also a Bayes rule. $\square$

11.3 Binary Minimax Detection

Recall from Section 11.1.5 that the minimax decision rule δ_m minimizes the worst case risk:

$$\delta_m = \arg \min_{\delta \in \Delta} R_{\max}(\delta).$$

where $R_{\max}(\delta) = \max\{R_0(\delta), R_1(\delta)\}$.

11.3.1 Bayes Risk Line and Minimum Risk Curve

We find δ_m indirectly by using the solution to Bayesian detection problem as follows. Since the prior on the states is not specified in the minimax setting, we allow the prior π_0 ($= 1 - \pi_1$) to be a variable over which we can optimize. We begin with the following definitions.

Definition 11.3.1. *Bayes Risk Line.* For any $\delta \in \Delta$,

$$r(\pi_0; \delta) = \pi_0 R_0(\delta) + (1 - \pi_0) R_1(\delta).$$

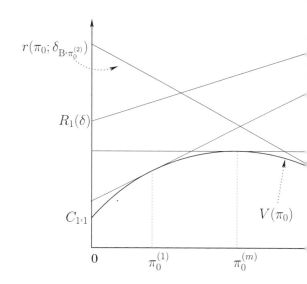

Figure 11.1: Bayes risk lines and mini

Definition 11.3.2. *Bayes Minimum Risk Curve.*

$$V(\pi_0) = \min_{\delta \in \Delta} r(\pi_0; \delta) = r(\pi_0; \delta_{\mathrm{B},\pi_0})$$

where $\delta_{\mathrm{B},\pi_0}$ is a Bayes rule for prior π_0.

Bayes risk lines and the minimum risk curve are ill
following result states some useful properties of $V(\pi$

Lemma 11.3.1. V is a concave (continuous) funct
and $V(1) = C_{0,0}$.

Proof: The minimum of concave functions is cond
of V follows from the fact that each of the risk line
concave) in π_0. As for the end point properties,

$$V(0) = \min_{\delta \in \Delta} R_1(\delta) = \min_{\delta \in \Delta} C_{0,1} \mathrm{P}_1(\mathcal{Y}_0) + C$$

where the minimizing rule is $\delta^*(y) = 1$, for all $y \in \mathcal{Y}$
We can write $V(\pi_0)$ in terms of the likelihood ratio

$$V(\pi_0) = \pi_0 [C_{1,0} \mathrm{P}_0\{L(Y) \geq \tau\} + C_{0,0} \mathrm{P}_0\{L$$
$$+ (1 - \pi_0)[C_{1,1} \mathrm{P}_1\{L(Y) \geq \tau\} +$$

If $L(y)$ has no point masses[2] under P_0 or P_1, then V
τ is differentiable in π_0).

---

[2]This condition typically holds for continuous observation
with the same support, but not necessarily even in this case.

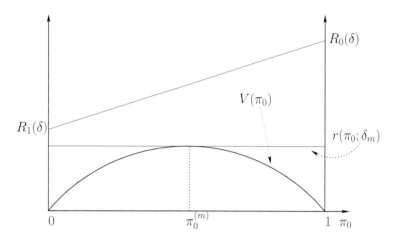

Figure 11.2: Minimax (equalizer) rule when V is differentiable at $\pi_0^{(m)}$.

11.3.2 Equalizer Rule

Let us first consider the case where V is indeed differentiable for all π_0. Then $V(\pi_0)$ achieves its maximum value at either the end points $\pi_0 = 0$ or $\pi_0 = 1$ or within the interior $\pi_0 \in (0,1)$. If we assume uniform costs, then $V(0) = V(1) = 0$, and the maximum cannot be attained at the end points. Therefore, we further restrict our analysis to the case of uniform costs (the more general setting is considered in [2]).

Theorem 11.3.1. *If $C_{0,0} = C_{1,1} = 0$ and V is differentiable on $[0,1]$, then*

$$\delta_m = \delta_{B,\pi_0^m}$$

where $\pi_0^m = \arg\max_{\pi_0} V(\pi_0)$, obtained by solving $dV(\pi_0)/d\pi_0 = 0$, i.e., δ_m is a Bayes rule for the worst case prior. Furthermore, δ_m is a Bayes equalizer rule, i.e., $R_0(\delta_m) = R_1(\delta_m)$. Note that randomization cannot improve the minimax rule in this case.

Proof: The proof follows from Figure 11.2 using the following steps:

1. For any $\delta \in \Delta$, the risk line $r(\pi_0; \delta)$ cannot intersect with $V(\pi_0)$.

2. For fixed $\pi_0^{(1)}$, the risk line $r(\pi_0; \delta_{B,\pi_0^{(1)}})$ is tangent to V at $\pi = \pi_0^{(1)}$.

3. Any rule with risk line that is not tangential to V cannot be minimax because one can always find a rule with risk line that has the same slope and is tangential to V with smaller $R_{\max}$.

4. Among all Bayes rules, the one that has $R_0 = R_1$ is minimax.

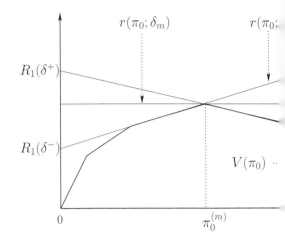

Figure 11.3: Minimax rule when V is *not* c

Since the tangent to V at any fixed prior π_0 is uniqu
ministic Bayes rule for that prior, randomization c‹
rule.

If V is *not* differentiable for all π_0, then the arg
Theorem 11.3.1 can still be used as long as V is d
and the minimax rule is still the unique Bayes rule fc
not differential at its maximum, then we have the sce
Note that δ^- and δ^+ are deterministic Bayes rules
and since they are likelihood ratio tests with δ^- hav

$$\delta^- = \begin{cases} 1 & \text{if } L(y) \geq \tau(\pi_0^{\mathrm{m}}) \\ 0 & \text{if } L(y) < \tau(\pi_0^{\mathrm{m}}) \end{cases}, \qquad \delta^+ = \begin{cases} 1 \\ (\end{cases}$$

where $\tau(\pi_0^{\mathrm{m}}) = \pi_0^{\mathrm{m}}/(1 - \pi_0^{\mathrm{m}})$. For δ^- and δ^+ to be
point mass at $\tau(\pi_0^{\mathrm{m}})$, i.e., $\mathrm{P}_j\{L(Y) = \tau(\pi_0^{\mathrm{m}})\} \neq 0$, ‹
that V is *not* differentiable at π_0^{m}. Also, if δ^- and δ^-
them can be an equalizer rule.

Finding the minimax rule within the set of deter›
in this case, since step 2 in the proof of Theorem 1
possible for a rule that has risk line that is not ta›
within Δ. We may need to resort to brute force ‹
rules within Δ as we did in Example 11.1.2. Fortur
problem by allowing for randomized decision rules.

It should be clear from Figure 11.3 that if an equalizer rule exists in $\tilde{\Delta}$, which is tangential to V at π_0^m, then it must be minimax within the class $\tilde{\Delta}$. Now, consider

$$\tilde{\delta}_{B,\pi_0^m} = \begin{cases} \delta^- & \text{with probability } q \\ \delta^+ & \text{with probability } (1-q) \end{cases}$$

The conditional risks of this randomized decision rule are given by

$$R_0(\tilde{\delta}_{B,\pi_0^m}) = qR_0(\delta^-) + (1-q)R_0(\delta^+)$$
$$R_1(\tilde{\delta}_{B,\pi_0^m}) = qR_1(\delta^-) + (1-q)R_1(\delta^+)$$

Thus, setting

$$q = \frac{R_1(\delta^+) - R_0(\delta^+)}{(R_1(\delta^+) - R_0(\delta^+)) + (R_0(\delta^-) - R_1(\delta^-))} \overset{\Delta}{=} q_m \qquad (11.14)$$

produces an equalizer rule.

Theorem 11.3.2. *If $C_{0,0} = C_{1,1} = 0$ and V is not differentiable at its maximum, then the minimax solution within the set of randomized decision rules $\tilde{\Delta}$ is given by the equalizer rule:*

$$\tilde{\delta}_m = \tilde{\delta}_{B,\pi_0^m} = \begin{cases} 1 & \text{if } L(y) > \tau(\pi_0^m) \\ 1 \text{ w.p. } q_m & \text{if } L(y) = \tau(\pi_0^m) \\ 0 & \text{if } L(y) < \tau(\pi_0^m) \end{cases}$$

where $\pi_0^m = \arg\max_{\pi_0} V(\pi_0)$ and q_m is given in (11.14).

11.3.3 Examples

Example 11.3.1. *Signal Detection in Gaussian Noise (continued).* In this example we study the minimax solution to the detection problem described in Example 11.2.1. We assume uniform costs. We can compute the minimum Bayes risk curve as:

$$V(\pi_0) = \pi_0 P_0\{Y \geq \tau'\} + (1-\pi_0)P_1\{Y < \tau'\}$$
$$= \pi_0 Q\left(\frac{\tau' - \mu_0}{\sigma}\right) + (1-\pi_0)\Phi\left(\frac{\tau' - \mu_1}{\sigma}\right)$$

with

$$\tau' = \frac{\sigma^2}{\mu_1 - \mu_0}\log\left(\frac{\pi_0}{1 - \pi_0}\right) + \frac{\mu_1 + \mu_0}{2}.$$

Clearly V is a differentiable function, and therefore the deterministic equalizer rule is minimax. We can solve for the equalizer rule without explicitly maximizing V. In particular, if we denote the LRT with threshold τ' (see (11.12)) by $\delta_{\tau'}$, then

$$R_0(\delta_{\tau'}) = Q\left(\frac{\tau' - \mu_0}{\sigma}\right), \quad R_1(\delta_{\tau'}) = \Phi\left(\frac{\tau' - \mu_1}{\sigma}\right) = Q\left(\frac{\mu_1 - \tau'}{\sigma}\right).$$

Setting $R_0(\delta_{\tau'}) = R_1(\delta_{\tau'})$ yields

$$\tau'_{\mathrm{m}} = \frac{\mu_1 + \mu_0}{2}$$

from which we can conclude that $\tau_{\mathrm{m}} = 1$ and $\pi_0^{\mathrm{m}} =$
Thus the minimax decision rule is given by

$$\delta_{\mathrm{m}} = \delta_{\mathrm{B},0.5} = \begin{cases} 1 & \text{if } y \geq \mu \\ 0 & \text{otherwis} \end{cases}$$

and the minimax risk is given by

$$r(\delta_{\mathrm{m}}) = V(0.5) = Q\left(\frac{\mu_1 -}{2\sigma}\right.$$

Example 11.3.2. *Discrete Observations (continued*
the minimax solution to the detection problem descri
that $L(a) = 0$, $L(b) = 1$, and $L(c) = \infty$. Assuming
prior π_0 (randomized and deterministic) are given b

$$\tilde{\delta}_{\mathrm{B},\pi_0}(y) = \begin{cases} 1 & \text{if } L(y) \\ 1 \text{ w.p. } q & \text{if } L(y) \\ 0 & \text{if } L(y) \end{cases}$$

where $\tau(\pi_0) = \pi_0/(1 - \pi_0)$ and $q \in [0, 1]$.
For $\pi_0 \in (0, 0.5)$, $\tau(\pi_0) \in (0, 1)$, and thus all the
to the single deterministic rule:

$$\delta^-(y) = \begin{cases} 1 & \text{if } y = b, c \\ 0 & \text{if } y = a \end{cases}$$

Similarly, for $\pi_0 \in (0.5, 1)$, $\tau(\pi_0) \in (1, \infty)$, and thus
collapse to the single deterministic rule:

$$\delta^+(y) = \begin{cases} 1 & \text{if } y = c \\ 0 & \text{if } y = a, b \end{cases}$$

For $\pi_0 = 0.5$, the following set of randomized decisio

$$\tilde{\delta}_{\mathrm{B},0.5}(y) = \begin{cases} 1 & \text{if } y \\ 1 \text{ w.p. } q & \text{if } y \\ 0 & \text{if } y \end{cases}$$

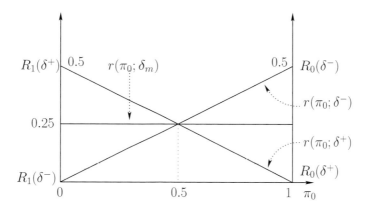

Figure 11.4: Minimax rule for Example 11.3.2.

and these rules can be obtained by randomizing between δ^+ and δ^-. From the above discussion it is clear that the minimum Bayes risk curve V is as shown in Figure 11.4, with the worst case prior $\pi_0^m = 0.5$. Furthermore, it is easy to check that $R_1(\delta^-) = R_0(\delta^+) = 0$, and $R_0(\delta^-) = R_1(\delta^+) = 0.5$. Therefore, from (11.14), $q_m = 0.5$, and the minimax decision rule is given by:

$$
\tilde{\delta}_m = \begin{cases} 1 & \text{if } y = c \\ 1 \text{ w.p. } 0.5 & \text{if } y = b \\ 0 & \text{if } y = a \end{cases}
$$

with minimax risk $r(\tilde{\delta}_m) = V(0.5) = 0.25$.

It is interesting to note that δ_2 and δ_4 in Example 11.1.2 are the same as δ^+ and δ^-, respectively, and that randomizing between these rules with equal probability is indeed the minimax solution within $\tilde{\Delta}$. □

11.4 Binary Neyman-Pearson Detection

For binary detection problems without a prior on the state, a commonly used alternative to minimax formulation is the Neyman-Pearson formulation, which is based on trading off the following two types of error probabilities:

$$
\begin{aligned}
\text{Probability of False Alarm} &\triangleq P_F(\tilde{\delta}) = P_0\{\tilde{\delta}(Y) = 1\} \\
\text{Probability of Miss} &\triangleq P_M(\tilde{\delta}) = P_1\{\tilde{\delta}(Y) = 0\}
\end{aligned}
\tag{11.16}
$$

The goal is to minimize P_M subject to the constraint $P_F \leq \alpha$, for $\alpha \in (0, 1)$.

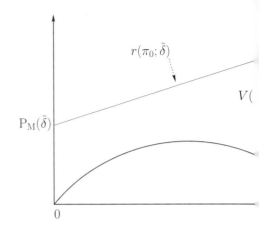

Figure 11.5: Risk line and Bayesian minimum ris

An alternative measure of performance that is
surveillance applications is:

$$\text{Probability of Detection} \triangleq P_D(\tilde{\delta}) = P_1\{\tilde{\delta}(Y$$

$P_D(\tilde{\delta})$ is also called the *power* of the decision rule $\tilde{\delta}$
problem is generally stated in terms P_D and P_F as:

$$\tilde{\delta}_{NP} = \arg \max_{\substack{\tilde{\delta} \in \tilde{\Delta}: \\ P_F(\tilde{\delta}) \leq \alpha}} P_D(\tilde{\delta}) \quad \text{for } \alpha$$

Note that unlike the Bayesian and minimax opti
formulated in terms of conditional risks, the N-P opt
terms of conditional error probabilities. In particula
uniform costs, which means $P_D(\tilde{\delta}) = 1 - R_1(\tilde{\delta})$ and
optimization is to minimize $R_1(\tilde{\delta})$ subject to $R_0(\tilde{\delta})$:

11.4.1   Solution to the N-P Optimizat

To solve the N-P optimization problem, we once agai
and the minimum risk curve $V(\pi_0)$ with uniform cost
the risk line $r(\pi_0; \tilde{\delta})$ for any rule $\tilde{\delta} \in \tilde{\Delta}$ lies above the
intersects the $\pi_0 = 0$ line at level $P_M(\tilde{\delta})$ and the $\pi_0 =$
all decision rules with risk lines that have intersecti
level less than or equal to α, we are interested in th
intersection with the $\pi_0 = 0$ line. As in the solution t
first consider the case where V is differentiable for al

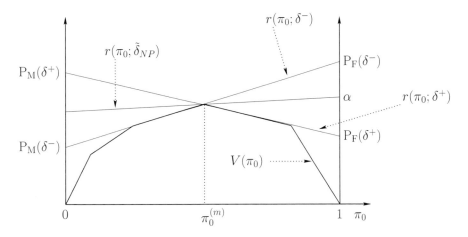

Figure 11.6: N-P optimization when V is not differentiable for all $\pi_0 \in [0, 1]$.

decision rule that solves the N-P problem has a risk line that is tangential to V and intersects the $\pi_0 = 1$ line at a level exactly equal to α. Such a rule is *deterministic* Bayes rule (LRT) that compares the likelihood ratio $L(y)$ to a threshold η that satisfies the P_F constraint.

Theorem 11.4.1. *If V is differentiable on $[0, 1]$, then*

$$\tilde{\delta}_{NP}(y) = \delta_\eta = \begin{cases} 1 & \text{if } L(y) \geq \eta \\ 0 & \text{otherwise} \end{cases}$$

where η is chosen so that $P_0\{L(Y) \geq \eta\} = \alpha$.

Now consider the case where V is not differentiable, and we have the scenario depicted in Figure 11.6. The decision rule δ^+ is the deterministic LRT that has the largest value of P_F satisfying the constraint $P_F \leq \alpha$, and the decision rule δ^- is the other deterministic LRT for the same prior. By randomizing between δ^+ and δ^- we can produce a decision rule that has $P_F = \alpha$, and is hence a solution to (11.17).

Theorem 11.4.2. *If V is not differentiable for all $\pi_0 \in [0, 1]$, then*

$$\tilde{\delta}_{NP}(y) = \tilde{\delta}_{\eta,\gamma} = \begin{cases} 1 & \text{if } L(y) > \eta \\ 1 \text{ w.p. } \gamma & \text{if } L(y) = \eta \\ 0 & \text{if } L(y) < \eta \end{cases}$$

where η and γ are chosen so that $P_0\{L(Y) > \eta\} + \gamma P_0\{L(Y) = \eta\} = \alpha$.

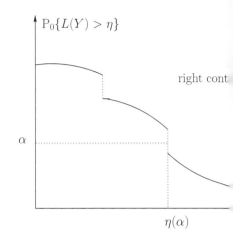

Figure 11.7: Complementary CDF of the li

11.4.2   N-P Rule and Receiver Opera

The procedure for finding the parameters η and γ of
is illustrated in Figure 11.7, where we plot $P_0\{L(y)$
seen in Figure 11.7, $P_0\{L(y) > \eta\}$ is a right contin
constraint α, we first choose $\eta(\alpha)$ as:

$$\eta(\alpha) = \min\{\eta \geq 0 : P_0\{L(y) >$$

If $P_0\{L(y) > \eta(\alpha)\} = \alpha$, then we do not need t
$\gamma(\alpha) = 0$. If $P_0\{L(y) > \eta(\alpha)\} < \alpha$, then we pick $\gamma(\alpha)$

$$\alpha = P_0\{L(y) > \eta(\alpha)\} + \gamma(\alpha)P_0\{L$$

which implies that

$$\gamma(\alpha) = \frac{\alpha - P_0\{L(y) > \eta(}{P_0\{L(y) = \eta(\alpha)}$$

The probability of detection (power) of $\tilde{\delta}_{NP}$ for P_F

$$P_D(\tilde{\delta}_{NP}) = P_1\{L(y) > \eta(\alpha)\} + \gamma(\alpha)P_1$$

A plot of $P_D(\tilde{\delta}_{NP})$ versus $P_F(\tilde{\delta}_{NP}) = \alpha$ is called the r
tics (ROC) of the Neyman-Pearson decision rule (see
of the ROC are discussed in Exercise 11. In particula
tion that lies above the 45° line, i.e., $P_D(\tilde{\delta}_{NP}) \geq P_F$

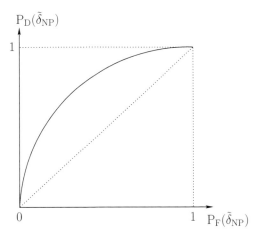

Figure 11.8: Receiver operating characteristic (ROC).

11.4.3 Examples

Example 11.4.1. *Signal Detection in Gaussian Noise (continued).* In this example we study the N-P solution to the detection problem described in Example 11.2.1. As in the Bayesian setting of this problem, we can simplify the form of the LRT by noting that

$$L(y) > \eta \iff y > \eta' = \frac{\sigma^2}{\mu_1 - \mu_0} \log \eta + \frac{\mu_1 + \mu_0}{2}$$

Thus

$$\tilde{\delta}_{NP}(y) = \begin{cases} 1 & \text{if } y > \eta' \\ 1 \text{ w.p. } \gamma & \text{if } y = \eta' \\ 0 & \text{if } y < \eta' \end{cases}.$$

Randomization is not needed since $P_0\{Y = \eta'\} = P_1\{Y = \eta'\} = 0$ for all $\eta' \in \mathbb{R}$, and therefore

$$\tilde{\delta}_{NP}(y) = \delta_{\eta'}(y) = \begin{cases} 1 & \text{if } y \geq \eta' \\ 0 & \text{if } y < \eta' \end{cases}.$$

Now

$$P_F(\delta_{\eta'}) = P_0\{Y \geq \eta'\} = Q\left(\frac{\eta' - \mu_0}{\sigma}\right).$$

Therefore, we can meet a P_F constraint of α by setting

$$\eta'(\alpha) = \sigma Q^{-1}(\alpha) + \mu_0.$$

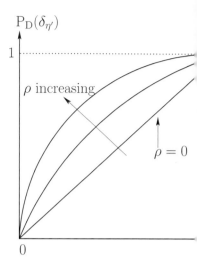

Figure 11.9: ROC for Example

The power of $\delta_{\eta'}$ is given by:

$$P_D(\delta_{\eta'}) = P_1\{Y \geq \eta'\} = Q\left(\frac{\eta'(\alpha) - \mu_1}{\sigma}\right)$$

where $\rho = (\mu_1 - \mu_0)/\sigma$ is a measure of the signal-to
is plotted in Figure 11.9. As ρ increases, the P_D inc
□

Example 11.4.2. *Discrete Observations (continue*
the N-P solution to the detection problem describe
the fact that $L(a) = 0$, $L(b) = 1$, and $L(c) = \infty$, we

$$P_0\{L(Y > \eta\} = \begin{cases} 0.5 & \text{if } \eta \in \\ 0 & \text{if } \eta \in \end{cases}$$

Thus, for $\alpha \in (0, 0.5)$, $\eta(\alpha) = 1$ and $\gamma(\alpha) = \frac{\alpha - 0}{0.5} = 2$

$$\tilde{\delta}_{NP}(y) = \begin{cases} 1 & \text{if } y \\ 1 \text{ w.p. } 2\alpha & \text{if } y \\ 0 & \text{if } y \end{cases}$$

and

$$P_D(\tilde{\delta}_{NP}) = p_1(c) + 2\alpha p_1(b) =$$

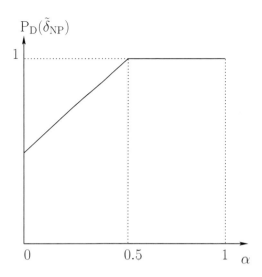

Figure 11.10: ROC for Example 11.4.2.

For $\alpha \in [0.5, 1)$, $\eta(\alpha) = 0$ and $\gamma(\alpha) = \frac{\alpha - 0.5}{0.5} = 2\alpha - 1$, which yields

$$\tilde{\delta}_{\mathrm{NP}}(y) = \begin{cases} 1 & \text{if } y = c, b \\ 1 \text{ w.p. } 2\alpha - 1 & \text{if } y = a \end{cases}$$

and $\mathrm{P_D}(\tilde{\delta}_{\mathrm{NP}}) = 1$. The ROC is plotted in Figure 11.10.   □

11.5 Bayesian Composite Detection

So far we have assumed that conditional densities p_0 and p_1 are specified completely. Under this assumption, we saw that all three formulations of the binary detection problem (Bayes, minimax, Neyman-Pearson) led to the same solution structure, LRT, which is a comparison of the likelihood ratio $L(y)$ to an appropriately chosen threshold. We now study the situation where p_0 and p_1 are not specified explicitly, but we are told that they come from a parametrized family of densities $\{p_\theta, \theta \in \Lambda\}$, with Λ being a discrete set or a subset of a Euclidean space. The hypothesis H_j corresponds to $\theta \in \Lambda_j$, $j = 0, 1$, and $\Lambda_0 \cup \Lambda_1 = \Lambda$, $\Lambda_0 \cap \Lambda_1 = \emptyset$.

We can consider composite binary detection (hypothesis testing) as a statistical decision theory problem where the set of states $\mathcal{S} = \Lambda$ is nonbinary, but the set of decisions $\mathcal{D} = \{0, 1\}$ is still binary, and the cost function relating the decisions and states is of the form:

$$C(i, \theta) = C_{i,j} \text{ for all } \theta \in \Lambda_j, \ i, j = 0, 1. \tag{11.18}$$

In this section we consider a Bayesian formulation of
that the state θ is a realization of a random variab
given by $\pi(\theta)$. From (11.8), we immediately have t
hypothesis testing is given by:

$$\delta_{\mathrm{B}}(y) = \arg\min_{i \in \mathcal{D}} C(i|y)$$

where, using the notation introduced in (11.2),

$$C(i|y) = \int_{\theta \in \Lambda} C(i, \theta) p(\theta|y)\mu(d\theta), \quad \text{with}$$

Using (11.18), we can expand $C(i|y)$ as:

$$C(i|y) = C_{i,0} \int_{\theta \in \Lambda_0} p(\theta|y)\mu(d\theta) \ + \ C_{i,1}$$

from which we can easily conclude that:

$$C(1|y) \le C(0|y) \quad \Longleftrightarrow \quad \frac{\int_{\theta \in \Lambda_1} p_\theta(y)\pi(\theta)\mu(d\theta)}{\int_{\theta \in \Lambda_0} p_\theta(y)\pi(\theta)\mu(d\theta)}$$

Now, if we define the priors on the hypotheses as

$$\pi_j \overset{\Delta}{=} \int_{\theta \in \Lambda_j} \pi(\theta)\mu(d\theta), \ j =$$

and the conditional densities for the hypotheses as

$$p(y|\Lambda_j) \overset{\Delta}{=} \frac{1}{\pi_j} \int_{\theta \in \Lambda_j} p_\theta(y)\pi(\theta$$

then we can see that

$$C(1|y) \le C(0|y) \quad \Longleftrightarrow \quad L($$

with τ as defined in (11.11) and $L(y) = p(y|\Lambda_1)/p(y$
Therefore, we can conclude that Bayes rule for c
but a LRT for the (simple) binary detection problem

$$H_0 : Y \sim p(y|\Lambda_0) \quad \text{versus} \quad H_1 : Y$$

with priors π_0 and π_1 as defined in (11.20).

Example 11.5.1. Consider the composite detection
$\Lambda_0 = [0, 1)$, and $\Lambda_1 = [1, \infty)$, with uniform costs, and

$$p_\theta(y) = \theta e^{-\theta y} \, \mathbb{I}_{\{y \ge 0\}}, \quad \pi(\theta) = \epsilon$$

where $\mathbb{I}$ is the indicator function. To compute the Bayes rule for this problem, we first compute

$$\int_{\theta \in \Lambda_1} p_\theta(y) \pi(\theta) \mu(d\theta) = \int_1^\infty \theta e^{-\theta(y+1)} d\theta = \frac{(y+2)e^{-(y+1)}}{(y+1)^2}$$

and

$$\int_{\theta \in \Lambda_0} p_\theta(y) \pi(\theta) \mu(d\theta) = \int_0^1 \theta e^{-\theta(y+1)} d\theta = \frac{1 - (y+2)e^{-(y+1)}}{(y+1)^2}.$$

Then, from (11.19), we get that

$$\delta_{\mathrm{B}} = \begin{cases} 1 & \text{if } (y+2) \geq 0.5 e^{(y+1)} \\ 0 & \text{otherwise} \end{cases}$$

which can be simplified to

$$\delta_{\mathrm{B}} = \begin{cases} 1 & \text{if } 0 \leq y \leq \tau' \\ 0 & \text{if } y > \tau' \end{cases}$$

where τ' is a solution to the transcendental equation $(y+2) = 0.5 e^{(y+1)}$. $\qquad \square$

11.6 Neyman-Pearson Composite Detection

We now consider the more interesting setting for the composite detection problem where there is no prior on the state. A common way to pose the optimization problem in this setting is a generalization of the Neyman-Pearson formulation (see (11.16)). We define the probabilities of false alarm and detection of a test $\tilde{\delta} \in \tilde{\Delta}$ by:

$$P_F(\tilde{\delta}; \theta) = P_\theta\{\tilde{\delta}(Y) = 1\}, \quad \theta \in \Lambda_0$$
$$P_D(\tilde{\delta}; \theta) = P_1\{\tilde{\delta}(Y) = 1\}, \quad \theta \in \Lambda_1$$

The goal in UMP detection is to constrain $P_F(\tilde{\delta}; \theta) \leq \alpha$, for all $\theta \in \Lambda_0$, and to simultaneously maximize $P_D(\tilde{\delta}; \theta)$, for all $\theta \in \Lambda_1$. If such a test exists, it is called UMP.

11.6.1 UMP Detection with One Composite Hypothesis

We begin by studying the special case where only H_1 is composite, i.e., Λ_0 is the singleton set equal to $\{\theta_0\}$. The UMP optimization problem can be stated as:

Maximize $P_D(\tilde{\delta}; \theta)$, for all $\theta \in \Lambda_1$, subject to $P_F(\tilde{\delta}; \theta_0) \leq \alpha$.

For fixed $\theta_1 \in \Lambda_1$, we can compute the likelihood rati

$$L_{\theta_1}(y) = \frac{p_{\theta_1}(y)}{p_{\theta_0}(y)}$$

and the corresponding Neyman-Pearson test is given

$$\tilde{\delta}_{\mathrm{NP}}(y; \theta_1) = \begin{cases} 1 & \text{if } L_{\theta_1} \\ 1 \text{ w.p. } \gamma_\alpha(\theta_1) & \text{if } L_{\theta_1} \\ 0 & \text{if } L_{\theta_1} \end{cases}$$

with $\gamma_\alpha(\theta_1)$ and $\gamma_\alpha(\theta_1)$ satisfying

$$\mathrm{P}_{\theta_0}\{L(Y) > \eta_\alpha(\theta_1)\} + \gamma_\alpha(\theta_1)\mathrm{P}_{\theta_0}\{L(Y)$$

Now, if it turns out that $\tilde{\delta}_{\mathrm{NP}}(y; \theta_1)$ is independent
since it is the N-P solution for all $\theta_1 \in \Lambda_1$. Otherwise
the following, we provide some illustrative examples.

Example 11.6.1. *Detection of One-Sided Composi*
This detection problem arises in communications and
signal amplitude is unknown but the phase is know
described by:

$$H_0 : Y = Z \text{ versus } H_1 : Y =$$

where $\theta > 0$ is an unknown parameter (signal amplitu
is a composite detection problem with $\theta_0 = 0$, and Λ
    For fixed $\theta > 0$, $L_\theta(y) = p_\theta(y)/p_0(y)$ has no point
therefore $\tilde{\delta}_{\mathrm{NP}}(y; \theta)$ is deterministic LRT:

$$\delta_{\mathrm{NP}}(y; \theta) = \begin{cases} 1 & \text{if } L_\theta(y) \geq \eta(\theta) \\ 0 & \text{if } L_\theta(y) < \eta(\theta) \end{cases} =$$

where (see Example 11.4.1) $\eta'(\theta)$ is given by

$$\eta'(\theta) = \frac{\sigma^2 \log \eta(\theta)}{\theta} + \frac{\theta}{2}.$$

For an α-level test we need to find $\eta'_\alpha(\theta)$ such that P_0
the fact that $Y \sim \mathcal{N}(0, \sigma^2)$ under P_0, we get

$$Q\left(\frac{\eta'_\alpha(\theta)}{\sigma}\right) = \alpha \implies \eta'_\alpha(\theta) =$$

Note that $\eta'_\alpha(\theta)$ is independent of θ, and therefore th

$$\delta_{\mathrm{UMP}} = \begin{cases} 1 & \text{if } y \geq \sigma Q^{-1}(\\ 0 & \text{if } y < \sigma Q^{-1}(\end{cases}$$

Note that while the test δ_{UMP} is independent of the θ, the performance of the test in terms of the P_D depends strongly on θ. In particular,

$$P_D(\delta_{\text{UMP}}; \theta) = P_\theta\{Y \geq \sigma Q^{-1}(\alpha)\} = Q(Q^{-1}(\alpha) - \theta/\sigma).$$

$\square$

Example 11.6.2. *Detection of Two-Sided Composite Signal in Gaussian Noise.* This detection problem arises in communications and radar applications where the signal amplitude and phase are both unknown. The two hypotheses are as described in Example 11.6.1, except that $\theta \in \mathbb{R}$, i.e., θ can be both positive and negative. There is no UMP test for this problem. This can be seen as follows.

First consider $\theta = 1$. Then following the same steps as in Example 11.6.1, we can show that the α-level N-P test is given by:

$$\delta_{\text{NP}}(y; 1) = \begin{cases} 1 & \text{if } y \geq \sigma Q^{-1}(\alpha) \\ 0 & \text{if } y < \sigma Q^{-1}(\alpha) \end{cases}.$$

Now, consider $\theta = -1$. Then, it is not difficult to see that $L_{-1}(y) \geq \eta$ iff $y \leq \eta'$ in this case. Therefore the α-level N-P test is given by:

$$\delta_{\text{NP}}(y; -1) = \begin{cases} 1 & \text{if } y \leq \sigma \Phi^{-1}(\alpha) \\ 0 & \text{if } y > \sigma \Phi^{-1}(\alpha) \end{cases}.$$

Since the most powerful tests for $\theta = -1$ and $\theta = 1$ are not the same, there is no uniformly most powerful test. $\square$

Example 11.6.3. *Detection of One-Sided Composite Signal in Cauchy Noise.* From Examples 11.6.1 and 11.6.2, we may be tempted to conclude that for problems involving signal detection in noise, UMP tests exist as long as H_1 is one-sided. To see that this is not true in general, we consider the example where the noise has a Cauchy distribution, i.e.,

$$p_\theta(y) = \frac{1}{\pi[1 + (y - \theta)^2]}$$

and we are testing $H_0 : \theta = 0$ against the one-sided composite hypothesis $H_1 : \theta > 0$. Then

$$L_\theta(y) = \frac{1 + y^2}{1 + (y - \theta)^2}.$$

It is easy to check that the α-level N-P tests for $\theta = 1$ and $\theta = 2$ are different, and hence there is no UMP solution. $\square$

11.6.2   UMP Detection with Both Co[

We now consider the more general case where both t[
The UMP optimization problem can be stated as:

$$\text{Maximize } \mathrm{P_D}(\tilde{\delta}; \theta), \text{ for all } \theta \in \Lambda_1, \text{ subject to } \underset{\theta \in}{s}$$

If a UMP test $\tilde{\delta}_{\mathrm{UMP}}$ exists, then it must satisfy the f[

$$\sup_{\theta_0 \in \Lambda_0} \mathrm{P_F}(\tilde{\delta}_{\mathrm{UMP}}; \theta_0) \leq \alpha.$$

Second, for any $\tilde{\delta} \in \tilde{\Delta}$ that satisfies $\sup_{\theta_0 \in \Lambda_0} \mathrm{P_F}(\tilde{\delta}; \theta_0$

$$\mathrm{P_D}(\tilde{\delta}; \theta_1) \leq \mathrm{P_D}(\tilde{\delta}_{\mathrm{UMP}}; \theta_1) \text{ for all }$$

The following example illustrates a case where a UM[
see Exercises 11.9.12 and 11.9.13.

Example 11.6.4. *Testing Between Two One-Sided C*
Noise. This is an extension of Example 11.6.1 in which
with $Z \sim \mathcal{N}(0, \sigma^2)$, and we are testing

$$H_0 : \theta \in \Lambda_0 = [0, 1] \quad \text{versus} \quad H_1 : \theta \in$$

For fixed $\theta_0 \in \Lambda_0$ and $\theta_1 \in \Lambda_1$, $L_{\theta_0, \theta_1}(y)$ has no poi[
and therefore $\tilde{\delta}_{\mathrm{NP}}(y; \theta_0, \theta_1)$ is a deterministic LRT:

$$\delta_{\mathrm{NP}}(y; \theta_0, \theta_1) = \begin{cases} 1 & \text{if } L_{\theta_0, \theta_1}(y) \geq \eta(\theta_0, \theta_1) \\ 0 & \text{if } L_{\theta_0, \theta_1}(y) < \eta(\theta_0, \theta_1) \end{cases} =$$

where $\eta'(\theta_0, \theta_1)$ is given by

$$\eta'(\theta_0, \theta_1) = \frac{\sigma^2 \log \eta(\theta_0, \theta_1)}{\theta_1 - \theta_0} + \frac{\theta}{}$$

Now in order to set the threshold η' to meet the const[
we first compute:

$$\mathrm{P_F}(\delta_{\eta'}; \theta_0) = \mathrm{P}_{\theta_0}\{Y \geq \eta'\} = Q\left(\right.$$

and note that this probability is an increasing functic[

$$\sup_{\theta_0 \in [0,1]} \mathrm{P_F}(\delta_{\eta'}; \theta_0) = Q\left(\frac{\eta' -}{\sigma}\right.$$

and we can meet the P_F constraint with equality by setting η' such that:

$$Q\left(\frac{\eta'-1}{\sigma}\right) = \alpha \quad \Longrightarrow \quad \eta'_\alpha = \sigma Q^{-1}(\alpha) + 1.$$

Note that η'_α is independent of θ_0 and θ_1. Define the test

$$\delta_{\eta'_\alpha}(y) = \begin{cases} 1 & \text{if } y \geq \eta'_\alpha \\ 0 & \text{if } y < \eta'_\alpha \end{cases}.$$

We will now establish that $\delta_{\eta'_\alpha}$ is a UMP test, by showing that conditions (11.22) and (11.23) hold. By construction,

$$\sup_{\theta_0 \in [0,1]} P_F(\delta_{\eta'_\alpha}; \theta_0) = P_F(\delta_{\eta'_\alpha}; 1) = \alpha$$

and so (11.22) holds. Also, $\delta_{\eta'_\alpha}$ is an α-level N-P test between the simple hypotheses $H_0 : \theta = 1$ and $H_1 : \theta = \theta_1$, and being independent of θ_1, it is an α-level N-P test between these hypotheses for all $\theta_1 \in (1, \infty)$. Now, consider any test $\tilde{\delta} \in \tilde{\Delta}$ that satisfies $\sup_{\theta \in [0,1]} P_F(\tilde{\delta}; \theta) \leq \alpha$. Then clearly it is also true that $P_F(\tilde{\delta}; 1) \leq \alpha$. This means that $\tilde{\delta}$ is an α-level test for testing the simple hypotheses $H_0 : \theta = 1$ versus $H_1 : \theta = \theta_1$, and it cannot be more powerful than $\delta_{\eta'_\alpha}$, i.e.,

$$P_D(\tilde{\delta}; \theta_1) \leq P_D(\delta_{\eta'_\alpha}; \theta_1) \text{ for all } \theta_1 \in (1, \infty).$$

Therefore (11.23) holds and we have:

$$\delta_{\text{UMP}}(y) = \delta_{\eta'_\alpha}(y) = \begin{cases} 1 & \text{if } y \geq \sigma Q^{-1}(\alpha) + 1 \\ 0 & \text{if } y < \sigma Q^{-1}(\alpha) + 1 \end{cases}.$$

Again, while the test δ_{UMP} is independent of the θ_1, the performance of the test in terms of the P_D depends on θ_1. In particular

$$P_D(\delta_{\text{UMP}}; \theta_1) = P_\theta\{Y \geq \sigma Q^{-1}(\alpha) + 1\} = Q\left(Q^{-1}(\alpha) - \frac{\theta_1 - 1}{\sigma}\right).$$

$\square$

11.6.3 Generalized Likelihood Ratio (GLR) Detection

While it is always desirable to have a UMP solution to the composite hypothesis testing problem, such solutions rarely exist in practice, especially in situations where both hypotheses are composite. One approach to generating a good test when UMP solutions do not exist is through the use of a "GLR" defined by

$$T_{\text{GLR}}(y) = \frac{\sup_{\theta_1 \in \Lambda_1} p_{\theta_1}(y)}{\sup_{\theta_0 \in \Lambda_0} p_{\theta_0}(y)}.$$

It is important to note that the maximization over θ_0
for each realization of the observation y, and so this
more complex that the LRT. Also the result of the m.
a PDF (or PMF) in the numerator and denominate
$T_{\mathrm{GLR}}(y)$ to produce a test, which is called the "gen
(GLRT)":

$$\tilde{\delta}_{\mathrm{GLRT}}(y) = \begin{cases} 1 & \text{if } T_{\mathrm{GLR}} \\ 1 \text{ w.p. } \gamma & \text{if } T_{\mathrm{GLR}} \\ 0 & \text{if } T_{\mathrm{GLR}} \end{cases}$$

The use of the GLRT can be justified via an asymptot
independent and identically distributed (i.i.d.) observ.
where it can be shown to have certain optimality prop
numerator and denominator in $T_{\mathrm{GLR}}(y)$ can also be
of maximum likelihood parameter estimation [2].

Example 11.6.5. *Detection of One-Sided Composite*
tinued). This problem was introduced in Example 11
given by

$$p_\theta(y) = \frac{1}{\pi[1 + (y - \theta)^2]}$$

and we are testing $H_0 : \theta = 0$ against the one-sided co
0. As we saw in Example 11.6.3, there is no UMP so
GLR statistic is given by

$$T_{\mathrm{GLR}}(y) = \frac{\sup_{\theta > 0} p_\theta(y)}{p_0(y)}$$

with

$$\sup_{\theta > 0} p_\theta(y) = \sup_{\theta > 0} \frac{1}{\pi[1 + (y - \theta)^2]} = \begin{cases} \frac{1}{\pi} \\ \frac{1}{\pi} \end{cases}$$

Thus

$$T_{\mathrm{GLR}}(y) = \begin{cases} 1 + y^2 & \text{if } y \geq \\ 1 & \text{if } y < \end{cases}$$

To find an α-level test we need to evaluate $P_0\{T_{\mathrm{GLR}}($

$$P_0\{T_{\mathrm{GLR}}(Y) \geq \eta\} = 1 \text{ for } 0 \leq$$

For $\eta \geq 1$

$$P_0\{T_{\mathrm{GLR}}(Y) \geq \eta\} = \int_{\sqrt{\eta - 1}}^{\infty} \frac{1}{\pi} \frac{1}{1 + y^2} \, dy = 0$$

There is a point of discontinuity in $P_0\{T_{\mathrm{GLR}}(Y) \geq \eta\}$
from 1 to the left to 0.5 to the right. For $\alpha \in (0.5, 1],$

to meet the P_F constraint with equality. For $\alpha \in (0, 0.5]$, which would be more relevant in practice, the GLRT is a deterministic test:

$$\delta_{\mathrm{GLRT}}(y) = \begin{cases} 1 & \text{if } T_{\mathrm{GLR}}(y) \geq \eta_\alpha \\ 0 & \text{if } T_{\mathrm{GLR}}(y) < \eta_\alpha \end{cases}$$

where

$$\eta_\alpha = [\tan(\pi(0.5 - \alpha))]^2 + 1.$$

□

11.6.4 Locally Most Powerful (LMP) Detection

Another approach to finding good detectors in cases where UMP tests do not exist is via a local optimization approach, which works when only one of the hypotheses is composite. Consider the scenario where $Y \sim P_\theta$, we are interested in testing $H_0 : \theta = \theta_0$ versus $H_1 : \theta > \theta_0$, and there is no UMP solution. Also, suppose that θ takes values close to θ_0 under H_1; this might occur in practice in the detection of weak signals with unknown amplitude in noise.

Fix $\theta > \theta_0$ and let $\tilde{\delta}_\theta$ be an α-level N-P test between θ and θ_0. Then assuming that $P_D(\tilde{\delta}_\theta; \theta)$ is differentiable with respect to θ, we can write the Taylor series approximation:

$$P_D(\tilde{\delta}_\theta; \theta) = P_D(\tilde{\delta}_{\theta_0}; \theta_0) + (\theta - \theta_0) \left.\frac{\partial}{\partial \theta} P_D(\tilde{\delta}_\theta; \theta)\right|_{\theta=\theta_0} + o(\theta - \theta_0)$$

$$\approx \alpha + (\theta - \theta_0) \left.\frac{\partial}{\partial \theta} P_D(\tilde{\delta}_\theta; \theta)\right|_{\theta=\theta_0}.$$

The locally optimal criterion can described as:

$$\text{Maximize } \left.\frac{\partial}{\partial \theta} P_D(\tilde{\delta}_\theta; \theta)\right|_{\theta=\theta_0} \text{ subject to } P_F(\tilde{\delta}; \theta_0) \leq \alpha \qquad (11.24)$$

the idea being that maximizing P_D should be approximately the same as maximizing the slope of P_D at $\theta = \theta_0$ for values of θ close to θ_0. Now

$$P_D(\tilde{\delta}_\theta; \theta) = \int_{\mathcal{Y}} \mathbb{I}_{\{\tilde{\delta}(y)=1\}} \, p_\theta(y) \, \mu(dy).$$

Assuming that $p_\theta(y)$ is differentiable in θ

$$\left.\frac{\partial}{\partial \theta} P_D(\tilde{\delta}_\theta; \theta)\right|_{\theta=\theta_0} = \int_{\mathcal{Y}} \mathbb{I}_{\{\tilde{\delta}(y)=1\}} \left.\frac{\partial}{\partial \theta} p_\theta(y)\right|_{\theta=\theta_0} \mu(dy).$$

Therefore, the solution to the locally optimal detecti●
seen as being equivalent to N-P testing between $p_{\theta_0}($

$$\frac{\partial}{\partial \theta} p_\theta(y)\bigg|_{\theta=\theta_0}.$$

Even though the latter quantity is not necessarily a P●
we followed in deriving the N-P solution in Section ●
that the solution to (11.24) has the form:

$$\tilde{\delta}_{\mathrm{LMP}}(y) = \begin{cases} 1 & \text{if } T_{\mathrm{lo}}(● \\ 1 \text{ w.p. } \gamma & \text{if } T_{\mathrm{lo}}(● \\ 0 & \text{if } T_{\mathrm{lo}}(● \end{cases}$$

where

$$T_{\mathrm{lo}}(y) = \frac{\frac{\partial}{\partial \theta} p_\theta(y)\big|_{\theta=\theta_0}}{p_{\theta_0}(y)}.$$

Example 11.6.6. *Detection of One-Sided Composite*
tinued). This problem was introduced in Example 1●
was no UMP solution. We studied the GLRT in ●
examine the LMP solution.

$$p_\theta(y) = \frac{1}{\pi[1 + (y - \theta)^2]} \quad \Longrightarrow \quad \frac{\partial}{\partial \theta} p_\theta(y)\bigg|$$

Thus

$$T_{\mathrm{lo}}(y) = \frac{2y}{1 + y^2}$$

and

$$\tilde{\delta}_{\mathrm{LMP}}(y) = \begin{cases} 1 & \text{if } T_{\mathrm{lo}}(y) \geq \\ 0 & \text{if } T_{\mathrm{lo}}(y) < \end{cases}$$

Randomization is not needed since $T_{\mathrm{lo}}(y)$ does not ha●

11.7   Binary Detection with Vec●

In the detection problems we have studied so far, w●
assumptions about the observation space, although th●
scalar observations. The theory that we have develo●
and vector observations. Nevertheless, it is useful ●
observations in more detail as such a study reveals a●
that are useful in applications.

Consider the detection problem:

$$H_0 : \boldsymbol{Y} \sim p_0(\boldsymbol{y}) \quad \text{versus} \quad H_1 : \boldsymbol{Y} \sim p_1(\boldsymbol{y})$$

where $\boldsymbol{Y} = [Y_1 \ Y_2 \cdots Y_n]^\top$ and $\boldsymbol{y} = [y_1 \ y_2 \cdots y_n]^\top$. The optimum detector for this problem, no matter which criterion (Bayes, Neyman-Pearson, minimax) we choose, is of the form

$$\tilde{\delta}_{\mathrm{OPT}}(y) = \begin{cases} 1 & \text{if } \log L(\boldsymbol{y}) > \eta \\ 1 \text{ w.p. } \gamma & \text{if } \log L(\boldsymbol{y}) = \eta \\ 0 & \text{if } \log L(\boldsymbol{y}) < \eta \end{cases} \qquad (11.25)$$

where $L(\boldsymbol{y}) = p_1(\boldsymbol{y})/p_0(\boldsymbol{y})$ is the likelihood ratio, and taking the log of $L(\boldsymbol{y})$ does not affect the structure of the test since log is a monotonic function. The threshold η and randomization parameter γ are chosen based on the criterion used for detection. Of course, in the Bayesian setting, $\eta = \log \tau$, with τ given in (11.11), and $\gamma = 0$.

11.7.1 Conditionally Independent Observations

Consider the special case where the observations are (conditionally) independent under each hypothesis. In this case

$$p_j(\boldsymbol{y}) = \prod_{k=1}^{n} p_{j,k}(y_k)$$

and the log likelihood ratio in (11.25) can be written as

$$\log L(\boldsymbol{y}) = \sum_{k=1}^{n} \log L_k(y_k)$$

where $L_k(y_k) = p_{1,k}(y_k)/p_{0,k}(y_k)$.

Example 11.7.1. *Deterministic signals in i.i.d. noise.* Here, the hypotheses are given by:

$$H_0 : \boldsymbol{Y} = \boldsymbol{s}_0 + \boldsymbol{Z} \quad \text{versus} \quad H_1 : \boldsymbol{Y} = \boldsymbol{s}_1 + \boldsymbol{Z}$$

where $\boldsymbol{s}_0$ and $\boldsymbol{s}_1$ are deterministic vectors (signals) and $Z_1, Z_2, \ldots, Z_n$ are i.i.d. random variables with zero mean and density given by p_Z. Hence, the log likelihood ratio in (11.25) can be written as:

$$\log L(\boldsymbol{y}) = \sum_{k=1}^{n} \log \frac{p_Z(y_k - s_{1,k})}{p_Z(y_k - s_{0,k})}.$$

A special case of this example is one where $\boldsymbol{Z}$ is a vector of i.i.d. $\mathcal{N}(0, \sigma^2)$ random variables, in which case (based on the more general result derived in the following section), we can show that the optimum detector str▪

$$\delta_{\text{OPT}}(y) = \begin{cases} 1 & \text{if } (s_1 - s_0)^\top \\ 0 & \text{if } (s_1 - s_0)^\top \end{cases}$$

11.7.2 Deterministic Signals in Correl

In general, the detection problem with vector observ▪
dependent, given the hypothesis, does not admit any s▪
is described in (11.25). However, in some special ca▪
pression for the log likelihood ratio to obtain some n▪
structure. In this section, we consider the example o▪
nals in correlated Gaussian noise, for which the hypo▪

$$H_0 : \boldsymbol{Y} = \boldsymbol{s}_0 + \boldsymbol{Z} \quad \text{versus} \quad H_1 : \boldsymbol{Y}$$

with $\boldsymbol{s}_0$ and $\boldsymbol{s}_1$ being deterministic signals as in Exam▪
sian vector with zero mean and covariance matrix Σ,
this case

$$p_j(\boldsymbol{y}) = \frac{1}{\sqrt{(2\pi)^n |\Sigma|}} \exp\left\{ -\frac{1}{2}(\boldsymbol{y} - \boldsymbol{s}_j)^\top \right.$$

where $|\Sigma|$ is the absolute value of the determinant of

$$\log L(\boldsymbol{y}) = \log \frac{p_1(\boldsymbol{y})}{p_0(\boldsymbol{y})} = (\boldsymbol{s}_1 - \boldsymbol{s}_0)^\top \Sigma^{-1} \Big($$

Since $\log L(\boldsymbol{y})$ does not have any point masses under
mum detector is deterministic and has the form:

$$\delta_{\text{OPT}}(\boldsymbol{y}) = \begin{cases} 1 & \text{if } T(\boldsymbol{y}) \geq \\ 0 & \text{if } T(\boldsymbol{y}) < \end{cases}$$

where $T(\boldsymbol{y}) = (\boldsymbol{s}_1 - \boldsymbol{s}_0)^\top \Sigma^{-1} \boldsymbol{y}$ and the η is chosen bas▪
In the special case of Bayesian detection,

$$\eta = \log \tau + \frac{1}{2}(\boldsymbol{s}_1 - \boldsymbol{s}_0)^\top \Sigma^{-1}(\boldsymbol{s}_1$$

with τ given in (11.11).

If we define the *pseudosignal* $\tilde{\boldsymbol{s}}$ by

$$\tilde{\boldsymbol{s}} \overset{\Delta}{=} \Sigma^{-1}(\boldsymbol{s}_1 - \boldsymbol{s}_0)$$

then the test statistic $T(\boldsymbol{y})$ can be written as:

$$T(\boldsymbol{y}) = \tilde{\boldsymbol{s}}^{\top} \boldsymbol{y} = \sum_{k=1}^{n} \tilde{s}_k y_k.$$

We see that the optimum detector is a *correlation* detector or matched filter [2].

Note that $T(\boldsymbol{y})$ is linear in $\boldsymbol{Y}$ and hence has a Gaussian PDF under both H_0 and H_1. In particular,

$$E_j[T(\boldsymbol{Y})] = \tilde{\boldsymbol{s}}^{\top} \tilde{\boldsymbol{s}}_j \stackrel{\Delta}{=} \tilde{\mu}_j$$

and

$$\mathrm{Var}_j[T(\boldsymbol{Y})] = \mathrm{Var}(\tilde{\boldsymbol{s}}^{\top} \boldsymbol{Z}) = \tilde{\boldsymbol{s}}^{\top} \Sigma \tilde{\boldsymbol{s}} = \tilde{\mu}_1 - \tilde{\mu}_0 \stackrel{\Delta}{=} d^2$$

where d^2 is called the *Mahalanobis distance* between the signals $\boldsymbol{s}_1$ and $\boldsymbol{s}_0$.

Based on the above characterization of $T(\boldsymbol{y})$, we can conclude that the problem of deterministic signal detection in correlated Gaussian noise is equivalent to the following detection problem involving the scalar observation $T(\boldsymbol{y})$:

$$H_0 : T(\boldsymbol{y}) \sim \mathcal{N}(\tilde{\mu}_0, d^2) \quad \text{versus} \quad H_1 : T(\boldsymbol{y}) \sim \mathcal{N}(\tilde{\mu}_1, d^2).$$

11.7.3 Gaussian Signals in Gaussian Noise

In this section we consider another important example involving dependent observations, that of detecting Gaussian signals in Gaussian noise. The hypotheses are described by:

$$H_0 : \boldsymbol{Y} = \boldsymbol{S}_0 + \boldsymbol{Z} \quad \text{versus} \quad H_1 : \boldsymbol{Y} = \boldsymbol{S}_1 + \boldsymbol{Z}$$

where $\boldsymbol{S}_0$, $\boldsymbol{S}_1$, and $\boldsymbol{Z}$ are jointly Gaussian random vectors. It is easy to see that this problem is equivalent to the following detection problem:

$$H_0 : \boldsymbol{Y} \sim \mathcal{N}(\boldsymbol{\mu}_0, \Sigma_0) \quad \text{versus} \quad H_1 : \boldsymbol{Y} \sim \mathcal{N}(\boldsymbol{\mu}_1, \Sigma_1) \tag{11.26}$$

for some vectors $\boldsymbol{\mu}_0$, $\boldsymbol{\mu}_1$, and covariance matrices Σ_0 and Σ_1. Note that

$$p_j(\boldsymbol{y}) = \frac{1}{\sqrt{(2\pi)^n |\Sigma_j|}} \exp\left\{ -\frac{1}{2}(\boldsymbol{y} - \boldsymbol{\mu}_j)^{\top} \Sigma_j^{-1} (\boldsymbol{y} - \boldsymbol{\mu}_j) \right\}$$

and therefore the log likelihood ratio is given by:

$$\log L(\boldsymbol{y}) = \frac{1}{2} \boldsymbol{y}^{\top} (\Sigma_0^{-1} - \Sigma_1^{-1}) \boldsymbol{y} + \left(\boldsymbol{\mu}_1^{\top} \Sigma_1^{-1} - \boldsymbol{\mu}_0^{\top} \Sigma_0^{-1} \right) \boldsymbol{y}$$

$$+ \frac{1}{2} \left[\log \frac{|\Sigma_0|}{|\Sigma_1|} + \boldsymbol{\mu}_0^{\top} \Sigma_0^{-1} \boldsymbol{\mu}_0 - \boldsymbol{\mu}_1^{\top} \Sigma_1^{-1} \boldsymbol{\mu}_1 \right].$$

Thus, the optimum detector in general involves both a *quadratic* term as well as a *linear* term in $\boldsymbol{y}$. If $\Sigma_0 = \Sigma_1$ and $\boldsymbol{\mu}_0 \neq \boldsymbol{\mu}_1$, then the quadratic term vanishes and we have the detector structure we saw earlier for the detection of deterministic signals in Gaussian noise. If $\boldsymbol{\mu}_0 = \boldsymbol{\mu}_1 = \boldsymbol{0}$ and $\Sigma_1 \neq \Sigma_0$, then the linear term vanishes and we have a purely quadratic detector.

Example 11.7.2. *Signaling over Rayleigh Fading (*
The following detection problem arises in the contex
systems, when the carrier phase is not known at the

$$H_0 : \boldsymbol{Y} = \boldsymbol{Z} \quad \text{versus} \quad H_1 : \boldsymbol{Y} = \begin{bmatrix} A \\ A \end{bmatrix}$$

where $\boldsymbol{Z} \sim \mathcal{N}(\boldsymbol{0}, \sigma^2 I)$, A is the fading amplitude that
ϕ is the random phase that is uniformly distributed
given by:

$$p_A(a) = \frac{a}{\nu^2} \exp\left[-\frac{a^2}{2\nu^2}\right] \mathbb{I}_{\{a}$$

If we define the fading signal vector $\boldsymbol{S}$ to have co
$S_2 = A\sin\phi$, then it is not difficult to show that
$\mathcal{N}(0, \nu^2)$ random variables. Thus the hypothesis test

$$H_0 : \boldsymbol{Y} \sim \mathcal{N}(\boldsymbol{0}, \sigma^2 I) \quad \text{versus} \quad H_1 : \boldsymbol{Y} \sim \mathcal{N}$$

This is a special case of (11.26) with $\boldsymbol{\mu}_0 = \boldsymbol{\mu}_1 = 0$, an
Thus the log likelihood ratio has the form:

$$\log L(\boldsymbol{y}) = (\text{constant}) \, \boldsymbol{y}^\top \boldsymbol{y} + (\text{co}$$

from which we can conclude that the optimum detec

$$\delta_{\mathrm{OPT}}(\boldsymbol{y}) = \begin{cases} 1 & \text{if } \boldsymbol{y}^\top \boldsymbol{y} \geq \\ 0 & \text{if } \boldsymbol{y}^\top \boldsymbol{y} < \end{cases}$$

The test statistic $\boldsymbol{Y}^\top \boldsymbol{Y} = Y_1^2 + Y_2^2$ has an exponentia
under H_0, and an exponential distribution with mean
if we are interested in N-P detection, for an α-level
$\eta_\alpha\} = \alpha$ by setting

$$\exp\left[-\frac{\eta_\alpha}{2\sigma^2}\right] = \alpha \quad \Longrightarrow \quad \eta_\alpha = -2$$

The corresponding power of the test is given by:

$$\mathrm{P_D}(\delta_{\mathrm{OPT}}) = \mathrm{P}_1\{\boldsymbol{Y}^\top \boldsymbol{Y} \geq \eta_\alpha\} = \exp\left[-\frac{}{2(\sigma^2}\right.$$

11.8   Summary and Further Rea

This chapter covered the fundamentals of detection the
nary detection problems. In Section 11.1, we provided

theory framework for detection problems. In Sections 11.2–11.4, we introduced the three basic formulations for the binary detection problem: Bayesian, minimax, and Neyman-Pearson. We saw that in all cases the optimum detection rule is a LRT with possible randomization. In Sections 11.5–11.6, we studied composite detection problems where the distributions of the observations are not completely specified. In particular, we saw that Bayesian composite detection can be reduced to an equivalent simple detection problem. The Neyman-Pearson version of the composite detection problem is more interesting, and we studied various approaches to this problem, including UMP detection, GLR detection, and LMP detection. Finally, we examined the detection problem with vector observations in more detail, and discussed optimum detector structures for both the cases where the observations are conditionally independent and dependent, under each hypothesis.

This chapter was inspired by the textbook on detection and estimation theory by Poor [2]. While we focused almost exclusively on binary detection problems, extension to M-ary detection is straightforward at least in the Bayesian setting (see Exercise 11.9.6). More details on M-ary detection can be found in the books by Van Trees [3], Levy [4] and Kay [5]. An alternative formulation to the detection problem with incompletely specified distributions is the robust formulation of Huber [6]. Other extensions of detection theory include sequential [7] and quickest change detection [8], where observations are taken sequentially in time and decisions about the hypothesis need to be made online. Asymptotic performance analysis and design of detection procedures for large number of observations using tools from large deviations theory has been an active area of research (see, e.g., [9]). Finally, distributed sensor networks have generated interesting new directions for research in detection theory [10].

Acknowledgments

The writing of this chapter was supported in part by the U.S. National Science Foundation, under grant CCF-0830169, through the University of Illinois at Urbana-Champaign. The author would also like to thank Taposh Banerjee for help with the figures.

11.9   Exercises

Exercise 11.9.1. Consider the binary statistical decision theory problem for which $\mathcal{S} = \mathcal{D} = \{0, 1\}$. Suppose the cost function is given by

$$C(i, j) = \begin{cases} 0 & \text{if } i = j \\ 1 & \text{if } j = 0, i = 1 \\ 10 & \text{if } j = 1, i = 0 \end{cases}$$

The observation Y takes values in the set $\Gamma = \{a, b, c$
of Y are:

$$p_0(a) = p_0(b) = 0.5 \qquad p_1(a) = p_1(b) = 0$$

1. Is there a best decision rule based on condition

2. Find Bayes (for equal priors) and minimax rule
   istic decision rules.

3. Now consider the set of randomized decision r
   equal priors). Also construct a randomized r
   smaller than that of the minimax rule of part (

Exercise 11.9.2. For the binary hypothesis testing
and $C_{1,1} < C_{0,1}$, show there is no "best" rule based o
the trivial case case where $p_0(y)$ and $p_1(y)$ have disjo

Exercise 11.9.3. Let $\mathcal{S} = \{0, 1\}$, and $\mathcal{D} = \{0, 1, e$
binary communication with erasures. Now suppose

$$p_j(y) = \frac{1}{\sqrt{2\pi\sigma^2}} \exp\left[-\frac{(y - (-1)^{j+1})^2}{2\sigma^2}\right], \quad j =$$

That is, Y has distribution $\mathcal{N}(-1, \sigma^2)$ when the state
$\mathcal{N}(1, \sigma^2)$ when the state is 1. Assume a cost structur

$$C_{i,j} = \begin{cases} 0 & \text{if } i = 0, j = 0 \text{ or } i = \\ 1 & \text{if } i = 1, j = 0 \text{ or } i = \\ c & \text{if } i = e \end{cases}$$

Furthermore, assume that the two states are equally

1. First assume that $c < 0.5$. Show that the Bayes
   form:

$$\delta_B(y) = \begin{cases} 0 & y \leq -t \\ e & -t < y \\ 1 & y \geq t \end{cases}$$

   Also give an expression for t in terms of the pa

2. Now find $\delta_B(y)$ when $c \geq 0.5$.

Exercise 11.9.4. Consider the binary detection pro

$$p_1(y) = \begin{cases} 1/4 & \text{if } y \in [0, 4 \\ 0 & \text{otherwise} \end{cases}$$

and

$$p_0(y) = \begin{cases} (y+3)/18 & \text{if } y \in [\\ 0 & \text{otherw} \end{cases}$$

1. Find a Bayes rule for uniform costs and equal priors and the corresponding minimum Bayes risk.

2. Find a minimax rule for uniform costs, and the corresponding minimax risk.

Exercise 11.9.5. For Exercise 11.9.2 above, find the minimum Bayes risk function $V(\pi_0)$, and then find a minimax rule in the set of randomized decision rules using $V(\pi_0)$.

Exercise 11.9.6. In this chapter, we formulated and solved the general Bayesian binary detection problem. We may generalize this formulation to M-ary detection ($M > 2$) as follows:

- $\mathcal{S} = \{0, \ldots, M-1\}$, with *a priori* probability of state j being π_j.

- $\mathcal{D} = \{0, \ldots, M-1\}$

- $C(i,j) = C_{ij} \geq 0$, for $i, j = 0, \ldots, M-1$.

- $\mathcal{Y}$, the observation space being continuous/discrete with conditional density (PDF/PMF) $p_j(y)$, $j = 0, \ldots, M-1$.

- $\delta \in \Delta$, δ partitions $\mathcal{Y}$ into M regions $\mathcal{Y}_0, \ldots, \mathcal{Y}_{M-1}$, where $\delta(y) = i$ when $y \in \mathcal{Y}_i$.

Find $\delta_B(y)$ by specifying the Bayes decision regions $\mathcal{Y}_i$, $i = 0, \ldots, M-1$. Simplify as much as possible.

Exercise 11.9.7. Consider the 5-ary detection problem in which the hypotheses are given by
$$H_j \ : \ Y = (j-2) + Z, \quad j = 0, 1, 2, 3, 4,$$
where $Z \sim \mathcal{N}(0,1)$. Assume that the hypotheses are equally likely.

1. Find the decision rule with minimum probability of error (i.e., Bayes rule with uniform costs).

2. Also find the corresponding minimum Bayes risk.

   Hint: Find the probability of correct decision making first.

Exercise 11.9.8. Consider the binary detection problem with

$$p_0(y) = \frac{1}{2}e^{-|y|} \quad \text{and} \quad p_1(y) = e^{-2|y|}, \quad y \in \mathbb{R}$$

1. Find the Bayes rule for equal priors and a cost structure of the form $C_{00} = C_{11} = 0$, $C_{10} = 1$, and $C_{01} = 2$.

2. Find the Bayes risk for the Bayes rule of part (a). (Note that the costs are not uniform.)

3. Find a Neyman-Pearson rule for $\alpha = 1/4$.

4. Find the probability of detection for the rule o█

Exercise 11.9.9. Consider the detection problem fc
the PMF's of the observations under the two hypoth█

$$p_0(y) = (1 - \beta_0)\beta_0^y, \; y = 0, 1,$$

and

$$p_1(y) = (1 - \beta_1)\beta_1^y, \; y = 0, 1, ░$$

Assume that $0 < \beta_0 < \beta_1 < 1$.

1. Find the Bayes rule for uniform costs and equa█

2. Find the Neyman-Pearson rule with false-alarn
   find the corresponding probability of detection

Exercise 11.9.10. Consider a binary detection pr█
minimize the following risk measure

$$\rho(\tilde{\delta}) = [\mathrm{P_F}(\tilde{\delta})]^2 + \mathrm{P_M}(\tilde{\delta})$$

1. Show that the optimal solution is a (possibly
   test.

2. Find the optimal solution for the observation n

$$p_0(y) = \begin{cases} 1 & \text{if } y \in [0█ \\ 0 & \text{otherwi█} \end{cases}$$

and

$$p_1(y) = \begin{cases} 2y & \text{if } y \in [█ \\ 0 & \text{otherwi█} \end{cases}$$

Exercise 11.9.11. Consider the detection probler
masses under either hypothesis. Let δ_η denote the lik█

$$\delta_\eta(y) = \begin{cases} 1 & \text{if } L(y) \geq \eta \\ 0 & \text{if } L(y) < \eta \end{cases}$$

As discussed in Section 11.4.2, a plot of $\mathrm{P_D}(\delta_\eta)$ vers█
of η is called the ROC. This plot is a concave functi█
responding to $\eta = \infty$, and the point $(1, 1)$ correspc
following properties of ROC's:

1. $\mathrm{P_D}(\delta_\eta) \geq \mathrm{P_F}(\delta_\eta)$ for all η. (Hint: consider cases

2. The slope of the ROC at a particular point is equal to the value of the threshold η required to acheive the P_D and P_F at that point, i.e.,

$$\frac{dP_D}{dP_F} = \eta.$$

(Hint: Use the fact that $L(Y)$ has a density under each hypothesis.)

Exercise 11.9.12. Consider the following composite detection problem with $\Lambda = \mathbb{R}$:

$$H_0 : \theta \le \tilde{\theta} \quad \text{versus} \quad H_1 : \theta > \tilde{\theta}$$

where $\tilde{\theta}$ is a fixed real number. Now suppose that for each fixed $\theta_0 \le \tilde{\theta}$ and each fixed $\theta_1 > \tilde{\theta}$, we have

$$\frac{p_{\theta_1}(y)}{p_{\theta_0}(y)} = g_{\theta_0,\theta_1}(T(y))$$

where the function T does not depend on θ_1 or θ_0, and the function g_{θ_0,θ_1} is *strictly increasing* in its argument.

Show that for any level α, a UMP test between H_0 and H_1 exists.

Exercise 11.9.13. Consider the composite binary detection problem in which

$$p_\theta(y) = \begin{cases} \theta e^{-\theta y} & \text{if } y \ge 0 \\ 0 & \text{if } y < 0 \end{cases}$$

1. For $\alpha \in (0,1)$, show that a UMP test of level α exists for testing the hypotheses

$$H_0 : \Lambda_0 = [1,2] \quad \text{versus} \quad H_1 : \Lambda_1 = (2, \infty).$$

Find this UMP test as a function of α.

2. Find the structure of the generalized likelihood ratio test.

Exercise 11.9.14. (UMP testing with Laplacian Observations) Consider the composite binary detection problem in which

$$p_\theta(y) = \frac{1}{2} e^{-|y-\theta|}, \ y \in \mathbb{R}.$$

and we are testing:

$$H_0 : \theta = 0 \quad \text{versus} \quad H_1 : \theta > 0$$

1. Does a UMP test exist? If so, find it for level α and derive its power P_D. If not, find the generalized likelihood ratio test for level α.

2. Find a locally most powerful α-level test and derive its power P_D.

Exercise 11.9.15. Consider the detection problem:

$$H_0: \quad \boldsymbol{Y} = \begin{bmatrix} -a \\ 0 \end{bmatrix} + \boldsymbol{Z} \quad \text{versus} \quad H_1:$$

where $\boldsymbol{Z} \sim \mathcal{N}(0, \Sigma)$ with

$$\Sigma = \begin{bmatrix} 1 & \rho \\ \rho & 1 + \rho^2 \end{bmatrix}.$$

Assume that $a > 0$ and $\rho \in (0, 1)$.

1. For equal priors show that the minimum-probab
   by

$$\delta_B(\boldsymbol{y}) = \begin{cases} 1 & \text{if } y_1 - by \\ 0 & \text{if } y_1 - by \end{cases}$$

   where $b = \rho/(1 + \rho^2)$ and $\tau = 0$.

2. Determine the minimum probability of error.

3. Consider the test of part (a) in the limit as $\rho -$
   dence on y_2 goes away in this limit.

4. Now suppose the observations $\boldsymbol{Y} \sim \mathcal{N}([a\ 0]^T, \Sigma$
   unknown parameter, and we wish to test betwe

$$H_0: 0 < a < 1 \quad \text{versus} \quad H_1$$

   Show that a UMP test exists for this problem, a
   $\alpha \in (0, 1)$.

Exercise 11.9.16. Consider the detection problem
tions:

$$H_0: \boldsymbol{Y} = \boldsymbol{Z} \quad \text{versus} \quad H_1: \boldsymbol{Y} =$$

where the components of $\boldsymbol{Z}$ are zero mean correlated

$$E[Z_k Z_\ell] = \sigma^2 \rho^{|k-\ell|}, \quad \text{for all } 1 \leq$$

where $|\rho| < 1$.

1. Show that the N-P test for this problem has th

$$\delta_\eta(y) = \begin{cases} 1 & \text{if } \sum_{k=1}^n b_k \\ 0 & \text{if } \sum_{k=1}^n b_k \end{cases}$$

   where $b_1 = s_1/\sigma$, $x_1 = y_1/\sigma$, and

$$b_k = \frac{s_k - \rho s_{k-1}}{\sigma\sqrt{1 - \rho^2}}, \quad x_k = \frac{y_k - \rho y_{k-1}}{\sigma\sqrt{1 - \rho^2}}$$

Hint: Note that $\Sigma_Z^{-1} = A/(\sigma^2(1-\rho^2))$, where A is a tridiagonal matrix with main diagonal $(1 \quad 1+\rho^2 \quad 1+\rho^2 \quad \ldots \quad 1+\rho^2 \quad 1)$ and superdiagonal and subdiagonal entries all being $-\rho$.

2. Find the α-level N-P test, δ_{η_α}.

3. Find the ROC for the above detector, i.e., find $P_D(\delta_{\eta_\alpha})$ as a function of α.

Exercise 11.9.17. Consider the composite detection problem with two-dimensional observations:

$$H_0 : \boldsymbol{Y} = \boldsymbol{Z} \quad \text{versus} \quad H_1 : \boldsymbol{Y} = \theta \boldsymbol{s} + \boldsymbol{Z}$$

where Z_1 and Z_2 are independent $\mathcal{N}(0,1)$ random variables, and $s_1 = 1$ and $s_2 = -1$.

The parameter θ is a deterministic but unknown parameter that takes one of *two* possible values $+1$ or -1.

1. Is there a UMP test for this problem? If so, find it for level α. If not, explain why not.

2. Show that an α-level GLRT for this problem is given by:

$$\delta_{\mathrm{GLRT}}(y) = \begin{cases} 1 & \text{if } |y_1 - y_2| \geq \eta_\alpha \\ 0 & \text{otherwise} \end{cases}$$

with $\eta_\alpha = \sqrt{2}\, Q^{-1}(\frac{\alpha}{2})$.

3. Give a clear argument to establish that the probability of detection for the GLRT of part (b) is independent of θ.

4. Now find the probability of detection for the GLRT as a function of η_α.

References

[1] Ferguson, T.S., *Mathematical Statistics: A Decision Theoretic Approach.* Academic Press, 1967.

[2] Poor, H.V., *An Introduction to Signal Detection and Estimation*, second edition. Springer-Verlag, 1994.

[3] Van Trees, H.L., *Detection, Estimation and Modulation Theory, Part 1.* Wiley, 1968.

[4] Levy, B.C., *Principles of Signal Detection and Parameter Estimation.* Springer-Verlag, 2008.

410

[5] Kay, S.M., *Fundamentals of Statistical Signal P*
Prentice Hall, 1998.

[6] Huber, P.J., *Robust Statistics*. Wiley, 1981.

[7] Wald, A., *Sequential Analysis*. Wiley, 1947.

[8] Poor, H.V. and Hadjiliadis, O., *Quickest Dete*
Press, 2009.

[9] Dembo, A. and Zeitouni, O., *Large Deviations*
Second Edition. Springer-Verlag, 1998.

[10] Varshney, P.K., *Distributed Detection and Data F*

Chapter 12 Monte Carlo Methods for Statistical Signal Processing

Xiaodong Wang[‡]
[‡]Columbia University, New York, USA

12.1 Introduction

In many problems encountered in signal processing, it is possible to describe accurately the underlying statistical model using probability distributions. Statistical inference can then theoretically be performed based on the relevant likelihood function or posterior distribution in a Bayesian framework. However, most problems encountered in applied research require non-Gaussian and/or nonlinear models in order to correctly account for the observed data. In these cases, it is typically impossible to obtain the required statistical estimates of interest, e.g., maximum likelihood, conditional expectation, in closed form as it requires integration and/or maximization of complex multidimensional functions. A standard approach consists of making model simplifications or crude analytic approximations in order to obtain algorithms that can be easily implemented. With the recent availability of high-powered computers, numerical simulation based approaches can now be considered and the full complexity of real problems can be addressed.

These integration and/or optimization problems could be tackled using analytic approximation techniques or deterministic numerical integration/optimization methods. These classical methods are often either not precise and robust enough or are too complex to implement. An attractive alternative consists of Monte Carlo algorithms. These algorithms are remarkably flexible and extremely powerful. The basic idea is to draw a large number of samples distributed according to some probability distribution(s) of interest so as to obtain simulation-based consistent estimates. These methods first became popular in physics [1] before literally revolutionizing applied statistics and related fields such as bioinformatics and econometrics in the 1990s [2–5].

Despite their ability to allow statistical inference
complex models, these flexible and powerful method
signal processing. This chapter provides a simple
methods in a signal processing context. We describe
which can be used to perform statistical inference i
contexts. We illustrate their applications in solving
digital communications and bioinformatics.

12.1.1   Model-Based Signal Processing

In statistical signal processing, many problems can b
is interested in obtaining an estimate of an unobserve
values in $\mathcal{X}$ given the realization of some statistically
In a model-based context, one has access to the li
probability or PDF $p(y|x)$ of $Y = y$ given $X = x$.
estimate of X is given by the maximum likelihood es

$$x_{ML} = \arg\max_{x \in \mathcal{X}} p(y|x)$$

For simple models, it is possible to compute $p(y|x)$
imization of the probability distribution/PDF can be
when the model includes latent variables, some non-C
ements, it is often impossible to compute in closed-f
is difficult to maximize it as it is a multimodal and
function. This severely limits the applications of ma
for complex models.

The problem appears even more clearly when on
Bayesian inference [6, 7]. In this context, one sets a
$p(x)$, and all (Bayesian) inference relies on the po
Bayes' theorem

$$p(x|y) = \frac{p(y|x)\,p(x)}{p(y)}$$

where

$$\int p(y|x)\,p(x)\,dx = p(y)$$

For example the MMSE estimate of X given $Y = y$ i

$$x_{MMSE} = \int x p(x|y)\,dx$$

To be able to compute this estimate, it is necessar
It is only feasible to perform these calculations anal
models.

12.1.2 Examples

To illustrate these problems, we discuss a few standard signal processing applications here. For the sake of simplicity, we do not distinguish random variables and their realizations from now on. We will use the notation $z_{i:j} = (z_i, z_{i+1}, \ldots, z_j)^{\mathrm{T}}$ for any sequence $\{z_n\}$.

Spectral Analysis

Consider the problem of estimating some sinusoids in noise. Let $y_{1:T}$ be an observed vector of T real data samples. The elements of $y_{1:T}$ may be represented by different models $\mathcal{M}_k$ corresponding either to samples of noise only ($k = 0$) or to the superposition of k ($k \geq 1$) sinusoids corrupted by noise, more precisely

$$
\begin{aligned}
\mathcal{M}_0 &: \ y_n = v_{n,k} & k = 0 \\
\mathcal{M}_k &: \ y_n = \sum_{j=1}^{k} \left(a_{c_{j,k}} \cos [\omega_{j,k} n] + a_{s_{j,k}} \sin [\omega_{j,k} n] \right) + v_{n,k} & k \geq 1
\end{aligned}
$$

where $\omega_{j_1,k} \neq \omega_{j_2,k}$ for $j_1 \neq j_2$ and $a_{c_{j,k}}, a_{s_{j,k}}, \omega_{j,k}$ are respectively the amplitudes and the radial frequency of the jth sinusoid for the model with k sinusoids. The noise sequence $v_{1:T,k}$ is assumed zero-mean white Gaussian of variance σ_k^2. In vector-matrix form, we have

$$
y_{1:T} = D\left(\omega_k\right) a_k + v_{k,1:T}
$$

where $a_k = \left(a_{c_{1,k}}, a_{s_{1,k}}, \ldots, a_{c_{k,k}}, a_{s_{k,k}}\right)^{\mathrm{T}}$ and $\omega_k = \left(\omega_{1,k}, \ldots, \omega_{k,k}\right)^{\mathrm{T}}$. The $T \times 2k$ matrix $D\left(\omega_k\right)$ is defined as

$$
\begin{aligned}
\left[D\left(\omega_k\right)\right]_{i,2j-1} &= \cos\left[\omega_{j,k} i\right], \ (i = 1, \ldots, T, \ j = 1, \ldots, k) \\
\left[D\left(\omega_k\right)\right]_{i,2j} &= \sin\left[\omega_{j,k} i\right], \ (i = 1, \ldots, T, \ j = 1, \ldots, k)
\end{aligned}
$$

We assume here that the number k of sinusoids and their parameters $\left(a_k, \omega_k, \sigma_k^2\right)$ are unknown. Given $y_{1:T}$, our objective is to estimate $\left(k, a_k, \omega_k, \sigma_k^2\right)$. It is standard in signal processing to perform parameter estimation and model selection using a (penalized) ML approach. First, an approximate ML estimate of the parameters is found; we emphasize that unfortunately the likelihood is highly nonlinear in its parameters ω_k and admits typically severe local maxima. Model selection is then performed by maximizing an information criterion (IC) such as AIC (Akaike), BIC (Bayes) or MDL (Minimum Description Length). Note that when the number of observations is small, these criteria can perform poorly. In this problem, a Bayesian approach is considered; see [8] for a motivation of this model. One has

$$
a_k | \sigma_k^2 \sim \mathcal{N}\left(0, \sigma_k^2 \delta^2 \left(D^{\mathrm{T}}\left(\omega_k\right) D\left(\omega_k\right)\right)^{-1}\right), \quad \sigma_k^2 \sim \mathcal{IG}\left(\frac{v_0}{2}, \frac{\gamma_0}{2}\right) \tag{12.1}
$$

and the frequencies ω_k are independent and uniformly distributed over $(0, \pi)$. Finally, we assume that the prior distribution $p(k)$ is a truncated Poisson distribution

of intensity Λ where $k_{\max} \triangleq \lfloor (N-1)/2 \rfloor$ (this const
wise the columns of $D(\omega_k)$ would be linearly depend
be respectively interpreted as an expected signal-to-
number of sinusoids.

In this case, it can easily be established that the m
of the frequencies ω_k is proportional on $\Omega = \{0, 1, \ldots$

$$p(\omega_k, k | y_{1:T}) \propto \left(\gamma_0 + y_{1:T}^{\mathrm{T}} P_k y_{1:T} \right)^{-\frac{T+v_0}{2}} (\Lambda/$$

where

$$M_k^{-1} = \left(1 + \delta^{-2} \right) D^{\mathrm{T}}(\omega_k) D(\omega_k), \quad m_k$$
$$P_k = I_T - D(\omega_k) M_k D^{\mathrm{T}}(\omega_k).$$

This posterior distribution is highly nonlinear in the p
cannot compute explicitly its normalizing constant
to compute the Bayes factors to perform model se
integration techniques could be used but they are ty
the dimension of the space of interest is high.

Optimal Filtering in State-Space Models

Consider an unobserved Markov process $\{x_n\}_{n \geq 1}$ of
tion density $x_n | x_{n-1} \sim f(\cdot | x_{n-1})$. The observatio
independent given $\{x_n\}_{n \geq 1}$ of marginal density $y_n |$
models is extremely wide. For example, it includes

$$x_n = \varphi(x_{n-1}, v_n), \quad y_n = \Psi(x$$

where φ and Ψ are two nonlinear deterministic mappin
are two independent and mutually independent seque

All inference on $x_{1:n}$ based on $y_{1:n}$ is based on th

$$p(x_{1:n} | y_{1:n}) = \frac{p(y_{1:n} | x_{1:n}) p}{\int p(y_{1:n} | x_{1:n}) p(x}$$

where

$$p(x_{1:n}) = p(x_1) \prod_{k=2}^{n} f(x_k | x_{k-1}), \quad p(y_{1:n} | x$$

This posterior distribution satisfies the following recu

$$p(x_{1:n} | y_{1:n}) = \frac{f(x_n | x_{n-1}) g(y_n | x_n)}{p(y_n | y_{1:n-1})} p($$

Unfortunately, except in the case where $\{x_n\}_{n\geq 1}$ takes values in a finite state-space (Hidden Markov model techniques) or the model is linear Gaussian (Kalman filtering techniques) then it is impossible to come up with a closed-form expression for this sequence of posterior distributions. Many suboptimal methods have been proposed to approximate this sequence; e.g., Extended Kalman filter, Gaussian sum approximations. However these methods tend to be unreliable as soon as the model includes strong nonlinear and/or non-Gaussian models. Deterministic numerical integration methods have been proposed but they are complex to implement, not flexible, and realistically can only be applied to models where $\{x_n\}_{n\geq 1}$ takes values in $\mathbb{R}$ or $\mathbb{R}^2$.

DNA Sequence Motif Discovery

Efforts by various genomic projects have steadily expanded the pool of sequenced deoxyribonucleic acid (DNA) data. Motifs, or DNA patterns found in different locations within the genome, are often of interest to biologists. By seeking out these similarities exhibited in sequences, we can further our knowledge on the functions and evolutions of these sequences [9]. Let $S = \{s_1, s_2, \cdots, s_T\}$, with $s_t = [s_{t1}, \cdots, s_{tL}]$, be the set of DNA sequences of length L where we wish to find a common motif. Let us assume that a motif of length w is present in each one of the sequences. The distribution of the motif is described by the $4 \times w$ position weight matrix (PWM) $\boldsymbol{\Theta} = [\boldsymbol{\theta}_1, \boldsymbol{\theta}_2, \cdots, \boldsymbol{\theta}_w]$, where the column vector $\boldsymbol{\theta}_j = [\theta_{j1}, \cdots, \theta_{j4}]^T, j = 1, \cdots, w$, is the probability distribution of the nucleotides $\{A, C, G, T\}$ at the j-th position of the PWM. The remaining non-motif nucleotides are assumed to follow a Markovian distribution with probabilities given by $\boldsymbol{\Theta}_0$.

To formulate the motif-finding problem we use the state space model, where the states represent the locations of the first nucleotides of the different occurrences of the motif in the sequence, whereas the observation for the state at step t is the entire nucleotide sequence, s_t. Since the ending $w - 1$ nucleotides in a sequence are not valid locations for the beginning of a motif with length w, at step $t, t = 1, \cdots, T$, the state, denoted as x_t, takes value from the set $\mathcal{X} = \{1, 2, \cdots, L_m\}$, where $L_m = L - w + 1$.

Let $\boldsymbol{a}_{t,x_t}$ be a sequence fragment of length w from s_t starting from position x_t in s_t, and denote $\boldsymbol{a}_{t,x_t}^c$ as the remaining fragment from s_t with $\boldsymbol{a}_{t,x_t}$ removed. For example, for $s_t = [AAAAGGGGAAAA]$ and $x_t = 5$ with $w = 4$, $\boldsymbol{a}_{t,x_t} = [GGGG]$ and $\boldsymbol{a}_{t,x_t}^c = [AAAAAAAA]$. Let us further define a vector $\boldsymbol{n}(\boldsymbol{a}) = [n_1, n_2, n_3, n_4]$ where $n_i, i = 1, \cdots, 4$, denotes the number of different nucleotides in the sequence fragment $\boldsymbol{a}$. Given the vectors $\boldsymbol{\theta} = [\theta_1, \cdots, \theta_4]$ and $\boldsymbol{n} = [n_1, \cdots, n_4]$, we define

$$\boldsymbol{\theta}^{\boldsymbol{n}} \triangleq \prod_{j=1}^{4} \theta_j^{n_j}. \tag{12.3}$$

In DNA sequences, a nucleotide is often influenced by the surrounding nucleotides. We assume for our system model a 3rd order Markov model for the

non-motif nucleotides in the sequence. Let us deno
a^c_{t,x_t}. For example, if $a^c_{t,x_t} = [ATAAG]$, the probabi

$$P^3_{t,x_t} = p(A)p(T \mid A)p(A \mid A,T)p(A \mid A,T,A$$

In general, the 0-th to 3-rd order Markov chain pro
non-motif nucleotides can be averaged over a large
sumed to be known, which we denote as $\boldsymbol{\Theta}_0$.   To
can be given as a known parameter by the user or
Since the nucleotides being located in the motif are in
nucleotides and non-motif nucleotides, given the PW
bution $\boldsymbol{\Theta}_0$, and the state at time t, the distribution c
then given as follows:

$$p(\boldsymbol{s}_t \mid x_t = i, \boldsymbol{\Theta}) = P^3_{t,x_t} \prod_{k=1}^{w} \boldsymbol{\theta}_k^{\boldsymbol{n}(\boldsymbol{a}_{t,i}(\boldsymbol{k}))}$$

where $\boldsymbol{a}_{t,i}(k)$ is the k-th element of the sequence frag
1×4 vector of zeros except at the position correspond
where it is a one.

From the discussion above, we formulate the infer
us denote the state realizations up to time T as $\boldsymbol{x} \triangleq$
the sequences up to time T as $\boldsymbol{S} \triangleq [\boldsymbol{s}_1, \boldsymbol{s}_2, \cdots, \boldsymbol{s}_T]$, v
$\boldsymbol{\Theta}$, the position weight matrix. Given the sequences
motif nucleotide distribution $\boldsymbol{\Theta}_0$, we wish to estim
which are the starting locations of the motif in eacl
weight matrix $\boldsymbol{\Theta}$, which describes the statistics of the

Remark: All problems described above require co
high-dimensional probability distributions. It is poss
ministic techniques to approximate these distributio
problems get very complex, the performance of these i
quickly. In this chapter, we advocate that Monte C
set of techniques which can provide satisfactory answ

12.2   Monte Carlo Methods

Let us consider the probability distribution or PDF
assume from now on that $\pi(x)$ is known pointwise u
i.e.,

$$\pi(x) = Z^{-1}\widetilde{\pi}(x)$$

where $\widetilde{\pi}(x)$ is known pointwise but the normalizing c

$$Z = \int_{\mathcal{X}} \widetilde{\pi}(x)\, dx$$

is unknown. Note this assumption is satisfied in all the examples discussed in the previous section if x corresponds to all the unknown variables/parameters.

In most applications of interest, the space $\mathcal{X}$ is typically high-dimensional; say $\mathcal{X} = \mathbb{R}^{1000}$ or $\mathcal{X} = \{0, 1\}^{1000}$. We are interested in the following generic problems.

- *Computing integrals.* For any test function $\varphi : \mathcal{X} \to \mathbb{R}$, we want to compute

$$E_\pi(\varphi) = \int_{\mathcal{X}} \varphi(x) \pi(x) \, dx. \tag{12.6}$$

- *Marginal distributions.* Assume $x = (x_1, x_2) \in \mathcal{X}_1 \times \mathcal{X}_2$, then we want to compute the marginal distribution

$$\pi(x_1) = \int_{\mathcal{X}_2} \pi(x_1, x_2) \, dx_2. \tag{12.7}$$

- *Optimization.* Given $\pi(x)$, we are interested in finding

$$\arg\max_{x \in \mathcal{X}} \pi(x) = \arg\max_{x \in \mathcal{X}} \tilde{\pi}(x). \tag{12.8}$$

- *Integration/Optimization.* Given the marginal distribution (12.7), we want to compute

$$\arg\max_{x_1 \in \mathcal{X}_1} \pi(x_1) = \arg\max_{x_1 \in \mathcal{X}_1} \tilde{\pi}(x_1) \tag{12.9}$$

Assume it is possible to obtain a large number of N independent random samples $\{x^{(i)}\}$ $(i = 1, \ldots, N)$ distributed according to π. The Monte Carlo method approximates π by the following point-mass measure

$$\hat{\pi}(x) = \frac{1}{N} \sum_{i=1}^{N} \delta\left(x - x^{(i)}\right). \tag{12.10}$$

It follows that an estimate of (12.6) is given by

$$\hat{E}_\pi(\varphi) = \int_{\mathcal{X}} \varphi(x) \hat{\pi}(x) \, dx = \frac{1}{N} \sum_{i=1}^{N} \varphi\left(x^{(i)}\right). \tag{12.11}$$

Marginal distributions can also be estimated straightforwardly as

$$\hat{\pi}(x_1) = \int_{\mathcal{X}_2} \hat{\pi}(x_1, x_2) \, dx_2$$

$$= \int_{\mathcal{X}_2} \frac{1}{N} \sum_{i=1}^{N} \delta\left(x_1 - x_1^{(i)}, x_2 - x_2^{(i)}\right) dx_2$$

$$= \frac{1}{N} \sum_{i=1}^{N} \delta\left(x_1 - x_1^{(i)}\right). \tag{12.12}$$

The samples $\{x^{(i)}\}$ being distributed according to π
proportion of them will be in the vicinity of the mod
(12.8) is

$$\arg\max_{\{x^{(i)}\}} \widetilde{\pi}\left(x^{(i)}\right).$$

Optimizing marginal distribution is more difficult, one

as the marginal distribution cannot be computed even
If the scenario where $\pi\left(x_1 | x_2\right)$ is known analytically,
is

$$\widehat{\pi}\left(x_1\right) = \int_{\mathcal{X}_2} \pi\left(x_1 | x_2\right) \widehat{\pi}\left(x_2\right) dx_2$$

$$= \int_{\mathcal{X}_2} \pi\left(x_1 | x_2\right) \left(\frac{1}{N} \sum_{i=1}^{N} \delta\left(x_2\right)\right)$$

$$= \frac{1}{N} \sum_{i=1}^{N} \pi\left(x_1 | x_2^{(i)}\right).$$

It is then possible to estimate (12.9) by $\arg\max_{\{x_1^{(i)}\}} \widehat{\pi}\left(x\right)$
tional complexity of this algorithm is unfortunately
(12.14) pointwise involves $N \gg 1$ terms. Alternative
later.

   A natural question to ask is why the Monte Carl
typical answer is that if one considers (12.11), then
erties; i.e., it is clearly unbiased and one can easily sh

$$\operatorname{var}\left\{\widehat{E}_\pi\left(\varphi\right)\right\} = \frac{\int \varphi^2\left(x\right) \pi\left(x\right) dx -}{N}$$

The truly remarkable property of this estimate is tha
zero of its variance is independent of the space $\mathcal{X}$ (
whereas all deterministic integration methods have
approximation error decreasing severely as the dime
Note however that it does not imply that Monte Ca
perform deterministic methods as the numerator of (1
Monte Carlo tends to be much more flexible and pow

   Nevertheless, they rely on the assumption that we
$\{x^{(i)}\}$ from π. The next question is how we obtain s

12.3 Markov Chain Monte Carlo (MCMC) Methods

12.3.1 General MCMC Algorithms

MCMC is a class of algorithms that allow one to draw (pseudo) random samples from an arbitrary target probability distribution, $p(\boldsymbol{x})$, known up to a normalizing constant. The basic idea behind these algorithms is that one can achieve the sampling from p by running a Markov chain whose equilibrium distribution is exactly p. Two basic types of MCMC algorithms, the Metropolis algorithm and the Gibbs sampler, have been widely used in diverse fields. The validity of the both algorithms can be proved by the basic Markov chain theory.

Metropolis-Hastings Algorithm

Let $p(\boldsymbol{x}) = c \exp\{-f(\boldsymbol{x})\}$ be the target probability distribution from which we want to simulate random draws. The normalizing constant c may be unknown to us. Metropolis et al. [1] first introduced the fundamental idea of evolving a Markov process in Monte Carlo sampling, which was later generalized by Hastings [10]. Starting with any configuration $\boldsymbol{x}^{(0)}$, the algorithm evolves from the current state $\boldsymbol{x}^{(t)} = \boldsymbol{x}$ to the next state $\boldsymbol{x}^{(t+1)}$ as follows:

Algorithm 12.3.1. [Metropolis-Hastings algorithm]

- *Propose a random "perturbation" of the current state, i.e., $\boldsymbol{x} \to \boldsymbol{x}'$, where $\boldsymbol{x}'$ is generated from a transition function $T(\boldsymbol{x}^{(t)} \to \boldsymbol{x}')$, which is nearly arbitrary (of course, some are better than others in terms of efficiency) and is completely specified by the user.*

- *Compute the Metropolis ratio*

$$r(\boldsymbol{x}, \boldsymbol{x}') = \frac{p(\boldsymbol{x}')T(\boldsymbol{x}' \to \boldsymbol{x})}{p(\boldsymbol{x})T(\boldsymbol{x} \to \boldsymbol{x}')}. \qquad (12.16)$$

- *Generate a random number $u \sim uniform(0,1)$. Let $\boldsymbol{x}^{(t+1)} = \boldsymbol{x}'$ if $u \leq r(\boldsymbol{x}, \boldsymbol{x}')$, and let $\boldsymbol{x}^{(t+1)} = \boldsymbol{x}^{(t)}$ otherwise.*

It is easy to prove that the M-H transition rule results in an "actual" transition function $A(\boldsymbol{x}, \boldsymbol{y})$ (it is different from T because a acceptance/rejection step is involved) that satisfies the detailed balance condition

$$p(\boldsymbol{x})A(\boldsymbol{x}, \boldsymbol{y}) = p(\boldsymbol{y})A(\boldsymbol{y}, \boldsymbol{x}), \qquad (12.17)$$

which necessarily leads to a *reversible* Markov chain with $p(\boldsymbol{x})$ as its invariant distribution.

The Metropolis algorithm has been extensively uς
the past forty years and is the cornerstone of all
adopted and generalized in the statistics communit
algorithms, the Gibbs sampler [11], differs from the
it uses conditional distributions based on $p(\boldsymbol{x})$ to con

Gibbs Sampler

Suppose $\boldsymbol{x} = (x_1, \cdots, x_d)$, where x_i is either a sca
sampler, one systematically or randomly chooses a
updates its value with a new sample x_i' drawn from
$p(\cdot \mid \boldsymbol{x}_{[-i]})$. Algorithmically, the Gibbs sampler can b

Algorithm 12.3.2. [Gibbs sampler]
Let the current state be $\boldsymbol{x}^{(t)} = \left(x_1^{(t)}, \cdots, x_d^{(t)} \right)$.

For $i = 1, \cdots, d$, *we draw* $x_i^{(t+1)}$ *from the cond*

$$p\left(x_i \mid x_1^{(t+1)}, \cdots, x_{i-1}^{(t+1)}, x_{i+1}^{(t)} \right.$$

Alternatively, one can randomly scan the coordin
currently $\boldsymbol{x}^{(t)} = (x_1^{(t)}, \cdots x_d^{(t)})$. Then one can rand
set $\{1, \cdots, d\}$ according to a given probability vector
$x_i^{(t+1)}$ from the conditional distribution $p\left(\cdot \mid \boldsymbol{x}_{[-i]}^{(t)} \right)$,
It is easy to check that *every* individual condition
Suppose currently $\boldsymbol{x}^{(t)} \sim p$. Then $\boldsymbol{x}_{[-i]}^{(t)}$ follows its m
Thus,

$$p\left(x_i^{(t+1)} \mid \boldsymbol{x}_{[-i]}^{(t)} \right) \cdot p\left(\boldsymbol{x}_{[-i]}^{(t)} \right) = p\left(x_i^{(t+1)} \right.$$

which implies that the joint distribution of $\left(\boldsymbol{x}_{[-i]}^{(t)}, x_i^{(*} \right.$
one update.

The Gibbs sampler's popularity in statistics comn
sive use of *conditional distributions* in each iteratic
method [12] first linked the Gibbs sampling structure
and the EM-type algorithms. The Gibbs sampler wa
where it was pointed out that the conditionals needed
monly available in many Bayesian and likelihood con
conditions, one can show that the Gibbs sampler chain
its convergence rate is related to how the variables cc
Therefore, grouping highly correlated variables toget.
greatly speed up the sampler.

Other techniques - A main problem with all the MCMC algorithms is that they may, for some problems, move very slowly in the configuration space or may be trapped in a local mode. This phenomenon is generally called *slow-mixing* of the chain. When the chain is slow-mixing, estimation based on the resulting Monte Carlo samples becomes very inaccurate. Some recent techniques suitable for designing more efficient MCMC samplers include parallel tempering [15], multiple-try method [16], and evolutionary Monte Carlo [17].

12.3.2 Applications of MCMC in Digital Communications

In this section, we discuss MCMC-based receiver signal processing algorithms for several typical communication channels, when the channel conditions are unknown *a priori*.

MCMC Detectors in AWGN Channels

We start with the simplest channel model in digital communications – the additive white Gaussian noise (AWGN) channel. After filtering and sampling of the continuous-time received waveform, the discrete-time received signal in such a channel is given by

$$y_t = \phi x_t + v_t, \quad t = 1, 2, \ldots, n, \tag{12.19}$$

where y_t is the received signal at time t; $x_t \in \{+1, -1\}$ is the transmitted binary symbol at time t; $\phi \in \mathbb{R}$ is the received signal amplitude; and v_t is an independent Gaussian noise sample with zero-mean and variance σ^2, i.e., $v_t \sim \mathcal{N}(0, \sigma^2)$. Denote $\boldsymbol{X} \triangleq [x_1, \ldots, x_n]$ and $\boldsymbol{Y} \triangleq [y_1, \ldots, y_n]$. Our problem is to estimate the *a posteriori* probability distribution of each symbol based on the received signal $\boldsymbol{Y}$, without knowing the channel parameters (ϕ, σ^2). The solution to this problem based on the Gibbs sampler is as follows. Assuming a uniform prior for ϕ, a uniform prior for $\boldsymbol{X}$ (on $\{-1, +1\}^n$) and an inverse χ^2 prior for σ^2, $\sigma^2 \sim \chi^{-2}(\nu, \lambda)$, the complete posterior distribution is given by

$$p\left(\boldsymbol{X}, \phi, \sigma^2 \mid \boldsymbol{Y}\right) \propto p\left(\boldsymbol{Y} \mid \boldsymbol{X}, \phi, \sigma^2\right) p(\phi) p\left(\sigma^2\right) p(\boldsymbol{X}). \tag{12.20}$$

The Gibbs sampler starts with arbitrary initial values of $\boldsymbol{X}^{(0)}$ and for $k = 0, 1, \ldots$, iterates between the following two steps.

Algorithm 12.3.3. [Two-component Gibbs detector in AWGN channel]

- *Draw a sample $\left(\phi^{(k+1)}, \sigma^{2(k+1)}\right)$ from the conditional distribution (given $\boldsymbol{X}^{(k)}$)*

$$p\left(\phi, \sigma^2 \mid \boldsymbol{X}^{(k)}, \boldsymbol{Y}\right) \quad \propto \quad \left(\sigma^2\right)^{-\frac{n}{2}} \exp\left[-\frac{}{2}\right.$$

$$\cdot \left(\sigma^2\right)^{-\frac{\nu+2}{2}} \exp\Big($$

$$\propto \quad \pi^{(k+1)}\left(\phi \mid \sigma^2\right)$$

where

$$\pi^{(k+1)}\left(\sigma^2\right) \sim \chi^{-2}\left(\nu + n - 1, \; \frac{1}{\nu + n - 1}\left[\nu\lambda + \right.\right.$$

and

$$\pi^{(k+1)}\left(\phi \mid \sigma^2\right) \sim \mathcal{N}\left(\frac{1}{n}\sum_{t=1}^{n} y_t\right.$$

- *Draw a sample* $\boldsymbol{X}^{(k+1)}$ *from the following co* $\left(\phi^{(k+1)}, \sigma^{2(k+1)}\right)$,

$$p\left(\boldsymbol{X} \mid \phi^{(k+1)}, \sigma^{2(k+1)}, \boldsymbol{Y}\right) \quad = \quad \prod_{t=1}^{n} p\left(x_t \mid y_t\right.$$

$$\propto \quad \prod_{t=1}^{n} \exp\left[-\frac{}{2\sigma}\right.$$

That is, for $t = 1, \ldots, n$ *and* $b \in \{+1, -1\}$, *draw*

$$P\left(x_t^{(k+1)} = b\right) \quad = \quad \left[1 + \exp\left(-\frac{2}{}\right.\right.$$

It is worthwhile to note that one can integrate out cally to get the marginal target distribution of $\boldsymbol{X}$, wh insight. More precisely, we have

$$\pi(\boldsymbol{X}) \propto \left[\nu\lambda + \sum_{t=1}^{n} y_t^2 - \frac{1}{n}\left(\sum_{t=1}^{n} x_t y_t\right)^2\right]$$

This defines a distribution on the space of a n-dimens distribution is clearly at $\tilde{\boldsymbol{X}}$, and $-\tilde{\boldsymbol{X}}$, where $\tilde{\boldsymbol{X}} = $ sig "obvious solution" in this simple setting but it is not (12.26), we can derive another Gibbs sampling algori

Algorithm 12.3.4. [One-component Gibbs detector in AWGN channel]

- *Choose t from $1, \ldots, n$ by either the random scan (i.e., the t is chosen at random) or the deterministic scan (i.e., one cycles t from 1 to n systematically). Update $\boldsymbol{X}^{(k)}$ to $\boldsymbol{X}^{(k+1)}$, where $x_s^{(k)} = x_s^{(k+1)}$ for $s \neq t$ and $x_t^{(k+1)}$ is drawn from the conditional distribution*

$$\pi\left(x_t = b \mid \boldsymbol{X}_{[-t]}^{(k)}\right) = \frac{\pi\left(x_t = b, \boldsymbol{X}_{[-t]}^{(k)}\right)}{\pi\left(x_t = b, \boldsymbol{X}_{[-t]}^{(k)}\right) + \pi\left(x_t = -b, \boldsymbol{X}_{[-t]}^{(k)}\right)}, \quad (12.27)$$

where $\pi(\boldsymbol{X})$ is as in (12.26). When the variance σ^2 is known,

$$\pi(\boldsymbol{X}) \propto \exp\left\{\frac{1}{2n\sigma^2}\left(\sum_{t=1}^{n} x_t y_t\right)^2\right\}. \quad (12.28)$$

Besides the two Gibbs samplers just described, an attractive alternative is the Metropolis algorithm applied directly to (12.26). Suppose $\boldsymbol{X}^{(k)} = \left(x_1^{(k)}, \ldots, x_n^{(k)}\right)$. At step $k+1$, the Metropolis algorithm proceeds as follows:

Algorithm 12.3.5. [Metropolis detector in AWGN channel]

- *Choose $t \in \{1, \ldots, n\}$ either by the random scan or by the deterministic scan. Define $\boldsymbol{Z} = (z_1, \ldots, z_n)$ where $z_t = -x_t^{(k)}$ and $z_s = x_s^{(k)}$ for $s \neq t$. Generate independently $U \sim \text{uniform}(0, 1)$. Let $\boldsymbol{X}^{(k+1)} = \boldsymbol{Z}$ if*

$$U \leq \min\left\{1, \frac{\pi(\boldsymbol{Z})}{\pi(\boldsymbol{X}^{(k)})}\right\}. \quad (12.29)$$

and let $\boldsymbol{X}^{(k+1)} = \boldsymbol{X}^{(k)}$ otherwise.

This Metropolis algorithm differs from the one-component Gibbs detector only slightly in the way of updating $x_t^{(k)}$ to $x_t^{(k+1)}$. That is, the Metropolis algorithm always forces the change (to $-x_t^{(k)}$) unless it is rejected, whereas the Gibbs sampler "voluntarily" selects whether to make the change so that no rejection is incurred. It is known that when the random scan is used, the Metropolis rule always results in a smaller second-largest eigenvalue (not in absolute value) than the corresponding Gibbs sampler [4]. Thus, when the target distribution is relatively peaked (high signal–to–noise ratio (SNR)) the Metropolis algorithm is slightly preferable. However, the Metropolis algorithm may have a large (in absolute value) negative eigenvalue when the target distribution is flatter (low SNR). In practice, however, the large negative eigenvalue is not a serious concern. No clear theory is available when a deterministic scan is used for updating. Simulations suggest that a similar result to that of the random scan samplers seems to hold well.

To overcome the phase ambiguity, one can eith
or, alternatively, use differential encoding.   Let t
$s_t \in \{+1, -1\}$, $t = 2, \ldots, n$. In differential coding.
ted sequence $x_t \in \{+1, -1\}$, $t = 1, \ldots, n$, such that x
Carlo draws from the posterior distribution of $p(S,$
MCMC algorithms to generate a Markov chain on $(X$
samples of X to S using $s_t^{(k)} = x_t^{(k)} x_{t-1}^{(k)}$, $t = 2, \ldots,$
and $-X$ result in the same S. Since $\left\{ X^{(k)} \right\}$ is a Mar
transition probability from $S^{(k)}$ to $S^{(k+1)}$ is given by

$$P\left(S^{(k+1)} \mid S^{(k)}\right) \;=\; P\left(X^{(k+1)} \mid X^{(k)}\right) + I$$

where both $X^{(k+1)}$ and $-X^{(k+1)}$ result in $S^{(k+1)}$, ar
that, both $X^{(k)}$ and $-X^{(k)}$ result in $S^{(k)}$, but sin
$P\left(X^{(k+1)} \mid X^{(k)}\right)$, either one can be used.

By denoting $s_1 = x_1$ and $S \triangleq [s_1, s_2, \ldots s_n]$, we
rise to the marginal target distribution for the s_t:

$$\pi(s_1, \ldots, s_n) \propto \left\{ \nu\lambda + \sum_{t=1}^{n} y_t^2 - \frac{1}{n}\left(\sum_{t=1}^{n} y_t \prod_{i=1}^{t}\right.\right.$$

Clearly, s_1 is independent of all the other s and has a u
It is trickier to implement an efficient Gibbs sam
based on (12.31). For example, the single-site updat
s_t at a time) may be inefficient because when we p
all the signs on $y_t, y_{t+1}, \ldots$ have to be changed. Thi
acceptance rate. Since a single update from x_t to -
(s_t, s_{t+1}) $(-s_t, -s_{t+1})$, we can employ proposals

$$(s_t, s_{t+1}) \propto (-s_t, -s_{t+1}), \quad t$$

and $s_n \to -s_n$ for distribution (12.31).

MCMC Equalizers in ISI Channels

Next we consider the Gibbs sampler for blind equali
terference (ISI) channel [18, 19]. After filtering and s
received waveform, the discrete-time received signal i

$$y_t \;=\; \sum_{s=0}^{q} \phi_s x_{t-s} + v_t, \quad t = 1,$$

where $(q + 1)$ is the channel order; $\phi_i \in \mathbb{R}$ is the value of the i-th channel tap, $i = 0, \ldots, q$; $x_t \in \{+1, -1\}$ is the transmitted binary symbol at time t; and $v_t \sim \mathcal{N}(0, \sigma^2)$ is an independent Gaussian noise sample at time t.

Let $\boldsymbol{X} \triangleq [x_{1-q}, \ldots, x_n]$, $\boldsymbol{Y} \triangleq [y_1, \ldots, y_n]$, $\boldsymbol{\phi} \triangleq [\phi_0, \ldots, \phi_q]^T$. With a uniform prior for $\boldsymbol{\phi}$, a uniform prior for $\boldsymbol{X}$, and an inverse χ^2 prior for σ^2 (e.g., $\sigma^2 \sim \chi^{-2}_{\nu, \lambda}$), the complete posterior distribution is

$$p\left(\boldsymbol{X}, \boldsymbol{\phi}, \sigma^2 \mid \boldsymbol{Y}\right) \sim p\left(\boldsymbol{Y} \mid \boldsymbol{X}, \boldsymbol{\phi}, \sigma^2\right) p(\boldsymbol{\phi}) p\left(\sigma^2\right) p(\boldsymbol{X}). \tag{12.33}$$

The Gibbs sampler approach to this problem starts with an arbitrary initial value of $\boldsymbol{X}^{(0)}$ and iterates between the following two steps:

Algorithm 12.3.6. [Two-component Gibbs equalizer in ISI channel]

- *Draw a sample $\left(\boldsymbol{\phi}^{(k+1)}, \sigma^{2(k+1)}\right)$ from the conditional distribution (given $\boldsymbol{X}^{(k)}$)*

$$p\left(\boldsymbol{\phi}, \sigma^2 \mid \boldsymbol{X}^{(k)}, \boldsymbol{Y}\right) \propto (\sigma^2)^{-\frac{n}{2}} \exp\left[-\frac{1}{2\sigma^2} \sum_{t=1}^{n} \left(y_t - \boldsymbol{\phi}^T \boldsymbol{x}_t^{(k)}\right)^2\right] (\sigma^2)^{-\frac{\nu+2}{2}}$$

$$\exp\left(-\frac{\nu\lambda}{2\sigma^2}\right)$$

$$\propto \pi^{(k+1)}\left(\boldsymbol{\phi} \mid \sigma^2\right) \pi^{(k+1)}\left(\sigma^2\right), \tag{12.34}$$

where $\boldsymbol{x}_t^{(k)} \triangleq \left[x_t^{(k)}, \ldots, x_{t-q}^{(k)}\right]^T$ for $k = 0, 1, \ldots,$ and

$$\pi^{(k+1)}\left(\sigma^2\right) \sim \chi^{-2}\left(\nu + n - 1, \frac{\nu\lambda + W^{(k+1)}}{\nu + n - 1}\right), \tag{12.35}$$

$$\pi^{(k+1)}\left(\boldsymbol{\phi} \mid \sigma^2\right) \sim \mathcal{N}\left(\boldsymbol{\mu}^{(k+1)}, \boldsymbol{\Sigma}^{(k+1)}\right), \tag{12.36}$$

$$W^{(k+1)} = \sum_{t=1}^{n} y_t^2 - \left[\sum_{t=1}^{n} \boldsymbol{x}_t^{(k)} y_t\right]^T \left[\sum_{t=1}^{n} \boldsymbol{x}_t^{(k)} \boldsymbol{x}_t^{(k)T}\right]^{-1} \left[\sum_{t=1}^{n} \boldsymbol{x}_t^{(k)} y_t\right], \tag{12.37}$$

$$\boldsymbol{\Sigma}^{(k+1)} = \left[\frac{1}{\sigma^2} \sum_{t=1}^{n} \boldsymbol{x}_t \boldsymbol{x}_t^T\right]^{-1}, \tag{12.38}$$

$$\boldsymbol{\mu}^{(k+1)} = \boldsymbol{\Sigma}^{(k+1)} \left(\frac{1}{\sigma^2} \sum_{t=1}^{n} \boldsymbol{x}_t^{(k)} y_t\right). \tag{12.39}$$

- *Draw a sample $\boldsymbol{X}^{(k+1)}$ from the conditional distribution, given $\left(\boldsymbol{\phi}^{(k+1)}, \sigma^{2(k+1)}\right)$*

through the following iterations. For $t = 1 - q$,

$$p\left(x_t \mid \phi^{(k+1)}, \sigma^{2(k+1)}, \boldsymbol{Y}, \boldsymbol{X}_{[-t]}^{(k)}\right)$$

$$\exp\left[-\frac{1}{2\sigma^{2(k+1)}} \sum_{j=1}^{n} \left(y_j\right.\right.$$

where $\boldsymbol{X}_{[-t]}^{(k)} \triangleq \left[x_{1-q}^{(k+1)}, \ldots, x_{t-1}^{(k+1)}, x_{t+1}^{(k)}, \ldots, x_M^{(k}\right.$

Another interesting Gibbs sampling scheme is base
In particular, a forward-backward algorithm can be e
conditional on $\boldsymbol{Y}$ and the parameters. This scheme
forms a Gaussian Markov model or a Markov chain
only a few values. In the ISI channel equalization pr
a priori, but they are correlated *a posteriori* because
the relationship (12.32). The induced correlation am
q. More precisely, instead of using formula (12.40)
can draw $\boldsymbol{X}$ altogether:

Algorithm 12.3.7. [Grouping-Gibbs equalizer in I

- *The first few steps are identical to the previous*

- *The last step is replaced by the forward-backwa*
  and σ (we suppress the superscript for iteration
  distribution of $\boldsymbol{X}$:

$$p\left(\boldsymbol{X} \mid \phi, \sigma, \boldsymbol{Y}\right) \quad \propto \quad \exp\left[-\frac{1}{2\sigma^2} \sum_{j=}^{n}\right.$$

$$\equiv \quad \exp\left\{g_1(\boldsymbol{x}_1) + \right.$$

*where $\boldsymbol{x}_j = (x_{j-q}, \ldots, x_j)$. Thus, each $\boldsymbol{x}_j$ can t
following two steps produce a sample $\boldsymbol{X}$ from p*

- *Forward summation. Define $f_1(\boldsymbol{x}_1) = \exp$*
  sively

$$f_{j+1}(\boldsymbol{x}_{j+1}) = \sum_{x_{j-q}=-1}^{1} [f_j(\boldsymbol{x}_j) \exp$$

     – *Backward sampling. First draw $\boldsymbol{x}_n = (x_{n-q}, \dots, x_n)$ from distribution $P(\boldsymbol{x}_n) \propto f_n(\boldsymbol{x}_n)$. Then, for $j = n - q - 1, \dots, 1$, draw $P(x_j \mid x_{j+1}, \dots, x_n) \propto f_{j+q}(x_j, \dots, x_{j+q})$.*

Although the grouping idea is attractive for overcoming the channel memory problem, the additional computation cost may offset its advantages. More precisely, the forward-backward procedure needs about 2^q times more memory and about 2^q times more basic operations.

Similar to the previous section, we can integrate out the continuous parameters and write down the marginal target distribution of $\boldsymbol{X}$:

$$\pi(\boldsymbol{X}) \propto [\nu\lambda + W]^{-(n+\nu)/2} \tag{12.43}$$

where

$$W = \sum y_t^2 - \left[\sum \boldsymbol{x}_t y_t\right]^T \left[\sum \boldsymbol{x}_t \boldsymbol{x}_t^T\right]^{-1} \left[\sum \boldsymbol{x}_t y_t\right]. \tag{12.44}$$

We can then derive the one-component Gibbs and Metropolis algorithms accordingly. The phase ambiguity (i.e., likelihood unchanged when $\boldsymbol{X}$ is changed to $-\boldsymbol{X}$) can be clearly seen from this joint distribution.

Algorithm 12.3.8. [One-component Gibb/Metropolis equalizer in ISI channel]

- *Choose t from $1, \dots, n$ by either the random scan or the systematic scan. Let $\boldsymbol{X}^{(k+1)} = \boldsymbol{Z}$, where $z_s = x_s^{(k)}$ for $s \neq t$ and $z_t = -x_t^{(k)}$, with probability*

$$\frac{\pi(\boldsymbol{Z})}{\pi\left(\boldsymbol{X}^{(k)}\right) + \pi(\boldsymbol{Z})}, \tag{12.45}$$

for the Gibbs equalizer, or with probability

$$\min\left\{1, \frac{\pi(\boldsymbol{Z})}{\pi(\boldsymbol{X}^{(k)})}\right\} \tag{12.46}$$

for the Metropolis equalizer, where $\pi(\boldsymbol{X})$ is as in (12.43). Otherwise let $\boldsymbol{X}^{(k+1)} = \boldsymbol{X}^{(k)}$. When the variance σ^2 is known,

$$\pi(\boldsymbol{X}) \propto \left\{\frac{1}{|\sum \boldsymbol{x}_t \boldsymbol{x}_t^T|^{q/2}}\right\} \exp\left(\frac{1}{2\sigma^2} \left[\sum \boldsymbol{x}_t y_t\right]^T \left[\sum \boldsymbol{x}_t \boldsymbol{x}_t^T\right]^{-1} \left[\sum \boldsymbol{x}_t y_t\right]\right).$$

To overcome the phase ambiguity, we use differential coding in all of our algorithms. Denote $\boldsymbol{S} \triangleq [s_2, \dots, s_n]$ as the information bits. Let $s_t^{(k)} = x_t^{(k)} x_{t-1}^{(k)}$, $t = 2, \dots, n$. Since $\boldsymbol{X}^{(k)}$ forms a Markov chain, $\boldsymbol{S}^{(k)}$ is a Markov chain too. The transition probability from $\boldsymbol{S}^{(k)}$ to $\boldsymbol{S}^{(k+1)}$ is

$$P\left(\boldsymbol{S}^{(k+1)} \mid \boldsymbol{S}^{(k)}\right) = P\left(\boldsymbol{X}^{(k+1)} \mid \boldsymbol{X}^{(k)}\right) + P\left(-\boldsymbol{X}^{(k+1)} \mid \boldsymbol{X}^{(k)}\right), \tag{12.47}$$

where both $\boldsymbol{X}^{(k+1)}$ and $-\boldsymbol{X}^{(k+1)}$ result in $\boldsymbol{S}^{(k+1)}$ and $\boldsymbol{X}^{(k)}$ results in $\boldsymbol{S}^{(k)}$.

12.4   Sequential Monte Carlo (S

12.4.1   General SMC Algorithms

Sequential Importance Sampling

Importance sampling is perhaps one of the most versatile Monte Carlo techniques. Suppose we want respect to p), using Monte Carlo method. Since di difficult, we want to find a *trial* distribution, $q(\boldsymbol{x})$, w but is easy to draw samples from. Because of the sin

$$
E\{h(\boldsymbol{x})\} \;=\; \int h(\boldsymbol{x})\,p(\boldsymbol{x})\,\mathrm{d}\boldsymbol{x}
$$

$$
\;=\; \int h(\boldsymbol{x})\,w(\boldsymbol{x})\,q(
$$

where

$$
w(\boldsymbol{x}) \;\triangleq\; \frac{p(\boldsymbol{x})}{q(\boldsymbol{x})},
$$

is the importance weight, we can approximate (12.48

$$
E\{h(\boldsymbol{x})\} \;\cong\; \frac{1}{W}\sum_{j=1}^{\nu} h\left(\boldsymbol{x}^{(j)}\right) u
$$

where $\boldsymbol{x}^{(1)}, \boldsymbol{x}^{(2)}, \cdots, \boldsymbol{x}^{(\nu)}$ are random samples from q using this method, we only need to know the expressio constant, which is the case for many processing prob nications. Each $\boldsymbol{x}^{(j)}$ is said to be properly weighted p.

However, it is usually difficult to design a good t dimensional problems. One of the most useful strat build up the trial density sequentially. Suppose we can where each of the x_j may be multidimensional. Th constructed as

$$
q(\boldsymbol{x}) = q_1(x_1)q_2(x_2 \mid x_1)\cdots q_d(x_d \mid x_1,
$$

by which we hope to obtain some guidance from the up the trial density. Corresponding to the decomposi target density as

$$
p(\boldsymbol{x}) = p(x_1)p(x_2 \mid x_1)\cdots p(x_d \mid x_1,
$$

and the importance weight as

$$w(\boldsymbol{x}) = \frac{p(x_1)p(x_2 \mid x_1) \cdots p(x_d \mid x_1, \cdots, x_{d-1})}{q_1(x_1)q_2(x_2 \mid x_1) \cdots q_d(x_d \mid x_1, \cdots, x_{d-1})}. \tag{12.53}$$

Equation (12.53) suggests a recursive way of computing and monitoring the importance weight. That is, by denoting $\boldsymbol{x}_t = (x_1, \cdots, x_t)$ (thus, $\boldsymbol{x}_d \equiv \boldsymbol{x}$), we have

$$w_t(\boldsymbol{x}_t) = w_{t-1}(\boldsymbol{x}_{t-1})\frac{p(x_t \mid \boldsymbol{x}_{t-1})}{q_t(x_t \mid \boldsymbol{x}_{t-1})}. \tag{12.54}$$

Then w_d is equal to $w(\boldsymbol{x})$ in (12.53). Potential advantages of this recursion and (12.52) are: (a) We can stop generating further components of $\boldsymbol{x}$ if the *partial weight* derived from the sequentially generated *partial sample* is too small; and (b) we can take advantage of $p(x_t|\boldsymbol{x}_{t-1})$ in designing $q_t(x_t|\boldsymbol{x}_{t-1})$. In other words, the marginal distribution $p(\boldsymbol{x}_t)$ can be used to guide the generation of $\boldsymbol{x}$.

Although the "idea" sounds interesting, the trouble is that expressions (12.52) and (12.53) are not useful at all! The reason is that in order to get (12.52), one needs to have the marginal distribution

$$p(\boldsymbol{x}_t) = \int p(x_1, \cdots, x_d)\mathrm{d}x_{t+1} \cdots \mathrm{d}x_d, \tag{12.55}$$

which is perhaps more difficult than the original problem.

In order to carry out the sequential sampling idea, we need to find a sequence of "auxiliary distributions," $\pi_1(x_1), \pi_2(\boldsymbol{x}_2), \cdots, \pi_d(\boldsymbol{x})$, so that $\pi_t(\boldsymbol{x}_t)$ is a reasonable approximation to the marginal distribution $p(\boldsymbol{x}_t)$, for $t = 1, \cdots, d-1$, and $\pi_d = p$. We want to emphasize that the π_t are only required to be known up to a normalizing constant and they *only* serve as "guides" to our construction of the whole sample $\boldsymbol{x} = (x_1, \cdots, x_d)$. The *sequential importance sampling* (SIS) method can then be defined as the following recursive procedure.

Algorithm 12.4.1. [Sequential importance sampling (SIS)]
  For $t = 2, \cdots, d$:

- *Draw x_t from $q_t(x_t|\boldsymbol{x}_{t-1})$, and let $\boldsymbol{x}_t = (\boldsymbol{x}_{t-1}, x_t)$.*

- *Compute*

$$u_t = \frac{\pi_t(\boldsymbol{x}_t)}{\pi_{t-1}(\boldsymbol{x}_{t-1})q_t(x_t \mid \boldsymbol{x}_{t-1})}, \tag{12.56}$$

  and let $w_t = w_{t-1}u_t$. Here u_t is called an incremental weight.

It is easy to show that $\boldsymbol{x}_t$ is properly weighted by w_t with respect to π_t provided that $\boldsymbol{x}_{t-1}$ is properly weighted by w_{t-1} with respect to π_{t-1}. Thus, the whole sample $\boldsymbol{x}$ obtained by SIS is properly weighted by w_d with respect to the target density $p(\boldsymbol{x})$. The "auxiliary distributions" can also be used to help construct a more efficient trial distribution:

- We can build q_t in light of π_t. For example, on

$$q_t(x_t \mid x_{t-1}) = \pi_t(x_t \mid x_{t-})$$

  Then the incremental weight becomes

$$u_t = \frac{\pi_t(x_t)}{\pi_{t-1}(x_{t-1})}.$$

  In the same token, we may also want q_t to be latter involves integrating out x_{t+1}.

- When we observe that w_t is getting too small sample half way and restart. In this way, we avo: samples that are deemed to have little effect in t as an outright rejection incurs bias, techniques are needed [21].

- Another problem with the SIS is that the resu often very skewed, especially when d is large. A sequential Monte Carlo to address this problem [21–23].

SMC for Dynamic Systems

Consider the following dynamic system modeled i

$$\text{state equation} \qquad z_t = f_t(z_{t-}$$
$$\text{observation equation} \qquad y_t = g_t(z_{t-}$$

where z_t, y_t, u_t and v_t are, respectively, the state v state noise, and the observation noise at time t. T vectors.

Let $Z_t = (z_0, z_1, \cdots, z_t)$ and let $Y_t = (y_0, y_1, \cdots, z$ ence of Z_t is of interest; that is, at current time t we w of a function of the state variable Z_t, say $h(Z_t)$, bas observation, Y_t. With the Bayes theorem, we reali to this problem is $E\{h(Z_t)|Y_t\} = \int h(Z_t)p(Z_t|Y_t)$ evaluation of this expectation is analytically intractat of such a dynamic system. Monte Carlo methods pr native to the required computation. Specifically, supp $\left\{Z_t^{(j)}\right\}_{j=1}^{\nu}$ is generated from the trial distribution q weight

$$w_t^{(j)} = \frac{p\left(Z_t^{(j)} \mid Y_t\right)}{q\left(Z_t^{(j)} \mid Y_t\right)}$$

to the sample $\mathbf{Z}_t^{(j)}$, we can approximate the quantity of interest, $E\{h(\mathbf{Z}_t)|\mathbf{Y}_t\}$, as

$$E\{h(\mathbf{Z}_t) \mid \mathbf{Y}_t\} \cong \frac{1}{W_t} \sum_{j=1}^{\nu} h\left(\mathbf{Z}_t^{(j)}\right) w_t^{(j)}, \tag{12.61}$$

where $W_t = \sum_{j=1}^{\nu} w_t^{(j)}$. The pair $\left(\mathbf{Z}_t^{(j)}, w_t^{(j)}\right)$, is a *properly weighted sample* with respect to distribution $p(\mathbf{Z}_t|\mathbf{Y}_t)$. A trivial but important observation is that $z_t^{(j)}$ (one of the components of $\mathbf{Z}_t^{(j)}$) is also properly weighted by $w_t^{(j)}$ with respect to the marginal distribution $p(z_t|\mathbf{Y}_t)$.

To implement Monte Carlo techniques for a dynamic system, a set of random samples properly weighted with respect to $p(\mathbf{Z}_t|\mathbf{Y}_t)$ is needed for any time t. Because the state equation in system (12.59) possesses a Markovian structure, we can implement a SMC strategy [21]. Suppose a set of properly weighted samples $\left\{(\mathbf{Z}_{t-1}^{(j)}, w_{t-1}^{(j)})\right\}_{j=1}^{\nu}$ (with respect to $p(\mathbf{Z}_{t-1}|\mathbf{Y}_{t-1})$) is given at time $(t-1)$. A sequential Monte Carlo filter generates from the set a new one, $\left\{\mathbf{Z}_t^{(j)}, w_t^{(j)}\right\}_{j=1}^{\nu}$, which is properly weighted at time t with respect to $p(\mathbf{Z}_t|\mathbf{Y}_t)$, according to the following algorithm.

Algorithm 12.4.2. [Sequential Monte Carlo filter for dynamic systems]
 For $j = 1, \cdots, \nu$:

- *Draw a sample $z_t^{(j)}$ from a trial distribution $q(z_t|\mathbf{Z}_{t-1}^{(j)}, \mathbf{Y}_t)$ and let $\mathbf{Z}_t^{(j)} = \left(\mathbf{Z}_{t-1}^{(j)}, z_t^{(j)}\right)$;*

- *Compute the importance weight*

$$w_t^{(j)} = w_{t-1}^{(j)} \cdot \frac{p\left(\mathbf{Z}_t^{(j)} \mid \mathbf{Y}_t\right)}{p\left(\mathbf{Z}_{t-1}^{(j)} \mid \mathbf{Y}_{t-1}\right) q\left(z_t^{(j)} \mid \mathbf{Z}_{t-1}^{(j)}, \mathbf{Y}_t\right)}. \tag{12.62}$$

The algorithm is initialized by drawing a set of i.i.d. samples $z_0^{(1)}, \cdots, z_0^{(m)}$ from $p(z_0|y_0)$. When y_0 represents the "null" information, $p(z_0|y_0)$ corresponds to the prior of z_0.

A useful choice of the trial distribution $q\left(z_t \mid \mathbf{Z}_{t-1}^{(j)}, \mathbf{Y}_t\right)$ for the state space model (12.59) is of the form

$$q\left(z_t \mid \mathbf{Z}_{t-1}^{(j)}, \mathbf{Y}_t\right) = p\left(z_t \mid \mathbf{Z}_{t-1}^{(j)}, \mathbf{Y}_t\right)$$

$$= \frac{p\left(y_t \mid z_t\right) p\left(z_t \mid z_{t-1}^{(j)}\right)}{p\left(y_t \mid z_{t-1}^{(j)}\right)}. \tag{12.63}$$

For this trial distribution, the importance weight is ι

$$w_t^{(j)} \;\propto\; w_{t-1}^{(j)} \cdot p\left(\boldsymbol{y}_t \mid \boldsymbol{z}_{t-}^{(j)}\right)$$

Mixture Kalman Filter

Many dynamic system models belong to the class models (CDLM) of the form

$$
\begin{aligned}
\boldsymbol{x}_t &= F_{\lambda_t}\boldsymbol{x}_{t-1} + G_{\lambda_t}\boldsymbol{u}_t, \\
\boldsymbol{y}_t &= H_{\lambda_t}\boldsymbol{x}_t + K_{\lambda_t}\boldsymbol{v}_t,
\end{aligned}
$$

where $\boldsymbol{u}_t \sim \mathcal{N}_c(0, I)$, $\boldsymbol{v}_t \sim \mathcal{N}_c(0, I)$ (here I denotes an random indicator variable. The matrices F_{λ_t}, G_{λ_t}, H λ_t. In this model, the "state variable" $\boldsymbol{z}_t$ correspond

We observe that for a given trajectory of the indica is both linear and Gaussian, for which the Kalman statistical characterization of the system dynamics. (MKF) [24] can be employed for on-line filtering an exploits the conditional Gaussian property and utilize to improve the algorithmic efficiency. Instead of de MKF draws Monte Carlo samples only in the indicat of Gaussian distributions to approximate the target the generic SMC method, MKF is substantially mor accurate results with the same computing resources).

Let $\boldsymbol{Y}_t = (\boldsymbol{y}_0, \boldsymbol{y}_1, \cdots, \boldsymbol{y}_t)$ and let $\boldsymbol{\Lambda}_t = (\lambda_0, \lambda_1,$ erating a set of properly weighted random samples $\{$ $p(\boldsymbol{\Lambda}_t|\boldsymbol{Y}_t)$, the MKF approximates the target distrib mixture of Gaussian distributions

$$\frac{1}{W_t}\sum_{j=1}^{\nu} w_t^{(j)}\mathcal{N}_c\left(\boldsymbol{\mu}_t^{(j)}, \boldsymbol{\Sigma}_t^{(j)}\right),$$

where $\boldsymbol{\kappa}_t^{(j)} \triangleq \left[\boldsymbol{\mu}_t^{(j)}, \boldsymbol{\Sigma}_t^{(j)}\right]$ is obtained by implemen given indicator trajectory $\boldsymbol{\Lambda}_t^{(j)}$ and $W_t = \sum_{j=1}^{\nu} w_t^{(j)}$. production at time t of a weighted sample of indica

based on the set of samples, $\left\{(\boldsymbol{\Lambda}_{t-1}^{(j)}, \boldsymbol{\kappa}_{t-1}^{(j)}, w_{t-1}^{(j)})\right\}_{j=1}^{\nu}$ according to the following algorithm.

Algorithm 12.4.3. [Mixture Kalman filter]
   For $j = 1, \cdots, \nu$:

- *Draw a sample $\lambda_t^{(j)}$ from a trial distribution $q\left(\lambda_t \mid \boldsymbol{\Lambda}_{t-1}^{(j)}, \boldsymbol{\kappa}_{t-1}^{(j)}, \boldsymbol{Y}_t\right)$.*

- *Run a one-step Kalman filter based on $\lambda_t^{(j)}$, $\boldsymbol{\kappa}_{t-1}^{(j)}$, and $\boldsymbol{y}_t$ to obtain $\boldsymbol{\kappa}_t^{(j)}$.*

- *Compute the weight*

$$
w_t^{(j)} \quad \propto \quad w_{t-1}^{(j)} \cdot \frac{p\left(\boldsymbol{\Lambda}_{t-1}^{(j)}, \lambda_t^{(j)} \mid \boldsymbol{Y}_t\right)}{p\left(\boldsymbol{\Lambda}_{t-1}^{(j)} \mid \boldsymbol{Y}_{t-1}\right) q\left(\lambda_t^{(j)} \mid \boldsymbol{\Lambda}_{t-1}^{(j)}, \boldsymbol{\kappa}_{t-1}^{(j)}, \boldsymbol{Y}_t\right)}. \quad (12.67)
$$

12.4.2 Resampling Procedures

The importance sampling weight $w_t^{(j)}$ measures the "quality" of the corresponding imputed signal sequence $\boldsymbol{Z}_t^{(j)}$. A relatively small weight implies that the sample is drawn far from the main body of the posterior distribution and has a small contribution in the final estimation. Such a sample is said to be ineffective. If there are too many ineffective samples, the Monte Carlo procedure becomes inefficient. This can be detected by observing a large *coefficient of variation* in the importance weight. Suppose $\{w_t^{(j)}\}_{j=1}^m$ is a sequence of importance weights. Then the coefficient of variation, v_t is defined as

$$
v_t^2 \quad = \quad \frac{\sum_{j=1}^m \left(w_t^{(j)} - \bar{w}_t\right)^2 / m}{\bar{w}_t^2} = \frac{1}{m} \sum_{j=1}^m \left(\frac{w_t^{(j)}}{\bar{w}_t} - 1\right)^2, \quad (12.68)
$$

where $\bar{w}_t = \sum_{j=1}^m w_t^{(j)}/m$. Note that if the samples are drawn exactly from the target distribution, then all the weights are equal, implying that $v_t = 0$. It is shown in [25] that the importance weights resulting from a sequential Monte Carlo filter form a martingale sequence. As more and more data are processed, the coefficient of variation of the weights increases — that is, the number of ineffective samples increases — rapidly.

A useful method for reducing ineffective samples and enhancing effective ones is *resampling* [23]. Roughly speaking, resampling allows those "bad" samples (with small importance weights) to be discarded and those "good" ones (with large importance weights) to replicate so as to accommodate the dynamic change of the system. Specifically, let $\{(\boldsymbol{Z}_t^{(j)}, w_t^{(j)})\}_{j=1}^m$ be the original properly weighted samples at time t. A *residual resampling* strategy forms a new set of weighted samples $\{(\tilde{\boldsymbol{Z}}_t^{(j)}, \tilde{w}_t^{(j)})\}_{j=1}^m$ according to the following algorithm (assume that $\sum_{j=1}^m w_t^{(j)} = m$):

Algorithm 12.4.4. [Resampling algorithm]

- *For $j = 1, \cdots, m$, retain $k_j = \lfloor w_t^{(j)} \rfloor$ copies of the sample $\boldsymbol{Z}_t^{(j)}$. Denote $K_r = m - \sum_{j=1}^m k_j$.*

- *Obtain K_r i.i.d. draws from the original samp*
  bilities proportional to $(w_t^{(j)} - k_j), j = 1, \cdots, m$

- *Assign equal weight, i.e., set $\tilde{w}_t^{(j)} = 1$, for each*

The samples drawn by the above residual resam
weighted with respect to $p(\mathbf{Z}_t | \mathbf{Y}_t)$, provided that m is
when small to modest m is used the resampling procec
between bias and variance. That is, the new samples
from the resampling procedure are only approximat
small bias in Monte Carlo estimation. On the othe
greatly reduces Monte Carlo variance for the future s

Resampling can be done at any time. However
computational burden and decreases "diversities" of
it decreases the number of distinctive filters and lose
hand, resampling too rarely may result in a loss of e
to give guidance on when to do resampling. A me
importance sampling scheme is the *effective sample s*

$$\bar{m}_t \triangleq \frac{m}{1 + v_t^2}.$$

Heuristically, $\bar{m}_t$ reflects the equivalent size of a set
of m weighted ones. It is suggested in [21] that resa
when the effective sample size becomes small, e.g.,
can conduct resampling at every fixed-length time in
    Instead of the previous resampling scheme sugges
implement a more flexible resampling scheme as follov
m):

For $j = 1, \cdots, m$,

(a) For $w_t^{(j)} \geq 1$,

- Retain k_j copies of the sample $\mathbf{Z}_t^{(j)}$,
  (see below);
- Assign weight $\tilde{w}_t^{(j)} = w_t^{(j)}/k_j$ for each

(b) For $w_t^{(j)} < 1$,

- Kill the sample with probability $1 - f$
- Assign weight $w_t^{(j)}/f_j$ to the survived

The advantage of this new resampling method is
of choosing a proper resampling size k_j as we wish.
eliminate those hopeless samples and emphasize thos
other hand, however, we do not want to throw away the

prove important later on (as the dynamical system moves towards their way). An empirical choice of the resample size formula is $k_j = \lfloor \sqrt{w_t^{(j)}} \rfloor$ and $f_j = \sqrt{w_t^{(j)}}$. The intuition behind this choice is that it effectively removes those hopeless samples with small weights but still maintains the diversity of the Monte Carlo sample.

12.4.3 Applications of SMC in Bioinformatics

In this section we illustrate the application of SMC in solving the DNA sequence motif discovery problem described in Section 12.1.2.

SMC Motif Discovery Algorithm

For the system states up to time t, $\boldsymbol{x}_t = [x_1, \cdots, x_t]$, and the corresponding sequences $\boldsymbol{S}_t = [\boldsymbol{s}_1, \cdots, \boldsymbol{s}_t]$, we will first present their prior distributions and their conditional posterior distributions, and then describe the steps of the SMC motif discovery algorithm.

Prior Distributions: Denote $\boldsymbol{\theta}_j \triangleq [\theta_{j1}, \cdots, \theta_{j4}]^T$, $j = 1, \cdots, w$, as the j-th column of the position weight matrix $\boldsymbol{\Theta}$. In Monte Carlo methods, the prior distribution is often chosen so that the posterior and the prior are conjugate pairs, i.e., they belong to the same functional family. It can be seen that for all of the motifs in the dataset $\boldsymbol{S}$, the nucleotide counts at each motif location are drawn from multinomial distributions. It is well known that the Dirichlet distribution provides conjugate pairs for such distribution. Therefore, we use a multivariate Dirichlet distribution as the prior for $\boldsymbol{\theta}$. The prior distribution for the i-th column of the PWM is then given by

$$\boldsymbol{\theta}_i \sim \mathcal{D}(\rho_{i1}, \cdots, \rho_{i4}), \quad i = 1, 2, \cdots, w. \tag{12.70}$$

Denote $\boldsymbol{\rho}_i \triangleq [\rho_{i1}, \cdots, \rho_{i4}]$. Assuming independent priors, then the prior distribution for the PWM $\boldsymbol{\Theta}$ is the product Dirichlet distribution

$$\boldsymbol{\Theta} \sim \prod_{i=1}^{w} \mathcal{D}(\boldsymbol{\rho}_i). \tag{12.71}$$

Conditional Posterior Distributions: Here we describe the conditional posterior distributions that are used in the SMC algorithm:

1. The conditional posterior distribution of the PWM $\boldsymbol{\Theta}$:

$$
\begin{aligned}
p\left(\boldsymbol{\Theta} \mid \boldsymbol{S}_t, \boldsymbol{x}_{t-1}, x_t = i\right) \;&\propto\; p\left(\boldsymbol{s}_t \mid \boldsymbol{\Theta}, \boldsymbol{x}_{t-1}, x_t = i, \boldsymbol{S}_{t-1}\right) p\left(\boldsymbol{\Theta} \mid \boldsymbol{x}_{t-1}, \boldsymbol{S}_{t-1}\right) \\
&\propto\; \prod_{j=1}^{w} \theta_j^{n(a_{t,i}(j))} \prod_{\ell=1}^{w} \theta_\ell^{\rho_\ell(t-1)-1} \\
&\propto\; \Lambda_w\Big(\boldsymbol{\Theta}; \boldsymbol{\rho}_1(t-1) + n\left(a_{t,i}(1)\right), \cdots, \\
&\qquad\qquad \boldsymbol{\rho}_w(t-1) + n\left(a_{t,i}(w)\right)\Big),
\end{aligned}
\tag{12.72}
$$

where we denote $\boldsymbol{\Lambda}_w(\boldsymbol{\Theta}; \boldsymbol{\rho}_1, \cdots, \boldsymbol{\rho}_w)$ as the pr
$\boldsymbol{\rho}_i(t) \triangleq [\rho_{i1}(t), \cdots, \rho_{i4}(t)], i = 1, \cdots, w$, as the p
of $\boldsymbol{\Theta}$ at time t, and $\boldsymbol{\theta}_k^{\boldsymbol{\rho}_k(t)-1} \triangleq \prod_{\ell=1}^4 \theta_{k\ell}^{(\rho_{k\ell}(t)-}$
distribution of $\boldsymbol{\Theta}$ depends only on the sufficien
$i \le w, 1 \le j \le 4\}$, which is easily updated based
by (12.72), i.e., $\boldsymbol{T}_t = \boldsymbol{T}_t(\boldsymbol{T}_{t-1}, x_t, \boldsymbol{s}_t)$.

2. The conditional posterior distribution of state

$$p\left(x_t = i \mid \boldsymbol{S}_t, \boldsymbol{\Theta}\right) = p\left(x_t = i \mid \boldsymbol{s}_t, \boldsymbol{\Theta}\right) \propto \mathcal{B}\left(\boldsymbol{s}_t\right)$$

SMC Estimator: We now outline the SMC algorith
the PWM is unknown, assuming that there is only on
present in each of the sequences in the dataset. At tin
of $x_t^{(k)}$ we use the optimal proposal distribution

$$q_2\left(x_t = i \mid \boldsymbol{x}_{t-1}^{(k)}, \boldsymbol{S}_t, \boldsymbol{\Theta}\right) = p\left(x_t = i \mid \boldsymbol{x}_{t-1}^{(k)}, \boldsymbol{S}_t, \boldsymbol{\Theta}\right)$$

To sample $\boldsymbol{\Theta}$, we use the following proposal distribut

$$q_1\left(\boldsymbol{\Theta} \mid \boldsymbol{x}_{t-1}^{(k)}, \boldsymbol{S}_t\right) \propto \sum_{i=1}^{L_m} p\left(\boldsymbol{s}_t \mid x_t = i, \boldsymbol{\Theta}, \boldsymbol{x}_{t-1},$$

$$\propto \sum_{i=1}^{L_m} P_{t,x_t}^3 \prod_{k=1}^w \boldsymbol{\theta}_k^{\boldsymbol{\rho}_k(t-1)+\boldsymbol{n}(\boldsymbol{a}}$$

$$\propto \sum_{i=1}^{L_m} \lambda_{i,t} \boldsymbol{\Lambda}_w\left(\boldsymbol{\Theta}; \boldsymbol{\rho}_1(t-1) + \right.$$

$$\left. + \boldsymbol{n}(\boldsymbol{a}_{t,i}(w))\right).$$

where

$$\lambda_{i,t} \triangleq P_{t,x_t}^3 \prod_{\ell=1}^w \boldsymbol{\rho}_\ell(t-1)^{\boldsymbol{n}(\boldsymbol{a}_{t,}}$$

with $\boldsymbol{\rho}_\ell(t)^{\boldsymbol{n}(\boldsymbol{a}_{t,i}(\ell))} \triangleq \prod_{j=1}^4 \rho_{\ell j}(t)^{\mathbb{I}(s_{t,i+\ell-1}-j)}$. The we

$$w_t \propto w_{t-1} \frac{\sum_{i=1}^{L_m} \lambda_{i,t}}{\prod_{k=1}^w \sum_{j=1}^4 \rho_{kj}(t -}$$

We are now ready to give the SMC motif discover

Algorithm 12.4.5. [SMC motif discovery algorithm for single motif present in all sequences]

- *For $k = 1, \cdots, K$*

    - *Sample $\boldsymbol{\Theta}^{(k)}$ from the mixture Dirichlet distribution given by (12.75).*
    - *Sample $x_t^{(k)}$ from (12.74).*
    - *Update the sufficient statistics $\boldsymbol{T}_t^{(k)} = \boldsymbol{T}_t(\boldsymbol{T}_{t-1}^{(k)}, x_t^{(k)}, \boldsymbol{s}_t)$ from (12.72).*

- *Compute the new weights according to (12.77).*

- *Compute $\widehat{K_{eff}} = \left(\sum_{k=1}^{K} (w_t^{(k)})^2 \right)^{-1}$. If $\widehat{K_{eff}} \leq \frac{K}{10}$ perform resampling.*

Motif Scores: When searching for motifs in a dataset, it is often necessary to assign confidence scores to the estimated motif locations. A natural choice in this case would be to use the a posteriori probability

$$p(x_t \mid \boldsymbol{s}_t) \propto p(\boldsymbol{s}_t \mid x_t) p(x_t), \tag{12.78}$$

as the confidence score for our estimation, where $p(x_t)$, the prior probability of the starting location of the motif in sequence t is assumed to be uniformly distributed. Note that

$$p(\boldsymbol{s}_t \mid x_t) = \int p(\boldsymbol{s}_t \mid x_t, \boldsymbol{\Theta}) p(\boldsymbol{\Theta}) d\boldsymbol{\Theta}. \tag{12.79}$$

From [26], [27], (12.79) can be approximated by

$$p(\boldsymbol{s}_t \mid x_t) \approx p\left(\boldsymbol{s}_t \mid x_t, \hat{\boldsymbol{\Theta}}\right) p\left(\hat{\boldsymbol{\Theta}}\right) = \mathcal{B}\left(\boldsymbol{s}_t; x_t, \hat{\boldsymbol{\Theta}}\right) \Lambda_w\left(\hat{\boldsymbol{\Theta}}; \boldsymbol{\rho}_{1,t}, \cdots, \boldsymbol{\rho}_{w,t}\right), \tag{12.80}$$

and we denote (12.80) as the Bayesian score. Extensions of the above basic SMC motif discovery algorithm can be found in [9].

12.5 Conclusions and Further Readings

Monte Carlo techniques rely on random number generation to calculate more efficiently deterministic variables and functions, to solve complicated optimization and estimation problems, and to simulate complex phenomena and systems. They found applicability in a wide variety of fields including engineering, bioinformatics, statistics, and physical sciences (phsyics, astronomy, chemistry, etc.). In the areas of signal processing, communications and networking, Monte Carlo techniques combined with Bayesian statistics proved to be very powerful tools for solving complex estimation, detection, optimization and simulation problems (see e.g., [28–30]). Recently, the class of sequential Monte Carlo techniques helped to design efficient recursive algorithms for diverse estimation and detection applications (see e.g., [4, 31–33], as well as the tutorials [34–37]). For a more comprehensive treatment of Monte Carlo techniques and Bayesian statistics, we recommend the excellent references [3–7].

12.6   Exercises

Exercise 12.6.1 (Variance of MCMC sampler)
a Markov Chain Monte Carlo sampler are given by
according to π. Let us further assume that the pr
enough and has achieved the equilibrium distributior

$$N\text{var}\left\{\sum_{i=1}^{N}\frac{\phi(x^{(i)})}{N}\right\} = \sigma^2\left[1 + 2\sum_{i=1}^{N-}\right.$$

where $\sigma^2 = var[\phi(x)]$ and $\rho_i = E[\phi(x^j)\phi(x^{j+i})]$.

Exercise 12.6.2 (Two-component Gibbs sample
resulting from a two-component Gibbs sampler which

$$\text{cov}\left\{\phi(x_1^{(0)}), \phi(x_1^{(1)})\right\} = var\left\{E\left\{\phi\right.\right.$$

holds for any function ϕ.

Exercise 12.6.3 (Computational efficiency of N
have Monte Carlo samples from data augmentation
$E[\phi(x_1)]$. Which one of following estimators should t

$$\hat{I} = \frac{1}{m}\{\phi(x_1^{(1)}) + ... + \phi(x_1^{(m)})\}$$

$$I' = \frac{1}{m}\{E[\phi(x_1^{(1)})|x_2^{(1)}] + ... + E[\phi(x_1^{(m)})|x_2^{(1)}]... \}$$

Justify your result by finding the variances for the tv

Exercise 12.6.4 (Mean of the importance weigh
density is $p(x)$ and the trial density is $q(x)$. We draw
from q and the sum of weights is given by $W = \sum$
expectation of W is equal to n.

Exercise 12.6.5 (Normalizing the importance w
the weights have been normalized to sum one and
according to the normalized weights. Prove the follov

$$E\left\{\frac{1}{\nu}\sum\phi(\hat{x}^{(j)})\right\} = E\left\{\sum_{j=1}^{\nu}w_j\phi\right.$$

Exercise 12.6.6 (Importance sampling estimat
importance sampling estimator given by

$$\frac{1}{W}\sum_{j=1}^{\nu}h(x^{(j)})w(x^{(j)})$$

in terms of $K(x) = h(x)w(x)$, where W is the sum of

Exercise 12.6.7 (More analytical work is good in importance sampling).
Prove that
$$\mathrm{var}\left\{\frac{f_{X_1 X_2}(x_1, x_2)}{g_{X_1 X_2}(x_1, x_2)}\right\} \geq \mathrm{var}\left\{\frac{f_{X_1}(x_1)}{g_{X_1}(x_1)}\right\}$$
where the variance is calculated with respect to the density g.

References

[1] N. Metropolis, A. Rosenbluth, A. Teller, and E. Teller, "Equations of state calculations by fast computing machines," *J. Chemical Physics*, vol. 21, pp. 1087–1091, 1953.

[2] J. Besag, P. Green, D. Higdon, and K. Mengersen, "Bayesian computation and stochastic systems (with discussion)," *Statist. Sci.*, vol. 10, pp. 3–66, 1995.

[3] W. Gilks, S. Richardson, and D. Spiegelhalter, *Markov Chain Monte Carlo in Practice*. Chapman & Hall, New York, 1995.

[4] J. Liu, *Monte Carlo Methods for Scientific Computing*. Springer-Verlag, New York, 2001.

[5] C. Robert and G. Casella, *Monte Carlo Statistical Methods*. Springer-Verlag, New York, 1999.

[6] J. Bernardo and A. Smith, *Bayesian Theory*. Wiley, New York, 1995.

[7] C. P. Robert, "Mixtures of distributions: inference and estimation," in *Markov Chain Monte Carlo in Practice*. Chapman & Hall, New York, 1996, ch. 24, pp. 441–464.

[8] C. Andrieu and A. Doucet, "Joint Bayesian detection and estimation of noisy sinusoids via reversible jump MCMC," *IEEE Transactions on Signal Processing*, vol. 47, pp. 2667–2676, 1999.

[9] K.-C. Liang, X. Wang, and D. Anastassiou, "A sequential Monte Carlo method for motif discovery," *IEEE Transactions on Signal Processing*, vol. 56, no. 9, pp. 4486–4495, September 2008.

[10] W. Hastings, "Monte Carlo sampling methods using Markov chains and their applications," *Biometrika*, vol. 57, pp. 97–109, 1970.

[11] S. Geman and D. Geman, "Stochastic relaxation, Gibbs distribution, and the Bayesian restoration of images," *IEEE Trans. Pattern Anal. Machine Intell.*, vol. PAMI-6, no. 11, pp. 721–741, Nov. 1984.

[12] M. Tanner and W. Wong, "The calculation of posterior distribution by data augmentation (with discussion)," *J. Amer. Statist. Assoc.*, vol. 82, pp. 528–550, 1987.

440

[13] A. Gelfand and A. Smith, "Sampling-based appro densities," *J. Amer. Stat. Assoc.*, vol. 85, pp. 39

[14] J. Liu, "The collapsed Gibbs sampler with app problem," *J. Amer. Statist. Assoc*, vol. 89, pp. 9

[15] C. Geyer, "Markov chain Monte Carlo maximu *Science and Statistics: Proceedings of the 23rd* E. Keramigas, Ed. Fairfax: Interface Foundatic

[16] J. Liu, F. Ling, and W. Wong, "The use of m optimization in Metropolis sampling," *J. Amer* 121–134, 2000.

[17] F. Liang and W. Wong, "Evolutionary Monte Ca sampling and change point problem," *Statistica* 2000.

[18] R. Chen and T. Li, "Blind restoration of linearly Gibbs sampling," *IEEE Trans. Sig. Proc.*, vol. 4: 1995.

[19] X. Wang and R. Chen, "Blind turbo equalizatic noise," *IEEE Transactions Vehicular Technology*, July 2001.

[20] C. Carter and R. Kohn, "On Gibbs samplin *Biometrika*, vol. 81, pp. 541–553, 1994.

[21] J. Liu and R. Chen, "Sequential Monte Carlo me *Journal of the American Statistical Association*,

[22] N. Gordon, D. Salmon, and A. Smith, "A novel Gaussian Bayesian state estimation," *IEE Proc* pp. 107–113, 1993.

[23] J. Liu and R. Chen, "Blind deconvolution via sequ *of the American Statistical Association*, vol. 90,

[24] R. Chen and J. Liu, "Mixture Kalman filters," *J.* no. 3, pp. 493–509, 2000.

[25] A. Kong, J. Liu, and W. Wong, "Sequential impu data problems," *J. Amer. Statist. Assoc*, vol. 89,

[26] X. Zhou, X. Wang, R. Pal, I. Ivanov, M. Bittne Bayesian connectivity-based approach to constru ulatory networks," *Bioinformatics*, vol. 20, no. 1

[27] C. Andrieu, J. Freitas, and A. Doucet, "Robust full Bayesian learning from neural networks," *Neural Computation*, vol. 13, pp. 2359–2407, 2001.

[28] X. Wang and V. Poor, *Wireless Communication Systems: Advanced Techniques for Signal Reception.* US: Prentice Hall, 2003.

[29] X. Wang and A. Doucet, "Monte Carlo methods for signal processing: a review in the statistical signal processing context," *IEEE Signal Processing Magazine*, vol. 22, no. 6, pp. 152–170, November 2005.

[30] X. Wang, R. Chen, and J. Liu, "Monte Carlo signal processing for wireless communications," *J. VLSI Sig. Proc.*, vol. 30, no. 1-3, pp. 89–105, Jan.-Mar. 2002.

[31] O. Cappe, E. Moulines, and T. Ryden, *Inference in Hidden Markov Models.* Berlin: Springer, 2005.

[32] A. Doucet, N. D. Freitas, and N. Gordon, *Sequential Monte Carlo Methods in Practice.* Berlin: Springer, 2001.

[33] B. Ristic, S. Arulampalam, and N. Gordon, *Beyond the Kalman Filter: Particle Filters for Tracking Applications.* Artech House, Norwood, MA, 2004.

[34] M. S. Arulampalam, S. Maskell, N. Gordon, and T. Clapp, "A tutorial on particle filters for online nonlinear/non-Gaussian Bayesian tracking," *IEEE Transactions on Signal Processing*, vol. 50, no. 2, p. 174–188, February 2002.

[35] O. Cappe, S. Godsill, and E. Moulines, "An overview of existing methods and recent advances in sequential Monte Carlo," *Proceedings of IEEE*, vol. 95, no. 5, pp. 899–924, April 2007.

[36] P. M. Djuric, J. H. Kotecha, J. Zhang, Y. Huang, T. Ghirmai, M. F. Bugallo, and J. Miguez, "Particle Filtering," *IEEE Signal Processing Magazine*, vol. 20, no. 5, pp. 19–38, September 2003.

[37] A. Doucet, S. Godsill, and C. Andrieu, "On Sequential Monte Carlo Methods for Bayesian Filtering," *Statistics and Computing*, vol. 10, no. 3, pp. 197–208, 2000.

Chapter 13 Factor Graphs and Message Passing Algorithms

Aitzaz Ahmad[‡], Erchin Serpedin[‡], and Khalid A. Qaraqe[♯]
[‡]Texas A&M University, College Station, USA
[♯]Texas A&M University at Qatar

13.1 Introduction

Complex modern day systems are often characterized by the presence of many interacting variables that govern the dynamics of the system. Statistical inference in such systems requires efficient algorithms that offer ease of implementation while delivering the prespecified performance guarantees. In developing an algorithm for a sophisticated system, accurate and representative modeling of the underlying system is often the first step. The use of graphical models to explain the working of complex systems has gained a lot of attention in recent years. Stochastic models are often represented by a Bayesian network or a Markov random field. The graphical representation not only provides a better understanding of the system model but also offers numerous exciting opportunities to develop new and improved algorithms. Factor graphs belong to the class of graphical models that serve to explain the dependencies between several interacting variables. They can be used to model a wide variety of systems and are increasingly applied in statistical learning, signal processing, and artificial intelligence.

13.1.1 Why Factor Graphs?

A fundamental problem in statistical inference is to estimate certain unknown variables in a model based on a given set of noisy observations. Often in sophisticated systems, this task becomes prohibitively complex due to multidimensional integrations which are either too complex or do not produce closed form solutions. Hence, not much insight is gained about the structure of the system that is described by a multidimensional global function composed of the system variables. Factor graphs provide a natural graphical description of systems that is easier for understanding and development. While other algorithms may be computationally demanding,

factor graphs circumvent this problem by exploiting
ate function factors into a product of several *local*
represents dependencies between different variables
a factor graph. Statistical inference can be performe
the edges representing variables in a factor graph. T
implementation of the task at hand, but a more gene
of the problem enabling even simpler algorithms. Sor
factor graph representation are as follows:

- Perhaps the most important benefit of factor gra
  a universal approach to understanding several
  estimation theory, and artificial intelligence. M:
  these fields may be viewed as specific instances
  graphs.

- They exploit the manner in which a global fu:
  factors into a product of functions involving a su
  in much simpler computations that offer severa
  in implementation.

- Factor graphs provide a simple graphical inte
  hand. It allows an engineer greater flexibility t
  for statistical inference in a much shorter span

- Because of the modular structure inherent in f
  sions and changes can be easily carried out.

- New algorithms can often be produced in fac
  tables that provide local computation rules for
  the system model. It also enables one to mix a
  without additional derivations and computatior

- When a factor graph representation of a system
  can be made. Even in applications where the sys
  having cycles, promising approximations to the

- Factor graphs can unify the tasks of modeling a
  tems. For example, by using a factor graph rep
  wireless communication system, tasks such as c
  and decoding can be expressed in a unified mar

13.1.2 Literature Survey

Factor graphs were originally developed for problems i
matical foundation of factor graphs is the "generalize

---

[1]A closed path in the graph is called a *cycle*. The length of
the cycle and the length of the smallest cycle is defined as the

by Aji and McEliece in [1]. The idea of factor graphs as a generalization of Tanner graphs was put forward by Wiberg in [2]. While Tanner primarily employed graphs for LDPC applications [3], Wiberg et al. suggested applications beyond coding. Forney introduced the normal graphs called Forney-style factor graphs (FFG) in [4] and this will be our choice representation for factor graphs throughout this chapter. An excellent tutorial on factor graphs is [5]. FFG's are also introduced in [6]. The use of factor graphs in model-based signal processing is discussed in [7]. As mentioned previously, an important advantage of factor graphs is their ability to provide a unified view of several algorithms in such diverse fields as coding theory, estimation theory, and machine learning. In particular, many algorithms in these fields can be viewed as specific instances of message passing in factor graphs. Algorithms such as BCJR [8], Viterbi algorithm [9], Kalman filtering [10], Expectation-Maximization algorithm [11], etc. have been shown to be instances of message passing in factor graph. Pearl's belief propagation algorithm in Bayesian networks [12] has a natural interpretation as message passing on a factor graph that expresses a factorization of the underlying Bayesian network. Steepest descent, particle filtering, and FFT can also be recovered through inference in factor graphs [5], [13], [14]. Factor graphs with cycles have also been used to explain the iterative decoding of turbo codes and LDPC codes. Although the inference in this case is only approximate, extensive simulations have proved this inference to be quite close to optimal [15–17].

Factor graphs find numerous applications in modeling and inference in systems. Some of these applications include iterative decoding of turbo codes [17], autoregressive model parameter estimation [18], electromyographic (EMG) signal separation [19], receiver design [20] and joint iterative detection and decoding [21]. Factor graphs have also been employed in linear equalization [22], LMMSE turbo equalization [23] and adaptive equalization [24]. Iterative detection schemes for ISI channels based on the sum-product inference algorithm in factor graphs representing the joint *a posteriori* probability of transmitted symbols are proposed in [25]. Factor graphs have also found applications in joint decoding and phase estimation [26], [27]. Since factor graphs offer an opportunity for modeling and inference simultaneously, they are also finding increasing applications in channel modeling and statistical inference.

13.1.3 Organization of the Chapter

This chapter provides a tutorial introduction to factor graphs and associated message passing algorithms. We adopt Forney-style factor graphs as our choice notation throughout this chapter. In FFG, all variables in the system are represented by edges. The global multivariate function factorizes into a product of simpler "local" functions and these functions are expressed by factor nodes. Section 13.2 introduces the basic terms associated with factor graphs and some rules that help in constructing factor graphs with the help of illustrative examples. A discussion on modeling systems using factor graphs is presented in Section 13.3 where two

specific modeling regimes, *behavioral modeling* and *f*
scribed. Factor graphs bear close resemblance to other
namely, Markov random fields and Bayesian networ
these graphical models are explained in Section 13.4
rithm for inference in factor graphs are presented in
the sum-product algorithm and the max-product alg
graphs with cycles and some general comments are br
and Section 13.7, respectively. Various algorithms :
rithm, Kalman filtering, and EM algorithm are prese
passing in factor graphs in Section 13.8. An applicat
rithm on factor graphs to jointly estimate the carrie
noise in OFDM transmission is described in Section
with some directions for future research in factor gra

13.2   Factor Graphs

A *factor graph* is a bipartite graph[2] that represent
function f of many variables which factors as a pro
factor being dependent on a subset of variables. Fac
produce a graphical model of a collection of interacti
terms associated with factor graphs are briefly descri

Configuration Space: The domain S of the global
lection of variables is called *configuration space*.

Configuration: A *configuration* ω is defined as a pa
to all variables of the global function and a configu
$f(\omega) \neq 0$.

Nodes and Edges: Every local function is represe
every variable has an edge or a half-edge. An edge c
to a factor node if and only if it is an argument of th
the factor node. An edge is connected to no more tha
local functions are functions of the same variable, an
be used to *clone* the variables [19].

For a given factorization of a global function of se
rules summarize the construction of a factor graph whi

---

[2]A *bipartite graph* is a graph G whose vertices can be pla
G_2 such that every edge in G connects a vertex in G_1 to a ve
vertices in G_1 are painted in one color and those in G_2 are pain
by edges are of separate colors. A direct consequence of this p
cannot contain any odd-length cycles. See [28] for more details

relationship [6].

1. Every factor has a unique node.

2. There is a unique edge (or half edge) for every variable.

3. The node representing a local function is connected with an edge (or half edge) representing some variable if and only if it is an argument of the local function.

Factor graphs based on such construction are also sometimes referred to as *Forney-style factor graphs*. The following examples serve to explain these terms and subsequent construction of a factor graph.

Example 13.2.1. Consider a global function $f(x_1, x_2, x_3, x_4, x_5, x_6)$ that factorizes as

$$f(x_1, x_2, x_3, x_4, x_5, x_6) = f_1(x_1, x_2) f_2(x_2, x_3) f_3(x_3, x_4, x_6) f_4(x_5, x_6), \quad (13.1)$$

where each x_i is considered discrete. The configuration space for this global function is $\Omega = \mathbb{Z}^6$ and a valid configuration ω such that $f(\omega) \neq 0$ could be $\omega = (x_1, x_2, x_3, x_4, x_5, x_6) = (1, 4, 3, 2, 3, 1)$. The local functions, expressed by the factors f_1, f_2, f_3 and f_4, constitute the nodes in the factor graph representation. The variables x_2, x_3 and x_6 are denoted by edges, whereas there are half edges for the variables x_1, x_4 and x_5, since they are arguments of only one local function.

Using the aforementioned rules of construction, the resulting factor graph is shown in Figure 13.1.

Example 13.2.2. Multivariate Probability Distribution. Factor graphs are often employed to represent factorizations of arbitrary multivariate functions that describe a particular system model. A common example is a probabilistic model

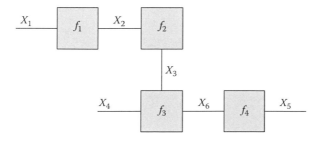

Figure 13.1: Factor graph for Example 13.2.1.

where the multivariate function and the corresponding
ity distributions. The configuration space, in this case
set of all possible outcomes of an experiment. For ex
$\mathcal{X}, \mathcal{Y}$ and $\mathcal{Z}$, a possible node function is their joint p

$$f(x, y, z) = p_{\mathcal{X}\mathcal{Y}\mathcal{Z}}(x, y, z)$$

The resulting factor graph consists of a single node,
random variables $\mathcal{X}, \mathcal{Y}$ and $\mathcal{Z}$, and is shown in Figure
tion of the multivariate global function is obtained by
distributions

$$f(x, y, z) = p_{\mathcal{X}\mathcal{Y}\mathcal{Z}}(x, y, z) = p_{\mathcal{X}}(x)\, p_{\mathcal{Y}|\mathcal{X}}(y|x)$$

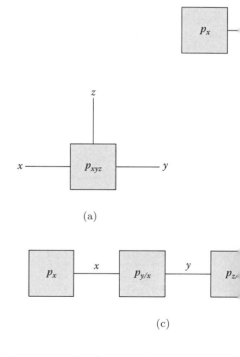

(a)

(c)

Figure 13.2: Factor graphs for a multivariate proba
probability distribution. (b) Conditional probabilit
chain relationship.

Clearly, three local functions are functions of the
previously, an equality constraint node is added to *clo*
is best described by Figure 13.3. Hence, two new va

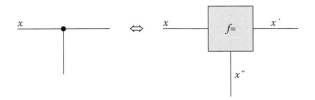

Figure 13.3: Equality constraint node.

a new factor (or local function), $f_= (x, x', x'') \triangleq \delta (x - x') \delta (x - x'')$, is produced. A factor graph representation, after the addition of an equality constraint node, is depicted in Figure 13.4.

For the case when the random variables $\mathcal{X}, \mathcal{Y}$ and $\mathcal{Z}$ form a *Markov chain*, the multivariate probability distribution function has a reduced factorization

$$f (x, y, z) = p_{\mathcal{X}\mathcal{Y}\mathcal{Z}} (x, y, z) = p_{\mathcal{X}} (x) \, p_{\mathcal{Y}|\mathcal{X}} (y|x) \, p_{\mathcal{Z}|\mathcal{Y}} (z|y) \ . \qquad (13.4)$$

Figure 13.2(c) shows the corresponding factor graph, where given $\mathcal{Y}$, $\mathcal{X}$ and $\mathcal{Z}$ are independent, a property attributed to the Markov chain relationship.

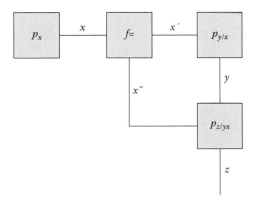

Figure 13.4: Factor graph after addition of equality constraint node.

13.3   Modeling Systems Using Factor Graphs

An important application of factor graphs is to model complex systems and highlight the way a complex global function factorizes into a product of simpler local functions. We have already seen how factor graphs are used to represent the factorization of a multivariate probability distribution in a probabilistic model. Broadly

speaking, the modeling of systems using factor graphs
egories: *behavioral modeling* and *probabilistic model*
a system in each of the two categories is briefly discu

13.3.1 Behavioral Modeling

In behavioral modeling of systems, set-theoretic conce
behavior by specifying the valid configurations of in
graph is ideally suited to represent the characteristic
[5]. Let $x_1, x_2, \ldots, x_n$ represent the variables of a sys
$S = A_1 \times A_2 \times \ldots \times A_n$. A subset B of S is called ε
are *valid configurations*. The indicator function for be
given by

$$I_B(x_1, \ldots, x_n) = \begin{cases} 1 & \text{if } (x_1, \ldots, \\ 0 & \text{if } (x_1, \ldots, \end{cases}$$

The determination of validity of a particular configu
applying a series of checks, each involving a subset of
the global indicator function for the behavior B in
of local functions, each of which is a link in the chai
declared valid if it passes all tests, i.e., each link in
local functions are, hence, expressed by factor nodes a
resented by a suitably constructed factor graph. The
through a factor graph depiction of linear codes is ne

Example 13.3.1. The characteristic function of a li
parity check matrix $\mathbf{H}$ can be represented by a factor
r factor nodes. For example, consider the parity chec

$$\mathbf{H} = \begin{bmatrix} 1 & 1 & 0 & 0 & 0 & 0 \\ 0 & 1 & 1 & 0 & 0 & 0 \\ 0 & 0 & 1 & 1 & 0 & 1 \\ 0 & 0 & 0 & 0 & 1 & 1 \end{bmatrix}$$

The binary linear code C is the set of all binary 6-
satisfy the matrix equation $\mathbf{H}\mathbf{x}^T = 0$. In this case, we
functions, to be described using appropriate factor
completely determined by checking *each* of the four e
necessary check for each link in the chain. The global

$$I_C(x_1, \ldots, x_6) = [(x_1, \ldots, x_6) \in C]$$
$$= [x_1 \oplus x_2 = 0][x_2 \oplus x_3 = 0][x_3 \oplus x$$

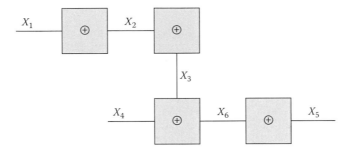

Figure 13.5: A factor graph representation of the linear code in Example 13.3.1.

The corresponding factor graph is shown in Figure 13.5. Notice that the factor graph is exactly the same as the one in Example 13.2.1, the only difference being the definition of the local functions, i.e., the local functions, herein, standing for the sum in binary *Galois Field GF*(2). A factor graph obtained in this way is often referred to as a *Tanner graph* [3].

13.3.2 Probabilistic Modeling

Factor graphs find a host of applications in representing systems that are inherently probabilistic in nature. As described earlier, factor graphs are used to express the factorization of a global multivariate probability distribution in such systems. This process was depicted earlier in Example 13.2.2. We look at another example where factor graphs are employed to describe a probabilistic model.

Example 13.3.2. Hidden Markov Model. A *Hidden Markov model* is a statistical model in which the system being modeled is a Markov process with an unobserved state. In such a model, only the output, depending on the current state, is visible, but the state is not directly visible, i.e., we only get to observe Y_i, the output of a memoryless channel with X_i as input. Hidden Markov models find a number of applications in bioinformatics and signal processing [29]. The factorization of the joint probability distribution function for a hidden Markov model is given by

$$p\left(x_0, x_1, x_2, \ldots, x_n, y_1, y_2, \ldots, y_n\right) = p\left(x_0\right) \prod_{i=1}^{n} p\left(x_i | x_{i-1}\right) p\left(y_i | x_i\right) . \qquad (13.8)$$

The resulting factor graph is shown in Figure 13.6.

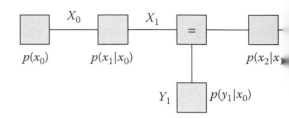

Figure 13.6: Factor graph representation of a

13.4 Relationship with Other P
ical Models

Factor graphs are widely used to represent complex
inherent simplicity. As shown in Example 13.2.2, o
bilistic model. In this scenario, factor graphs are use
probability distributions and bear close resemblance
based on undirected graphs (Markov random fields) a
acyclic graphs (Bayesian networks) [30].

13.4.1 Markov Random Fields

Consider an undirected graph $G = (V, E)$, where the
to a collection of random variables connected throu
graph G is a *Markov random field* if the distributic
local Markov property: $(\forall v \in V)\ p\,(v|V \setminus \{v\}) = p\,($
the set of neighbors of v. Hence, G is a Markov rand
values of its neighbors, every variable $v \in V$ is indep
Markov random fields are frequently used in statistics
   A set X is often called a *clique*, if every two nodes
edge. Such a clique is *maximal* if it is not containe
joint probability mass function of an MRF can be e:
collection of clique potential functions[3], defined on t

---

[3]Each clique in the graphical model representation is associ
clique potential is only a function of the *local variables* in the

in the MRF, i.e.,

$$p\left(v_1,\ldots,v_n\right) = \frac{1}{Z} \prod_{E \in C} \psi_E\left(V_E\right), \tag{13.9}$$

where Z is a normalization constant. It is clear that the joint probability distribution in (13.9) is suitable for representation by a factor graph where the local functions are the clique potential functions. The relationship between factor graphs and Markov random fields is to be highlighted further in Example 13.4.1.

13.4.2 Bayesian Networks

Bayesian networks are a class of directed acyclic graphs (DAGs)[4] that are closely related to factor graphs. The nodes in a DAG represent the random variables in a Bayesian sense: they may be observable quantities, latent variables, unknown parameters or hypotheses, while the edges highlight the conditional dependencies. Unconnected nodes in a Bayesian network represent the random variables that are conditionally independent of each other. Bayesian networks find extensive applications in modeling systems for gene regulatory networks, gene expression analysis, image processing and inference in engineering [33].

Let $\mathbf{a}\left(v\right)$ denote the set of *parents* of v, i.e., the set of vertices having an edge incident on v, then the distribution expressed by the Bayesian network is given by

$$p\left(v_1,\ldots,v_n\right) = \prod_{i=1}^{n} p\left(v_i | \mathbf{a}\left(v_i\right)\right). \tag{13.10}$$

This factorization of the multivariate probability distribution is a natural fit for representation by a factor graph with the local functions in this case being the conditional probability distributions. The following example shows the close resemblance between factor graphs, MRFs, and Bayesian networks.

Example 13.4.1. Consider the following factorization of a multivariate probability distribution function

$$p\left(x_1,x_2,x_3,x_4,x_5\right) = p\left(x_1\right)p\left(x_2|x_1\right)p\left(x_3|x_1\right)p\left(x_4|x_2\right)p\left(x_5|x_3\right).$$

The various graphical representations of this factorization in terms of FFGs, MRF, and Bayesian network are depicted in Figure 13.7.

13.5 Message Passing in Factor Graphs

In a variety of applications, one may be interested in determining the individual local functions in the factorization of a global function. For example, associated

---

[4]A *directed acyclic graph* is a graph formed by a collection of vertices connected by directed edges such that there is no way to start from a particular vertex, move around the edges and end at the same vertex after some hops.

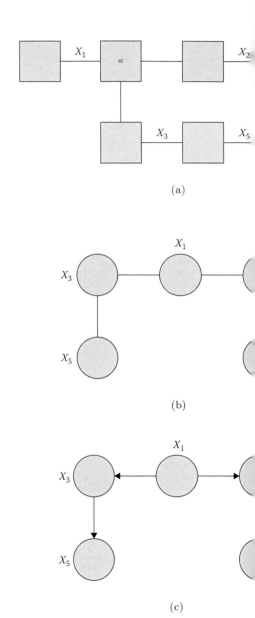

Figure 13.7: A probabilistic model using different gr
FFG. (b) Markov random field. (c) Bayesian network

with a function $f(x_1, \ldots, x_n)$ are n marginal functions $f_i(x_i)$. For each $a \in A_i$, the value of the marginal function $f_i(a)$ is obtained by summing the value of $f(x_1, \ldots, x_n)$ over all configurations of variables that have $x_i = a$. We call this process of finding the marginal associated with x_i as *summary for x_i* [5]. In cases where the global function is a probability distribution, summary for x_i represents the marginal probability distribution for each x_i. We define the summary for x_i using the following special notation

$$f_i(x_i) = \sum_{\backslash\{x_i\}} f(x_1, \ldots, x_n), \tag{13.11}$$

i.e., a summary of x_i is depicted by showing the variable not being summed over, for ease of representation.

Message passing algorithms constitute a class of algorithms where objects in a system can send or receive messages and subsequently, conclusions about reliability or synchronization can be made. In factor graphs, these algorithms correspond to efficient ways of calculating the marginal function associated with each variable. Two of the most popular message passing algorithms are *sum-product algorithm* and *max-product algorithm*, which differ in the type of operator used for summary propagation. These algorithms are described in detail below.

13.5.1 Sum-Product Algorithm

One of the most widely used message passing algorithm in factor graphs is the sum-product algorithm, which takes its name due to the fact that a summation operator is used as a summary propagation operator. We can determine either a single, or all marginal functions, associated with the variables using this algorithm [5].

Computing a Single Marginal Function

The computation of a single marginal function $f_i(x_i)$ begins at the leaves of the factor graph, i.e., variables connected through a half edge. Each leaf variable node sends a trivial *unit* message to its parent and each leaf factor node f_i sends the description of its local function to its parent. Both the variable and factor nodes wait for all messages to arrive from their children before delivering a final message to their parent. In case of a variable node, it simply sends a product of all messages received from its children, while a factor node f_i, with a parent x_i, forms the product of its local function f_i with the messages received from its children, and propagates this message to its parent after applying a summary operator $\sum_{\backslash\{x_i\}}$

on the result. After all messages have been passed
graph in a similar manner, the computation terminate
corresponding marginal function $f_i(x_i)$ is obtained a
messages on x_i.

It should be mentioned here that a message emana
out to x, is a single-argument function of x, since a
summed out using the summary operator. Similarly
variable node are functions of that variable, and so is a

Computing All Marginal Functions

Quite often, we are interested in computing either
marginal functions associated with a given global mu
a recursive application of the algorithm described a
marginal function associated with each variable, this
subcomputations which are similar for a number of
cient to determine $f_i(x_i)$ for each x_i separately. Si
marginal functions can be achieved by "overlaying"
single-i sum-product algorithm on a factor graph [5].

In this case, we do not prespecify a parent/child re
of s serves as the parent of s at some stage of the algo
m acts as the parent, a message from s to m is exactl
product algorithm i.e., once s receives messages from
as children (for the moment), it sends the resulting
parent m. After sending this message, s returns to
return message from m, which will now be regarded a
arrives, s is able to compute and send messages to eac
m), each being regarded, in turn, as parent [5]. Once
message has been passed on each edge, the algorithm
function $f_i(x_i)$ for a variable x_i is then the product
The sum-product algorithm can be summarized as

The message sent from a node s on an edge e is t
tion at s (or the unit function if s is a variable node)
s on edges other than e, summarized for the variable

Some notations are now specified to describe the wo
gorithm. A message from a variable x to a factor no
and a message from f to x is given by $m_{f \to x}(x)$. Let
the graph be designated as $n(x)$. The computations
are expressed as follows [5]:

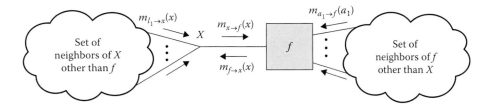

Figure 13.8: Update rules of the sum-product algorithm in a factor graph.

Variable to local function:

$$m_{x \to f}(x) = \prod_{l \in n(x) \setminus f} m_{l \to x}(x) \; . \tag{13.12}$$

Local function to variable:

$$m_{f \to x}(x) = \sum_{\setminus \{x\}} \left(f(X) \prod_{a \in n(f) \setminus \{x\}} m_{a \to f}(a) \right), \tag{13.13}$$

where $X = n(f)$ is the set of arguments of the local function f. These computations are depicted pictorially in Figure 13.8. It is clear that a message emanating from a variable is simple to compute since, a *summary* for x of a product of functions of x is the product itself. For an outgoing message from a factor node, function multiplications are involved, followed by a summation. We describe the working of the aforementioned algorithm in detail in the following example.

Example 13.5.1. Consider the following factorization of a global function of six variables:

$$f(x_1, x_2, x_3, x_4, x_5, x_6) = f_1(x_1, x_2) f_2(x_2, x_3) f_3(x_3, x_4, x_6) f_4(x_5, x_6) f_5(x_5) \; .$$

Notice that this factor graph is a slight modification of the one in Example 13.2.1, and the modification has been introduced to explain the working of the sum-product algorithm in greater detail. The flow of messages takes place in nine steps and is illustrated in Figure 13.9.

Step 1:

$$m_{x_1 \to f_1}(x_1) = 1$$
$$m_{x_4 \to f_3}(x_4) = 1$$
$$m_{f_5 \to x_5}(x_5) = \sum_{\setminus \{x_5\}} f_5(x_5) = f_5(x_5) \; .$$

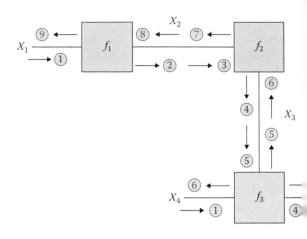

Figure 13.9: Sum-product algorithm messaging steps
13.5.1.

Step 2:

$$m_{f_1 \to x_2}(x_2) = \sum_{\backslash \{x_2\}} m_{x_1 \to f_1}(x_1) \, f$$

$$m_{x_5 \to f_4}(x_5) = m_{f_5 \to x_5}(x_5) \, .$$

Step 3:

$$m_{x_2 \to f_2}(x_2) = m_{f_1 \to x_2}(x_2)$$

$$m_{f_4 \to x_6}(x_6) = \sum_{\backslash \{x_6\}} m_{x_5 \to f_4}(x_5) \, f_4$$

Step 4:

$$m_{f_2 \to x_3}(x_3) = \sum_{\backslash \{x_3\}} m_{x_2 \to f_2}(x_2) \, f$$

$$m_{x_6 \to f_3}(x_6) = m_{f_4 \to x_6}(x_6) \, .$$

Step 5:

$$m_{x_3 \to f_3}(x_3) = m_{f_2 \to x_3}(x_3)$$

$$m_{f_3 \to x_3}(x_3) = \sum_{\setminus \{x_3\}} m_{x_6 \to f_3}(x_6)\, m_{x_4 \to f_3}(x_4)\, f_3(x_3, x_4, x_6)\ .$$

Step 6:

$$m_{x_3 \to f_2}(x_3) = m_{f_3 \to x_3}(x_3)$$

$$m_{f_3 \to x_4}(x_4) = \sum_{\setminus \{x_4\}} m_{x_3 \to f_3}(x_3)\, f_3(x_3, x_4, x_6)$$

$$m_{f_3 \to x_6}(x_6) = \sum_{\setminus \{x_6\}} m_{x_3 \to f_3}(x_3)\, f_3(x_3, x_4, x_6)\ .$$

Step 7:

$$m_{f_2 \to x_2}(x_2) = \sum_{\setminus \{x_2\}} m_{x_3 \to f_2}(x_3)\, f_2(x_2, x_3)$$

$$m_{x_6 \to f_4}(x_6) = m_{f_3 \to x_6}(x_6)\ .$$

Step 8:

$$m_{x_2 \to f_1}(x_2) = m_{f_2 \to x_2}(x_2)$$

$$m_{f_4 \to x_5}(x_5) = \sum_{\setminus \{x_5\}} m_{x_6 \to f_4}(x_6)\, f_4(x_5, x_6)\ .$$

Step 9:

$$m_{f_1 \to x_1}(x_1) = \sum_{\setminus \{x_1\}} m_{x_2 \to f_1}(x_2)\, f_1(x_1, x_2)$$

$$m_{x_5 \to f_5}(x_5) = m_{f_4 \to x_5}(x_5)\ .$$

Marginal functions computed at the termination of the algorithm:

$$f(x_1) = m_{f_1 \to x_1}(x_1)$$
$$f(x_2) = m_{f_1 \to x_2}(x_2)\, m_{f_2 \to x_2}(x_2)$$
$$f(x_3) = m_{f_2 \to x_3}(x_3)\, m_{f_3 \to x_3}(x_3)$$
$$f(x_4) = m_{f_3 \to x_4}(x_4)$$
$$f(x_5) = m_{f_4 \to x_5}(x_5)\, m_{f_5 \to x_5}(x_5)$$
$$f(x_6) = m_{f_3 \to x_6}(x_6)\, m_{f_4 \to x_6}(x_6)\ .$$

As mentioned previously, the algorithm terminates when a message has been sent in either direction on each edge. Marginal functions for all variables are obtained by taking a product of all incoming messages on a variable.

13.5.2   Max-Product Algorithm

The *max-product algorithm* derives its name from the
tion, as in the sum-product algorithm, the operator
operation. The working of the max-product algorithm
algorithm described earlier. In much the same way,
tween variables and factor nodes and vice versa. Reca
computations involved in a max-product algorithm a

Variable to local function:

$$m_{x \to f}(x) = \prod_{h \in n(x) \backslash f} m_{h \to x} ($$

Local function to variable:

$$m_{f \to x}(x) = \max_{\backslash \{x\}} \left(f(X) \prod_{y \in n(f) \backslash \{x\}} r \right.$$

where $X = n(f)$ is the set of arguments of the local
algorithm differs from the sum-product algorithm by
operator in (13.15). Once a message has been passe
edge, the algorithm terminates. Marginal functions
are obtained as a product of all incoming messages o

$$f(x_i) = \prod_{h \in n(x)} m_{h \to x_i}(x_i)$$

13.6   Factor Graphs with Cycles

It is clear that the summary propagation algorithms d
only if the factor graph does not contain any cycles, i
Hence, for a factor graph with cycles, it no longer re
the leaves since, messages sent through the update ru
result in an endless propagation of messages around
and max-product algorithm can be viewed as an *itero*
that is terminated after a predefined time period ha
has been met. After the algorithm terminates, the m
are, at best, only an approximation of the true margi

13.6.1   Message Passing Schedules

A message passing *schedule* in a factor graph amount
sages are passed at each time step. So far, we have on

schedules that specify messages to be initiated at the leaves of the factor graph. This is frequently described as the *two-way schedule*. Clearly, several other message passing schedules are possible depending on whether messages are passed serially, all at once, etc., in a factor graph. Two important message passing schedules are:

Serial Schedule In a *serial schedule*, only one message is passed along any edge in the factor graph at each time step. The two-way schedule is ideally suited to a serial implementation, as it involves the smallest number of messages needed; namely, $2E$, where E is the number of edges in the spanning tree [30].

Flooding Schedule The *flooding schedule* involves sending a message along all edges in each direction at a time step. In some cases, it may lead to a faster convergence when a factor graph has cycles.

The interested reader is referred to [17] for a detailed description of different message passing schedules and how they may lead to efficient message passing algorithms in factor graphs with cycles.

13.6.2   Iterative Message Passing

The summary propagation algorithms operating on a factor graph with cycles, in their natural form, proceed endlessly without any natural termination. This resembles an *iterative* process with messages propagated multiple times. Factor graphs with cycles are not uncommon. In fact, two important applications of message passing in factor graphs with cycles are in decoding of turbo codes and LDPC codes. In these instances, the *thickness*[5] of some edges renders the exact marginalization computationally infeasible. It has been shown through extensive simulation in [15] and [16] that if we proceed with the sum-product algorithm as if the factor graph contained no cycles, these decoding algorithms can remarkably achieve a performance quite close to the Shannon limit.

13.7   Some General Remarks on Factor Graphs

This section highlights some general comments on factor graphs that may find applications in diverse areas. The information in this section is based largely on the contributions reported by Loeliger [34].

13.7.1   Mappers

It is a common practice in communication systems to map arbitrary data to discrete constellations useful for transmitting data over a channel. Consider two binary

---

[5]Thickness of an edge is usually associated with the number of variables to be carried along that edge.

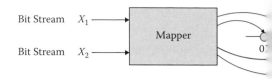

Figure 13.10: Bits-to-symbol m

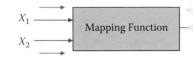

Figure 13.11: Messages through th

symbols, X_1 and X_2, that are to be mapped to a 4-A.
ure 13.10. This mapping is expressed by the function
and x_1 is mapped to the most significant bit. Such
FFG can be represented by the local function

$$\phi_f(x_1, x_2, y) \triangleq \begin{cases} 1, & \text{if } f(x_1, x_2) \\ 0, & \text{otherw} \end{cases}$$

where y is the output of the mapper. Recall from
emanating out of a variable is the product of all mes
message out of a factor node is the product of all inco
function of the factor node and summarized over the
propagated. The various messages in the factor grap
For example, we can write

$$m_Y(y) = \sum_{x_1, x_2} \phi_f(x_1, x_2, y) \, m_{X_1}(x_1$$

13.7.2 Hybrid Equality Constraint

A *hybrid* equality constraint arises between a variable
set $\mathcal{X}$ and a variable Y defined on the set of real nur
factor graph is shown in Figure 13.12(a) where the
$\delta(x - y)$. This is to be interpreted as a Kronecker delt
Using the previously described message update rules

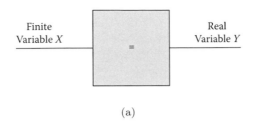

Finite
Variable X

Real
Variable Y

(a)

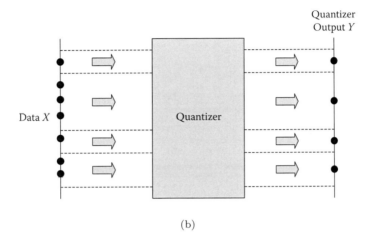

Quantizer
Output Y

Data X

Quantizer

(b)

Figure 13.12: (a) Hybrid equality node. (b) Quantizer.

the message out of the X-edge is

$$m_X(x) = \int_y \delta(x - y) m_Y(y) \, dy \tag{13.19}$$
$$= m_Y(x) ,$$

which can be obtained by sampling the incoming density for different values of X in $\mathcal{X}$. Similarly, the message out of the Y-edge is a sum of weighted Dirac deltas given by

$$m_Y(y) = \sum_{x \in \mathcal{X}} \delta(y - x) m_X(x) . \tag{13.20}$$

13.7.3 Quantizers

It happens quite often that we need to quantize the incoming data into a finite subset of $\mathbb{R}$ or a set of (finite) intervals in $\mathbb{R}$. Such an operation is depicted in

Figure 13.12(b). The messages through a quantizer (

$$m_Y(y) = \sum_{x:q(x)=y} m_X(x$$

$$m_X(x) = m_Y(q(x)),$$

where $q(x)$ is the quantizer function.

13.7.4 Continuous Variables

The variables in a factor graph, considered thus far
values in a finite set $\mathcal{X}$. There are certain adjustme
the case of continuous variables. For example, in the
summation operator in (13.13) has to be replaced v
However, proceeding as in the sum-product algorit
may lead to intractable integrals. Some suitable mes
to circumvent this problem. A *constant* message ca
estimate to represent the continuous variable. We ca
using a quantizer such as the one in Section 13.7.3.
choices based on the type of problem at hand [10], [3

13.8 Some Important Algorithr
Message Passing in Factor

The notion of message passing in factor graphs has c
of a large number of algorithms in coding theory, est;
intelligence. Several popular algorithms in these field;
of message passing on the appropriate factor graph
a universal framework to explain the working of a la
rithms, the message passing approach on factor grap
extensions of these algorithms. Similarly, new algor
may also be derived by using summary propagation i;
ples of algorithms viewed as instances of message pas;
the forward/backward algorithm, the Viterbi algorith
algorithm and iterative turbo decoding. Pearl's bel
networks [17], steepest descent [13], particle filtering [
be obtained by message passing in factor graphs. We
coding theory (the forward/backward algorithm and
estimation theory (the Kalman filter and the EM alg
algorithms may be viewed as specific instances of mess
The material in this section is based largely on the cc

13.8.1 Forward/Backward Algorithm

An important algorithm in coding theory is the *forward/backward algorithm*, sometimes called the BCJR algorithm after its inventors Bahl, Cocke, Jelinek and Raviv [8]. It computes the *a posteriori* probabilities of all the *hidden* state variables using a given set of observations that represent a hidden Markov model. The algorithm derives its name from the fact that for each state, it computes a *forward probability*, the probability of landing in a particular state based on the observations up to a certain time, and a *backward probability*, the probability of observing the remaining observations based on the current state.

The computation of these probabilities in the forward/backward algorithm can be viewed as an instance of message passing on the corresponding factor graph representing the underlying hidden Markov model. Such a factor graph is shown in Figure 13.13 where the variables u_i, x_i and s_i represent the input, output, and state in the Markov model at state i. In the hidden Markov model, we do not get to observe the output directly. Instead, we make inference based on the available observation y. The global multivariate function in this case is the global trellis function that factors into a product of *local* trellis functions. Each of these local trellis functions, represented by T_i in Figure 13.13, is like an indicator function that checks for valid configurations of s_{i-1}, s_i, u_i and x_i. Recalling Example 13.3.2, the joint probability density function of u, s and x given y in this case is expressed as [5]

$$h\left(u, s, x \mid y\right) = \prod_{i=1}^{n} T_i\left(s_{i-1}, u_i, x_i, s_i\right) \prod_{i=1}^{n} f\left(y_i | x_i\right). \tag{13.23}$$

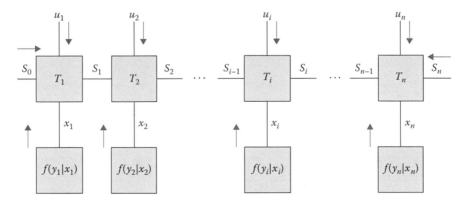

Figure 13.13: The factor graph corresponding to the hidden Markov model in (13.23). The arrows show the initial messages passed by the leaves of the factor graph.

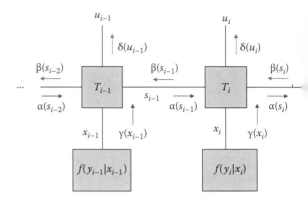

Figure 13.14: Message updates of the sum-product al
recursions.

The marginal a posteriori probabilities $p(u_i|y)$ can
the density function in (13.23) over all u_k, $k \neq i$

$$p(u_i|y) = \sum_{\backslash \{u_i\}} h(u, s, x | y$$

This a posteriori probability can be obtained wit
sum-product algorithm on the factor graph of Figur
with the input variables u_i (leaves of the factor grapl
sage to the local check functions T_i. Similarly, the
to the output variables x_i. As explained in Section
forwards the product of all messages incident on it, wl
the product of the incoming messages and the local fur
variable node. The details of the computations involv
by considering a fragment of the hidden Markov mode
notation from [8], we denote:

- The message $m_{x_i \rightarrow T_i}(x_i)$ by $\gamma(x_i)$.

- The message $m_{T_i \rightarrow u_i}(u_i)$ by $\delta(u_i)$.

- The message $m_{s_i \rightarrow T_{i+1}}(s_i)$ by $\alpha(s_i)$. $\alpha(s_i)$ is t
  and should be interpreted as the probability of l
  of observations $y_1, \ldots, y_i$.

- The message $m_{s_i \rightarrow T_i}(s_i)$ by $\beta(s_i)$. $\beta(s_i)$ is the l
  ing the probability of s_i given the future observat

As before, the marginal function associated with each node is the product of all messages incident on it from its neighboring nodes. With the above notation, the two recursions, forward and backward, that make up the process of the forward/backward algorithm can be expressed as [5]:

Forward Recursion

$$\alpha\left(s_{i}\right)=\sum_{\backslash\{s_{i}\}} T_{i}\left(s_{i-1}, x_{i}, u_{i}, s_{i}\right) \gamma\left(s_{i}\right) \alpha\left(s_{i-1}\right) . \tag{13.25}$$

The recursion in (13.25) stems from the usual operation of the sum-product algorithm expressed in (13.13). The probability of landing in state s_i, $\alpha\left(s_i\right)$, is given by the product of the incoming messages and the local function T_i and summarized over the variable s_i.

Backward Recursion

$$\beta\left(s_{i-1}\right)=\sum_{\backslash\{s_{i-1}\}} T_{i}\left(s_{i-1}, x_{i}, u_{i}, s_{i}\right) \gamma\left(s_{i}\right) \beta\left(s_{i}\right) . \tag{13.26}$$

Using (13.25) and (13.26), the marginal probabilities, $\alpha\left(s_i\right)$ and $\beta\left(s_i\right)$, can be recursively computed. The algorithm terminates once a message has been passed in either direction on each edge of the factor graph. After termination, the a posteriori probabilities, $p\left(u_i|y\right)$, can be determined as

$$\delta\left(u_{i}\right)=\sum_{\backslash\{u_{i}\}} T_{i}\left(s_{i-1}, x_{i}, u_{i}, s_{i}\right) \gamma\left(s_{i}\right) \alpha\left(s_{i-1}\right) \beta\left(s_{i}\right) . \tag{13.27}$$

These computations show that the forward/backward algorithm is an instance of message passing in factor graph using the sum-product algorithm. The basic operations in the working of the forward and backward recursions can be delineated as *sums of products*. The a posteriori probabilities of the variables x_i and s_i can be obtained by the sum-product algorithm in a similar manner.

13.8.2 The Viterbi Algorithm

The Viterbi algorithm, proposed in [9], is a recursive algorithm to determine the most likely state sequence that resulted in the given set of observations. In the context of the hidden Markov model of Figure 13.13, the Viterbi algorithm finds the configuration that has the largest a posteriori probability. For equally likely codes, this problem is recast as *maximum-likelihood sequence detection* (MLSD) [5].

Before establishing the computations of the Viterbi algorithm as being an instance of message passing on factor graphs, we describe two important concepts.

Max-Product and Min-Sum Semirings

A *semiring* is a set S equipped with two binary op
satisfies the following conditions.

1. Addition is commutative: $\forall x, y \in S, \ x + y = y$

2. Addition is associative: $\forall x, y, z \in S, \ x + (y + z$

3. Multiplication is associative: $\forall x, y, z \in S, \ x * ($

4. Distributive property: $\forall x, y, z \in S, \ x * (y + z)$

Up to this point, we have considered the set of real
of any global multivariate function whose factorizati
graph. The distributive law with the usual operati
defined as

$$\forall a, b, c \in \mathbb{R} \qquad a \cdot (b + c) = (a \cdot b)$$

In a *max-product semiring*, the notion of summatic
operation and "." is distributed over max for non-ne
as

$$a \left(\max (b, c) \right) = \max \left(ab, ac \right.$$

Hence, in a max-product semiring, the max operator
ator. The complete summary of a multivariate funct
[5]

$$\max f (x_1, \ldots, x_n) = \max_{x_1} \left(\max_{x_2} \left(\ldots \left(\max_{x_n} f (\right. \right. \right.$$

$$= \sum_{\backslash \{\}} f (x_1, \ldots, x_n) \ .$$

Another semiring that finds many applications in dec
ing. Compared to distributive law of the max-prod
corresponding distributive law for the case of min-su
as

$$a + \min (b, c) = \min (a + b, a$$

where the "max" and "product" operators are replac
spectively.

The Viterbi algorithm aims to determine the seque
have produced the current set of observations. It calc
branch metrics, to specify the costs associated with
resenting the hidden Markov model. In Gaussian c
amounts to computing the squared Euclidean distance
put states in a trellis. The sequence of states throu

lowest accumulation of branch metrics is declared as the *survivor sequence*, i.e., the sequence of states most likely to have resulted in the observations at hand. Specific details of the algorithm can be found in [35]. The computations of the branch metrics needed in the Viterbi algorithm can be done by message passing in an appropriate factor graph by employing the sum-product algorithm in the min-sum semiring where products become sums. The Viterbi algorithm in its most general sense operates only in the forward direction in the factor graph of a hidden Markov model, like the one in Figure 13.13. Using the min-sum algorithm, the forward recursion of (13.25) can be written as

$$\alpha\left(s_i\right) = \min_{\backslash\{s_i\}} \left(T_i\left(s_{i-1}, x_i, u_i, s_i\right) + \gamma\left(s_i\right) + \alpha\left(s_{i-1}\right)\right). \tag{13.33}$$

The calculation of these branch metrics in this case is akin to taking "minimum of sums." These recursions can be used to determine the most likely sequence of states that produced the given set of observations. Hence, the conventional Viterbi algorithm, operating in the forward direction only, may also be recovered by a message passing approach on the factor graph of a hidden Markov model.

13.8.3 Kalman Filter

The Kalman filter, proposed by Rudolf E. Kalman in [36], is a recursive method to estimate the true measurements (states) of a linear system using a set of noisy observations. The algorithm gradually averages the estimate by assigning a weight to every new coming observation based on its level of uncertainty. The Kalman filter has found widespread uses in control theory, signal processing, telecommunications, econometrics, weather forecasting and navigation systems. For a detailed look at the applications of Kalman filtering in various fields of interest, see [37]. The computations in a Kalman filter as a representation of message passing on a factor graph are described below [5].

The Kalman filter considers the discrete-time linear system:

$$x\left(i+1\right) = Ax\left(i\right) + Bu\left(i\right) \tag{13.34}$$
$$y\left(i\right) = Cx\left(i\right) + Dw\left(i\right),$$

where $x\left(i\right)$ is the state of the system at time step i. We limit ourselves to the case where A, B, C and D are time-invariant scalars, although generalization to the vector and time-variant case is straightforward. Output of the system at time i is denoted by $y\left(i\right)$ and we assume that the input $u\left(i\right)$ and noise $w\left(i\right)$ are independently distributed as standard Gaussian random variables. The inherent structure of the system is similar to the hidden Markov model described in Example 13.3.2. Therefore, the conditional joint density function of the states $x\left(1\right), \ldots, x\left(k\right)$ is given by

$$f\left(x\left(1\right), \ldots, x\left(k\right) \middle| y\left(1\right), \ldots, y\left(k\right)\right) = \prod_{i=1}^{k} f\left(x\left(i\right) \middle| x\left(i-1\right)\right) f\left(y\left(i\right) \middle| x\left(i\right)\right), \tag{13.35}$$

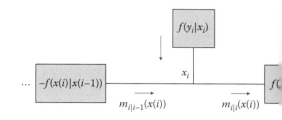

Figure 13.15: Message updates in the factc

where the system is initialized with an initial state
for this joint density was shown in Figure 13.6. A f
is depicted in Figure 13.15 along with the messages 1
propagation algorithm. We know that a linear combin
dom variables produces another Gaussian random va
$x(i)$ and $y(i)$ are jointly Gaussian random variables.
conditional densities $f(x(i)|x(i-1))$ and $f(y(i)|x$
as $\mathcal{N}(Ax(i-1), B^2)$ and $\mathcal{N}(Cx(i), D^2)$, respectiv
we know that the MMSE estimator of current state x
$y(1), \ldots, y(k)$ is given by the conditional expectatio

$$\hat{x}_{k|k} = E\{x(k)|y(1), \ldots, y($$

where $\hat{x}_{k|k}$ is to be interpreted as the estimate of t
the observations up to time k. In order to determi
need the marginal conditional density $f(x(k)|y(1),$
by marginalizing the joint density function $f(x(1), .$

$$f(x(k)|y(1), \ldots, y(k)) = \int_{\backslash x(k)} f(x(1), \ldots x(k)|$$

It is precisely this density function that can be obtai
sage passing schedule on a factor graph representing
usual notation, we denote the message passed by .
$m_{i|i-1}(x(i))$. Clearly, such a message is a scalar m
where $\hat{x}_{i|i-1}$ is the MMSE estimate of $x(i)$ given obse
Gaussian densities $\mathcal{N}(\mu_1, \sigma_1^2)$ and $\mathcal{N}(\mu_2, \sigma_2^2)$, their p

$$\mathcal{N}(\mu_1, \sigma_1^2)\,\mathcal{N}(\mu_2, \sigma_2^2) \propto \mathcal{N}(\mu_3$$

where

$$\mu_3 = \frac{\mu_1\sigma_2^2 + \mu_2\sigma_1^2}{\sigma_1^2 + \sigma_2^2},$$

and

$$\sigma_3^2 = \frac{\sigma_1^2 \sigma_2^2}{\sigma_1^2 + \sigma_2^2} .$$

Using this property, the message emanating from the variable $x\,(i)$ to the factor node $f\,(x\,(i+1)\,|x\,(i))$ can be computed as the product of messages received on $x\,(i)$ [5]

$$m_{i|i}\,(x\,(i)) = m_{i|i-1}\,(x\,(i))\,f\,(y\,(i)\,|x\,(i)) \tag{13.39}$$

$$= \mathcal{N}_{x(i)}\left(\hat{x}_{i|i-1}, \sigma_{i|i-1}^2\right)\mathcal{N}_{y(i)}\left(Cx\,(i), D^2\right).$$

By a linear transformation on the Gaussian density, we can equivalently write

$$m_{i|i}\,(x\,(i)) \propto \mathcal{N}_{x(i)}\left(\hat{x}_{i|i-1}, \sigma_{i|i-1}^2\right)\mathcal{N}_{x(i)}\left(y\,(i)\,/C, D^2/C^2\right) \tag{13.40}$$

$$\propto \mathcal{N}_{x(i)}\left(\hat{x}_{i|i}, \sigma_{i|i}^2\right),$$

where

$$\hat{x}_{i|i} = \frac{D^2\hat{x}_{i|i-1} + Cy\,(i)\,\sigma_{i|i-1}^2}{C^2\sigma_{i|i-1}^2 + D^2} \tag{13.41}$$

$$= \hat{x}_{i|i-1} + \frac{C\sigma_{i|i-1}^2}{C^2\sigma_{i|i-1}^2 + D^2}\left(y\,(i) - C\hat{x}_{i|i-1}\right), \tag{13.42}$$

and

$$\sigma_{i|i}^2 = \frac{D^2\sigma_{i|i-1}^2}{C^2\sigma_{i|i-1}^2 + D^2} . \tag{13.43}$$

We see from (13.42) that the MMSE estimate of the state at time i is determined by the previous estimate using observations up to time $i-1$, the observation at time i and scale factors using (13.43). In order to compute the message passed from the factor node $f\,(x\,(i+1)\,|x\,(i))$ to the variable $x\,(i+1)$, we make use of another property of the Gaussian distribution:

$$\int_{-\infty}^{+\infty}\mathcal{N}_x\left(\mu_1, \sigma_1^2\right)\mathcal{N}_y\left(\alpha x, \sigma_2^2\right)dx \propto \mathcal{N}_y\left(\alpha\mu_1, \alpha^2\sigma_1^2 + \sigma_2^2\right). \tag{13.44}$$

The message emanating from the factor node $f\,(x\,(i+1)\,|x\,(i))$, denoted by $m_{i+1|i}\,(x\,(i+1))$, is given by

$$m_{i+1|i}\,(x\,(i+1)) = \int m_{i|i}\,(x\,(i))\,\mathcal{N}_{x(i+1)}\left(Ax\,(i), B^2\right)dx\,(i). \tag{13.45}$$

Using (13.41) and (13.44), (13.45) can be stated as

$$m_{i+1|i}\,(x\,(i+1)) \propto \mathcal{N}_{x(i+1)}\left(\hat{x}_{i+1|i}, \sigma_{i+1|i}^2\right), \tag{13.46}$$

where

$$\hat{x}_{i+1|i} = A\hat{x}_{i|i}$$
$$= A\hat{x}_{i|i-1} + K_i\left(y\left(i\right) - C\right.$$

and

$$\sigma^2_{i+1|i} = A^2\sigma^2_{i|i} + B^2$$
$$= \frac{A^2 D^2 \sigma^2_{i|i-1}}{C^2\sigma^2_{i|i-1} + D^2} + I$$

The Kalman filter gain K_i is given by

$$K_i = \frac{AC\sigma^2_{i|i-1}}{C^2\sigma^2_{i|i-1} + D^2}.$$

Equation (13.48) can be interpreted as the predictic based on the observation at time i and the predicte i. These are exactly the updates involved in a stan computations in the above-mentioned steps can be product algorithm on the corresponding factor graph replacing the summation as the summary operator The marginal density function $f\left(x\left(i\right)|y\left(1\right), \ldots, y\left(i\right.\right.$ estimator of (13.36) is given by the product of messag

$$f\left(x\left(i\right)|y\left(1\right), \ldots, y\left(i\right)\right) = m_{i|i-1}\left(x\left(i\right)\right.$$
$$\propto \mathcal{N}_{x(i)}\left(\hat{x}_{i|i}, \sigma\right.$$

Hence, the recursive updates in a Kalman filter can be in factor graphs. For additional results on the update factor graph of Figure 13.15 is replaced by an equa resulting computational rules, see problems at the en

13.8.4   Expectation Maximization (EM

The Expectation Maximization (EM) algorithm, orig iterative method for finding the maximum likelihood of a given distribution from incomplete data. The E either the data is indeed incomplete due to limitatior nism, or when optimization of an otherwise intractab simplified by assuming additional hidden parameters tion. Due to its analytical tractability, the EM algorit in various fields such as signal processing, data clust tern recognition [39–41]. EM algorithm has also four estimation in graphical models where it is often used to break cycles in the factor graph [42]. We describe here how the EM algorithm may be viewed as an instance of message passing in an appropriate factor graph.

Basics of the EM algorithm

Consider a set of observations $\mathbf{x}$, whose density function is parameterized by a certain parameter θ. In a maximum likelihood (ML) estimation scenario, we determine the ML estimate of θ by computing

$$\theta_{ML} = \arg\max_{\theta} f(\mathbf{x}; \theta), \tag{13.54}$$

i.e., the value of θ for which the observations $\mathbf{x}$ are most likely to appear. In many cases, this maximization is not analytically tractable. The EM algorithm strives to ease the computational complexity by augmenting the data $\mathbf{x}$ with variables $\mathbf{h}$, which we call *hidden*. The complete data set is, therefore, given by $(\mathbf{x}, \mathbf{h})$. The two steps in the EM algorithm can be expressed as:

1. Expectation Step

$$Q(\theta; \theta^*) \triangleq E\{\log f(\mathbf{x}, \mathbf{h}; \theta) \,|\, \mathbf{x}, \theta = \theta^*\} \tag{13.55}$$

$$= \int_{\mathbf{h}} \log f(\mathbf{x}, \mathbf{h}; \theta) \, f(\mathbf{h}|\mathbf{x}; \theta = \theta^*) \, d\mathbf{h}. \tag{13.56}$$

where θ^* is the current estimate of θ.

2. Maximization Step
   A new estimate $\theta^{*'}$ is obtained by computing

$$\theta^{*'} = \arg\max_{\theta} Q(\theta; \theta^*). \tag{13.57}$$

The two steps are iterated by assigning $\theta^* = \theta^{*'}$ and the iterations are continued till a satisfactory estimate is obtained or a prespecified time has elapsed. After each iteration, we are guaranteed to move towards a local maximum [38].

Both the expectation and maximization steps in (13.56) and (13.57) can be efficiently computed via appropriate message passing on a factor graph [11]. We describe this process below.

Expectation Step on a Factor Graph

Suppose the global multivariate probability density function factorizes as

$$f(\mathbf{x}, \mathbf{h}, \theta) = \prod_{i \in \mathcal{I}} f_i(\mathbf{x}_i, \mathbf{h}_i, \theta_i), \tag{13.58}$$

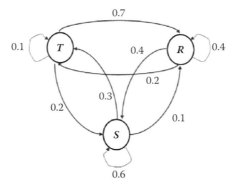

Figure 7.3: Illustration of the transitions between various states of a discrete Markov chain.

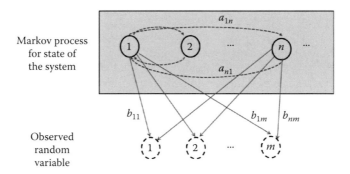

Figure 7.4: Structure of a general Hidden Markov Model.

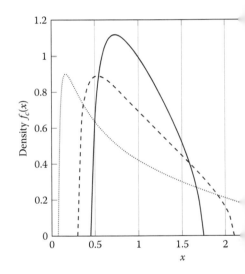

Figure 8.1: Marčenko-Pastur law for different li

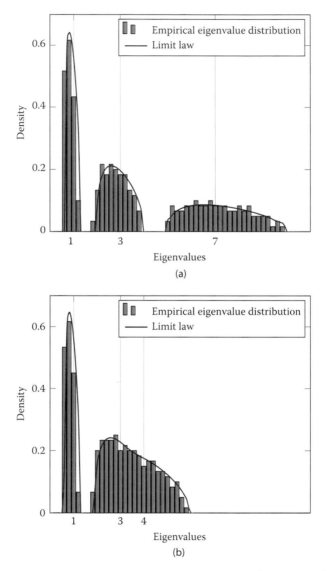

Figure 8.2: Histogram of the eigenvalues of $\mathbf{B}_N = \mathbf{T}_N^{\frac{1}{2}} \mathbf{X}_N \mathbf{X}_N^{\dagger} \mathbf{T}_N^{\frac{1}{2}}$, $N = 300$, $n = 3000$, with $\mathbf{T}_N$ diagonal composed of three evenly weighted masses in (a) 1, 3, and 7, (b) 1, 3, and 4.

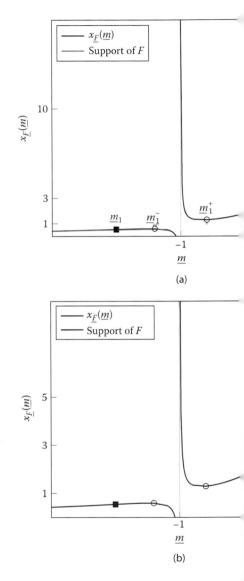

Figure 8.3: $x_{\underline{F}}(\underline{m})$ for $\mathbf{T}_N$ diagonal composed of thre 1, 3, and 10 (a) and 1, 3, and 5 (b), $c = 1/10$ in bo marked in circles, inflexion points are marked in squar read on the right vertical axes.

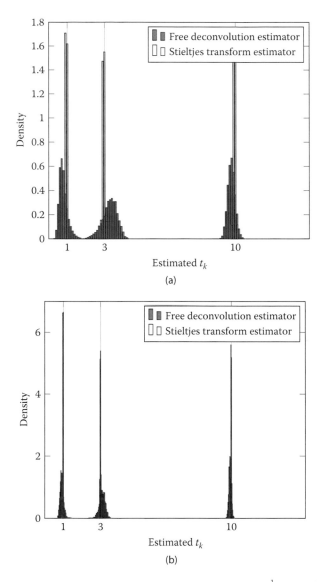

Figure 8.4: Estimation of t_1, t_2, t_3 in the model $\mathbf{B}_N = \mathbf{T}_N^{\frac{1}{2}} \mathbf{X}_N \mathbf{X}_N^{\dagger} \mathbf{T}_N^{\frac{1}{2}}$ based on the first three empirical moments of $\mathbf{B}_N$ and Newton-Girard inversion, see [32], for $N_1/N = N_2/N = N_3/N = 1/3$, $N/n = 1/10$, for 100,000 simulation runs; (a) $N = 30$, $n = 90$; (b) $N = 90$, $n = 270$. Comparison is made against the Stieltjes transform estimator of Theorem 8.4.1.

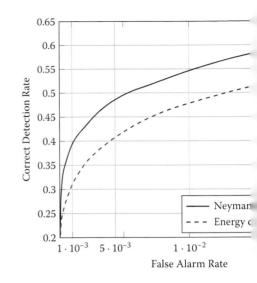

Figure 8.5: ROC curve for single-source detection, SNR = −3 dB, FAR range of practical interest.

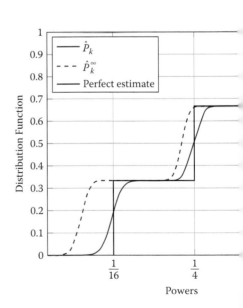

Figure 8.6: Distribution function of the estimators $\hat{P}$
$P_1 = 1/16$, $P_2 = 1/4$, $P_3 = 1$, $n_1 = n_2 = n_3 = 4$
sensors, $M = 128$ samples, and SNR = 20 dB. Opti
dashed lines.

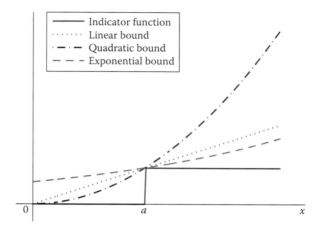

Figure 9.1: Bounds of the indicator function $\mathbf{1}_{[a,\infty)}(x)$.

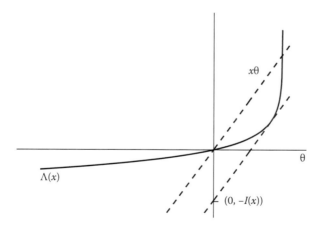

Figure 9.2: Geometric relation between the CGF and rate function.

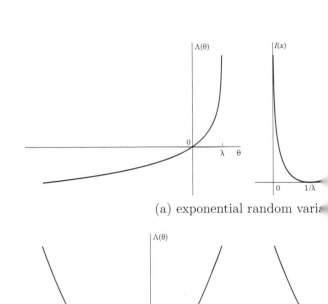

(a) exponential random varia

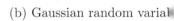

(b) Gaussian random varial

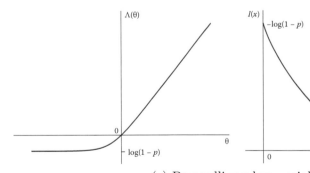

(c) Bernoulli random variab

Figure 9.3: Cumulant generating function (CGF) Λ

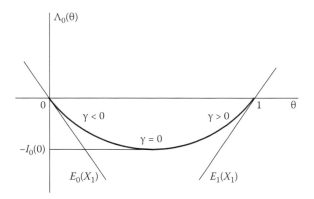

Figure 9.4: Geometric illustration of the CGF $\Lambda_0(\theta)$, threshold γ, and Chernoff's information $I_0(0)$ (adapted from [1, p.93]).

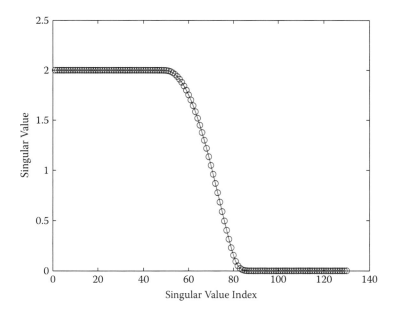

Figure 10.5: An example of the singular values distribution of the ill-conditioned matrix $\mathbf{A}_\epsilon^H$.

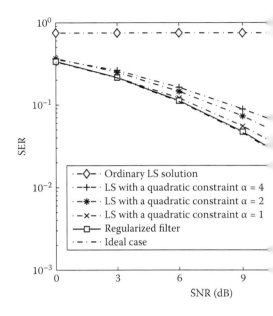

Figure 10.7: Symbol Error Rate (SER) performance
designed by regularized LS against the performance
solution and the LS with a quadratic constraint wi
modulation.

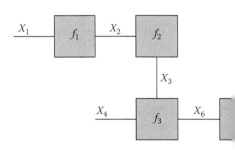

Figure 13.1: Factor graph for Exam

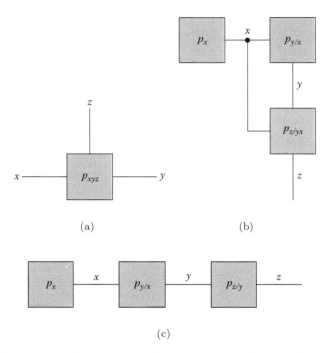

(a)

(b)

(c)

Figure 13.2: Factor graphs for a multivariate probability distribution: (a) Joint probability distribution. (b) Conditional probability distribution. (c) Markov chain relationship.

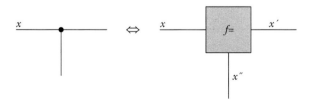

Figure 13.3: Equality constraint node.

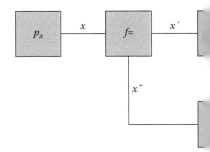

Figure 13.4: Factor graph after addition of equ

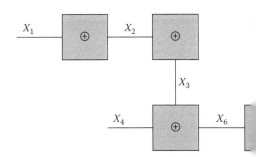

Figure 13.5: A factor graph representation of the lin

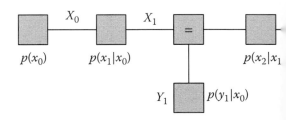

Figure 13.6: Factor graph representation of a H

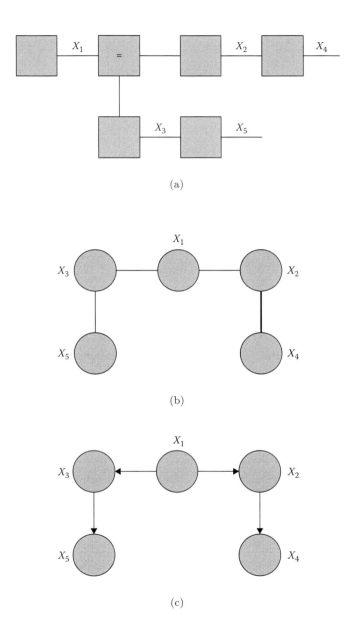

Figure 13.7: A probabilistic model using different graphical representations: (a) FFG. (b) Markov random field. (c) Bayesian network.

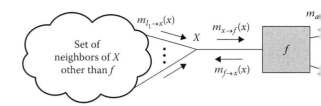

Figure 13.8: Update rules of the sum-product alg

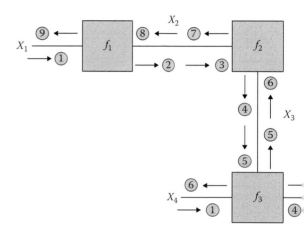

Figure 13.9: Sum-product algorithm messaging steps
13.5.1.

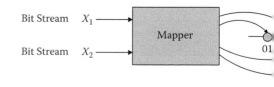

Figure 13.10: Bits-to-symbol m

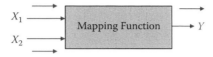

Figure 13.11: Messages through the mapper.

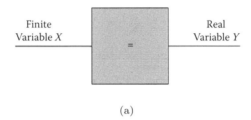

(a)

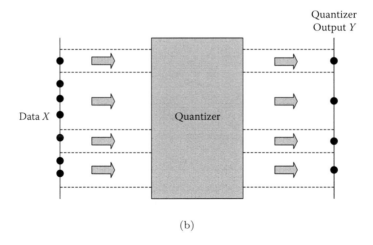

(b)

Figure 13.12: (a) Hybrid equality node. (b) Quantizer.

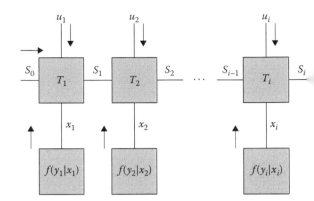

Figure 13.13: The factor graph corresponding to th[...]
(13.23). The arrows show the initial messages passe[...]
graph.

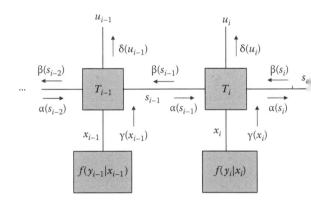

Figure 13.14: Message updates of the sum-product alg[...]
recursions.

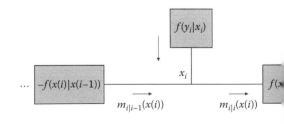

Figure 13.15: Message updates in the facto[...]

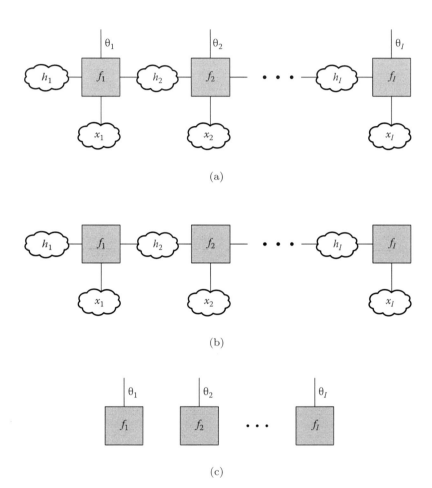

Figure 13.16: (a) Factor graph for (13.58). (b) Expectation step factor graph. (c) Maximization step factor graph.

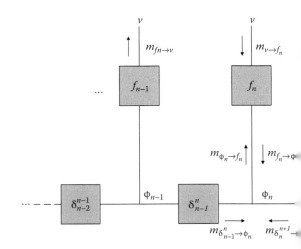

Figure 13.17: Message updates in factor graph for j
and phase noise estimation.

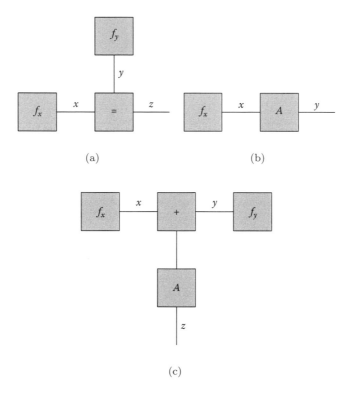

(a) (b)

(c)

Figure 13.18: Some common nodes: (a) Equality constraint node. (b) Multiplier node. (c) Additive and multiplier node.

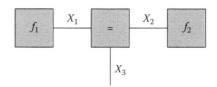

Figure 13.19: Factor graph for pointwise multiplication.

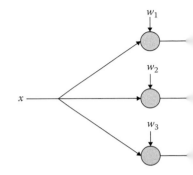

Figure 13.20: A noisy communicati

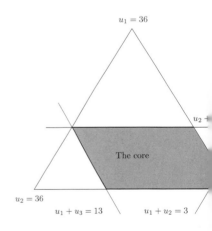

Figure 19.12: Illustration of the core of a three-player

where $\mathcal{I}$ is set of factor nodes (local functions). Equa
as

$$Q\left(\theta;\theta^*\right) = \sum_{i\in\mathcal{I}} E\left\{\log f_i\left(\mathbf{x}_i,\mathbf{h}_i,\theta_i\right)\right| $$

The ith term in the above summation can be express

$$E\left\{\log f_i\left(\mathbf{x}_i,\mathbf{h}_i,\theta_i\right)|\mathbf{x},\theta=\theta^*\right\} = \int_{\mathbf{h}_i} f_i\left(\mathbf{h}_i|\mathbf{x};\theta=\right.$$

The message passing approach on a factor graph allow
function $f_i\left(\mathbf{h}_i|\mathbf{x};\theta=\theta^*\right)$ using the sum-product alge
factor graph of $f\left(\mathbf{x},\mathbf{h},\theta=\theta^*\right)$ can be obtained from
by removing the edges corresponding to θ. The marg
can then be computed by an appropriate message
tor graph. The process of forming the subgraph f (
Figure 13.16(b). The expectation step on the factor

1. Form the subgraph for $f\left(\mathbf{x},\mathbf{h},\theta=\theta^*\right)$ from the

2. Using the sum-product algorithm together with
   ing schedule, determine $f_i\left(\mathbf{h}_i|\mathbf{x};\theta=\theta^*\right)$ for eac

3. Use this marginal density to obtain the expecta

The marginal density in Step 2 is obtained by taking
messages on each $\mathbf{h}_i$ independently.

Maximization Step on a Factor Graph

Notice that (13.59) can be written as:

$$Q\left(\theta;\theta^*\right) = \sum_{i=1}^{|\mathcal{I}|} Q_i\left(\theta_i;\theta_i^*\right)$$

where $Q_i\left(\theta_i;\theta_i^*\right) \triangleq E\left\{\log f_i\left(\mathbf{x}_i,\mathbf{h}_i,\theta_i\right)|\mathbf{x},\theta=\theta^*\right\}$.
of $\exp\left(.\right)$ and invariance of the ML estimate, we
$\exp(Q\left(\theta;\theta^*\right))$ w.r.t θ. Hence, (13.61) can be restated

$$\exp(Q\left(\theta;\theta^*\right)) = \prod_{i\in\mathcal{I}} \exp(Q_i\left(\theta_i\right.$$

The factor graph of $\exp(Q\left(\theta;\theta^*\right))$ in the maximization
the factors containing θ in the original factor graph and
edges) corresponding to θ. The edges representing $\mathbf{h}$ an
θ can be removed from the factor graph to obtain the
This process is shown in Figure 13.16(c). The maxi
graph can be summarized as [11]:

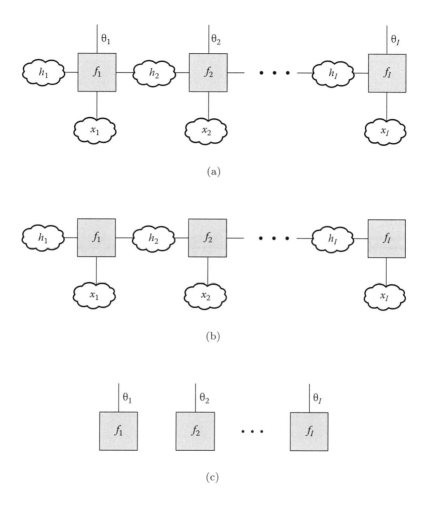

Figure 13.16: (a) Factor graph for (13.58). (b) Expectation step factor graph. (c) Maximization step factor graph.

1. Form the maximization step factor graph for e
   graph of the factor graph of $f(\mathbf{x}, \mathbf{h}, \theta)$.

2. Determine $\theta^{*\prime}$ which maximizes $\exp(Q(\theta; \theta^*)$
   semiring on the maximization step factor graph

3. Pass $\theta^{*\prime}$ to the expectation step.

It is clear that the marginal density required in the ex
determined as long as the subgraph of $f(\mathbf{x}, \mathbf{h}, \theta = \theta$
maximization step subgraph is a tree, the maximiz
computed. These facts suggest an interesting propert
EM algorithm on a factor graph. Even if the graph c
choice of θ such that the subgraphs in the expectatio
trees, the EM algorithm can be exactly implemented
the cycles.

More details on how the EM algorithm may be
along with some applications in linear Gaussian sta
identification, etc., can be found in [43, and ref. ther

13.9 Applications of Message
Graphs

Factor graphs have proved very useful in modeling of
applications in such diverse fields as coding theory,
machine learning, etc., and offer low complexity solu
Some specific applications of factor graphs were highl
describe an application of message passing in factor gr
signals transmitted over a wireless fading channel [4
of how practical systems are implemented and infere
graphs.

13.9.1 Detection of OFDM Signals in
rier Frequency Offset and Phas

We describe the system model very briefly. The s
in [44, and ref. therein]. Notation has been borrowed
tion similar and enable the reader to understand the
system, the input data stream is mapped onto a seque
The sequence is divided into non-overlapping blocks of
subcarriers. After the addition of N_{vc} null carriers, we
vector. Let the lth OFDM symbol be denoted by $\mathbf{a}^{(l}$
other length N vector $\mathbf{b}^{(l)} = [b_0^l, b_1^l, \ldots, b_{N-1}^l]^T$ is cre
Fourier transform (IDFT) of vector $\mathbf{a}$. A cyclic prefi
mented with the vector $\mathbf{y}^{(l)}$ to guard against the ISI a

N_T. The OFDM symbols are to be transmitted over a *WSS uncorrelated scattering* (WSS-US) Rayleigh fading channel which is deemed static over the duration of each OFDM symbol interval. The channel impulse response is given by

$$h\left(t, \tau\right) \cong h^{l}\left(\tau\right) = \sum_{n=0}^{L-1} h_n\left[l\right] \delta\left(\tau - nT_s\right), \tag{13.63}$$

where L is the number of multipaths and $h_n\left[l\right]$ is the value of the nth tap gain $h_n\left(t\right)$ for $n = 0, \ldots, L-1$. $h_n\left(t\right)$ are mutually independent, WSS Gaussian complex random processes. After down-conversion and low-pass filtering have been performed, the received signal is expressed as [44]:

$$r\left(t\right) = s\left(t - lN_T T_s, \mathbf{x}^{(l)}, \mathbf{H}^{(l)}, \nu^{(l)}, \phi\left(t\right)\right) + w\left(t\right), \tag{13.64}$$

where $t \in \left[lN_T T_s, \left(lN_T + N\right) T_s\right]$ and

$$
\begin{aligned}
s\left(t, \mathbf{x}^{(l)}, \mathbf{H}^{(l)}, \nu^{(l)}, \phi\left(t\right)\right) &= \exp\left(j2\pi\nu^{(l)}t/NT_s + j\phi\left(t\right)\right) \\
&\quad \cdot \frac{1}{\sqrt{NT_s}} \sum_{k=0}^{N-1} x_k^{(l)} H_k^{(l)} \exp\left(j2\pi f_k t\right), \tag{13.65}
\end{aligned}
$$

is the signal component in $r\left(t\right)$. The noise $w\left(t\right)$ is complex white Gaussian with a two-sided power spectral density $2N_o$. $\nu^{(l)}$ and $\phi\left(t\right)$ are residual carrier frequency offset and phase noise affecting the demodulated signal in the lth symbol, respectively, and are considered independent of each other. $\phi\left(t\right)$ is a zero mean continuous Brownian motion process with variance $2\pi\beta t$ [44]. After sampling, the received signal is expressed as

$$\mathbf{r}^{(l)} = \mathbf{E}^{(l)} \mathbf{F}_N^{(l)} \mathbf{A}^{(l)} \mathbf{H}^{(l)} + \mathbf{w}^{(l)}, \tag{13.66}$$

where $\mathbf{E}^{(l)} = \mathrm{diag}\left(\mathbf{e}^{(l)}\right), \mathbf{e}^{(l)} = \left[e_0^{(l)}, e_1^{(l)}, \ldots, e_{N-1}^{(l)}\right]$ with $e_n^{(l)} = \exp\left(j2\pi n\nu^{(l)} + j\phi_n^{(l)}\right)$ and $\phi_n^{(l)} = \phi\left(t_n^{(l)}\right)$. $\mathbf{F}_N$ is an $N \times N$ IDFT matrix. The phase noise process $\phi\left(t\right)$ can be modeled as a discrete time Markov process

$$\phi_n^{(l)} = \phi_{n-1}^{(l)} + \Delta_n^{(l)}, \tag{13.67}$$

where $\Delta_n^{(l)}$ is a zero mean white Gaussian noise process with variance $\sigma^2 = 2\pi\beta T_s$. The cumbersome details of construction of (13.66) are omitted and can be found in [44]. Since the signal processing operations are performed on a block-by-block basis, the index l in (13.66) is dropped in the remainder of the discussion. The received signal vector can be equivalently written as

$$\mathbf{r} = \mathbf{EZ} + \mathbf{w}, \tag{13.68}$$

where $\mathbf{Z} = \mathbf{F}_N \mathbf{A} \mathbf{F}_{N,L} \mathbf{h}$, $\mathbf{A}$ being the data matrix, $\mathbf{F}_I$ and $\mathbf{h}$ represents the channel coefficients. We confine ML estimates $\hat{\nu}$ and $\hat{\phi}$ by operating the sum-product of $\mathbf{r}$ expressed in (13.66). Assuming that the data ma known, we can use the sum-product algorithm to obt $f(\nu, \phi | \mathbf{r}, \mathbf{A}, \mathbf{h})$. Thereafter, the marginal density for summing over $\phi(\nu)$. The ML estimate is derived l marginal density. Notice that given $\mathbf{r}$ and $\mathbf{Z}$, the jo factored as

$$f(\nu, \phi | \mathbf{r}, \mathbf{Z}) = f(\nu, \phi) f(\mathbf{r} | \mathbf{Z}, \nu, \phi)$$

$$= f(\nu) f(\phi) \prod_{n=0}^{N-1} f_n(r_n$$

where (1.70) follows by the independence of ν and ϕ and ϕ, $\mathbf{r}$ consists of statistically independent Gaussia

$$f_n(r_n | z_n, \nu, \phi_n) = \frac{1}{\sqrt{2\pi \sigma_w^2}} \exp\left(-\frac{|r_n|}{}\right)$$

where $\tilde{e} = \exp(j2\pi n\nu + j\phi_n)$. Owing to the discrete N it follows that

$$f(\phi) = f(\phi_0) \prod_{n=1}^{N-1} f(\phi_n | \phi_{n-})$$

where $f(\phi_n | \phi_{n-1}) = f_\Delta(\phi_n - \phi_{n-1})$ and $f_\Delta(x) = \frac{}{\sqrt{}}$ of the increment Δ_n. Equation (13.72) can be expres

$$f(\phi) = \delta^0 \prod_{n=1}^{N-1} \delta_{n-1}^n,$$

where $\delta^0 = f(\phi_0)$ and $\delta_{n-1}^n = f_\Delta(\phi_n - \phi_{n-1})$. This expressed as

$$f(\nu, \phi | \mathbf{r}, \mathbf{Z}) = f(\nu) \delta^0 \prod_{n=1}^{N-1} \delta_{n-1}^n \prod_{n=0}^{N-1} f_n$$

The factor graph corresponding to (13.74) is show can be performed by passing messages along the edg downward messages can be computed in parallel. Us rules of the sum-product algorithm, it follows that

$$m_{f_n \to \nu}(\nu) = \int_{\Gamma_{\phi_n}} f_n(r_n | z_n, \nu, \phi_n) m_{\phi_n \to}$$

$$m_{f_n \to \phi_n}(\phi_n) = \int_{\Gamma_\nu} f_n(r_n | z_n, \nu, \phi_n) m_{\nu \to f_n}$$

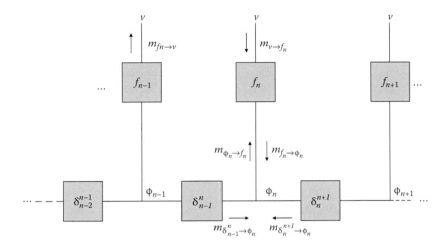

Figure 13.17: Message updates in factor graph for joint carrier frequency offset and phase noise estimation.

The final step in the downward message passing is the recursive evaluation of the *forward* and *backward* message:[6]

$$m_{\delta_{n-1}^n \to \phi_n}(\phi_n) = \int_{\Gamma_{\phi_{n-1}}} m_{\phi_{n-1} \to \delta_{n-1}^n} \cdot \delta_{n-1}^n d\phi_{n-1}, \tag{13.77}$$

where $m_{\phi_{n-1} \to \delta_{n-1}^n} = m_{f_{n-1} \to \phi_{n-1}} \cdot m_{\delta_{n-2}^{n-1} \to \phi_{n-1}}$ and

$$m_{\delta_n^{n+1} \to \phi_n}(\phi_n) = \int_{\Gamma_{\phi_{n+1}}} m_{\phi_{n+1} \to \delta_n^{n+1}} \cdot \delta_{n+1}^n d\phi_{n+1}, \tag{13.78}$$

where $m_{\phi_{n+1} \to \delta_n^{n+1}} = m_{f_{n+1} \to \phi_{n+1}} \cdot m_{\delta_{n+1}^{n+2} \to \phi_{n+1}}$. The upward messages in the final step of each iteration are given by

$$m_{\phi_n \to f_n} = m_{\delta_{n-1}^n \to \phi_n} \cdot m_{\delta_n^{n+1} \to \phi_n}. \tag{13.79}$$

$$m_{\nu \to f_n} = \prod_{k=0, k \neq n}^{N-1} m_{f_k \to \nu}. \tag{13.80}$$

An estimate of the marginalized PDF of ν can be obatined by taking the product of all incoming messages on the variable ν in the factor graph

$$\hat{f}(\nu) = \prod_{n=0}^{N-1} m_{f_n \to \nu}. \tag{13.81}$$

---

[6]This process is similar to the forward/backward recursions described in Section 13.8.1.

The PDF of phase noise ϕ can be determined in a si
for this PDF is

$$\hat{f}(\phi) = m_{\delta_{n-1}^n \to \phi_n} \cdot m_{\delta_n^{n+1} \to \phi_n} \cdot r$$

The ML estimates $\hat{\nu}$ and $\hat{\phi}$ can then be obtained by s
the PDF's expressed in (13.81) and (13.82). The ap
algorithm on the factor graph representing the joint
produces accurate estimates of the carrier frequency o
discussion on the computational complexity of the m
in [44].

13.10 Exercises

Exercise 13.10.1 (Factor Graph of a Binary F
the following parity check matrix for a binary Hamm

$$\mathbf{H} = \begin{bmatrix} 1 & 1 & 0 & 1 & 1 & 0 \\ 0 & 1 & 0 & 0 & 1 & 1 \\ 1 & 1 & 0 & 0 & 0 & 1 \\ 0 & 0 & 1 & 1 & 1 & 0 \end{bmatrix}$$

(a) Show how the global indicator function I_C fa
functions, each of which is another indicator function
(b) Draw the factor graph corresponding to this c
(c) A factor graph of a *dual code* is obtained by in
nodes and the equality check nodes. Using equality
that are an argument of two or more local functions,
dual code in (b).

Exercise 13.10.2 (Representation of Systems
Suppose that a global multivariate function factorize

$$f(x_1, x_2, x_3, x_4, x_5, x_6) = f_1(x_1) f_2(x_2) f_3(x_3|x_1, x$$

Represent this factorization in terms of the three graph
style factor graph, Markov random field and Bayesian

Exercise 13.10.3 (Computing Marginal Functi
ing). For the factorization of a function given by

$$f(x_1, x_2, x_3, x_4, x_5, x_6, x_7) = f_1(x_1) f_2(x_1,$$
$$\times f_4(x_5) f_5(x)$$

(a)   Compute the marginal function $f(x_3)$. Show all the messages passed in this computation.

(b)   Compute the marginal functions for all variables showing the details of messages passed along all the edges.

Exercise 13.10.4 (Inference Using Max-Product Algorithm). Compute all marginal functions in the factor graph of Example 13.5.1 using the max-product algorithm.

Exercise 13.10.5 (Some Common Node Computations). (a)   For the equality node shown in Figure 13.18(a), determine the message out of the variable $\mathcal{Z}$. Assume $\mathcal{X} \sim \mathcal{N}\left(\mu_x, \sigma_x^2\right)$ and $\mathcal{Y} \sim \mathcal{N}\left(\mu_y, \sigma_y^2\right)$.

(b)   Compute the marginal function for the variable $\mathcal{Y}$ in Figure 13.18(b). Take $\mathcal{X}$ to be distributed as in (a).

(c)   Compute the marginal function for the variable $\mathcal{Z}$ in Figure 13.18(c) with the distributions of $\mathcal{X}$ and $\mathcal{Y}$ the same as in (a).

Exercise 13.10.6 (Fourier Transform as Message Passing). As mentioned in the text, it is possible to determine the Fourier transform of a function through a factor graph. The rules can be summarized as [6]

1. Every variable is replaced by its dual (frequency) variable.

2. Each factor is replaced by its Fourier transform.

3. For each edge, a minus sign is introduced into one of the two adjacent factors.

These rules work only when the variable of interest is represented by a half edge. We know that multiplication in time domain is equal to convolution of Fourier transforms in frequency domain. Consider the factor graph in Figure 13.19, showing the pointwise multiplication of X_1 and X_2 to obtain X_3. Using the above mentioned rules and the sum-product algorithm, show how the Fourier transform of this pointwise multiplication may be obtained as the convolution of respective Fourier transforms.

Exercise 13.10.7 (a posteriori Probability Distribution). Consider the transmission of a random input $\mathcal{X}$ through a multipath additive Gaussian noise channel. Such a system can be represented as in Figure 13.20 where $\mathcal{W}_i \sim \mathcal{N}\left(\mu_i, \sigma_i^2\right)$ are independent of each other. The outputs $\mathcal{Y}_i$ are also conditionally independent given $\mathcal{X}$. We would like to determine the a posteriori probability $f(x|y_1, y_2, y_3)$ using the observations y_1, y_2 and y_3.

(a)   Compute the factorization of the joint density function into a product of local functions, each of which is a probability distribution.

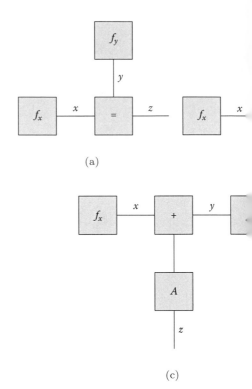

(a)

(c)

Figure 13.18: Some common nodes: (a) Equality con
node. (c) Additive and multiplier node.

(b)   Form the complete factor graph, showing an eq
variable which is an argument of more than two func
(c)   Determine the a posteriori probability $f(x|y$
product algorithm on the factor graph obtained in (b

Exercise 13.10.8 (MAP Estimation in a Multipa
output of a multipath fading channel with additive G

$$y_k' = \sum_{l=0}^{L} h_l x_{k-l} + w_k$$

where L is the number of multipath components, $\mathbf{h} =$
alent discrete time channel impulse response and w_k

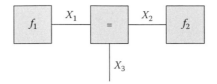

Figure 13.19: Factor graph for pointwise multiplication.

noise at discrete time k. We assume that the fading coefficients are completely known at the receiver. We would like to determine the a posteriori probabilities $p(x_k|y)$ and form a MAP estimate.

(a)   Show that the above equation may be written in the state space form

$$\boldsymbol{\Theta}_k = \mathbf{A}\boldsymbol{\Theta}_{k-1} + \mathbf{B}x_k$$
$$y_k = \mathbf{C}\boldsymbol{\Theta}_k$$

where $y_k = y_k' - w_k$.

(b)   Form a factor graph of the resulting state space model.

(c)   Describe the process of obtaining a MAP estimate from a noisy observation y'.

Exercise 13.10.9 (The Kalman Filter). Verify Equations (13.41) and (13.48) in the Kalman filter message updates.

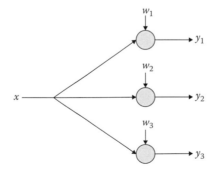

Figure 13.20: A noisy communication system.

484

References

[1] S. M. Aji and R. J. McEliece, "The generalized di
    actions on Information Theory, vol. 46, pp. 325-

[2] N. Wiberg, "Codes and decoding on general gr
    Lonkoping University, Linkoping, Sweden, 1996.

[3] R. M. Tanner, "A recursive approach to low con
    actions on Information Theory, vol. 27, pp. 533-

[4] G. D. Forney, "Codes on graphs: Normal realis
    on Information Theory, vol. 47, no. 2, pp. 520-5

[5] F. Kschischang and H. A. Loeliger, "Factor grapl
    rithm," IEEE Transactions on Information Theo
    Feb. 2001.

[6] H. A. Loeliger, "An introduction to factor grap
    Magazine, pp. 28-41, Jan. 2004.

[7] H. A. Loeliger, J. Dauwels, J. Hu, S. Korl, L. F
    "The factor graph approach to model-based signa
    the IEEE, vol. 95, no. 6, pp. 1295-1322, 2007.

[8] L. R. Bahl, J. Cocke, F. Jelinek, and J. Raviv,
    codes for minimizing symbol error rate," IEEE
    Theory, vol. 20, pp. 284-287, 1974.

[9] A. Viterbi, "Error bounds for convolutional c
    optimum decoding algorithm," IEEE Transactr
    vol. 13, pp. 260-269, 1967.

[10] H. A. Loeliger, "Least squares and Kalman filt
    Codes, Graphs and Systems, Kluwer, 2002, pp. 1

[11] A. W. Eckford, "The factor graph EM algoritl
    codes," in Proc. 6th Workshop Signal Proc. Adv
    cations, June 2005, pp. 910-914.

[12] J. Pearl, Probabilistic Reasoning in Intelligent
    Francisco, CA, 1988.

[13] J. Dauwels, S. Korl, and H. A. Loeliger, "Steepest
    Proc. IEEE ITSOC Information Theory Worksho
    2005, pp. 42-46.

[14] J. Dauwels, S. Korl, and H. A. Loeliger, "Partic
    ing," in Proc. IEEE Int. Symp. on Information '

[15] C. Berrou, A. Glavieux, and P. Thitimajshima, "Near Shannon limit error-correcting codes and decoding: Turbo codes," in *Proc. IEEE Int. Conf. Communications*, May 1993, pp. 1064–1070.

[16] D. J. C. Mackay, "Good error-correcting codes based on very sparse matrices," *IEEE Transactions on Information Theory*, vol. 45, pp. 399–431, 1999.

[17] F. R. Kschischang and B. J. Frey, "Iterative decoding of compound codes by probability propagation in graphical models," *IEEE J. Select. Areas Commun.*, vol. 16, pp. 219–230, Feb. 1998.

[18] S. Korl, "A factor graph approach to signal modeling, system identification and filtering," in *PhD Dissertation No. 16170*, ETH, Zurich, Switzerland, 2005.

[19] V. M. Koch, "A factor graph approach to model-based signal separation," in *PhD Dissertation No. 17038*, ETH, Zurich, Switzerland, 2007.

[20] A. P. Worthen and W. E. Stark, "Unified design of iterative receivers using factor graphs," *IEEE Transactions on Information Theory*, vol. 47, no. 2, pp. 843–849.

[21] A. Barbieri, G. Colavolpe, and G. Caire, "Joint iterative detection and decoding in the presence of phase noise and frequency offset," *IEEE Transactions on Communications*, vol. 55, no. 1, pp. 171–179.

[22] R. J. Drost and A. C. Singer, "Linear equalization via factor graphs," in *Proc. IEEE Int. Symp. on Information Theory*, 2004, p. 132.

[23] Q. Guo and L. Ping, "LMMSE turbo equalization based on factor graphs," *IEEE J. Select. Areas Commun.*, vol. 26, no. 2, pp. 219–230, Feb. 2008.

[24] H. A. Loeliger, "On hybrid factor graphs and adaptive equalization," in *Proc. IEEE Int. Symp. on Information Theory*, 2001, p. 268.

[25] G. Colavolpe and G. Germi, "On the application of factor graphs and the sum-product algorithm to ISI channels," *IEEE Transactions on Communications*, vol. 53, no. 5, pp. 818–825, May 2005.

[26] J. Dauwels and H. A. Loeliger, "Joint decoding and phase estimation: an exercise using factor graphs," in *Proc. IEEE Int. Symp. on Information Theory*, 2003, p. 231.

[27] J. Dauwels and H. A. Loeliger, "Phase estimation by message passing," in *Proc. IEEE Int. Conf. Communications*, June 2004, pp. 523–527.

[28] G. Chartrand, *Introductory Graph Theory*. Dover, NY, 1985.

[29] L. Rabiner, "A tutorial on hidden Markov models and selected applications in speech recognition," in *Proc. EEE*, Feb. 1989, pp. 257–286.

486

[30] B. J. Frey, F. R. Kschischang, H.-A. Loeliger, an
and algorithms," in *Proc. 35th Allerton Conf. c
and Computing*, Sept. 1997, pp. 666–680.

[31] R. Kindermann and J. L. Snell, *Markov Random
American Mathematical Society, Providence, Rl

[32] D. Koller and N. Friedman, *Probabilistic Grap*
Techniques. MIT Press, Cambridge, Massachus

[33] R. E. Neapolitan, *Learning Bayesian Networks.*

[34] H. A. Loeliger, "Some remarks on factor graphs,
Turbo Codes and Related Topics, 2003, pp. 111–

[35] J. G. Proakis, *Digital Communications.* McGra

[36] R. E. Kalman, "A new approach to linear filterin
Journal of Basic Engineering, pp. 35–45, 1960.

[37] C. K. Chui and G. Chen, *Kalman Filtering u*
Springer-Verlag, NY, 1998.

[38] A. P. Dempster, N. M. Laird, and D. B. Rubin, "
complete data via the EM algorithm," *Journal of*
vol. 39, Series B, pp. 1–38, 1977.

[39] K. J. Molnar and J. W. Modestino, "Applicati
the multitarget/multisensor tracking problem,"
Processing, vol. 46, pp. 115–129, Jan. 1998.

[40] J. M. Ollinger, "Maximum-likelihood reconstruct
emission computed tomography via the EM alg
on Medical Imaging, vol. 13, pp. 89–101, 1994.

[41] M. H. C. Law, M. A. T. Figueiredo, and A. K.
selection and clustering using mixture models," *I*
Analysis and Machine Intelligence, vol. 26, pp. 1

[42] A. W. Eckford and S. Pasupathy, "Iterative mul
ical modeling," in *IEEE Int. Conf. on Persona*
Hyderabad, India, 2000.

[43] J. Dauwels, A. Eckford, S. Korl, and H. A. Loeli
tion as message passing - part i: Principles and Ga
http://arxiv.org/abs/0910.2832, 2009.

[44] F. Z. Merli and G. M. Vitetta, "Factor graph a
tection of OFDM signals in the presence of carrie
noise," *IEEE Transactions on Wireless Commu*
868–877, 2008.

Chapter 14 Unconstrained and Constrained Optimization Problems

Shuguang Cui[‡], Anthony Man-Cho So[♯], and Rui Zhang[♮]
[‡]Texas A&M University, College Station, USA
[♯]The Chinese University of Hong Kong, Hong Kong, China
[♮]National University of Singapore, Singapore

In the first section of this chapter, we will give an overview of the basic mathematical tools that are useful for analyzing both unconstrained and constrained optimization problems. In order to allow the readers to focus on the applications of these tools and not to be burdened with too many technical details, we shall state most of the results without proof. However, the readers are strongly encouraged to refer to the texts [1–4] for expositions of these results and other further developments. In the second section, we provide three application examples to illustrate how we could apply the optimization techniques to solve real-world problems, with a focus on communications, networking, and signal processing. In the last section, several exercise questions are given to help the audience gain a deeper understanding of the material.

14.1 Basics of Convex Analysis

The notion of convexity plays a very important role in both the theoretical and algorithmic aspects of optimization. Before we discuss the relevance of convexity in optimization, let us first introduce the notions of convex sets and convex functions and state some of their properties.

Definition 14.1.1. Let $S \subset \mathbb{R}^n$ be a set. We say that

1. S is **affine** if $\alpha \mathbf{x} + (1 - \alpha)\mathbf{y} \in S$ whenever $\mathbf{x}, \mathbf{y} \in S$ and $\alpha \in \mathbb{R}$;

2. S is **convex** if $\alpha \mathbf{x} + (1 - \alpha)\mathbf{y} \in S$ whenever $\mathbf{x}, \mathbf{y} \in S$ and $\alpha \in [0, 1]$.

Given $\mathbf{x}, \mathbf{y} \in \mathbb{R}^n$ and $\alpha \in \mathbb{R}$, the vector $\mathbf{z} = \alpha\mathbf{x} +$ **combination** of $\mathbf{x}$ and $\mathbf{y}$. If $\alpha \in [0, 1]$, then $\mathbf{z}$ is ca of $\mathbf{x}$ and $\mathbf{y}$.

Geometrically, when $\mathbf{x}$ and $\mathbf{y}$ are distinct points i

$$L = \{\mathbf{z} \in \mathbb{R}^n : \mathbf{z} = \alpha\mathbf{x} + (1 - \alpha)\mathbf{y}$$

of all affine combinations of $\mathbf{x}$ and $\mathbf{y}$ is simply the *line* the set

$$S = \{\mathbf{z} \in \mathbb{R}^n : \mathbf{z} = \alpha\mathbf{x} + (1 - \alpha)\mathbf{y},$$

is the *line segment* between $\mathbf{x}$ and $\mathbf{y}$. By convention,

It is clear that one can generalize the notion of affir of two points to any finite number of points. In partic the points $\mathbf{x_1}, \ldots, \mathbf{x_k} \in \mathbb{R}^n$ is a point $\mathbf{z} = \sum_{i=1}^{k} \alpha_i\mathbf{x_i}$, v a convex combination of the points $\mathbf{x_1}, \ldots, \mathbf{x_k} \in \mathbb{R}^n$ is $\sum_{i=1}^{k} \alpha_i = 1$ *and* $\alpha_1, \ldots, \alpha_k \geq 0$.

Here are some sets in Euclidean space whose conve by first principles:

Example 14.1.1. (Some Examples of Convex S

1. **Non-negative Orthant**: $\mathbb{R}_+^n = \{\mathbf{x} \in \mathbb{R}^n : \mathbf{x} \geq$

2. **Hyperplane**: $H(\mathbf{s}, c) = \{\mathbf{x} \in \mathbb{R}^n : \mathbf{s}^T\mathbf{x} = c\}$.

3. **Halfspaces**: $H^+(\mathbf{s}, c) = \{\mathbf{x} \in \mathbb{R}^n : \mathbf{s}^T\mathbf{x} \leq c\}$, H

4. **Euclidean Ball**: $B(\bar{\mathbf{x}}, r) = \{\mathbf{x} \in \mathbb{R}^n : \|\mathbf{x} - \bar{\mathbf{x}}\|_2$

5. **Ellipsoid**: $E(\bar{\mathbf{x}}, \mathbf{Q}, r) = \{\mathbf{x} \in \mathbb{R}^n : (\mathbf{x} - \bar{\mathbf{x}})^T Q($ $n \times n$ symmetric, positive definite matrix (i.e., $\mathbf{x}^T$ and is denoted by $\mathbf{Q} \succ \mathbf{0}$.

6. **Simplex**: $\Delta = \left\{ \sum_{i=0}^{n} \alpha_i\mathbf{x_i} : \sum_{i=0}^{n} \alpha_i = 1, \alpha \right.$ where $\mathbf{x_0}, \mathbf{x_1}, \ldots, \mathbf{x_n}$ are vectors in $\mathbb{R}^n$ su $\mathbf{x_0}, \mathbf{x_2} - \mathbf{x_0}, \ldots, \mathbf{x_n} - \mathbf{x_0}$ are linearly independe $\mathbf{x_0}, \mathbf{x_1}, \ldots, \mathbf{x_n}$ are *affinely* independent).

7. **Positive Semidefinite Cone**: $\mathcal{S}_+^n = \{\mathbf{A} \in \mathbb{R}^{n \times n} : \mathbf{A}$ is symmetric and $\mathbf{x}^T\mathbf{A}\mathbf{x}$ metric matrix $\mathbf{A} \in \mathbb{R}^{n \times n}$ is said to be positive all $\mathbf{x} \in \mathbb{R}^n$, and is denoted by $\mathbf{A} \succeq \mathbf{0}$).

Let us now turn to the notion of a convex functio

Definition 14.1.2. Let $S \subset \mathbb{R}^n$ be a nonempty convex set, and let $f : S \to \mathbb{R}$ be a real-valued function.

1. We say that f is **convex** on S if

$$f(\alpha\mathbf{x_1} + (1 - \alpha)\mathbf{x_2}) \le \alpha f(\mathbf{x_1}) + (1 - \alpha)f(\mathbf{x_2}) \qquad (14.1)$$

for all $\mathbf{x_1}, \mathbf{x_2} \in S$ and $\alpha \in [0, 1]$. We say that f is **concave** if $-f$ is convex.

2. We say that f is **strictly convex** on S if

$$f(\alpha\mathbf{x_1} + (1 - \alpha)\mathbf{x_2}) < \alpha f(\mathbf{x_1}) + (1 - \alpha)f(\mathbf{x_2})$$

for all $\mathbf{x_1}, \mathbf{x_2} \in S$ and $\alpha \in (0, 1)$.

3. The **epigraph** of f is the set $\text{epi}(f) = \{(\mathbf{x}, r) \in S \times \mathbb{R} : f(\mathbf{x}) \le r\}$.

The relationship between convex sets and convex functions can be summarized as follows:

Proposition 14.1.1. Let f be as in Definition 14.1.2. Then, f is convex (as a function) iff $\text{epi}(f)$ is convex (as a set in $S \times \mathbb{R}$).

Let $r \in \mathbb{R}$ be arbitrary. A set closely related to the epigraph is the so-called r-**level set** of f, which is defined as $L(r) = \{\mathbf{x} \in \mathbb{R}^n : f(\mathbf{x}) \le r\}$. It is clear that if f is convex, then $L(r)$ is convex for all $r \in \mathbb{R}$. However, the converse is *not* true, as illustrated by the function $x \mapsto x^3$. A function $f : S \to \mathbb{R}$ whose domain is convex and whose r-level sets are convex for all $r \in \mathbb{R}$ is called **quasi-convex**.

One of the most desirable features of convexity is the following:

Proposition 14.1.2. Consider the optimization problem:

$$\begin{aligned} \text{minimize} \quad & f(\mathbf{x}) \\ \text{subject to} \quad & \mathbf{x} \in S, \end{aligned}$$

where $S \subset \mathbb{R}^n$ is a convex set and $f : S \to \mathbb{R}$ is convex. Then, any local minimum of f is also a global minimum[1].

Now, let $S \subset \mathbb{R}^n$ be an open convex set, and let $f : S \to \mathbb{R}$ be an arbitrary function. When f has suitable degree of differentiability, we can characterize its convexity by its gradient or Hessian. Specifically, we have the following:

Theorem 14.1.1. *Let $S \subset \mathbb{R}^n$ be an open convex set, and let $f : S \to \mathbb{R}$ be a differentiable function on S. Then, f is convex on S iff*

$$f(\mathbf{x_1}) \ge f(\mathbf{x_2}) + (\nabla f(\mathbf{x_2}))^T (\mathbf{x_1} - \mathbf{x_2})$$

for all $\mathbf{x_1}, \mathbf{x_2} \in S$. Furthermore, if f is twice continuously differentiable function on S, then f is convex on S iff $\nabla^2 f(\mathbf{x})$ is positive semidefinite for all $\mathbf{x} \in S$.

---

[1]Recall that for a generic optimization problem $\min_{\mathbf{x} \in S \subset \mathbb{R}^n} f(\mathbf{x})$, a point $\mathbf{x}^* \in S$ is called a **global minimum** if $f(\mathbf{x}^*) \le f(\mathbf{x})$ for all $\mathbf{x} \in S$. On the other hand, if there exists an $\epsilon > 0$ such that the point $\mathbf{x}^* \in S$ satisfies $f(\mathbf{x}^*) \le f(\mathbf{x})$ for all $\mathbf{x} \in S \cap B^\circ(\mathbf{x}^*, \epsilon)$, then it is called a **local minimum**. Here, $B^\circ(\bar{\mathbf{x}}, \epsilon) =$ denotes the *open* ball centered at $\bar{\mathbf{x}} \in \mathbb{R}^n$ of radius $\epsilon > 0$.

Sometimes it may be difficult to verify directly f
given function is convex or not. However, a functio
composition of several, more elementary functions. W
functions is convex, it is natural to ask whether thei
In general, the answer is no. On the other hand, he
that preserve convexity.

Theorem 14.1.2. *Let $S \subset \mathbb{R}^n$ be a nonempty con*
hold:

1. **(Non-negative Combinations)** *Let $f_1, \ldots,$*
   tions, and let $\alpha_1, \ldots, \alpha_m \geq 0$. Then, the func
   S.

2. **(Pointwise Supremum)** *Let $\{f_i\}_{i \in I}$ be an ar*
   tions on S. Then, the pointwise supremum $f =$

3. **(Affine Composition)** *Let $f : \mathbb{R}^n \to \mathbb{R}$ be a cc*
   $\mathbb{R}^n$ be an affine mapping^2. Then, the function
   $(f \circ A)(\mathbf{x}) = f(A(\mathbf{x}))$ is convex on $\mathbb{R}^m$.

4. **(Composition with an Increasing Convex**
   be a convex function, and let $g : \mathbb{R} \to \mathbb{R}$ be ar
   Then, the function $g \circ f : S \to \mathbb{R}$ defined by $(g$
   on S.

5. **(Restriction on Lines)** *Let $f : S \to \mathbb{R}$ be a*
   $\mathbf{h} \in \mathbb{R}^n$, define the function $\tilde{f}_{\mathbf{x_0}, \mathbf{h}} : \mathbb{R} \to \mathbb{R} \cup \{+$

$$\tilde{f}_{\mathbf{x_0}, \mathbf{h}}(t) = \begin{cases} f(\mathbf{x_0} + t\mathbf{h}) & \text{if } \mathbf{x} \\ +\infty & \text{othe} \end{cases}$$

Then, f is convex on S iff $\tilde{f}_{\mathbf{x_0}, \mathbf{h}}$ is convex on $\mathbb{R}$

Let us now illustrate an application of Theorem 1

Example 14.1.2. Let $f : \mathbb{R}^{m \times n} \to \mathbb{R}_+$ be given b
denotes the **spectral norm** or **largest singular va**
By the Courant–Fischer theorem (see, e.g., [5]), we h

$$f(\mathbf{X}) = \sup \{\mathbf{u}^T \mathbf{X} \mathbf{v} : \|\mathbf{u}\|_2 = 1, \ \|$$

Now, for each $\mathbf{u} \in \mathbb{R}^m$ and $\mathbf{v} \in \mathbb{R}^n$ with $\|\mathbf{u}\|_2 = \|\mathbf{v}$
$f_{\mathbf{u}, \mathbf{v}} : \mathbb{R}^{m \times n} \to \mathbb{R}$ by

$$f_{\mathbf{u}, \mathbf{v}}(\mathbf{X}) = \mathbf{u}^T \mathbf{X} \mathbf{v}.$$

---

^{2}A map $A : \mathbb{R}^m \to \mathbb{R}^n$ is said to be **affine** if there exists
$\mathbf{d} \in \mathbb{R}^n$ such that $A(\mathbf{x}) = \mathbf{Bx} + \mathbf{d}$ for all $\mathbf{x} \in \mathbb{R}^m$.

Note that $f_{\mathbf{u},\mathbf{v}}$ is a convex (in fact, linear) function of $\mathbf{X}$ for each $\mathbf{u}, \mathbf{v}$. Hence, it follows from (14.2) that f is a pointwise supremum of a family of linear functions of $\mathbf{X}$. By Theorem 14.1.2, this implies that f is convex. $\qquad\square$

14.2 Unconstrained vs. Constrained Optimization

14.2.1 Optimality Conditions for Unconstrained Optimization

One of the most fundamental problems in optimization is to derive conditions for identifying potential optimal solutions to an optimization problem. Typically, such conditions, which are known as **optimality conditions**, would enable us to reduce the original optimization problem to that of checking the validity of certain geometric conditions, or to that of checking the consistency of certain system of inequalities. As an illustration and to motivate our discussion, let us first consider a univariate, twice continuously differentiable function $f : \mathbb{R} \to \mathbb{R}$. Recall from basic calculus that if $\bar{x} \in \mathbb{R}$ is a local minimum of f, then we must have

$$\frac{df(x)}{dx}\bigg|_{x=\bar{x}} = 0. \tag{14.3}$$

In other words, condition (14.3) is a *necessary condition* for $\bar{x}$ to be a local minimum. However, it is *not* a sufficient condition, as an $\bar{x} \in \mathbb{R}$ that satisfies (14.3) can be a local maximum or just a stationary point. In order to certify that $\bar{x}$ is indeed a local minimum, one could check, in addition to (14.3), whether

$$\frac{d^2 f(x)}{dx^2}\bigg|_{x=\bar{x}} > 0. \tag{14.4}$$

In particular, condition (14.4) is a *sufficient condition* for $\bar{x}$ to be a local minimum.

In the above discussion, conditions (14.3) and (14.4) together yield a system of inequalities whose solutions are local minima of the function f. Alternatively, they can be viewed as stating the geometric fact that there is no descent direction in a neighborhood of a local minimum. In particular, the former is an algebraic interpretation of local optimality, while the latter is a geometric interpretation. It is worth noting that each interpretation has its own advantage. Indeed, the geometric interpretation can often help us gain intuitions about the problem at hand, and the algebraic interpretation would help to make those intuitions precise. Thus, it is good to keep both interpretations in mind.

To derive optimality conditions for the local minima of a *multivariate* twice continuously differentiable function $f : \mathbb{R}^n \to \mathbb{R}$, we first recall that $\nabla f(\mathbf{x})$, the gradient of f at $\mathbf{x} \in \mathbb{R}^n$, is the direction of steepest ascent at $\mathbf{x}$. Thus, if $\nabla f(\mathbf{x}) \neq \mathbf{0}$, then starting at $\mathbf{x}$, we can proceed in the direction $-\nabla f(\mathbf{x})$ and achieve a smaller function value. More specifically, we have the following

Proposition 14.2.1. Suppose that $f : \mathbb{R}^n \to \mathbb{R}$ is $\bullet$ $\bar{\mathbf{x}} \in \mathbb{R}^n$. If there exists a $\mathbf{d} \in \mathbb{R}^n$ such that $\nabla \left(f(\bar{\mathbf{x}}) \right)$ $\alpha_0 > 0$ such that $f(\bar{\mathbf{x}} + \alpha \mathbf{d}) < f(\bar{\mathbf{x}})$ for all $\alpha \in (0,$ **descent direction** of f at $\bar{\mathbf{x}}$.

Using Proposition 14.2.1, we can establish the fol

Corollary 14.2.1. (First Order Necessary Con Optimization) Suppose that $f : \mathbb{R}^n \to \mathbb{R}$ is continuc If $\bar{\mathbf{x}}$ is a local minimum, then we have $\nabla f(\bar{\mathbf{x}}) = \mathbf{0}$. In

Similar to the univariate case, even if $\bar{\mathbf{x}} \in \mathbb{R}^n$ sat conclude that $\bar{\mathbf{x}}$ is a local minimum. For instance, con given by $f(x_1, x_2) = -x_1^2 - (x_1 - x_2)^2$. Then, we hav

$$\nabla f(\mathbf{x}) = -2(2x_1 - x_2, x_2 -$$

In particular, the (unique) solution to $\nabla f(\mathbf{x}) = \mathbf{0}$ is $\mathbf{a}$ be easily verified, the point $(\bar{x}_1, \bar{x}_2) = (0,0)$ is a glob
The above example shows that some extra condi guarantee that a solution to the equation $\nabla f(\mathbf{x}) = \mathbf{0}$ instance, we have the following proposition, which st then the necessary condition in Corollary 14.2.1 is al

Proposition 14.2.2. Suppose that $f : \mathbb{R}^n \to \mathbb{R}$ is cc convex at $\bar{\mathbf{x}}$. Then, $\bar{\mathbf{x}}$ is a global minimum iff $\nabla f(\bar{\mathbf{x}})$

Alternatively, if $\nabla f(\bar{\mathbf{x}}) = \mathbf{0}$ and $\nabla^2 f(\bar{\mathbf{x}})$, the H definite, then $\bar{\mathbf{x}}$ is a local minimum. Specifically, we ha which generalizes the corresponding result for the u (14.4)).

Proposition 14.2.3. (Second Order Sufficien strained Optimization) Suppose that $f : \mathbb{R}^n \to$ ferentiable at $\bar{\mathbf{x}} \in \mathbb{R}^n$. If $\nabla f(\bar{\mathbf{x}}) = \mathbf{0}$ and $\nabla^2 f(\bar{\mathbf{x}})$ is local minimum.

Let us now illustrate the above results with an ex

Example 14.2.1. Let $f : \mathbb{R}^n \to \mathbb{R}$ be defined by f $\mathbf{Q} \in \mathbf{S}^n$ and $\mathbf{c} \in \mathbb{R}^n$ are given. Then, f is continu have $\nabla f(\mathbf{x}) = \mathbf{Q}\mathbf{x} + \mathbf{c}$ and $\nabla^2 f(\mathbf{x}) = \mathbf{Q}$. Now, if if $\mathbf{Q} \succeq \mathbf{0}$, then by Proposition 14.2.2, any $\bar{\mathbf{x}} \in \mathbb{R}^n$ will be a global minimum of f. Note that in this ca

---

[3]Let S be a nonempty convex subset of $\mathbb{R}^n$. We say that if $f(\alpha\bar{\mathbf{x}} + (1 - \alpha)\mathbf{x}) \leq \alpha f(\bar{\mathbf{x}}) + (1 - \alpha)f(\mathbf{x})$ for all $\alpha \in (0,1)$ $f : S \to \mathbb{R}$ can be convex at a particular point $\bar{\mathbf{x}} \in S$ without t

from Proposition 14.2.3 that $\bar{\mathbf{x}}$ is a local minimum of f, since we only have $\mathbf{Q} \succeq \mathbf{0}$. On the other hand, suppose that $\mathbf{Q} \succ \mathbf{0}$. Then, $\mathbf{Q}$ is invertible, and by Proposition 14.2.3, the point $\bar{\mathbf{x}} = -\mathbf{Q}^{-1}\mathbf{c}$ is a local minimum of f. However, since f is convex, Proposition 14.2.2 allows us to draw a stronger conclusion, namely, the point $\bar{\mathbf{x}} = -\mathbf{Q}^{-1}\mathbf{c}$ is in fact the unique global minimum. $\qquad\square$

14.2.2 Optimality Conditions for Constrained Optimization

After deriving optimality conditions for unconstrained optimization problems, let us turn our attention to *constrained* optimization problems of the form

$$\min_{\mathbf{x} \in S} f(\mathbf{x}), \qquad (14.5)$$

where S is a nonempty subset of $\mathbb{R}^n$. Note that due to the constraint $\mathbf{x} \in S$, even if $\bar{\mathbf{x}} \in \mathbb{R}^n$ satisfies $\nabla f(\bar{\mathbf{x}}) = \mathbf{0}$ and $\nabla^2 f(\bar{\mathbf{x}}) \succ \mathbf{0}$, it may not be a solution to (14.5), since $\bar{\mathbf{x}}$ need not lie in S. Similarly, a local minimum $\bar{\mathbf{x}}$ of f over S need not satisfy $\nabla f(\bar{\mathbf{x}}) = \mathbf{0}$, since all the descent directions of f at $\bar{\mathbf{x}}$ may lead to points that do not lie in S. Thus, in order to derive optimality conditions for (14.5), we need to consider not only the **set of descent directions** at $\bar{\mathbf{x}}$, i.e.,

$$D = \left\{ \mathbf{d} \in \mathbb{R}^n : \nabla f(\bar{\mathbf{x}})^T \mathbf{d} < 0 \right\}, \qquad (14.6)$$

but also the **set of feasible directions** at $\bar{\mathbf{x}}$, i.e.,

$$F = \{\mathbf{d} \in \mathbb{R}^n \setminus \{\mathbf{0}\} : \text{there exists an } \alpha_0 > 0 \text{ such that } \bar{\mathbf{x}} + \alpha\mathbf{d} \in S \text{ for all } \alpha \in (0, \alpha_0)\}. \qquad (14.7)$$

We emphasize that in order for $\mathbf{d} \in F$, the *entire* open line segment $\{\bar{\mathbf{x}} + \alpha\mathbf{d} : \alpha \in (0, \alpha_0)\}$ must belong to S. This is to ensure that whenever $\mathbf{d} \in D$, one can find a feasible solution $\bar{\mathbf{x}}' \in S$ with $f(\bar{\mathbf{x}}') < f(\bar{\mathbf{x}})$ by proceeding from $\bar{\mathbf{x}}$ in the direction $\mathbf{d}$. Indeed, by Proposition 14.2.1, if $\mathbf{d} \in D$, then there exists an $\alpha_1 > 0$ such that $f(\bar{\mathbf{x}} + \alpha\mathbf{d}) < f(\bar{\mathbf{x}})$ for all $\alpha \in (0, \alpha_1)$. However, if $\bar{\mathbf{x}} + \alpha\mathbf{d} \notin S$ for any $\alpha \in (0, \alpha_1)$, then we cannot rule out the local minimality of $\bar{\mathbf{x}}$, even if $\bar{\mathbf{x}} + \alpha\mathbf{d} \in S$ for some $\alpha > \alpha_1$.

As the following proposition shows, the sets D and F provide a necessary, and under some additional assumptions, even sufficient condition for optimality.

Proposition 14.2.4. Consider Problem (14.5). Suppose that $f : \mathbb{R}^n \to \mathbb{R}$ is continuously differentiable at $\bar{\mathbf{x}} \in S$. If $\bar{\mathbf{x}}$ is a local minimum, then we have $D \cap F = \emptyset$. Conversely, suppose that (i) $D \cap F = \emptyset$, (ii) f is convex at $\bar{\mathbf{x}}$, and (iii) there exists an $\epsilon > 0$ such that $\mathbf{d} = \mathbf{x} - \bar{\mathbf{x}} \in F$ for any $\mathbf{x} \in S \cap B^\circ(\bar{\mathbf{x}}, \epsilon)$. Then, $\bar{\mathbf{x}}$ is a local minimum of f over S.

REMARKS: Condition (iii) is to ensure that the entire line segment $\{\bar{\mathbf{x}} + \alpha(\mathbf{x} - \bar{\mathbf{x}}) : \alpha \in [0, 1]\}$ lies in S for any $\mathbf{x} \in S \cap B^\circ(\bar{\mathbf{x}}, \epsilon)$, so that $\mathbf{d} = \mathbf{x} - \bar{\mathbf{x}} \in F$; see the remark after (14.7).

So far we have only discussed optimality conditio
optimization problems, i.e., problems of the form (14
a necessary condition for local optimality in terms c
that $D \cap F = \emptyset$. However, such a condition is large
easy to manipulate as algebraic conditions (e.g., a sy
other hand, as we will show below, if the feasible reg
one can circumvent such difficulty and derive algebr
begin, let us consider the following class of optimizat

$$
\begin{aligned}
\text{minimize} \quad & f(\mathbf{x}) \\
\text{subject to} \quad & g_i(\mathbf{x}) \leq 0 \quad \text{for } i = 1 \\
& \mathbf{x} \in X,
\end{aligned}
$$

where $f : \mathbb{R}^n \to \mathbb{R}$ and $g_i : \mathbb{R}^n \to \mathbb{R}$ are continuously
X is a nonempty open subset of $\mathbb{R}^n$ (usually we take
following:

Proposition 14.2.5. Let $S = \{\mathbf{x} \in X : g_i(\mathbf{x}) \leq 0$ for
region of problem (14.8), and let $\bar{\mathbf{x}} \in S$. Define

$$
I = \{i \in \{1, \ldots, m\} : g_i(\bar{\mathbf{x}}) =
$$

to be the index set for the **active** or **binding** constr

$$
\begin{aligned}
G &= \left\{\mathbf{d} \in \mathbb{R}^n : \nabla g_i(\bar{\mathbf{x}})^T \mathbf{d} < 0 \text{ for} \right. \\
\overline{G} &= \left\{\mathbf{d} \in \mathbb{R}^n \setminus \{\mathbf{0}\} : \nabla g_i(\bar{\mathbf{x}})^T \mathbf{d} \leq \right.
\end{aligned}
$$

Then, we have $G \subset F \subset \overline{G}$, where F is defined in (14.
g_i, where $i \in I$, are strictly convex (resp. concave) at $\bar{\mathbf{x}}$

Using Proposition 14.2.4 and Proposition 14.2.5, v
geometric optimality condition for (14.8):

Corollary 14.2.2. Let S be the feasible region of pr
define $I = \{i \in \{1, \ldots, m\} : g_i(\bar{\mathbf{x}}) = 0\}$. If $\bar{\mathbf{x}}$ is a loca
where D is defined in (14.6) and G is defined in (14.

The intuition behind Corollary 14.2.2 is quite strai
that $\mathbf{d} \in D \cap G$. Then, by Proposition 14.2.1, there
$f(\bar{\mathbf{x}} + \alpha \mathbf{d}) < f(\bar{\mathbf{x}})$ and $g_i(\bar{\mathbf{x}} + \alpha \mathbf{d}) < g_i(\bar{\mathbf{x}}) = 0$ for
Moreover, by the continuity of the functions $g_1, \ldots, g_m$
we have $g_i(\bar{\mathbf{x}} + \alpha \mathbf{d}) < 0$ for all $i \notin I$. It follows that
that $\bar{\mathbf{x}} + \alpha \mathbf{d} \in S$ and $f(\bar{\mathbf{x}} + \alpha \mathbf{d}) < f(\bar{\mathbf{x}})$ for all $\alpha \in (0,$
a local minimum.

The upshot of Corollary 14.2.2 is that it allows
ditions for (14.8) that is more algebraic in nature. S
together with Farkas' lemma, yields the following:

Theorem 14.2.1. (Karush–Kuhn–Tucker Necessary Conditions) *Let $\bar{\mathbf{x}} \in S$ be a local minimum of problem (14.8), and let $I = \{i \in \{1, \ldots, m\} : g_i(\bar{\mathbf{x}}) = 0\}$ be the index set for the active constraints. Suppose that the family $\{\nabla g_i(\bar{\mathbf{x}})\}_{i \in I}$ of vectors is linearly independent. Then, there exist $\bar{u}_1, \ldots, \bar{u}_m \in \mathbb{R}$ such that*

$$
\begin{aligned}
\nabla f(\bar{\mathbf{x}}) + \sum_{i=1}^{m} \bar{u}_i \nabla g_i(\bar{\mathbf{x}}) &= \mathbf{0}, \\
\bar{u}_i g_i(\bar{\mathbf{x}}) &= 0 \qquad for\ i = 1, \ldots, m, \\
\bar{u}_i &\geq 0 \qquad for\ i = 1, \ldots, m.
\end{aligned}
\tag{14.10}
$$

We say that $\bar{\mathbf{x}} \in \mathbb{R}^n$ is a **KKT point** if (i) $\bar{\mathbf{x}} \in S$ and (ii) there exist **Lagrange multipliers** $\bar{u}_1, \ldots, \bar{u}_m$ such that $(\bar{\mathbf{x}}, \bar{u}_1, \ldots, \bar{u}_m)$ satisfies the system (14.10).

Note that if the gradient vectors of the active constraints are not linearly independent, then the KKT conditions are *not* necessary for local optimality, *even when the optimization problem is convex*. This is demonstrated in the following example.

Example 14.2.2. Consider the following optimization problem:

$$
\begin{aligned}
\text{minimize} \quad & x_1 \\
\text{subject to} \quad & (x_1 - 1)^2 + (x_2 - 1)^2 \leq 1, \\
& (x_1 - 1)^2 + (x_2 + 1)^2 \leq 1.
\end{aligned}
\tag{14.11}
$$

Since there is only one feasible solution (i.e., $(x_1, x_2) = (1, 0)$), it is naturally optimal. Besides the primal feasibility condition, the KKT conditions of (14.11) are given by

$$
\begin{aligned}
\begin{bmatrix} 1 \\ 0 \end{bmatrix} + 2u_1 \begin{bmatrix} x_1 - 1 \\ x_2 - 1 \end{bmatrix} + 2u_2 \begin{bmatrix} x_1 - 1 \\ x_2 + 1 \end{bmatrix} &= \mathbf{0}, \\
u_1 \left((x_1 - 1)^2 + (x_2 - 1)^2 - 1 \right) &= 0, \\
u_2 \left((x_1 - 1)^2 + (x_2 + 1)^2 - 1 \right) &= 0. \\
u_1, u_2 &\geq 0
\end{aligned}
$$

However, it is clear that there is no solution $(u_1, u_2) \geq \mathbf{0}$ to the above system when $(x_1, x_2) = (1, 0)$. $\qquad \square$

Let us now illustrate Theorem 14.2.1 with an example.

Example 14.2.3. (Optimization of a Matrix Function) Let $\mathbf{A} \succ \mathbf{0}$ and $b > 0$ be given. Consider the following problem:

$$
\begin{aligned}
\text{minimize} \quad & -\log \det(\mathbf{Z}) \\
\text{subject to} \quad & \operatorname{tr}(\mathbf{AZ}) \leq b, \\
& \mathbf{Z} \succ \mathbf{0}.
\end{aligned}
\tag{14.12}
$$

Note that (14.12) is of the form (14.8), since we may

$$
\begin{aligned}
\text{minimize} \quad & -\log\det(\mathbf{Z}) \\
\text{subject to} \quad & \operatorname{tr}(\mathbf{AZ}) \le b, \\
& \mathbf{Z} \in \mathbf{S}^n_{++},
\end{aligned}
$$

and $\mathbf{S}^n_{++} \subset \mathbb{R}^{n(n+1)/2}$ is an open set. Now, it is know

$$
\nabla \log\det(\mathbf{X}) = \mathbf{X}^{-1}, \quad \nabla \operatorname{tr}(\mathbf{A}\cdot
$$

see, e.g., [6]. Hence, the KKT conditions associated $\mathbf{v}$

$$
\begin{aligned}
\operatorname{tr}(\mathbf{AZ}) &\le b, \quad \mathbf{Z} \succ \\
-\mathbf{Z}^{-1} + u\mathbf{A} &= \mathbf{0}, \quad u \ge \\
u\,(\operatorname{tr}(\mathbf{AZ}) - b) &= 0.
\end{aligned}
$$

Condition (a) is simply primal feasibility. Condition (
tarity. As we shall see later, condition (b) can be i
respect to a certain dual of (14.12).

Note that Theorem 14.2.1 applies only to inequal
problems of the form (14.8). However, by extending t.
to prove Corollary 14.2.2, one can establish similar ne
for optimization problems of the form

$$
\begin{aligned}
\text{minimize} \quad & f(\mathbf{x}) \\
\text{subject to} \quad & g_i(\mathbf{x}) \le 0 \quad \text{for } i = 1, \\
& h_j(\mathbf{x}) = 0 \quad \text{for } j = 1 \\
& \mathbf{x} \in X,
\end{aligned}
$$

where $f, g_1, \ldots, g_{m_1}, h_1, \ldots, h_{m_2} : \mathbb{R}^n \to \mathbb{R}$ are cont
tions, and X is a nonempty open subset of $\mathbb{R}^n$. Speci

Theorem 14.2.2. (Karush–Kuhn–Tucker Neces
the feasible region of Problem (14.13). *Suppose that*
of problem (14.13), *with* $I = \{i \in \{1, \ldots, m_1\} : g_i($
for the active constraints. Furthermore, suppose that
$\{\nabla g_i(\bar{\mathbf{x}})\}_{i \in I} \cup \{\nabla h_j(\bar{\mathbf{x}})\}_{j=1}^{m_2}$ *of vectors is linearly inc*
$\bar{v}_1, \ldots, \bar{v}_{m_1} \in \mathbb{R}$ *and* $\bar{w}_1, \ldots, \bar{w}_{m_2} \in \mathbb{R}$ *such that*

$$
\begin{aligned}
\nabla f(\bar{\mathbf{x}}) + \sum_{i=1}^{m_1} \bar{v}_i \nabla g_i(\bar{\mathbf{x}}) + \sum_{j=1}^{m_2} \bar{w}_j \nabla h_j(\bar{\mathbf{x}}) &= \mathbf{0}, \\
\bar{v}_i g_i(\bar{\mathbf{x}}) &= 0 \\
\bar{v}_i &\ge 0
\end{aligned}
$$

As demonstrated in Exercise 14.2.2, the linear independence of the gradient vectors of the active constraints is generally needed to guarantee the existence of Lagrange multipliers. However, such a regularity condition is not always easy to check. As it turns out, there are other forms of regularity conditions, a more well-known of which is the following:

Theorem 14.2.3. *Suppose that in Problem (14.13), the functions $g_1, \ldots, g_{m_1}$ are convex and $h_1, \ldots, h_{m_2}$ are linear. Let $\bar{\mathbf{x}} \in S$ be a local minimum, and let $I = \{i \in \{1, \ldots, m_1\} : g_i(\bar{\mathbf{x}}) = 0\}$. If the Slater condition is satisfied, i.e., if there exists an $\mathbf{x}' \in S$ such that $g_i(\mathbf{x}') < 0$ for all $i \in I$, then $\bar{\mathbf{x}}$ satisfies the KKT conditions (14.14).*

Another setting in which the existence of Lagrange multipliers is guaranteed is the following:

Theorem 14.2.4. *Suppose that in Problem (14.13), the functions $g_1, \ldots, g_{m_1}$ are concave and $h_1, \ldots, h_{m_2}$ are linear. Let $\bar{\mathbf{x}} \in S$ be a local minimum. Then, $\bar{\mathbf{x}}$ satisfies the KKT conditions (14.14).*

In particular, Theorem 14.2.4 implies that when all the constraints in problem (14.13) are linear, one can always find Lagrange multipliers for any local minimum of problem (14.13).

So far we have only discussed necessary optimality conditions for constrained optimization problems. Let us now turn our attention to sufficient conditions. The following theorem can be viewed as an extension of the first order sufficient condition in Proposition 14.2.2 to the constrained setting.

Theorem 14.2.5. *Suppose that in Problem (14.13), the functions $f, g_1, \ldots, g_{m_1}$ are convex, $h_1, \ldots, h_{m_2}$ are linear, and $X = \mathbb{R}^n$. Let $\bar{\mathbf{x}} \in \mathbb{R}^n$ be feasible for (14.13). If there exist vectors $\bar{\mathbf{v}} \in \mathbb{R}^{m_1}$ and $\bar{\mathbf{w}} \in \mathbb{R}^{m_2}$ such that $(\bar{\mathbf{x}}, \bar{\mathbf{v}}, \bar{\mathbf{w}})$ satisfies the KKT conditions (14.14), then $\bar{\mathbf{x}}$ is a global minimum.*

To demonstrate the usage of the above results, let us consider the following example:

Example 14.2.4. (Linear Programming) Consider the standard form linear programming (LP):

$$
\begin{aligned}
\text{minimize} \quad & f(\mathbf{x}) \equiv \mathbf{c}^T \mathbf{x} \\
\text{subject to} \quad & h_j(\mathbf{x}) \equiv \mathbf{a_j}^T \mathbf{x} - b_j = 0 \quad \text{for } j = 1, \ldots, m, \\
& g_i(\mathbf{x}) \equiv -x_i \leq 0 \quad \text{for } i = 1, \ldots, n,
\end{aligned} \quad (14.15)
$$

where $\mathbf{a_1}, \ldots, \mathbf{a_m}, \mathbf{c} \in \mathbb{R}^n$ and $b_1, \ldots, b_m \in \mathbb{R}$. Since

$$
\begin{aligned}
\nabla f(\mathbf{x}) &= \mathbf{c}, \\
\nabla g_i(\mathbf{x}) &= -\mathbf{e_i} \quad \text{for } i = 1, \ldots, n, \\
\nabla h_j(\mathbf{x}) &= \mathbf{a_j} \quad \text{for } j = 1, \ldots, m,
\end{aligned}
$$

the KKT conditions associated with (14.15) are give

$$\mathbf{c} - \sum_{i=1}^{n} v_i \mathbf{e_i} + \sum_{j=1}^{m} w_j \mathbf{a_j} \;=\; \mathbf{0},$$

$$
\begin{aligned}
x_i v_i &= 0 \quad \text{for} \\
v_i &\geq 0 \quad \text{for} \\
\mathbf{a_j}^T \mathbf{x} &= b \quad \text{for} \\
x_i &\geq 0 \quad \text{for}
\end{aligned}
$$

The above system can be written more compactly as

$$
\begin{aligned}
\mathbf{A}\mathbf{x} &= \mathbf{b}, \quad \mathbf{x} \geq \mathbf{0}, \\
\mathbf{A}^T \mathbf{w} + \mathbf{c} &= \mathbf{v}, \quad \mathbf{v} \geq \mathbf{0}, \\
\mathbf{x}^T \mathbf{v} &= 0,
\end{aligned}
$$

where $\mathbf{A}$ is an $m \times n$ matrix whose j–th row is $\mathbf{a_j}$,
who are familiar with the theory of linear programmin
that (a) is primal feasibility, (b) is dual feasibility,
In particular, when we apply Theorem 14.2.4 to Pro
strong duality theorem of linear programming.

14.2.3   Lagrangian Duality

Given an optimization problem $\mathcal{P}$ (the primal proble
a dual problem whose properties are closely related
investigation, consider the following primal problem:

$$
(P) \qquad
\begin{aligned}
v_p^* \;=\; & \quad \inf \quad f(\mathbf{x}) \\
& \text{subject to} \quad g_i(\mathbf{x}) \leq 0 \\
& \qquad\qquad\quad\; h_j(\mathbf{x}) = 0 \\
& \qquad\qquad\quad\; \mathbf{x} \in X.
\end{aligned}
$$

Here, $f, g_1, \ldots, g_{m_1}, h_1, \ldots, h_{m_2} : \mathbb{R}^n \to \mathbb{R}$ are *arbi*
arbitrary nonempty subset of $\mathbb{R}^n$. For the sake of br
two sets of constraints in (P) as $\mathbf{g}(\mathbf{x}) \leq \mathbf{0}$ and $\mathbf{h}(\mathbf{x})$
is given by $\mathbf{g}(\mathbf{x}) = (g_1(\mathbf{x}), \ldots, g_{m_1}(\mathbf{x}))$ and $\mathbf{h} : \mathbb{R}^n$
$(h_1(\mathbf{x}), \ldots, h_{m_2}(\mathbf{x}))$.

Now, the **Lagrangian dual problem** associate
problem:

$$
(D) \qquad
\begin{aligned}
v_d^* \;=\; & \quad \sup \quad \theta(\mathbf{u}, \mathbf{v}) \equiv \text{in} \\
& \text{subject to} \quad \mathbf{u} \geq \mathbf{0}.
\end{aligned}
$$

Here, $L : \mathbb{R}^n \times \mathbb{R}^{m_1} \times \mathbb{R}^{m_2} \to \mathbb{R}$ is the Lagrangian function given by

$$L(\mathbf{x}, \mathbf{u}, \mathbf{v}) = f(\mathbf{x}) + \sum_{i=1}^{m_1} u_i g_i(\mathbf{x}) + \sum_{j=1}^{m_2} v_j h_j(\mathbf{x}) = f(\mathbf{x}) + \mathbf{u}^T \mathbf{g}(\mathbf{x}) + \mathbf{v}^T \mathbf{h}(\mathbf{x}). \quad (14.16)$$

Observe that the above formulation can be viewed as a penalty function approach, in the sense that we incorporate the primal constraints $\mathbf{g}(\mathbf{x}) \leq \mathbf{0}$ and $\mathbf{h}(\mathbf{x}) = \mathbf{0}$ into the objective function of (D) using the Lagrange multipliers $\mathbf{u}$ and $\mathbf{v}$. Also, since the set X is arbitrary, there can be many different Lagrangian dual problems for the same primal problem, depending on which constraints are handled as $\mathbf{g}(\mathbf{x}) \leq \mathbf{0}$ and $\mathbf{h}(\mathbf{x}) = \mathbf{0}$, and which constraints are treated by X. However, different choices of the Lagrangian dual problem will in general lead to different outcomes, both in terms of the dual optimal value as well as the computational efforts required to solve the dual problem.

Let us now investigate the relationship between (P) and (D). For any $\bar{\mathbf{x}} \in X$ and $(\bar{\mathbf{u}}, \bar{\mathbf{v}}) \in \mathbb{R}_+^{m_1} \times \mathbb{R}^{m_2}$, we have

$$\inf_{\mathbf{x} \in X} L(\mathbf{x}, \bar{\mathbf{u}}, \bar{\mathbf{v}}) \leq f(\bar{\mathbf{x}}) + \bar{\mathbf{u}}^T \mathbf{g}(\bar{\mathbf{x}}) + \bar{\mathbf{v}}^T \mathbf{h}(\bar{\mathbf{x}}) \leq \sup_{\mathbf{u} \geq \mathbf{0}} L(\bar{\mathbf{x}}, \mathbf{u}, \mathbf{v}).$$

This implies that

$$\sup_{\mathbf{u} \geq \mathbf{0}} \inf_{\mathbf{x} \in X} L(\mathbf{x}, \mathbf{u}, \mathbf{v}) \leq \inf_{\mathbf{x} \in X} \sup_{\mathbf{u} \geq \mathbf{0}} L(\mathbf{x}, \mathbf{u}, \mathbf{v}). \quad (14.17)$$

In particular, we have the following weak duality theorem, which asserts that the dual objective value is always a lower bound on the primal objective value:

Theorem 14.2.6. (Weak Duality) *Let $\bar{\mathbf{x}}$ be feasible for (P) and $(\bar{\mathbf{u}}, \bar{\mathbf{v}})$ be feasible for (D). Then, we have $\theta(\bar{\mathbf{u}}, \bar{\mathbf{v}}) \leq f(\bar{\mathbf{x}})$. In particular, if $v_d^* = +\infty$, then (P) has no feasible solution.*

Given the primal–dual pair of problems (P) and (D), the **duality gap** between them is defined as $\Delta = v_p^* - v_d^*$. By Theorem 14.2.6, we always have $\Delta \geq 0$. It would be nice to have $\Delta = 0$ (i.e., zero duality gap). However, as the following example shows, this is not true in general.

Example 14.2.5. Consider the following problem from [1, Example 6.2.2]:

$$\begin{aligned} \text{minimize} \quad & f(\mathbf{x}) \equiv -2x_1 + x_2 \\ \text{subject to} \quad & h(\mathbf{x}) \equiv x_1 + x_2 - 3 = 0, \\ & \mathbf{x} \in X, \end{aligned} \quad (14.18)$$

where $X \subset \mathbb{R}^2$ is the following discrete set:

$$X = \{(0,0), (0,4), (4,4), (4,0), (1,2), (2,1)\} .$$

By enumeration, we see that the optimal value of (
point $(x_1, x_2) = (2, 1)$. Now, one can verify that the
by

$$\theta(v) \quad = \quad \min_{x \in X} \{-2x_1 + x_2 + v(x_1$$

$$= \quad \begin{cases} -4 + 5v & \text{for } v \leq -\mathbf{1} \\ -8 + v & \text{for } -1 \leq \\ -3v & \text{for } v \geq 2. \end{cases}$$

It follows that $\max_v \theta(v) = -6$, which is attained at
gap in this example is $\Delta = -3 - (-6) = 3 > 0$.

The above example raises the important questio:
zero. It turns out that there is a relatively simple an
we proceed, let us introduce the following definition:

Definition 14.2.1. We say that $(\bar{\mathbf{x}}, \bar{\mathbf{u}}, \bar{\mathbf{v}})$ is a **sadd**
function L defined in (14.16) if the following conditic

1. $\bar{\mathbf{x}} \in X$,

2. $\bar{\mathbf{u}} \geq \mathbf{0}$, and

3. for all $\mathbf{x} \in X$ and $(\mathbf{u}, \mathbf{v}) \in \mathbb{R}^{m_1} \times \mathbb{R}^{m_2}$ with $\mathbf{u} \geq$

$$L(\bar{\mathbf{x}}, \mathbf{u}, \mathbf{v}) \leq L(\bar{\mathbf{x}}, \bar{\mathbf{u}}, \bar{\mathbf{v}}) \leq L$$

In particular, observe that $(\bar{\mathbf{x}}, \bar{\mathbf{u}}, \bar{\mathbf{v}})$ is a saddle poi
X when $(\mathbf{u}, \mathbf{v})$ is fixed at $(\bar{\mathbf{u}}, \bar{\mathbf{v}})$, and that $(\bar{\mathbf{u}}, \bar{\mathbf{v}})$ m
$\mathbb{R}^{m_1} \times \mathbb{R}^{m_2}$ with $\mathbf{u} \geq \mathbf{0}$ when $\mathbf{x}$ is fixed at $\bar{\mathbf{x}}$.
We are now ready to state the following theorem:

Theorem 14.2.7. (Saddle Point Optimality Cor
with $\bar{\mathbf{x}} \in X$ and $\bar{\mathbf{u}} \geq \mathbf{0}$ is a saddle point of L iff

1. $L(\bar{\mathbf{x}}, \bar{\mathbf{u}}, \bar{\mathbf{v}}) = \min_{\mathbf{x} \in X} L(\mathbf{x}, \bar{\mathbf{u}}, \bar{\mathbf{v}})$,

2. $\mathbf{g}(\bar{\mathbf{x}}) \leq \mathbf{0}$ and $\mathbf{h}(\bar{\mathbf{x}}) = \mathbf{0}$, and

3. $\bar{\mathbf{u}}^T \mathbf{g}(\bar{\mathbf{x}}) = 0$.

Moreover, the point $(\bar{\mathbf{x}}, \bar{\mathbf{u}}, \bar{\mathbf{v}})$ is a saddle point of L iff
solutions to (P) and (D), respectively, with $f(\bar{\mathbf{x}}) = \theta($
gap.

In other words, the existence of a saddle point $(\bar{\mathbf{x}}, \bar{\mathbf{u}}, \bar{\mathbf{v}})$ of L implies that

$$\inf_{\mathbf{x} \in X} L(\mathbf{x}, \bar{\mathbf{u}}, \bar{\mathbf{v}}) = L(\bar{\mathbf{x}}, \bar{\mathbf{u}}, \bar{\mathbf{v}}) = \sup_{\mathbf{u} \geq \mathbf{0}} L(\bar{\mathbf{x}}, \mathbf{u}, \mathbf{v}),$$

which in turn implies that

$$\sup_{\mathbf{u} \geq \mathbf{0}} \inf_{\mathbf{x} \in X} L(\mathbf{x}, \mathbf{u}, \mathbf{v}) = \inf_{\mathbf{x} \in X} \sup_{\mathbf{u} \geq \mathbf{0}} L(\mathbf{x}, \mathbf{u}, \mathbf{v}),$$

i.e., inequality (14.17) holds with equality, and $v_p^* = v_d^*$.

Now, if we want to apply Theorem 14.2.7 to certify that the duality gap between (P) and (D) is zero, we need to produce a saddle point of the Lagrangian function L, which is not always an easy task. The following theorem, which is an application of Sion's minimax theorem [7] (see [8] for an elementary proof), provides an easy-to-check sufficient condition for certifying zero duality gap.

Theorem 14.2.8. *Let L be the Lagrangian function defined in* (14.16). *Suppose that*

1. *X is a compact convex subset of $\mathbb{R}^n$,*

2. *$(\mathbf{u}, \mathbf{v}) \mapsto L(\mathbf{x}, \mathbf{u}, \mathbf{v})$ is continuous and concave on $\mathbb{R}_+^{m_1} \times \mathbb{R}^{m_2}$ for each $\mathbf{x} \in X$, and*

3. *$\mathbf{x} \mapsto L(\mathbf{x}, \mathbf{u}, \mathbf{v})$ is continuous and convex on X for each $(\mathbf{u}, \mathbf{v}) \in \mathbb{R}_+^{m_1} \times \mathbb{R}^{m_2}$.*

Then, we have

$$\sup_{\mathbf{u} \geq \mathbf{0}} \inf_{\mathbf{x} \in X} L(\mathbf{x}, \mathbf{u}, \mathbf{v}) = \inf_{\mathbf{x} \in X} \sup_{\mathbf{u} \geq \mathbf{0}} L(\mathbf{x}, \mathbf{u}, \mathbf{v}).$$

Let us now illustrate some of the above results with an example.

Example 14.2.6. (Semidefinite Programming) Consider the following standard form semidefinite programming (SDP):

$$\begin{aligned} \inf \quad & f(\mathbf{Z}) \equiv \operatorname{tr}(\mathbf{CZ}), \\ \text{subject to} \quad & h_j(\mathbf{Z}) \equiv b_j - \operatorname{tr}(\mathbf{A_jZ}) = 0 \quad \text{for } j = 1, \ldots, m, \qquad (14.19) \\ & \mathbf{Z} \in X \equiv \mathbf{S}_+^n, \end{aligned}$$

where $\mathbf{C}, \mathbf{A_1}, \ldots, \mathbf{A_m} \in \mathbb{R}^{n \times n}$ are symmetric matrices, $b_1, \ldots, b_m \in \mathbb{R}$ and $\mathbf{S}_+^n$ is the set of $n \times n$ symmetric positive semidefinite matrices. The Lagrangian dual associated with (14.19) is given by

$$\sup \quad \theta(\mathbf{v}) \equiv \inf_{\mathbf{Z} \in \mathbf{S}_+^n} \left\{ \operatorname{tr}(\mathbf{CZ}) + \sum_{j=1}^m v_j (b_j - \operatorname{tr}(\mathbf{A_jZ})) \right\}. \qquad (14.20)$$

Now, for any fixed $\mathbf{v} \in \mathbb{R}^m$, we have

$$\theta(\mathbf{v}) = \begin{cases} \mathbf{b}^T\mathbf{v} & \text{if } \mathbf{C} - \sum_{j=1}^{m} v_j \mathbf{A}_{} \\ -\infty & \text{otherwise.} \end{cases}$$

To see this, let $\mathbf{U}\mathbf{\Lambda}\mathbf{U}^T$ be the spectral decomposition
pose that $\Lambda_{ii} < 0$ for some $i = 1, \ldots, n$. Consider th
Clearly, we have $\mathbf{Z}(\alpha) \in \mathbf{S}_+^n$ for all $\alpha > 0$. Moreover,

$$\text{tr}\left(\left(\mathbf{C} - \sum_{j=1}^{m} v_j \mathbf{A_j}\right)\mathbf{Z}(\alpha)\right) = \alpha \cdot \text{tr}\left((\mathbf{U}\right.$$

$$= \alpha \cdot \text{tr}\left(\mathbf{\Lambda}\mathbf{e}\right.$$

whence

$$\theta(\mathbf{v}) = \mathbf{b}^T\mathbf{v} + \inf_{\mathbf{Z} \in \mathbf{S}_+^n} \text{tr}\left(\left(\mathbf{C} - \sum_{j=1}^{m} v_j \mathbf{A}\right.\right.$$

On the other hand, if $\mathbf{C} - \sum_{j=1}^{m} v_j \mathbf{A_j} \in \mathbf{S}_+^n$, then we ha
0 for any $\mathbf{Z} \in \mathbf{S}_+^n$. It follows that $\theta(\mathbf{v}) = \mathbf{b}^T\mathbf{v}$ in this
Now, using (14.21), we see that (14.20) is equival

$$\sup \quad \mathbf{b}^T\mathbf{v}$$

$$\text{subject to} \quad \mathbf{C} - \sum_{j=1}^{m} v_j \mathbf{A_j} \in$$

which is known as a dual standard form SDP.

14.3 Application Examples

In the past decade optimization techniques, especial
niques, have been widely used in various engineering
gineering, mechanical engineering, and electrical eng
neering in particular, optimization techniques have be
in communications [9–14], networking [15–19], signal
circuit design [23]. In this section, we briefly go throu
munications, networking, and signal processing to ill
the results introduced in the previous section to solve

Example 14.3.1. (Power Allocation Optimiza
Channels) Consider the transmission over n paralle

channel, $i \in \{1, \ldots, n\}$, is characterized by the channel power gain, $h_i \geq 0$, and the additive Gaussian noise power, $\sigma_i > 0$. Let the transmit power allocated to the ith channel be denoted by $p_i \geq 0$. The maximum information rate that can be reliably transmitted over the ith channel is given by [24]

$$r_i = \log\left(1 + \frac{h_i p_i}{\sigma_i}\right). \tag{14.23}$$

Given a constraint P on the total transmit power over n channels, i.e., $\sum_{i=1}^{n} p_i \leq P$, we want to optimize the allocated power $p_1, \ldots, p_n$ such that the sum rate of n channels, $\sum_{i=1}^{n} r_i$, is maximized. This problem is thus formulated as

$$
\begin{aligned}
\text{maximize} \quad & \sum_{i=1}^{n} \log\left(1 + \frac{h_i p_i}{\sigma_i}\right) \\
\text{subject to} \quad & \sum_{i=1}^{n} p_i \leq P, \\
& p_i \geq 0 \quad \text{for } i = 1, \ldots, n.
\end{aligned} \tag{14.24}
$$

For convenience, we rewrite the above problem equivalently as

$$
\begin{aligned}
\text{minimize} \quad & f(\mathbf{p}) \equiv -\sum_{i=1}^{n} \log\left(1 + \frac{h_i p_i}{\sigma_i}\right) \\
\text{subject to} \quad & h(\mathbf{p}) \equiv \sum_{i=1}^{n} p_i - P \leq 0, \\
& g_i(\mathbf{p}) \equiv -p_i \leq 0 \quad \text{for } i = 1, \ldots, n, \\
& \mathbf{p} \in \mathbb{R}^n,
\end{aligned} \tag{14.25}
$$

where $\mathbf{p} = [p_1, \ldots, p_n]^T$. It is easy to verify that $f(\mathbf{p})$ is convex, and $h(\mathbf{p})$, $g_1(\mathbf{p}), \ldots, g_n(\mathbf{p})$ are all affine and thus convex. According to Theorem 14.2.5, if we can find a set of feasible solutions $\bar{\mathbf{p}} = [\bar{p}_1, \ldots, \bar{p}_n]^T \in \mathbb{R}^n$ for the above constrained minimization problem as well as a set of $u \geq 0$ and $v_i \geq 0, i = 1, \ldots, n$ such that the following KKT conditions are satisfied,

$$
\begin{aligned}
\nabla f(\bar{\mathbf{p}}) + u\nabla h(\bar{\mathbf{p}}) + \sum_{i=1}^{n} v_i \nabla g_i(\bar{\mathbf{p}}) &= \mathbf{0}, & \text{(a)} \\
u h(\bar{\mathbf{p}}) &= 0, & \text{(b)} \\
v_i g_i(\bar{\mathbf{p}}) &= 0 \quad \text{for } i = 1, \ldots, n, & \text{(c)}
\end{aligned} \tag{14.26}
$$

then we can claim that $\bar{\mathbf{p}}$ is a global minimum for this problem. Suppose that $u > 0$. From (b), it follows that $h(\bar{\mathbf{p}}) = 0$, i.e., $\sum_{i=1}^{n} \bar{p}_i = P$. From (a), it follows that

$$\bar{p}_i = \frac{1}{u - v_i} - \frac{\sigma_i}{h_i} \quad \text{for } i = 1, \ldots, n. \tag{14.27}$$

Suppose that $\bar{p}_i > 0$. From (c), it follows that $v_i = 0$. Then from (14.27), it follows that $\bar{p}_i = \frac{1}{u} - \frac{\sigma_i}{h_i} > 0$. Clearly, if this inequality holds, the corresponding $\bar{p}_i$ will satisfy both (a) and (c). Otherwise, the preassumption of $\bar{p}_i > 0$ cannot be true and the only feasible value for $\bar{p}_i$ is $\bar{p}_i = 0$. In this can always find a $v_i \geq 0$ such that $\bar{p}_i = 0$ holds in (1- $u > 0$, the set of feasible values for $\bar{p}_i$ that satisfy bo

$$\bar{p}_i = \left(\frac{1}{u} - \frac{\sigma_i}{h_i} \right)^+ \quad \text{for } i = 1, .$$

where $(x)^+ = \max(0, x)$ for $x \in \mathbb{R}$. Furthermore, rec to satisfy $\sum_{i=1}^n \bar{p}_i = P$, i.e.,

$$\sum_{i=1}^n \left(\frac{1}{u} - \frac{\sigma_i}{h_i} \right)^+ = P.$$

Note that for any $P > 0$, in the above equation there of u (which can be found numerically by a simple bise $0 < u < \max_i(h_i/\sigma_i)$. With the root of u, the corres satisfy all the KKT conditions in (a), (b), and (c), an solutions for Problem (14.25). It is worth noting that power allocation in (14.28) is known as the "water-fil

**Example 14.3.2. (Transmit Optimization for]
with Per-Antenna Power Constraints)** Consid
MIMO AWGN channel with n transmitting antenna
The propagation channel from the transmitter to tl
a real matrix, $\mathbf{H} \in \mathbb{R}^{m \times n}$, in which all the colum
empty", i.e., there is at least one element in each c
additive noises at m receiving antennas are assumed
variables with zero mean and unit variance. The t1
antenna, $i \in \{1, \ldots, n\}$, are denoted by $x_i(t) \in \mathbb{R}, t$
to a per-antenna average power constraint P_i, i.e., E
denotes the expectation. Let $\mathbf{Z} \in \mathbf{S}_+^n$ denote the trai
$\mathbf{Z} = E\left\{ \mathbf{x}(t) \left(\mathbf{x}(t) \right)^T \right\}$, where $\mathbf{x}(t) = [x_1(t), \ldots, x_n(t]$
transmit power constraints can then be expressed as

$$\text{tr}\left(\mathbf{A}_i \mathbf{Z} \right) \leq P_i \quad \text{for } i = 1, \ldots$$

where $\mathbf{A}_i \in \mathbb{R}^{n \times n}$ is a matrix with all zero elements element being one.

For any transmit covariance matrix $\mathbf{Z} \in \mathbf{S}_+^n$, the 1 the MIMO AWGN channel is given by [25]

$$r = \log \det \left(\mathbf{I} + \mathbf{H} \mathbf{Z} \mathbf{H}^T \right)$$

where $\mathbf{I}$ denotes an identity matrix. The problem of ou the rate r over $\mathbf{Z} \in \mathbf{S}_+^n$ subject to the set of per-antenna which can be equivalently formulated as

$$
\begin{aligned}
v_p^* \quad = \quad &\text{minimize} \quad f(\mathbf{Z}) \equiv -\log\det\left(\mathbf{I} + \mathbf{HZH}^T\right) \\
&\text{subject to} \quad g_i(\mathbf{Z}) \equiv \text{tr}\left(\mathbf{A}_i \mathbf{Z}\right) - P_i \le 0 \ \text{ for } i = 1,\dots,n, \\
&\qquad\qquad \mathbf{Z} \in \mathbf{S}_+^n.
\end{aligned}
\tag{14.32}
$$

In the following, we apply the Lagrangian duality to solve the above problem. The Lagrangian function for this problem is given by

$$
L(\mathbf{Z},\mathbf{u}) = f(\mathbf{Z}) + \sum_{i=1}^n u_i g_i(\mathbf{Z}) = -\log\det\left(\mathbf{I} + \mathbf{HZH}^T\right) + \sum_{i=1}^n u_i(\text{tr}\left(\mathbf{A}_i \mathbf{Z}\right) - P_i),
\tag{14.33}
$$

where $\mathbf{u} = [u_1,\dots,u_n]^T \in \mathbb{R}_+^n$. The Lagrangian dual problem associated with problem (14.32) is then given by

$$
\begin{aligned}
v_d^* \quad = \quad &\text{maximize} \quad \theta(\mathbf{u}) \equiv \min_{\mathbf{Z}\in\mathbf{S}_+^n} L(\mathbf{Z},\mathbf{u}) \\
&\text{subject to} \quad \mathbf{u} \ge \mathbf{0}.
\end{aligned}
\tag{14.34}
$$

It can be verified that the conditions listed in Theorem 14.2.8 are all satisfied for the Lagrangian function $L(\mathbf{Z},\mathbf{u})$ given in (14.33). We thus conclude that $v_p^* = v_d^*$, i.e., the duality gap for Problem (14.32) is zero. Accordingly, we can solve this problem equivalently by solving its dual problem (14.34), as shown next.

First, we solve the minimization problem in (14.34) to obtain the dual function $\theta(\mathbf{u})$ for any given $\mathbf{u} \ge \mathbf{0}$. Observe that $\theta(\mathbf{u})$ can be explicitly written as

$$
\theta(\mathbf{u}) = \min_{\mathbf{Z}\in\mathbf{S}_+^n} -\log\det\left(\mathbf{I} + \mathbf{HZH}^T\right) + \text{tr}\left(\mathbf{A}_u \mathbf{Z}\right) - \sum_{i=1}^n u_i P_i
\tag{14.35}
$$

where $\mathbf{A}_u = \sum_{i=1}^n u_i \mathbf{A}_i$ is a diagonal matrix with the ith diagonal element equal to $u_i, i = 1,\dots,n$. Note that for the minimization problem in the above, the optimal solution for $\mathbf{Z}$ is independent of the term $\sum_{i=1}^n u_i P_i$, which thus can be ignored. To solve this minimization problem, we first observe that if any diagonal element in $\mathbf{A}_u$, say, $u_i, i \in \{1,\dots,n\}$, is equal to zero, then the minimum value for this problem becomes $-\infty$, which is attained by, e.g., taking $\mathbf{Z} = \alpha\mathbf{1}_i\mathbf{1}_i^T$, where $\mathbf{1}_i$ denotes an $n \times 1$ vector with all zero elements except for the ith element being one, and letting $\alpha \to \infty$. Next, we consider the case where all u_i's are greater than zero. In this case, $\mathbf{A}_u$ is full-rank and thus its inverse exists. By defining a new variable $\bar{\mathbf{Z}} = \mathbf{A}_u^{1/2}\mathbf{Z}\mathbf{A}_u^{1/2} \in \mathbf{S}_+^n$ and using the fact that $\text{tr}\left(\mathbf{AB}\right) = \text{tr}\left(\mathbf{BA}\right)$, the minimization problem in (14.35) can be rewritten as

$$
\min_{\bar{\mathbf{Z}}\in\mathbf{S}_+^n} -\log\det\left(\mathbf{I} + \mathbf{HA}_u^{-1/2}\bar{\mathbf{Z}}\mathbf{A}_u^{-1/2}\mathbf{H}^T\right) + \text{tr}\left(\bar{\mathbf{Z}}\right).
\tag{14.36}
$$

Let the SVD of $\mathbf{HA}_u^{-1/2}$ be denoted by

$$
\mathbf{HA}_u^{-1/2} = \mathbf{U}\mathbf{\Lambda}\mathbf{V}^T,
\tag{14.37}
$$

where $\mathbf{U} \in \mathbb{R}^{m \times m}$ and $\mathbf{V} \in \mathbb{R}^{n \times n}$ are unitary matrices
matrix with the diagonal elements being denoted b
and $\lambda_i \geq 0, i = 1, \ldots, k$. Substituting (14.37) into (1
$\log \det (\mathbf{I} + \mathbf{AB}) = \log \det (\mathbf{I} + \mathbf{BA})$ yield

$$\min_{\bar{\mathbf{Z}} \in \mathbf{S}_+^n} - \log \det \left(\mathbf{I} + \boldsymbol{\Lambda} \mathbf{V}^T \bar{\mathbf{Z}} \mathbf{V} \boldsymbol{\Lambda}^T \right)$$

By letting $\hat{\mathbf{Z}} = \mathbf{V}^T \bar{\mathbf{Z}} \mathbf{V}$ and using the fact that tr
equivalent problem of (14.38) as

$$\min_{\hat{\mathbf{Z}} \in \mathbf{S}_+^n} - \log \det \left(\mathbf{I} + \boldsymbol{\Lambda} \hat{\mathbf{Z}} \boldsymbol{\Lambda}^T \right) + $$

Recall the Hadamard's inequality [24], which states th
$\prod_{i=1}^{m} \mathbf{X}_{ii}$, iff $\mathbf{X}$ is a diagonal matrix, where $\mathbf{X}_{ii}$ denc
of $\mathbf{X}$, $i = 1, \ldots, m$. Applying this result to Problem
minimum value for this problem is attained iff $\hat{\mathbf{Z}}$ is
diagonal elements of $\hat{\mathbf{Z}}$ be denoted by $p_1, \ldots, p_n$. Sin
can be simplified as

$$\begin{aligned} \text{minimize} \quad & - \sum_{i=1}^{n} \log(1 + \lambda_i^2 p_i) - \\ \text{subject to} \quad & p_i \geq 0 \ \text{ for } i = 1, \ldots, n \end{aligned}$$

Note that in the above problem, for convenience we
for $i = k + 1, \ldots, n$. Similar to Exercise 14.3.1, the gl
problem can be shown to be the following water-fillir

$$p_i = \left(1 - \frac{1}{\lambda_i^2} \right)^+ \quad \text{for } i = 1, .$$

To summarize, for any given $\mathbf{u} > \mathbf{0}$, the optimal sc
problem in (14.35) is given by

$$\mathbf{Z}_u = \mathbf{A}_u^{-1/2} \mathbf{V} \hat{\mathbf{Z}} \mathbf{V}^T \mathbf{A}_u^{-1/2}$$

where $\hat{\mathbf{Z}}$ is a diagonal matrix with the diagonal element
the dual function $\theta(\mathbf{u})$ in (14.35) can be simplified to

$$\theta(\mathbf{u}) = \begin{cases} - \sum_{i=1}^{k} \left(\log(\lambda_i^2) \right)^+ + \sum_{i=1}^{k} \left(1 - \frac{1}{\lambda_i^2} \right)^+ - \\ -\infty \end{cases}$$

where $\lambda_1, \ldots, \lambda_k$ are related to $\mathbf{u}$ via (14.37).

Next, we solve the dual problem (14.34) by maximizing the dual function $\theta(\mathbf{u})$ in (14.43) over $\mathbf{u} \geq \mathbf{0}$. The corresponding dual optimal solution of $\mathbf{u}$ then leads to the optimal solution of $\mathbf{Z}_u$ in (14.42) for the primal problem (14.32). Since $v_d^* = v_p^* \geq 0$, in fact we only need to consider the maximization of $\theta(\mathbf{u})$ over $\mathbf{u} > \mathbf{0}$ in (14.43). However, due to the coupled structure of λ_i's and u_i's shown in (14.37), it is not evident whether $\theta(\mathbf{u})$ in (14.43) is differentiable over u_i's for $\mathbf{u} > \mathbf{0}$. As a result, conventional decent methods to find the global minimum for differentiable convex functions such as Newton's method are ineffective for our problem at hand. Thus, we resort to an alternative method, known as *subgradient* based method, to handle the non-differentiable function $\theta(\mathbf{u})$. First, we introduce the definition of subgradient for an arbitrary real-valued function $z(\mathbf{x})$ defined over a nonempty convex set $S \subset \mathbb{R}^n$. We assume that $z(\mathbf{x})$ has a finite maximum. However, $z(\mathbf{x})$ need not be continuously differentiable nor have an analytical expression for its differential. In this case, a vector $\mathbf{v} \in \mathbb{R}^n$ is called the subgradient of $z(\mathbf{x})$ at point $\mathbf{x} = \mathbf{x}_0$ if for any $\mathbf{x} \in S$, the following inequality holds:

$$z(\mathbf{x}) \leq z(\mathbf{x}_0) + \mathbf{v}^T(\mathbf{x} - \mathbf{x}_0). \tag{14.44}$$

If at any point $\mathbf{x} \in S$ a corresponding subgradient $\mathbf{v}$ for $z(\mathbf{x})$ is attainable, then the maximum of $z(\mathbf{x})$ can be found via an iterative search over $\mathbf{x} \in S$ based on $\mathbf{v}$ (see, e.g., the ellipsoid method [26]). Since $\theta(\mathbf{u})$ is defined over a convex set $\mathbf{u} > \mathbf{0}$ and has a finite maximum, the dual problem (14.34) can thus be solved by a subgradient based method. Next, we show that the subgradient of $\theta(\mathbf{u})$ at any point $\mathbf{u} > \mathbf{0}$ is given by $[\mathrm{tr}\,(\mathbf{A}_1\mathbf{Z}_u) - P_1, \ldots, \mathrm{tr}\,(\mathbf{A}_n\mathbf{Z}_u) - P_n]^T$, where $\mathbf{Z}_u$ is given in (14.42). Suppose that at any two points $\mathbf{u} > \mathbf{0}$ and $\mathbf{u}' > \mathbf{0}$, $\theta(\mathbf{u})$ and $\theta(\mathbf{u}')$ are attained by $\mathbf{Z} = \mathbf{Z}_u$ and $\mathbf{Z} = \mathbf{Z}'_u$, respectively. Then, we have the following inequalities:

$$
\begin{aligned}
\theta(\mathbf{u}') &= L(\mathbf{Z}'_u, \mathbf{u}') \\
&= \min_{\mathbf{Z} \in \mathbf{S}_+^n} L(\mathbf{Z}, \mathbf{u}') \\
&\leq L(\mathbf{Z}_u, \mathbf{u}') \\
&= -\log\det\left(\mathbf{I} + \mathbf{H}\mathbf{Z}_u\mathbf{H}^T\right) + [\mathrm{tr}\,(\mathbf{A}_1\mathbf{Z}_u) - P_1, \ldots, \mathrm{tr}\,(\mathbf{A}_n\mathbf{Z}_u) - P_n]\mathbf{u}' \\
&= -\log\det\left(\mathbf{I} + \mathbf{H}\mathbf{Z}_u\mathbf{H}^T\right) + [\mathrm{tr}\,(\mathbf{A}_1\mathbf{Z}_u) - P_1, \ldots, \mathrm{tr}\,(\mathbf{A}_n\mathbf{Z}_u) - P_n]\mathbf{u} \\
&\quad + [\mathrm{tr}\,(\mathbf{A}_1\mathbf{Z}_u) - P_1, \ldots, \mathrm{tr}\,(\mathbf{A}_n\mathbf{Z}_u) - P_n](\mathbf{u}' - \mathbf{u}) \\
&= L(\mathbf{Z}_u, \mathbf{u}) + [\mathrm{tr}\,(\mathbf{A}_1\mathbf{Z}_u) - P_1, \ldots, \mathrm{tr}\,(\mathbf{A}_n\mathbf{Z}_u) - P_n](\mathbf{u}' - \mathbf{u}) \\
&= \theta(\mathbf{u}) + [\mathrm{tr}\,(\mathbf{A}_1\mathbf{Z}_u) - P_1, \ldots, \mathrm{tr}\,(\mathbf{A}_n\mathbf{Z}_u) - P_n](\mathbf{u}' - \mathbf{u}),
\end{aligned}
$$

from which the subgradient of $\theta(\mathbf{u})$ follows.

Last, we can verify that the optimal primal and dual solutions, $\mathbf{Z}_u$ given in

(14.42) and the corresponding $\mathbf{u} > 0$ satisfy (a) of th

$$\nabla f(\mathbf{Z}_u) + \sum_{i=1}^{n} u_i \nabla g_i(\mathbf{Z}_u) \;=\; \mathbf{0},$$

$$u_i g_i(\mathbf{Z}_u) \;=\; 0 \quad \text{for } i = 1,$$

while since $\mathbf{u} > 0$, from (b) it follows that $g_i(\mathbf{Z}_u) =$
hold for $i = 1, \ldots, n$. Thus, all transmit antennas
maximum power levels with the optimal transmit co
consistent with the observation that the subgradient
the optimal dual solution of $\mathbf{u}$ should vanish to $\mathbf{0}$.

Example 14.3.3. (Power Efficient Beamformin
work via SDP Relaxation) In this example, we
nonconvex problem can be solved via convex techniqu
we consider a two-way relay channel (TWRC) consis
and S2, each with a single antenna and a relay node, R
$M \geq 2$. It is assumed that the transmission protoco
utive equal-duration time slots for one round of infor
and S2 via R. During the first time slot, both S1 and
R, which linearly processes the received signal and t
signal to S1 and S2 during the second time slot. It is a
chronization has been established among S1, S2, and
The received baseband signal at R in the first time s

$$\mathbf{y}_R(n) = \mathbf{h}_1 \sqrt{p_1} s_1(n) + \mathbf{h}_2 \sqrt{p_2} s_2(n) \cdot$$

where $\mathbf{y}_R(n) \in \mathbb{C}^M$ is the received signal vector at sy
with N denoting the total number of transmitted sy
$\mathbf{h}_1 \in \mathbb{C}^M$ and $\mathbf{h}_2 \in \mathbb{C}^M$ represent the channel vector
to R, respectively, which are assumed to be constan
$s_1(n)$ and $s_2(n)$ are the transmitted symbols from S
$E\{|s_1(n)|\} = 1$, $E\{|s_2(n)|\} = 1$, and $|\cdot|$ denoting the
number; p_1 and p_2 denote the transmit powers of S
$\mathbf{z}_R(n) \in \mathbb{C}^M$ is the receiver noise vector, independe
of generality, it is assumed that $\mathbf{z}_R(n)$ has a circula
sian (CSCG) distribution with zero mean and identit
by $\mathbf{z}_R(n) \sim \mathcal{CN}(\mathbf{0}, \mathbf{I}), \forall n$. Upon receiving the mixe
processes it with amplify-and-forward (AF) relay ope
analogue relaying, and then broadcasts the processed s
second time slot. Mathematically, the linear processi
at the relay can be concisely represented as

$$\mathbf{x}_R(n) = \mathbf{A} \mathbf{y}_R(n), \quad n = 1, \ldots, N$$

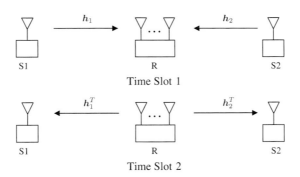

Figure 14.1: The two-way multi antenna relay channel.

where $\mathbf{x}_R(n) \in \mathbb{C}^M$ is the transmitted signal at R, and $\mathbf{A} \in \mathbb{C}^{M \times M}$ is the relay processing matrix.

Note that the transmit power of R can be shown equal to

$$\begin{aligned}
p_R(\mathbf{A}) &= \mathrm{E}\left[\mathrm{tr}\left(\mathbf{x}_R(n)\mathbf{x}_R^H(n)\right)\right] \\
&= \|\mathbf{A}\mathbf{h}_1\|_2^2\, p_1 + \|\mathbf{A}\mathbf{h}_2\|_2^2\, p_2 + \mathrm{tr}(\mathbf{A}\mathbf{A}^H). \quad (14.48)
\end{aligned}$$

We can assume w.l.o.g. that channel reciprocity holds for TWRC during uplink and downlink transmissions, i.e., the channels from R to S1 and S2 during the second time slot are given as $\mathbf{h}_1^T$ and $\mathbf{h}_2^T$, respectively. Thus, the received signals at S1 can be written as

$$\begin{aligned}
y_1(n) &= \mathbf{h}_1^T \mathbf{x}_R(n) + z_1(n) \\
&= \mathbf{h}_1^T \mathbf{A}\mathbf{h}_1 \sqrt{p_1} s_1(n) + \mathbf{h}_1^T \mathbf{A}\mathbf{h}_2 \sqrt{p_2} s_2(n) + \mathbf{h}_1^T \mathbf{A}\mathbf{z}_R(n) + z_1(n) \quad (14.49)
\end{aligned}$$

for $n = 1, \ldots, N$, where $z_1(n)$'s are the independent receiver noise samples at S1, and it is assumed that $z_1(n) \sim \mathcal{CN}(0,1), \forall n$. Note that on the right-hand side of (14.49), the first term is the self-interference of S1, while the second term contains the desired message from S2. Assuming that both $\mathbf{h}_1^T \mathbf{A}\mathbf{h}_1$ and $\mathbf{h}_1^T \mathbf{A}\mathbf{h}_2$ are perfectly known at S1 via training-based channel estimation prior to data transmission, S1 can first subtract its self-interference from $y_1(n)$ and then coherently demodulate $s_2(n)$. The above practice is known as analogue network coding (ANC). From (14.49), subtracting the self-interference from $y_1(n)$ yields

$$\tilde{y}_1(n) = \tilde{h}_{21}\sqrt{p_2} s_2(n) + \tilde{z}_1(n), \quad n = 1, \ldots, N \quad (14.50)$$

where $\tilde{h}_{21} = \mathbf{h}_1^T \mathbf{A}\mathbf{h}_2$, and $\tilde{z}_1(n) \sim \mathcal{CN}(0, \left\|\mathbf{A}^H \mathbf{h}_1^*\right\|_2^2 + 1)$, where $*$ denotes the complex conjugate. From (14.50), for a given $\mathbf{A}$, the maximum achievable SNR for the end-to-end link from S2 to S1 via R, denoted

$$\gamma_{21} = \frac{\left|\mathbf{h}_1^T \mathbf{A} \mathbf{h}_2\right|^2 p_2}{\left\|\mathbf{A}^H \mathbf{h}_1^*\right\|_2^2 + 1}$$

Similarly, it can be shown that the maximum SNR γ
via R is given as

$$\gamma_{12} = \frac{\left|\mathbf{h}_2^T \mathbf{A} \mathbf{h}_1\right|^2 p_1}{\left\|\mathbf{A}^H \mathbf{h}_2^*\right\|_2^2 + 1}.$$

Now we minimize the relay transmission power
constraints that the achievable SNRs γ_{21} and γ_{12} over
two target values, $\bar{\gamma}_1$ and $\bar{\gamma}_2$. As such, the optimizati

$$\text{minimize}_{\mathbf{A}} \quad p_R := \|\mathbf{A}\mathbf{h}_1\|_2^2 p_1 + \|\mathbf{A}\mathbf{h}_2\|_2^2$$

$$\text{subject to} \quad \left|\mathbf{h}_1^T \mathbf{A} \mathbf{h}_2\right|^2 \geq \frac{\bar{\gamma}_1}{p_2} \left\|\mathbf{A}^H \mathbf{h}_1^*\right\|_2^2$$

$$\left|\mathbf{h}_2^T \mathbf{A} \mathbf{h}_1\right|^2 \geq \frac{\bar{\gamma}_2}{p_1} \left\|\mathbf{A}^H \mathbf{h}_2^*\right\|_2^2$$

For the convenience of analysis, we further modify t
First, let $\text{Vec}(\mathbf{Q})$ be a $K^2 \times 1$ vector associated w
$\mathbf{Q} = [\mathbf{q}_1, \ldots, \mathbf{q}_K]^T$, where $\mathbf{q}_k \in \mathbb{C}^K, k = 1, \ldots, K$
$\left[\mathbf{q}_1^T, \ldots, \mathbf{q}_K^T\right]^T$. Next, with $\mathbf{b} = \text{Vec}(\mathbf{A})$ and $\boldsymbol{\Theta} = $
can express p_R in the objective function of (14.53) as
with $\boldsymbol{\Phi} = (\text{diag}(\boldsymbol{\Theta}^T, \boldsymbol{\Theta}^T))^{\frac{1}{2}}$, where $\text{diag}(\mathbf{A}, \mathbf{B})$ denc
with $\mathbf{A}$ and $\mathbf{B}$ as the diagonal square matrices. Sir
and $\mathbf{f}_2 = \text{Vec}\left(\mathbf{h}_2 \mathbf{h}_1^T\right)$. Then, from (14.53) it follows t
$\left|\mathbf{h}_2^T \mathbf{A} \mathbf{h}_1\right|^2 = \left|\mathbf{f}_2^T \mathbf{b}\right|^2$. Furthermore, by defining

$$\mathbf{h}_i = \begin{bmatrix} \mathbf{h}_i(1,1) & 0 & \mathbf{h}_i(2,1) & 0 \\ 0 & \mathbf{h}_i(1,1) & 0 & \mathbf{h}_i(2,1 \end{bmatrix}$$

we have $\left\|\mathbf{A}^H \mathbf{h}_i^*\right\|_2^2 = \|\mathbf{h}_i \mathbf{b}\|_2^2, i = 1, 2$. Using the ab
can be rewritten as

$$\text{minimize}_{\mathbf{b}} \quad p_R := \|\boldsymbol{\Phi}\mathbf{b}\|_2^2$$

$$\text{subject to} \quad \left|\mathbf{f}_1^T \mathbf{b}\right|^2 \geq \frac{\bar{\gamma}_1}{p_2} \|\mathbf{h}_1 \mathbf{b}\|$$

$$\left|\mathbf{f}_2^T \mathbf{b}\right|^2 \geq \frac{\bar{\gamma}_2}{p_1} \|\mathbf{h}_2 \mathbf{b}\|$$

The above problem can be shown to be still nonconvex. However, in the following, we show that the exact optimal solution could be obtained via a relaxed SDP problem.

We first define $\mathbf{E}_0 = \mathbf{\Phi}^H \mathbf{\Phi}$, $\mathbf{E}_1 = \frac{p_2}{\tilde{\gamma}_1} \mathbf{f}_1^* \mathbf{f}_1^T - \mathbf{h}_1^H \mathbf{h}_1$, and $\mathbf{E}_2 = \frac{p_1}{\tilde{\gamma}_2} \mathbf{f}_2^* \mathbf{f}_2^T - \mathbf{h}_2^H \mathbf{h}_2$. Since standard semidefinite programming (SDP) formulations only involve real variables and constants, we introduce a new real matrix variable as $\mathbf{X} = [\mathbf{b}_R; \mathbf{b}_I] \times [\mathbf{b}_R; \mathbf{b}_I]^T$, where $\mathbf{b}_R = Re(\mathbf{b})$ and $\mathbf{b}_I = Im(\mathbf{b})$ are the real and imaginary parts of $\mathbf{b}$, respectively. To rewrite the norm representations at (14.54) in terms of $\mathbf{X}$, we need to rewrite $\mathbf{E}_0$, $\mathbf{E}_1$, and $\mathbf{E}_2$, as expanded matrices $\mathbf{F}_0$, $\mathbf{F}_1$, and $\mathbf{F}_2$, respectively, in terms of their real and imaginary parts. Specifically, to write out $\mathbf{F}_0$, we first define the short notations $\mathbf{\Phi}_R = Re(\mathbf{\Phi})$ and $\mathbf{\Phi}_I = Im(\mathbf{\Phi})$; then we have

$$\mathbf{F}_0 = \begin{bmatrix} \mathbf{\Phi}_R^T \mathbf{\Phi}_R + \mathbf{\Phi}_I^T \mathbf{\Phi}_I & \mathbf{\Phi}_I^T \mathbf{\Phi}_R - \mathbf{\Phi}_R^T \mathbf{\Phi}_I \\ \mathbf{\Phi}_R^T \mathbf{\Phi}_I - \mathbf{\Phi}_I^T \mathbf{\Phi}_R & \mathbf{\Phi}_R^T \mathbf{\Phi}_R + \mathbf{\Phi}_I^T \mathbf{\Phi}_I \end{bmatrix}.$$

The expanded matrices $\mathbf{F}_1$ and $\mathbf{F}_2$ can be generated from $\mathbf{E}_1$ and $\mathbf{E}_2$ in a similar way, where the two terms in $\mathbf{E}_1$ or $\mathbf{E}_2$ could first be expanded separately then summed together.

As such, problem (14.54) can be equivalently rewritten as

$$\begin{aligned} \text{minimize}_{\mathbf{X}} \quad & p_R := \text{tr}(\mathbf{F}_0 \mathbf{X}) \\ \text{subject to} \quad & \text{tr}(\mathbf{F}_1 \mathbf{X}) \geq 1, \; \text{tr}(\mathbf{F}_2 \mathbf{X}) \geq 1, \; \mathbf{X} \succeq 0, \\ & \text{rank}(\mathbf{X}) = 1. \end{aligned} \tag{14.55}$$

The above problem is still not convex given the last rank-one constraint. However, if we remove such a constraint, this problem is relaxed into a convex SDP problem as shown below.

$$\begin{aligned} \text{minimize}_{\mathbf{X}} \quad & p_R := \text{tr}(\mathbf{F}_0 \mathbf{X}) \\ \text{subject to} \quad & \text{tr}(\mathbf{F}_1 \mathbf{X}) \geq 1, \; \text{tr}(\mathbf{F}_2 \mathbf{X}) \geq 1, \; \mathbf{X} \succeq 0. \end{aligned} \tag{14.56}$$

Given the convexity of the above SDP problem, the optimal solution could be efficiently found by various convex optimization methods. Note that SDP relaxation usually leads to an optimal $\mathbf{X}$ for problem (14.56) that is of rank r with $r \geq 1$, which makes it impossible to reconstruct the exact optimal solution for Problem (14.54) when $r > 1$. A commonly adopted method in the literature to obtain a feasible rank-one (but in general suboptimal) solution from the solution of SDP relaxation is via "randomization" (see, e.g., [27] and references therein). Fortunately, we show in the following that with the special structure in Problem (14.56), we could efficiently reconstruct an optimal rank-one solution from its optimal solution that could be of rank r with $r > 1$, based on some elegant results derived for SDP relaxation in [28]. In other words, we could obtain the exact optimal solution for the nonconvex problem in (14.55) without losing any optimality, and as efficiently as solving a convex problem.

Theorem 14.3.1. *Assume that an optimal solution* *found for Problem* (14.56), *we could efficiently const* *solution* $\mathbf{X}^{\star\star}$ *of rank one, i.e.,* $\mathbf{X}^{\star\star}$ *is the optimal* (14.56).

Proof: Please refer to [21].

Note that the above proof is self-constructive, ba obtain a routine to obtain an optimal rank-one solut $\mathbf{X}^{\star}$. Then we could map the solution back to obtain problem in (14.53).

14.4 Exercises

Exercise 14.4.1. Please indicate whether the follow

1. $\left\{\mathbf{x} : \frac{\mathbf{a}^T\mathbf{x}-\mathbf{b}}{\mathbf{c}^T\mathbf{x}+\mathbf{d}} \leq 1; \mathbf{c}^T\mathbf{x} + \mathbf{d} < 0\right\}$;

2. $\{\mathbf{x} : \mathbf{A}\mathbf{x} = \mathbf{b}, \|\mathbf{x}\|_2 = 1\}$;

3. $\{\mathbf{X} : X_{11}\mathbf{a_0} + X_{22}\mathbf{a_1} \succeq 0, \mathbf{a_0} \in \mathbf{S}^n, \mathbf{a_1} \in \mathbf{S}^n\}$; ($X$ in matrix $\mathbf{X}$)

4. $\{\mathbf{X} : \mathbf{a}^T\mathbf{X}\mathbf{a} = 1\}$

Exercise 14.4.2. Please indicate whether the follow concave or neither.

1. $f(\mathbf{x}) = \sup_w \left\{\log \sum_{i=1}^n e^{\frac{x_i}{w}}\right\}$;

2. $f(\mathbf{x}) = -(x_1x_2x_3)^{1/3}$, $\mathbf{x} > \mathbf{0}$;

3. $f(\mathbf{X}) = \log\det\left(\mathbf{A}^T\mathbf{X}\mathbf{A}\right)$; $\mathbf{X} \succ \mathbf{0}$;

4. $f(\mathbf{x}) = \mathbf{x}^T\mathbf{A}\mathbf{x} + 2\mathbf{x} - 5$, $\mathbf{A} = \begin{bmatrix} 0 & 1 \\ 1 & 0 \end{bmatrix}$.

Exercise 14.4.3. With the following problem form questions.

$$\begin{array}{ll} \text{minimize}_\mathbf{x} & -(x_1 + x_2) \\ \text{subject to} & \|\mathbf{a_1}\mathbf{x}\|_2 \leq 1, \\ & \|\mathbf{a_2}\mathbf{x} - \mathbf{b_2}\|_2 \end{array}$$

where $\mathbf{x} = [x_1, x_2]^T$, $\mathbf{a_1} = \mathbf{a_2} = \begin{bmatrix} 1 & 0 \\ 0 & 1 \end{bmatrix}$, and $\mathbf{b_2} =$

1. Is this problem convex?

2. Does Slater's constraint condition hold?

3. What is the optimal solution for this problem? (Hint: Try to solve this problem graphically if the KKT conditions are hard to solve.)

4. What is the optimal objective value for the dual problem?

5. What is the optimal value for the dual variable associated with the second constraint?

Exercise 14.4.4. Given the optimization problem shown in Exercise 14.4.3, please reformulate it as a semidefinite programming (SDP) problem, then derive the dual problem of the resulting SDP problem.

Exercise 14.4.5. With the following optimization problem, answer the followup questions.

$$\text{maximize}_\mathbf{P} \quad \sum_{i=1}^{n} \log\left(1 + \frac{P_i}{\delta_i}\right)$$

$$\text{subject to} \quad \sum_{i=1}^{n} P_i = P_{total},$$

$$\mathbf{P} \geq \mathbf{0},$$

where $\mathbf{P} = [P_1, \ldots, P_n]^T$, and $\delta_i > 0, i = 1, \ldots, n$.

1. Is KKT sufficient for us to get the optimal solution for the above problem?

2. Is KKT necessary for the optimal solution?

3. Please write out the KKT conditions for this problem.

4. Please solve the general form of optimal P_i's.

5. If $n = 3$, $\delta_1 = 2$, $\delta_2 = 10$, $\delta_3 = 5$, and $P_{total} = 10$, what are the optimal P_i values?

Exercise 14.4.6. Let $1 \leq m \leq n$ be integers, and let $\mathbf{A}$ be an $m \times n$ matrix with full row rank. Furthermore, let $\mathbf{c} \in \mathbb{R}^n$ and $\mathbf{Q} \in \mathbf{S}_{++}^n$ be given. Consider the following optimization problem:

$$\text{minimize} \quad \frac{1}{2}\mathbf{x}^T\mathbf{Q}\mathbf{x} + \mathbf{c}^T\mathbf{x} \tag{14.57}$$

$$\text{subject to} \quad \mathbf{A}\mathbf{x} = \mathbf{0}.$$

1. Explain why the KKT conditions are necessary

2. Write down the KKT conditions associated wit optimal solution to (14.57) in closed form.

Exercise 14.4.7. Let $f : \mathbb{R}^n \to \mathbb{R}$ be a differentiab the following problem:

$$\text{minimize} \quad f(\mathbf{x})$$
$$\text{subject to} \quad \mathbf{x} \geq \mathbf{0}.$$

Show that $\bar{\mathbf{x}} \in \mathbb{R}^n$ is an optimal solution to (14.58 system:

$$\nabla f(\bar{\mathbf{x}}) \geq \mathbf{0},$$
$$\bar{\mathbf{x}} \geq \mathbf{0},$$
$$\bar{\mathbf{x}}^T \nabla f(\bar{\mathbf{x}}) = 0.$$

Exercise 14.4.8. This problem is concerned with f enclosing ellipsoid of a set of vectors.

1. Let $\mathbf{u} \in \mathbb{R}^n$ be fixed, and define the function $g :$ Find $\nabla g(\mathbf{X})$.

2. Let $V = \{\mathbf{v^1}, \dots, \mathbf{v^m}\} \subset \mathbb{R}^n$ be a set of vectors following problem:

$$\inf \quad -\log \det(\mathbf{X})$$
$$\text{subject to} \quad \left\| \mathbf{X}\mathbf{v^i} \right\|_2^2 \leq 1 \quad i =$$
$$\mathbf{X} \in \mathbf{S}^n_{++}.$$

Let $\bar{\mathbf{X}}$ be an optimal solution to (14.59) (it ca exists). Write down the KKT conditions that $\bar{\mathbf{X}}$

3. Suppose that $m = n$ and $\mathbf{v^i} = \mathbf{e_i}$ for $i = 1,$ standard basis vector. Using the above result, de to (14.59) and find the corresponding Lagrange

Exercise 14.4.9. Let $\mathbf{a} \in \mathbb{R}^n$, $b \in \mathbb{R}$ and $\mathbf{c} \in \mathbb{R}^n$ be Consider the following problem:

$$\text{minimize} \quad \sum_{i=1}^{n} \frac{c_i}{x_i}$$
$$\text{subject to} \quad \sum_{i=1}^{n} a_i x_i = b,$$
$$\mathbf{x} \geq \mathbf{0}.$$

1. Let $u_1 \in \mathbb{R}$ and $\mathbf{u_2} \in \mathbb{R}^n$ be the Lagrange multipliers associated with the equality and inequality constraints, respectively. Write down the KKT conditions associated with (14.60).

2. Give explicit expressions for $\bar{\mathbf{x}} \in \mathbb{R}^n$, $\bar{u}_1 \in \mathbb{R}$ and $\bar{\mathbf{u}}_2 \in \mathbb{R}^n$ such that $(\bar{\mathbf{x}}, \bar{u}_1, \bar{\mathbf{u}}_2)$ satisfies the KKT conditions above.

3. Is the solution $\bar{\mathbf{x}} \in \mathbb{R}^n$ found above an optimal solution to (14.60)? Explain.

References

[1] M. S. Bazaraa, H. D. Sherali, and C. M. Shetty, *Nonlinear Programming: Theory and Algorithms*, 2nd ed., ser. Wiley-Interscience Series in Discrete Mathematics and Optimization. New York: John Wiley & Sons, Inc., 1993.

[2] D. P. Bertsekas, *Nonlinear Programming*, 2nd ed. Belmont, Massachusetts: Athena Scientific, 1999.

[3] S. Boyd and L. Vandenberghe, *Convex Optimization*. Cambridge: Cambridge University Press, 2004, available online at `http://www.stanford.edu/~boyd/cvxbook/`.

[4] D. G. Luenberger and Y. Ye, *Linear and Nonlinear Programming*, 3rd ed., ser. International Series in Operations Research and Management Science. New York: Springer Science+Business Media, LLC, 2008, vol. 116.

[5] R. A. Horn and C. R. Johnson, *Matrix Analysis*. Cambridge: Cambridge University Press, 1985.

[6] M. Brookes, "The Matrix Reference Manual," 2005, available online at `http://www.ee.ic.ac.uk/hp/staff/dmb/matrix/intro.html`.

[7] M. Sion, "On General Minimax Theorems," *Pacific Journal of Mathematics*, vol. 8, no. 1, pp. 171–176, 1958.

[8] H. Komiya, "Elementary Proof for Sion's Minimax Theorem," *Kodai Mathematical Journal*, vol. 11, no. 1, pp. 5–7, 1988.

[9] S. Cui, A. J. Goldsmith, and A. Bahai, "Energy-constrained modulation optimization," *IEEE Transactions on Wireless Communications*, vol. 4, no. 5, pp. 2349–2360, September 2005.

[10] S. Cui, M. Kisialiou, Z.-Q. Luo, and Z. Ding, "Robust blind multiuser detection against signature waveform mismatch based on second order cone programming," *IEEE Transactions on Wireless Communications*, vol. 4, no. 4, pp. 1285–1291, July 2005.

516

[11] R. Zhang and Y.-C. Liang, "Exploiting multi-ant
trum sharing in cognitive radio networks," *IEE*
in Signal Processing, vol. 2, no. 1, pp. 88–102, F

[12] W. Yu and R. Lui, "Dual methods for nonconv
multicarrier systems," *IEEE Transactions on Cc*
pp. 1310–1322, July 2006.

[13] L. Zhang, R. Zhang, Y.-C. Liang, Y. Xin, an
ship between the multi-antenna secrecy commu
communications," *IEEE Transactions on Comm*
1877–1886, June 2010.

[14] R. Zhang, S. Cui, and Y.-C. Liang, "On ergodic
nitive multiple-access and broadcast channels,"
mation Theory, vol. 55, no. 11, pp. 5161–5178, N

[15] M. Chiang, "Balancing transport and physical la
works: Jointly optimal congestion control and p
on Selected Areas in Communications, vol. 23,
2005.

[16] J. Xiao, S. Cui, Z. Q. Luo, and A. J. Goldsmith
networks under energy constraint," *IEEE Trans*
vol. 54, no. 2, pp. 413–422, February 2005.

[17] A. So and Y. Ye, "Theory of semidefinite prog
localization," *Mathematical Programming*, vol. 1

[18] S. Cui and A. J. Goldsmith, "Cross-layer desig
works using cooperative MIMO techniques," *E*
Journal, Special Issue on Advances in Signal Pro
signs, vol. 86, pp. 1804–1814, August 2006.

[19] R. Madan, S. Cui, S. Lall, and A. Goldsmit
tion of transmission schemes in energy-constrain
IEEE/ACM Transactions on Networking, vol. 1
cember 2007.

[20] Z. Quan, S. Cui, H. V. Poor, and A. Sayed, "C
ing for cognitive radios," *IEEE Signal Processir*
cognitive radios, vol. 25, no. 6, pp. 60–73, Janua

[21] R. Zhang, Y.-C. Liang, C.-C. Chai, and S. Cui
two-way multi-antenna relay channel with analc
Journal on Selected Areas of Communications,
June 2009.

[22] R. Zhang and S. Cui, "Cooperative interference management with miso beam-forming," *IEEE Transactions on Signal Processing*, vol. 58, no. 10, pp. 5450 –5458, October 2010.

[23] S. P. Boyd, S.-J. Kim, D. D. Patil, and M. A. Horowitz, "Digital Circuit Optimization via Geometric Programming," *Operations Research*, vol. 53, no. 6, pp. 899–932, November 2005.

[24] T. Cover and J. Thomas, *Elements of Information Theory.* New York: Wiley, 1991.

[25] I. E. Telatar, "Capacity of multi-antenna Gaussian channels," *Eur. Trans. Telecommun.*, vol. 10, no. 6, pp. 585–595, November 1999.

[26] R. G. Bland, D. Goldfarb, and M. J. Todd, "The ellipsoid method: a survey," *Operations Research*, vol. 29, no. 6, pp. 1039–1091, November 1981.

[27] Z.-Q. Luo and W. Yu, "An introduction to convex optimization for communications and signal processing," *IEEE Journal of Selected Topics in Signal Processing*, vol. 24, no. 8, pp. 1426–1438, August 2006.

[28] Y. Ye and S. Zhang, "New results on quadratic minimization," *SIAM J. Optim.*, vol. 14, pp. 245–267, 2003.

Chapter 15 Linear Programming and Mixed Integer Programming

Bogdan Dumitrescu[‡]
[‡]Tampere University of Technology, Finland

Linear programming (LP) is the class of optimization problems whose criterion and constraints are linear. Such problems appeared first in economy, for example in activity planning or resource allocation, but are now ubiquitous also in engineering. Although being apparently the simplest type of optimization problem that has no analytic solution, LP has interesting properties and its study has led to significant developments and generalizations for the whole optimization field, both in terms of theory and algorithms.

Mixed integer programming (MIP) problems also have linear criterion and constraints, but some of their variables are constrained to integer values. Despite their resemblance to LP, they have distinct properties and are solved by dedicated algorithms, using LP as a fundamental tool, but much more complex.

This chapter will present the basic theory, the main algorithmic approaches and a few typical signal processing problems that can be modeled through LP or MIP.

15.1 Linear Programming

Having linear criterion and constraints, LP belongs to convex optimization and shares all the good properties given by convexity. Some of these properties have particular forms that can be used for designing algorithms tailored specifically for LP. Besides an introduction to the standard LP forms and transformations that lead to them, this section contains characterizations of optimality and, in particular, duality, and a short presentation of the two main algorithmic solutions to LP: the simplex method and interior-point methods.

15.1.1   General Presentation

An LP problem, in standard (named also primal, or

$$
\begin{aligned}
\mu = \quad &\min \quad && \mathbf{c}^T\mathbf{x} \\
&\text{s.t.} \quad && \mathbf{A}\mathbf{x} = \mathbf{b} \\
& && \mathbf{x} \geq 0
\end{aligned}
$$

By $\mathbf{x} \geq 0$ we understand that each element of the v
We name μ the value of the LP problem; we denote
i.e., a vector for which $\mathbf{c}^T\mathbf{x}^\star = \mu$, $\mathbf{A}\mathbf{x}^\star = \mathbf{b}$, $\mathbf{x}^\star \geq 0$
and n columns. Typically, the number of rows is sm.
happen that the system $\mathbf{A}\mathbf{x} = \mathbf{b}$ has no solution at
independent rows, i.e., rank $(\mathbf{A}) = m$; if this is not t
of $\mathbf{A}$ can be eliminated using the QR factorization be
Even with these assumptions, Problem (15.1) may be
of $\mathbf{A}\mathbf{x} = \mathbf{b}$ satisfies the positivity constraint; in this
$\mu = \infty$. Due to the linearity of the constraints, the f
that may be bounded or unbounded. Since the grad
hence constant, if (15.1) has a finite solution, then i
feasible polytope, i.e., it is a vertex, an edge or even a
is unbounded in the direction $-\mathbf{c}$, then the criterion
LP problem is *unbounded* and $\mu = -\infty$.

Example 15.1.1. The LP problem

$$
\begin{aligned}
&\min \quad && x_2 + x_3 \\
&\text{s.t.} \quad && 0.8x_1 + 0.5x_2 + x_3 \\
& && x_1 \geq 0, \ x_2 \geq 0, \ x_3
\end{aligned}
$$

is a particular case of (15.1), with $\mathbf{A} = [0.8 \ 0.5 \ 1]$
feasible domain is the gray triangle shown on the left
decreases along the direction $-\mathbf{c} = [0 \ -1 \ -1]^T$ and
when x_2 and x_3 have their least possible values; the s
vertex of the feasible triangle.

All types of linear constraints can be brought to t
inequality $\mathbf{A}\mathbf{x} \leq \mathbf{b}$ can be transformed into the equa
a vector $\mathbf{s} \geq 0$ of *slack* variables. Similarly, the inequ
into $\mathbf{A}\mathbf{x} - \mathbf{s} = \mathbf{b}$, now $\mathbf{s} \geq 0$ being named *surplus*.
unrestricted, it can be expressed as $\mathbf{x} = \mathbf{x}_+ - \mathbf{x}_-$, w
have negative values but it has a lower bound $\mathbf{x} \geq \mathbf{x}$
$\mathbf{x} - \mathbf{x}_0 \geq 0$, which changes the equality constraint of
Concerning the criterion, if the LP problem would imp
this is equivalent to minimizing $-\mathbf{c}^T\mathbf{x}$, i.e., the sign c
getting the standard form. Note that several of the ab

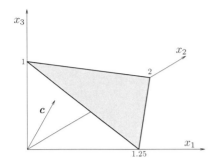 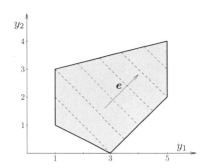

Figure 15.1: Feasible domains of the LP problems from Examples 15.1.1 (left) and 15.1.2 (right).

the number of variables when bringing the problem to standard form; however, the standard form has the advantage of efficient specialized algorithms.

Another common formulation of LP problems is in the inequality (named also dual) form

$$\begin{aligned} \max \quad & \mathbf{e}^T \mathbf{y} \\ \text{s.t.} \quad & \mathbf{D}\mathbf{y} \le \mathbf{f} \end{aligned} \tag{15.3}$$

The variable $\mathbf{y} \in \mathbb{R}^m$ is unrestricted and the constraints are only in inequality form, with all variables at the left of the "$\le$" sign and the free term at the right. Typically, the matrix $\mathbf{D} \in \mathbb{R}^{n \times m}$ has more rows than columns. Again, one can bring any LP problem to form (15.3), if so desired. The inequality $\mathbf{D}\mathbf{y} \ge \mathbf{f}$ is transformed by reversing the sign into $-\mathbf{D}\mathbf{y} \le -\mathbf{f}$. Equality constraints are transformed by variable elimination. Consider a single equality $\mathbf{g}^T\mathbf{y} = \alpha$; the vector $\mathbf{g}$ has at least a non-zero component, say, g_k; it results that $y_k = \left(\alpha - \sum_{i \neq k} g_i y_i\right)/g_k$; this expression of y_k is substituted in all the other constraints of the optimization problem and hence the variable y_k is eliminated.

Example 15.1.2. The LP problem

$$\begin{aligned} \max \quad & y_1 + y_2 \\ \text{s.t.} \quad -y_1 \quad\quad & \le -1 \\ y_1 \quad\quad & \le 5 \\ -y_1 - 2y_2 & \le -3 \\ y_1 - y_2 & \le 3 \\ -y_1 + 4y_2 & \le 11 \end{aligned} \tag{15.4}$$

is a particular case of (15.3). Its feasible domain is the gray polygon shown on the right of Figure 15.1; the reader is invited to identify the correspondence between the five edges of the polygon and the five inequality constraints. The level curves are lines for which $y_1 + y_2$ is constant and are drawn with dashed lines; they are

orthogonal to the gradient $\mathbf{e} = [1\ 1]^T$. The criterion in
the gradient. The optimum is attained in $\mathbf{y}^\star = [5\ 4]^T$,
polygon. If the criterion were defined by $\mathbf{e} = [1\ 0]^T$
the whole rightmost edge of the polygon, defined b
Finally, if the last constraint would be removed, the
would disappear; the polygon would become unbou
would grow indefinitely since y_2 has no upper boune
unbounded.

15.1.2   Transformations of the Standar

The data of Problem (15.1), namely $\mathbf{A}$, $\mathbf{b}$ and $\mathbf{c}$, ca
an equivalent problem is obtained (in the sense that
the same solution as (15.1)), also in standard form.
the transformed problem by $\tilde{\mathbf{A}}$, $\tilde{\mathbf{b}}$ and $\tilde{\mathbf{c}}$, but in gen
performed in place. The two important transformatic
  (i) Reordering the variables: $\tilde{\mathbf{x}} = \mathbf{Px}$, where $\mathbf{P} \in \mathbb{R}$
Since $\mathbf{Ax} = \mathbf{AP}^T\mathbf{Px}$, this is equivalent to the transfo
which amounts to rearrange the *columns* of $\mathbf{A}$ and th
the permutation of the variable. The vector $\mathbf{b}$ is unch
that the variables are conveniently ordered, since thi
of the solution.
  (ii) Multiplication with a nonsingular matrix $\mathbf{M}$
$\mathbf{MAx} = \mathbf{Mb}$ has the same solution as $\mathbf{Ax} = \mathbf{b}$, the tr
$\mathbf{Mb}$ is valid. The vector $\mathbf{c}$ is not affected. In particul
matrix, in which case the transformation consists of
the system $\mathbf{Ax} = \mathbf{b}$.
  An example of transformed problem is the canon
describe all transformations as performed in place)

$$\tilde{\mathbf{A}} \to \mathbf{A} = \begin{bmatrix} 1 & 0 & \cdots & 0 & a_{1,m+1} \\ 0 & 1 & \cdots & 0 & a_{2,m+1} \\ & & \vdots & & \\ 0 & 0 & \cdots & 1 & a_{m,m+1} \end{bmatrix}$$

We assume that the first columns of $\mathbf{A}$ are linearly in
case, permutation of columns can be used. The abo
Gauss-Jordan elimination. The process has m steps a
is transformed into the i-th unit vector $\mathbf{e}_i$ (whose eler
i-th, which is equal to 1). To this purpose, we first
by $a_{i,i}$, which means that the i-th row of $\mathbf{A}$ and b_i are
then pivoting is necessary: equation i of $\mathbf{Ax} = \mathbf{b}$ is pe
for which $a_{k,i} \neq 0$. A better numerical choice is to al
$|a_{k,i}|$. Also, complete pivoting can be used, searching

$\ell \geq i$; a column permutation, i.e., transformation of type (i) may be necessary.)
Then, we apply of transformation of type (ii), with

$$\mathbf{M}_i = \mathbf{I} - \mathbf{m}_i \mathbf{e}_i^T, \quad \text{with } \mathbf{m}_i = [a_{1,i} \ \cdots \ a_{i-1,i} \ 0 \ a_{i+1,i} \ \cdots \ a_{i,m}]^T.$$

These operations amount to multiply equation i of $\mathbf{Ax} = \mathbf{b}$ with $a_{k,i}$ and subtract
it from equation k, for each $k \neq i$. Note that the first $i-1$ columns of $\mathbf{A}$ are not
affected, since the i-th row of $\mathbf{A}$ has zeros in its first $i-1$ positions.

Complete pivoting has the advantage that it always finds m independent
columns if $\text{rank}(\mathbf{A}) = m$; if the rank is smaller, then the last $m - \text{rank}(\mathbf{A})$ rows
of the transformed matrix are zero and hence can be eliminated or, if the corre-
sponding elements of $\mathbf{b}$ are non-zero, infeasibility is detected. However, orthogonal
transformations are numerically better to detect rank deficiency.

The canonical form (15.5) is used in the implementation of the simplex method,
discussed in Section 15.1.5.

15.1.3 Optimality Characterization

We look now at the LP standard form (15.1) and present some properties related
to optimality. The next result is the most significant.

Theorem 15.1.3. *If the LP problem (15.1) has finite value, then it has a solution*
$\mathbf{x}^\star$ *with at most m non-zero elements.*

Proof. If (15.1) has finite value, then the optimum is attained. Assume that $\mathbf{x}$
is a solution of (15.1) having the minimum number $p > m$ of non-zero elements.
Without loss of generality, we can assume that these are $x_1, x_2, \ldots, x_p$. Since
$\text{rank}(\mathbf{A}) < p$, it results that the first p columns of $\mathbf{A}$ are linearly dependent and
so there exists a vector $\mathbf{u} \neq 0$, with $u_i = 0$ for $i > p$, such that $\mathbf{Au} = 0$. We can
assume that $\mathbf{c}^T \mathbf{u} \geq 0$, otherwise we take $-\mathbf{u}$ instead of $\mathbf{u}$.

Suppose first that $\mathbf{c}^T \mathbf{u} > 0$. Since $x_i > 0$ for $i = 1 : p$, it results that for some
$\varepsilon > 0$ small enough, it is true that $x_i - \varepsilon u_i \geq 0$. Putting, $\tilde{\mathbf{x}} = \mathbf{x} - \varepsilon \mathbf{u}$, it results
that $\mathbf{A}\tilde{\mathbf{x}} = \mathbf{b}$, $\tilde{\mathbf{x}} \geq 0$ and $\mathbf{c}^T \tilde{\mathbf{x}} < \mathbf{c}^T \mathbf{x}$. Hence, $\mathbf{x}$ is not optimal.

Suppose now that $\mathbf{c}^T \mathbf{u} = 0$. Since $\mathbf{u} \neq 0$, we can vary ε (no longer constrained
to be positive) until one of the first p components of $\tilde{\mathbf{x}} = \mathbf{x} - \varepsilon \mathbf{u}$ becomes zero,
the others remaining positive. It results that $\tilde{\mathbf{x}}$ is also a solution of (15.1), but has
$p - 1$ non-zero elements.

Hence the assumption $p > m$ leads to a contradiction, and so the theorem is
proved. ∎

Let $\mathcal{I}$ be a subset of $1 : n$ having m elements. If $\mathbf{x}$ satisfies the constraints of
(15.1), the elements x_i are zero for $i \notin \mathcal{I}$ and the matrix $\mathbf{B}$ formed by the columns
of $\mathbf{A}$ with indices in $\mathcal{I}$ is nonsingular, then $\mathbf{x}$ is named a *basic feasible point*. If
$x_i > 0$, $\forall i \in \mathcal{I}$, then the basic feasible point is called *nondegenerate*. The set of
variables x_i with $i \in \mathcal{I}$ is called a *base* (equivalently, $\mathcal{I}$ may be called the base).

Theorem 15.1.4. *A vector* $\mathbf{x}$ *is a basic feasible poir of the feasible polytope of the LP problem (15.1).*

Proof.   Assume that $\mathbf{x}$ is a basic feasible point and th $\mathbf{v}$ and $\mathbf{w}$, both different from $\mathbf{x}$, and $\lambda \in (0,1)$ such tha not a vertex of the feasible polytope. Let $\mathbf{v}_1 = [v_i]_{i \in \mathcal{I}}$, $\mathbf{w}_1, \mathbf{w}_2, \mathbf{x}_1, \mathbf{x}_2$ and note that $\mathbf{x}_2 = 0$. Since $\lambda \mathbf{v}_2 +$ $\mathbf{w}_2 \geq 0$ ($\mathbf{v}$ and $\mathbf{w}$ are feasible) and $\lambda > 0$, $1 - \lambda > 0$, It follows that $\mathbf{Av} = \mathbf{Bv}_1 = \mathbf{b}$, $\mathbf{Aw} = \mathbf{Bw}_1 = \mathbf{b}$ and nonsingular, we conclude that $\mathbf{v}_1 = \mathbf{w}_1$, which is im; that $\mathbf{v} = \mathbf{w} = \mathbf{x}$, contrary to the initial assumptions.
Conversely, assume that $\mathbf{x}$ is a vertex and has ; $\mathcal{I} = \{i \in 1 : n \mid x_i \neq 0\}$. Let $\mathbf{B} = [\mathbf{a}_i]_{i \in \mathcal{I}}$, i.e., the s columns corresponding to the non-zero elements of $\mathbf{x}$ i.e., the columns of $\mathbf{B}$ are not linearly independen Theorem 15.1.3, there exists $\mathbf{u} \neq 0$, with $u_i = 0$ fo Then, for ε small enough, $\mathbf{v} = \mathbf{x} + \varepsilon \mathbf{u}$ and $\mathbf{w} = \mathbf{x}$ $\mathbf{x} = (\mathbf{v} + \mathbf{w})/2$, hence $\mathbf{x}$ is not a vertex. This contr rank of $\mathbf{B}$. It results then that rank $(\mathbf{B}) = p \leq m$ (cannot be linearly independent). If $p = m$, then $\mathbf{x}$ $p < m$, since rank $(\mathbf{A}) = m$, there are $m - p$ colum of $\mathbf{B}$ make a nonsingular matrix, hence fulfilling the c point.

Corollary 15.1.5. If the LP problem (15.1) has finit that is a basic feasible point.

Proof.   Theorem 15.1.3 says that there exists a s zero elements and its proof shows that the correspc maximum rank (otherwise we can decrease p, like in proof of Theorem 15.1.4 shows that in this case $\mathbf{x}$ is ;

So, we have proved rigorously that, as suggested I a vertex of the feasible polytope is always a solutio (this does not exclude the existence of other solutions property and the notion of basic feasible point are es of the simplex method.

15.1.4   Duality Aspects

The Lagrangian associated with the LP standard pro

$$L(\mathbf{x}, \mathbf{y}, \mathbf{s}) = \mathbf{c}^T \mathbf{x} + \mathbf{y}^T (\mathbf{b} - \mathbf{Ax})$$

where $\mathbf{y} \in \mathbb{R}^m$ is the vector of Lagrange multiplier: ity constraints and $\mathbf{s} \in \mathbb{R}^n$, $\mathbf{s} \geq 0$, the vector of mu inequality $\mathbf{x} \geq 0$. The Lagrange dual of the optimization problem (15.1) is

$$\max_{\mathbf{y},\mathbf{s}} \inf_{\mathbf{x}} L(\mathbf{x},\mathbf{y},\mathbf{s}). \tag{15.7}$$

The Lagrangian $L(\mathbf{x},\mathbf{y},\mathbf{s}) = \left(\mathbf{c} - \mathbf{A}^T\mathbf{y} - \mathbf{s}\right)^T \mathbf{x} + \mathbf{y}^T\mathbf{b}$ is linear in $\mathbf{x}$, so it is unbounded from below when the term multiplying $\mathbf{x}$ is non-zero, hence producing

$$\inf_{\mathbf{x}} L(\mathbf{x},\mathbf{y},\mathbf{s}) = \begin{cases} \mathbf{b}^T\mathbf{y}, & \text{if } \mathbf{A}^T\mathbf{y} + \mathbf{s} = \mathbf{c} \\ -\infty, & \text{otherwise} \end{cases} \tag{15.8}$$

Taking into account that $\mathbf{s} \geq 0$, it results that the dual of (15.1) is

$$\nu = \begin{array}{ll} \max & \mathbf{b}^T\mathbf{y} \\ \text{s.t.} & \mathbf{A}^T\mathbf{y} \leq \mathbf{c} \end{array} \tag{15.9}$$

This is an LP problem in inequality form (15.3), with $\mathbf{D} = \mathbf{A}^T$, $\mathbf{e} = \mathbf{b}$, $\mathbf{f} = \mathbf{c}$; this justifies the name *dual* associated with (15.3). Conversely, the dual of (15.3) is an LP problem in standard form (15.1), see Exercise 15.4.

Since LP is a convex optimization problem, strong duality holds, i.e., the primal and dual problems have the same optimal value. For LP, unlike for other convex problems, no special assumption (like the Slater condition) is required for strong duality: feasibility is enough.

Theorem 15.1.6. *If one of the problems (15.1) and (15.9) is feasible, then their optimal values are equal, namely: (a) if one of the problems has finite value, then the other has the same value, i.e., $\mu = \nu$, $\mathbf{c}^T\mathbf{x}^\star = \mathbf{b}^T\mathbf{y}^\star$, where $\mathbf{x}^\star$ and $\mathbf{y}^\star$ are solutions of (15.1) and (15.9), respectively; (b) if one of the problems is unbounded, then the other is infeasible.*

Proof.   Since $\inf_{\mathbf{x}} L(\mathbf{x},\mathbf{y},\mathbf{s}) \leq \mathbf{c}^T\mathbf{x}^\star$, the construction (15.7) of the dual problem implies that $\mathbf{b}^T\mathbf{y} \leq \mathbf{c}^T\mathbf{x}$ for any feasible $\mathbf{x}$, $\mathbf{y}$, i.e., $\nu \leq \mu$. So, if the primal is unbounded ($\mu = -\infty$), then the dual is infeasible; if the dual is unbounded ($\nu = \infty$), then the primal is infeasible. We will prove that if the primal has finite value, then the dual has the same value; the reverse implication can be proved similarly. We present two proofs. The first is simple, but treats only the typical case when $\mathbf{x}^\star$ is a nondegenerate basic feasible point; hence, the proof is incomplete. The second is general, but more technical; for brevity, we use Farkas' lemma.

Proof 1. Corollary 15.1.5 says that $\mathbf{x}^\star$ is a basic feasible point. Assume, without loss of generality, that the first m variables form the base and $\mathbf{A} = [\mathbf{B}\,\mathbf{N}]$, with $\mathbf{B} \in \mathbb{R}^{m \times m}$ nonsingular. Split $\mathbf{x}^T = [\mathbf{x}_1^T\,\mathbf{x}_2^T]$, $\mathbf{c}^T = [\mathbf{c}_1^T\,\mathbf{c}_2^T]$, where the first component has length m. Since $\mathbf{x}_2^\star = 0$, it results that $\mathbf{x}_1^\star = \mathbf{B}^{-1}\mathbf{b}$ and $\mu = \mathbf{c}^T\mathbf{x}^\star = \mathbf{c}_1^T\mathbf{B}^{-1}\mathbf{b}$. For any feasible $\mathbf{x}$, it follows from $\mathbf{A}\mathbf{x} = \mathbf{b}$ that

$$\mathbf{B}\mathbf{x}_1 + \mathbf{N}\mathbf{x}_2 = \mathbf{b} \;\Rightarrow\; \mathbf{x}_1 = \mathbf{B}^{-1}\left(\mathbf{b} - \mathbf{N}\mathbf{x}_2\right) = \mathbf{x}_1^\star - \mathbf{B}^{-1}\mathbf{N}\mathbf{x}_2 \tag{15.10}$$

and so

$$\mathbf{c}^T\mathbf{x} = \mathbf{c}_1^T\mathbf{x}_1 + \mathbf{c}_2^T\mathbf{x}_2 = \mathbf{c}_1^T\mathbf{B}^{-1}\mathbf{b} + (\mathbf{c}_2^T - \mathbf{c}_1^T\mathbf{B}^{-1}\mathbf{N}$$

where we have denoted $\mathbf{r} = \mathbf{c}_2 - \mathbf{N}^T\mathbf{B}^{-T}\mathbf{c}_1$. Assume r
i.e., $\mathbf{x}_1^\star > 0$ (remind that this assumption makes the
that the vector $\mathbf{r}$ has at least one negative element. T
that $\mathbf{r}^T\mathbf{x}_2 < 0$. Since $\|\mathbf{x}_2\|$ can be taken arbitrarily s
by (15.10) can stay into any neighborhood of $\mathbf{x}_1^\star$, suc
for such an $\mathbf{x}_2$, we have obtained $\mathbf{x} \geq 0$ such that A
contradicts the optimality of $\mathbf{x}^\star$. So, the inequality $\mathbf{r}$
　　The constraint of the dual (15.9) has the form

$$\begin{bmatrix} \mathbf{B}^T \\ \mathbf{N}^T \end{bmatrix} \mathbf{y} \leq \begin{bmatrix} \mathbf{c}_1 \\ \mathbf{c}_2 \end{bmatrix}.$$

We build $\mathbf{y}^\star$ by forcing the first m inequality cons
become equalities, which implies that $\mathbf{y}^\star = \mathbf{B}^{-T}\mathbf{c}_1$. I

$$\mathbf{b}^T\mathbf{y}^\star = \mathbf{b}^T\mathbf{B}^{-T}\mathbf{c}_1 = \mathbf{c}_1^T\mathbf{x}_1^\star$$

and

$$\mathbf{r} \geq 0 \ \Rightarrow \ \mathbf{N}^T\mathbf{B}^{-T}\mathbf{c}_1 \leq \mathbf{c}_2 \ \Rightarrow \ \mathbf{N}$$

so $\mathbf{y}^\star$ is feasible and optimal.

　　Proof 2. Given a matrix $\boldsymbol{\Phi} \in \mathbb{R}^{p \times q}$ and a vector $\mathbf{f}$
lemma says that exactly one of the following affirmati
is often named a "theorem of alternatives").
　　(i) $\exists \mathbf{u} \in \mathbb{R}^q$ such that $\boldsymbol{\Phi}\mathbf{u} = \mathbf{f}$ and $\mathbf{u} \geq 0$.
　　(ii) $\exists \mathbf{v} \in \mathbb{R}^p$ such that $\boldsymbol{\Phi}^T\mathbf{v} \geq 0$ and $\mathbf{f}^T\mathbf{v} < 0$.
　　The proof of Farkas' lemma is based on the const
separates two convex sets and we omit it. However,
very intuitive. If (i) holds, then $\mathbf{f}$ lies in the cone form
of the columns of $\boldsymbol{\Phi}$. Then, a vector $\mathbf{v}$ cannot make si
with all columns of $\boldsymbol{\Phi}$ (i.e., $\boldsymbol{\Phi}^T\mathbf{v} \geq 0$) and an obtuse
　　Given an arbitrary $\varepsilon > 0$, it results from the optim
is no $\mathbf{x}$ such that $\mathbf{A}\mathbf{x} = \mathbf{b}$, $\mathbf{x} \geq 0$, $\mathbf{c}^T\mathbf{x} = \mu - \varepsilon - \alpha$,
$\mathbf{u} = [\mathbf{x}^T \ \alpha]^T$ such that

$$\begin{bmatrix} \mathbf{A} & 0 \\ \mathbf{c}^T & 1 \end{bmatrix} \mathbf{u} = \begin{bmatrix} \mathbf{b} \\ \mu - \varepsilon \end{bmatrix},$$

Hence the second alternative of Farkas' lemma hold
that

$$\begin{bmatrix} \mathbf{A}^T & \mathbf{c} \\ 0 & 1 \end{bmatrix} \mathbf{v} \geq 0, \quad [\mathbf{b}^T \ \mu - \varepsilon]$$

Putting $\mathbf{v} = [-\mathbf{w}^T \ \beta]$, it results from the above that

$$-\mathbf{A}^T\mathbf{w} + \beta\mathbf{c} \geq 0 \tag{15.12a}$$

$$\beta \geq 0 \tag{15.12b}$$

$$-\mathbf{b}^T\mathbf{w} + \beta(\mu - \varepsilon) < 0 \tag{15.12c}$$

If $\beta = 0$, then (15.12a) gives $\mathbf{A}^T\mathbf{w} \leq 0$; multiplying this inequality to the left with $\mathbf{x}^T$, where $\mathbf{x}$ is a feasible variable, gives $\mathbf{b}^T\mathbf{w} \leq 0$; this contradicts (15.12c), which is $\mathbf{b}^T\mathbf{w} > 0$. Hence $\beta > 0$. Putting $\mathbf{y}^\star = \mathbf{w}/\beta$, it results from (15.12a) that $\mathbf{A}^T\mathbf{y}^\star \leq \mathbf{c}$, hence $\mathbf{y}^\star$ is dual feasible, and from (15.12c) that $\mathbf{b}^T\mathbf{y}^\star > \mu - \varepsilon$. Since this is true for any $\varepsilon > 0$, it follows that $\mathbf{b}^T\mathbf{y}^\star \geq \mu$. Since the dual value cannot be large than μ, we conclude that $\mathbf{b}^T\mathbf{y}^\star = \mu$ and so the primal and dual LP problems have the same value. ∎

Note that Theorem 15.1.6 does not exclude the possibility that both the primal (15.1) and dual (15.9) are infeasible; see Exercise 15.3 for an example.

The KKT optimality conditions associated with the primal (15.1) and the dual (15.9) have the form

$$\mathbf{A}\mathbf{x} = \mathbf{b} \tag{15.13a}$$

$$\mathbf{A}^T\mathbf{y} + \mathbf{s} = \mathbf{c} \tag{15.13b}$$

$$\mathbf{x} \geq 0 \tag{15.13c}$$

$$\mathbf{s} \geq 0 \tag{15.13d}$$

$$\mathbf{s}^T\mathbf{x} = 0 \tag{15.13e}$$

The first four conditions are simply the constraints of the primal and dual problems, with the multiplier $\mathbf{s}$ considered. Condition (15.13e) is called *complementarity slackness*; since all the products $s_i x_i$ are non-negative, this condition can be also expressed as: for all $i = 1 : n$, at least one of the equalities $x_i = 0$, $s_i = 0$ holds. Otherwise said, an inequality constraint of (15.1) is active ($x_i = 0$) or the corresponding Lagrange multiplier is zero ($s_i = 0$), which means that the i-th inequality of the constraint $\mathbf{A}^T\mathbf{y} \leq \mathbf{c}$ is active in the dual (15.9). In the nondegenerate case, there are m non-zero elements in $\mathbf{x}^\star$, see Theorem 15.1.3 and the subsequent discussion, so it results that (at least) m constraints of the dual problem must be active; typically, exactly one of $x_i^\star$ and $s_i^\star$ is zero. Note that Proof 1 of Theorem 15.1.6 actually enforces complementarity slackness.

Theorem 15.1.7. *Conditions* (15.13a)–(15.13e) *hold if and only if both LP problems* (15.1) *and* (15.9) *are feasible and* $\mathbf{x}$ *and* $\mathbf{y}$ *are optimal.*

Proof. Conditions (15.13a)–(15.13d) are equivalent to the feasibility of both the primal and the dual LP problems. By transposing (15.13a) and multiplying at the right with $\mathbf{y}$, we obtain $\mathbf{b}^T\mathbf{y} = \mathbf{x}^T\mathbf{A}^T\mathbf{y}$. By transposing (15.13b), multiplying at the right with $\mathbf{x}$ and making use of the previous equality, we end up with

$$\mathbf{b}^T\mathbf{y} + \mathbf{s}^T\mathbf{x} = \mathbf{c}^T\mathbf{x}.$$

Since $\mathbf{s}^T\mathbf{x} \geq 0$ and $\mathbf{b}^T\mathbf{y} = \mathbf{c}^T\mathbf{x}$ holds only at optimal
equivalent to optimality.

15.1.5   The Simplex Method

The simplex method, invented by Dantzig in 1947, is t
finding the solution of the LP primal problem (15.1).
builds a path of vertices of the feasible polytope. A
which shares an edge with the current vertex, such
smaller. When this is no longer possible, the minin
present here only the main ideas of the algorithm. F
ignore that the polygon from Figure 15.1 (right) con
assume that we start the search of the optimum fro
a vertex of the feasible domain. There are two pat
from a vertex to one of its better neighbors: $(1,1) -$
$(3,0) \rightarrow (5,2) \rightarrow (5,4)$. In general, there are many su
followed by the simplex method.

Returning to the general case of the LP primal pr
have a basic feasible point $\mathbf{x}$, which is a vertex of the
by Theorem 15.1.4; finding such a point is not trivia
We can transform the matrix $\mathbf{A}$ into the canonical for
m variables defining the basic feasible point into the
pursuing Gauss-Jordan elimination as described in
Jordan elimination process with row pivoting is guara
non-zero diagonal elements, since the m columns of
point are linearly independent. We assume that all tra
in place.

Note that, since $\mathbf{x}$ is a basic feasible point and s
form (15.5) of the matrix $\mathbf{A}$ it results that $x_i = b_i$, i
the criterion is

$$z = \mathbf{c}^T\mathbf{x} = \sum_{i=1}^{m} b_i c_i.$$

Since $\mathbf{x}$ is feasible, it necessarily results that $b_i \geq 0$ a
We present now a step of the simplex method. Tr
feasible polytope to a neighbor vertex is equivalent to
$k \leq m$, with a non-basic variable x_ℓ, $\ell > m$, such that
obtained and the criterion decreases; this operation is
be confused with pivoting in the elimination process
first m variables is the *base*, x_k is named *leaving* varia
In what follows, the notations for the new basic feasib
a tilde. The new solution has $\tilde{x}_k = 0$, $\tilde{x}_j = 0$ for $j > n$
the system $\mathbf{A}\tilde{\mathbf{x}} = \mathbf{b}$ reads $\tilde{x}_i + a_{i,\ell}\tilde{x}_\ell = b_i$, while equa

Hence, the non-zero elements of the new solution can be found by substitution and are

$$\tilde{x}_\ell = \frac{b_k}{a_{k,l}},$$
$$\tilde{x}_i = b_i - a_{i,\ell}\frac{b_k}{a_{k,\ell}}, \quad i = 1:m. \tag{15.15}$$

Note that we can obtain $\tilde{x}_\ell \geq 0$ only if $a_{k,\ell} > 0$. Also, remark that for $i = k$ the above formula gives the correct result $\tilde{x}_k = 0$.

Improvement. The entering variable x_ℓ is chosen first, such that the criterion is decreased. The value of the criterion for the new solution (15.15) is

$$\tilde{z} = \mathbf{c}^T\tilde{\mathbf{x}} = \sum_{i=1}^m c_i\left(b_i - a_{i,\ell}\frac{b_k}{a_{k,\ell}}\right) + c_\ell\frac{b_k}{a_{k,l}} = z + \left(c_\ell - \sum_{i=1}^m c_i a_{i,\ell}\right)\frac{b_k}{a_{k,\ell}} \stackrel{\text{def}}{=} z + s_\ell\frac{b_k}{a_{k,\ell}}. \tag{15.16}$$

The vector $\mathbf{s}$ whose ℓ-th element is defined in the rightmost equality is named *reduced cost* and is

$$\mathbf{s} = \mathbf{c} - \mathbf{A}^T[c_1 \ c_2 \ \dots \ c_m]^T. \tag{15.17}$$

Note that its first m elements are equal to zero, due to the form (15.5) of $\mathbf{A}$. Moreover, once ℓ has been chosen, the vector $\mathbf{s}$ can be obtained in the Gauss-Jordan elimination process as if $\mathbf{c}$ would be appended to $\mathbf{A}$ as the $m + 1$-th row; this is how the simplex method is actually implemented.

It results from (15.16) that the criterion decreases if and only if there exists an index ℓ such that $s_\ell < 0$. We also conclude that, if $s_i \geq 0$, for all $i > m$, then the optimum has been attained: no feasible direction of decrease exists. In the standard simplex method, the new variable is chosen such that $s_\ell = \min\{s_i \mid i = m+1 : n\}$; this is the greedy choice that hopes, but is not guaranteed, to maximize descent.

Feasibility. The leaving variable x_k is chosen such that the new solution (15.15) is feasible, namely $\tilde{\mathbf{x}} \geq 0$. Since $b_k \geq 0$ and $a_{k,\ell} > 0$, we have to care about the sign of $\tilde{x}_i$ only if $a_{i,\ell}$ is positive. By choosing k such that

$$\frac{b_k}{a_{k,\ell}} = \min_i\left\{\frac{b_i}{a_{i,\ell}}\ \middle|\ a_{i,\ell} > 0, \ i = 1:m\right\}, \tag{15.18}$$

feasibility is ensured. Indeed, if $\tilde{x}_i < 0$ for some i with $a_{i,\ell} > 0$, then

$$\tilde{x}_i \stackrel{(15.15)}{=} b_i - a_{i,\ell}\frac{b_k}{a_{k,\ell}} < 0 \implies \frac{b_i}{a_{i,\ell}} < \frac{b_k}{a_{k,\ell}},$$

which contradicts the choice (15.18).

Unboundedness. A distinct possibility is that $a_{i,\ell} \leq 0$, for all $i = 1:m$. In this case, by *appending* x_ℓ to the basis (hence allowing $m + 1$ variables to be positive) and giving it an arbitrary positive value ξ, feasibility is preserved for the new solution $\tilde{x}_i = b_i - a_{i,\ell}\xi$, $i = 1:m$, no matter how large is ξ. Since the criterion is

Table 15.1: Basic form of the simp

Input data: **A**, **b**, **c** of the LP primal (15.1).

Initialization. Find a basic feasible point. Trans
form (15.5) by Gauss-Jordan elimination (and
Compute the reduced costs **s** as in (15.17).

Step 1. Find entering variable index ℓ such that s_ℓ
If $s_\ell \geq 0$, then the optimum has been attained a
variables are $x_i = b_i$, $i = 1 : m$. Stop.

Step 2. If $a_{i,\ell} \leq 0$, $\forall i = 1 : m$, then the prob
Otherwise, find the leaving variable index k accor

Step 3. Interchange variables k and ℓ (i.e., th
elements of **c**). Restore **A** to canonical form (15.5
costs. Go to step 1.

---

linear in ξ and decreases as ξ increases, the criterion
So, in such a case, the problem is unbounded.

A sketch of the simplex method is presented in T
further comments. After permuting columns k and ℓ
k is not in canonical form. Since $a_{k,k}$ (former $a_{k,\ell}$) i
a pivot to zero all the other elements of the column in
see again the paragraph after (15.5). The reduced cc
being the $m + 1$-th row of **A**.

Initialization. Unless some a priori information
feasible point can be found by solving, with the sin
problem

$$\min \quad \mathbf{1}^T \mathbf{w}$$
$$\text{s.t.} \quad \mathbf{Ax} + \mathbf{Dw} = \mathbf{b}$$
$$\mathbf{x} \geq 0, \ \mathbf{w} \geq 0$$

where $\mathbf{1} = [1\ 1\ \ldots\ 1]^T$ and $\mathbf{D} = \text{diag}(b_1, b_2, \ldots, b_m)$.
problem is $\mathbf{x} = 0$, $\mathbf{w} = \mathbf{1}$, so the simplex method for
Since $\mathbf{1}^T \mathbf{w} \geq 0$, $\forall \mathbf{w} \geq 0$, if there exists a feasible $\mathbf{x}$ for
then the optimal value of problem (15.19) is zero,
method will find it. If the final base contains only v
be used as initialization for solving (15.1). Howeve
variables from $\mathbf{w}$ are still in the base; in this case, s
simplex method solve (15.1) with these extra variabl
zero.

Termination. In principle, since the criterion dec
path built by the simplex method does not contain loop

once. Since the number of vertices of the feasible polytope is finite, termination is guaranteed. This is actually true only if all basic feasible points are nondegenerate, i.e., they have exactly m non-zero elements, not less. A degenerate basic feasible point may cause the simplex method to advance along a zero-length "edge," into another degenerate point, and possibly loop through such points. However, looping can be prevented by diverse modifications of the basic form of the simplex method from Table 15.1, not discussed here.

Complexity. Since the simplex method may go through all vertices of the feasible polytope—and an example [1] was produced that this can happen for a polytope with 2^n vertices—the complexity is exponential in the worst case. Despite this, in practice, the simplex method has very good average behavior. Extensive empirical evidence suggests that most often only $2m$ to $3m$ iterations are sufficient [2]. However, there are LP problems for which the simplex method is slow and other (interior-point) methods are superior.

Variations. There are many variations of the simplex method. The entering variable can be chosen in several ways, besides using the most negative reduced cost coefficient; steepest descent (through the edge making the smallest angle with the negative gradient) or best-neighbor (neighbor vertex for which criterion is minimum) may give better advance towards the optimum, but their computational cost per iteration is higher; also, they still have exponential complexity in the worst case. Better numerical behavior (at the expense of extra operations) is obtained if an upper triangular matrix is kept and updated at the left of the canonical form (15.5), instead of the unit matrix. This modification, proposed by Bartels-Golub [3], is equivalent to computing an LU factorization of the current basic feasible matrix made by the basic columns of $\mathbf{A}$, instead of the inverse.

15.1.6   Interior-Point Methods

While the simplex method deals only with LP problems, interior-point methods are based on ideas suited to many categories of convex optimization problems. We will present here only the basics of a single type of interior-point algorithm, using the central path. This algorithm is called primal-dual, because it solves simultaneously the primal LP problem (15.1) and the dual (15.9), written in the form

$$\max \quad \mathbf{b}^T \mathbf{y} \tag{15.20}$$
$$\text{s.t.} \quad \mathbf{A}^T \mathbf{y} + \mathbf{s} = \mathbf{c}$$
$$\mathbf{s} \geq 0$$

to stress the connection with the KKT conditions (15.13a)–(15.13e). For any feasible $\mathbf{x}, \mathbf{s} \geq 0$ and $\mathbf{y}$, the quantity (see the proof of Theorem 15.1.7)

$$\mathbf{s}^T \mathbf{x} = \mathbf{c}^T \mathbf{x} - \mathbf{b}^T \mathbf{y} \tag{15.21}$$

is non-negative and is called *duality gap.*

We start by associating with (15.1) a logarithmic l

$$\min \quad \mathbf{c}^T\mathbf{x} - \lambda \sum_{i=1}^n \ln .$$
$$\text{s.t.} \quad \mathbf{Ax} = \mathbf{b}$$
$$(\mathbf{x} > 0)$$

where $\lambda > 0$ is a parameter. Since the logarithm is de
feasible domain of (15.22) coincides with the feasibl
(15.1), with the exception of those points $\mathbf{x}$ with a ze
$\mathbf{x} > 0$ needs not be posed explicitly. The value of th
as some x_i approaches zero, hence the solution of (
the feasible domain of (15.1); as the parameter λ ap
(15.22) approaches from the interior the LP solution,
feasible domain.

For a given λ, we attempt to solve (15.22) toge
corresponding KKT conditions. The Lagrangian asso

$$L(\mathbf{x}, \mathbf{y}) = \mathbf{c}^T\mathbf{x} + \mathbf{y}^T (\mathbf{b} - \mathbf{Ax}) - \lambda$$

and its gradient with respect to $\mathbf{x}$ is

$$\nabla_{\mathbf{x}} L(\mathbf{x}, \mathbf{y}) = \mathbf{c} - \mathbf{A}^T\mathbf{y} - \lambda \mathbf{X}$$

where $\mathbf{X} = \text{diag}(x_1, \ldots, x_n)$ and $\mathbf{1}$ is the vector w
Denoting

$$\mathbf{s} = \lambda \mathbf{X}^{-1}\mathbf{1} > 0,$$

forcing the gradient to zero and adding the explicit co
conditions are

$$\mathbf{Ax} = \mathbf{b}$$
$$\mathbf{A}^T\mathbf{y} + \mathbf{s} = \mathbf{c}$$
$$\mathbf{Xs} = \lambda \mathbf{1}$$

together with the implicit constraints $\mathbf{x}, \mathbf{s} > 0$.

Note that (15.26c) is equivalent to $x_i s_i = \lambda$, $i =$
feasible for the LP problems (15.1) and (15.20), it re
duality gap (15.21) is $\mathbf{x}^T\mathbf{s} = \lambda n$; hence, it depends ex
The points $\mathbf{x}$ and $(\mathbf{y}, \mathbf{s})$, defined for all values of $\lambda > 0$
primal and dual LP problems, respectively. The centr.
domains of the problems; as λ approaches zero, the
solution of the LP problems, since the duality gap te

Most primal-dual interior-point methods try to fol
ing the KKT system (15.26a)-(15.26c) for decreasing

(15.26c) is nonlinear, an iterative method is necessary. Newton's method is the simplest for solving systems of equations and very successful in this case. At iteration k, given some approximate solution $\mathbf{x}^{(k)}$, $\mathbf{y}^{(k)}$, $\mathbf{s}^{(k)}$ of system (15.26a)-(15.26c), the next approximation is found using the search direction given by the solution of

$$\begin{bmatrix} \mathbf{A} & 0 & 0 \\ 0 & \mathbf{A}^T & \mathbf{I} \\ \mathbf{S}^{(k)} & 0 & \mathbf{X}^{(k)} \end{bmatrix} \begin{bmatrix} \delta\mathbf{x}^{(k)} \\ \delta\mathbf{y}^{(k)} \\ \delta\mathbf{s}^{(k)} \end{bmatrix} = \begin{bmatrix} \mathbf{b} - \mathbf{A}\mathbf{x}^{(k)} \\ \mathbf{c} - \mathbf{A}^T\mathbf{y}^{(k)} - \mathbf{s}^{(k)} \\ \lambda^{(k)}\mathbf{1} - \mathbf{X}^{(k)}\mathbf{s}^{(k)} \end{bmatrix}, \qquad (15.27)$$

where the left matrix and the right vector are the Jacobian and the approximation error of (15.26a)-(15.26c), respectively; we have denoted $\mathbf{S} = \mathrm{diag}(s_1, \ldots, s_n)$; note that $\mathbf{Xs} = \mathbf{Sx}$, which explains the last block row of the Jacobian.

The overall algorithm is summarized in Table 15.2. The algorithm takes an aggressive approach in the choice of the parameter λ, by not attempting to solve exactly the KKT equations (15.26a)-(15.26c) for a given value λ. Instead, the parameter λ is modified at each iteration, with the aim of a faster convergence; hence, each Newton iteration is performed for another system. The typical choice of λ takes into account its relation with the duality gap and is

$$\lambda^{(k)} = \tau \frac{\mathbf{x}^{(k)T}\mathbf{s}^{(k)}}{n},$$

where $0 < \tau < 1$; the target gap is a fraction of the current gap (if $\mathbf{x}^{(k)}$ and $\mathbf{s}^{(k)}$ are feasible). The step length α is usually taken almost equal (e.g., 99%) to the value that zeroes an element of $\mathbf{x}^{(k+1)}$ or $\mathbf{s}^{(k+1)}$, while the other elements remain positive.

The algorithm shown in Table 15.2 belongs to the category of infeasible interior-point methods, because $\mathbf{x}^{(k)}$, $\mathbf{y}^{(k)}$, and $\mathbf{s}^{(k)}$ do not satisfy exactly the feasibility conditions (15.26a) and (15.26b) at each iteration. In practice, infeasible methods appear to work better than feasible ones, in which (15.26a) and (15.26b) are enforced by starting with a feasible point $\mathbf{x}^{(0)}$, $\mathbf{y}^{(0)}$, and $\mathbf{s}^{(0)}$ (which is not trivial); feasibility is maintained because the first two (vector) elements of the right side of (15.27) are zero at each iteration; the search direction always lies in the feasible domain.

The key feature of interior-point methods is that not only they converge, but they are guaranteed to approximate the LP solution to the desired accuracy in polynomial time; convergence analysis is outside the scope of this book. So, unlike the simplex method, the worst case behavior is manageable. In fact, interior-point methods are less sensitive to the data and solve the LP problems in an almost constant number of iterations, not depending on the size n of the problem. Of course, the complexity of an iteration depends on n; the most time consuming step consists of finding the solution of the linear system (15.27); its special structure is exploited by efficient algorithms.

Table 15.2: Sketch of primal-dual interior-point

Input data: $\mathbf{A}, \mathbf{b}, \mathbf{c}$ defining the LP primal (15.1) a

Initialization. Take $\mathbf{x}^{(0)} > 0$, $\mathbf{s}^{(0)} > 0$, $\mathbf{y}^{(0)}$. Put k

Step 1. Choose target duality gap $\lambda^{(k)}$.

Step 2. Solve the linear system (15.27) to find dir

Step 3. Choose a step length α and compute the

$$
\begin{bmatrix} \mathbf{x}^{(k+1)} \\ \mathbf{y}^{(k+1)} \\ \mathbf{s}^{(k+1)} \end{bmatrix} = \begin{bmatrix} \mathbf{x}^{(k)} \\ \mathbf{y}^{(k)} \\ \mathbf{s}^{(k)} \end{bmatrix} + \alpha \begin{bmatrix} \\ \\ \end{bmatrix}
$$

The step length α is chosen such that $\mathbf{x}^{(k+1)} > 0$,

Step 4. If $\mathbf{x}^{(k+1)}$, $\mathbf{y}^{(k+1)}$, $\mathbf{s}^{(k+1)}$ approximate sat

mality conditions (15.13a), (15.13b), and (15.13e

put $k = k + 1$ and go to step 1.

15.2   Modeling Problems via Lin

This section shows, for a few typical problems, how the
or constraints are transformed to linear form; since th
will not make special efforts to bring the resulting LP
however, the reader is encouraged to do this exercise.

15.2.1   Optimization with 1-norm and

In many optimization problems, the aim is to collectiv
(in absolute value) the elements of a vector, typicall
in a given way on some variables. This is realized b
vector, whose choice is dictated by the desired relati
If the chosen norm is the 1-norm or the ∞-norm, and
on the variables, then an LP problem results. Let $\mathbf{e} \in$
$\mathbf{x} \in \mathbb{R}^n$ the variable vector, their dependence being

$$\mathbf{e} = \mathbf{A}\mathbf{x} - \mathbf{b} \quad \Leftrightarrow \quad e_i = \mathbf{a}_i^T \mathbf{x} - b_i,$$

where $\mathbf{a}_i^T$ denotes the i-th row of $\mathbf{A}$. In this context,
a linear regression problem. The 1-norm and ∞-norm

$$\|\mathbf{e}\|_1 = \sum_{i=1}^m |e_i|, \qquad \|\mathbf{e}\|_\infty = \max_{1 \le i \le}$$

The ∞-norm is minimized when the largest error element has to be as small as possible; as a result, the optimized error has often several elements equal in absolute value to the largest one and many elements having non-negligible values. On the contrary, 1-norm optimization tends to produce error vectors with few large elements and many others approaching zero. We consider optimization problems minimizing one of the norms (15.29), subject to (15.28) and possibly other linear constraints involving $\mathbf{x}$, ignored in the presentation below but trivial to incorporate. So, the problem under scrutiny is

$$\mu_1 \text{ or } \infty = \begin{array}{ll} \min & \|\mathbf{e}\|_{1 \text{ or } \infty} \\ \text{s.t.} & e_i = \mathbf{a}_i^T\mathbf{x} - b_i, \ i = 1 : m \end{array} \tag{15.30}$$

The ∞-norm optimization can be put in LP form by adding a variable representing a bound on the maximum value of an error element, thus obtaining

$$\mu_\infty = \begin{array}{ll} \min & \xi \\ \text{s.t.} & |\mathbf{a}_i^T\mathbf{x} - b_i| \le \xi, \ i = 1 : m \end{array} \tag{15.31}$$

It is clear that, at optimality, $\xi^\star$ is equal to the largest $|e_i^\star|$ and so equal to $\mu_\infty = \|\mathbf{e}^\star\|_\infty$. Each absolute value inequality from (15.31) can be transformed into two inequalities, obtaining the LP problem (in inequality form)

$$\mu_\infty = \begin{array}{ll} \min & \xi \\ \text{s.t.} & \mathbf{a}_i^T\mathbf{x} - \xi \le +b_i, \ i = 1 : m \\ & -\mathbf{a}_i^T\mathbf{x} - \xi \le -b_i, \ i = 1 : m \end{array} \tag{15.32}$$

The 1-norm minimization can be treated similarly, but now a variable is needed for each $|e_i|$. Problem (15.30) is equivalent to

$$\mu_1 = \begin{array}{ll} \min & \sum_{i=1}^m z_i \\ \text{s.t.} & |\mathbf{a}_i^T\mathbf{x} - b_i| \le z_i, \ i = 1 : m \end{array} \tag{15.33}$$

At optimality, all inequalities become in fact the equalities $z_i^\star = |e_i^\star|$, $i = 1 : m$. The corresponding LP problem is

$$\mu_1 = \begin{array}{ll} \min & \sum_{i=1}^m z_i \\ \text{s.t.} & \mathbf{a}_i^T\mathbf{x} - z_i \le b_i, \ i = 1 : m \\ & -\mathbf{a}_i^T\mathbf{x} - z_i \le -b_i, \ i = 1 : m \end{array} \tag{15.34}$$

Since z_i is forced to be non-negative by the nature of the constraints, it is not needed to impose explicitly the constraint $\mathbf{z} \ge 0$. Note that, having $n+m$ variables, the 1-norm LP problem (15.34) has a higher complexity than the ∞-norm problem (15.32), which has only $n + 1$ variables.

Example 15.2.1. Let us study first a linear model identification problem. Assume that some physical process is supposed to be described by the linear relation $v = \sum_{k=1}^n c_k u_k$, where $\mathbf{u} \in \mathbb{R}^n$ is the input vector and $v \in \mathbb{R}$ is the output. Making

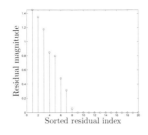
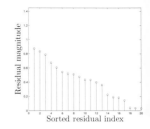

Figure 15.2: Sorted absolute values of the residuals
system, obtained by minimizing their 1-norm (left), 2
(right); see Example 15.2.1.

m measurements of the input and output, v_i and u_{ik}
want estimate the values of the coefficients c_k (which
errors

$$e_i = \sum_{k=1}^{n} u_{ik}c_k - v_i, \quad i = 1:$$

are as small as possible. We have thus a problem (1?
and $x_k = c_k$. The number of measurements is larger tl
otherwise in general we obtain $\mathbf{e} = 0$. We can inter
pseudo-solution to the overdetermined linear system
residuals of this system. If the true process is indeed
noise is Gaussian, then the best solution is the least-sc
from the minimization of $\|\mathbf{e}\|_2$. In other conditions, c
     Let us illustrate the results obtained for different
ated data $\mathbf{A}$ and $\mathbf{b}$, with $m = 20$, $n = 12$. Figure
of the actual residuals $|e_i|$, when the estimation pro
2-, and ∞-norms. Since the LP problem (15.34) is i
complementary slackness condition (15.13e) and Thee
cally, $n + m$ of the $2m$ constraints are active at optir
simultaneously $\mathbf{a}_i^T \mathbf{x}^\star - z_i^\star = b_i$ and $-\mathbf{a}_i^T \mathbf{x}^\star - z_i^\star = -$
results that usually only $2m - (n + m) = m - n$ resi
when the 1-norm is minimized. Note that in Figure
non-zero residuals. Similarly, in the ∞-norm minimiz
are usually $n + 1$ active constraints at optimality.
unlikely situation when all measurements are perfec
that the equality $|\mathbf{a}_i^T \mathbf{x}^\star - b_i| = \xi^\star$ holds for $n + 1$ ind
residuals are equal to their maximum value, which i
by Figure 15.2 (right), where 13 residuals are equal t
least-squares residuals are in between these extreme l

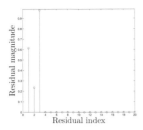 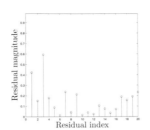 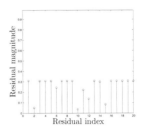

Figure 15.3: Residuals of an overdetermined linear system, obtained by minimizing their 1-norm (left), 2-norm, (center), and ∞-norm (right); see Example 15.2.2.

Example 15.2.2. Keeping the same setup as in the previous example, let us assume that only a few measurements are affected by sufficiently large errors, while the others are perfect. Our purpose is to detect those measurements and to eliminate them in order to get the correct solution using the remaining measurements. Otherwise put, we want to find the outliers. The 1-norm optimization can be used to this purpose; we give here a simplified discussion; the interested reader should consult [4] for details.

Assume that there are p outliers and $p \ll m$, $p < n$; the number of outliers is unknown. We want to find $\mathbf{x}$ such that $\mathbf{e} = \mathbf{Ax} - \mathbf{b}$ has the smallest number of non-zeros, which is p. The number of non-zeros of a vector is usually called its 0-norm and denoted $\|\mathbf{e}\|_0$, despite not being a norm. In other words, we want the sparsest residual. This is not a convex optimization problem—in fact it is a hard one. However, if the number of non-zeros is sufficiently small, the 1-norm minimization problem (15.34) has the same solution as the 0-norm minimization; this is an example of degenerate optimum, since usually $p < m-n$. Even for a larger number of non-zeros, the 1-norm minimization may give valuable information, by producing a residual with large values for the outliers and small or zero values elsewhere. The case where the other measurements are affected by (a smaller) noise can be also accommodated.

Figure 15.3 shows the residuals resulting from 1-, 2-, and ∞-norm minimization for a linear system with $m = 20$, $n = 12$ and $p = 3$. The outliers are the first p measurements. It is visible that the 1-norm minimization produces an $\mathbf{x}$ for which the first three residuals are non-zero; so, not only that the outliers are found, but also this is practically the true solution, since all the other equations are satisfied (almost) exactly. The 2-norm minimization results are more difficult to interpret; the figure may suggest that the first and third residuals belong to outliers. The ∞-norm minimization result is shown only for completing the picture; we should not expect any relevant information about the outliers—on the contrary.

15.2.2   Chebyshev Center of a Polytop

Given the polytope $\mathcal{P} = \left\{ \mathbf{x} \in \mathbb{R}^m \mid \mathbf{A}^T\mathbf{x} \leq \mathbf{b} \right\}$, wi
Chebyshev center $\mathbf{z}$ is its innermost point, that for
exterior of $\mathcal{P}$ is maximum. Geometrically, $\mathbf{z}$ is the cen
inscribed in the polytope and r is the radius of th
illustrates such a construction in 2D ($m = 2$), for the
domain of (15.4); the hypersphere (circle in 2D) is ta
in 2D).

For a polytope, the distance from an inner point t
from that point to the nearest face. Let $\mathbf{a}_i^T\mathbf{x} = b_i$
face of the polytope, where $\mathbf{a}_i$ is the i-th column of
to this hyperplane is

$$d_i = (b_i - \mathbf{a}_i^T\mathbf{z})/\left\|\mathbf{a}_i\right\|_2.$$

To prove this relation, let $\mathbf{x}_i$ be the point on the hype
Then $\mathbf{z} - \mathbf{x}_i$ is orthogonal to the hyperplane, i.e., par
value of the scalar product of these vectors is equal t

$$\left| \mathbf{a}_i^T(\mathbf{z} - \mathbf{x}_i) \right| = \left\|\mathbf{a}_i\right\|_2 \left\| \mathbf{z} - \mathbf{x}_i \right.$$

Since $d_i = \left\| \mathbf{z} - \mathbf{x}_i \right\|_2$, $\mathbf{a}_i^T\mathbf{x}_i = b_i$ and $\left| \mathbf{a}_i^T\mathbf{z} - b_i \right| = b_i -$

The Chebyshev center is the $\mathbf{z}$ maximizing $\min d_i$
found by solving the LP problem

$$\begin{aligned} r = \quad &\max \quad \Delta \\ &\text{s.t.} \quad b_i - \mathbf{a}_i^T\mathbf{z} \geq \Delta \left\|\mathbf{a}_i\right\|_2, \end{aligned}$$

Indeed, this amounts to maximize Δ, with $\Delta \leq d_i$, t
to the nearest face. The problem has $m + 1$ variabl
only inequality constraints, so the complementarity
says that typically there are at least $m + 1$ active con
means that the hypersphere centered in $\mathbf{z}$ and with r
$m + 1$ faces of the polytope. This is actually the gec
circle inscribed in a triangle is tangent to the edges,
draw a circle tangent to all edges of a quadrangle.
for which the Chebyshev center is not unique, for exa
(15.36) may lead to a circle that is tangent to less th

15.2.3   Classification with Separating H

A basic classification problem is the following: given
function f such that $f(\mathbf{x}) < 0$ for $\mathbf{x} \in \mathcal{X}$, and $f(\mathbf{y})$
and $\mathcal{Y}$ are finite sets containing observations $\mathbf{x}_i$, $i =$
respectively, of a phenomenon; the decision to assign a

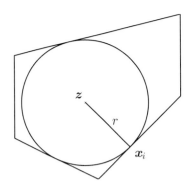

Figure 15.4: Chebyshev center $\mathbf{z}$ for the polygon defined by (15.4).

by an expert. The discriminating function f is sought for automatically classifying future observation and sometimes for gaining some insight of the phenomenon.

The simplest discriminating function is affine: $f(\mathbf{x}) = \mathbf{a}^T\mathbf{x} + b$, with $\mathbf{a} \in \mathbb{R}^n$, $b \in \mathbb{R}$. So, the two sets are separated by the hyperplane $\mathbf{a}^T\mathbf{x} + b = 0$. If it exists, the separating hyperplane is generally not unique, so in a first instance we can simply solve the following feasibility problem for finding one:

$$
\begin{aligned}
&\text{find} \quad \mathbf{a}, b \\
&\text{s.t.} \quad \mathbf{a}^T\mathbf{x}_i + b < 0, \ i = 1 : N_x \\
&\qquad\ \ \mathbf{a}^T\mathbf{y}_i + b > 0, \ i = 1 : N_y
\end{aligned}
\tag{15.37}
$$

Because of the strict inequalities, this is not an LP problem, but it can be transformed immediately into one by noticing that $\mathbf{a}$ and b can be multiplied with the same positive constant without changing the classification or the hyperplane. So, instead of (15.37), we solve

$$
\begin{aligned}
&\text{find} \quad \mathbf{a}, b \\
&\text{s.t.} \quad \mathbf{a}^T\mathbf{x}_i + b \leq -1, \ i = 1 : N_x \\
&\qquad\ \ \mathbf{a}^T\mathbf{y}_i + b \geq 1, \ i = 1 : N_y
\end{aligned}
\tag{15.38}
$$

Of course, any positive number can replace 1 in the above problem. If the two sets can be separated by a hyperplane, then solving (15.38) provides one; if not, the LP problem has no solution.

Example 15.2.3. We have generated two separable sets of $N_x = N_y = 50$ points in $\mathbb{R}^2$. A separating line has been found by solving (15.38). The result is illustrated on the left of Figure 15.5. Since the feasibility problem finds just one of the many separating lines, it may not find the best one. In this case, we might not like that the line is too close to the points in the "circles" set.

An optimality criterion can result if we impose the minimal distance from a point in $\mathcal{X}$ or $\mathcal{Y}$ to the separating hyperplane to be as large as possible. Adapting

the formula (15.35) of the distance from a point to
hand, this results in

$$
\begin{aligned}
\max \quad & \Delta \\
\text{s.t.} \quad & \mathbf{a}^T \mathbf{x}_i + b \leq -\Delta \left\| \mathbf{a} \right\|_2, \ i \\
& \mathbf{a}^T \mathbf{y}_i + b \geq \Delta \left\| \mathbf{a} \right\|_2, \ i =
\end{aligned}
$$

Besides not belonging to LP, this formulation is aga
nation in the size of $\mathbf{a}$ and b. In order to bound th
norm of $\mathbf{a}$, obtaining

$$
\begin{aligned}
\max \quad & \Delta \\
\text{s.t.} \quad & \mathbf{a}^T \mathbf{x}_i + b \leq -\Delta, \ i = \\
& \mathbf{a}^T \mathbf{y}_i + b \geq \Delta, \ i = 1 \\
& \left\| \mathbf{a} \right\|_2 \leq 1
\end{aligned}
$$

At optimality, the equality $\left\| \mathbf{a} \right\|_2 = 1$ always holds; if it
each constraint with $\left\| \mathbf{a} \right\|_2$ we will get a larger value
exactly the distance from the line to the nearest poir
optimization problem (15.40) is not LP, but convex (li
and can be solved with algorithms discussed in the pr
lower complexity, it was proposed in [5] to replace $\left\| \mathbf{a} \right.$
the LP problem (remind the definition (15.29) of the

$$
\begin{aligned}
\max \quad & \Delta \\
\text{s.t.} \quad & \mathbf{a}^T \mathbf{x}_i + b \leq -\Delta, \ i = \\
& \mathbf{a}^T \mathbf{y}_i + b \geq \Delta, \ i = 1 \\
& -1 \leq a_k \leq 1, \ k = 1 :
\end{aligned}
$$

Although only an approximation of (15.40), this LP
good results in practice.

Example 15.2.4. We solved (15.40) and (15.41) for t.
The results are virtually the same, in the sense that tl
(the optimal $\mathbf{a}$ and b are different, but can be scaled t
The separating line is shown on the right of Figure 15.
dashed lines, parallels to the separating line through
$\mathcal{Y}$. The distances from these lines to the separating li

15.2.4   Linear Fractional Programming

Linear fractional programming (LFP) is an example
actually) optimization problem that can be transform
problem has the form

$$
\begin{aligned}
\mu_0 = \quad \min \quad & \frac{\mathbf{c}^T \mathbf{x} + d}{\mathbf{e}^T \mathbf{x} + f} \\
\text{s.t.} \quad & \mathbf{A}_e \mathbf{x} = \mathbf{b}_e \\
& \mathbf{A}_i \mathbf{x} \leq \mathbf{b}_i
\end{aligned}
$$

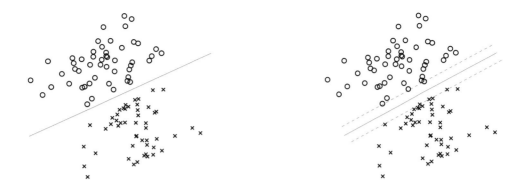

Figure 15.5: Separating lines obtained by solving the feasibility problem (15.38) (left) and the optimization problems (15.40) and (15.41) (right).

So, it has linear constraints, but the criterion function is linear fractional. It is assumed that the feasible set is such that the denominator is always positive: $\mathbf{e}^T\mathbf{x} + f > 0$ for any feasible $\mathbf{x}$. Since the multiplication of both the denominator and numerator of the criterion with a positive number does not change its value (and keeps the denominator positive), the denominator could be eliminated by multiplying with a new variable t, such that $(\mathbf{e}^T\mathbf{x} + f)t = 1$, which means that $t = 1/(\mathbf{e}^T\mathbf{x} + f)$. In order to preserve linearity, we replace $\mathbf{x}$ by

$$\mathbf{y} = t\mathbf{x} = \mathbf{x}/(\mathbf{e}^T\mathbf{x} + f).$$

So, the criterion of (15.42) becomes $\mathbf{c}^T\mathbf{y} + dt$ and is linear. The new variables are obviously not independent, since $\mathbf{e}^T\mathbf{y} + ft = 1$; so this constraint, also linear, should be imposed explicitly. Transforming also the constraints of (15.42) by multiplication with t, we end up with the LP problem

$$
\begin{aligned}
\mu_1 = \quad \min \quad & \mathbf{c}^T\mathbf{y} + dt && (15.43)\\
\text{s.t.} \quad & \mathbf{A}_e\mathbf{y} - \mathbf{b}_e t = 0 \\
& \mathbf{A}_i\mathbf{y} - \mathbf{b}_i t \le 0 \\
& \mathbf{e}^T\mathbf{y} + ft = 1 \\
& t \ge 0
\end{aligned}
$$

The constraint $t \ge 0$ is added since strict inequalities like $t > 0$ are not permitted in LP. We show now that the LP formulation (15.43) is indeed equivalent to the initial LFP problem (15.42).

If some $\mathbf{x}$ is feasible for (15.42), then $t = 1/(\mathbf{e}^T\mathbf{x} + f)$ and $\mathbf{y} = \mathbf{x}/(\mathbf{e}^T\mathbf{x} + f)$ are feasible for (15.43) and

$$\frac{\mathbf{c}^T\mathbf{x} + d}{\mathbf{e}^T\mathbf{x} + f} = \mathbf{c}^T\mathbf{y} + dt,$$

i.e., the criteria of (15.42) and (15.43) are equal, so $\mu_1 \le \mu_0$.

Conversely, if $t > 0$ and $\mathbf{y}$ are feasible, then $\mathbf{x} =$
criteria of (15.42) and (15.43) are equal. The only
feasible $t = 0$. Assume that some feasible $\mathbf{x}_0$ exist
feasible $\mathbf{y}$ satisfies $\mathbf{A}_e\mathbf{y} = 0$, $\mathbf{A}_i\mathbf{y} \leq 0$, it results tha
any non-negative α. Moreover, since $\mathbf{e}^T\mathbf{y} = 1$, it resu

$$\frac{\mathbf{c}^T\mathbf{x} + d}{\mathbf{e}^T\mathbf{x} + f} = \frac{\mathbf{c}^T\mathbf{x}_0 + \alpha\mathbf{c}^T\mathbf{y} + d}{\mathbf{e}^T\mathbf{x}_0 + \alpha + f} \; \alpha \text{-}$$

i.e., the criterion of (15.42) has values arbitrarily clo
(15.43). We can thus conclude that $\mu_0 \leq \mu_1$.

15.2.5   Continuous Constraints and Di

Continuous constraints produce infinite dimensional
like all the LP problems previously discussed in this
number of constraints. Consider the constraint $f(\omega$
an uncountable set, for example an interval $[\alpha, \beta]$; f
consider a scalar function, depending on a single para
that the function has the form

$$f(\omega) = \phi_0(\omega) + \sum_{k=1}^{N} x_k\phi_k(\omega$$

where x_k, $k = 1 : N$, are the variables of an optir
some elementary functions. (Note that we reverse th
variables and ω parameter, adopting the viewpoint of
say, function theory.) For example, $f(\omega)$ could be a p
have to be optimized; in this case, the functions ϕ,
members of a non-canonic polynomial basis. Essentia
linearity of $f(\omega)$ in the variables x_k, although the fu
optimization problem studied here is

$$\begin{aligned} \mu = \quad &\min \quad \mathbf{c}^T\mathbf{x} \\ &\text{s.t.} \quad f(\omega) \leq 0, \; \forall\omega \in \Omega, \quad f \; \varepsilon \\ &\qquad \text{possibly other linear co} \end{aligned}$$

Such a problem belongs to the class of *semi-infinite* L
is infinite, but the number of variables is finite, hen
the variables appear linearly, hence LP.

In a few cases, e.g., that of polynomials or trig
constraint $f(\omega) \leq 0$, $\forall\omega \in \Omega$, has an *equivalent* finite
polynomial constraints are equivalent with linear ma
not true in general, hence the need of algorithms for th
LP. These algorithms are generalized simplex method

and their presentation is beyond the purpose of this chapter; the interested reader can consult [7, 8].

We discuss here only the poor man's approach to semi-infinite LP, which is *discretization*. Despite only *approximating* (15.45), discretization is used in many engineering applications, due to its sheer simplicity. Discretization consists of replacing the continuous constraint of (15.45) by the finite version

$$f(\omega_\ell) \leq 0, \quad \ell = 1 : L, \tag{15.46}$$

where ω_ℓ are L points from Ω, named the discretization set or grid. If $\Omega = [\alpha, \beta]$, then ω_ℓ could belong to an equidistant grid covering the interval, i.e. $\omega_\ell = \alpha + (\ell - 1)\Delta$, with $\Delta = (\beta - \alpha)/(L - 1)$ being the distance between two successive points. Due to the form (15.44) of the function, the discretized form (15.46) can be written as

$$\mathbf{Ax} \leq \mathbf{b}, \quad \text{with } \mathbf{A} \in \mathbb{R}^{L \times N}, \ a_{\ell,k} = \phi_k(\omega_\ell), \ \mathbf{b} \in \mathbb{R}^L, \ b_\ell = -\phi_0(\omega_\ell). \tag{15.47}$$

This is a linear constraint and so the semi-infinite problem (15.45) is approximated with the LP problem

$$\mu = \begin{array}{ll} \min & \mathbf{c}^T \mathbf{x} \\ \text{s.t.} & \mathbf{Ax} \leq \mathbf{b} \\ & \text{possibly other linear constraints} \end{array} \tag{15.48}$$

The discretized problem (15.48) can be equivalent to (15.45) only if the discretization set contains the points ω that are active in the inequality constraint of (15.45) at optimality (the number of active points is finite—this property is intensely used in semi-infinite programming). However, it is impossible to know these points in advance. So, no matter how large is L, it is always possible to have $f(\omega) > 0$ for some $\omega \in \Omega$ outside the discretization grid; see Figure 15.7 (explanations in Example 15.2.5 below).

To cope with this problem, it is advisable to take a dense enough grid, typically with at least $L = 10N$ points if an equidistant grid is used, and to impose $f(\omega_\ell) \leq -\varepsilon$ instead of (15.46), where ε is a small positive constant, introduced with the purpose of ensuring (15.46) between the discretization points. Once (15.48) is solved, the constraint (15.46) can be checked on a finer grid and, if no significant violation occurs, the solution is deemed acceptable. Of course, the solution is only near-optimal, even if the constraint is satisfied $\forall \omega \in \Omega$.

Example 15.2.5. We illustrate here discretization as used for designing FIR linear-phase filters; it is not the best approach for the problem posed below, but we aim to show its advantages and drawbacks. A linear-phase filter with even degree $2n$ has the transfer function

$$\begin{aligned} H(z) \quad = \quad & h_0 + h_1 z^{-1} + \ldots + h_{n-1} z^{-(n-1)} + h_n z^{-n} + h_{n-1} z^{-(n+1)} \\ & + \ldots + h_1 z^{-(2n-1)} + h_0 z^{-2n}. \end{aligned} \tag{15.49}$$

On the unit circle ($z = e^{j\omega}$), the transfer function
sponse)

$$H(e^{j\omega}) = e^{-j\omega n} \left(h_n + 2h_{n-1} \cos\omega + \ldots + 2h_0 \text{ cc} \right.$$

where $\mathbf{h} \in \mathbb{R}^{n+1}$ contains the coefficients of the filter

$$\mathbf{a}(\omega) = [2\cos(n\omega) \; 2\cos(n-1)\omega \ldots$$

Note that the magnitude of the frequency response
value $|H(e^{j\omega})|$ shows how much a sinusoidal signal wit
the filter, hence it determines the behavior of the filt
design problem, that of a lowpass filter, whose fre
magnitude 1 in the passband $[0, \omega_p]$ and 0 in the stopl
are given. Since this is not possible, we allow a toler
γ_s in the stopband. Let us assume that γ_p is given
the resulting filter is called Chebyshev or minimax fi
problem is

$$\begin{aligned}
\min \quad & \gamma_s \\
\text{s.t.} \quad & 1 - \gamma_p \le |\mathbf{a}(\omega)^T \mathbf{h}| \le 1 + \gamma_p, \\
& |\mathbf{a}(\omega)^T \mathbf{h}| \le \gamma_s, \; \forall \omega \in [\omega_s, \pi]
\end{aligned}$$

We assume that $\mathbf{a}(\omega)^T \mathbf{h}$ is positive in the passband (
replace $\mathbf{h}$ with $-\mathbf{h}$) and attempt to solve (15.50) by
LP problem

$$\begin{aligned}
\min \quad & \gamma_s \\
\text{s.t.} \quad & 1 - \gamma_p \le \mathbf{a}(\omega_\ell)^T \mathbf{h} \le 1 + \gamma_p, \; \ell = \\
& -\gamma_s \le \mathbf{a}(\omega_\ell)^T \mathbf{h} \le \gamma_s, \; \ell = L_p + 1
\end{aligned}$$

The discretization grid has L_p points in the passban
band.

We solve this problem for $n = 15$, $\omega_p = 0.33\pi$,
equidistant grids in the passband and stopband. Wi
optimal value of (15.50) is $\gamma_s = 0.0517$ and the magnit
is shown on the left of Figure 15.6; the dashed lines
for the response: $1 + \gamma_p$ and $1 - \gamma_s$ in the passband,
is fine enough for the violation of the continuous cc
If a sparse grid is used, with $L_p = 11$, $L_s = 21$ (we
points), the value of (15.50) becomes $\gamma_s = 0.0469$, i.
grid. This gain is obtained on the expense of larger
between the discretization points, that can be notice
from Figure 15.6 (right), detailed in Figure 15.7. The
the discretization points; obviously, in these points the
some are active, as it should be expected at optimali

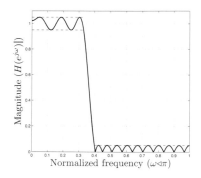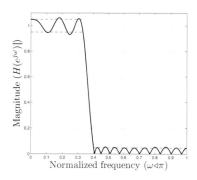

Figure 15.6: Magnitude of frequency responses of linear-phase FIR filters design via discretization and LP, using a fine grid (left) and a sparse grid (right).

constraints are violated between the points. This happens regardless of the density of the grid. To conclude, discretization is a simple to implement method that can give quickly approximate designs. Moreover, the linear constraints from (15.51) can be easily combined with other constraints on the filter to form new optimization problems.

15.3 Mixed Integer Programming

Mixed integer linear programming, also named mixed integer programming (MIP), consists of LP problems like (15.1) or (15.3) in which some or all the variables have integer values. Although typically each integer variable can take only a finite number of values—particularly when the variable is binary and hence takes only the values 0 or 1—MIP problems are more difficult to solve, since they are nonconvex. Moreover, they are NP-hard, which basically means that, in the worst case, only exhaustive search over all the values of the integer variables can guarantee the optimality of the computed solution. Using heuristics based on the branch and bound technique presented in this section, optimal or nearly-optimal solutions can be found quite often.

15.3.1 Problem Statement and LP Relaxation

For the purpose of presentation and without losing generality from an algorithmic viewpoint, as it will be clear later, we assume that all variables are integer, i.e., the MIP problems have one of the standard forms

$$
\begin{array}{llll}
\min & \mathbf{c}^T\mathbf{x} & \max & \mathbf{e}^T\mathbf{y} \\
\text{s.t.} & \mathbf{A}\mathbf{x} = \mathbf{b} & \text{s.t.} & \mathbf{D}\mathbf{y} \leq \mathbf{f} \\
& \mathbf{x} \geq 0, \quad \mathbf{x} \in \mathbb{Z}^n & & \mathbf{y} \in \mathbb{Z}^m
\end{array} \qquad (15.52)
$$

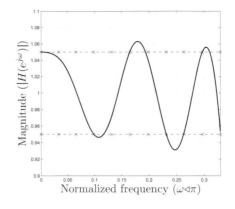

Figure 15.7: Details of the frequency response obta
sparse grid: passband (left), stopband (right).

Such problems appear when the variables have natura
or when they represent logical conditions. We start
sources of such problems are discussed in Section 15.

Example 15.3.1. Let us consider the following MIP

$$
\begin{aligned}
\max \quad & -\epsilon y_1 + y_2 \\
\text{s.t.} \quad -y_1 \qquad\quad &\leq -1 \\
y_1 + \; y_2 &\leq 8.2 \\
-y_1 - 2y_2 &\leq -3 \\
y_1 - \; y_2 &\leq 3 \\
-y_1 + 5y_2 &\leq 14 \\
y_1, y_2 &\in \mathbb{Z}
\end{aligned}
$$

with ϵ a positive constant that is smaller than 0.2.
feasible domain, the gradient, and the level curves,
is the polygon defined by the inequality constraints
represent the points with integer coordinates inside the
feasibility domain of the MIP problem; infeasible poi
are marked with $\times$. We see from the level curves t
criterion is obtained for $y_1 = 1$, $y_2 = 3$; so, this is the
   For a problem of such small size, the solution can
enumeration; for example, by realizing that $1 \leq y_1 \leq$
covering the feasible set with a rectangle, we would
points with integer coordinates in this rectangle (some
problems with n variables, this exhaustive approach
2^n points, in the favorable case where the variables a
is computationally prohibitive in general.

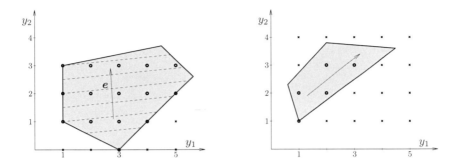

Figure 15.8: Left: feasible points of the MIP problem from Example 15.3.1. Right: any rounding of the LP solution may produce an infeasible point.

Another naive approach would be to remove the integrality constraints, solve the relaxed resulting LP instead of the MIP, and then round the solution to integer values. In Example 15.3.1, the solution of the LP problem is $y_1 = 4.5$, $y_2 = 3.7$. Rounding to the nearest integer gives $\mathbf{y} = (4, 4)$ or $\mathbf{y} = (5, 4)$, which are not feasible points. Truncating gives $\mathbf{y} = (4, 3)$, which is feasible, but not optimal. In general, a $\mathbf{y} \in \mathbb{R}^n$ is in a box with 2^n vertices with integer coordinates (either $\lfloor y_i \rfloor$ or $\lceil y_i \rceil$, for each $i = 1 : n$), so the search of a feasible vertex is by itself difficult. Moreover, it is possible that none of the vertices is a feasible point, as suggested on the right of Figure 15.8; the solution of the LP problem is the rightmost vertex of the corresponding feasible polygon; none of the neighboring points with integer coordinates is a feasible point for the MIP problem. However, the idea of solving the LP problem is not at all worthless, as we will see next.

For some problems, it can be told by simple inspection of the problem that the LP solution is optimal for the MIP problem. Consider the assignment problem, in which ℓ agents have to execute simultaneously ℓ different tasks; so, an agent executes a single task and a task is executed by a single agent. The cost of executing task i by agent j is $c_{i,j}$. The purpose is to find the assignment with minimum cost. The problem can be modeled by introducing binary variables $x_{i,j}$, $i, j = 1 : \ell$, with the meaning that $x_{i,j} = 1$ if task i is executed by agent j and $x_{i,j} = 0$ otherwise. The resulting MIP optimization problem is

$$
\begin{aligned}
\min \quad & \sum_{i=1}^{\ell} \sum_{j=1}^{\ell} c_{i,j} x_{i,j} \\
\text{s.t.} \quad & \sum_{i=1}^{\ell} x_{i,j} = 1, \quad j = 1 : \ell \\
& \sum_{j=1}^{\ell} x_{i,j} = 1, \quad i = 1 : \ell \\
& x_{i,j} \in \{0, 1\}, \quad i, j = 1 : \ell
\end{aligned} \tag{15.54}
$$

The first constraint ensures that agent j executes exactly one task; the second constraint ensures that task i is executed by a single agent. The problem (15.54) can be relaxed to LP by replacing the binary constraints with $0 \leq x_{i,j} \leq 1$.

Hence, instead of searching the solution among the n
($n = \ell^2$), whose coordinates are either 0 or 1, the sear
hypercube. Since the solution of the LP problem is a
and all vertices have binary coordinates, the solution
of the MIP problem.

In degenerate cases, a numerical solver may retur
ample, if for some i_1, i_2, j_1, j_2 it happens that $c_{i_1,j_1} =$
there is no preference in assigning tasks i_1 and i_2 to a
could be obtained from an LP solver for the correspc
x_{i_2,j_1}, x_{i_2,j_2}, but any rounding to $\{0,1\}$ that satisfi
gives an optimal solution to the MIP problem.

In general, the LP relaxation gives the optimal MI
of its feasible polytope have integer coordinates. Thi
instance, assume that, in the assignment problem, age
to execute task i; bounding the total resources by a cor
ρ will destroy the nice properties of problem (15.54),
probably "cut" a part of the hypercube and introduce
coordinates to the feasible polytope of the LP proble

Still, the LP relaxation is at the core of the bra
hence is extremely useful.

15.3.2   Branch and Bound

Branch and bound is a general algorithmic idea bas
strategy. It is the background algorithm for the mos
this section, we will describe its basics and point out
details can be found in the specialized literature. A
algorithmic approaches to MIP, such as the cutting p

In MIP, branch and bound consists of solving LP p
constraints that restrain the feasibility domain, with t
integer solutions. The LP problems are seen as the nc
Figure 15.9. Once a problem is solved, two complem
to it, producing two subproblems; this is called bra
the information available from the LP solutions is u:
branching on a node cannot give the optimum, and
the node without exploring its subtree; this is called
MIP the common expression is that the node was *fat*

We will describe the algorithm for the maximizatio
right of (15.52) and illustrate it for the problem (15.53
the integrality constraints from these problems we ob
(P), whose solution and optimal value are $\mathbf{y}^\star$ and z, r
are $\mathbf{y}^\star = (4.5, 3.7)$ and $z = 3.25$. Since the MIP pro
results that its optimal value cannot be larger than
algorithm updates an upper bound $z_{\max}$ and a lower t

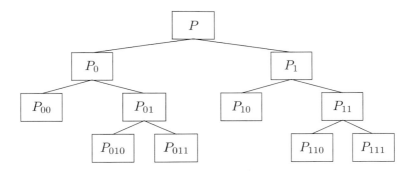

Figure 15.9: Branch and bound tree.

after solving (P), we can set $z_{\max} = z$. However, there is no information on the lower bound, so we can only say that $z_{\min} = -\infty$, i.e., the value for an infeasible MIP problem.

Typically, the solution of (P) is not integer, and this is indeed the case in our example, so further exploration is needed. Branching is performed by selecting one variable, say y_1, and adding to (P) the constraints $y_1 \le \lfloor y_1^\star \rfloor$ and $y_1 \ge \lceil y_1^\star \rceil$ to obtain two subproblems, denoted (P_0) and (P_1), respectively. Formally, we will write $(P_0) = (P, y_1 \le \lfloor y_1^\star \rfloor)$ and $(P_1) = (P, y_1 \ge \lceil y_1^\star \rceil)$. This choice is guided not only by the purpose of getting subproblems with smaller feasible domains, but also by the hope to obtain an integer optimal value for y_1 when solving the subproblems, although as we have seen above this will not necessarily happen. In our example, the subproblems are $(P_0) = (P, y_1 \le 4)$ and $(P_1) = (P, y_1 \ge 5)$; their feasible domains are shown on the left of Figure 15.10; since it is no longer possible to have $4 < y_1 < 5$, a part of the initial feasible domain has been eliminated. Since no feasible points with integer coordinates have been eliminated, the solution of the initial MIP problem is the solution of the MIP version of one of (P_0) or (P_1).

We go on by solving the subproblems. The solution of (P_0) is $\mathbf{y}_0^\star = (4, 3.6)$, and the optimal value $z_0 = 3.2$. Although $z_0 < z_{\max}$, we cannot yet lower the upper bound of the MIP value to z_0, because it is possible that (P_1) gives a value larger than z_0. The solution is still noninteger, so we continue the branching, this time using the variable y_2 (however, in general one can branch several times on the same variable). The corresponding subproblems are $(P_{00}) = (P_0, y_2 \le \lfloor (\mathbf{y}_0^\star)_2 \rfloor)$ and $(P_{01}) = (P_0, y_2 \ge \lceil (\mathbf{y}_0^\star)_2 \rceil)$. For our example, they are $(P_{00}) = (P, y_1 \le 4, y_2 \le 3)$ (see right of Figure 15.10) and $(P_{01}) = (P, y_1 \le 4, y_2 \ge 4)$.

We have now three active (not solved) subproblems: (P_1), (P_{00}), (P_{01}). Any of them can be solved now, following diverse strategies of exploring the branch and bound tree. If (P_1) is solved first, the solution is $\mathbf{y}_1^\star = (5, 3.2)$ and the optimal value $z_1 = 2.7$. Branching continues with $(P_{10}) = (P_1, y_2 \le \lfloor (\mathbf{y}_1^\star)_2 \rfloor)$ and $(P_{11}) = (P_1, y_2 \ge \lceil (\mathbf{y}_1^\star)_2 \rceil)$. For our example, the subproblems are $(P_{10}) = (P, y_1 \ge 5, y_2 \le 3)$ and $(P_{11}) = (P, y_1 \ge 5, y_2 \ge 4)$. We can set $z_{\max} = \max(z_0, z_1) = 3.2$, since

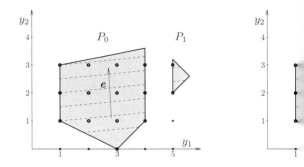

Figure 15.10: Branch and bound subproblems. Left:
Right: feasible problems after second branching, on y

none of the subproblems of (P_0) and (P_1) can have a v
parents.

The LP problem (P_{00}) has the integer solution $\mathbf{y}$
2.9. Since we have found a feasible point for the i
set $z_{\min} = z_{00}$. For the same reason, further branch
Although we actually have found the optimum, we c
indeed the case without examining the remaining activ
P_{01} is infeasible, hence again there is no reason to b
update the upper bound as $z_{\max} = \max(z_{00}, z_{01}, z_1)$
results that $\mathbf{y}_{00}^{\star}$ is the solution of the initial MIP prob
to explore (P_{10}) and (P_{11}).

If (P_{00}) and (P_{01}) are solved before (P_1), i.e., the s
then, when solving (P_1), we notice that $z_1 < z_{\min} =$
solution of (P_1) is worse than a currently available int
can come from branching (P_1); so, the search on that

From the discussion above, we conclude that there
node is considered fathomed:

- the solution of the corresponding LP problem i
  LP problem is larger than $z_{\min}$, then $z_{\min}$ is set

- the LP problem is infeasible;

- the value of the LP problem is smaller than th
  of an already found integer solution.

Based on the above discussion, a general form of
rithm for solving MIP problems is given in Table 1$
sponding to a node of the branch and bound tree is
nodes are discovered in steps 2 and 3. Branching is pϵ

Table 15.3: Basic form of the branch and bound algorithm for solving MIP problems.

Input data: $\mathbf{A}$, $\mathbf{b}$, $\mathbf{c}$ of the MIP problem (15.52) in inequality form.

Initialization. Create list $\mathcal{L}$ of active LP subproblems and initialize it with (P), the initial MIP problem without integrality constraints. Initialize lower and upper bounds to the optimal value: $z_{\max} = \infty$, $z_{\min} = -\infty$. Create empty list $\mathcal{M}$ of solved subproblems whose value may be $z_{\max}$.

Step 1. Choose a subproblem (S) from $\mathcal{L}$ and find its solution $\mathbf{y}^\star$ and value z. Remove (S) from $\mathcal{L}$. Append (S) to $\mathcal{M}$.

Step 2. If $z \leq z_{\min}$, then either (S) is infeasible $(z = -\infty)$ or its value is too small for branching to produce good candidates to the optimum. Go to step 5.

Step 3. If $\mathbf{y}^\star \in \mathbb{Z}^m$, then (since $z > z_{\min}$), set $z_{\min} = z$ and $\mathbf{y}^\star_{MIP} = \mathbf{y}^\star$. Go to step 5.

Step 4. If $\mathbf{y}^\star \notin \mathbb{Z}^m$, then choose a variable y_i on which to branch. Create subproblems $(S_0) = (S, y_i \leq \lfloor y_i^\star \rfloor)$, $(S_1) = (S, y_i \geq \lceil y_i^\star \rceil)$ and add them to $\mathcal{L}$.

Step 5. Let (S_{up}) be the parent of (S). If both subproblems of (S_{up}) are in $\mathcal{M}$ (i.e., are both solved), remove (S_{up}) from $\mathcal{M}$. Update $z_{\max}$ as the maximum value of a problem from $\mathcal{M}$.

Step 6. If $z_{\min} = z_{\max}$, then the MIP optimum has been found, stop. If the list $\mathcal{L}$ is empty, then the MIP problem is infeasible, stop. Otherwise, go back to step 1.

Output: the MIP solution $\mathbf{y}^\star_{MIP}$ and the corresponding optimal value $z_{\min}$.

---

the value of $z_{\max}$ is updated using the following remark: if both LP subproblems of a node are solved, their values are less than or equal to the value of the LP problem corresponding to the node; the latter becomes an obsolete upper bound and can be ignored from now on. Finally, step 6 contains the conclusions; if the lower and upper bounds are equal, then the optimum has been found and the other active subproblems can remain unsolved; if there are no more active subproblems, then the only possibility is that $z_{\min} = -\infty$, i.e., no feasible integer point has been found for the MIP problem, which is hence infeasible.

The fine details that make the difference between the branch and bound methods are in the choice of the next LP problem to be solved (step 1) and in the choice of the variable on which to branch (step 4). Both are guided by heuristics and can never guarantee a certain complexity of the algorithm. However, there are classes of problems that are favored by some algorithms. So, although MIP is a

hard problem in general, branch and bound can give
if the number of variables m is relatively small. If m
complexity can easily explode, although current progr
with thousands of variables. Anyway, the branch a
stopped with a suboptimal solution, if $z_{\min}$ is deemed
hence can give useful information.

To give a taste of the different choices in step 1, let
exploration of the tree from Figure 15.9 is valid, as long
its subnodes. A depth-first search has the advantag
nodes with feasible integer solutions; they can give
lower bound, hence $z_{\min}$ can grow rapidly. Anothe
subproblems of a node are very similar with the pro
one constraint is added to it; the simplex algorithm
initial guess and it will have a very low complexity. A l
explores all nodes at the same level in the tree, has th
upper bound $z_{\max}$. Since these search strategies have
practical algorithms combine them, with the purpose
between $z_{\min}$ and $z_{\max}$.

If the variables are binary, the branch and boun
refined, and its complexity is lower than that of a
number of variables and the same constraints. That
that the variables are binary can be used algorithmic

Finally, note that if only some of the variables of ti
values, then the branch and bound algorithm works
optimality is detected with respect to integer variable:
have optimal values.

15.3.3   Examples of Mixed Integer Pro

We describe here some problems whose modeling leac

Logical conditions on LP constraints. The constra
lem are implicitly connected by a logical AND. All of t
solution. A natural question is how to model other
first the case of an exclusive OR operation: an LP p
constraints $\mathbf{A}\mathbf{x} \leq \mathbf{b}$ and either one of the constraints

$$\mathbf{A}_1\mathbf{x} \leq \mathbf{b}_1,$$
$$\mathbf{A}_2\mathbf{x} \leq \mathbf{b}_2.$$

If these are the only alternatives, then one can imp
LP problems; the solution giving the best criterion
if several pairs of alternatives like (15.55) are presen
solve an LP problem for each combination of constra
general way of dealing with (15.55) is to introduce t

ξ_1, ξ_2 and to impose simultaneously the constraints

$$\begin{aligned}
\mathbf{A}_1\mathbf{x} &\leq \mathbf{b}_1 + \xi_1\mathbf{u}_1, \\
\mathbf{A}_2\mathbf{x} &\leq \mathbf{b}_2 + \xi_2\mathbf{u}_2, \\
\xi_1 + \xi_2 &= 1, \quad \xi_1, \xi_2 \in \{0, 1\},
\end{aligned} \tag{15.56}$$

where $\mathbf{u}_1$, $\mathbf{u}_2$ are vectors with positive elements, chosen big enough such that the constraints $\mathbf{A}_1\mathbf{x} \leq \mathbf{b}_1 + \mathbf{u}_1$, $\mathbf{A}_2\mathbf{x} \leq \mathbf{b}_2 + \mathbf{u}_2$ are satisfied for all desirable values of $\mathbf{x}$; typically, all elements of $\mathbf{u}_1$, $\mathbf{u}_2$ are taken equal to a large constant M, and building (15.56) is called the "big-M" technique. The constraint $\xi_1 + \xi_2 = 1$ of the resulting MIP problem forces only one of the variables ξ_1 or x_2 to be zero, hence imposing only one of the constraints (15.55). In this case, since $\xi_2 = 1 - \xi_1$, one can use a single binary variable; for a generalization, see Exercise 15.10.

Other logical conditions make little sense. Imposing that at most one of the conditions (15.55) is satisfied can be modeled by $\xi_1 + \xi_2 \geq 1$, but the optimal solution of the MIP problem will likely give $\xi_1 = \xi_2 = 1$, i.e., the solution of the LP problem without any of the constraints (15.55). Imposing that at least one of the conditions (15.55) is satisfied can be modeled by $\xi_1 + \xi_2 \leq 1$, but the solution should be the same as that obtained with (15.56), since the optimization criterion should be improved by giving up one of the constraints. Note also that, in practice, it is recommended to take M as small as possible, since very large values, although theoretically desirable, can affect the numerical accuracy of the solution.

Sparse solutions to LP feasibility problems. Assume that we seek a vector $\mathbf{x} \in \mathbb{R}^n$ satisfying some linear constraints $\mathbf{A}\mathbf{x} \leq \mathbf{b}$ (the exact form of these constraints is not important), such that its number of non-zero elements (namely $\|\mathbf{x}\|_0$) is minimum. For example, we want to design an FIR filter like in Example 15.2.5, satisfying the constraints of (15.51), but, instead of optimizing the stopband error bound γ_s, we keep it fixed to a preset value and minimize the number of non-zero elements. Such an optimization makes sense for hardware devices where the filter implementation cost depends on the number of coefficients, while the performance requirements are known. There are many heuristics for finding suboptimal solutions, like minimizing $\|\mathbf{x}\|_1$ instead of $\|\mathbf{x}\|_0$ (see a similar idea in Example 15.2.2) or using greedy approaches that find the non-zero elements one by one [9]. Using the big-M technique, the minimization of $\|\mathbf{x}\|_0$, subject to $\mathbf{A}\mathbf{x} \leq \mathbf{b}$, can be transformed into the equivalent MIP problem with binary variables

$$\begin{aligned}
\min \quad & \sum_{i=1}^{n} \xi_i \\
\text{s.t.} \quad & \mathbf{A}\mathbf{x} \leq \mathbf{b} \\
& -\xi_i M \leq x_i \leq \xi_i M, \quad i = 1 : n \\
& \xi_i \in \{0, 1\}, \quad i = 1 : n
\end{aligned} \tag{15.57}$$

Taking M big enough, if the binary variable ξ_i is zero, then x_i is also zero. If $\xi_i = 1$, then x_i is practically unrestricted. Hence, the optimization criterion is the number of non-zero elements of $\mathbf{x}$.

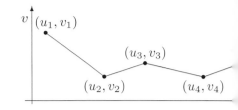

Figure 15.11: Piecewise linear fr

Piecewise linear constraint. Consider two scalar v.
piecewise linear relation like in Figure 15.11, with gi
To model such a nonconvex constraint in MIP style,
segment is made by all convex combinations of its enc
the segment i connecting (u_i, v_i) and (u_{i+1}, v_{i+1}), the

$$u = \xi u_i + \zeta u_{i+1},$$
$$v = \xi v_i + \zeta v_{i+1},$$

with $\xi + \zeta = 1$ and $\xi, \zeta \geq 0$. Since the variable point
segment, we have to introduce real variables ξ_i, ζ_i fo
and model the piecewise linear relation through

$$u = \sum_{i=1}^{p-1} [\xi_i u_i + \zeta_i u_{i+1}],$$
$$v = \sum_{i=1}^{p-1} [\xi_i v_i + \zeta_i v_{i+1}],$$

where all pairs ξ_i, ζ_i are zero, except one, since $(u$
segment. This constraint can be modeled with binar
by imposing

$$\xi_i + \zeta_i = \eta_i,$$
$$\xi_i \geq 0, \ \zeta_i \geq 0, \quad i = 1 : p -$$
$$\sum_{i=1}^{p-1} \eta_i = 1,$$
$$\eta_i \in \{0, 1\}, \quad i = 1 : p - 1.$$

So, only one of the binary variables η_i can be one, the
one pair of variables ξ_i, ζ_i is non-zero and obeys to $\xi_i +$
The constraints (15.59) and (15.60) are linear and de
v and ξ_i, ζ_i, $i = 1 : p - 1$, and the binary variables
be indeed inserted in a MIP problem. This model is
number of real variables can be reduced to almost a
suggested in Exercise 15.11; however, the complexity
is dictated mostly by the number of integer variables.

Design of FIR filters with signed power-of-two co
the FIR filter design problem in Example 15.2.5 from
ing its hardware implementation complexity, which is

specialized circuits. An important part of the implementation cost is dictated by the physical realization of the coefficients of the filter (15.49), which are implemented with finite precision. We have seen above that one can minimize the number of non-zero coefficients, but here we consider a more detailed description of the coefficients. In fixed point arithmetic, a coefficient h_k can be represented as

$$h_k = \sum_{i=-\alpha}^{\beta} h_{ik} 2^i, \tag{15.61}$$

where $h_{ik} \in \{-1, 0, 1\}$ are signed digits in a binary representation with $p = \alpha + \beta + 1$ digits, that covers $\beta + 1$ digits in the integer part and α digits in the fractional part. The value of β dictates the maximum magnitude and can have small values like 1 or 2 and the value of α sets the precision of the implementations and can be 8–10 or even larger. The relation (15.61) is called signed power-of-two (SPT) representation. Note that, unlike the canonical representation that admits only the digits 0 and 1 and has a separate representation of the sign, the SPT representation is not unique; for example, the decimal number 0.375 can be represented as $2^{-2} + 2^{-3}$ or $2^{-1} - 2^{-3}$.

Using the notations from Example 15.2.5, the frequency response of the filter is essentially given by

$$\mathbf{a}(\omega)^T \mathbf{h} = \sum_{k=0}^{n} h_k a_k(\omega) = \sum_{k=0}^{n} \sum_{i=-\alpha}^{\beta} h_{ik} 2^i a_k(\omega) = \mathbf{b}(\omega)^T \mathbf{y}$$

where $a_k(\omega) = 2 \cos(n - k)\omega$, $k = 0 : n - 1$, $a_n(\omega) = 1$, and

$$\begin{aligned}
\mathbf{y}^T &= [\ h_{-\alpha,0} \quad \cdots \quad h_{\beta,0} \quad h_{-\alpha,1} \quad \cdots \quad h_{\beta,n} \ ], \\
\mathbf{b}(\omega)^T &= [2^{-\alpha} a_0(\omega) \quad \cdots \quad 2^\beta a_0(\omega) \quad 2^{-\alpha} a_1(\omega) \quad \cdots \quad 2^\beta a_n(\omega)].
\end{aligned}$$

So, the frequency response depends linearly on the unknown vector $\mathbf{y} \in \mathbb{R}^{p(n+1)}$. Hence, by imposing tolerances γ_p and γ_s and minimizing the number of nonzero bits of the SPT representation of the filter coefficients, the discretized LP design problem (15.51) can be transformed into the MIP problem

$$
\begin{aligned}
\min \quad & \|\mathbf{y}\|_1 && (15.62) \\
\text{s.t.} \quad & 1 - \gamma_p \leq \mathbf{b}(\omega_\ell)^T \mathbf{y} \leq 1 + \gamma_p, \ \ell = 1 : L_p, \\
& -\gamma_s \leq \mathbf{b}(\omega_\ell)^T \mathbf{y} \leq \gamma_s, \ \ell = L_p + 1 : L_p + L_s \\
& \mathbf{y} \in \{-1, 0, 1\}^{p(n+1)}
\end{aligned}
$$

Other linear constraints can be added to this problem. For example, if the number of non-zero bits per coefficient has to be bounded to $\tilde{p}$, the constraint is

$$\sum_{i=-\alpha}^{\beta} |h_{ik}| \leq \tilde{p}, \quad k = 0 : n.$$

The redundancy of the SPT representation is elimin
never consecutive, which can be written as

$$|h_{ik}| + |h_{i+1,k}| \leq 1, \quad i = -\alpha : \beta - 1$$

Although this increases the number of constraints, it a
time in many cases. The above condition also sligh
coefficients: the larger value is no longer $2^{\beta} + 2^{\beta-1} +$
$2^{\beta} + 2^{\beta-2} + 2^{\beta-4} + \ldots$; however, inside the new rang
are attainable.

Alternatively to the SPT representation, one co
standard fixed point representation

$$h_k = x_k \cdot 2^{-\alpha}, \quad x_k \in \mathbb{Z}, \ |x_k| \leq$$

Although the result of the corresponding MIP optin
ily converted to SPT representation, some constrai:
representation (15.63), for example those regarding
However, (15.63) is useful when the coefficients are op
mentation of the filter in fixed point arithmetic, wher
the range and precision of the coefficients, not the pa

15.4   Historical Notes and Furth

Although there are earlier descriptions, the first treatm
a distinct optimization problem belonged to L.V. Ka:
with production planning. In 1947, G.B. Dantzig in
the first efficient tool for solving LP problems on a c
brought different improvements to the simplex metho
principles, together with the understanding of its expe
L.G. Khachiyan [11] presented the ellipsoid method, w
time algorithm, with a $O(n^4)$ complexity; however, i:
always worse than the simplex, as its average comple
one. The first efficient interior-point algorithm, with
$O(n^{3.5})$, but usually much better, was given by N. K:
a huge impulse on the research of interior-point alg
several classes of convex optimization. Currently, th
methods are considered of practically equal importanc
for LP problems with different characteristics and pro

More historical information, with emphasis on in
found in [13], [2], and [14]. An early simple prese:
method and comparisons with the simplex method
reader interested in more detailed presentations of line
textbooks like [10] and [16]. LP belongs to the mor
optimization, treated extensively in [17].

Mixed integer programming belongs essentially to combinatorial optimization, where branch and bound is the preferred algorithmic tool. As such, solving MIP problems via branch and bound was not especially innovative in principle; however, there are many methods that cleverly take advantage of the particular connection with LP. For a broader view on the field, see for example [18].

There are many software packages for solving LP and MIP problems. The reader is encouraged to try noncommercial software like LP_SOLVE, GPLK or MINTO. Semidefinite programming solvers can be easily used for LP problems; packages like SeDuMi, SDPT3, CVX, Yalmip are all free and the latter has also a MIP module.

15.5 Exercises

Exercise 15.1. Consider the problem of minimizing $\mathrm{Re}\left(\mathbf{c}^T\mathbf{x}\right)$, with the constraint $\mathbf{A}\mathbf{x} = \mathbf{b}$, where the variable $\mathbf{x}$ and the data $\mathbf{A}$, $\mathbf{b}$, $\mathbf{c}$ are *complex*. Assuming that the real and imaginary parts of $\mathbf{x}$ are non-negative, transform it into the standard form (15.1).

Exercise 15.2. Is there a value of $\mathbf{c}$ such that all feasible points of problem (15.2) are also solutions? In general, show that if $\mathbf{c} \in \mathcal{I}m\left(\mathbf{A}^T\right)$, then all feasible points are optimal, i.e., have the same value $\mathbf{c}^T\mathbf{x}$. Hint: express the solutions of the linear system $\mathbf{A}\mathbf{x} = \mathbf{b}$ as $\mathbf{x} = \mathbf{x}_0 + \mathbf{u}$, with $\mathbf{x}_0$ a particular solution and $\mathbf{u} \in \mathcal{K}er\left(\mathbf{A}\right)$.

Exercise 15.3. Consider the LP primal problem (15.1) with $\mathbf{A} = [-1\ 0]$, $\mathbf{b} = 1$, $\mathbf{c} = [0\ -1]^T$. Show that both the primal and its dual (15.9) are infeasible.

Exercise 15.4. Show that the dual of the optimization problem (15.3) is

$$\begin{aligned} \min \quad & \mathbf{f}^T\mathbf{x} \\ \text{s.t.} \quad & \mathbf{D}^T\mathbf{x} = \mathbf{e} \\ & \mathbf{x} \geq 0 \end{aligned}$$

which means that the dual of the dual of the primal LP problem (15.1) is the primal itself.

Exercise 15.5. An LP problem in the form

$$\begin{aligned} \max \quad & \mathbf{c}^T\mathbf{x} \\ \text{s.t.} \quad & \mathbf{A}\mathbf{x} \leq \mathbf{b} \\ & \mathbf{x} \geq 0 \end{aligned}$$

is also called standard. Show that its dual is

$$\begin{aligned} \min \quad & \mathbf{b}^T\mathbf{y} \\ \text{s.t.} \quad & \mathbf{A}^T\mathbf{y} \geq \mathbf{c} \\ & \mathbf{y} \geq 0 \end{aligned}$$

If $\mathbf{b} > 0$, after introducing slack variables to transfor
(15.1), find a basic feasible point.

Exercise 15.6. Strong duality of LP is basically equ
page 526 for its statement) used in the proof of Th
lemma by applying Theorem 15.1.6 to the LP progra
    (i) Primal (feasibility problem): $\min 0$, subject to
    Hint: show that its dual is
    (ii) $\max \mathbf{f}^T \mathbf{v}$, subject to $\mathbf{\Phi}^T \mathbf{v} \leq 0$ (note that the i
on page 526).
    Use the facts that: (a) if the primal is feasible, t
(b) if there is a feasible $\mathbf{v}$ such that $\mathbf{f}^T \mathbf{v} > 0$, then t
the primal is infeasible.

Exercise 15.7. Express as LP in standard equality f
ing $\|\mathbf{x}\|_1$ or $\|\mathbf{x}\|_\infty$ subject to the equality constraint
the matrix of the standard LP problem (15.1) shou
1-norm minimization and $(m+n) \times (2n+1)$ for the c
for the 1-norm, use the substitution $\mathbf{x} = \mathbf{x}_+ - \mathbf{x}_-$ a
$\mathbf{x}_+$ and $\mathbf{x}_-$ cannot have non-zero elements in the san

Exercise 15.8. Assume that the positions (x_i, y_i) c
Express as LP the problem of finding a polynomial
the maximum error $|y_i - f(x_i)|$, $i = 1 : N$, is minimu

Exercise 15.9. Consider an LP problem which, tog
constraints, is also constrained to exactly one of the
$\mathbf{b}_2$. Model this problem as MIP. (Hint: transform ea
then apply the big-M technique.)

Exercise 15.10. Consider an LP problem in which e
$\mathbf{A}_i \mathbf{x} \leq \mathbf{b}_i$, $i = 1 : N$, should be satisfied, with $\mathbf{A}_i$, $\mathbf{b}$
of appropriate sizes. How can it be modeled as a MII

Exercise 15.11. Alternatively to (15.59), the piecew
ure 15.11 can be modeled via only p variables θ_i thro

$$u = \sum_{i=1}^p \theta_i u_i,$$
$$v = \sum_{i=1}^p \theta_i v_i,$$

where $\theta_i \geq 0$, $\sum_{i=1}^p \theta_i = 1$; each variable θ_i is associat
the variables (u, v) belong to a single segment i, the
θ_{i+1} must be non-zero (their sum being equal to one)

Prove that, together with the above, the conditions

$$
\begin{aligned}
&\theta_1 \leq \eta_1 \\
&\theta_i \leq \eta_{i-1} + \eta_i, \quad i = 2 : p - 1 \\
&\theta_p \leq \eta_{p-1} \\
&\sum_{i=1}^{p-1} \eta_i = 1 \\
&\eta_i \in \{0, 1\}, \quad i = 1 : p - 1
\end{aligned}
$$

describe correctly the piecewise linear dependence. (Hint: note that variable η_i corresponds to segment i and $\eta_{i-1} + \eta_i = 1$ only for two consecutive values i.)

Exercise 15.12. We want to place N objects on the floor of a rectangular room. Each object has a rectangular basis, so we can model the problem in 2D; the length ℓ_i and width w_i of object i are given; the height is not relevant, since the objects cannot be superposed. The position of object i is given by the coordinates (x_i, y_i) of one of its corners, the same for all objects.

What is the condition that objects i and j occupy distinct positions in space (their basis rectangles do not superpose)?

How can this condition be modeled in the constraints of an optimization problem (the criterion is not relevant now)?

References

[1] V. Klee and G. Minty, "How good is the simplex method," in *Inequalities III*, O. Shisha, Ed. Academic Press, 1972, pp. 159–175.

[2] M. Todd, "The many facets of linear programming," *Math. Program. Ser. B*, vol. 91, no. 3, pp. 417–436, Feb. 2002.

[3] R. Bartels and G. Golub, "The simplex method of linear progamming using LU decomposition," *Comm. ACM*, vol. 12, no. 5, pp. 266–268, May 1969.

[4] J. Picard and A. Weiss, "Bounds on the Number of Identifiable Outliers in Source Localization by Linear Programming," *IEEE Trans. Signal Proc.*, vol. 58, no. 5, pp. 2884–2895, May 2010.

[5] W. Zhou, L. Zhang, and L. Jiao, "Linear programming support vector machines," *Pattern Recognition*, vol. 35, pp. 2927–2936, 2002.

[6] B. Dumitrescu, *Positive Trigonometric Polynomials and Signal Processing Applications*. Springer, Dordrecht, The Netherlands, 2007.

[7] R. Reemtsen and J.-J. Rückman, Eds., *Semi-Infinite Programming*. Kluwer Academic Publishers, Dordrecht, The Netherlands, 1998.

[8] M. Goberna and M. Lopez, "Semi-infinite programming theory: an updated survey," *Eur. J. Operational Research*, vol. 143, pp. 390–405, 2002.

[9] T. Baran, D. Wei, and A. Oppenheim, "Linear sparse filter design," *IEEE Trans. Signal Proc.*, March 2010.

[10] D. Bertsimas and J. Tsitsiklis, *Introduction to* Scientific, Nashua, NH, USA, 1997.

[11] L. Khachiyan, "A polynomial algorithm in linear *Dokl.*, vol. 20, pp. 191–194, 1979.

[12] N. Karmarkar, "A new polynomial-time algoritl *Combinatorica*, vol. 4, no. 4, pp. 373–395, 1984.

[13] R. Freund and S. Mizuno, "Interior point metl ture directions," in *High Performance Optimizat* Academics, pp. 441–466, 2000.

[14] M. Wright, "The interior-point revolution in c developments, and lasting consequences," *Bull.* no. 1, pp. 39–56, 2005.

[15] G. Astfalk, I. Lustig, R. Marsten, and D. Shanno for linear programming," *IEEE Software*, vol. 9,

[16] R. Vanderbei, *Linear Programming: Foundatic* Springer, New York, NY, USA, 2001.

[17] S. Boyd and L. Vandenberghe, *Convex Optimiza* Press, Cambridge, UK, 2004.

[18] L. Wolsey and G. Nemhauser, *Integer and Combi* Interscience, New York, NY, USA, 1999.

Chapter 16 Majorization Theory and Applications

Jiaheng Wang[‡]and Daniel Palomar[♯]
[‡]KTH Royal Institute of Technology, Stockholm, Sweden
[♯]Hong Kong University of Science and Technology, Hong Kong

In this chapter we introduce a useful mathematical tool, namely Majorization Theory, and illustrate its applications in a variety of scenarios in signal processing and communication systems. Majorization is a partial ordering and precisely defines the vague notion that the components of a vector are "less spread out" or "more nearly equal" than the components of another vector. Functions that preserve the ordering of majorization are said to be Schur-convex or Schur-concave. Many problems arising in signal processing and communications involve comparing vector-valued strategies or solving optimization problems with vector- or matrix-valued variables. Majorization theory is a key tool that allows us to solve or simplify these problems.

The goal of this chapter is to introduce the basic concepts and results on majorization that serve mostly the problems in signal processing and communications, but by no means to enclose the vast literature on majorization theory. A complete and superb reference on majorization theory is the book by Marshall and Olkin [1]. The building blocks of majorization can be found in [2], and [3] also contains significant material on majorization. Other textbooks on matrix and multivariate analysis, e.g., [4] and [5], may also include a part on majorization. Recent applications of majorization theory to signal processing and communication problems can be found in two good tutorials [6] and [7].

The chapter contains two parts. The first part is devoted to building the framework of majorization theory. The second part focuses on applying the concepts and results introduced in the first part to several problems arising in signal processing and communication systems.

16.1   Majorization Theory

16.1.1   Basic Concepts

To explain the concept of majorization, let us first d
for increasing and decreasing orders of a vector.

Definition 16.1.1. For any vector $\mathbf{x} \in \mathbb{R}^n$, let

$$x_{[1]} \geq \cdots \geq x_{[n]}$$

denote its components in decreasing order, and le

$$x_{(1)} \leq \cdots \leq x_{(n)}$$

denote its components in increasing order.

Majorization[1] defines a partial ordering between t
precisely describes the concept that the components
"more nearly equal" than the components of $\mathbf{y}$.

Definition 16.1.2. (**Majorization** [1, 1.A.1]) For a
say $\mathbf{x}$ is majorized by $\mathbf{y}$ (or $\mathbf{y}$ majorizes $\mathbf{x}$), denoted

$$\sum_{i=1}^{k} x_{[i]} \ \leq \ \sum_{i=1}^{k} y_{[i]}, \qquad 1 \leq$$

$$\sum_{i=1}^{n} x_{[i]} \ = \ \sum_{i=1}^{n} y_{[i]}.$$

Alternatively, the previous conditions can be rewritte

$$\sum_{i=1}^{k} x_{(i)} \ \geq \ \sum_{i=1}^{k} y_{(i)}, \qquad 1 \leq$$

$$\sum_{i=1}^{n} x_{(i)} \ = \ \sum_{i=1}^{n} y_{(i)}.$$

There are several equivalent characterizations of t
$\mathbf{y}$ in addition to the conditions given in Definition 16.1.
of majorization given in [2] is that $\mathbf{x} \prec \mathbf{y}$ if

$$\sum_{i=1}^{n} \phi\left(x_{i}\right) \leq \sum_{i=1}^{n} \phi\left(y_{i}\right)$$

---

[1]The majorization ordering given in Definition 16.1.2 is als
distinguish it from multiplicative majorization (or log-majoriza

for all continuous convex functions ϕ. Another interesting characterization of $\mathbf{x} \prec \mathbf{y}$, also from [2], is that $\mathbf{x} = \mathbf{P}\mathbf{y}$ for some doubly stochastic matrix[2] $\mathbf{P}$. In fact, the latter characterization implies that the set of vectors $\mathbf{x}$ that satisfy $\mathbf{x} \prec \mathbf{y}$ is the convex hull spanned by the $n!$ points formed from the permutations of the elements of $\mathbf{y}$.[3] Yet another interesting definition of $\mathbf{y} \succ \mathbf{x}$ is given in the form of waterfilling as

$$\sum_{i=1}^{n} (x_i - a)^+ \leq \sum_{i=1}^{n} (y_i - a)^+ \tag{16.2}$$

for any $a \in \mathbb{R}$ and $\sum_{i=1}^{n} x_i = \sum_{i=1}^{n} y_i$, where $(u)^+ \triangleq \max(u, 0)$. The interested reader is referred to [1, Ch. 4] for more alternative characterizations.

Observe that the original order of the elements of $\mathbf{x}$ and $\mathbf{y}$ plays no role in the definition of majorization. In other words, $\mathbf{x} \prec \boldsymbol{\Pi}\mathbf{x}$ for all permutation matrices $\boldsymbol{\Pi}$.

Example 16.1.1. The following are simple examples of majorization:

$$\left(\frac{1}{n}, \frac{1}{n}, \ldots, \frac{1}{n}\right) \prec \left(\frac{1}{n-1}, \frac{1}{n-1}, \ldots, \frac{1}{n-1}, 0\right)$$

$$\prec \cdots \prec \left(\frac{1}{2}, \frac{1}{2}, 0, \ldots, 0\right) \prec (1, 0, \ldots, 0).$$

More generally

$$\left(\frac{1}{n}, \frac{1}{n}, \ldots, \frac{1}{n}\right) \prec (x_1, x_2, \ldots, x_n) \prec (1, 0, \ldots, 0)$$

whenever $x_i \geq 0$ and $\sum_{i=1}^{n} x_i = 1$. $\qquad\square$

It is worth pointing out that majorization provides only a partial ordering, meaning that there exist vectors that cannot be compared within the concept of majorization. For example, given $\mathbf{x} = (0.6, 0.2, 0.2)$ and $\mathbf{y} = (0.5, 0.4, 0.1)$, we have neither $\mathbf{x} \prec \mathbf{y}$ nor $\mathbf{x} \succ \mathbf{y}$.

To extend Definition 16.1.2, which is only applicable to vectors with the same sum, the following definition provides two partial orderings between two vectors with different sums.

Definition 16.1.3. (Weak majorization [1, 1.A.2]) For any two vectors $\mathbf{x}, \mathbf{y} \in \mathbb{R}^n$, we say $\mathbf{x}$ is weakly submajorized by $\mathbf{y}$ (or $\mathbf{y}$ submajorizes $\mathbf{x}$), denoted by $\mathbf{x} \prec_w \mathbf{y}$ (or $\mathbf{y} \succ_w \mathbf{x}$), if

---

[2] A square matrix $\mathbf{P}$ is said to be stochastic if either its rows or columns are probability vectors, i.e., if its elements are all nonnegative and either the rows or the columns sums are one. If both the rows and columns are probability vectors, then the matrix is called doubly stochastic. Stochastic matrices can be considered representations of the transition probabilities of a finite Markov chain.

[3] The permutation matrices are doubly stochastic and, in fact, the convex hull of the permutation matrices coincides with the set of doubly stochastic matrices [1, 2].

$$\sum_{i=1}^{k} x_{[i]} \leq \sum_{i=1}^{k} y_{[i]}, \qquad 1 \leq k$$

We say $\mathbf{x}$ is weakly supermajorized by $\mathbf{y}$ (or $\mathbf{y}$ su
$\mathbf{x} \prec^{w} \mathbf{y}$ (or $\mathbf{y} \succ^{w} \mathbf{x}$), if

$$\sum_{i=1}^{k} x_{(i)} \geq \sum_{i=1}^{k} y_{(i)}, \qquad 1 \leq k$$

In either case, we say $\mathbf{x}$ is weakly majorized by $\mathbf{y}$

For nonnegative vectors, weak majorization can b
in terms of linear transformation by doubly substocha
trices (see [1, Ch. 2]). Note that $\mathbf{x} \prec \mathbf{y}$ implies $\mathbf{x}$
inverse does not hold. In other words, majorization
tion than weak majorization. A useful connection bet
majorization is given as follows.

Lemma 16.1.1. ([1, 5.A.9, 5.A.9.a]) If $\mathbf{x} \prec_{w} \mathbf{y}$, then
such that

$$\mathbf{x} \leq \mathbf{u} \text{ and } \mathbf{u} \prec \mathbf{y}, \qquad \mathbf{v} \leq \mathbf{y} \text{ and}$$

If $\mathbf{x} \prec^{w} \mathbf{y}$, then there exist vectors $\mathbf{u}$ and $\mathbf{v}$ such 1

$$\mathbf{x} \geq \mathbf{u} \text{ and } \mathbf{u} \prec \mathbf{y}, \qquad \mathbf{u} \geq \mathbf{y} \text{ and}$$

The notation $\mathbf{x} \leq \mathbf{u}$ means the componentwise or
for all entries of vectors $\mathbf{x}, \mathbf{u}$.

16.1.2   Schur-Convex/Concave Functic

Functions that are monotonic with respect to the c
called Schur-convex or Schur-concave functions. This
ticular importance in this chapter, as it turns out th
signal processing and communication systems are Sc
functions.

Definition 16.1.4. (Schur-convex/concave funct
function ϕ defined on a set $\mathcal{A} \subseteq \mathbb{R}^{n}$ is said to be Schu

$$\mathbf{x} \prec \mathbf{y} \text{ on } \mathcal{A} \quad \Rightarrow \quad \phi(\mathbf{x}) \leq \phi$$

If, in addition, $\phi(\mathbf{x}) < \phi(\mathbf{y})$ whenever $\mathbf{x} \prec \mathbf{y}$ but $\mathbf{x}$ is
ϕ is said to be strictly Schur-convex on $\mathcal{A}$. Similarly,
on $\mathcal{A}$ if

$$\mathbf{x} \prec \mathbf{y} \text{ on } \mathcal{A} \quad \Rightarrow \quad \phi(\mathbf{x}) \geq \phi$$

and ϕ is strictly Schur-concave on $\mathcal{A}$ if strict inequal
$\mathbf{x}$ is not a permutation of $\mathbf{y}$.

Clearly, if ϕ is Schur-convex on $\mathcal{A}$, then $-\phi$ is Schur-concave on $\mathcal{A}$, and vice versa.

It is important to remark that the sets of Schur-convex and Schur-concave functions do no form a partition of the set of all functions from $\mathcal{A} \subseteq \mathbb{R}^n$ to $\mathbb{R}$. In fact, neither are the two sets disjoint (i.e., the intersection is not empty), unless we consider strictly Schur-convex/concave functions, nor do they cover the entire set of all functions as illustrated in Figure 16.1.

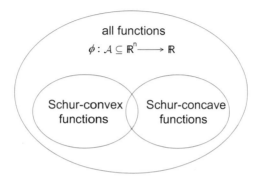

Figure 16.1: Illustration of the sets of Schur-convex and Schur-concave functions within the set of all functions $\phi : \mathcal{A} \subseteq \mathbb{R}^n \to \mathbb{R}$.

Example 16.1.2. The simplest example of a Schur-convex function, according to the definition, is $\phi(\mathbf{x}) = \max_k \{x_k\} = x_{[1]}$, which is also strictly Schur-convex. $\square$

Example 16.1.3. The function $\phi(\mathbf{x}) = \sum_{i=1}^{n} x_i$ is both Schur-convex and Schur-concave since $\phi(\mathbf{x}) = \phi(\mathbf{y})$ for any $\mathbf{x} \prec \mathbf{y}$. However, it is neither strictly Schur-convex nor strictly Schur-concave. $\square$

Example 16.1.4. The function $\phi(\mathbf{x}) = x_1 + 2x_2 + x_3$ is neither Schur-convex nor Schur-concave, as can be seen from the counterexample that for $\mathbf{x} = (2, 1, 1)$, $\mathbf{y} = (2, 2, 0)$ and $\mathbf{z} = (4, 0, 0)$, we have $\mathbf{x} \prec \mathbf{y} \prec \mathbf{z}$ but $\phi(\mathbf{x}) < \phi(\mathbf{y}) > \phi(\mathbf{z})$. $\square$

To distinguish Schur-convexity/concavity from common monotonicity, we also define increasing and decreasing functions that will be frequently used later.

Definition 16.1.5. (Increasing/decreasing functions) A function $f : \mathbb{R}^n \to \mathbb{R}$ is said to be increasing if it is increasing in each argument, i.e.,

$$\mathbf{x} \leq \mathbf{y} \quad \Rightarrow \quad f(\mathbf{x}) \leq f(\mathbf{y}),$$

and to be decreasing if it is decreasing in each argument, i.e.,

$$\mathbf{x} \leq \mathbf{y} \quad \Rightarrow \quad f(\mathbf{x}) \geq f(\mathbf{y}).$$

Using directly Definition 16.1.4 to check Schur-co
tion may not be easy. In the following, we presen
determine whether a function is Schur-convex or Sch

Theorem 16.1.1. *([1, 3.A.3]) Let the function ϕ*
$\mathcal{D}_n \triangleq \{\mathbf{x} \in \mathbb{R}^n : x_1 \geq \cdots \geq x_n\}$ *and continuously*
of $\mathcal{D}_n$. Then ϕ is Schur-convex (Schur-concave) o
decreasing (increasing) in $i = 1, \ldots, n$.

Theorem 16.1.2. *(**Schur's condition** [1, 3.A.4])*
and the function $\phi : \mathcal{I}^n \to \mathbb{R}$ be continuously differentia
on $\mathcal{I}^n$ if and only if ϕ is symmetric[4] on $\mathcal{I}^n$ and

$$(x_i - x_j)\left(\frac{\partial \phi}{\partial x_i} - \frac{\partial \phi}{\partial x_j}\right) \geq 0, \qquad 1$$

ϕ is Schur-concave on on $\mathcal{I}^n$ if and only if ϕ is symme
is reversed.

In fact, to prove Schur-convexity/concavity of a fu
and Theorem 16.1.2, one can take $n = 2$ without
i.e., check only the two-argument case [1, 3.A.5]. B
Theorem 16.1.2, it is possible to obtain some suffic
Schur-convexity/concavity of different composite func

Proposition 16.1.1. (Monotonic composition [
posite function $\phi(\mathbf{x}) = f(g_1(\mathbf{x}), \ldots, g_k(\mathbf{x}))$, where
defined on $\mathbb{R}^k$. Then, it follows that

- f is increasing and g_i is Schur-convex $\Rightarrow \phi$ is S
- f is decreasing and g_i is Schur-convex $\Rightarrow \phi$ is S
- f is increasing and g_i is Schur-concave $\Rightarrow \phi$ is S
- f is decreasing and g_i is Schur-concave $\Rightarrow \phi$ is S

Proposition 16.1.2. (Convex[5] composition [1, 3.
function $\phi(\mathbf{x}) = f(g(x_1), \ldots, g(x_n))$, where f is a rea
$\mathbb{R}^n$. Then, it follows that

- f is increasing Schur-convex and g convex $\Rightarrow \phi$
- f is decreasing Schur-convex and g concave $\Rightarrow$

---

[4]A function is said to be symmetric if its arguments can
changing the function value.
[5]A function $f : \mathcal{X} \to \mathbb{R}$ is convex if $\mathcal{X}$ is a convex set and
$f(\alpha x + (1 - \alpha)y) \leq \alpha f(x) + (1 - \alpha)f(y)$. f is concave if $-f$ i

For some special forms of functions, there exist simple conditions to check whether they are Schur-convex or Schur-concave.

Proposition 16.1.3. (Symmetric convex functions [1, 3.C.2]) If ϕ is symmetric and convex (concave), then ϕ is Schur-convex (Schur-concave).

Corollary 16.1.1. ([1, 3.C.1]) Let $\phi(\mathbf{x}) = \sum_{i=1}^{n} g(x_i)$, where g is convex (concave). Then ϕ is Schur-convex (Schur-concave).

Proposition 16.1.3 can be generalized to the case of quasi-convex functions.[6]

Proposition 16.1.4. (Symmetric quasi-convex functions [1, 3.C.3]) If ϕ is symmetric and quasi-convex, then ϕ is Schur-convex.

Schur-convexity/concavity can also be extended to weak majorization through the following fact.

Theorem 16.1.3. (*[1, 3.A.8]) A real-valued function ϕ defined on a set $\mathcal{A} \subseteq \mathbb{R}^n$ satisfies*

$$\mathbf{x} \prec_w \mathbf{y} \text{ on } \mathcal{A} \quad \Rightarrow \quad \phi(\mathbf{x}) \leq \phi(\mathbf{y})$$

if and only if ϕ is increasing and Schur-convex on $\mathcal{A}$. Similarly, ϕ satisfies

$$\mathbf{x} \prec^w \mathbf{y} \text{ on } \mathcal{A} \quad \Rightarrow \quad \phi(\mathbf{x}) \leq \phi(\mathbf{y})$$

if and only if ϕ is decreasing and Schur-convex on $\mathcal{A}$.

By using the above results, we are now able to find various Schur-convex/concave functions. Several such examples are provided in the following, while the interested reader can find more Schur-convex/concave functions in [1].

Example 16.1.5. Consider the l_p norm $|\mathbf{x}|_p = (\sum_i |x_i|^p)^{1/p}$, which is symmetric and convex when $p \geq 1$. Thus, from Proposition 16.1.3, $|\mathbf{x}|_p$ is Schur-convex for $p \geq 1$. $\square$

Example 16.1.6. Suppose that $x_i > 0$. Since x^a is convex when $a \geq 1$ and $a \leq 0$ and concave when $0 \leq a < 1$, from Corollary 16.1.1, $\phi(\mathbf{x}) = \sum_i x_i^a$ is Schur-convex for $a \geq 1$ and $a \leq 0$, and Schur-concave for $0 \leq a < 1$. Similarly, $\phi(\mathbf{x}) = \sum_i \log x_i$ and $\phi(\mathbf{x}) = -\sum_i x_i \log x_i$ are both Schur-concave, since $\log x$ and $-x \log x$ are concave. $\square$

Example 16.1.7. Consider $\phi : \mathbb{R}_+^2 \to \mathbb{R}$ with $\phi(\mathbf{x}) = -x_1 x_2$, which is symmetric and quasi-convex. Thus, from Proposition 16.1.4, it is Schur-convex. $\square$

---

[6]A function $f : \mathcal{X} \to \mathbb{R}$ is quasi-convex if $\mathcal{X}$ is a convex set and for any $x, y \in \mathcal{X}$ and $0 \leq \alpha \leq 1$, $f(\alpha x + (1-\alpha)y) \leq \max\{f(x), f(y)\}$. A convex function is also quasi-convex, but the converse is not true.

16.1.3   Relation to Matrix Theory

There are many interesting results that connect m
theory, among which a crucial finding by Schur is th
Hermitian matrix are majorized by its eigenvalues.
used to simplify optimization problems with matrix-

Theorem 16.1.4. *(Schur's inequality [1, 9.B.1])*
with diagonal elements denoted by the vector **d** *an*
vector $\boldsymbol{\lambda}$. *Then* $\boldsymbol{\lambda} \succ \mathbf{d}$.

Theorem 16.1.4 provides an "upper bound" on th
mitian matrix in terms of the majorization orderin
natural "lower bound" of a vector $\mathbf{x} \in \mathbb{R}^n$ would b
note the vector with equal elements given by $1_i \triangleq \sum$
Hermitian matrix we have

$$1 \prec \mathbf{d} \prec \boldsymbol{\lambda}$$

which is formally described in the following corollary.

Corollary 16.1.2. Let $\mathbf{A}$ be a Hermitian matrix an

$$\mathbf{1}(\mathbf{A}) \prec \mathbf{d}\left(\mathbf{U}^{\dagger}\mathbf{A}\mathbf{U}\right) \prec \boldsymbol{\lambda}(.$$

where $\mathbf{1}(\mathbf{A})$ denotes the vector of equal elements who
is the vector of the diagonal elements of $\mathbf{A}$, and $\boldsymbol{\lambda}($
values of $\mathbf{A}$.

Proof: It follows directly from (16.4), as well as the
$$\text{and } \boldsymbol{\lambda}\left(\mathbf{U}^{\dagger}\mathbf{A}\mathbf{U}\right) = \boldsymbol{\lambda}(\mathbf{A}).$$

Corollary 16.1.2 "bounds" the diagonal element
matrix $\mathbf{U}$. However, it does not specify what can be a
will be instrumental for that purpose.

Theorem 16.1.5. *([1, 9.B.2]) For any two vector*
y, *there exists a real symmetric (and therefore Her*
elements given by **x** *and eigenvalues given by* **y**.

Corollary 16.1.3. For any vector $\boldsymbol{\lambda} \in \mathbb{R}^n$, there $\in$
therefore Hermitian) matrix with equal diagonal ele
by $\boldsymbol{\lambda}$.

Corollary 16.1.4. Let $\mathbf{A}$ be a Hermitian matrix and
$\mathbf{x} \prec \boldsymbol{\lambda}(\mathbf{A})$. Then, there exists a unitary matrix $\mathbf{U}$ su

$$\mathbf{d}\left(\mathbf{U}^{\dagger}\mathbf{A}\mathbf{U}\right) = \mathbf{x}.$$

Proof: The proofs of Corollary 16.1.3 and Corollary 16.1.4 are straightforward from Corollary 16.1.2 and Theorem 16.1.5. □

Theorem 16.1.5 is the converse of Theorem 16.1.4 (in fact it is stronger than the converse since it guarantees the existence of a real symmetric matrix instead of just a Hermitian matrix). Now, we can provide the converse of Corollary 16.1.2.

Corollary 16.1.5. Let $\mathbf{A}$ be a Hermitian matrix. There exists a unitary matrix $\mathbf{U}$ such that

$$\mathbf{d}\left(\mathbf{U}^\dagger \mathbf{A} \mathbf{U}\right) = \mathbf{1}\left(\mathbf{A}\right),$$

and also another unitary matrix $\mathbf{U}$ such that

$$\mathbf{d}\left(\mathbf{U}^\dagger \mathbf{A} \mathbf{U}\right) = \boldsymbol{\lambda}\left(\mathbf{A}\right).$$

We now turn to the important algorithmic aspect of majorization theory which is necessary, for example, to compute a matrix with given diagonal elements and eigenvalues. The following definition is instrumental in the derivation of transformations that relate vectors that satisfy the majorization relation.

Definition 16.1.6. (*T-transform* [1, p. 21]) A *T-transform* is a matrix of the form
$$\mathbf{T} = \alpha \mathbf{I} + (1 - \alpha)\,\mathbf{\Pi} \tag{16.5}$$
for some $\alpha \in [0,1]$ and some $n \times n$ permutation matrix $\mathbf{\Pi}$ with $n-2$ diagonal entries equal to 1. Let $[\mathbf{\Pi}]_{ij} = [\mathbf{\Pi}]_{ji} = 1$ for some indices $i < j$, then

$$\mathbf{\Pi}\mathbf{y} = [y_1,\,\ldots,\,y_{i-1},\,y_j,\,y_{i+1},\,\ldots,\,y_{j-1},\,y_i,\,y_{j+1},\ldots,\,y_n]^T$$

and hence

$$\mathbf{T}\mathbf{y} = [y_1,\,\ldots,\,y_{i-1},\,\alpha\,y_i + (1-\alpha)\,y_j,\,y_{i+1},\,\ldots,$$
$$y_{j-1},\,\alpha\,y_j + (1-\alpha)\,y_i,\,y_{j+1},\,\ldots,\,y_n]^T.$$

Lemma 16.1.2. ([1, 2.B.1]) For any two vectors $\mathbf{x}, \mathbf{y} \in \mathbb{R}^n$ satisfying $\mathbf{x} \prec \mathbf{y}$, there exists a sequence of T-transforms $\mathbf{T}^{(1)}, \ldots, \mathbf{T}^{(K)}$ such that $\mathbf{x} = \mathbf{T}^{(K)} \cdots \mathbf{T}^{(1)}\mathbf{y}$ and $K < n$.

An algorithm to obtain such a sequence of T-transforms is introduced next.

Algorithm 16.1.1. (*[1, 2.B.1]*) *Algorithm to obtain a sequence of T-transforms such that* $\mathbf{x} = \mathbf{T}^{(K)} \cdots \mathbf{T}^{(1)}\mathbf{y}$.
  Input: Vectors $\mathbf{x}, \mathbf{y} \in \mathbb{R}^n$ *satisfying* $\mathbf{x} \prec \mathbf{y}$ *(it is assumed that the components of* $\mathbf{x}$ *and* $\mathbf{y}$ *are in decreasing order and that* $\mathbf{x} \neq \mathbf{y}$*).*
  Output: Set of T-transforms $\mathbf{T}^{(1)}, \ldots, \mathbf{T}^{(K)}$.

  0. Let $\mathbf{y}^{(0)} = \mathbf{y}$ *and* $k = 1$ *be the iteration index.*

1. *Find the largest index i such that $y_i^{(k-1)} > x_i$ an*
   than i such that $y_j^{(k-1)} < x_j$.

2. *Let $\delta = \min\left(x_j - y_j^{(k-1)}, y_i^{(k-1)} - x_i\right)$ and $\alpha =$*

3. *Use α to compute $\mathbf{T}^{(k)}$ as in (16.5) and let $\mathbf{y}^{(k}$*

4. *If $\mathbf{y}^{(k)} \neq \mathbf{x}$, then set $k = k+1$ and go to step 1,*

Recursive algorithms to obtain a matrix with giv
elements are provided in [1, 9.B.2] and [8]. Here, we
simple method proposed in [8] as follows.

Algorithm 16.1.2. *([8]) Algorithm to obtain a re
diagonal values given by $\mathbf{x}$ and eigenvalues given by $\mathbf{y}$*
   Input: *Vectors $\mathbf{x}, \mathbf{y} \in \mathbb{R}^n$ satisfying $\mathbf{x} \prec \mathbf{y}$ (it is*
of $\mathbf{x}$ and $\mathbf{y}$ are in decreasing order and that $\mathbf{x} \neq \mathbf{y}$).
   Output: *Matrix $\mathbf{A}$.*

1. *Using Algorithm 16.1.1, obtain a sequence of*
   $\mathbf{T}^{(K)} \cdots \mathbf{T}^{(1)} \mathbf{y}.$

2. *Define the Givens rotation $\mathbf{U}^{(k)}$ as*

$$
\left[\mathbf{U}^{(k)}\right]_{ij} = \begin{cases} \sqrt{\left[\mathbf{T}^{(k)}\right]_{ij}}, \\ -\sqrt{\left[\mathbf{T}^{(k)}\right]_{ij}}, \end{cases}
$$

3. *Let $\mathbf{A}^{(0)} = \mathrm{diag}(\mathbf{y})$ and $\mathbf{A}^{(k)} = \mathbf{U}^{(k)T} \mathbf{A}^{(k-1)}$*
   given by $\mathbf{A} = \mathbf{A}^{(K)}$. Define the unitary matrix
   desired matrix is given by $\mathbf{A} = \mathbf{U}^T \mathrm{diag}(\mathbf{y}) \mathbf{U}$.

Algorithm 16.1.2 obtains a real symmetric matrix
diagonal elements. For the interesting case in which
be equal and the desired matrix is allowed to be com
an alternative much simpler solution in closed form a

Lemma 16.1.3. *([9]) Let $\mathbf{U}$ be a unitary matrix satis*
$|[\mathbf{U}]_{il}| \; \forall i, k, l$. *Then, the matrix $\mathbf{A} = \mathbf{U}^H \mathrm{diag}(\lambda) \mathbf{U}$*
(and eigenvalues given by λ). Two examples of $\mathbf{U}$
and the Hadamard matrix (when the dimensions are
of two).

Nevertheless, Algorithm 16.1.2 has the nice proper
$\mathbf{U}$ is real-valued and can be naturally decomposed (b
uct of a series of rotations. This simple structure p

implementation. Interestingly, an iterative approach to construct a matrix with equal diagonal elements and with a given set of eigenvalues was obtained in [10], based also on a sequence of rotations.

16.1.4 Multiplicative Majorization

Parallel to the concept of majorization introduced in Section 16.1.1, which is often called additive majorization, is the notion of multiplicative majorization (also termed log-majorization) defined as follows.

Definition 16.1.7. The vector $\mathbf{x} \in \mathbb{R}_+^n$ is multiplicatively majorized by $\mathbf{y} \in \mathbb{R}_+^n$, denoted by $\mathbf{x} \prec_\times \mathbf{y}$, if

$$\prod_{i=1}^{k} x_{[i]} \leq \prod_{i=1}^{k} y_{[i]}, \qquad 1 \leq k < n$$

$$\prod_{i=1}^{n} x_{[i]} = \prod_{i=1}^{n} y_{[i]}.$$

To differentiate the two types of majorization, we sometimes use the symbol $\prec_+$ rather than $\prec$ to denote (additive) majorization. It is easy to see the relation between additive majorization and multiplicative majorization: $\mathbf{x} \prec_+ \mathbf{y}$ if and only if $\exp(\mathbf{x}) \prec_\times \exp(\mathbf{y})$.[7]

Example 16.1.8. Given $\mathbf{x} \in \mathbb{R}_+^n$, let $\mathbf{g}$ denote the vector of equal elements given by $g_i \triangleq (\prod_{j=1}^{n} x_j)^{1/n}$, i.e., the geometric mean of $\mathbf{x}$. Then, $\mathbf{g} \prec_\times \mathbf{x}$.

Similar to the definition of Schur-convex/concave functions, it is also possible to define multiplicatively Schur-convex/concave functions.

Definition 16.1.8. A function $\phi : \mathcal{A} \to \mathbb{R}$ is said to be multiplicatively Schur-convex on $\mathcal{A} \in \mathbb{R}^n$ if

$$\mathbf{x} \prec_\times \mathbf{y} \text{ on } \mathcal{A} \quad \Rightarrow \quad \phi(\mathbf{x}) \leq \phi(\mathbf{y}),$$

and multiplicatively Schur-concave on $\mathcal{A}$ if

$$\mathbf{x} \prec_\times \mathbf{y} \text{ on } \mathcal{A} \quad \Rightarrow \quad \phi(\mathbf{x}) \geq \phi(\mathbf{y}).$$

However, considering the correspondence between additive and multiplicative majorization, it may not be necessary to use the notion of multiplicatively Schur-convex/concave functions. Instead, the so-called multiplicatively Schur-convex/concave functions in Definition 16.1.8 can be equivalently referred to as

---

[7]Indeed, using the language of group theory, we say that the groups $(\mathbb{R}, +)$ and $(\mathbb{R}_+, \times)$ are isomorphic since there is a bijection function $\exp : \mathbb{R} \to \mathbb{R}_+$ such that $\exp(x+y) = \exp(x) \times \exp(y)$ for $\forall x, y \in \mathbb{R}$.

functions such that $\phi \circ \exp$ is Schur-convex and Schu
the composite function is defined as $\phi \circ \exp(\mathbf{x}) \triangleq \phi(\epsilon$

The following two lemmas relate Schur-convexity/
that of the composite function $\phi \circ \exp$.

Lemma 16.1.4. If ϕ is increasing and Schur-convex,

Proof: It is an immediate result from Proposition 1

Lemma 16.1.5. If the composite function $\phi \circ \exp$ is
$\mathbb{R}^n : x_1 \geq \cdots \geq x_n\}$, then ϕ is Schur-concave on $\mathcal{D}_n$ i

Proof: It can be easily proved using Theorem 16.1.1

The following two examples show that the implic
16.1.5 does not hold in the opposite direction.

Example 16.1.9. The function $\phi(\mathbf{x}) = \prod_{i=1}^{n} x_i$ is
$\frac{\partial \phi(\mathbf{x})}{\partial x_i} = \frac{\phi(\mathbf{x})}{x_i}$ is increasing in i on $\mathcal{D}_n$ (see Theorem 16.
function $\phi \circ \exp(\mathbf{x}) = \exp(\sum_i x_i)$ is Schur-convex (an

Example 16.1.10. The function $\phi(\mathbf{x}) = \sum_{i=1}^{n} \alpha_i x_i$
concave on $\mathcal{D}_n$. The composite function is $\phi \circ \exp(\mathbf{x})$
α_{i+1}, the derivative $\frac{\partial \phi \circ \exp(\mathbf{x})}{\partial x_i} = \alpha_i \exp(x_i)$ is not alwa
for any $\mathbf{x} \in \mathcal{D}_n$. Hence according to Theorem 16.1.1,
$\phi \circ \exp$ is not Schur-concave (neither Schur-convex) c

In contrast with (additive) majorization that leads
between eigenvalues and diagonal elements of a He
16.1.3), multiplicative majorization also brings some i
mostly on the relation between singular values and ei
following, we introduce one recent result that will be

Theorem 16.1.6. *(Generalized triangular decor*
$\mathbf{H} \in \mathbb{C}^{m \times n}$ *be a matrix with rank k and singular valu*
exists an upper triangular matrix $\mathbf{R} \in \mathbb{C}^{k \times k}$ and sen
such that $\mathbf{H} = \mathbf{Q}\mathbf{R}\mathbf{P}^H$ if and only if the diagonal el
where $|\mathbf{r}|$ is a vector with the absolute values of $\mathbf{r}$ ele
stand for the singular values of $\mathbf{H}$ and diagonal entri

The GTD is a generic form including many well-k
such as the SVD, the Schur decomposition, and the Q
16.1.6 implies that, given $|\mathbf{r}| \prec_{\times} \sigma$, there exists a matr
and eigenvalues being $\mathbf{r}$ and σ, respectively. A recurs
matrix was proposed in [11].

16.1.5   Stochastic Majorization

A comparison of some kind between two random variables X and Y is called stochastic majorization if the comparison reduces to the ordinary majorization $x \prec y$ in case X and Y are degenerate at x and y, i.e., $\Pr(X = x) = 1$ and $\Pr(Y = y) = 1$. Random vectors to be compared by stochastic majorization often have distributions belonging to the same parametric family, where the parameter space is a subset of $\mathbb{R}^n$. In this case, random variables X and Y with corresponding distributions F_θ and $F_{\theta'}$ are ordered by stochastic majorization if and only if the parameters θ and θ' are ordered by ordinary majorization.

Specifically, let $\mathcal{A} \subseteq \mathbb{R}^n$ and $\{F_\theta : \theta \in \mathcal{A}\}$ be a family of n-dimensional distribution functions indexed by a vector-valued parameter θ. Let

$$E\{\phi(X)\} = \int_{\mathbb{R}^n} \phi(x) dF_\theta(x) \tag{16.6}$$

denote the expectation of $\phi(X)$ when X has distribution F_θ, and let

$$\Pr(\phi(X) \le t) = \int_{\phi(x) \le t} dF_\theta(x) \tag{16.7}$$

denote the tail probability that $\phi(X)$ is less than or equal to t when X has distribution F_θ. We are particularly interested in investigating whether or in what conditions $E\{\phi(X)\}$ and $\Pr(\phi(X) \le t)$ are Schur-convex/concave in θ.

The following results provide the conditions in which $E\{\phi(X)\}$ is Schur-convex in θ for exchangeable random variables.[8]

Proposition 16.1.5. ([1, 11.B.1]) Let $X_1, \ldots, X_n$ be exchangeable random variables and suppose that $\Phi : \mathbb{R}^{2n} \to \mathbb{R}$ satisfies: (i) $\Phi(\mathbf{x}, \theta)$ is convex in θ for each fixed $\mathbf{x}$; (ii) $\Phi(\mathbf{\Pi x}, \mathbf{\Pi}\theta) = \Phi(\mathbf{x}, \theta)$ for all permutations $\mathbf{\Pi}$; (iii) $\Phi(\mathbf{x}, \theta)$ is Borel measurable in $\mathbf{x}$ for each fixed θ. Then,

$$\psi(\theta) = E\{\Phi(X_1, \ldots, X_n, \theta)\}$$

is symmetric and convex (and thus Schur-convex).

Corollary 16.1.6. ([1, 11.B.2, 11.B.3]) Let $X_1, \ldots, X_n$ be exchangeable random variables and $\phi : \mathbb{R}^n \to \mathbb{R}$ be symmetric and convex. Then,

$$\psi(\theta) = E\{\phi(\theta_1 X_1, \ldots, \theta_n X_n)\}$$

and

$$\psi(\theta) = E\{\phi(X_1 + \theta_1, \ldots, X_n + \theta_n)\}$$

are symmetric and convex (and thus Schur-convex).

---

[8]$X_1, \ldots, X_n$ are exchangeable random variables if the distribution of $X_{\pi(1)}, \ldots, X_{\pi(n)}$ does not depend on the permutation π. In other words, the joint distribution of $X_1, \ldots, X_n$ is invariant under permutations of its arguments. For example, independent and identically distributed random variables are exchangeable.

Corollary 16.1.7. ([1, 11.B.2.c]) Let $X_1, \ldots, X_n$ b
ables and g be a continuous convex function. Then,

$$\psi(\theta) = E\left[g\left(\sum_i \theta_i X_i\right)\right]$$

is symmetric and convex (and thus Schur-convex).

Compared to the expectation form, there are only
on the Schur-convexity/concavity of the tail probabi
θ, which are usually given in some specific form of ϕ
following is one important result concerning linear c
variables.

Theorem 16.1.7. (*[12]*) Let $\theta_i \geq 0$ for all i, and X
identically distributed (iid) random variables followin
the density

$$f(x) = \frac{x^{k-1}\exp(-x)}{\Gamma(k)}$$

where $\Gamma(k) = (k-1)!$. *Suppose that* $g : \mathbb{R} \to \mathbb{R}$ *is*
function g^{-1} *exists. Then,*

$$P\left\{g\left(\sum_i \theta_i X_i\right) \leq t\right\}$$

is Schur-concave in θ *for* $t \geq g(2)$ *and Schur-convex*

Corollary 16.1.8. Let $\theta_i \geq 0$ for all i, and $X_1, \ldots$
dom variables with the density $f(x) = \exp(-x)$. The
concave in θ for $t \geq 2$ and Schur-convex in θ for $t \leq 1$

For more examples of stochastic majorization in
tail probabilities, we refer the interested reader to [1]

16.2 Applications of Majorizati

16.2.1 CDMA Sequence Design

The code division multiple access (CDMA) system is a
technique in wireless networks. In a CDMA system, a
width and they are distinguished from each other by
codes. A fundamental problem in CDMA systems is t
sequences so that the system performance, such as the

Consider the uplink of a single-cell synchronous CDMA system with K users and processing gain N. In the presence of additive white Gaussian noise, the sampled baseband received signal vector in one symbol interval is

$$\mathbf{r} = \sum_{i=1}^{K} \mathbf{s}_i \sqrt{p_i} b_i + \mathbf{n} \tag{16.8}$$

where, for each user i, p_i is the received power, b_i is the transmitted symbol, and $\mathbf{s}_i \in \mathbb{R}^N$ is the unit-energy signature sequence, i.e., $\|\mathbf{s}_i\| = 1$, and $\mathbf{n}$ is a zero-mean Gaussian random vector with covariance matrix $\sigma^2 \mathbf{I}_N$, i.e., $\mathbf{n} \sim \mathcal{N}(\mathbf{0}, \sigma^2 \mathbf{I}_N)$. Introduce an $N \times K$ signature sequence matrix $\mathbf{S} \triangleq [\mathbf{s}_1, \ldots, \mathbf{s}_K]$ and let $\mathbf{P}^{1/2} \triangleq \operatorname{diag}\{\sqrt{p_1}, \ldots, \sqrt{p_K}\}$ and $\mathbf{b} \triangleq [b_1, \ldots, b_K]^T$. Then (16.8) can be compactly expressed as

$$\mathbf{r} = \mathbf{S}\mathbf{P}^{1/2}\mathbf{b} + \mathbf{n}. \tag{16.9}$$

There are different criteria to measure the performance of a CDMA system, among which the most commonly used one may be the sum capacity given by [13]

$$C_{\text{sum}} = \frac{1}{2} \log \det \left(\mathbf{I}_N + \sigma^{-2} \mathbf{S}\mathbf{P}\mathbf{S}^T \right). \tag{16.10}$$

In practice, the system performance may also be measured by the total MSE of all users, which, assuming that each uses a LMMSE filter at his receiver, is given by [14]

$$\text{MSE} = K - \operatorname{Tr}\left[\mathbf{S}\mathbf{P}\mathbf{S}^T \left(\mathbf{S}\mathbf{P}\mathbf{S}^T + \sigma^2 \mathbf{I}_N \right)^{-1} \right]. \tag{16.11}$$

Another important global quantity that measures the total interference in the CDMA system is the total weighted square correlation (TWSC), which is given by [15]

$$\text{TWSC} = \sum_{i=1}^{K} \sum_{j=1}^{K} p_i p_j \left(\mathbf{s}_i^T \mathbf{s}_j \right) = \operatorname{Tr}\left[\left(\mathbf{S}\mathbf{P}\mathbf{S}^T \right)^2 \right]. \tag{16.12}$$

The goal of the sequence design problem is to optimize the system performance, e.g., maximize C_{sum} or minimize MSE or TWSC, by properly choosing the signature sequences for all users or, equivalently, by choosing the optimal signature sequence matrix $\mathbf{S}$.

Observe that the aforementioned three performance measures are all determined by the eigenvalues of the matrix $\mathbf{S}\mathbf{P}\mathbf{S}^T$. To be more exact, denoting the eigenvalues of $\mathbf{SPS}^T$ by $\lambda \triangleq (\lambda_i)_{i=1}^N$, it follows that

$$C_{\text{sum}} = \frac{1}{2}\sum_{i=1}^N \log\left(1 + \right.$$

$$\text{MSE} = K - \sum_{i=1}^N \frac{\lambda_i}{\lambda_i + \sigma}$$

$$\text{TWSC} = \sum_{i=1}^N \lambda_i^2.$$

Now, we can apply majorization theory. Indeed, sinc
function, it follows from Corollary 16.1.1 that C_{sum}
with respect to λ. Similarly, given that $-x/(x + \sigma^2)$
both MSE and TWSC are Schur-convex in λ. Therefo
sequence matrix yielding the Schur-minimal eigenvalu
that are majorized by all other feasible eigenvalues,
sequences will not only maximize C_{sum} but also minir
same time.

To find the optimal signature sequence matrix $\mathbf{S}$,
all feasible $\mathbf{S}$

$$\mathcal{S} \triangleq \{\mathbf{S} \in \mathbb{R}^{N \times K} : \|\mathbf{s}_i\| = 1, \; i = 1$$

and correspondingly the set of all possible λ

$$\mathcal{L} \triangleq \left\{\lambda(\mathbf{SPS}^T) : \mathbf{S} \in \mathcal{S}\right\}$$

Now the question is how to find a Schur-minimal vecto
not easy to answer given the form of $\mathcal{L}$ in (16.14). To
transform $\mathcal{L}$ to a more convenient equivalent form t
relation.

Lemma 16.2.1. When $K \leq N$, $\mathcal{L}$ is equal to

$$\mathcal{M} \triangleq \{\lambda \in \mathbb{R}^N : (\lambda_1, \ldots, \lambda_K) \succ (p_1, \ldots, p_K), \; \lambda$$

When $K > N$, $\mathcal{L}$ is equal to

$$\mathcal{N} \triangleq \{\lambda \in \mathbb{R}^N : (\lambda_1, \ldots, \lambda_N, \underbrace{0, \ldots, 0}_{K-N}) \succ$$

Proof: Consider the case $K \leq N$. We first show that
$K \leq N$, $\lambda(\mathbf{SPS}^T)$ has at most K non-zero element
Let $\mathbf{p} \triangleq (p_i)_{i=1}^K$. Observe that, for $\mathbf{S} \in \mathcal{S}$, $\mathbf{p}$ and λ_a a
the eigenvalues of the matrix $\mathbf{P}^{1/2}\mathbf{S}^T\mathbf{SP}^{1/2}$, respectiv
have $\lambda_a \succ \mathbf{p}$, implying that $\lambda \in \mathcal{M}$.

To see the other direction, let $\lambda \in \mathcal{M}$ and thus $\lambda_a \succ \mathbf{p}$. According to Theorem 16.1.5, there exists a symmetric matrix $\mathbf{Z}$ with eigenvalues λ_a and diagonal elements $\mathbf{p}$. Denote the eigenvalue decomposition (EVD) of $\mathbf{Z}$ by $\mathbf{Z} = \mathbf{U}_z \boldsymbol{\Lambda}_z \mathbf{U}_z^T$ and introduce $\boldsymbol{\Lambda} = \mathrm{diag}\{\boldsymbol{\Lambda}_z, \mathbf{0}_{(N-K)\times(N-K)}\}$ and $\mathbf{U} = [\mathbf{U}_z \, \mathbf{0}_{K\times(N-K)}]$. Then we can choose $\mathbf{S} = \boldsymbol{\Lambda}^{1/2} \mathbf{U}^T \mathbf{P}^{-1/2}$. It is easy to check that the eigenvalues of $\mathbf{SPS}^T = \boldsymbol{\Lambda}$ are λ, and $\|\mathbf{s}_i\|^2$, $i = 1, \ldots, K$, coincide with the diagonal elements of $\mathbf{S}^T \mathbf{S} = \mathbf{P}^{-1/2} \mathbf{Z} \mathbf{P}^{-1/2}$ and thus are all ones, so we have $\lambda \in \mathcal{L}$.

The equivalence between $\mathcal{L}$ and $\mathcal{N}$ when $K > N$ can be obtained in a similar way, for which a detailed proof was provided in [8]. □

When $K \leq N$, the Schur-minimal vector in $\mathcal{L}$ (or $\mathcal{M}$), from Lemma 16.2.1, is $\lambda^\star = (p_1, \ldots, p_L, 0, \ldots, 0)$, which can be achieved by choosing arbitrary K orthonormal sequences, i.e., $\mathbf{S}^T \mathbf{S} = \mathbf{I}_K$.

When $K > N$, the problem of finding a Schur-minimal vector in $\mathcal{L}$ (or $\mathcal{N}$) is, however, not straightforward. It turns out that in this case the Schur-minimal vector is given in a complicated form based on the following definition.

Definition 16.2.1. (Oversized users [8]) User i is defined to be oversized if

$$p_i > \frac{\sum_{i=1}^{K} p_j 1_{\{p_i > p_j\}}}{N - \sum_{j=1}^{K} 1_{\{p_j \geq p_i\}}}$$

where $1_{\{\cdot\}}$ is the indication function. Intuitively, a user is oversized if his power is large relative to those of the others.

Theorem 16.2.1. *([8]) Assume w.l.o.g. that the users are ordered according to their powers $p_1 \geq \cdots \geq p_K$, and the first L users are oversized. Then, the Schur-minimal vector in $\mathcal{L}$ (or $\mathcal{N}$) is given by*

$$\lambda^\star = \left(p_1, \ldots, p_L, \frac{\sum_{j=L+1}^{K} p_j}{N - L}, \ldots, \frac{\sum_{j=L+1}^{K} p_j}{N - L} \right).$$

The left question is how to find an $\mathbf{S} \in \mathcal{S}$ such that the eigenvalues of $\mathbf{SPS}^T$ are $\lambda^\star$. Note that the constraint $\mathbf{S} \in \mathcal{S}$ is equivalent to saying that the diagonal elements of $\mathbf{S}^T \mathbf{S}$ are all equal to 1. Therefore, given the optimal $\mathbf{S}$, the matrix $\mathbf{P}^{1/2} \mathbf{S}^T \mathbf{S} \mathbf{P}^{1/2}$ has the diagonal elements $\mathbf{p} = (p_1, \ldots, p_K)$ and the eigenvalues $\lambda_b = (\lambda^\star, \mathbf{0})$. From Theorem 16.1.5, there exists a $K \times K$ symmetric matrix $\mathbf{M}$ such that its diagonal elements and eigenvalues are given by $\mathbf{p}$ and λ_b, respectively. Denote the (EVD) of $\mathbf{M}$ by $\mathbf{M} = \mathbf{U} \boldsymbol{\Lambda} \mathbf{U}^T$, where $\boldsymbol{\Lambda} = \mathrm{diag}\{\lambda^\star\}$ and $\mathbf{U} \in \mathbb{R}^{K \times N}$ contains the N eigenvectors corresponding to $\lambda^\star$. Then, the optimal signature sequence matrix can be obtained as $\mathbf{S} = \boldsymbol{\Lambda}^{1/2} \mathbf{U}^T \mathbf{P}^{-1/2}$. It can be verified that the eigenvalues of $\mathbf{SPS}^T$ are $\lambda^\star$ and $\mathbf{S} \in \mathcal{S}$.

Finally, to construct the symmetric matrix $\mathbf{M}$ with the diagonal elements $\mathbf{p}$ and the eigenvalues λ_b (provided $\mathbf{p} \prec \lambda_b$), one can exploit Algorithm 16.1.2 introduced in Section 16.1.3. Interestingly, an iterative algorithm was proposed in [14, 15]

to generate the optimal signature sequences. This a
signature sequence in a sequential way, and was prove
solution.

16.2.2   Linear MIMO Transceiver Desi

MIMO channels, usually arising from using multiple
wireless link, have been well recognized as an effective
and reliability of wireless communications [16].   A
harvest the benefits of MIMO channels is to exploit lir
precoder at the transmitter and a linear equalizer at t
transceivers for MIMO channels has a long history,
specific measure of the global performance. It has r
that the design of linear MIMO transceivers can be u
into a general framework that embraces a wide rar
criteria. In the following we briefly introduce this un

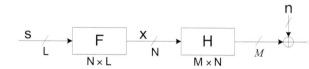

Figure 16.2: Linear MIMO transceiver consisting of a
equalizer.

Consider a communication link with N transmit a
signal model of such a MIMO channel is

$$\mathbf{y} = \mathbf{Hx} + \mathbf{n}$$

where $\mathbf{x} \in \mathbb{C}^N$ is the transmitted signal vector, $\mathbf{H} \in \mathbb{C}$
$\mathbf{y} \in \mathbb{C}^M$ is the received signal vector, and $\mathbf{n} \in \mathbb{C}^M$ is a z
ric complex Gaussian random vector with covariance
In the linear transceiver scheme as illustrated in Figur
$\mathbf{x}$ results from the linear transformation of a symbol ve
precoder $\mathbf{F} \in \mathbb{C}^{N \times L}$ and is given by $\mathbf{x} = \mathbf{Fs}$. Assume t
and $E\left\{\mathbf{ss}^H\right\} = \mathbf{I}$. The total average transmit power

$$P_T = E\left\{\|\mathbf{x}\|^2\right\} = \text{Tr}(\mathbf{FF}^H$$

At the receiver is a linear equalizer $\mathbf{G}^H \in \mathbb{C}^{L \times M}$ used t
in $\hat{\mathbf{s}} = \mathbf{G}^H\mathbf{y}$. Therefore, the relation between the t

---

[9]If the noise is not white, say with covariance matrix $\mathbf{R}_n$,
$\mathbf{R}_n^{-1/2}\mathbf{y} = \tilde{\mathbf{H}}\mathbf{x} + \bar{\mathbf{n}}$, where $\tilde{\mathbf{H}} = \mathbf{R}_n^{-1/2}\mathbf{H}$ is the equivalent char

estimated symbols can be expressed as

$$\hat{\mathbf{s}} = \mathbf{G}^H \mathbf{H} \mathbf{F} \mathbf{s} + \mathbf{G}^H \mathbf{n}. \tag{16.17}$$

An advantage of MIMO channels is the support of simultaneously transmitting multiple data streams, leading to significant capacity improvement.[10] Observe from (16.17) that the estimated symbol at the ith data stream is given by

$$\hat{s}_i = \mathbf{g}_i^H \mathbf{H} \mathbf{f}_i s_i + \mathbf{g}_i^H \mathbf{n}_i \tag{16.18}$$

where $\mathbf{f}_i$ and $\mathbf{g}_i$ are the ith columns of $\mathbf{F}$ and $\mathbf{G}$, respectively, and $\mathbf{n}_i = \sum_{j \neq i} \mathbf{H} \mathbf{f}_j s_j + \mathbf{n}$ is the equivalent noise seen by the ith data stream with covariance matrix $\mathbf{R}_{n_i} = \sum_{j \neq i} \mathbf{H} \mathbf{f}_j \mathbf{f}_j^H \mathbf{H}^H + \mathbf{I}$. In practice, the performance of a data stream can be measured by the MSE, signal-to-interference-plus-noise ratio (SINR), or bit error rate (BER), which according to (16.18) are given by

$$\mathrm{MSE}_i \triangleq E\left\{ |\hat{s}_i - s_i|^2 \right\} = \left| \mathbf{g}_i^H \mathbf{H} \mathbf{f}_i - 1 \right|^2 + \mathbf{g}_i^H \mathbf{R}_{n_i} \mathbf{g}_i$$

$$\mathrm{SINR}_i \triangleq \frac{\text{desired component}}{\text{undesired component}} = \frac{\left| \mathbf{g}_i^H \mathbf{H} \mathbf{f}_i \right|^2}{\mathbf{g}_i^H \mathbf{R}_{n_i} \mathbf{g}_i}$$

$$\mathrm{BER}_i \triangleq \frac{\text{\# bits in error}}{\text{\# transmitted bits}} \approx \varphi_i(\mathrm{SINR}_i)$$

where φ_i is a decreasing function relating the BER to the SINR at the ith stream [6, 9]. Any properly designed system should attempt to minimize the MSEs, maximize the SINRs, or minimize the BERs.

Measuring the global performance of a MIMO system with several data streams is tricky as there is an inherent tradeoff among the performance of the different streams. Different applications may require a different balance on the performance of the streams, so there are a variety of criteria in the literature, each leading to a particular design problem (see [6] for a survey). However, in fact, all these particular problems can be unified into one framework using the MSEs as the nominal cost. Specifically, suppose that the system performance is measured by an arbitrary global cost function of the MSEs $f_0\left(\{\mathrm{MSE}_i\}_{i=1}^L \right)$ that is increasing in each argument.[11] The linear transceiver design problem is then formulated as

$$\begin{aligned} \underset{\mathbf{F}, \mathbf{G}}{\text{minimize}} \quad & f_0\left(\{\mathrm{MSE}_i\} \right) \\ \text{subject to} \quad & \mathrm{Tr}(\mathbf{F}\mathbf{F}^H) \leq P \end{aligned} \tag{16.19}$$

where $\mathrm{Tr}(\mathbf{F}\mathbf{F}^H) \leq P$ represents the transmit power constraint.

---

[10]This kind of improvement is often called the multiplexing gain [18].

[11]The increasingness of f is a mild and reasonable assumption: if the performance of one stream improves, the global performance should improve too.

To solve (16.19), we first find the optimal $\mathbf{G}$ for
the optimal equalizer is the LMMSE filter, also term

$$\mathbf{G}^\star = (\mathbf{H}\mathbf{F}\mathbf{F}^H\mathbf{H}^H + \mathbf{I})^{-1}\mathbf{H}$$

To see this, let us introduce the MSE matrix

$$
\begin{aligned}
\mathbf{E}(\mathbf{F}, \mathbf{G}) &\triangleq E\left[(\hat{\mathbf{s}} - \mathbf{s})(\hat{\mathbf{s}} - \mathbf{s})^H\right] \\
&= (\mathbf{G}^H\mathbf{H}\mathbf{F} - \mathbf{I})(\mathbf{F}^H\mathbf{H}^H\mathbf{G}
\end{aligned}
$$

from which the MSE of the ith data stream is given
difficult to verify that

$$\mathbf{E}(\mathbf{F}, \mathbf{G}^\star) = (\mathbf{I} + \mathbf{F}^H\mathbf{H}^H\mathbf{H}\mathbf{F})^{-1} \preceq$$

for any $\mathbf{G}$, meaning that $\mathbf{G}^\star$ simultaneously minimiz
or all MSEs. At the same time, one can verify tha
$\mathbf{G}^\star$, also maximizes SINR_i (or equivalently minimiz
Wiener filter is optimal in the sense of both minimizi
all SINRs (or minimizing all BERs). Observe that t
the particular choice of the cost function f_0 in (16.19
Using the Wiener filter as the equalizer, we can e

$$
\begin{aligned}
\mathrm{MSE}_i &= [(\mathbf{I} + \mathbf{F}^H\mathbf{H}^H\mathbf{H}\mathbf{F} \\
\mathrm{SINR}_i &= \frac{1}{\mathrm{MSE}_i} - 1 \\
\mathrm{BER}_i &= \varphi_i(\mathrm{MSE}_i^{-1} - 1).
\end{aligned}
$$

This means that different performance measures based
the BERs can be uniformly represented by the MSE-b
the generality of the problem formulation (16.19).
    Now, the transceiver design problem (16.19) redu
design problem:

$$
\begin{aligned}
&\underset{\mathbf{F}}{\text{minimize}} && f_0\left(\{[(\mathbf{I} + \mathbf{F}^H\mathbf{H}^H\mathbf{H}\mathbf{F} \right. \\
&\text{subject to} && \mathrm{Tr}(\mathbf{F}\mathbf{F}^H) \le P.
\end{aligned}
$$

Solving such a general problem is very challenging
majorization theory.

Theorem 16.2.2. *([6, Theorem 3.13]) Suppose that*
is increasing in each argument. Then, the optimal so

$$\mathbf{F}^\star = \mathbf{V}_h\mathrm{diag}(\sqrt{\mathbf{p}})\mathbf{\Omega}$$

where

(i) $\mathbf{V}_h \in \mathbb{C}^{N \times L}$ *is a semi-unitary matrix with columns equal to the right singular vectors of* $\mathbf{H}$ *corresponding to the* L *largest singular values in increasing order;*

(ii) $\mathbf{p} \in \mathbb{R}_+^L$ *is the solution to the following power allocation problem:*[12]

$$
\begin{aligned}
\underset{\mathbf{p}, \rho}{\text{minimize}} \quad & f_0 \left(\rho_1, \ldots, \rho_L \right) \\
\text{subject to} \quad & \left(\frac{1}{1 + p_1 \gamma_1}, \ldots, \frac{1}{1 + p_L \gamma_L} \right) \succ^w \left(\rho_1, \ldots, \rho_L \right) \\
& \mathbf{p} \geq \mathbf{0}, \ \mathbf{1}^T \mathbf{p} \leq P
\end{aligned}
\tag{16.24}
$$

where $\{\gamma_i\}_{i=1}^L$ *are the* L *largest eigenvalues of* $\mathbf{H}^H \mathbf{H}$ *in increasing order;*

(iii) $\mathbf{\Omega} \in \mathbb{C}^{L \times L}$ *is a unitary matrix such that* $[(\mathbf{I} + \mathbf{F}^{\star H} \mathbf{H}^H \mathbf{H} \mathbf{F}^{\star})^{-1}]_{ii} = \rho_i$ *for all* i, *which can be computed with Algorithm 16.1.2.*

Proof: We start by rewriting (16.23) into the equivalent form

$$
\begin{aligned}
\underset{\mathbf{F}, \rho}{\text{minimize}} \quad & f_0 \left(\rho \right) \\
\text{subject to} \quad & \mathbf{d} \left((\mathbf{I} + \mathbf{F}^H \mathbf{H}^H \mathbf{H} \mathbf{F})^{-1} \right) \leq \rho \\
& \text{Tr}(\mathbf{F} \mathbf{F}^H) \leq P.
\end{aligned}
\tag{16.25}
$$

Note that, given any $\mathbf{F}$, we can always find another $\tilde{\mathbf{F}} = \mathbf{F} \mathbf{\Omega}^H$ with a unitary matrix $\mathbf{\Omega}$ such that $\tilde{\mathbf{F}}^H \mathbf{H}^H \mathbf{H} \tilde{\mathbf{F}} = \mathbf{\Omega} \mathbf{F}^H \mathbf{H}^H \mathbf{H} \mathbf{F} \mathbf{\Omega}^H$ is diagonal with diagonal elements in increasing order. The original MSE matrix is given by $(\mathbf{I} + \mathbf{F}^H \mathbf{H}^H \mathbf{H} \mathbf{F})^{-1} = \mathbf{\Omega}^H (\mathbf{I} + \tilde{\mathbf{F}}^H \mathbf{H}^H \mathbf{H} \tilde{\mathbf{F}})^{-1} \mathbf{\Omega}$. Thus we can rewrite (16.25) in terms of $\tilde{\mathbf{F}}$ and $\mathbf{\Omega}$ as

$$
\begin{aligned}
\underset{\tilde{\mathbf{F}}, \mathbf{\Omega}, \rho}{\text{minimize}} \quad & f_0 \left(\rho \right) \\
\text{subject to} \quad & \tilde{\mathbf{F}}^H \mathbf{H}^H \mathbf{H} \tilde{\mathbf{F}} \quad \text{diagonal} \\
& \mathbf{d} \left(\mathbf{\Omega}^H (\mathbf{I} + \tilde{\mathbf{F}}^H \mathbf{H}^H \mathbf{H} \tilde{\mathbf{F}})^{-1} \mathbf{\Omega} \right) \leq \rho \\
& \text{Tr}(\tilde{\mathbf{F}} \tilde{\mathbf{F}}^H) \leq P.
\end{aligned}
\tag{16.26}
$$

It follows from Lemma 16.1.1 and Corollary 16.1.2 that, for a given $\tilde{\mathbf{F}}$, we can always find a feasible $\mathbf{\Omega}$ if and only if

$$
\lambda \left((\mathbf{I} + \tilde{\mathbf{F}}^H \mathbf{H}^H \mathbf{H} \tilde{\mathbf{F}})^{-1} \right) \succ^w \rho.
$$

Therefore, using the diagonal property of $\tilde{\mathbf{F}}^H \mathbf{H}^H \mathbf{H} \tilde{\mathbf{F}}$, (16.26) is equivalent to

$$
\begin{aligned}
\underset{\tilde{\mathbf{F}}, \rho}{\text{minimize}} \quad & f_0 \left(\rho \right) \\
\text{subject to} \quad & \tilde{\mathbf{F}}^H \mathbf{H}^H \mathbf{H} \tilde{\mathbf{F}} \quad \text{diagonal} \\
& \mathbf{d} \left((\mathbf{I} + \tilde{\mathbf{F}}^H \mathbf{H}^H \mathbf{H} \tilde{\mathbf{F}})^{-1} \right) \succ^w \rho \\
& \text{Tr}(\tilde{\mathbf{F}} \tilde{\mathbf{F}}^H) \leq P.
\end{aligned}
\tag{16.27}
$$

---

[12] $\succ^w$ denotes weak supermajorization (see Definition 16.1.3).

Given that $\tilde{\mathbf{F}}^H\mathbf{H}^H\mathbf{H}\tilde{\mathbf{F}}$ is diagonal with diagonal ele
can invoke [9, Lemma 12] or [6, Lemma 3.16] to conclu
written as $\tilde{\mathbf{F}} = \mathbf{V}_h\mathrm{diag}(\sqrt{\mathbf{p}})$, implying that $\mathbf{F}^\star = \mathbf{V}$
the weak supermajorization relation as well as the
(16.27) can be expressed as (16.24).

If, in addition, f_0 is minimized when the argume
order,[13] then (16.24) can be explicitly written as

$$\begin{array}{ll}
\underset{\mathbf{p},\rho}{\text{minimize}} & f_0\left(\rho_1,\ldots,\rho_L\right) \\
\text{subject to} & \sum_{j=i}^{L}\frac{1}{1+p_i\gamma_i} \le \sum_{j=i}^{L}\rho_j, \\
& \rho_i \ge \rho_{i+1}, \quad 1 \le i \le L - \\
& \mathbf{p} \ge \mathbf{0}, \ \mathbf{1}^T\mathbf{p} \le P
\end{array}$$

which is a convex problem if f_0 is a convex function
solved in polynomial time [20]. In fact, the optimal
plified or even obtained in closed form, when the ob
of Schur-convex/concave functions.

Corollary 16.2.1. ([9, Theorem 1]) Suppose that th
is increasing in each argument.

(i) If f_0 is Schur-concave, then the optimal solutio

$$\mathbf{F}^\star = \mathbf{V}_h\mathrm{diag}(\sqrt{\mathbf{p}})$$

where $\mathbf{p}$ is the solution to the following power a

$$\begin{array}{ll}
\underset{\mathbf{p}}{\text{minimize}} & f_0\left(\{(1 + p_i\gamma_i)\right. \\
\text{subject to} & \mathbf{p} \ge \mathbf{0}, \ \mathbf{1}^T\mathbf{p} \le
\end{array}$$

(ii) If f_0 is Schur-convex, then the optimal solution

$$\mathbf{F}^\star = \mathbf{V}_h\mathrm{diag}(\sqrt{\mathbf{p}})\mathbf{\Omega}$$

where the power allocation $\mathbf{p}$ is given by

$$p_i = (\mu\gamma_i^{-1/2} - \gamma_i^{-1})^+, \quad 1$$

with μ chosen to satisfy $\mathbf{1}^T\mathbf{p} = P$, and $\mathbf{\Omega}$ is
$(\mathbf{I} + \mathbf{F}^{\star H}\mathbf{H}^H\mathbf{H}\mathbf{F}^\star)^{-1}$ has equal diagonal eleme
matrix satisfying $|[\mathbf{\Omega}]_{ik}| = |[\mathbf{\Omega}]_{il}|, \forall i, k, l$, such as
the unitary Hadamard matrix (see Lemma 16.1

---

[13]In practice, most cost functions are minimized when the ar
(if not, one can always use instead the function $\tilde{f}_0(\mathbf{x}) = \min_{\mathbf{P} \in}$
permutation matrices) and, hence, the decreasing order can be

Although Schur-convex/concave functions do not form a partition of all L-dimensional functions, they do cover most of the frequently used global performance measures. An extensive account of Schur-convexity/concavity of common performance measures was provided in [6] and [9] (see also Exercise 16.4.3). For Schur-concave functions, a nice property is that the MIMO channel is fully diagonalized by the optimal transceiver, whereas for Schur-convex functions, the channel is diagonalized subject to a specific rotation $\mathbf{\Omega}$ on the transmit symbols.

16.2.3   Nonlinear MIMO Transceiver Design

In this section, we introduce another paradigm of MIMO transceivers, consisting of a linear precoder and a nonlinear decision feedback equalizer (DFE). The DFE differs from the linear equalizer in that the DFE exploits the finite alphabet property of digital signals and recovers signals successively. Thus, the nonlinear decision feedback (DF) MIMO transceivers usually enjoy superior performance than the linear transceivers. Using majorization theory, the DF MIMO transceiver designs can also be unified, mainly based on the recent results in [11, 21, 22], into a general framework covering diverse design criteria, as was derived independently in [6, 23, 24]. Different from the linear transceiver designs that are based on additive majorization, the DF transceiver designs rely mainly on multiplicative majorization (see Section 16.1.4).

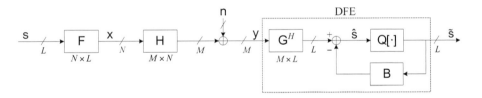

Figure 16.3: Nonlinear MIMO transceiver consisting of a linear precoder and a decision feedback equalizer (DFE).

Considering the MIMO channel in (16.15), we use a linear precoder $\mathbf{F} \in \mathbb{C}^{N \times L}$ at the transmitter to generate the transmitted signal $\mathbf{x} = \mathbf{Fs}$ from a symbol vector $\mathbf{s}$ satisfying $E\left\{\mathbf{ss}^H\right\} = \mathbf{I}$. For simplicity, we assume that $L \leq \operatorname{rank}(\mathbf{H})$. The receiver exploits, instead of a linear equalizer, a DFE that detects the symbols successively with the Lth symbol (s_L) detected first and the first symbol (s_1) detected last. As shown in Figure 16.3, a DFE consists of two components: a feed-forward filter $\mathbf{G}^H \in \mathbb{C}^{L \times M}$ applied to the received signal $\mathbf{y}$, and a feedback filter $\mathbf{B} \in \mathbb{C}^{L \times L}$ that is a strictly upper triangular matrix and feeds back the previously detected symbols. The block $Q[\cdot]$ represents the mapping from the "analog" estimated $\hat{s}_i$ to the closest "digital" point in the signal constellation. Assuming no error propagation,[14] the "analog" estimated $\hat{s}_i$ can be writ

$$\hat{s}_i = \mathbf{g}_i^H \mathbf{y} - \sum_{j=i+1}^{L} b_{ij} x_j, \quad 1 \leq$$

where $\mathbf{g}_i$ is the ith column of $\mathbf{G}$ and $b_{ij} = [\mathbf{B}]_{ij}$. Com
vector can be written as

$$\hat{\mathbf{s}} = \mathbf{G}^H \mathbf{y} - \mathbf{Bs} = (\mathbf{G}^H \mathbf{HF} - \mathbf{B})\mathbf{s}$$

Let $\mathbf{f}_i$ be the ith column of $\mathbf{F}$. The performance (
measured by the MSE or the SINR as

$$
\begin{aligned}
\mathrm{MSE}_i \quad &\triangleq \quad E\left\{|\hat{s}_i - s_i|^2\right\} \\
&= \quad \left|\mathbf{g}_i^H \mathbf{Hf}_i - 1\right|^2 + \sum_{j=i+1}^{L} \left|\mathbf{g}_i^H \mathbf{Hf}_j - b_{ij}\right|^2 \\
\mathrm{SINR}_i \quad &\triangleq \quad \frac{\text{desired component}}{\text{undesired component}} \\
&= \quad \frac{\left|\mathbf{g}_i^H \mathbf{Hf}_i\right|^2}{\sum_{j=i+1}^{L} \left|\mathbf{g}_i^H \mathbf{Hf}_j - b_{ij}\right|^2 + \sum_{j=1}^{i-1} \left|\mathbf{g}_i^H\right.}
\end{aligned}
$$

Alternatively, the performance can also be measured
a decreasing function φ_i. Similar to the linear tr
that the system performance is measured by a globa
$f_0\left(\{\mathrm{MSE}_i\}_{i=1}^{L}\right)$ that is increasing in each argument. T
transceiver design is formulated as the following prob

$$
\begin{aligned}
\underset{\mathbf{F}, \mathbf{G}, \mathbf{B}}{\text{minimize}} \quad & f_0\left(\{\mathrm{MSE}_i\}\right) \\
\text{subject to} \quad & \mathrm{Tr}(\mathbf{FF}^H) \leq P
\end{aligned}
$$

where $\mathrm{Tr}(\mathbf{FF}^H) \leq P$ denotes the transmit power con
It is easily seen that to minimize MSE_i, the DF
$\mathbf{g}_i^H \mathbf{Hf}_j$, $1 \leq i < j \leq L$, or, equivalently,

$$\mathbf{B} = \mathcal{U}(\mathbf{G}^H \mathbf{HF})$$

where $\mathcal{U}(\cdot)$ stands for keeping the strictly upper tria
while setting the others zero. To obtain the optima
$\mathbf{W} \triangleq \mathbf{HF}$ be the effective channel, and denote by

---

[14]Error propagation means that if the detection is erroneous
subsequent detections. By using powerful coding techniques,
can be made negligible.

consisting of the first i columns of $\mathbf{W}$ and by $\mathbf{w}_i$ the ith column of $\mathbf{W}$. Then, with $b_{ij} = \mathbf{g}_i^H \mathbf{H} \mathbf{f}_j$, the feed-forward filter minimizing MSE_i is given by [6, Sec. 4.3]

$$\mathbf{g}_i = (\mathbf{W}_i \mathbf{W}_i^H + \mathbf{I})^{-1} \mathbf{w}_i, \quad 1 \le i \le L. \tag{16.33}$$

In fact, there is a more computationally efficient expression of the optimal DFE given as follows.

Lemma 16.2.2. ([25]) Let the QR decomposition of the augmented matrix be

$$\mathbf{W}_a \triangleq \begin{bmatrix} \mathbf{W} \\ \mathbf{I}_L \end{bmatrix}_{(M+L) \times L} = \mathbf{QR}$$

and partition $\mathbf{Q}$ into

$$\mathbf{Q} = \begin{bmatrix} \bar{\mathbf{Q}} \\ \underline{\mathbf{Q}} \end{bmatrix}$$

where $\bar{\mathbf{Q}} \in \mathbb{C}^{M \times L}$ and $\underline{\mathbf{Q}} \in \mathbb{C}^{L \times L}$. The optimal feed-forward and feedback matrices that minimize the MSEs are

$$\mathbf{G}^\star = \bar{\mathbf{Q}} \mathbf{D}_R^{-1} \quad \text{and} \quad \mathbf{B}^\star = \mathbf{D}_R^{-1} \mathbf{R} - \mathbf{I} \tag{16.34}$$

where $\mathbf{D}_R$ is a diagonal matrix with the same diagonal elements as $\mathbf{R}$. The resulting MSE matrix is diagonal:

$$\mathbf{E} \triangleq E\left[(\hat{\mathbf{s}} - \mathbf{s})(\hat{\mathbf{s}} - \mathbf{s})^H\right] = \mathbf{D}_R^{-2}.$$

By using the optimal DFE in (16.34), the MSE and the SINR at the ith data stream are related by

$$\mathrm{SINR}_i = \frac{1}{\mathrm{MSE}_i} - 1$$

which is the same as in the linear equalizer case. Therefore, we can focus w.l.o.g. on the MSE-based performance measures, which, according to Lemma 16.2.2, depend on the diagonal elements of $\mathbf{R}$. The optimal precoder is then given by the solution to the following problem:

$$
\begin{aligned}
\underset{\mathbf{F}}{\text{minimize}} \quad & f_0\left(\{[\mathbf{R}]_{ii}^{-2}\}\right) \\
\text{subject to} \quad & \begin{bmatrix} \mathbf{HF} \\ \mathbf{I}_L \end{bmatrix} = \mathbf{QR} \\
& \mathrm{Tr}(\mathbf{FF}^H) \le P.
\end{aligned}
\tag{16.35}
$$

This complicated optimization can be simplified by using multiplicative majorization.

Theorem 16.2.3. ([6, Theorem 4.3]) Suppose that the cost function $f_0 : \mathbb{R}^L \mapsto \mathbb{R}$ is increasing in each argument. Then, the optimal solution to (16.35) is given by

$$\mathbf{F}^\star = \mathbf{V}_h \mathrm{diag}(\sqrt{\mathbf{p}}) \mathbf{\Omega}$$

where

(i) $\mathbf{V}_h \in \mathbb{C}^{N \times L}$ *is a semi-unitary matrix with co*
gular vectors of matrix $\mathbf{H}$ *corresponding to the*
increasing order;

(ii) $\mathbf{p} \in \mathbb{R}_+^L$ *is the solution to the following power a*

$$\begin{array}{ll} \underset{\mathbf{p},\mathbf{r}}{\text{minimize}} & f_0\left(r_1^{-2},\ldots,r_L^{-2}\right) \\ \text{subject to} & \left(r_1^2,\ldots,r_L^2\right) \prec_\times (1+p_1\gamma_1, \\ & \mathbf{p} \geq 0, \ \mathbf{1}^T\mathbf{p} \leq P \end{array}$$

where $\{\gamma_i\}_{i=1}^L$ *are the* L *largest eigenvalues of* $\blacksquare$

(iii) $\mathbf{\Omega} \in \mathbb{C}^{L \times L}$ *is a unitary matrix such that the ma*
tion

$$\begin{bmatrix} \mathbf{HF}^\star \\ \mathbf{I}_L \end{bmatrix} = \mathbf{QR}$$

has diagonal elements $\{r_i\}_{i=1}^L$. *To obtain* $\mathbf{\Omega}$, *it*
eralized triangular decomposition (GTD) [11]

$$\begin{bmatrix} \mathbf{HV}_h\text{diag}(\sqrt{\mathbf{p}}) \\ \mathbf{I}_L \end{bmatrix} = \mathbf{Q}$$

and then set $\mathbf{\Omega} = \mathbf{P}_J$.

Proof: The proof is involved, so we provide only
interested reader to [6, Appendix 4.C] for the detaile
Denote the diagonal elements of $\mathbf{R}$ by $\{r_i\}_{i=1}^L$, a
effective channel $\mathbf{W}$ and the augmented matrix $\mathbf{W}_a$
in decreasing order, respectively. One can easily see t

$$\sigma_{w_a,i} = \sqrt{1+\sigma_{w,i}^2} \quad 1 \leq i \leq$$

Consider the SVD $\mathbf{F} = \mathbf{U}_f\text{diag}(\sqrt{\mathbf{p}})\mathbf{\Omega}$. By using Thec
can prove that there exists an $\mathbf{\Omega}$ such that $\mathbf{W}_a = \mathbf{QR}$ ▮
[6, Lemma 4.9]. Therefore, the constraint

$$\begin{bmatrix} \mathbf{HF} \\ \mathbf{I}_L \end{bmatrix} = \mathbf{QR}$$

can be equivalently replaced by

$$\left(r_1^2,\ldots,r_L^2\right) \prec_\times \left(\sigma_{w_a,1}^2,\ldots,\sigma_u^2\right.$$

Next, by showing that

$$\prod_{i=1}^k \sigma_{w_a,i}^2 = \prod_{i=1}^k(1+\sigma_{w,i}^2) \leq \prod_{i=1}^k(1+\gamma_i p_i$$

where the equality holds if and only if $\mathbf{U}_f = \mathbf{V}_h$, one can conclude that the optimal $\mathbf{F}$ occurs when $\mathbf{U}_f = \mathbf{V}_h$. $\qquad\square$

Theorem 16.2.3 shows the solution to the general problem with an arbitrary cost function has a nice structure. In fact, when the composite objective function

$$f_0 \circ \exp(\mathbf{x}) \triangleq f_0(e^{x_1}, \ldots, e^{x_L}) \tag{16.37}$$

is either Schur-convex or Schur-concave, the nonlinear DF transceiver design problem admits a simpler or even closed-form solution.

Corollary 16.2.2. ([6, Theorem 4.4]) Suppose that the cost function $f_0 : \mathbb{R}^L \mapsto \mathbb{R}$ is increasing in each argument.

(i) If $f_0 \circ \exp$ is Schur-concave, then the optimal solution to (16.35) is given by

$$\mathbf{F}^\star = \mathbf{V}_h \mathrm{diag}(\sqrt{\mathbf{p}})$$

where $\mathbf{p}$ is the solution to the following power allocation problem:

$$\begin{aligned} \underset{\mathbf{p}}{\text{minimize}} \quad & f_0\left(\{(1+\gamma_i p_i)^{-1}\}_{i=1}^L\right) \\ \text{subject to} \quad & \mathbf{p} \geq \mathbf{0}, \ \mathbf{1}^T \mathbf{p} \leq P. \end{aligned}$$

(ii) If $f_0 \circ \exp$ is Schur-convex, then the optimal solution to (16.35) is given by

$$\mathbf{F}^\star = \mathbf{V}_h \mathrm{diag}(\sqrt{\mathbf{p}})\mathbf{\Omega}$$

where the power allocation $\mathbf{p}$ is given by

$$p_i = (\mu - \gamma_i^{-1})^+, \quad 1 \leq i \leq L$$

with μ chosen to satisfy $\mathbf{1}^T \mathbf{p} = P$, and $\mathbf{\Omega}$ is a unitary matrix such that the QR decomposition

$$\begin{bmatrix} \mathbf{H}\mathbf{F}^\star \\ \mathbf{I}_L \end{bmatrix} = \mathbf{QR}$$

yields $\mathbf{R}$ with equal diagonal elements.

It is interesting to relate the linear and nonlinear DF transceivers by the Schur-convexity/concavity of the cost function. From Lemma 16.1.5, $f_0 \circ \exp$ being Schur-concave implies that f_0 is Schur-concave, but not vice versa. From Lemma 16.1.4, if f_0 is Schur-convex, then $f_0 \circ \exp$ is also Schur-convex, but not vice versa. The examples of the cost function for which $f_0 \circ \exp$ is either Schur-concave or Schur-convex were provided in [6] (see also Exercise 16.4.4 and a recent survey [26]).

16.2.4   Impact of Correlation

A Measure of Correlation

Consider two n-dimensional random vectors $\mathbf{x}$ an
ily/class of distributions with zero means and cova
respectively. One question arising in many practical
$\mathbf{x}$ and $\mathbf{y}$ in terms of the degree of correlation. Majoriz
to measure correlation of a random vector.

Definition 16.2.2. ([7, Sec. 4.1.2]) Let $\lambda(\mathbf{A})$ denote
semidefinite matrix $\mathbf{A}$. Then, we say $\mathbf{x}$ is more correla
matrix $\mathbf{R}_x$ is more correlated than $\mathbf{R}_y$, if $\lambda(\mathbf{R}_x) \succ \lambda$

Note that comparing $\mathbf{x}$ and $\mathbf{y}$ (or equivalently $\mathbf{R}$
jorization ordering imposes an implicit constraint o
$\sum_{i=1}^{n} \lambda_i(\mathbf{R}_x) = \sum_{i=1}^{n} \lambda_i(\mathbf{R}_y)$, or equivalently, $\mathrm{Tr}(\mathbf{R}$
ment is actually quite reasonable. If we consider $E\{$
ith element of $\mathbf{x}$, then $\mathrm{Tr}(\mathbf{R}_x) = \sum_{i=1}^{n} E\{|x_i|^2\}$ is th
of $\mathbf{x}$. Therefore, the comparison is conducted in a fair
the two vectors is equal. Nevertheless, Definition 16.
case where $\mathrm{Tr}(\mathbf{R}_x) \neq \mathrm{Tr}(\mathbf{R}_y)$ by using weak majoriza

From Example 16.1.1, the most uncorrelated covar
values, whereas the most correlated covariance matrix
value. In the next, we demonstrate through several ex
tion theory along with Definition 16.2.2 to analyze
communication systems.

Colored Noise in CDMA Systems

Consider the uplink of a single-cell synchronous (
and processing gain N similar to the one that has beer
but with colored noise. More exactly, with the receiv
given by (16.8), the zero-mean noise $\mathbf{n}$ is now correlate
$\mathbf{R}_n$. In this case, the sum capacity of the CDMA sys

$$C_{\mathrm{sum}} = \frac{1}{2} \log \det \left(\mathbf{I}_N + \mathbf{R}_n^{-1} \mathbf{S}$$

where $\mathbf{S}$ is the signature sequence matrix and $\mathbf{P} = \mathrm{d}$
received power of each user. The maximum sum cap
$\max_{\mathbf{S} \in \mathcal{S}} C_{\mathrm{sum}}$, where $\mathcal{S}$ is defined in (16.13).

Denote the EVD of $\mathbf{R}_n$ by $\mathbf{R}_n = \mathbf{U}_n \mathbf{\Lambda}_n \mathbf{U}_n^H$ with e
C_{opt} can be characterized as follows.

Lemma 16.2.3. ([27, Lemma 2.2]) The maximum sun
CDMA system with colored noise is given by

$$C_{\mathrm{opt}} = \max_{\mathbf{S} \in \mathcal{S}} \frac{1}{2} \sum_{i=1}^{N} \log \left(1 + \frac{\lambda_i(\mathbf{S}}{\sigma}$$

Proposition 16.2.1. C_{opt} obtained in (16.39) is Schur-convex in $\sigma^2 \triangleq (\sigma_1^2, \ldots, \sigma_N^2)$.

Proof: Let $\phi(\sigma^2) = \frac{1}{2} \sum_{i=1}^{N} \log \left(1 + \lambda_i/\sigma_i^2\right)$. Since $g(x_i) = \log \left(1 + \lambda_i/x_i\right)$ is a convex function and $f(\mathbf{x}) = \frac{1}{2} \sum_{i=1}^{N} x_i$ is increasing and Schur-convex, it follows from Proposition 16.1.2 that $\phi(\sigma^2) = f(g(\sigma_1^2), \ldots, g(\sigma_N^2))$ is Schur-convex. Therefore, given $\sigma_a^2 \prec \sigma_b^2$, we have

$$C_{\text{opt}}(\sigma_a^2) = \max_{\mathbf{S} \in \mathcal{S}} \phi(\sigma_a^2) \leq \max_{\mathbf{S} \in \mathcal{S}} \phi(\sigma_b^2) = C_{\text{opt}}(\sigma_b^2).$$

$\square$

Proposition 16.2.1 indicates that the more correlated (according to Definition 16.2.2) the noise is, the higher the sum capacity could be. Intuitively, if one of the noise variances, say σ_N^2, is much larger than the rest, the users can avoid using signals in the direction of $\mathbf{R}_n$ corresponding to σ_N^2 and benefit from a reduced average noise variance (since the sum of all variances keeps unchanged). Apparently, white noise with equal $\sigma_i^2 = \sigma^2$, $i = 1, \ldots, N$, is one of the worst cases that lead to the minimum C_{opt}.

Spatial Correlation in MISO Channels

A multiple-input single-output (MISO) channel usually arises in using multiple transmit antennas and a single receive antenna in a wireless link. Consider a block-flat-fading[15] MISO channel with N transmit antennas. The channel model is given by

$$y = \mathbf{x}^H \mathbf{h} + n \tag{16.40}$$

where $\mathbf{x} \in \mathbb{C}^N$ is the transmitted signal, $y \in \mathbb{C}$ is the received signal, the complex Gaussian noise n has zero mean and variance σ^2, and the channel $\mathbf{h} \in \mathbb{C}^N$ is a circular symmetric Gaussian random vector with zero-mean and covariance matrix $\mathbf{R}_h$, i.e., $\mathbf{h} \sim \mathcal{CN}(\mathbf{0}, \mathbf{R}_h)$.

In MISO (as well as MIMO) channels, the transmit strategy is determined by the transmit covariance matrix $\mathbf{Q} = E\{\mathbf{x}\mathbf{x}^H\}$. Denote the EVD of $\mathbf{Q}$ by $\mathbf{Q} = \mathbf{U}_q \mathbf{\Lambda}_q \mathbf{U}_q^H$ with the diagonal matrix $\mathbf{\Lambda}_q = \text{diag}\{p_1, \ldots, p_N\}$. Then, the eigenvectors of $\mathbf{Q}$, i.e., the columns of $\mathbf{U}_q$, can be regarded as the transmit directions, and the eigenvalue p_i represents the power allocated to the ith data stream or eigenmode. Assuming that the receiver knows the channel perfectly and the transmitter uses a Gaussian codebook with zero mean and covariance matrix $\mathbf{Q}$, the maximum average mutual information, also termed the ergodic capacity, of the MISO channel is given by [16]

$$C = \max_{\mathbf{Q} \in \mathcal{Q}} E \left[\log(1 + \gamma \mathbf{h}^H \mathbf{Q} \mathbf{h})\right] \tag{16.41}$$

where γ is the signal-to-noise ratio, $\mathcal{Q} \triangleq \{\mathbf{Q} : \mathbf{Q} \succeq \mathbf{0}, \text{Tr}(\mathbf{Q}) = 1\}$ represents the normalized transmit power constraint, and the expectation is taken over $\mathbf{h}$.

---

[15] Block flat-fading means that the channel keeps unchanged for a block of T symbols, and then the channel changes to an uncorrelated channel realization.

The ergodic capacity depends on what kind of cha
transmitter (CSIT) is available. In the following, we

- No CSIT. Neither $\mathbf{h}$ nor its statistics are known
  usually assumed that $\mathbf{h} \sim \mathcal{CN}(\mathbf{0}, \mathbf{I})$, i.e., $\mathbf{R}_h =$

- Perfect CSIT. That is, $\mathbf{h}$ is perfectly known by

- Imperfect CSIT with covariance feedback. In
  $\mathbf{h} \sim \mathcal{CN}(\mathbf{0}, \mathbf{R}_h)$ with $\mathbf{R}_h$ known by the transmi

Denote the EVD of $\mathbf{R}_h$ by $\mathbf{R}_h = \mathbf{U}_h \mathbf{\Lambda}_h \mathbf{U}_h^H$ with
sorted w.l.o.g. in decreasing order, and let $w_1, \ldots, w$
iid random variables. In the case of no CSIT, the
matrix is given by $\mathbf{Q} = \frac{1}{N}\mathbf{I}$ [16], which results in

$$C_{\mathrm{noCSIT}}(\mu) = E\left[\log\left(1 + \frac{\gamma}{N}\sum_{i=1}^{N}\right.\right.$$

With perfect CSIT, the optimal $\mathbf{Q}$ is given by $\mathbf{Q} = \mathbf{h}$

$$C_{\mathrm{pCSIT}}(\mu) = E\left[\log\left(1 + \gamma\sum_{i=1}^{N}\mu\right.\right.$$

For imperfect CSIT with covariance feedback (i.e., $\mathbf{I}$
given in the form $\mathbf{Q} = \mathbf{U}_h \mathbf{\Lambda}_q \mathbf{U}_h^H$ [28], so the ergodic

$$C_{\mathrm{cfCSIT}}(\mu) = \max_{\mathbf{p} \in \mathcal{P}} E\left[\log\left(1 + \gamma\sum_{i=1}^{N}\right.\right.$$

where $\mathcal{P} \triangleq \{\mathbf{p} : \mathbf{p} \geq \mathbf{0}, \ \mathbf{1}^T\mathbf{p} = 1\}$ is the power
capacities of the three types of CSIT all depend on t
which is exactly characterized by the following result

Theorem 16.2.4. (*[29]*) *While* $C_{\mathrm{noCSIT}}(\mu)$ *and* C_{pCS}
in μ, $C_{\mathrm{cfCSIT}}(\mu)$ *is Schur-convex in* μ.

Proof: The Schur-concavity of $C_{\mathrm{noCSIT}}(\mu)$ and C_{p}
Corollary 16.1.6, since $f(\mathbf{x}) = \log(1 + a\sum_{i=1}^{N} x_i)$
function for $a > 0$ and $x_i \geq 0$. The proof of the Sch
based on Theorem 16.1.2 but quite involved. We refer
for more details.

Theorem 16.2.4 completely characterizes the imp
godic capacity of a MISO channel. To see this, assume
comparison under Definition 16.2.2) and the correlat

i.e., $\mu^1 \prec \mu^2$. We define the fully correlated vector $\psi = (N, 0, \ldots, 0)$ that majorizes all other vectors, and the least correlated vector $\chi = (1, 1, \ldots, 1)$ that is majorized by all other vectors. Then, according to Theorem 16.2.4, the impact of different types of CSIT and different levels of correlation on the MISO capacity is provided in the following inequality chain [29]:

$$
\begin{aligned}
C_{\text{noCSIT}}(\psi) &\leq C_{\text{noCSIT}}(\mu^2) \leq C_{\text{noCSIT}}(\mu^1) \leq C_{\text{noCSIT}}(\chi) \\
= C_{\text{cfCSIT}}(\chi) &\leq C_{\text{cfCSIT}}(\mu^1) \leq C_{\text{cfCSIT}}(\mu^2) \leq C_{\text{cfCSIT}}(\psi) \\
= C_{\text{pCSIT}}(\psi) &\leq C_{\text{pCSIT}}(\mu^2) \leq C_{\text{pCSIT}}(\mu^1) \leq C_{\text{pCSIT}}(\chi).
\end{aligned}
\tag{16.45}
$$

Simply speaking, correlation helps in the covariance feedback case, but degrades the channel capacity when there is either perfect or no CSIT. Nevertheless, the more amount of CSIT is available, the better the performance could be.

16.2.5 Robust Design

The performance of MIMO communication systems depends, to a substantial extent, on the channel state information (CSI) available at both ends of the communication link. While CSI at the receiver (CSIR) is usually assumed to be perfect, CSI at the transmitter (CSIT) is often imperfect due to many practical issues. Therefore, when devising MIMO transmit strategies, the imperfectness of CSIT has to be considered, leading to the so-called robust designs. A common philosophy of robust designs is to achieve worst-case robustness, i.e., to guarantee the system performance in the worst channel [30]. In this section, we use majorization theory to prove that the uniform power allocation is the worst-case robust solution for two kinds of imperfect CSIT.

Deterministic Imperfect CSIT

Consider the MIMO channel model in (16.15), where the transmit strategy is given by the transmit covariance matrix $\mathbf{Q}$. Indeed, assuming the transmit signal $\mathbf{x}$ is a Gaussian random vector with zero mean and covariance matrix $\mathbf{Q}$, i.e., $\mathbf{x} \sim \mathcal{CN}(\mathbf{0}, \mathbf{Q})$, the mutual information is given by [16]

$$
\Psi(\mathbf{Q}, \mathbf{H}) = \log \det \left(\mathbf{I} + \mathbf{H}\mathbf{Q}\mathbf{H}^H \right) = \log \det \left(\mathbf{I} + \mathbf{Q}\mathbf{H}^H \mathbf{H} \right).
\tag{16.46}
$$

If $\mathbf{H}$ is perfectly known by the transmitter, i.e., perfect CSIT, the channel capacity can be achieved by maximizing $\Psi(\mathbf{Q}, \mathbf{H})$ under the power constraint $\mathbf{Q} \in \mathcal{Q} \triangleq \{\mathbf{Q} : \mathbf{Q} \succeq \mathbf{0}, \ \text{Tr}(\mathbf{Q}) = P\}$.

In practice, however, the accurate channel value is usually not available, but belongs to a known set of possible values, often called an uncertainty region. Since $\Psi(\mathbf{Q}, \mathbf{H})$ depends on $\mathbf{H}$ through $\mathbf{R}_H = \mathbf{H}^H \mathbf{H}$, we can conveniently define an uncertainty region $\mathcal{H}$ as

$$
\mathcal{H} \triangleq \{\mathbf{H} : \mathbf{R}_H \in \mathcal{R}_H\}
\tag{16.47}
$$

where the set $\mathcal{R}_H$ could, for example, contain any constraints as

$$\mathcal{R}_H \triangleq \{\mathbf{R}_H : \{\lambda_i(\mathbf{R}_H)\} \in \mathcal{L}$$

where $\mathcal{L}_{R_H}$ denotes arbitrary eigenvalue constraints. $\mathbb{N}$ and (16.48) is an isotropic set in the sense that for each for any unitary matrix $\mathbf{U}$.

Following the philosophy of worst-case robustness, is obtained by optimizing $\Psi(\mathbf{Q}, \mathbf{H})$ in the worst cha region $\mathcal{H}$, thus resulting in a maximin problem

$$\max_{\mathbf{Q} \in \mathcal{Q}} \min_{\mathbf{H} \in \mathcal{H}} \Psi(\mathbf{Q}, \mathbf{H}).$$

The optimal value of this maximin problem is referred [31]. In the following, we show that the compound uniform power allocation.

Theorem 16.2.5. (*[32, Theorem 1]) The optimal so and the optimal value is*

$$C(\mathcal{H}) = \min_{\mathbf{H} \in \mathcal{H}} \log \det \left(\mathbf{I} + \frac{P}{N} \mathbf{H} \right.$$

Proof: Denote the eigenvalues of $\mathbf{Q}$ by $p_1 \geq \cdots$ order. From [32, Lemma 1], the optimal $\mathbf{Q}$ depend inner minimization of (16.49) is equivalent to

$$\underset{\{\lambda_i(\mathbf{R}_H)\} \in \mathcal{L}_{R_H}}{\text{minimize}} \sum_{i=1}^{N} \log\left(1 + p_i \lambda_i\right.$$

with $\lambda_1(\mathbf{R}_H) \leq \cdots \leq \lambda_N(\mathbf{R}_H)$ in increasing order. C $\sum_{i=1}^{N} g_i(x_i) = \sum_{i=1}^{N} \log(1 + a_i x_i)$ with $\{a_i\}$ in increas that $g_i'(x) \leq g_{i+1}'(y)$ whenever $x \geq y$. Thus, from The concave function, whose maximum is achieved by a ι power constraint $\sum_{i=1}^{N} p_i = P$, it follows that

$$\min_{\{\lambda_i(\mathbf{R}_H)\} \in \mathcal{L}_{R_H}} \sum_{i=1}^{N} \log\left(1 + p_i \lambda_i(\mathbf{R}_H)\right) \leq \min_{\{\lambda_i(\mathbf{R}_H)\} \in \mathcal{L}_{R}}$$

where the equality holds for the uniform power alloca

The optimality of the uniform power allocation is Due to the symmetry of the problem, if the transmi tribute power over the eigenvalues of $\mathbf{Q}$, then the worst singular value (or eigenvalue of $\mathbf{R}_H$) to the lowest eig avoid such a situation and achieve the best performa appropriate way is to use equal power on all eigenvalues of $\mathbf{Q}$, which is formally proved in Theorem 16.2.5.

Stochastic Imperfect CSIT

Tracking the instantaneous channel value may be difficult when the channel varies rapidly. The stochastic imperfect CSIT model assumes that the channel is a random quantity with its statistics such as mean or/and covariance known by the transmitter. Sometimes, even the channel statistics may not be perfectly known. The interests of this model would be on optimizing the average system performance using the channel statistics.

For simplicity, we consider the MISO channel in (16.40), where the channel $\mathbf{h}$ is a circular symmetric Gaussian random vector with zero-mean and covariance matrix $\mathbf{R}_h$, i.e., $\mathbf{h} \sim \mathcal{CN}(\mathbf{0}, \mathbf{R}_h)$. Mathematically, the channel can be expressed as

$$\mathbf{h} = \mathbf{R}_h^{1/2} \mathbf{z} \tag{16.50}$$

where $\mathbf{z} \sim \mathcal{CN}(\mathbf{0}, \mathbf{I})$. Different from the covariance feedback case where $\mathbf{R}_h$ is assumed to be known by the transmitter (see Section 16.2.4), here we consider an extreme case where the transmitter does not even know exactly $\mathbf{R}_h$. Instead, we assume that $\mathbf{R}_h \in \mathcal{R}_h$ with

$$\mathcal{R}_h \triangleq \{\mathbf{R}_h : \{\lambda_i(\mathbf{R}_h)\} \in \mathcal{L}_{R_h}\} \tag{16.51}$$

where $\mathcal{L}_{R_h}$ denotes arbitrary constraints on the eigenvalues of $\mathbf{R}_h$. In the case of no information on $\mathbf{R}_h$, we have $\mathcal{L}_{R_h} = \mathbb{R}_+^N$. To combat with the possible bad channels, the robust transmit strategy should take into account the worst channel covariance, thus leading to the following maximin problem

$$\max_{\mathbf{Q} \in \mathcal{Q}} \min_{\mathbf{R}_h \in \mathcal{R}_h} E\left[\log(1 + \mathbf{h}^H \mathbf{Q} \mathbf{h})\right] = E\left[\log(1 + \mathbf{z}^H \mathbf{R}_h^{1/2} \mathbf{Q} \mathbf{R}_h^{1/2} \mathbf{z})\right] \tag{16.52}$$

where $\mathcal{Q} \triangleq \{\mathbf{Q} : \mathbf{Q} \succeq \mathbf{0}, \ \mathrm{Tr}(\mathbf{Q}) = P\}$ and $\mathbf{z} \sim \mathcal{CN}(\mathbf{0}, \mathbf{I})$. The following result indicates that the uniform power allocation is again the robust solution.

Theorem 16.2.6. *The optimal solution to (16.52) is* $\mathbf{Q}^\star = \frac{P}{N}\mathbf{I}$ *and the optimal value is*

$$C(\mathcal{R}_h) = \min_{\mathbf{R}_h \in \mathcal{R}_h} E\left[\log\left(1 + \frac{P}{N}\sum_{i=1}^{N} \lambda_i(\mathbf{R}_h) w_i\right)\right]$$

where $w_1, \ldots, w_N$ *are standard exponentially iid random variables.*

Proof: Denote the eigenvalues of $\mathbf{Q}$ by $p_1 \geq \cdots \geq p_N$ w.l.o.g. in decreasing order. Considering that $\mathcal{Q}$ and $\mathcal{R}_h$ impose no constraint on the eigenvectors of $\mathbf{Q}$ and $\mathbf{R}_h$, respectively, and that $\mathbf{Uz}$ has the same distribution as $\mathbf{z}$ for any unitary matrix $\mathbf{U}$, the optimal $\mathbf{Q}$ should be a diagonal matrix depending on the eigenvalues $\{p_i\}$ (see, e.g., [28]) and thus (16.52) is equivalent to

$$\max_{\mathbf{Q} \in \mathcal{Q}} \min_{\mathbf{R}_h \in \mathcal{R}_h} E\left[\log\left(1 + \sum_{i=1}^{N} p_i \lambda_i(\mathbf{R}_h) w_i\right)\right] \tag{16.53}$$

with $w_i = |z_i|^2$, where z_i is the ith element of $\mathbf{z}$.

Given $\{p_i\}$ in decreasing order, the minimum c (16.53) must be achieved with $\{\lambda_i(\mathbf{R}_h)\}$ in increasir objective value can be obtained by changing the or lowing the similar steps in the proof of Theorem $E\left\{\log(1 + \sum_{i=1}^{N} p_i \lambda_i(\mathbf{R}_h)w_i)\right\}$ is a Schur-concave fur Hence, the maximum of (16.53) is achieved by a unif jorized by all other power vectors under the constrair

Another interesting problem is to investigate the all possible transmit strategies, which is given by t minimax problem:

$$\min_{\mathbf{R}_h \in \mathcal{R}_h} \max_{\mathbf{Q} \in \mathcal{Q}} E\left[\log(1 + \mathbf{h}^H \mathbf{Q}\right.$$

Through the similar steps in the proof of Theorem solution to (16.54) is proportional to an identity mat This provides a robust explanation for the assumpti case of no CSIT (see Section 16.2.4): $\mathbf{R}_h = \mathbf{I}$ is t $\mathcal{R}_h = \{\mathbf{R}_h : \text{Tr}(\mathbf{R}_h) = N\}$ [29].

16.3   Conclusions and Further F

This chapter introduced majorization as a partial orde vectors and described its main properties. This chapte of majorization theory in proving inequalities and problems in the fields of signal processing and wire more comprehensive treatment of majorization theo readers are directed to Marshall and Olkins book [1]. theory to signal processing and wireless communicati tutorials [6] and [7] .

16.4   Exercises

Exercise 16.4.1. Schur-convexity of sums of functio

a. Let $\phi(\mathbf{x}) = \sum_{i=1}^{n} g_i(x_i)$, where each g_i is dif Schur-convex on $\mathcal{D}_n$ if and only if

$$g_i'(a) \geq g_{i+1}'(b) \quad \text{whenever} \quad a \geq b,$$

b. Let $\phi(\mathbf{x}) = \sum_{i=1}^{n} a_i g(x_i)$, where $g(x)$ is decre $a_1 \leq \cdots \leq a_n$. Show that ϕ is Schur-convex on

Exercise 16.4.2. Schur-convexity of products of fun

a. Let $g : \mathcal{I} \to \mathbb{R}_+$ be continuous on the interval $\mathcal{I} \subseteq \mathbb{R}$. Show that $\phi(\mathbf{x}) = \prod_{i=1}^n g(x_i)$ is (strictly) Schur-convex on $\mathcal{I}^n$ if and only if $\log g$ is (strictly) convex on $\mathcal{I}$.

b. Show that $\phi(\mathbf{x}) = \prod_{i=1}^n \Gamma(x_i)$, where $\Gamma(x) = \int_0^\infty u^{x-1} e^{-u} du$ denotes the Gamma function, is strictly Schur-convex on $\mathbb{R}_{++}^n$.

Exercise 16.4.3. Linear MIMO Transceiver.

a. Prove Corollary 16.2.1, which shows that when the cost function f_0 is either Schur-concave or Schur-convex, the optimal linear MIMO transceiver admits an analytical structure.

b. Show that the following problem formulations can be rewritten as minimizing a Schur-concave cost function of MSEs:

- Minimizing $f(\{\text{MSE}_i\}) = \sum_{i=1}^L \alpha_i \text{MSE}_i$.[16]
- Minimizing $f(\{\text{MSE}_i\}) = \prod_{i=1}^L \text{MSE}_i^{\alpha_i}$.
- Maximizing $f(\{\text{SINR}_i\}) = \sum_{i=1}^L \alpha_i \text{SINR}_i$.
- Maximizing $f(\{\text{SINR}_i\}) = \prod_{i=1}^L \text{SINR}_i^{\alpha_i}$.
- Minimizing $f(\{\text{BER}_i\}) = \prod_{i=1}^L \text{BER}_i$.

c. Show that the following problem formulations can be rewritten as minimizing a Schur-convex cost function of MSEs:

- Minimizing $f(\{\text{MSE}_i\}) = \max_i\{\text{MSE}_i\}$.
- Maximizing $f(\{\text{SINR}_i\}) = \left(\prod_{i=1}^L \text{SINR}_i^{-1}\right)^{-1}$.
- Maximizing $f(\{\text{SINR}_i\}) = \min_i\{\text{SINR}_i\}$.
- Minimizing $f(\{\text{BER}_i\}) = \sum_{i=1}^L \text{BER}_i$.
- Minimizing $f(\{\text{BER}_i\}) = \max_i\{\text{BER}_i\}$.

Exercise 16.4.4. Nonlinear MIMO Transceiver.

a. Prove Corollary 16.2.2, which shows that the optimal nonlinear DF MIMO transceiver can also be analytically characterized if the composite cost function $f_0 \circ \exp$ is either Schur-concave or Schur-convex.

b. Show that the following problem formulations can be rewritten as minimizing a Schur-concave $f_0 \circ \exp$ of MSEs:

- Minimizing $f(\{\text{MSE}_i\}) = \prod_{i=1}^L \text{MSE}_i^{\alpha_i}$.

---

[16] Assume w.l.o.g. that $0 \leq \alpha_1 \leq \cdots \leq \alpha_L$.

596

- Maximizing $f(\{\mathrm{SINR}_i\}) = \sum_{i=1}^{L} \alpha_i \mathrm{SINR}_i$.

c. Show that, in addition to all problem formulat
following ones can also be rewritten as minimi
of MSEs:

- Minimizing $f(\{\mathrm{MSE}_i\}) = \sum_{i=1}^{L} \mathrm{MSE}_i$.
- Minimizing $f(\{\mathrm{MSE}_i\}) = \prod_{i=1}^{L} \mathrm{MSE}_i$.
- Maximizing $f(\{\mathrm{SINR}_i\}) = \prod_{i=1}^{L} \mathrm{SINR}_i$.

References

[1] A. W. Marshall and I. Olkin, *Inequalities: The*
Applications. New York: Academic Press, 1979

[2] G. H. Hardy, J. E. Littlewood, and G. Pólya, *In*
and New York: Cambridge University Press, 195

[3] R. Bhatia, *Matrix Analysis.* New York: Springe

[4] R. A. Horn and C. R. Johnson, *Matrix Analys*
University Press, 1985.

[5] T. W. Anderson, *An Introduction to Multivariat*
Hoboken, NJ: Wiley, 2003.

[6] D. P. Palomar and Y. Jiang, "MIMO transceiver
ory," *Foundations and Trends in Communicatic*
vol. 3, no. 4-5, pp. 331–551, 2006.

[7] E. A. Jorswieck and H. Boche, "Majorization and
in wireless communications," *Foundations and Tr*
Information Theory, vol. 3, no. 6, pp. 553–701, J

[8] P. Viswanath and V. Anantharam, "Optimal seq
synchronous CDMA systems," *IEEE Trans. Inf*
pp. 1984–1993, Sep. 1999.

[9] D. P. Palomar, J. M. Cioffi, and M. A. Lagunas,
design for multicarrier MIMO channels: A unifie
timization," *IEEE Trans. Signal Process.*, vol. 5
2003.

[10] C. T. Mullis and R. A. Roberts, "Synthesis of m
point digital filters," *IEEE Trans. on Circuits and*
pp. 551–562, Sept. 1976.

[11] Y. Jiang, W. Hager, and J. Li, "The generalized triangular decomposition," *Mathematics of Computation*, Nov. 2006.

[12] E. A. Jorswieck and H. Boche, "Outage probability in multiple antenna systems," *European Transactions on Telecommunications*, vol. 18, no. 3, pp. 217–233, Apr. 2007.

[13] S. Verdú, *Multiuser Detection*. New York, NY: Cambridge University Press, 1998.

[14] S. Ulukus and R. D. Yates, "Iterative construction of optimum signature sequence sets in synchronous CDMA systems," *IEEE Trans. Inform. Theory*, vol. 47, no. 5, pp. 1989–1998, Jul. 2001.

[15] C. Rose, S. Ulukus, and R. D. Yates, "Wireless systems and interference avoidance," *IEEE Trans. Wireless Commun.*, vol. 1, no. 3, pp. 415–428, Jul. 2002.

[16] I. E. Telatar, "Capacity of multi-antenna Gaussian channels," *European Trans. Telecommun.*, vol. 10, no. 6, pp. 585–595, Nov.-Dec. 1999.

[17] D. P. Palomar, M. A. Lagunas, and J. M. Cioffi, "Optimum linear joint transmit-receive processing for MIMO channels with QoS constraints," *IEEE Trans. Signal Process.*, vol. 52, no. 5, pp. 1179–1197, May 2004.

[18] L. Zheng and D. N. C. Tse, "Diversity and multiplexing: A fundamental tradeoff in multiple-antenna channels," *IEEE Trans. Inform. Theory*, vol. 49, no. 5, pp. 1073–1096, May 2003.

[19] S. M. Kay, *Fundamentals of Statistical Signal Processing: Estimation Theory*. Englewood Cliffs, NJ, USA: Prentice-Hall, 1993.

[20] S. Boyd and L. Vandenberghe, *Convex Optimization*. Cambridge, U.K.: Cambridge University Press, 2004.

[21] Y. Jiang, W. Hager, and J. Li, "The geometric mean decomposition," *Linear Algebra and Its Applications*, vol. 396, pp. 373–384, Feb. 2005.

[22] Y. Jiang, W. Hager, and J. Li, "Tunable channel decomposition for MIMO communications using channel state information," *IEEE Trans. Signal Process.*, vol. 54, no. 11, pp. 4405–4418, Nov. 2006.

[23] F. Xu, T. N. Davidson, J. K. Zhang, and K. M. Wong, "Design of block transceivers with decision feedback detection," *IEEE Trans. Signal Process.*, vol. 54, no. 3, pp. 965–978, Mar. 2006.

[24] A. A. D'Amico, "Tomlinson-Harashima precoding in MIMO systems: A unified approach to transceiver optimization based on multiplicative schur-convexity," *IEEE Trans. Signal Process.*, vol. 56, no. 8, pp. 3662–3677, Aug. 2008.

598

[25] B. Hassibi, "A fast square-root implementatic Thirty-Fourth Asilomar Conference on Signals, cific Grove, CA, USA, Nov. 2000.

[26] P. P. Vaidyanathan, S.-M. Phoong, and Y.-P. Li timization for Transceiver Systems. New York Press, 2010.

[27] P. Viswanath and V. Anantharam, "Optimal sec ored noise: A Schur-saddle function property," l vol. 48, no. 6, pp. 1295–1318, Jun. 2002.

[28] S. A. Jafar and A. Goldsmith, "Transmitter op beamforming for multiple antenna systems," IEE vol. 3, no. 4, pp. 1165–1175, Jul. 2004.

[29] E. A. Jorswieck and H. Boche, "Optimal transn of correlation in multiantenna systems with dif information," IEEE Trans. Signal Process., vol Dec. 2004.

[30] J. Wang and D. P. Palomar, "Worst-case robu imperfect channel knowledge," IEEE Trans. Si pp. 3086–3100, Aug. 2009.

[31] A. Lapidoth and P. Narayan, "Reliable commu certainty," IEEE Trans. Inform. Theory, vol. 44 1998.

[32] D. P. Palomar, J. M. Cioffi, and M. A. Lagunas, MIMO channels: A game-theoretic approach," l vol. 49, no. 7, pp. 1707–1727, Jul. 2003.

Chapter 17 Queueing Theory

Thomas Chen[‡]

[‡]Swansea University, Wales, UK

17.1 Introduction

Queueing theory has been successful for performance analysis of circuit-switched and packet-switched networks. Historically, queueing theory can be traced back to Danish mathematician A. K. Erlang who developed the well-known Erlang B and Erlang C formulas relating the capacity of telephone switches to the probability of call blocking. Since the emergence of packet switching in the late 1960s, queueing theory has become the primary mathematical apparatus for modeling and analysis of computer networks including the Internet. For packet-switched networks, the main performance metrics of interest are usually packet loss and packet delays.

Simple and intermediate queueing theory is centered mainly around queueing systems with Poisson arrivals, which can be treated relatively easily by Markov chains. For practical problems, the accuracy of these queueing models is often debatable. Unfortunately, more realistic queueing systems tend to become much more complicated and intractable (in advanced queueing theory).

17.2 Markov Chains

A brief review of Markov chains is a useful prologue to queueing theory. Recall from Chapter 6 that Markov processes are a class of stochastic processes that have the Markov (memoryless) property: the future evolution of a Markov process depends only on the current state and not on previous states. In other words, given the history of the process, the future depends only on the most recent state. For a Markov process, it is not necessary to know the entire history of the process in order to predict its future values; only the most recent point is sufficient.

Definition 17.2.1. For a discrete-time process X_n, the Markov property is

$$Pr(X_{n+1} \leq x | X_n = x_n, X_{n-1} = x_{n-1}, \ldots) = Pr(X_{n+1} \leq x | X_n = x_n) \quad (17.1)$$

For a continuous-time process X_t, the Markov property is

$$Pr(X_t \leq x | X_s, \ s \leq 0) = Pr(X_t \leq x | X_0) \quad (17.2)$$

for $t \geq 0$.

Markov chains are a subset of Markov processes associated with integers without loss of generality. M time making a transition at every time increment or c transitions at random times. The Markov property change depends only on the current state and not on got to that state).

17.2.1   Discrete-Time Markov Chains

A discrete-time Markov chain can be characterized by ability matrix

$$P = \begin{bmatrix} p_{00} & p_{01} & p_{02} & \cdots \\ p_{10} & p_{11} & p_{12} & \cdots \\ \vdots & \vdots & \vdots & \vdots \end{bmatrix}$$

where the matrix elements are the transition probabi

$$p_{ij} = P(X_{n+1} = j | X_n = i$$

for all states i, j. The transition (probability) matrix of possible states is infinite.

Queueing theory is usually interested in the stea Markov chain. Not every Markov chain has a stead Markov chains encountered in elementary queueing th that if the Markov chain goes for a long time, ther will approach some steady-state probability π_i that initial state. In another interpretation, if the Marko state distribution at time 0, it will continue indefini In addition, if the Markov chain is observed over a fraction of time spent in state i will be π_i.

Theorem 17.2.1. *If the steady-state probabilities* $[\pi_0 \ \pi_1 \ \cdots]$ *exist for a discrete-time Markov chain, the solution to the balance equations:*

$$\pi = \pi P$$

or

$$\pi_j = \sum_i \pi_i p_{ij}$$

under the constraint that the probabilities must sum t

17.2.2 Continuous-Time Markov Chains

Unlike discrete-time Markov chains, continuous-time Markov chains can make state transitions at random times. A discrete-time Markov chain can be found embedded in a continuous-time Markov chain.

Theorem 17.2.2. *For a continuous-time Markov chain X_t with transition times $\{t_1, t_2, \ldots\}$, the points $Y_n = X_{t_n}$ constitute an embedded discrete-time Markov chain. Given the current state $Y_n = i$, the time spent in the state $\tau_n \equiv t_{n+1} - t_n$ is exponentially distributed with some parameter a_i dependent only on the current state i.*

Definition 17.2.2. The transition rate from current state i to another state j is governed by the infinitessimal rates

$$a_{ij} = a_i P(Y_{n+1} = j | Y_n = i). \qquad (17.7)$$

The infinitessimal rate a_{ij} implies that the time spent in state i until a transition is made to state j is exponentially distributed with mean $1/a_{ij}$. The conditional rate matrix

$$A = \begin{bmatrix} a_{00} & a_{01} & a_{02} & \cdots \\ a_{10} & a_{11} & a_{12} & \cdots \\ \vdots & \vdots & \vdots & \vdots \end{bmatrix} \qquad (17.8)$$

is analogous to the transition probability matrix P for discrete-time Markov chain in that A completely characterizes a continuous-time Markov chain.

Theorem 17.2.3. *If a continuous-time Markov chain has steady-state or equilibrium probabilities $\{\pi_i\}$, then they are the unique solutions of the balance equations*

$$a_j \pi_j = \sum_{i \neq j} \pi_i a_{ij} \qquad (17.9)$$

under the constraint that the probabilities must sum to one, $\sum_i \pi_i = 1$.

The balance equations can be written compactly in matrix form as simply

$$0 = \pi Q \qquad (17.10)$$

where the Q matrix is defined as

$$Q \equiv \begin{bmatrix} -a_0 & a_{01} & a_{02} & a_{03} & \cdots \\ a_{10} & -a_1 & a_{12} & a_{13} & \cdots \\ a_{20} & a_{21} & -a_2 & a_{23} & \cdots \\ a_{30} & a_{31} & a_{32} & -a_3 & \cdots \\ \vdots & \vdots & \vdots & \vdots & \vdots \end{bmatrix} \qquad (17.11)$$

Poisson Counting Process

The Poisson counting process is a continuous-time M
to queueing theory. The conditional rate matrix is

$$A = \begin{bmatrix} 0 & \lambda & 0 & \cdots \\ 0 & 0 & \lambda & \ddots \\ 0 & 0 & 0 & \ddots \\ \vdots & \ddots & \ddots & \ddots \end{bmatrix}$$

Alternatively, the Poisson counting process can b
tion times. Imagine a series of arrivals (or "births")
starting from time 0, where interarrival times are in
distributed with parameter λ. The point process $\{t$
arrival process. The Poisson counting process X_t is
time t. It is called Poisson because X_t for any given t
with the probability mass function:

$$Pr(X_t = k) = \frac{(\lambda t)^k}{k!} e^{-\lambda t}$$

for $k = 0, 1, \ldots$. In fact, the number of arrivals in a
Poisson with mean λt.

A third way to look at the process is to consid
divide it into short intervals of length Δ. In each s
an arrival with probability $\lambda \Delta$ or no arrival with p
in each short interval is a Bernoulli trial, and assun
independent. It is evident that the number of arriv
a binomial distribution with mean λt. Shrinking th
$\Delta \to 0$ however, the number of trials becomes infi
of arrivals is fixed; the binomial distribution is kno
distribution in this case.

This third view of the Poisson arrival process is i
oryless property" (of particular importance to queu
Suppose that we know the last arrival occurred at tim
occurred up to time 0. When will the next arrival oc
struction, every short interval is independent, and th
arrival will be independent of the previous arrival tin
will be exponentially distributed with parameter λ re

This constructive view is also useful to show anoth
Poisson arrival process. Consider the aggregate of two
processes, one with rate λ_1 and the other with λ_2. A
$(0, t)$ divided into short intervals of length Δ. The p
an interval is $\lambda_1 \lambda_2 \Delta^2$ but this becomes negligible in

probability of one arrival, $(\lambda_1 + \lambda_2)\Delta$, is significant. Taking the limit, it becomes clear that the aggregate of two independent Poisson arrival processes is itself a Poisson arrival process with rate $\lambda_1 + \lambda_2$. Similarly, it can be shown by reversing the argument that random splitting of a Poisson arrival process will result in two independent Poisson arrival processes.

Birth-Death Process

Generally, a population can have random births (arrivals) and deaths (departures). A birth-death process is a continuous-time Markov chain often encountered in queueing theory. Its state transition diagram showing infinitessimal rates between consecutive states is shown in Figure 17.1. If the population is in state i, the time to the next birth is exponentially distributed with parameter λ_i while the time to the next death is exponentially distributed with parameter μ_i. The conditional rate matrix is

$$A = \begin{bmatrix} 0 & \lambda_0 & 0 & 0 & \cdots \\ \mu_1 & 0 & \lambda_1 & 0 & \ddots \\ 0 & \mu_2 & 0 & \lambda_2 & \ddots \\ \vdots & \ddots & \ddots & \ddots & \ddots \end{bmatrix} \qquad (17.14)$$

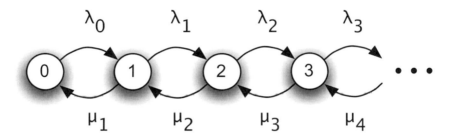

Figure 17.1: Birth-death process.

In order to find steady-state probabilities, one could use the balance equation $0 = \pi Q$, but consider an intuitive argument. In steady-state, we would expect that the total transition rates in opposite directions between states 0 and 1 should be equal, leading to the first balance equation

$$\lambda_0 \pi_0 = \mu_1 \pi_1 \qquad (17.15)$$

Likewise, considering the transition rates in opposite directions between states 1 and 2 leads to the next balance equation

$$\lambda_1 \pi_1 = \mu_2 \pi_2 \qquad (17.16)$$

In general, it can be seen that

$$\lambda_j \pi_j = \mu_{j+1} \pi_{j+1}$$

for all j. Along with the additional constraint π_0
solution exists to the balance equations, then the stea
birth-death process are

$$\pi_0 = \left(\sum_j \frac{\lambda_0 \cdots \lambda_{j-1}}{\mu_1 \cdots \mu_j} \right)^{-1}$$

$$\pi_j = \frac{\lambda_0 \cdots \lambda_{j-1}}{\mu_1 \cdots \mu_j} \pi_0 \quad , j >$$

17.3   Queueing Models

Queueing theory is a branch of applied probability
"jobs" that require service or otherwise wait in buffe
tomers are packets and service consists of the time to
The simplest system is a single-server queue as show
customer wants to occupy the server for a service tir
any new arrivals must join the queue. When the ser
from the queue enters into service. After a service tir
system. The operation of the system depends on the
the random service times.

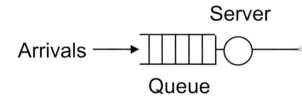

Figure 17.2: A single-server q

It is easy to imagine more complicated queueing
queues or multiple servers. The queues may use vario
priorities. The server might allow preemption (servi
Multiple queueing systems may work in parallel or se
In general, the analysis of packet networks involve
of interest:

- the delay through the system which is the sum
  queue and the service time;

- the number in the system which is the sum of the number waiting in the queue (queue length) and the number in service;

- the probability of losing customers if the buffer capacity is finite.

Typically, several assumptions are understood implicitly unless a different assumption is stated explicitly. These implicit assumptions include:

- customers arrive singly unless batch arrivals are specified;

- a server handles one customer at a time unless batch service is specified;

- the service discipline is FIFO (first in first out) also known as FCFS (first come first served);

- the buffer has infinite capacity;

- once queued, arrivals do not defect from the queue before receiving service or jump between queues (in the case of multiple queues);

- if the buffer is finite, arrivals are dropped from the tail of the queue;

- service is non-preemptive meaning that a customer in service must complete service without interruption;

- the system is work conserving meaning that a server cannot stand idle if any customers are waiting, and customers do not duplicate any completed service.

Traditional queueing models assume that arrivals are a renewal process, i.e., interarrival times are independent and identically distributed (i.i.d.) with a known probability distribution function $A(t)$. By convention, λ refers to the mean arrival rate implying that $1/\lambda$ is the mean interarrival time. In addition, service times are also i.i.d. with a known service time probability distribution function $B(t)$. By convention, μ is the service rate, or $1/\mu$ is the mean service time. The utilization or load factor is the ratio of mean service time to mean interarrival time, $\rho = \lambda/\mu$. For single-server queues, ρ is the fraction of time that the system is busy. Normally, performance analysis is interested in the stable case $\rho < 1$. When $\rho \geq 1$, the queue length grows without bounds.

Since a complete description of a queueing model would be lengthy, statistician David Kendall suggested a shorthand notation having the form A/B/C which can be expanded to a longer form A/B/C/D/E/F. The concise form of A/B/C is understood to mean that the other parameters have their default values. The parts of the notation have the meanings listed below.

A	Probability distribution function for intera:
B	Probability distribution function for service
C	Number of servers
D	Maximum capacity of system including ser¯
E	Total population of customers (infinite by ↑
F	Service discipline (FIFO by default)

Because certain probability distribution function
arrival times and service times, there is shorthand ↑
probability distributions:

M	Exponential
D	Deterministic
Ek	Erlang with paramete
G	General

17.4   M/M/1 Queue

The simplest queue is the M/M/1 queue, meaning a s
son arrivals and exponential service times. Poisson arr
that interarrival times are i.i.d. according to an exp
function with mean $1/\lambda$:

$$p(t) = \lambda e^{-\lambda t}$$

The service times are i.i.d. according to another exp
function with mean $1/\mu$:

$$p(s) = \mu e^{-\mu s}$$

The M/M/1 queue is the simplest to analyze beca
tured completely by the number in the system, X_t, v
process. However, some explanation is necessary to ju
system is sufficient to represent the entire system st
system, the complete system state would need to incl

- the number in the system, X_t;

- the time that the current customer has already

- the time since the last arrival, Z_t.

The next change in X_t will be either an arrival or de
current customer began service, Y_t, is important beca
ture time. Generally, the customer's remaining service
service has already been spent. The time since the l
because it affects the next arrival time. Hence, the tri
state of the system at time t sufficiently to determin
system.

For the M/M/1 queue, the Poisson arrival process is memoryless and exponential service times are memoryless. By the memoryless property of Poisson arrivals, the time to the next arrival will be exponentially distributed regardless of the time since the last arrival. Hence, Y_t is not relevant to the future behavior of X_t. Similarly, the current customer's remaining service time will be exponentially distributed regardless of how much service the customer has already received. It is unnecessary to examine Z_t to predict the next departure. Hence, knowledge of the number in the system X_t is sufficient to predict future changes in the M/M/1 queue.

17.4.1 Steady-State Probabilities for Number in System

The number in the system, X_t, will change by the next arrival or departure. The time to the next arrival will be exponentially distributed with parameter λ, and if $X_t > 0$, the time to the next departure will be exponentially distributed with parameter μ. Hence, $X(t)$ is a birth-death process with birth rates λ and death rates μ represented by the state transition diagram shown in Figure 17.3 and the conditional rate matrix

$$A = \begin{bmatrix} 0 & \lambda & 0 & 0 & \cdots \\ \mu & 0 & \lambda & 0 & \ddots \\ 0 & \mu & 0 & \lambda & \ddots \\ \vdots & \ddots & \ddots & \ddots & \ddots \end{bmatrix} \qquad (17.21)$$

The steady-state probabilities $\{\pi_i\}$ for X_t are found in the usual way from the

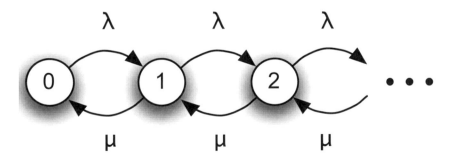

Figure 17.3: State transition diagram for $X(t)$.

balance equations:

$$\lambda \pi_0 = \mu \pi_1$$
$$\lambda \pi_1 = \mu \pi_2$$
$$\lambda \pi_2 = \mu \pi_3$$
$$\vdots$$

with the additional constraint that all probabilities mu
1. Putting all probabilities in terms of π_0, we get the

$$\pi_j = \rho^j \pi_0$$

By substitution, we find the solution is the geometric

$$\pi_j = \rho^j (1 - \rho)$$

for $j = 0, 1, \ldots$.

The stability condition $\rho < 1$ is easy to see. If
solution to the balance equations because the queue i
infinity. The non-obvious condition is $\rho = 1$. Intuitiv
to expect the queue to be stable if the arrival rate
However, the balance equations become $\pi_0 = \pi_1 =$
an infinite number of equal probabilities sum to one.
when $\rho = 1$.

The mean number in the system L can be found
bilities:

$$L = \frac{\rho}{1 - \rho}$$

Further, we can find the mean number waiting in
service) as

$$L_q = \sum_{j=1}^{\infty} (j - 1) \pi_j = \frac{\rho^2}{1 -}$$

The mean number in service is therefore $L - L_q = \rho$
number in service is the same as the utilization.

17.4.2   Little's Formula

In addition to the number in the system, the other pe
is the delay through the system. Fortunately, Little
provides a simple relation between the mean numbe
mean delay through the system W:

$$L = \lambda W$$

Little's formula is quite general and valid for any type of queue. Having found L for the M/M/1 queue, the mean delay through the system is immediately

$$W = \frac{1}{\mu - \lambda} \tag{17.28}$$

17.4.3 Probability Distribution of Delay Through System

The delay through the M/M/1 queue is exponentially distributed. This result follows from an introduction to PASTA (Poisson Arrivals See Time Averages).

Theorem 17.4.1. *According to PASTA, the probability that Poisson arrivals will see N_a customers in the system upon arrival is equal to the fraction of time that the system spends in state N_a.*

In particular for the M/M/1 queue, the probability that Poisson arrivals will see $N_a = j$ in the system upon arrival will be the steady-state probability π_j:

$$Pr(N_a = j) = \pi_j \tag{17.29}$$

PASTA has implications for the delay through the system. If an arrival finds j in the system, then it will have to wait for the j customers before it to finish their services, and then this arrival will receive service. Thus, the arrival will have a total delay through the system consisting of the sum of $j + 1$ service times.

The sum of service times suggests that it will be easier to work with characteristic functions. The characteristic function for an exponential random variable has the form $\frac{\mu}{\mu+s}$. Let N_a be the number in the system found by an arrival, and w be the arrival's delay through the system. Given that the arrival finds $N_a = j$ in the system, the conditional characteristic function for its delay is

$$E(e^{-sw}|N_a = j) = \left(\frac{\mu}{\mu + s}\right)^{j+1} \tag{17.30}$$

We can uncondition this characteristic function because we know the probability of finding $N_a = j$ in the system is π_j:

$$E(e^{-sw}) = \sum_{j=0}^{\infty} \left(\frac{\mu}{\mu + s}\right)^{j+1} \pi_j \tag{17.31}$$

$$= \frac{\mu - \lambda}{\mu - \lambda + s}$$

This characteristic function can be recognized as corresponding to the exponential probability distribution function, implying that w is exponentially distributed with parameter $\mu - \lambda$.

17.5   M/M/1/N Queue

In packet networks, buffers have finite capacities and
probability of packet loss is an important performance
delay. The M/M/1/N queue is the same as the M/M/
capacity of the system is N (including the one in serv
can be represented by the number in the system
process truncated at state N as shown in Figure 17.4

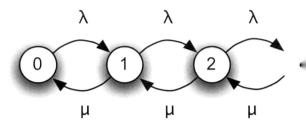

Figure 17.4: State transition diagra

The balance equations are now

$$\lambda\pi_0 = \mu\pi_1$$
$$\lambda\pi_1 = \mu\pi_2$$
$$\vdots$$
$$\lambda\pi_{N-1} = \mu\pi_N$$

with the additional constraint $\pi_0 + \pi_1 + \cdots + \pi_N = 1.$
state probabilities is

$$\pi_j = \frac{(1-\rho)\rho^j}{1-\rho^{N+1}}$$

for $j = 0, 1, \ldots, N$.

   If a finite sized buffer for packets is represented
probability that a packet will be lost is the same as th
will find the system full upon arrival:

$$Pr(loss) = \pi_N = \frac{(1-\rho)\rho}{1-\rho^{N+}}$$

For large N, it can be seen that the probability of loss
N:

$$Pr(loss) \approx (1-\rho)\rho^N$$

   For large N, one might guess that the loss probabi
could be approximated by the tail probability for the

$$Pr(loss) \approx P(X > N \text{ in } M/\Lambda$$

This conjecture can be easily checked by comparing

$$Pr(X > N \ in \ M/M/1) = \sum_{j=N+1}^{\infty} \rho^j(1-\rho) = \rho^{N+1} \qquad (17.37)$$

with the M/M/1/N loss probability in (17.35), which differs only by a factor of $\frac{\rho}{1-\rho}$. Thus, the M/M/1 queue may be an adequate approximation for the M/M/1/N queue for some purposes when N is very large.

17.5.1 Example: Queueing Model of a Packet Switch

The results from the performance analysis of the M/M/1/N queue are useful for analyzing the performance of the simple $K \times K$ output buffered packet switch shown in Figure 17.5. The assumptions are:

- Poisson arrivals at rate λ at each input port;

- packets require exponential service with rate μ at each server;

- packets can be transferred immediately through the fabric from inputs to output buffers;

- output buffers are finite capacity N;

- each incoming packet is addressed to the outputs with equal likelihood, i.e., probability $1/K$ to any output port.

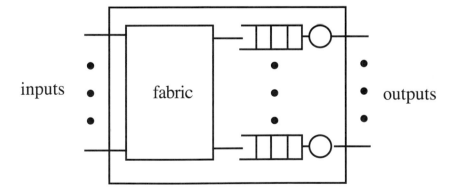

Figure 17.5: Output buffered packet switch.

The output buffers are statistically identical so one can examine any queue. A queue receives a flow of packets from input 1 with rate λ/K, a flow from input 2 with rate λ/K, and so on. The flow from each input port is Poisson because Poisson

arrivals have the property that random splitting of I
the Poisson character of the split processes (refer to
receives the aggregate of K Poisson arrival processe
aggregate of independent Poisson arrivals will be Pc
arrivals to the queue will be Poisson with rate λ.

Since packets receive exponential service, the ou
It was found in (17.35) that the probability of loss
approximately $(1-\rho)\rho^N$. A practical problem in que
the necessary buffer size N for a given loss probabilit
the necessary buffer capacity to meet a given loss pro

$$N = \frac{\log P_{loss} - \log(1-\rho)}{\log \rho}$$

17.6   M/M/N/N Queue

The M/M/N/N queue is an example of a queueing s
The arrival process is Poisson with rate λ. Each of
exponential service time with rate μ. An arrival can
a server is idle, or otherwise departs without service.
is a truncated birth-death process with $(N+1) \times (N$

$$A = \begin{bmatrix} 0 & \lambda & 0 & 0 & \cdots & 0 \\ \mu & 0 & \lambda & 0 & \ddots & 0 \\ 0 & 2\mu & 0 & \lambda & \ddots & 0 \\ \vdots & \ddots & \ddots & \ddots & \ddots & \lambda \\ 0 & 0 & \cdots & 0 & N\mu & 0 \end{bmatrix}$$

The balance equations are

$$\lambda \pi_{n-1} = n\mu \pi_n$$

for $n = 1, \ldots, N$ with the additional constraint $\pi_0 + \cdots$
the steady-state probabilities are

$$\pi_0 = \left(1 + \frac{\lambda}{\mu} + \cdots + \frac{\lambda}{N\mu}\right)$$

$$\pi_n = \frac{\lambda^n}{n!\mu^n}\pi_0 \quad , n = 1, \ldots$$

The probability of losing a customer is the steady-sta
system full, π_N. This is the Erlang B formula applied
to a telephone switch with the capacity to support N

17.7 M/M/1 Queues in Tandem

Burke's theorem is an important result related to the M/M/1 queue. The arrivals to an M/M/1 queue are Poisson with rate λ by definition. Burke's theorem is surprising because it states that the departures in steady-state are Poisson as well.

Theorem 17.7.1. *In steady-state, the departure process from an M/M/1 queue is a Poisson process, and the number in the queue at a time t is independent of the departure process prior to t.*

At first, the result may appear to contradict intuition. By intuition, we know that the queue is either busy or free. When it is busy, packets are being served at a rate μ, so departures are separated by exponential times with mean $1/\mu$. This implies that departures should be Poisson with rate μ. On the other hand, when the queue is free, it is waiting for the next packet arrival. The interarrival time is exponential with mean $1/\lambda$. The new arrival will be serviced immediately. Hence, its departure will be separated from the previous departure by the sum of one interarrival time and a service time. In either case, departures do not seem to be a Poisson process with rate λ.

A way to rationalize Burke's theorem is by a time reversal argument. Recall that the number in an M/M/1 queue is a continuous-time Markov chain X_t. Specifically, it is a birth-death process with birth rates λ and death rates μ. An increment in X_t corresponds to an arrival, and a decrement in X_t corresponds to a departure. As a Markov process, it is time reversible and the time-reversed process, $Y_t = X_{-t}$ is also a Markov process. It can be shown that the time-reversed process Y_t is also a birth-death process with birth rates λ and death rates μ. This implies that arrivals for Y_t are a Poisson process. But arrivals for Y_t are the same as departures for the forward-time process X_t, implying that departures from the M/M/1 queue are Poisson with rate λ.

17.7.1 Product Form

Burke's theorem is important because it allows simple analysis of a system of multiple queues. Consider the two queues in series shown in Figure 17.6. Arrivals to the first queue are Poisson with rate λ, and both servers are exponential and independent of each other. Departures from the first queue are arrivals to the second queue. Burke's theorem states that departures from the first queue are Poisson with rate λ, and therefore arrivals to the second queue are Poisson with rate λ. This implies that both queues are M/M/1. Both queues can be treated as independent M/M/1 queues.

Theorem 17.7.2. *Let $\pi_{i,j}$ denote the joint steady-state probability that there are i in the first queue and j in the second queue. The joint steady-state probabilities can be factored into*

$$\pi_{i,j} = \pi_i^{(1)} \pi_j^{(2)} \tag{17.42}$$

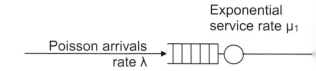

Figure 17.6: Two queues in s

*where $\pi_i^{(1)}$ is the steady-state probabilities for the fir:
the steady-state probabilities for the second M/M/1 q*

The system is said to have a product-form soluti
state probabilities can be shown to equal the product
ities for the individual queues. The product-form sol
in series can be treated as independent M/M/1 queue
Performance analysis is often concerned with th
(both queues). Because the queues are independent,
equals the sum $w = w_1 + w_2$ where w_1 is the exponentia
the first queue with parameter $\mu_1 - \lambda$ and w_2 is the ex
through the second queue with parameter $\mu_2 - \lambda$.

17.7.2   Jackson Networks

Jackson networks have product form which convenier
work of queues of any size by treating each queue as a
An example of an open Jackson network is shown in F
have a number of assumptions:

- Poisson arrivals enter the network at queue i w

- customers require exponential service with rate

- service times at each queue are independent of

- after service, customers are routed randomly w
  going from queue i to queue j (0 denotes leavin

Having product form, the Jackson network in the
a network of independent M/M/1 queues. Since the
given, it remains to find the aggregate arrival rate i
gate rates can be found by balancing the traffic rates
resulting in the system of equations:

$$\gamma_1 = \lambda_1 + r_{31}\gamma_3$$
$$\gamma_2 = \lambda_2 + r_{12}\gamma_1$$
$$\gamma_3 = \lambda_3 + r_{23}\gamma_2$$

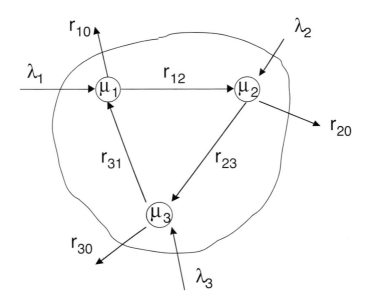

Figure 17.7: An open Jackson network.

These equations can be solved for the rates $\{\gamma_i\}$ in terms of the given routing probabilities r_{ij} and new arrival rates λ_i.

A practical question might be the delay along a given path, for instance, through queues 1 and 2. Because the queues are independent, the delay along this path will be the sum $w = w_1 + w_2$ where w_1 is the exponentially distributed delay through the first queue with parameter $\mu_1 - \gamma_1$ and w_2 is the exponentially distributed delay through the second queue with parameter $\mu_2 - \gamma_2$.

17.8 M/G/1 Queue

As mentioned before, the M/M/1 queue makes two assumptions that may be questionable for some applications. First, interarrival times are exponentially distributed which alleviates the need to account for the times of past arrivals. Second, service times are exponentially distributed. The memoryless property of the exponential probability distribution means it is unnecessary to account for how long the current packet has already spent in service.

In modeling packet networks, the second assumption of exponential service times is particularly questionable. The service time for a packet is the time to transmit it on a link, i.e., the ratio of its packet length to the link rate. Since transmission link rates are constant, the service times are proportional to packet lengths. The assumption of exponential service times implies that packet lengths

are exponentially distributed, which is not true in
packets in the Internet have shown that the packet l
ponential. There is a wide but limited range of packe
lengths occur frequently due to the nature of protocc

The M/G/1 queue keeps the assumption of Poi:
service times to have any given probability distributior
density function $b(y)$. The service times $\{y_1, y_2, \ldots\}$

A practically important special case is the M/D/1
deterministic $1/\mu$, that is, the service time probabili
is a step function at $y = 1/\mu$. The M/D/1 queue is
fixed-length packets such as ATM (asynchronous tr:
data link protocols.

17.8.1   Embedded Markov Chain

Unfortunately, the number in the system X_t is no
the entire state of an M/G/1 queue, and X_t is gene
departure time will depend on the previous departur
are not memoryless (unless $B(y)$ is the exponential
the system needs to keep track of the number in the
the current packet in service has already spent, Y_t. T
is Markovian but cumbersome to handle.

The analysis can be simplified to a single variable b
Markov chain. The idea is to carefully choose specific
process that will make up a discrete-time Markov cha
state probabilities for the embedded Markov chain, t
can be related back to the original process X_t. For
embedded Markov chain works well for the M/G/1 q

Suppose the chosen embedded points are $X_1, X_2,$
parture of each customer. Thus, X_n is the number of
system after the departure of the nth customer.

Theorem 17.8.1. X_n *is a discrete-time Markov*
steady-state probabilities of X_n *reflect the steady-stat*
queue.

In other words, the steady-state probabilities of
X_n, if they can be found, will completely describe the
why this result is true.

To show that X_n is Markovian, let us consider
$X_n = 0$ meaning that the nth packet leaves behind an
packet will arrive to an empty system and begin ser
one service time within the system before leaving. T
number of packets that arrived during its service time

of packets that arrive during the service time of the $(n+1)$th packet. If $X_n = 0$, then we have established that

$$X_{n+1} = A_n \tag{17.44}$$

In the second case, suppose that $X_n > 0$ meaning that the nth packet leaves behind a busy system. One of the X_n waiting packets enters into service immediately and leaves after one service time. Upon the departure of the $(n+1)$th packet, it leaves behind $X_n - 1$ of the previously waiting packets and the number of packets that arrived during its service time, A_n. Thus, if $X_n > 0$, we have

$$X_{n|1} = X_n - 1 + A_n \tag{17.45}$$

The two cases can be combined together in a single equation relating X_{n+1} and X_n:

$$X_{n+1} = max(X_n - 1, 0) + A_n \tag{17.46}$$

This relation shows that the future X_{n+1} depends directly on the current X_n, and implies that given X_n, the past $\{X_{n-1}, X_{n-2}, \ldots\}$ offers no additional information to X_{n+1}. The Markov property is expressed as

$$Pr(X_{n+1}|X_n, X_{n-1}, X_{n-2}, \ldots) = Pr(X_{n+1}|X_n) \tag{17.47}$$

As a discrete-time Markov chain, X_n is defined by its transition probabilities

$$p_{ij} = Pr(X_{n+1} = j | X_n = i) = \begin{cases} \alpha_j & \text{if } i = 0 \\ \alpha_{j-i+1} & \text{if } i > 0, \ j \geq i - 1 \\ 0 & \text{otherwise} \end{cases} \tag{17.48}$$

where $\alpha_i = Pr(A_n = i)$ is the probability mass function for A_n. Or equivalently, X_n is defined by its transition probability matrix

$$P = \begin{bmatrix} \alpha_0 & \alpha_1 & \alpha_2 & \cdots \\ \alpha_0 & \alpha_1 & \alpha_2 & \ddots \\ 0 & \alpha_0 & \alpha_1 & \ddots \\ \vdots & \ddots & \ddots & \ddots \end{bmatrix} \tag{17.49}$$

The next task is to find the probability mass function α_i. We note that given any service time Y, the number of arrivals during that service time will have a Poisson distribution:

$$Pr(A_n = i | Y = y) = \frac{(\lambda y)^i}{i!} e^{-\lambda y} \tag{17.50}$$

The conditional probability can be unconditioned with the service time probability density function as:

$$\alpha_i = Pr(A_n = i) = \int_0^\infty \frac{(\lambda y)^i}{i!} e^{-\lambda y} b(y) dy \tag{17.51}$$

It is not possible to be more specific without a given

Given the service time probability density funct
transition probabilities for X_n and then find its ste
The steady-state probability π_j of the embedded Mai
probability of a departure leaving behind j in the syst
the steady-state probabilities for the embedded Mark
probabilities for the queueing system.

We will make two arguments. First, the steady-st
hind j in the system, π_j, is the same as the steady-s
finding j in the system. Second, the steady-state pro
j in the system is the same as the steady-state proba
in state j. By these two arguments, the steady-state
ded Markov chain are the same as the steady-state ι
system.

To rationalize the first argument, consider the nur
stable queue, the process goes up and down infinitely
when X_t jumps from j to $j + 1$ correspond to an ar
whereas the transitions of $X(t)$ decreasing from $j+1$ t
leaving behind j in the system. Over a very long t
transitions downward from $j + 1$ to j, as a fraction
will be the steady-state probability that departure le
π_j. It will be the same as the number of upward trar
fraction of all downward transitions, which is the stez
arrival finds j in the system.

The second argument is justified by the PASTA pi
the M/G/1 queue. By the PASTA property, the prol
will find j in the system is the same as the steady-stat
will be in state j. We had just established that the ξ
an arrival finds j in the system will be equal to π_j. Th
π_j is the steady-state probability that the system wil

17.8.2   Mean Number in System

The most important result for the M/G/1 queue is tl
mean-value formula which gives the mean number or
queue (they are directly related through Little's form

Theorem 17.8.2. *The P-K mean value formula: the
queue is*

$$E(X) = \rho + \frac{\lambda^2 E(Y^2)}{2(1 - \rho)}$$

where $\rho = \lambda/\mu$ and $E(Y^2)$ is the second moment of ti

Interestingly, the P-K mean-value formula says tl
M/G/1 queue depends only on the utilization factor a:

service time. It does not depend on the entire service time probability distribution function.

The P-K mean-value formula can be rewritten in different forms. For example, substituting the coefficient of variation defined as the variance normalized by the squared mean:

$$C_y^2 = \frac{var(Y)}{[E(Y)]^2} \tag{17.53}$$

the P-K mean-value formula can be re-expressed as

$$E(X) = \rho + \rho^2 \frac{1 + C_y^2}{2(1 - \rho)} \tag{17.54}$$

In this form, the P-K mean-value formula is explicitly dependent only on the utilization and the coefficient of variation of service times, and increases linearly with the variance of service times.

If service times have more variability, the variability will cause longer queues. It reflects a general behavior of queueing systems: more randomness in the system (in either the arrival process or service times) tends to increase queueing.

The mean number in the system $E(X)$ is the sum of the number in service and the mean number waiting in the queue. The mean number in service is ρ, so the mean number waiting in the queue is

$$L_q = \rho^2 \frac{1 + C_y^2}{2(1 - \rho)} \tag{17.55}$$

One of the implications of the P-K mean-value formula is that the queue length is minimized when service times are deterministic and the coefficient of variation is zero. That is, the M/D/1 queue has the shortest queues among all types of M/G/1 queues. For the M/D/1 queue, the mean number in the system will be

$$E(X) = \rho + \frac{\rho^2}{2(1 - \rho)} \tag{17.56}$$

17.8.3   Distribution of Number in System

For the probability distribution of the number in the M/G/1 queue, the important result is the P-K transform formula, which gives the probabilities for the number in the system in terms of a probability generating function. If X is the steady-state number in the system, define the probability generating function for X as

$$G_X(z) = E(z^X) \tag{17.57}$$

which is essentially the z-transform of the probability mass function of X. We will use the service time in terms of its characteristic function

$$B^*(s) = E(e^{-sY}) \tag{17.58}$$

which is the Laplace transform of the service time probability density function.

Theorem 17.8.3. *The P-K transform formula: the p*
for X is

$$G_X(z) = B^*(\lambda - \lambda z)\frac{(1-z)(}{B^*(\lambda - \lambda}$$

If the service time characteristic function (17.5
probability generating function for X directly from
(17.59), but unfortunately, the formula leaves the p
z-transform of $G_X(z)$. For example, let us consider tr
characteristic function for the exponential service tim

$$B^*(s) = \frac{\mu}{s + \mu}$$

The P-K transform formula gives the probability gen

$$G_X(z) = \frac{1 - \rho}{1 - \rho z}$$

The inverse z-transform of $G_X(z)$ is the probability
nately, in the case of the M/M/1 queue, the invers
recognized as the geometric probabilities

$$Pr(X = j) = (1 - \rho)\rho^j$$

which agrees with our earlier results for the M/M/1

17.8.4   Mean Delay Through System

We know from Little's formula that the mean delay
mean number in the system L by $L = \lambda W$. Thus,
mean-value formula (17.52) is for the mean delay:

$$W = \frac{1}{\mu} + \frac{\lambda E(Y^2)}{2(1-\rho)}$$

Again, the mean delay is minimized when service
the M/D/1 queue. For the M/D/1 queue, the mea
system is

$$W = \frac{1}{\mu} + \frac{\lambda}{2\mu(\mu - \lambda)}$$

17.8.5   Distribution of Delay Through

Little's formula states an explicit relationship betwee
mean delay. A more general relationship exists bet
butions for the number in the system and delay thre

relationship, one would expect that an equivalent version of the P-K transform formula (17.59) gives the probability distribution of the delay through the M/G/1 queue.

Theorem 17.8.4. *Another version of the P-K transform formula is*

$$F^*(s) = B^*(s) \frac{s(1-\rho)}{s - \lambda + \lambda B^*(s)} \tag{17.65}$$

where $F^*(s) = E(e^{-sw})$ is the characteristic function for the delay through the system w.

The formula leaves the problem of finding the inverse Laplace transform of $F^*(s)$ which will be the probability density function of w, say $f(w)$.

As an example, consider the characteristic function for exponential service

$$B^*(s) = \frac{\mu}{s + \mu} \tag{17.66}$$

The P-K transform formula (17.65) gives

$$F^*(s) = \frac{\mu - \lambda}{s + \mu - \lambda} \tag{17.67}$$

Fortunately in this case, the inverse Laplace transform can be recognized easily as the exponential probability density function

$$f(w) = (\mu - \lambda)e^{-(\mu - \lambda)w} \tag{17.68}$$

which agrees with our earlier result for delay through the M/M/1 queue.

17.8.6   Example: Mixed Packets

Freedom from a restriction to exponential service times makes the M/G/1 queue generally more useful than the M/M/1 queue for analysis of packet networks. Consider the output buffer of a packet switch. We assume the usual Poisson arrivals, but there are three classes of packets, each requiring a different exponential service time (or in other words, each class has a different packet length distribution). The composition of traffic is listed in the table.

Class	Prob(class)	Mean service time
1	0.7	1
2	0.2	3
3	0.1	10

The first step in the analysis of an M/G/1 queue
time distribution. In this example, the overall service
exponential distributions:

$$B(y) = Pr(Y \le y|class\ 1)Pr(class\ 1) + Pr(Y \le$$
$$+Pr(Y \le y|class\ 3)Pr(class\ 3)$$
$$= 0.7(1 - e^{-y}) + 0.2(1 - e^{-y/3}) + 0.1(1 - e^{-})$$

Differentiating with respect to y, the service time pro

$$b(y) = 0.7e^{-y} + \frac{0.2}{3}e^{-y/3} + \frac{0.1}{10}$$

The P-K mean-value formula needs the first two
The mean service time is

$$E(Y) = 0.7E(Y|class\ 1) + 0.2E(Y|class\ 2)$$
$$= 2.3$$

The second moment of the service time is

$$E(Y^2) = 0.7E(Y^2|class\ 1) + 0.2E(Y^2|class\ 2$$
$$= 25$$

The P-K mean-value formula (17.63) states the mean

$$W = E(Y) + \frac{\lambda E(Y^2)}{2(1 - \rho)} = 2.3 + \frac{}{2(}$$

We can continue with the P-K transform formul
delay through this system. The P-K transform form
time characteristic function $B^*(s)$ which in this exan

$$B^*(s) = E(e^{-sY}|class\ 1)Pr(class\ 1) + E(e^{-sY}|$$
$$+E(e^{-sY}|class\ 3)Pr(class\ 3)$$
$$= 0.7\frac{1}{s + 1} + 0.2\frac{1/3}{s + 1/3} + 0.1\frac{1/10}{s + 1/10}$$
$$= \frac{0.7}{s + 1} + \frac{0.2}{3s + 1} + \frac{0.1}{10s + 1}$$

According to the P-K transform formula, the charact
through the system w is

$$F^*(s) = \left(\frac{0.7}{s + 1} + \frac{0.2}{3s + 1} + \frac{0.1}{10s + 1}\right)\frac{s}{s - \lambda + \lambda\left(\frac{0}{s-}\right)}$$

Unfortunately, the inverse Laplace transform of $F^*(s$
mediately obvious.

17.8.7 Example: Data Frame Retransmissions

For another example, suppose that a data frame is transmitted and stored in the buffer until a positive acknowledgement is received with probability p. The time from the start of frame transmission until an acknowledgement is a constant T. With probability $1 - p$, the acknowledgement will be negative, necessitating a retransmission. Each retransmission is an independent Bernoulli trial with probability p of a successful transmission. After a successful transmission, the data frame can be deleted from the buffer, and the next frame waiting in the buffer will be transmitted.

With the usual assumption of Poisson packet arrivals at rate λ, the system can be modeled by an M/G/1 queue. The "effective" service time is T with probability p, $2T$ with probability $(1 - p)p$, and generally nT with probability $(1 - p)^{n-1}p$. For this geometric service time, the mean is

$$E(Y) = \frac{T}{p} \tag{17.76}$$

and second moment is

$$E(Y^2) = \frac{T^2(2 - p)(1 - p)}{p^2} \tag{17.77}$$

The P-K mean-value formula (17.63) states the mean delay through the system is

$$W = \frac{T}{p} + \frac{\lambda T^2(2 - p)(1 - p)}{2p(p - \lambda T)} \tag{17.78}$$

17.9 Conclusions

This chapter has covered basic queueing theory encompassing the simple M/M/1 queue to the intermediate M/G/1 queue. The M/M/1 queue is particularly useful because its analysis can be extended to product-form networks. The M/G/1 queue is also fairly useful for networking problems because the service time can be general and the P-K formulas allow straightforward analysis. Unfortunately, the assumption of Poisson arrivals limits the applicability of these models.

This chapter has not covered the general G/G/1 queue which can be approached by means of Lindley's integral equation. Although general, the G/G/1 queue is quite difficult for analysis. It is often questionable whether an exact queueing analysis is worth the effort when queueing models are usually a conceptual approximation to real problems. Simpler models such as the stochastic fluid buffer have been successful alternatives to queueing models because they are more tractable, even though they appear at first to be more abstract.

Advanced queueing theory also covers a variety of complications such as service scheduling algorithms (essentially priorities), service preemption, multiple queues,

and buffer management or selective discarding algorit
ing theory and its applications to networking proble
possible variations of the basic queueing models.

17.10   Exercises

Exercise 17.10.1. Consider a discrete-time Mark
states, 0 and 1. The initial state is $X_0 = 0$. Its t
is

$$P = \begin{bmatrix} 0.3 & 0.7 \\ 0.6 & 0.4 \end{bmatrix}$$

(a) Find the steady-state probabilities. (b) What are
$n = 3$ and $n = 6$? Do they seem to be approaching t

Exercise 17.10.2. Sometimes a metric called "pow
throughput to mean delay. (a) Find the power for a st
throughput is the same as the arrival rate λ). (b) Fir
power.

Exercise 17.10.3. (a) For the M/M/1 queue, use t
to find the mean number waiting in the queue (exclu
L_q. (b) Find the mean waiting time in the queue I
the mean delay through the system and mean service
formula holds for the queue excluding the server, i.e.

Exercise 17.10.4. For an M/M/1 queue, find the all
that 99% of customers do not experience a delay thro
given T.

Exercise 17.10.5. Consider an M/M/1/2 queueing
tomer accepted into the system brings in a profit of $
results in a loss of $1. Find the profit for the system.

Exercise 17.10.6. Consider an output buffered pac
buffers are modeled as M/M/1/N queues. (a) With λ
size N is required to meet a loss probability of 10^-
both increased by a factor of 10 (representing a scal
would the answer to part (a) change?

Exercise 17.10.7. Consider two M/M/1 queues in
server has rate μ_1 and the second has rate μ_2. (a) Un
service rates is fixed to a constant, $\mu_1 + \mu_2 = \mu$, find
system (both queues). (b) Find μ_1 and μ_2 under the
fixed, to minimize the mean delay through the systen

Exercise 17.10.8. (a) Find the mean number in the M/D/1 queue as a function of ρ. (b) Find the ratio of the mean number in the M/D/1 queue to the mean number in the M/M/1 queue. Is the ratio less than one?

Exercise 17.10.9. Use the P-K transform formula for the M/D/1 queue. (a) Find the characteristic function for the delay through the system. (b) Find the characteristic function for the waiting time in the queue.

Exercise 17.10.10. Define X_n is the number of customers left behind in an M/D/1 system after the departure of the nth customer. (a) Describe the transition probabilities for this discrete-time Markov chain. (b) Set up the balance equations for the steady-state probabilities.

Exercise 17.10.11. Consider an M/G/1 queue with two classes of packets. Class 1 packets arrive as a Poisson process with rate λ_1 and require a constant service time μ_1. Class 2 packets arrive as a Poisson process with rate λ_2 and require a constant service time μ_2. Find the mean delay through the system.

References

[1] Bolch, G., Greiner, S., de Meer, H., and Trivedi, K. S., *Queueing Networks and Markov Chains: Modeling and Performance Evaluation with Computer Science Applications*, Wiley Interscience, Hoboken, NJ, 2006.

[2] Bose, S., *An Introduction to Queueing Theory*, Springer, New York, 2001.

[3] Burke, P. J., "The output of a queuing system," *Operations Research*, vol. 4, pp. 699–704, December 1956.

[4] Daigle, J., *Queueing Theory with Applications to Packet Telecommunication*, Springer, New York, 2010.

[5] Gelenbe, E., and Pujolle, G., *Introduction to Queueing Networks*, Wiley, Chichester, UK, 1998.

[6] Gross, D., Shortle, J., Thompson, J., and Harris, C., *Fundamentals of Queueing Theory*, 4th ed., Wiley Interscience, Hoboken, NJ, 2008.

[7] Jackson, J. R., "Jobshop-like queueing systems," *Management Science*, vol. 10, pp. 131-142, 1963.

[8] Kleinrock, L., *Queueing Systems Volume 1: Theory*, Wiley Interscience, New York, 1975.

[9] Little, J. D. C., "A proof of the queueing formula $L = \lambda W$," Operations Research, vol. 9, pp. 383-387, 1961.

626

[10] Medhi, J., *Stochastic Models in Queueing Theo* San Diego, 2002.

[11] Robertazzi, T., *Computer Networks and Systems* 2000.

[12] Wolff, R., *Stochastic Modeling and the Theory c* glewood Cliffs, NJ, 1989.

Chapter 18 Network Optimization Techniques

Michał Pióro[‡]

[‡]Lund University, Sweden, and Warsaw University of Technology, Poland

18.1 Introduction

This chapter is devoted to modeling and optimization techniques applicable to communication network design and planning. The main emphasis is put on the models dealing with optimization of the capacity of network resources and traffic routing that lead to tractable optimization problems. It is a common belief that the right means for such modeling are multicommodity flow networks (MFN). MFN form a field of operations research on its own, and are closely connected to integer programming. In consequence, the presented chapter aims at giving a systematic survey of basic MFN models and integer programming methods applicable to communication network design.

We start our presentation in Section 18.2 with basic ways of formulating MFN optimization problems related to the most important cases of link capacity and routing modeling. In particular, we introduce link-path and node-link formulations, different routing requirements, and several types of link dimensioning functions. In Section 18.3 we discuss notions of integer programming and present its fundamental techniques crucial for MFN optimization. This section is connected to Chapters 13 and 14 and presents two decomposition methods, namely path generation and Benders decomposition, and such approaches as cutting plane, branch-and-bound and its variants, and heuristics. Finally, in Section 18.4, we present a selected set of advanced MFN optimization problems related to multistate network design and hence to design of networks robust to equipment failures and traffic variations, and discuss how integer programming techniques are applicable to this important class of problems.

Intentionally, the chapter is not concentrated on technological background of communication network modeling. For this purpose the reader can for example use the handbooks [1–3].

18.2   Basic Multicommodity Flo
mization Models

In this section, after introducing notions and notatior
flow networks (MFN), we discuss basic kinds of MF
mulations and present representative yet fairly simpl
problems (NDP). These examples will be extended
chapter, covering a spectrum of NDPs relevant for cc

18.2.1   Notions and Notations

Network. A multicommodity flow network (MF
$\mathcal{N}(\mathcal{G}, \mathcal{D})$ where $\mathcal{G} = \mathcal{G}(\mathcal{V}, \mathcal{E})$ is the network graph
demands between the pairs of nodes of the graph. Gr
of nodes $\mathcal{V}$ and the set of links $\mathcal{E}$. For ease of exposi
the graph does not contain loops nor parallel links, t
the set of all two-element subsets of the set $\mathcal{V}$ (the
$\mathcal{E} \subseteq \mathcal{V}^2 \setminus \{(v, v) : v \in \mathcal{V}\}$ (the case of directed graph)
of link $e \in \mathcal{E}$ are denoted by $a(e)$ and $b(e)$. In the u
then $a(e) = v$ and $b(e) = w$ or $a(e) = w$ and $b(e) =$
In the directed case, if $e = (v, w)$ then $a(e) = v$ and
graph, $\delta(v) = \{e \in \mathcal{E} : v \in e\}$ is the set of links incic
rected graph, the sets $\delta^+(v) = \{e \in \mathcal{E} : a(e) = v\}$ an
represent, respectively, the sets of all links outgoing ε
The cost of realizing one unit of capacity (or one ur
context) on link $e \in \mathcal{E}$ is denoted by $\xi_e \geq 0$ and ξ
unit cost vector. Capacity of link $e \in \mathcal{E}$ is denoted ei
optimization variable) or by $c_e \geq 0$ (when it is given ar
capacity vectors are $y = (y_e : e \in \mathcal{E}) \in \mathbb{R}_+^{|\mathcal{E}|}$ and $c = ($
the quantities ξ, y, c are functions, as for example ξ
write ξ_e, y_e, c_e instead of $\xi(e), y(e), c(e)$, and identify
because the links (and the nodes for that matter) c
convention will be applied to other analogous quantit
    Consider the network graph depicted in Figure 18
to $\mathcal{V} = \{v_1, v_2, v_3, v_4\}$. In the undirected case, the
$\{e_1, e_2, \ldots, e_5\}$, and, for example, $e_1 = \{v_1, v_2\}$, $a(e_1)$
$\{e_1, e_2\}$. The entries of the vectors $\xi = (\xi_1, \xi_2, \ldots, \xi$
$(c_1, c_2, \ldots, c_5)$ correspond to links $e_1, e_2, \ldots, e_5$, respe
we have $\mathcal{E} = \{e_1, e_2, \ldots, e_6\}$ and, for example $e_1 = (u$
and $\delta^-(v_1) = \emptyset, \delta^+(v_1) = \{e_1, e_2\}$.

Demands.   The demands in set $\mathcal{D}$ can be either
notational convenience, we exclude parallel demands b

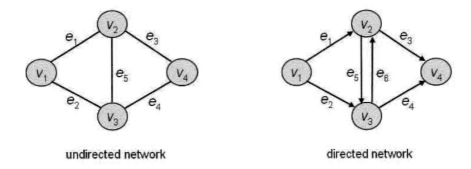

undirected network   directed network

Figure 18.1: A simple network example.

demands are undirected then $\mathcal{D} \subseteq \mathcal{V}^{|2|}$ and when demands are directed – $\mathcal{D} \subseteq \mathcal{V}^2 \setminus \{(v, v) : v \in \mathcal{V}\}$. The endpoints of a demand $d \in \mathcal{D}$ are denoted by $s(d)$ and $t(d)$ and are defined analogously to the endpoints of a link. By definition, they are different from each other. In the directed case, $s(d)$ is called the source of demand d while $t(d)$ – the sink of demand d. The demand value (volume) of demand $d \in \mathcal{D}$ is given by $h_d \geq 0$ and expresses the traffic generated by d to be realized by means of flows between its endpoints. The (traffic) demand vector is given by $h = (h_d : d \in \mathcal{D}) \in \mathbb{R}_+^{|\mathcal{D}|}$. In the sequel we will assume that both the network graph and the demand set are either undirected or directed.

For the network in Figure 18.1, we may consider all possible demands or a subset of them, for example $\mathcal{D} = \{d_1, d_2\}$ where $d_1 = \{v_1, v_4\}, s(d_1) = v_1, t(d_1) = v_4, d_2 = \{v_2, v_3\}, s(d_2) = v_2, t(d_2) = v_3$ in the undirected case, and $\mathcal{D} = \{d_1, d_2, d_3\}$ where $d_1 = (v_1, v_4), s(d_1) = v_1, t(d_1) = v_4, d_2 = (v_2, v_3), s(d_2) = v_2, t(d_2) = v_3, d_3 = (v_3, v_2), s(d_3) = v_3, t(d_3) = v_2$ in the directed case.

Network states. The network can be in different states $s \in \mathcal{S}$, where s is a state and $\mathcal{S}$ is the set of all states, that differ by availability of the links and/or by demand vectors. There are two main cases of the considered state sets, corresponding to survivable network design and to multihour design.

In the first case, the set $\mathcal{S}$ of network states is called the failure scenario (note that sometimes the notion of scenario is used for a single state, not for the set of all states). Every state $s \in \mathcal{S}$ identifies the subset of the links that are failed, i.e., $s \subseteq \mathcal{E}$. Observe that we assume that in each state a link is either fully available or is totally failed and that the normal, failure-less state (sometimes called the nominal state) corresponds to the empty set $s = \emptyset$. For failure scenarios we will also use the notation $\mathcal{S}_e = \{s \in \mathcal{S} : e \notin s\}$ (the set of all states with link e available, including the normal state $s = \emptyset$) and $\bar{\mathcal{S}}_e = \mathcal{S} \setminus \mathcal{S}_e$ (the set of all states with link e failed). Symmetrically, we will write $\mathcal{E}_s$ to denote the set of links surviving in state s, i.e., $\mathcal{E}_s = \{e \in \mathcal{E} : e \notin s\}$. The failure scenarios are used

to model the so-called survivable network design pro
Section 18.4.2). The failure states $\mathcal{S} \setminus \{\emptyset\}$ in a failu
called failure situations. In general, the failure situa
can contain several failed links. Such a situation is ca
set of affected links is referred to as the shared risk l
the most commonly considered failure scenarios ass
i.e., $\mathcal{S} \subseteq \{\{e\} : e \in \mathcal{E}\} \cup \{\emptyset\}$. In general, we assume
$(h^s = (h_d^s : d \in \mathcal{D}))$ for each $s \in \mathcal{S} \setminus \{\emptyset\}$ that has
situation s. Typically, $h^s \leq h^0 = h$ component-wise.
states, then we say that we require 100% demand pr

For the undirected network in Figure 18.1 we
the single-link failure scenario $\mathcal{S} = \{s_0, s_1, s_2, \ldots$
$s_i = \{e_i\}, i = 1, 2, \ldots, 5$. Another example of
$\mathcal{S} = \{s_0, s_1, s_2, \ldots, s_5\} \cup \{s_6, s_7\}$ where $s_0 = \{\emptyset\}$, s
$s_6 = \{e_1, e_4\}, s_7 = \{e_2, e_3\}$. Then, s_6 and s_7 represe
identifying the respective SRLGs.

For the directed network, the full single-link fail
$\{s_0, s_1, s_2, \ldots, s_6\}$ where $s_0 = \{\emptyset\}$ and $s_i = \{e_i\}, i =$

In our presentation, node failures will not be expl
be modeled through link failures. A way to do it,
is to substitute each node $v \in \mathcal{V}$ by two copies v'
$e_v = (v', v'')$ between them. All the links $e \in \delta^-$
incoming to node v' and all the links $e \in \delta^+(v)$ outgoi
from v''. In this way, a failure of node v can be model
e_v (see [2]).

The second case of network states is called the
states are denoted by τ and the state set by $\mathcal{T}$.
considered since in all states $\tau \in \mathcal{T}$ all links are availa
by their demand vectors h^τ, $\tau \in \mathcal{T}$. In a demand sc
the normal state – the states $\tau \in \mathcal{T}$ correspond to
different offered traffic. The demand scenarios are
multihour and multiperiod design problems (see Subs

Throughout this chapter, if explicitly not stated ot
the number of states in set $\mathcal{S}$ is polynomial with respe

Routing paths. We assume that each demand d
set $\mathcal{P}_d$ of admissible paths that can be used for rea
demand (the set of all admissible paths is denoted b
in the undirected case we consider undirected paths
in the directed case – directed paths from $s(d)$ to $t(d$
otherwise, $\mathcal{P}_d$ is a subset of all the elementary paths
the admissible paths do not traverse any node more
each path $p \in \mathcal{P}$ can simply be identified with the se
that $p \subseteq \mathcal{E}$. Note that since we have assumed at most

and elementary paths, the sets $\mathcal{P}_d$, $d \in \mathcal{D}$ are mutually disjoint. Further, $\mathcal{P}_e \subseteq \mathcal{P}$ is the set of all admissible paths containing link $e \in \mathcal{E}$ ($\mathcal{P}_e = \{p \in \mathcal{P} : p \ni e\}$), and $\mathcal{P}_{ed} \subseteq \mathcal{P}_d$ – the set of all admissible paths of demand $d \in \mathcal{D}$ containing link e ($\mathcal{P}_{ed} = \{p \in \mathcal{P}_d : p \ni e\}$).

The set $\mathcal{P}_d^s$ of admissible paths for demand $d \in \mathcal{D}$ available in state $s \in \mathcal{S}$ is equal to

$$\mathcal{P}_d^s = \{p \in \mathcal{P}_d : p \cap s = \emptyset\}.$$

Similarly, $\bar{\mathcal{P}}_d^s = \mathcal{P}_d \setminus \mathcal{P}_d^s$ denotes the complementary set of all admissible paths failing in state $s \in \mathcal{S}$. $\mathcal{P}_e^s = \mathcal{P}_e \cap (\bigcup_{d \in \mathcal{D}} \mathcal{P}_d^s)$ denotes the set of all paths containing link $e \in \mathcal{E}$ that are available in state $s \in \mathcal{S}$. Observe that by the definition of the normal state, $\mathcal{P}_d^\emptyset = \mathcal{P}_d$ and $\mathcal{P}_e^\emptyset = \mathcal{P}_e$. Also, $\mathcal{P}_{ed}^s = \{p \in \mathcal{P}_d^s : p \ni e\}$ is the set of all admissible paths of d that contain link e and are available in state s.

Notation

$$\mathcal{S}_p = \{s \in \mathcal{S} : p \cap s = \emptyset\}$$

and $\bar{\mathcal{S}}_p = \mathcal{S} \setminus \mathcal{S}_p$ refers to the sets of states $s \in \mathcal{S}$ in which path $p \in \mathcal{P}$ is available or unavailable, respectively. For the path protection/restoration mechanisms considered in Section 18.4, notation $\mathcal{Q}_p \subseteq \mathcal{P}$ refers to all admissible backup paths for protecting a particular path $p \in \mathcal{P}$ (in this context, path p is called a primary or basic path). These are the paths with the same end nodes as path p that never fail together with p (in other words, p and q are failure-disjoint). Hence, if $q \in \mathcal{Q}_p$ then for all $s \in \mathcal{S}$,

$$p \cap s \neq \emptyset \quad \Rightarrow \quad q \cap s = \emptyset.$$

Also, $\mathcal{Q}_{ep} := \{q \in \mathcal{Q}_p : q \ni e\}$ denotes the set of all paths protecting path p that contain a particular link $e \in \mathcal{E}$. The set of all admissible (failure-disjoint) primary-backup path pairs $r = (p, q)$ for demand $d \in \mathcal{D}$ will be denoted by $\mathcal{R}_d = \{r = (p, q) : p \in \mathcal{P}_d, q \in \mathcal{Q}_p\}$ ($\mathcal{R} = \bigcup_{d \in \mathcal{D}} \mathcal{R}_d$).

Finally, for each link $e \in \mathcal{E}$, the set of all pairs $r \in \mathcal{R}$ such that $e \in p$ will be denoted by $\mathcal{R}_e^1$, and the set of all pairs $r \in \mathcal{R}$ such that $e \in q$ will be denoted by $\mathcal{R}_e^2$. Similarly, the sets $\mathcal{R}_{ed}^1 = \{r = (p, q) \in \mathcal{R}_d : p \ni e\}$ and $\mathcal{R}_{ed}^2 = \{r = (p, q) \in \mathcal{R}_d : q \ni e\}$.

To illustrate the notions related to the routing paths, let us consider the undirected network from Figure 18.1 and the demand set $\mathcal{D} = \{d_1 = \{v_1, v_4\}, d_2 = \{v_2, v_3\}\}$. For the corresponding routing path lists we can assume the full sets of elementary paths, i.e., $\mathcal{P}_{d_1} = \mathcal{P}_1 = \{p_{11}, p_{12}, p_{13}, p_{14}\}$, where $p_{11} = \{e_1, e_3\}, p_{12} = \{e_2, e_4\}, p_{13} = \{e_1, e_5, e_4\}, p_{14} = \{e_2, e_5, e_3\}$, and $\mathcal{P}_{d_2} = \mathcal{P}_2 = \{p_{21}, p_{22}, p_{23}\}$, where $p_{21} = \{e_5\}, p_{22} = \{e_1, e_2\}, p_{23} = \{e_3, e_4\}$. Also, $\mathcal{P} = \mathcal{P}_1 \cup \mathcal{P}_2 = \{p_{11}, p_{12}, p_{13}, p_{14}, p_{21}, p_{22}, p_{23}\}$. For the full single-link failure scenario $\mathcal{S} = \{s_0, s_1, s_2, \ldots, s_5\}$, we have for example $\mathcal{P}_{d_1}^{s_0} = P_1^0 = \mathcal{P}_1, \mathcal{P}_{d_1}^{s_1} = P_1^1 = \{p_{12}, p_{14}\}$. Note also that $\mathcal{R}_1 = \{(p_{11}, p_{12}), (p_{12}, p_{11})\}$.

For the directed case the paths are directed. For $\mathcal{D} = \{d_1 = (v_1, v_4), d_2 = (v_2, v_3), d_3 = (v_3, v_2)\}$ the routing lists are as follows: $\mathcal{P}_1 = \{p_{11}, p_{12}, p_{13}, p_{14}\}$, where $p_{11} = \{e_1, e_3\}, p_{12} = \{e_2, e_4\}, p_{13} = \{e_1, e_5, e_4\}, p_{14} = \{e_2, e_6, e_3\}$, $\mathcal{P}_2 =$

$\{p_{21}\}$, where $p_{21} = \{e_5\}$, $\mathcal{P}_3 = \{p_{31}\}$, where $p_{31} = \{e$
of the rest of the notions to the reader.

Flows. As we will see in the next subsection, there
lations used to model the MFN optimization proble
link-path formulation (or path-flow formulation), use
paths (or path-pairs). In this case, the volume of a
to be realized by means of path-flows x_{dp} assigned t
More advanced optimization problems (taking into ac
mand scenarios) can involve path-flow variables relate
$(d \in \mathcal{D}, p \in \mathcal{P}_d^s, s \in \mathcal{S})$ or x_{dp}^τ $(d \in \mathcal{D}, p \in \mathcal{P}_d, \tau \in \mathcal{T})$
tion/restoration mechanisms, path-pair flow variable
used.

The second basic formulation, called node-link f
mulation), involves link-flow variables x_{ed} determinir
assigned to link $e \in \mathcal{E}$. In a modification of the nod
gregated node-link formulation, link flow variable x_e
the demands with the end node $t \in \mathcal{V}$ assigned to lin

Finally, if flow variables are binary rather than
u_{dp}, u_{dr}, u_{ed} instead of, respectively, x_{dp}, x_{dr}, x_{ed}. A
link-flows will be referred to as flow pattern.

Note that when links, demands, paths, and fa
example when $\mathcal{E} = \{e_1, e_2, \ldots, e_E\}$, $\mathcal{D} = \{d_1,$
$\{p_{k1}, p_{k2}, \ldots, p_{kP_k}\}$, $\mathcal{S} = \{s_0, s_1, \ldots, s_S\}$, then the
labels, i.e., $x_{dp}^s, d = 1, 2, \ldots, D$, $p = 1, 2, \ldots, P_d$, s
$1, 2, \ldots, D$, $e = 1, 2, \ldots, E$. For other flow cases the

18.2.2 Link-Path vs. Node-Link Formu location Problems

Flow Allocation – Link-Path Formulation

We start our presentation of NDPs with a simple optim
formulation (L-P formulation in short). Consider a
given link capacity reservations $c = (c_e, e \in \mathcal{E})$, flow
demand volumes $h = (h_d, d \in \mathcal{D})$, and predefined a
$\mathcal{D}$. The following optimization problem with non-ne
variables $x = (x_{dp}, d \in \mathcal{D}, p \in \mathcal{P}_d)$ is referred to as FA

$$\textbf{FAP(L-P):} \quad \min \ F(x) = \sum_{e \in \mathcal{E}} \xi_e \left(\sum_{d \in \mathcal{D}} \sum_{p \in \mathcal{P}_{ed}} x_{dp} \right) \tag{18.1a}$$

$$\sum_{p \in \mathcal{P}_d} x_{dp} = h_d \qquad\qquad d \in \mathcal{D} \tag{18.1b}$$

$$\sum_{d \in \mathcal{D}} \sum_{p \in \mathcal{P}_{ed}} x_{dp} \leq c_e \qquad\qquad e \in \mathcal{E} \tag{18.1c}$$

$$x_{dp} \in \mathbb{R}_+ \qquad\qquad d \in \mathcal{D}, \ p \in \mathcal{P}_d. \tag{18.1d}$$

Recall that $\mathcal{P}_{ed}$ denotes the set of all admissible paths of demand $d \in \mathcal{D}$ that contain link $e \in \mathcal{E}$. The demand constraint (18.1b) makes sure that the demand volumes are realized, and the capacity constraint (18.1c) – that for each link its (given) capacity is not exceeded by its load (specified by the left-hand side of (18.1c)). The objective (18.1a) of FAP is to minimize the total cost of the link loads. Certainly, (18.1) is a linear programming (LP) problem and can be solved as such using for example the simplex method (see Chapter 14). Observe that the above formulation is valid for undirected links (demands) as well as for directed links (demands) as the issue of direction is hidden in the definition of the paths.

L-P formulation (18.1) assumes predefined lists of admissible paths. Consequently, when all elementary paths are to be considered, then the full lists of paths $\mathcal{P}_d$, $d \in \mathcal{D}$ have to be used in the problem formulation. Because in general the number of elementary paths grows exponentially with the graph size, FAP requires an exponential number of path-flow variables x_{dp}, $d \in \mathcal{P}_d, p \in \mathcal{P}_d$. For this reason, L-P formulation (18.1) is called noncompact. A general way of dealing with such excessive (exponential) number of variables in noncompact LP formulations is column generation (called path generation in the context of MFN). Path generation is discussed in Subsection 18.3.1.

We note that objectives other than (18.1a) can be thought of – we will see examples of such other objectives further in this chapter. Finally, observe that in FAP, feasible flows are in general bifurcated as can be easily seen for a two-node network with one demand h and two parallel links with capacities $c_1 = c_2 = \frac{h}{2}$.

Flow Allocation – Node-Link Formulation

In fact, there is a way to consider all elementary paths for admissible paths in FAP using a compact LP formulation, i.e., a formulation using a polynomial number of variables. This is the so-called node-link formulation (N-L formulation in short). Formally, N-L formulations require directed graphs and directed demands, and use link-flow variables $x = (x_{ed}, e \in \mathcal{E}, d \in \mathcal{D})$. For FAP, an N-L formulation is as follows:

$$\textbf{FAP(N-L):} \quad \min F(x) = \sum_{e \in \mathcal{E}} \xi_e \left(\sum_{d \in \mathcal{D}} x_{ed} \right)$$

$$\sum_{e \in \delta^+(v)} x_{ed} - \sum_{e \in \delta^-(v)} x_{ed} = \begin{cases} 0, \\ h_d, \\ -h_d, \end{cases}$$

$$\sum_{d \in \mathcal{D}} x_{ed} \leq c_e \qquad e \in \mathcal{E}$$

$$x_{ed} \in \mathbb{R}_+ \qquad e \in \mathcal{E},\ d \in \mathcal{D}.$$

Observe that for any fixed $d \in \mathcal{D}$, one of the equat
on the rest and thus redundant – therefore we can o
formulation path-flows do not appear. Instead, link
demand on every link) are used. For every fixed dem
(demand) constraints (18.2b) make sure that the ent
from source $s(d)$ to destination $t(d)$. Note that the
is a bit simpler than (18.1c) in the L-P formulatio
solution to (18.2) does not explicitly define the pat
somehow retrieved from the link flows, for example u
algorithm [4]. It is important that due to objective (18
would use only elementary paths although nonoptima
L) can lead to path-flows assigned to nonelementary
path flows can appear because of loops like loop v_2
network in Figure 18.1. For the retrieved path-flows
each $e \in \mathcal{E}$ and each $d \in \mathcal{D}$ (where $\mathcal{P}_d$ denotes the s
demand d). Certainly, there may be more than one
given link-flow pattern (Exercise 18.6.1), but not vice

The N-L formulation of FAP consists of $E \cdot D$ va
constraints (not counting non-negativity constraints).
lation is compact as the number of variables (and con
with the size of the network.

N-L formulation can be applied to undirected gra
every undirected link $e = \{v, w\}$ is associated with
and $e'' = (w, u)$. The set of arcs is denoted by $\mathcal{A}$, whe
resulting graph $(\mathcal{V}, \mathcal{A})$ is called bi-directed. Certainly,
contain outgoing and incoming arcs $a \in \mathcal{A}$, respective
demand $d = \{s, t\} \in \mathcal{D}$ is made directed in one of
say $d = (s, t)$ (the direction is chosen arbitrarily).

corresponding to the original undirected network becomes as follows:

$$\textbf{FAP(N-L):} \quad \min \ F(x) = \sum_{e \in \mathcal{E}} \xi_e \left(\sum_{d \in \mathcal{D}} x_{ed} \right) \tag{18.3a}$$

$$\sum_{a \in \delta^+(v)} x'_{ad} - \sum_{a \in \delta^-(v)} x'_{ad} = \begin{cases} 0, & v \in \mathcal{V} \setminus \{s(d), t(d)\} \\ h_d, & v = s(d) \end{cases}$$

$$d \in \mathcal{D} \quad (18.3b)$$

$$x_{ed} = x'_{e'd} + x'_{e''d} \qquad e \in \mathcal{E} \tag{18.3c}$$

$$\sum_{d \in \mathcal{D}} x_{ed} \leq c_e \qquad e \in \mathcal{E} \tag{18.3d}$$

$$x'_{ad} \in \mathbb{R}_+ \qquad a \in \mathcal{A}, \ d \in \mathcal{D}. \tag{18.3e}$$

Note that typically, only at most one of the opposite arc flows $x'_{e'd}, x'_{e''d}$ on link e will be non-zero (think why).

Flow Allocation – Aggregated N-L Formulation

Observe that the number of variables in the compact N-L formulation discussed above is equal to $E \cdot D$, and although polynomial, can be very large for large networks. Assuming $V = 100$, $E = 300$ and $D = V(V-1)$, the number of variables in FAP(N-L) is of the order of 10^6. This number can be substantially reduced using the so-called aggregated node-link formulation where instead of link flows x_{ed} corresponding to the demands we consider link flows x_{et} corresponding to the destination nodes.

As FAP(N-L), the following aggregated node-link formulation (AG) assumes a directed graph and directed demands.

$$\textbf{FAP(AG):} \quad \min \ F(x) = \sum_{e \in \mathcal{E}} \xi_e \left(\sum_{t \in \mathcal{B} \setminus \{a(e)\}} x_{et} \right) \tag{18.4a}$$

$$\sum_{e \in \delta^+(v)} x_{et} = \sum_{e \in \delta^-(v)} x_{et} + \sum_{d \in \mathcal{D}(v,t)} h_d \quad v \in \mathcal{V}, t \in \mathcal{B} \setminus \{v\} \quad (18.4b)$$

$$\sum_{t \in \mathcal{B} \setminus \{a(e)\}} x_{et} \leq c_e \qquad e \in \mathcal{E} \tag{18.4c}$$

$$x_{et} \in \mathbb{R}_+ \qquad e \in \mathcal{E}, \ t \in \mathcal{B}. \tag{18.4d}$$

Above, $\mathcal{B}$ is the set of all nodes that are the end nodes for at least one demand ($\mathcal{B} = \{t \in \mathcal{V} : \exists \, d \in \mathcal{D}, \ t = t(d)\}$) and $\mathcal{D}(v,t)$ is the set of the demands with $s(d) = v$ and $t(d) = t$. The specific constraint for FAP(AG), i.e., (18.4b), assures that all flows destined to node t outgoing from node v are equal to all flows destined to t incoming to v plus all flows to t that originate at v. The number of variables

in the aggregated formulation FAP(AG) is equal to ⊙
our example considered above. It is, however, impor
the number of variables (and constraints) in the AG
reduced with respect to the ordinary N-L formulati⊙
have much more dense coefficient matrices than the
(for example using a simplex solver) the superiority
time required by each of the two formulations can va
problem in hand.

As for N-L, an AG formulation can as well be ⊔
after an analogous transformation.

Routing Restrictions

A simple routing restriction is the so-called hop-lim⊙
admissible paths have a limit on the number of transit
nodes can be traversed by any $p \in \mathcal{P}$ (i.e., $|p| \leq n +$
can easily be included to the L-P formulation (also w⊙
but there is no easy way to incorporate the hop-limit ⊔
general way (N-L formulations for a small number of

Another important kind of routing restriction is ⊙
too much of a demand volume on any path or on an⊙
link diversity). A link-path formulation of FAP with
adding constraint

$$x_{dp} \leq \frac{h_d}{n_d} \qquad d \in \mathcal{D}, \ p \in \mathcal{P}$$

to formulation (18.1), where n_d is a given demand-de
If for example $n_d = 3$ then the demand volume will be
Adding constraint

$$x_{ed} \leq \frac{h_d}{n_d} \qquad d \in \mathcal{D}, \ e \in \mathcal{E}$$

to formulation (18.2) results in an analogous node-li⊙
versity. Note that both extensions remain LP prob.
quirement can be easily expressed in the link-path no⊙
requirement cannot be expressed in the node-link not

So far we have assumed bifurcated routing, i.e., a⊙
volume between its admissible paths and presented the
problems. In many applications, however, bifurcatio⊙
bifurcated routing (also called single-path routing or ⊔
i.e., allocation of the whole demand volume h_d to on⊙
The single-path requirement calls for binary variable

mixed-integer programming (MIP) problem in the L-P formulation.

FAP/SP(L-P): $\quad \min F(u) = \sum_{e \in \mathcal{E}} \xi_e \left(\sum_{d \in \mathcal{D}} \sum_{p \in \mathcal{P}_d} h_d u_{dp} \right)$ $\qquad$ (18.7a)

$$\sum_{p \in \mathcal{P}_d} u_{dp} = 1 \qquad\qquad d \in \mathcal{D} \qquad\qquad (18.7b)$$

$$\sum_{d \in \mathcal{D}} \sum_{p \in \mathcal{P}_{ed}} h_d u_{dp} \leq c_e \qquad\qquad e \in \mathcal{E} \qquad\qquad (18.7c)$$

$$u_{dp} \in \{0, 1\} \qquad\qquad d \in \mathcal{D}, \ p \in \mathcal{P}_d. \qquad (18.7d)$$

In fact, the above problem is $\mathcal{NP}$-hard, as most of other network design problems involving unsplittable flows, since the bin-packing problem and other well-known combinatorial problems can be easily reduced to (18.7), see [6–9]. FAP/SP can also be written in N-L notation:

FAP/SP(N-L): $\quad \min F(u) = \sum_{e \in \mathcal{E}} \xi_e \left(\sum_{d \in \mathcal{D}} h_d u_{ed} \right)$ $\qquad$ (18.8a)

$$\sum_{e \in \delta^+(v)} u_{ed} - \sum_{e \in \delta^-(v)} u_{ed} = \begin{cases} 0, & v \in \mathcal{V} \setminus \{s(d), t(d)\} \\ 1, & v = s(d) \end{cases}$$
$$d \in \mathcal{D} \qquad (18.8b)$$

$$\sum_{d \in \mathcal{D}} h_d u_{ed} \leq c_e \qquad e \in \mathcal{E} \qquad\qquad (18.8c)$$

$$u_{ed} \in \{0, 1\} \qquad e \in \mathcal{E}, \ d \in \mathcal{D}. \qquad (18.8d)$$

Observe that the hop-limit requirement can be added to the above binary formulation by adding constraint $\sum_{e \in \mathcal{E}} u_{ed} \leq n$, $d \in \mathcal{D}$, although this was not possible for the LP version (18.2) of FAP(N-L). Note that FAP/SP can also be formulated in the aggregated notation by combining (18.4) and (18.8), using integer rather than binary variables u_{et} – the number of paths realized on link e for all demands terminating at t.

$\qquad$ Finally, we note that expressing the requirement of avoiding tiny flows (a path flow is either greater than a given threshold or equal to 0) requires binary variables. For example, FAP with such an requirement can be formulated as follows (LB means that a lower bound, $l_d, d \in \mathcal{D}$, is imposed on r

$$\textbf{FAP/LB:} \quad \min \; F(x) = \sum_{e \in \mathcal{E}} \xi_e \left(\sum_{d \in \mathcal{D}} \sum_{p \in \mathcal{P}_d} x_{dp} \right)$$

$$\sum_{p \in \mathcal{P}_d} x_{dp} = h_d$$

$$l_b u_{dp} \le x_{dp} \le h_d u_{dp}$$

$$\sum_{d \in \mathcal{D}} \sum_{p \in \mathcal{P}_{ed}} x_{dp} \le c_e$$

$$u_{dp} \in \{0, 1\}$$

$$x_{dp} \in \mathbb{R}_+$$

18.2.3 Dimensioning Problems

The problems considered in this subsection involve f
tions (on top of flow patterns) and minimizing their
traditionally called dimensioning problems (DP).

Simple Dimensioning Problems

In the simplest case, a flow allocation problem is turn
lem by removing the capacity constraint in FAP (i.e.,
(18.4c), (18.7c), (18.8c), and (18.9d) in the correspor
treating link loads $\sum_{d \in \mathcal{D}} \sum_{p \in \mathcal{P}_{ed}} x_{dp}$ and unit flow o
and unit link capacity costs ξ_e, respectively. Certainl
use auxiliary variables $y = (y_e : e \in \mathcal{E})$ to define the lir
link load and modify the FAP formulations according
(18.1) this leads to the following simple dimensioning

$$\textbf{SDP(L-P):} \quad \min \; F(y) = \sum_{e \in \mathcal{E}} \xi_e y_e$$

$$\sum_{p \in \mathcal{P}_d} x_{dp} = h_d \qquad\qquad\qquad d$$

$$\sum_{d \in \mathcal{D}} \sum_{p \in \mathcal{P}_{ed}} x_{dp} \le y_e \qquad\qquad e$$

$$x_{dp} \in \mathbb{R}_+ \qquad\qquad\qquad d$$

$$y_e \in \mathbb{R} \qquad\qquad\qquad e$$

Clearly, (18.10) is a linear programming problem
an optimal solution can be found directly withou
deed, in an optimal solution of (18.10) we can ass

skip constraint (18.10c), and transform the objective function into $F(x) = \sum_{d \in \mathcal{D}, p \in \mathcal{P}_d}(\sum_{e \in p} \xi_e)x_{dp}$. The quantity in the brackets is precisely the length of path $p \in \mathcal{P}_d$ with respect to link weights ξ and thus finding an optimal solution to SDP consists in selecting a shortest path $p(d)$ in each $\mathcal{P}_d$, and assigning the entire demand volume h_d as the flow to the selected path, i.e., $x_{dp(d)} = h_d, x_{dp} = 0,\ p \in \mathcal{P}_d \setminus \{p(d)\}, d \in \mathcal{D}$. In fact, the considered problem is totally unimodular and the above-described single path solution is an optimal vertex solution (see Subsection 18.3.1).

Certainly, if there is more than one shortest path among admissible paths $\mathcal{P}_d$ of a demand $d \in \mathcal{D}$, in an optimal solution the demand volume can be split arbitrarily between these paths. Still, in a vertex solution of LP (18.10) (found for example by the simplex method) only one shortest path for every demand will be used to realize the demand flow. Hence, we will get a nonbifurcated solution even though this is not forced by the constraints. Moreover, if we do not wish to predefine admissible paths and rather consider all possible elementary paths, an optimal solution can easily be found be applying a shortest path algorithm (like the classical Dijkstra algorithm, see for example Appendix C in [2]) instead of finding a shortest path by enumerating the path lists $\mathcal{P}_d$. We can thus say that SDP can be solved by "shortest-path allocation rule" (referred to as SPA in the sequel). Note that SPA resolves the issue of path generation in the case of SDP (see Exercise 18.6.2). This nice property is present also for hop-limit since generating a shortest path with a given hop-limit is a polynomial problem solved by a simple modification of the Dijkstra algorithm (see Appendix C in [2]).

Formulation (18.10) is straightforward and simple to solve using the SPA rule. However, deviations from the linear capacity reservation model considered in formulation (18.10) can greatly complicate the dimensioning problem. For example, this is the case when we require modular link capacity reservation, i.e., $y_e \in \{0, M, 2M, \ldots\}$ for a given capacity module M. The resulting modification, referred to as MDP (modular dimensioning problem), is a MIP problem:

$$\textbf{MDP(L-P):} \quad \min \ F(y) = \sum_{e \in \mathcal{E}} \xi_e y_e \tag{18.11a}$$

$$\sum_{p \in \mathcal{P}_d} x_{dp} = h_d \qquad\qquad d \in \mathcal{D} \tag{18.11b}$$

$$\sum_{d \in \mathcal{D}} \sum_{p \in \mathcal{P}_{ed}} x_{dp} \le M y_e \qquad\qquad e \in \mathcal{E} \tag{18.11c}$$

$$x_{dp} \in \mathbb{R}_+ \qquad\qquad d \in \mathcal{D},\ p \in \mathcal{P}_d \tag{18.11d}$$

$$y_e \in \mathbb{Z}_+ \qquad\qquad e \in \mathcal{E}. \tag{18.11e}$$

Certainly, this time inequality in (18.11c) is necessary since it may not be feasible to fill the installed link capacity modules. It is well known that MDP is $\mathcal{NP}$-hard as it can be reduced from the Steiner-tree problem [10]. Therefore, the SPA rule cannot apply. In fact, optimal solutions of MDP are in general bifurcated which can

be easily seen for a three-node symmetric network wi*
and the value of the module much larger than the d*
parameter M can be omitted by dividing all h_d by M
demand values.

As MDP is a MIP problem, path generation can*
within a branch-and-price process, see Subsection 18.3*
can be applied so that efficient path generation is nc
gated N-L formulation of MDP can be pretty effecti
MIP solvers.

Link Dimensioning Models

The case with just one capacity module M in MDP
facility case (see [11]). It can be easily extended b*
kind of link capacity module. In such a multifacility
modules M_k, $k \in \mathcal{K}$, where $\mathcal{K} = \{1, 2, \ldots, K\}$, with
ξ_{ek}, $e \in \mathcal{E}$. Then, the appropriate constraints in (18.*

$$\sum_{d \in \mathcal{D}} \sum_{p \in \mathcal{Q}_{ed}} x_{dp} \leq \sum_{k \in \mathcal{K}} M_k y_{ek} \qquad e \in$$

$$y_{ek} \in \mathbb{Z}_+ \qquad k \in$$

and the objective function becomes equal to $\sum_{e \in \mathcal{E}} \sum$

Assuming link modularity leads to a step-wise din
i.e., a function that specifies the link capacity rese.
link load, or the cost of a link for the given load.
more efficient in terms of lower bounds, way of mo*
by using module increments (see Paragraph 4.3.1 in
are K such modules $m_1, m_2, \ldots, m_K$ and associate,
variables $u_{e1}, u_{e2}, \ldots, u_{eK}$ determining whether or n
link e. Certainly, for each link we require that $u_{e1} \geq$
Writing down an DP with this incremental way of mo
function is left to the reader as Exercise 18.6.3 (see
incremental approach).

Modular link dimensioning models lead to mixed-*
seen earlier in this subsection. Still, in many appl
capacity, namely concave and convex models, are of i*

Let $\underline{y}_e$ denote the link load (as for example giv
(18.10c)) and consider a nondecreasing concave real
that $f(0) = 0$. (Recall that concavity of f means that
$0 < \alpha < 1$, $f(\alpha z_1 + (1 - \alpha)z_2) \geq \alpha f(z_1) + (1 - \alpha)f($
required to carry load $\underline{y}_e$ is defined by $y_e = f(\underline{y}_e)$ so
SDP-type problems becomes

$$F = \sum_{e \in \mathcal{E}} \xi_e f(\underline{y}_e).$$

Consider problem SDP (18.10) with objective function (18.13). It is well known that this problem is hard to solve [13], as in general it has a large number of local minima (at the vertices of the solution polyhedron) defined by constraints (18.10b)-(18.10e). A fast algorithm for finding a local minimum for such a concave version of SDP can be found in [14] (see also [2, 13]). A way to solve the problem in an exact way is to use a piecewise approximation of a concave function $f(z) = \min\{a_k z + b_k : k \in \mathcal{K}\}$ where $\mathcal{K} = \{1, 2, \ldots, K\}$, $a_1 > a_2 > \ldots > a_K > 0$ and $b_K > b_{K-1} > \ldots > b_1$ are the coefficients of the consecutive linear pieces of the approximation (see Exercise 18.6.4). Then an appropriate DP can be formulated as the following MIP problem (see also Paragraph 4.3.3 in [2]).

$$\textbf{SDP/CV:} \quad \min \ F(y) = \sum_{e \in \mathcal{E}} \xi_e \sum_{k \in \mathcal{K}} (a_k y_{ek} + b_k u_{ek}) \tag{18.14a}$$

$$\sum_{p \in \mathcal{P}_d} x_{dp} = h_d \qquad\qquad d \in \mathcal{D} \tag{18.14b}$$

$$\sum_{d \in \mathcal{D}} \sum_{p \in \mathcal{Q}_{ed}} x_{dp} \leq \underline{y}_e \qquad\qquad e \in \mathcal{E} \tag{18.14c}$$

$$\sum_{k \in \mathcal{K}} y_{ek} = \underline{y}_e \qquad\qquad e \in \mathcal{E} \tag{18.14d}$$

$$\sum_{k \in \mathcal{K}} u_{ek} = 1 \qquad\qquad e \in \mathcal{E} \tag{18.14e}$$

$$y_{ek} \leq \Delta u_{ek} \qquad\qquad e \in \mathcal{E}, \ k \in \mathcal{K} \tag{18.14f}$$

$$\underline{y}_e \in \mathbb{R} \qquad\qquad e \in \mathcal{E} \tag{18.14g}$$

$$x_{dp} \in \mathbb{R}_+ \qquad\qquad d \in \mathcal{D}, \ p \in \mathcal{P}_d \tag{18.14h}$$

$$y_{ek} \in \mathbb{R}_+ \qquad\qquad e \in \mathcal{E}, \ k \in \mathcal{K} \tag{18.14i}$$

$$u_{ek} \in \{0, 1\} \qquad\qquad e \in \mathcal{E}, \ k \in \mathcal{K}. \tag{18.14j}$$

Above, constraints (18.14d)-(18.14f) together with the objective function assure, for each link $e \in \mathcal{E}$, the proper value of $\min\{a_k \underline{y}_e + b_k : k \in \mathcal{K}\}$ for a given value of the link load $\underline{y}_e$. (Δ is a sufficiently large constant, e.g., $\Delta = \sum_{d \in \mathcal{D}} h_d$.) We note that in general MIPs involving piecewise linear approximations of concave functions are hard to solve, partly because of the presence of the "big M" in (18.14f), denoted by Δ in this case.

Dealing with a convex dimensioning function $y_e = f(\underline{y}_e)$ is much simpler. (Recall that convexity of f means that for any $0 \leq z_1 < z_2$ and any $0 < \alpha < 1$, $f(\alpha z_1 + (1 - \alpha)z_2) \leq \alpha f(z_1) + (1 - \alpha)f(z_2)$.) First of all, when dealing with the linear programs with a (linear) objective substituted by a convex one, we are still within the efficiently tractable area of convex programming (see Chapter 14). For example, for SDP with objective (18.13) involving a convex f instead of (18.10a) we may use standard algorithms for optimizing a convex objective subject to linear constraints (for example the Frank-Wolfe method [15] or the Rosen gradient projection method [16], see also Chapter 5 in [2]). In fact, the piece-

wise linear approximation of SDP with a convex di▮
a linear programming problem. The dimensioning
$f(z) = \max \{a_k z + b_k : k \in \mathcal{K}\}$ where $\mathcal{K} = \{1, 2, \ldots,$
and $b_K < b_{K-1} < \ldots < b_1$ are the coefficients of th▮
the approximation (Exercise 18.6.5). SDP becomes a
4.3.2 in [2]):

$$\textbf{SDP/CX:} \quad \min \ F(y) = \sum_{e \in \mathcal{E}} \xi_e z_e$$

$$\sum_{p \in \mathcal{P}_d} x_{dp} = h_d \qquad\qquad d$$

$$\sum_{d \in \mathcal{D}} \sum_{p \in \mathcal{Q}_{ed}} x_{dp} \leq \underline{y}_e \qquad\qquad e$$

$$z_e \geq a_k \underline{y}_e + b_k \qquad\qquad e$$

$$\underline{y}_e \in \mathbb{R} \qquad\qquad e$$

$$x_{dp} \in \mathbb{R}_+ \qquad\qquad d$$

$$z_e \in \mathbb{R}_+ \qquad\qquad e$$

Finally, we note that these are the concave dimension
urally in network design (as they reflect the natural ▮
scale in resource dimensioning) and that the convex
penalty functions.

18.3   Optimization Methods for
Flow Networks

As we have seen in the previous section, network d
cally take a form of linear programs (LP) or mixed-
problems. In this section we illustrate how optimizat▮
integer programming (see Chapter 15) are used for N▮
methods including decomposition methods (Subsecti
plane (CP) and the branch-and-bound (BB) approac▮
latter are the basic approaches to the MIP problems,
such as CPLEX [17]. Additionally, in Subsection 18.3▮
ods for NDP, but only briefly, as in this chapter heuri
as a means for finding upper bounds for the optima▮
with BB. Certainly, the LP algorithms, first of all
Subsection 15.1.5 in Chapter 15) and, to a less ext▮
(see Subsection 15.1.6 in Chapter 15) are the prereq▮
and BB. In the following, we assume that the reader
method.

18.3.1 Decomposition Methods

Path Generation

Path generation (PG), a basic method for the MFN linear programming problems, is similar to column generation in the revised simplex method. It can also be regarded as an application of the Dantzig-Wolfe decomposition (see for example [15, 18]). PG is best illustrated with the following example.

Consider the following version of the FAP problem (18.1) in the noncompact L-P formulation:

$$\min \ z \tag{18.16a}$$

$$[\lambda_d] \quad \sum_{p \in \mathcal{P}_d} x_{dp} = h_d \qquad\qquad d \in \mathcal{D} \tag{18.16b}$$

$$[\pi_e] \quad \sum_{d \in \mathcal{D}} \sum_{p \in \mathcal{P}_{ed}} x_{dp} \leq c_e + z \qquad\qquad e \in \mathcal{E} \tag{18.16c}$$

$$x_{dp} \in \mathbb{R}_+ \qquad\qquad d \in \mathcal{D}, \ p \in \mathcal{P}_d. \tag{18.16d}$$

Note that if for a feasible solution (z, x) we have that $z \leq 0$, then the induced link loads do not exceed link reservations c. Anyhow, our goal is to solve (18.16) when all elementary paths are considered for $\mathcal{P}$ – this can be done through PG.

Problem (18.16) is a linear programming problem and hence its dual can be derived by applying the standard formulae (see Subsection 15.1.4 in Chapter 15 and for example [15, 18, 19]). Using the dual variables given in the brackets on the left-hand side of (18.16b) and (18.16c) we obtain the following dual problem:

$$\max \ W(\lambda, \pi) = \sum_{d \in \mathcal{D}} h_d \lambda_d - \sum_{e \in \mathcal{E}} c_e \pi_e \tag{18.17a}$$

$$\lambda_d \leq \sum_{e \in p} \pi_e \qquad\qquad d \in \mathcal{D}, \ p \in \mathcal{P}_d \tag{18.17b}$$

$$\sum_{e \in \mathcal{E}} \pi_e = 1 \tag{18.17c}$$

$$\pi_e \geq 0 \qquad\qquad e \in \mathcal{E}, \tag{18.17d}$$

where $W(\lambda, \pi)$ is the dual function. It is easy to notice that in any optimal solution $\lambda^* = (\lambda_d^*, \ d \in \mathcal{D})$, $\pi^* = (\pi_e^*, \ e \in \mathcal{E})$ of the dual, for each demand $d \in \mathcal{D}$ the value of λ_d^* is equal to the length of the shortest path on the list $\mathcal{P}_d$ with respect to link metrics π^*. Now suppose that there exists a demand d and a path $p(d)$ between $s(d)$ and $t(d)$ outside the list $\mathcal{P}_d$ that is sharply shorter with respect to π^* than λ_d^*. Adding this path to the list, $\mathcal{P}_d := \mathcal{P}_d \cup \{p(d)\}$, and the corresponding constraint

$$\lambda_d \leq \sum_{e \in p(d)} \pi_e \tag{18.18}$$

to (18.17b) eliminates the current optimal dual solutic
ble solutions of (18.17), hence opening a possibility o
objective $W^* = W(\lambda^*, \pi^*)$, and hence of decreasing
z^* for the primal problem (18.16) with the so-extende
problems the primal and the dual optimal objectives
programming problem is convex.) This observation
generation algorithm.

Algorithm PG

Step 0: Form initial lists of admissible paths $\mathcal{P}_d$, d

Step 1: Solve the dual problem (18.17). Let $(\lambda^*, \pi$
the dual.

Step 2: For each $d \in \mathcal{D}$ find a shortest path, $p(d)$
$\mathcal{G}(\mathcal{V}, \mathcal{E})$ with respect to link metrics π^*. Let $L($
If $L(d) < \lambda^*$ then $\mathcal{P}_d := \mathcal{P}_d \cup \{p(d)\}$.

Step 3: If at least one path has been added, then g

Step 4: Current admissible path lists are sufficient
of (18.16) (the problem admits all elementary $_1$

Initial lists of admissible paths can for example be fo
path for each demand. Observe that if the goal of s
a feasible flow pattern for given h and c, then we ca
Step 3 when $W(\lambda^*, \pi^*) \leq 0$. Certainly, in the genera
of Step 3 must be used. The PG algorithm is further
It should be noted that in practice, when using an I
consider the dual problem explicitly and instead use t
variables corresponding to the primal optimal solution
15.1.4 of Chapter 15, such a dual solution is readil
basis. For example, in the LP problem min $\{cx : Ax \leq$
matrix B defines the optimal dual variables $\lambda^* = (\lambda^*_1$
to the m rows of the coefficient matrix A) by the for
Step 1 we may consider solving the primal problem ir
    Roughly speaking, PG is usually quite fast for dim
the number of the paths added in the process is low, w
ing for allocation problems (Exercise 18.6.7). In the la
process can be made more efficient by using an appro
Könemman [20]. The method of [20] is capable of g
fast way and getting a solution for the problem like t
termined accuracy. Hence, it can be used to generate
the optimum, and then switched to PG for generating

In Section 18.4 we will consider more applications of PG. In particular we will see that there are noncompact network optimization formulations that lead to $\mathcal{NP}$-hard path generation problems (called the pricing problems).

Benders Decomposition

Benders decomposition (BD) is a decomposition method for linear and mixed-integer programming [15, 18, 22]. The idea of BD consists in projecting out a subset of variables (as for example flow variables) from the problem and generating a set of new inequalities (called Benders inequalities) involving the subset of variables that are left (as for example link capacity reservations).

BD is well illustrated using the SDP problem (18.10). The idea is to generate inequalities for link capacity reservations $y = (y_e, \ e \in \mathcal{E})$ using the values $y^* = (y_e^*, \ e \in \mathcal{E})$ that are consecutively generated during the BD process by the so-called master problem. The inequalities are generated using the feasibility test (FT) based on the FAP problem (18.16). Suppose $y^* = (y_e^*, \ e \in \mathcal{E})$ is a given fixed link capacity vector and consider an optimal solution (λ^*, π^*) of the dual problem (18.17) with c substituted with y^*. If $W^* = W(\lambda^*, \pi^*) \leq 0$ then the test is positive: capacity reservations y^* are sufficient to realize demand volumes h (for the given sets of admissible paths $\mathcal{P}_d, \ d \in \mathcal{D}$). If not, inequality

$$\sum_{e \in \mathcal{E}} \pi_e^* y_e \geq \sum_{d \in \mathcal{D}} h_d \lambda_d^*$$

will eliminate the vector y^*. The BD algorithm is as follows.

Algorithm BD

Step 0: Form initial list of inequalities Ω determining the set $\mathcal{Y}(\Omega)$ of variables y.

Step 1: Solve MP:

$$\min \ F(y) = \sum_{e \in \mathcal{E}} \xi_e y_e \tag{18.19a}$$

$$y \in \mathcal{Y}(\Omega). \tag{18.19b}$$

Step 2: Let y^* be an optimal solution of (18.19). Perform FT by solving the dual problem:

$$\max \ W(\lambda, \pi) = \sum_{d \in \mathcal{D}} h_d \lambda_d - \sum_{e \in \mathcal{E}} y_e^* \pi_e$$

$$\lambda_d \leq \sum_{e \in p} \pi_e$$

$$\sum_{e \in \mathcal{E}} \pi_e = 1$$

$$\pi_e \geq 0$$

Step 3: Let (λ^*, π^*) be an optimal solution of (18.

$$\Omega := \ \Omega \cup \Big\{ \sum_{e \in \mathcal{E}} \pi_e^* y_e \geq \sum_{d \in \mathcal{I}}$$

and go to Step 1.

Step 4: Current y^* is an optimal solution of the SI

The inequalities generated in Step 3 are called m
set Ω can be simply set to $\Omega = \{y_e \geq 0, \ e \in \mathcal{E}\}$ bu
can be put there to decrease the number of iterations
process. An example of such inequalities are cut ineq

$$\sum_{e \in \mathcal{E}(\mathcal{V}^1, \mathcal{V}^2)} y_e \geq \sum_{d \in \mathcal{D}(\mathcal{V}^1, \mathcal{V}^2)} h$$

In (18.21), $(\mathcal{V}^1, \mathcal{V}^2)$ is a cut which means that the
partitioning of the set of nodes $\mathcal{V}$. Then, $\mathcal{E}(\mathcal{V}^1, \mathcal{V}^2) =$
$|\{b(e)\} \cap \mathcal{V}^2| \ = \ 1\}$ is the set of links forming
$\{d \in \mathcal{D} : \ |\{s(d)\} \cap \mathcal{V}^1| = 1, |\{t(d)\} \cap \mathcal{V}^2| = 1\}$ is the
use the links of the cut. In fact, as shown in [23], cut
planar graphs but in the general case BD may requir
inequalities which are not cut inequalities.

Example 1. Consider the planar network depicted in
between the nodes s and t with $h = 10$. The metric ine
a final master problem are as follows:

$$y_1 + y_2 \geq 10, \ \ y_1 + y_3 + y_4 \geq$$

Both relations are cut inequalities. $\square$

Note that in Step 2 we must use path generation
elementary paths. Alternatively, we can use a N-L f

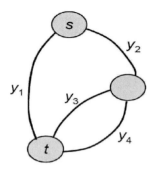

Figure 18.2: A planar network.

The BD algorithm and the above properties are further illustrated with Exercise 18.6.8.

It is important to emphasize that the master problem in BD can be a MIP as well. For example, we could use BD for the modular dimensioning case MDP (18.11) by considering the integer requirement on y in the master problem and using $c_e = My_e^*$ in the feasibility test and for generating the metric inequalities. Alternatively we could scale down the demand volumes $h_d := \frac{h_d}{M}$ and use exactly the same form of the feasibility test as in the algorithm given above.

We will discuss other (and more useful) applications of BG in Section 18.4.

18.3.2 Solving MIP Problems

In this section we will summarize the basic approaches of integer programming and give examples of how they apply to NDP. The presentation is based on a paper published by M. Padberg [24] (and in fact on his tutorial [25] that is unfortunately not accessible on the Internet anymore) and the excellent integer programming handbooks by L. Wolsey [26], A. Schrijver [19], G. Nemhauser and L. Wolsey [27], and M. Grötschel, L. Lovász and A. Schrijver [28].

MIP Formulations and Their LP Relaxations

Consider a mixed-integer programming (MIP) problem, i.e., the problem of the form

$$\textbf{MIP:} \quad \min z = cx + dy \tag{18.22a}$$

$$Ax + Dy \leq b \tag{18.22b}$$

$$x \in \mathbb{Z}_+^n, \ y \in \mathbb{R}_+^p \tag{18.22c}$$

where $x = (x_1, x_2, \ldots, x_n)$ is the vector of integer variables, $y = (y_1, y_2, \ldots, y_p)$ is the vector of continuous variables, $c = (c_1, c_2, \ldots, c_n)$ and $d = (d_1, d_2, \ldots, d_p)$

are the cost vectors, $A = [a_{ij}]_{i=1,2,\ldots,m,\ j=1,2,\ldots,n}$ and
are the coefficient matrices, and $b = (b_1, b_2, \ldots, b_m)$ is
sides of the constraints. When all integer variables x
binary, then we call the problem mixed-binary progra

If continuous variables do not appear in the proble
integer programming (IP) problem:

$$\textbf{IP:} \quad \min \ z = cx$$
$$Ax \leq b$$
$$x \in \mathbb{Z}_+^n.$$

When x are binary instead of integer, then problem (18
(BP) or 0-1 integer program.

The set of all feasible points (solutions) of (18.22)
X^{IP} for (18.23)), and referred to also as the feasible
set, optimization space, etc. The optimal objective
(and z^{IP} for (18.23a)). We will use the notation z^M
is unbounded and $z^{MIP} = +\infty$ when the problem is
avoid certain theoretical difficulties (see [24, 27]), we
data c, A, D, b contains only rational numbers.

The linear relaxation (LP relaxation in short) o
obtained by dropping the integrality requirement in (1
instead of $x \in \mathbb{Z}_+^n$. Let P^{LP} denote the polyhedro
b, $x \geq 0$, $y \geq 0\}$ defining the optimization space of t
$P^{LP} \supseteq X^{MIP}$ and thus $z^{LP} \leq z^{IP}$, that is, the LP rel
of the original MIP problem, imposing a lower bound
$z^{LP} = -\infty$ implies $z^{MIP} = -\infty$ (if the LP relaxa
MIP is unbounded) and $z^{LP} = +\infty$ implies $z^{MIP} =$
infeasible, then the MIP is infeasible).

Let $P^{MIP} = \text{conv}(X^{MIP})$ denote the convex hul
$X \subseteq \mathbb{R}_+^n$, $\text{conv}(X)$ is the set composed of all convex
of X, i.e., the set of all $x \in \mathbb{R}_+^n$ of the form $x = \sum_{i=}^{I}$
$\{x^1, x^2, \ldots, x^I\}$ of X and some scalars $\alpha_1, \alpha_2, \ldots, \alpha_I$
$\alpha_i \geq 0$ for $i = 1, 2, \ldots, I$. Alternatively, $\text{conv}(X)$ ca
convex set containing X.) It is known that P^{MIP} is
that the LP problem defined on P^{MIP}

$$\min \ z = cx$$
$$x \in P^{MIP}$$

is equivalent to the original MIP problem (18.22) since
of the polyhedron P^{MIP} are identical to the set X^M
the original MIP problem to an LP problem. How
of theoretical importance because in general it is vir

explicit characterization of P^{MIP} corresponding to (18.22), i.e., a characterization of the form $Gx + Hy \leq g$, $x \geq 0$, $y \geq 0$. This difficulty is implied by the fact that the number of inequalities characterizing P^{MIP} (i.e., the number of rows in matrices G and H) is typically exponential and the problem of generating these inequalities is, as discussed below, of the same complexity as the problem (18.22) itself.

It is clear that $P^{MIP} \subseteq P^{LP}$. In most cases the inclusion is proper ($P^{MIP} \subsetneq P^{LP}$) and it is common that the optimal objective of the LP relaxation is substantially smaller than the optimal objective of the original problem ($z^{LP} \ll z^{MIP}$). Still, it can happen (although very rarely) that $P^{LP} = P^{MIP}$. This is the case for the so-called totally unimodular IP problems discussed in the next paragraph.

Totally Unimodular Problems

A $m \times n$ matrix A is called totally unimodular if every square submatrix of A has determinant equal to $+1, -1$ or 0 [26, 29]. It is known that A is totally unimodular if and only if (A, I) is totally unimodular, where I is an $m \times m$ identity matrix. Consider the LP relaxation of an IP problem (18.23) with integral data and a totally unimodular coefficient matrix A. The form of the simplex basic feasible solutions ($x = B^{-1}b$) and Cramer's rule imply that in such a totally unimodular LP problem, the vertex solutions are integral and hence $P^{LP} = P^{IP}$. As already mentioned in Subsection 18.2.3, the L-P formulation of SDP

$$\min \left\{ F(x) = \sum_{e \in \mathcal{E}} \xi_e \left(\sum_{d \in \mathcal{D}} \sum_{p \in \mathcal{P}_{ed}} h_d x_{dp} \right) : \sum_{p \in \mathcal{P}_d} x_{dp} = 1, d \in \mathcal{D}; x \geq 0 \right\} \quad (18.25)$$

is totally unimodular. In fact, vertex solutions of (18.25) are binary.

Another well-known NDP example of a totally unimodular problem is the following FAP problem in the N-L formulation:

$$\min \ F(x) = \sum_{e \in \mathcal{E}} \xi_e \, x_e \quad (18.26a)$$

$$\sum_{e \in \delta^+(v)} x_e - \sum_{e \in \delta^-(v)} x_e = \begin{cases} 0, & v \in \mathcal{V} \setminus \{s, t\} \\ 1, & v = s \end{cases} \quad (18.26b)$$

$$0 \leq x_e \leq 1 \qquad\qquad e \in \mathcal{E}. \quad (18.26c)$$

Because of the totally unimodular coefficient matrix, the above problem solves the problem of finding a ξ-shortest path from s to t. The reader should note the similarity of (18.26) to the one-demand version of (18.8) with $h = 1$ and $c_e \equiv 1$.

Thus, IP problems with totally unimodular LP relaxations are easy as compared with general IP problems.

The Cutting Plane Method

As mentioned in the introduction to this section, the
proaches to MIPs. One is the cutting plane metho
branch-and-bound method using LP relaxations (BF
handbook of integer programming, for example [15, ?
series of LP relaxations of consecutive versions of (18.?
(cuts) that cut off the current solutions (x^{LP}, y^{LP}) v

Consider the MIP problem specified with (18.22).
(18.22), a discrete-mixed set in $\mathbb{Z}_+^n \times \mathbb{R}_+^p$ is denoted by
polyhedron in $\mathbb{R}_+^{n+p}$ – by $P^{MIP} = \text{conv}(X^{MIP})$. Rec:
solution set (a polyhedron) of the LP relaxation of t
$P^{MIP} \subseteq P^{LP}$ (usually $P^{MIP} \subsetneq P^{LP}$). An inequali
where $(f, g, f_0) \in \mathbb{R}^{n+p+1}$ is called a valid inequalit
$P^{MIP} \subseteq \{(x, y) \in \mathbb{R}_+^{n+p} : fx + gy \leq f_0\}$. The fami
the considered MIP will be denoted by $\hat{\mathcal{F}}$. A valid i:
plane (or simply a cut) for the considered MIP if

$$P^{LP} \cap \{(x, y) \in \mathbb{R}_+^{n+p} : fx + gy \leq$$

so that the cutting plane (f, g, f_0) actually cuts off a

Example 2. Consider an IP problem

$$\max\{5x_1 + 5x_2 : 2x_1 + x_2 \leq 10, \ x_1 + 2x_2 \leq 1$$

The graphical illustration of the problem reveals that
problem are the points $(2, 4), (3, 3), (4, 2)$ with the o[
Skipping the integrality constraint, i.e., considering tl
its optimal solution is equal to $(3\frac{1}{3}, 3\frac{1}{3})$ with the o[
This nonintegral LP relaxation solution can be cut o

$$x_1 + x_2 \leq 6$$

which is a facet of the polyhedron $P^{IP} = \text{conv}(X^I$
$2x_1 + x_2 \leq 10, \ x_1 + 2x_2 \leq 10, \ x_1 \in \mathbb{Z}_+, x_2 \in \mathbb{Z}_+\}$ is t
IP problem. $\square$

An example of a cutting plane is a lifted cut ine
problem (18.11) involving modular links:

$$\sum_{e \in \mathcal{E}(\mathcal{V}^1, \mathcal{V}^2)} y_e \geq \left\lceil \frac{\sum_{d \in \mathcal{D}(\mathcal{V}^1, \mathcal{V}^2)}}{M} \right.$$

Note that above, the notion of the cut is used in two
cutting plane, and cut as a cut in a graph.

The idea of CP is straightforward. We first solve the LP relaxation of the MIP. If the solution (x^{LP}, y^{LP}) is integral in x we stop as we have just solved the MIP. Otherwise, we find a family $\mathcal{F} \subseteq \hat{\mathcal{F}}$ of cuts cutting off the point (x^{LP}, y^{LP}) (i.e., $fx^{LP} + gy^{LP} > f_0$ for all $(f, g, f_0) \in \mathcal{F}$), add the cuts in $\mathcal{F}$ to the LP relaxation, and iterate. Let $\text{LP}(\mathcal{F})$ denote the so-extended LP relaxation

$$\textbf{LP}(\mathcal{F}): \quad \min \ z = cx + dy \tag{18.29a}$$

$$Ax + Dy \le b \tag{18.29b}$$

$$fx + gy \le f_0, \quad (f, g, f_0) \in \mathcal{F} \tag{18.29c}$$

$$x, y \ge 0 \tag{18.29d}$$

and let $(x^{\mathcal{F}}, y^{\mathcal{F}}) \in \mathbb{R}_+^{n+p}$ denote an optimal solution of (18.29). A generic CP algorithm is as follows.

Algorithm CP

Step 0: Initialize $\mathcal{F}$ as the empty set.

Step 1: Solve $\text{LP}(\mathcal{F})$. If $x^{\mathcal{F}} \in \mathbb{Z}_+^n$ (or $z^{\mathcal{F}} \in \{-\infty, +\infty\}$) then stop the algorithm: current $(x^{\mathcal{F}}, y^{\mathcal{F}})$ is an optimal solution of (18.22).

Step 2: Find a family of cuts $\mathcal{F}'$ for the MIP cutting off $(x^{\mathcal{F}}, y^{\mathcal{F}})$. $\mathcal{F} := \mathcal{F} \cup \mathcal{F}'$ and go to Step 1.

Observe that if we denote the polyhedrons of the consecutive problems $\text{LP}(\mathcal{F})$ solved in Step 1 by $P^0, P^1, P^2 \dots$ then $P^0 \supseteq P^1 \supseteq P^2 \supseteq \cdots \supseteq P^{MIP}$ and therefore $z^0 \le z^1 \le z^2 \le \cdots \le z^{MIP}$ where z^k is the solution for P^k.

The efficiency of the CP algorithm depends on the way the families $\mathcal{F}'$ are found in Step 2. A classical approach is to use the Gomory cuts [30, 31] (named after R. Gomory, the inventor of the CP method). The original Gomory's version of CP produces just one cut in Step 2 (i.e., $|\mathcal{F}'| = 1$). Moreover, a Gomory cut is obtained from the current basis B of the simplex algorithm [24, 26] and therefore this way of generating cuts is general, i.e., it uses problem-independent ready-to-use formulas. It is proved (under an essential assumption that the optimal z^{MIP} is an integer) that the Gomory method is convergent in a finite number of steps. Certainly, if the MIP treated by the CP algorithm is $\mathcal{NP}$-hard, we cannot expect a polynomial number of steps when the cut generation method is polynomial. Nevertheless, the Gomory method (and its further improvements) is regarded as impractical as it usually requires an enormous number of steps. In fact, the CP method as such is not used in today's MIP solvers (like for example CPLEX [17]) although the Gomory-like cuts are applied within the branch-and-cut algorithm (see below).

Branch-and-Bound Using LP Relaxations

The branch-and-bound (BB) method (described already in Subsection 15.3.2 of Chapter 15, see also any handbook of integer programming, e.g., [15, 26, 27])

generates a tree of linear subproblems related to t
solve. The consecutive subproblems have more and
values allowed for the integer variables. The root of
the LP relaxation of (18.22), and the leaves – to integ
the BB tree the corresponding LP subproblem is solv
a subproblem is defined by $Ax + Dy \leq b$, $x, y \geq 0$ a
bounding the values of integer variables. Then its so
current upper bound, i.e., the currently best feasibl
lower bound is greater or equal to the upper bound,
node and all its successors are discarded as the lea
cannot contain solutions that are better than the cur
   Consider the MIP problem given by (18.22). Tl
scribed below creates a tree of LP subproblems. E
corresponds to a specific additional set Ω of bounding
integer variables of (18.22), $\Omega = \{\alpha_j(\Omega) \leq x_j \leq \beta_j(\Omega$

$$
\textbf{LP}(\Omega): \quad \min \; z = cx + dy
$$

$$
Ax + Dy \leq b
$$

$$
\alpha_j(\Omega) \leq x_j \leq \beta_j(\Omega), \; j = 1, 2, \ldots,
$$

$$
y \geq 0.
$$

Each node of the BB tree is identified with the corres
and thus with Ω. The algorithm makes use of a set c
to be processed before terminating.

Algorithm BB

Step 0: Set the upper bound $z^{UB} = +\infty$. Form
(corresponding to the root of the BB tree) by
$+\infty$, $j = 1, 2, \ldots, n$. Let the set of active BB n

Step 1: Is there any active BB node remaining?
current (x^{UB}, y^{UB}) is an optimal solution of (1

Step 2: Select a BB node Ω in the set of active n
(18.30) for the current Ω to get its optimal soluti
go to Step 1 (bounding).

Step 3: If at least one x_j^{Ω} is not an integer, go to St

Step 4: Update the current best solution $z^{UB} :=$
and go to Step 1.

Step 5: Select a noninteger x_j^{Ω} and create two new (branching):

$$\Omega' := (\Omega \setminus \{\alpha_j(\Omega) \le x_j \le \beta_j(\Omega)\}) \cup \{\alpha_j(\Omega) \le x_j \le \lfloor x_j^\Omega \rfloor\}, \text{ i.e., } \beta_j(\Omega') = \lfloor x_j^\Omega \rfloor$$
instead of $\beta_j(\Omega)$

$$\Omega'' := (\Omega \setminus \{\alpha_j(\Omega) \le x_j \le \beta_j(\Omega)\}) \cup \{\lceil x_j^\Omega \rceil \le x_j \le \beta_j(\Omega)\}, \text{ i.e., } \alpha_j(\Omega'') = \lceil x_j^\Omega \rceil$$
instead of $\alpha_j(\Omega)$

and add them to the set of active BB nodes. Go to Step 1.

The BB algorithm creates a binary tree of the BB nodes. After processing, each active node Ω is either fathomed (when it cannot offer any solution that is better than the current upper bound, i.e., when $z^\Omega \ge z^{UB}$, or its solution (x^Ω, y^Ω) is integral in x) or it creates two new active nodes, each with the range of the branching variable x_j more limited than in Ω. Certainly, the way of selecting an active BB node (Step 2) and a variable x_j to branch on (Step 5) can strongly influence the efficiency of the algorithm.

We observe that efficiency of the BB algorithm heavily depends on the quality of the lower bounds delivered by $LP(\Omega)$. This is because of condition $z^\Omega \ge z^{UB}$ that allows for discarding the part of the BB tree below the current BB node, where $LP(\Omega)$ is the LP relaxation of the MIP subproblem specified by (18.30) and the condition $x \in \mathbb{Z}_+^n$. Certainly, we can expect that the better the quality of the original LP relaxation of the MIP (18.22), the better the LP relaxations $LP(\Omega)$. Although this is an important observation, in general a significant improvement of the lower bounds z^Ω may be achieved by applying a limited number of iterations of the CP method at the BB nodes; we will discuss this enhancement of the BB algorithm (called branch-and-cut, BC) later in this subsection.

We also note that in the BB process for a MBP (i.e., for a MIP involving only binary variables x), we could use the Lagrangean relaxation (LR) [15, 18, 26, 27] instead of the LP relaxation. Recall that the LR relaxation of a MIP problem gives a lower bound that is greater or equal to the lower bound provided by the LP relaxation, and that the LR solution can be substantially better than the LP solution. However, in many cases, also in network design, these two bounds are equal – we then say that the problem possesses the integrality property, see [26, 29]. Still, even though the LR lower bound can be better than the LP relaxation bound, calculating the value of LR involves subgradient maximization based on solving the so-called Lagrangean subproblems. In fact, the potential gains in BB from the better lower bounds can be overbalanced by excessive computational burden of LR (not mentioning implementation difficulty for an unexperienced user) with respect to the LP relaxation. For an application of LR to the topological design problem (18.33) see [32]; see also [33–36] for applications of LR to other multicommodity flow problems.

Finally, we note that the BB algorithm can be terminated when the gap $z^{UB} - z^{LP}$ between the currently best solution of MIP (18.22) and the solution of its LP

relaxation is less than some threshold, for example w
getting such approximate, suboptimal solutions can s
since, for example, it may happen in BB that the curr
but the process continues forever to prove that.

Strengthening the MIP Formulation

As the reader may have already noticed, a fundamer
ming is to construct an LP relaxation with the solutio
and hence giving the lower bound close to z^{MIP}. Sin
dard LP relaxation we observe that $P^{LP} \supsetneq P^{MIP}$
strong lower bound we typically need to reinforce (st
by generating additional valid inequalities. However
it is first of all important to use an appropriate fo
lem in hand because in general different (but equiva
LP relaxations of different quality. This issue is best
topological design problem (also called fixed-charge p
taking into account the fixed cost of installing links
in [2]). The problem can be stated as the following N

$$\textbf{TDP:} \quad \min F(y, u) = \sum_{e \in \mathcal{E}} \xi_e y_e + \sum_{e \in \mathcal{E}} \kappa_e u_e$$

$$\sum_{p \in \mathcal{P}_d} x_{dp} = h_d$$

$$\sum_{d \in \mathcal{D}} \sum_{p \in \mathcal{P}_{ed}} x_{dp} \leq y_e$$

$$y_e \leq \Delta_e u_e$$

$$x_{dp} \in \mathbb{R}_+$$

$$y_e \in \mathbb{R}, \ u_e \in \{0, 1\}$$

Above, κ_e is the given cost of installing link $e \in \mathcal{E}$, ar
number being an upper bound for the link load, for
all demand volumes $H = \sum_{d \in \mathcal{D}} h_d$. It is easy to see
the LP relaxation of (18.31), and hence a lower boun

$$\sum_{d \in \mathcal{D}} L(d) h_d$$

where $L(d)$ denotes the length of the shortest path of c
$\alpha_e = (\xi_e + \frac{\kappa_e}{\Delta_e})$. We notice that the "big M" paramete
influence the lower bound because each $\alpha_e, e \in \mathcal{E}$,
decreases as $\Delta_e \to \infty$. Thus, when κ_e are large with r
with respect to κ_e, the lower bound can be poor.

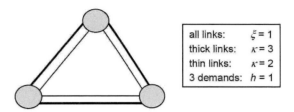

Figure 18.3: Network example for topological design.

Now let us consider another formulation of TDP [32]:

TDP/1: $\quad \min\ F(y,u) = \sum_{e \in \mathcal{E}} \xi_e y_e + \sum_{e \in \mathcal{E}} \kappa_e u_e$ (18.33a)

$$\sum_{p \in \mathcal{P}_d} x_{dp} = 1 \qquad\qquad\qquad d \in \mathcal{D} \qquad (18.33\text{b})$$

$$\sum_{d \in \mathcal{D}} \sum_{p \in \mathcal{P}_{ed}} h_d x_{dp} \le y_e \qquad\qquad e \in \mathcal{E} \qquad (18.33\text{c})$$

$$x_{dp} \le u_e \qquad\qquad e \in \mathcal{E},\ d \in \mathcal{D},\ e \in \mathcal{P}_{ed} \quad (18.33\text{d})$$

$$x_{dp} \in \mathbb{R}_+ \qquad\qquad d \in \mathcal{D},\ p \in \mathcal{P}_d \qquad (18.33\text{e})$$

$$y_e \in \mathbb{R},\ u_e \in \{0,1\} \qquad\qquad e \in \mathcal{E} \qquad (18.33\text{f})$$

Clearly, the two formulations (18.31) and (18.33) are equivalent in terms of optimal solutions. Although the LP relaxation of (18.33) requires an LP solver (its optimal solution is in general bifurcated and therefore cannot be solved by single path allocation, as the LP relaxation of (18.31)), it provides a better lower bound than the LP relaxation of (18.31), and hence for (18.31), as the two formulations are equivalent, and therefore is more effective for BB.

Example 3. The difference in the two lower bounds is illustrated by the example depicted in Figure 18.3. The (undirected) network consists of 3 nodes, and 3 "thick" links and 3 "thin" links; there is one thick and one thin link in parallel between each pair of nodes. The capacity dependent cost ξ_e is equal to 1 for all links, and the installation cost κ_e is equal to 3 for the thick links and to 2 for the thin links. There are also 3 demands corresponding to the three pairs of nodes, each with the demand volume $h = 1$. The value of the first LP relaxation (corresponding to (18.31) with $\Delta_e \equiv H = 3$) is equal to $\text{LB}_1 = 5$ while $\text{LB}_2 = 7.5$ where LB_2 is the value of the LP relaxation of (18.33). This is a substantial difference that significantly influences the number of nodes visited in the BB tree (Exercise 18.6.9). Thus, the lower bounds at the root of the BB tree are equal to 5 and 7.5, respectively. Considering another node, for example the BB node that excludes

thin links (forcing the values of u_e corresponding t
to 0 and leaving the values of u_e corresponding to
respective lower bounds are equal to 6 and 9. In fact,
that the optimal MIP solution F^* (i.e., the solution
be integer-valued, LB_2 can be lifted to 8 and hence
certainly a nice property for an LP relaxation. □

Problem-Dependent Cuts

Even with the best available formulation of a MIP, its
quality in terms of the gap $z^{MIP} - z^{LP}$. This, and t
CP methods are not focused on minimizing the gap
an approach using problem-dependent cuts is in plac
is how to find cuts that are reasonably effective to ge
substantially increase the value z^{LP}.

For convenience, the considered MIP problem can
programming problem (called LOP, linear optimizatio
(rational) polyhedron P in $\mathbb{R}^q$ and a rational cost vec

$$\textbf{LOP:} \quad \min\{z = cx : \ x \in$$

In LOP, the polyhedron P corresponds to the polyh
$x \in \mathbb{R}^q$ to variables $(x, y) \in \mathbb{R}_+^{n+p}$ in (18.22)) so th
system of inequalities of the form $Ex \le e$ is not know
number of inequalities). Hence, the fundamental qu
solved in polynomial time.

Let $\hat{\mathcal{F}}$ denote the family of all valid inequalities fo

$$\hat{\mathcal{F}} = \{(f, f_0) \in \mathbb{R}^{q+1} : \ P \subseteq \{x \in \mathbb{R}^q$$

and consider the following cutting plane procedure fo

Algorithm CP/LOP

Step 0: Set $k = 0$ and $P^0 = \{x \in \mathbb{R}^q : \ Ax \le b\}$
relaxation of (18.22)).

Step 1: Find $x^k \in P^k$ minimizing cx over P^k. If x^k
x^k is an optimal solution of LOP.

Step 2: Find a cut $(f, f_0) \in \hat{\mathcal{F}}$ with $fx^k > f_0$.

Step 3: Set $P^{k+1} = P^k \cap \{x \in \mathbb{R}^q : \ fx \le f_0\}$, $k :=$

Intuitively, the above procedure would be efficient if in
an "optimal separator" (Step 2 is called the constrain

step), that is, a cut that maximizes optimal $z = cx$ when added to P^k. This requirement is expressed as

$$\max_{(f,f_0) \in \hat{\mathcal{F}}_{x^k}} \left\{ \min \left\{ cx : \ x \in P^k \cap \{x \in \mathbb{R}^q : \ fx \le f_0\} \right\} \right\} \qquad (18.36)$$

where $\hat{\mathcal{F}}_{x^k}$ is a subfamily of cuts in $\hat{\mathcal{F}}$ that cut off the current solution x^k. Note that the so-obtained separator drives up the lower bound of LOP as much as possible. However, devising a reasonable procedure for finding such optimal or even suboptimal separators is not realistic and instead in practice we look for the separators (f, f_0) like:

$$\max \left\{ f x^k - f_0 : \ (f, f_0) \in \hat{\mathcal{F}}_\infty \right\} \qquad (18.37)$$

where $\hat{\mathcal{F}}_\infty = \left\{ (f, f_0) \in \hat{\mathcal{F}} : \max\{|f_i| : \ j = 1, 2, \ldots, q\} = 1 \right\}$. The cut resulting from (18.37) is called the most violated cut; if the maximum is nonpositive, then $x^k \in P$. Observe that in general the most violated cut does not define a facet of P (facet is a fully dimensional face of a polyhedron) – a highly desired feature.

Note that the CP/LOP procedure based on finding the most violated cuts has been used in Algorithm PG for path generation in Subsection (18.3.1) to solve a LOP problem corresponding to the polyhedron P of the dual problem (18.17) whose characterization through a system of inequalities involves an exponential number of constraints.

A fundamental result of integer programming states that the complexity of LOP is equivalent to the complexity of a procedure applied in the separation step (see [27]):

Theorem 1: For any rational polyhedron P and any rational c, problem (18.34) is solvable in polynomial time if and only if the polyhedral separation problem for any rational $x \in \mathbb{Q}^q$

$$\max \left\{ f x - f_0 : \ (f, f_0) \in \hat{\mathcal{F}}_\infty \right\} \qquad (18.38)$$

is solvable in polynomial time.

In fact, the theorem remains true for the following version of the polyhedral separation problem: for any rational $x \in \mathbb{Q}^q$

- conclude that $x \in P$

- if $x \notin P$, find any violated separator $(f, f_0) \in \hat{\mathcal{F}}$ such that $fx > f_0$.

Observe that Theorem 1 does not imply that *any* Algorithm CP/LOP is polynomial when the separation problem is polynomial since only special CP algorithms, as the ellipsoidal method [27], can guarantee this. Moreover, Theorem 1 does not imply that the problem represented by LOP is $\mathcal{NP}$-hard when the separation problem is

$\mathcal{NP}$-hard. This is simply because problem (18.34) c
in another way than by LOP, and then solvable in a
Theorem 1 really says in this case is that there exi
LOP is $\mathcal{NP}$-hard. Nevertheless, even if the LOP in
we can try to generate the most violated cuts despite
($\mathcal{NP}$-hard) as well. The reason is that solving the
practical instances, be much less time consuming that
it can be solved to a reasonable extent (i.e., to near-c

As far as MFN are concerned, considerable work h.
dependent cuts for the modular dimensioning [11, 37–
ing models [34, 42, 43]. We mention here that somet
called, somewhat ambiguously, valid inequalities.

In fact, instead of trying to solve a given LOP pr
CP/LOP, it can be reasonable to generate only som
enter the BB procedure. This approach is sometimes
(CB). In this case, there is no particular solution tha
just generate a reasonable number of cuts we consider
LP relaxation lower bound. For example, when solvir
problem (18.11), these could be cut inequalities (18
graph cuts. An example of using the CB approach c
specialized types of cuts (superadditive cuts) are use
mulation of the single-path FAP problem (18.8).

Branch-and-Cut

A crucial extension of BB that considerably impro
solvers is the so called branch-and-cut (BC), see [27
and CP techniques into one unified scheme. Consid
and the family $\hat{\mathcal{F}}$ of all its valid inequalities. For a g
given set Ω (see the description of BB above, and in
the following LP subproblem:

$$\mathbf{LP}(\Omega, \mathcal{F}): \quad \min \ z = cx + dy$$
$$Ax + Dy \leq b$$
$$\alpha_j(\Omega) \leq x_j \leq \beta_j(\Omega), \ j = 1, 2, \ldots$$
$$fx + gy \leq f_0, \quad (f, g, f_0) \in \mathcal{F}$$
$$y \geq 0.$$

An optimal solution of (18.39) will be denoted by z^Ω

Algorithm BC

Step 0: Set the upper bound $z^{UB} = +\infty$. Initializ
the initial active BB node (the root of the B
$0, \beta_j(\Omega) = +\infty, \ j = 1, 2, \ldots, n$. Make node Ω a

Step 1: Is there any active BB node remaining? If not, stop the algorithm: current (x^{UB}, y^{UB}) is an optimal solution of (18.22).

Step 2.1: Select a BB node Ω in the set of active nodes.

Step 2.2: Solve subproblem (18.39) for the current Ω and $\mathcal{F}$. If $z^{\Omega, \mathcal{F}} \geq z^{UB}$ go to Step 1 (bounding).

Step 3: If at least one $x_j^{\Omega, \mathcal{F}}$ is not an integer, go to Step 5.

Step 4: Update the current best solution $z^{UB} := z^{\Omega, \mathcal{F}}$, $(x^{UB}, y^{UB}) := (x^{\Omega, \mathcal{F}}, y^{\Omega, \mathcal{F}})$ and go to Step 1.

Step 5.1: Find a family of cuts $\mathcal{F}' \subseteq \hat{\mathcal{F}}$ for the MIP cutting off $(x^{\Omega, \mathcal{F}}, y^{\Omega, \mathcal{F}})$. If $\mathcal{F}'$ is not empty, $\mathcal{F} := \mathcal{F} \cup \mathcal{F}'$ and go to Step 2.2.

Step 5.2: Select a noninteger $x_j^{\Omega, \mathcal{F}}$ and create two new active BB nodes Ω' and Ω'' (branching):

$$\Omega' := (\Omega \setminus \{\alpha_j(\Omega) \leq x_j \leq \beta_j(\Omega)\}) \cup \{\alpha_j(\Omega) \leq x_j \leq \lfloor x_j^{\Omega, \mathcal{F}} \rfloor\}, \text{ i.e., } \beta_j(\Omega') = \lfloor x_j^{\Omega, \mathcal{F}} \rfloor$$
instead of $\beta_j(\Omega)$

$$\Omega'' := (\Omega \setminus \{\alpha_j(\Omega) \leq x_j \leq \beta_j(\Omega)\}) \cup \{\lceil x_j^{\Omega, \mathcal{F}} \rceil \leq x_j \leq \beta_j(\Omega)\}, \text{ i.e., } \alpha_j(\Omega'') = \lceil x_j^{\Omega, \mathcal{F}} \rceil$$
instead of $\alpha_j(\Omega)$

and add them to the set of active BB nodes. Go to Step 1.

Note the difference between Algorithm BB and Algorithm BC. To get BC, Step 2 of BB is divided to Step 2.1 and Step 2.2, Step 5 of BB becomes Step 5.2, and Step 5.1 is added.

BC can use both problem-dependent valid inequalities and general cuts (Gomory-like cuts, superadditve cuts, etc.), see [26, 27]. The latter way is useful for the mixed-binary problems, for which the mixed-integer Gomory cuts are, under some technical conditions, globally valid as shown by E. Balas et al. [46]. A survey of general cutting plane techniques can be found in [24, 46, 47] (also see the list of references there and in [48]).

Algorithm BC can be viewed as a combination of BB and CP. The idea is to generate a limited number of cuts while processing a BB node (in this way improving the current lower bound in a reasonable time) and let the cutting plane process continue after the branching. The BC algorithm is considered to be the most effective general exact approach in integer programming and therefore is commonly used in the MIP solvers. The technique has been extensively used in integer programming (see [24, 27] for applications for IP problems in general) and in particular for MIP problems in network design, see for example [33, 34, 42, 43, 49, 50].

Branch-and-Price

Branch-and-price (BP) is a way of dealing with noncor
52]. When integer variables x driving the BB proces
i.e., when pricing involves only continuous variables
straightforward as illustrated by the following examp
   Consider the noncompact modular dimensioning

$$\textbf{MDP/SP/HL(L-P):} \quad \min \ F(y) = \sum_{e \in \mathcal{E}} \xi_e y_e$$

$$[\lambda_d] \quad \sum_{p \in \mathcal{P}_d} x_{dp} = h_d$$

$$[\pi_e] \quad \sum_{d \in \mathcal{D}} \sum_{p \in \mathcal{P}_{ed}} x_{dp} \leq y_e$$

$$x_{dp} \in \mathbb{R}_+$$

$$y_e \in \mathbb{Z}_+$$

Note the above problem is the MDP (18.11) from Su
volumes normalized by M. In (18.40) the paths in
assumed hop-limit. As we have already noted at
18.2.2, there is no easy way of using a compact form
solving (18.40) makes sense. The BB process for this
the LP relaxations in the BB nodes through path gene
are noncompact. Observe that it is not correct to ge
relaxation of the root of the BB tree (i.e., LP relaxat
the so-obtained lists of admissible paths for all subse
the reason is that the paths generated for nonmodu
contain all the paths required for optimality of MDP/S
paths for a BB node Ω, we need to consider, besides
$\sigma^1 \geq 0$ and $\sigma^2 \geq 0$ corresponding to additional con
$y_e - \beta_e(\Omega) \leq 0$, respectively. Path generation consists
$d \in \mathcal{D}$, a shortest path (obeying the hop limit) betwee
to optimal π^* (where $\pi_e^* = \xi_e + \sigma_e^{2*} - \sigma_e^{1*}$, see Exercise
exists, adding it to the admissible path list $\mathcal{P}_d(\Omega)$.
   When path generation is used for the relaxed prob
and-bound tree, the two subsequently generated and
in general require different path-lists, still the path-lis
are usually very close to the final lists of the subpro
procedure is performed at every BB node to obtain th
lists of candidate paths are typically only extended v
BB tree.
   However, when pricing involves the integer variat
compromise the tractability of the pricing process w

the main LP relaxation is considered. To illustrate this important issue, consider the following nonbifurcated version of the noncompact L-P formulation (18.16):

$$\min \ z \tag{18.41a}$$

$$\sum_{p \in \mathcal{P}_d} u_{dp} = 1 \qquad\qquad d \in \mathcal{D} \tag{18.41b}$$

$$\sum_{d \in \mathcal{D}} \sum_{p \in \mathcal{P}_{ed}} h_d u_{dp} \le c_e + z \qquad\qquad e \in \mathcal{E} \tag{18.41c}$$

$$u_{dp} \in \{0, 1\} \qquad\qquad d \in \mathcal{D}, \ p \in \mathcal{P}_d. \tag{18.41d}$$

with a hop-limit imposed on the paths in $\mathcal{P}$. Note that this time (18.41) has its compact N-L counterpart (see the remark after formulation (18.8) in Subsection 18.2.2). Still, the N-L formulation provides a worse lower bound than the LP relaxation of (18.41), since the former actually admits paths longer than the hop-limit (think why).

As discussed in [53], path generation in the BB nodes generated for this problem imposes a difficulty. If we were to use BB directly for the L-P formulation (18.41), i.e., if we were to branch on variables u_{dp}, then setting a particular path flow u_{dp} to 0 would require that path $p \in \mathcal{P}_d$ is not used in the current BB solution. Still, this path can appear to be the shortest path generated as the result of solving the pricing problem. In such a case we would have to generate the second shortest path. Now, imagine that there are many u_{dp} for a demand d set to 0, say there are L such paths. Then we could in general have to generate $L + 1$ shortest paths for this demand to obtain a new path to be added to the current path-list $\mathcal{P}_d$. Since generation of L shortest paths is $\mathcal{NP}$-hard (with increasing L), path generation may become excessively time consuming.

Instead of branching on variables u_{dp} we could branch on binary variables x_{ed}, i.e., on link flows, where

$$x_{ed} = \sum_{p \in \mathcal{P}_{ed}} u_{dp}, \quad e \in \mathcal{E}, \ d \in \mathcal{D} \tag{18.42}$$

after adding equations (18.42) to the MIP formulation. Now, fixing x_{ed} to 0 makes no problem – we just forbid to use link e when we look for a shortest paths of demand d. Still, setting x_{ed} to 1 for a subset of links creates a difficulty because it is $\mathcal{NP}$-hard to find a shortest path that must traverse a given set of links.

To resolve this issue, we use a specialized branching scheme due to [53]. We add equations (18.42) to formulation (18.41) and define the branching rule as follows. If the vector $(x_{ed}, \ e \in \mathcal{E}, d \in \mathcal{D})$ in the current BB node is not integral, we select a demand $d \in \mathcal{D}$ and a node $v \in \mathcal{V}$ with two fractional flows $x_{e_1 d}$ and $x_{e_2 d}$ for some links $e_1, e_2 \in \delta^+(v)$ outgoing from node v (observe that such pair (d, v) exists). Then we divide the set $\delta^+(v)$ into two subsets $\mathcal{A}_1(v)$ and $\mathcal{A}_2(v)$ so that $e_1 \in \mathcal{A}_1(v)$ and $e_2 \in \mathcal{A}_2(v)$, and branch by creating two new BB nodes, one with additional constraint $\sum_{e \in \mathcal{A}_1(v)} x_{ed} = 0$, and the other with additional constraint

$\sum_{e \in \mathcal{A}_2(v)} x_{ed} = 0$, added to the LP subproblem of t
way we eliminate the bifurcation of the flow of deman
BB subtree below the current BB node. Note that
solution of the single-path allocation problem, the
either traverses node $v \in \mathcal{V}$ or does not. If it does,
the links outgoing from v (except for the end node t
belong to exactly one of the sets $\mathcal{A}_1(v), \mathcal{A}_2(v)$, say
flow variables assigned to the links belonging to the
0: $\sum_{e \in \mathcal{A}_2(v)} x_{ed} = 0$. This simple observation justif
Namely, we require that at each BB node only varia
of the two sets can obtain non-zero values.

For this particular branching rule, the pricing
not force using particular subsets of links by the d
subsets of links not be used. When looking for a s
dual variables π^* corresponding to constraints (18.4
Dijkstra algorithm. Since we know which links outgoi
used for selecting the shortest paths for a demand, we
the graph while searching for a shortest path. Certai
of a BB node is solved by PG, the new paths found
to the path lists $\mathcal{P}_d, d \in \mathcal{D}$ of the LP subproblem of
u_{dp} are added to constraints (18.41b)-(18.41d) and (1
Certainly, BP and BC can be combined in one sch
and-cut (BPC). This approach was for example applie
knapsack inequalities [54].

18.3.3   Heuristic Approaches

The role of heuristics in network design is two-fold.
methods discussed above, heuristic methods can hel
(sometimes even near-optimal) solutions of the MIP
tics can serve just as a means of finding a reasonable e
time and, what can be important, using simple to und
rithms. Looking at the literature of NDP, we may not
approaches applied to particular problems are either
approaches to combinatorial optimization called met
problem-dependent (specialized) heuristic algorithms.

Stochastic Meta-heuristics

Consider a combinatorial optimization problem

$$\min_{x \in X} F(x)$$

where X is a finite set and $F : X \mapsto \mathbb{R}_+$. Roughly
heuristics can be viewed as local search methods fo

mechanism for leaving the local minima at the expense of temporary increase of the cost function. There are several such methods used, including simulated annealing (SAN), evolutionary algorithm (EA), simulated allocation (SAL) or tabu search (TS). These methods are quite popular and are summarized in many handbooks, in the context of NDP for example in [2, 3].

Probably the oldest (at least among the commonly known methods) general stochastic heuristic method is called simulated annealing [55–57]. In SAN, having the current solution $x \in X$, we pick up a solution $y \in X$ in its neighborhood at random and proceed to y ($x := y$) when $F(y) \leq F(x)$. However, when $F(y) > F(x)$, we accept the move to y ($x := y$) with probability equal to $e^{\frac{-\Delta F}{T}}$ (where $\Delta F = F(y) - F(x)$ – this is the so-called Metropolis test). Once L steps of this kind are performed, the "temperature" parameter T is decreased and the search of the best solution in the optimization space X continued. A typical temperature reduction is $T := \tau T$, for some parameter τ from interval $(0, 1)$, e.g., $\tau = 0.99$. Note that for a fixed ΔF, the acceptance probability, decreases with T, so in the consecutive execution of the inner loop the uphill moves are more and more rare. The stopping criterion can be, for instance, the lack of significant improvement of the objective function.

An application of SAN requires first of all a proper specification of the neighborhood $\mathcal{N}(x)$. To illustrate this point consider for example the nonbifurcated flow allocation problem:

FAP/SP(L-P): min z (18.44a)

$$\sum_{p \in \mathcal{P}_d} u_{dp} = 1 \qquad d \in \mathcal{D} \qquad (18.44b)$$

$$\sum_{d \in \mathcal{D}} \sum_{p \in \mathcal{P}_{ed}} h_d u_{dp} \leq c_e + z \qquad e \in \mathcal{E} \qquad (18.44c)$$

$$u_{dp} \in \{0, 1\} \qquad d \in \mathcal{D}, \, p \in \mathcal{P}_d. \qquad (18.44d)$$

The set X is the set of all feasible routing vectors u and $F(u) = z$. The neighborhood of flow pattern $u \in X$ would be the set $\mathcal{N}(u)$ of flow patterns u' differing from u for exactly one demand:

$$u' \in \mathcal{N}(u) \Leftrightarrow \exists \, d' \in \mathcal{D}, \, u'_{dp} = u_{dp}, \, d \in \mathcal{D} \setminus \{d'\}, \, p \in \mathcal{P}_d. \qquad (18.45)$$

Another popular algorithms are evolutionary algorithm (see [58–60]), tabu search (see [3, 61, 62]), and simulated allocation (see [2, 63, 64]).

Feasibility Pump and Randomized Rounding

A straightforward general heuristic, different from the stochastic meta-heuristics described above is called "feasibility pump" (FP). FP is intended for finding feasible solutions of MIPs and was proposed by Fischetti et al. in [65]. The idea of FP is

explained below for pure binary programs (an extensi
is straightforward).

Let $X^{BP} \subseteq \{0,1\}^n$ denote the set of feasible solut
IP problem (18.23) with binary x, and let P^{LP} be th
LP relaxation of the BP. Suppose we are interested in
$x \in X^{BP}$. To do this, we first find a solution $x^* \in P$
then round off variables $x_1^*, x_2^*, \ldots, x_n^*$ to binary valu
off vector x^* by $[x^*] = ([x^*]_1, [x^*]_2, \ldots, [x^*]_n)$. If $[x$
Otherwise we solve the following LP:

$$\min \sum_{i \in I_0} x_i + \sum_{i \in I_1} (1 - x_i)$$

$$Ax \leq b$$

$$0 \leq x_i \leq 1, \ i = 1, 2, \ldots, n$$

where $I_0 = \{i : \ 1 \leq i \leq n, \ [x^*]_i = 0\}$ and $I_1 = \cdot$
Note that in effect we are finding a point $x \in P^{LP}$
form the current point $[x^*] \in \{0,1\}^n$. Now we der
(18.46) by x^* and repeat the above procedure. Certai
smart rounding-off (e.g., involving randomness) in or
general, we are not guaranteed that the procedure wi

In NDP, the feasibility pump approach is potent
like the pure flow allocation problem FAP/SP (see (
problem we may skip the objective function when a
is just what we are looking for. Finally, let us note th
does have an objective function $f(x)$ (like (18.7)) we n
to the LP problem (18.46), where f^* is related to the
reasonably bigger than the lower bound), and in th
optimal feasible solution.

In branch-and-bound it is important to systema
bounds for the optimal solutions in the nodes of the B
called "randomized rounding" and was proposed by P.
in [66]. Randomized rounding is a probabilistic me
procedure yields random approximate solutions of a C

Randomized rounding applied to linear relaxations
(18.7) of the single-path allocation problems goes
$d \in \mathcal{D}$ the current demand volume distribution (x_{dp}
as a probability distribution (as $\sum_{p \in \mathcal{P}_d} x_{dp} = 1, x_{dp}$
used to draw exactly one path to be used to carry th
$\text{Prob}\{x_{dp'} = 1; \ x_{dp} = 0, p = 1, 2, \ldots, \mathcal{P}_d, \ p \neq p'\} =$
any of the formulations with soft capacity constrain
(18.16)) any such drawing will yield a feasible solutio
each BB node the drawing is repeated sufficiently m
upper bounds.

18.4 Optimization Models for Multistate Networks

This section is devoted to optimization models of multistate networks. The presentation is concentrated on network design problems taking into consideration network survivability, i.e., robustness to failures. In particular, we consider problems related to protection and restoration mechanisms that can be used in survivable networks. On top of survivability, in the last part of this section, we briefly discuss models reflecting nonsimultaneous traffic matrices scenarios. Below we assume that, if not stated explicitly otherwise, all elementary paths are admissible for traffic demands.

18.4.1 Protection Models

Conceptually, the simplest way of protecting traffic against failing network components is by over-provisioning. The corresponding protection mechanisms are passive (static), and protect demands by realizing redundant (with respect to normal demand h_d, $d \in \mathcal{D}$) flows, out of which a sufficient subset survives any of the assumed failure states. Consider a network $\mathcal{N}(\mathcal{G}, \mathcal{D})$ and a failure scenario $\mathcal{S}$ (see Subsection 18.2.1).

Path Diversity

The protection concept based on path diversity (PD) follows the idea of over-provisioning by routing more demand volume than the specified values h_d, $d \in \mathcal{D}$ in the normal failure-less state $s = \emptyset$, and assuring that at least a specified fraction of the normal flow survives each failure situation s in the considered failure scenario $\mathcal{S}$. We note that several path diversity protection concepts similar to the one discussed below can be found in the literature, for example diversification [38] and its generalization demand-wise shared protection (DSP) [67–69].

The PD network design problem is given by the following noncompact linear programming formulation.

$$\textbf{PD(L-P):} \quad \min \ F(y) = \sum_{e \in \mathcal{E}} \xi_e y_e \qquad (18.47\text{a})$$

$$[\lambda_d^s] \quad \sum_{p \in \mathcal{P}_d^s} x_{dp} \geq h_d^s \qquad\qquad d \in \mathcal{D}, \ s \in \mathcal{S} \qquad (18.47\text{b})$$

$$\sum_{d \in \mathcal{D}} \sum_{p \in \mathcal{P}_{ed}} x_{dp} \leq y_e \qquad\qquad e \in \mathcal{E} \qquad (18.47\text{c})$$

$$x_{dp} \in \mathbb{R}_+ \qquad\qquad d \in \mathcal{D}, \ p \in \mathcal{P}_d \qquad (18.47\text{d})$$

$$y_e \in \mathbb{R} \qquad\qquad e \in \mathcal{E}. \qquad (18.47\text{e})$$

In essence, the formulation simply states that for eac
of flow must survive in every network state (recall th
paths of demand $d \in \mathcal{D}$ that survive in state $s \in \mathcal{S}$
some demand $d \in \mathcal{D}$ fails in state $s \in \mathcal{S}$ we assume
problem becomes infeasible due to $\mathcal{P}_d^s = \emptyset$. The symb
constraint (18.47b) denotes the corresponding dual v
the problem dual to PD(L-P). Notice that with cont
as assumed in (18.47), each inequality in (18.47c) cou
and then the problem would decompose into a set of s
demand $d \in \mathcal{D}$:

$$\min\ F(x) = \sum_{p \in \mathcal{P}_d} \left(\sum_{e \in p} \xi_e \right) x_p$$

$$[\lambda^s] \quad \sum_{p \in \mathcal{P}_d^s} x_p \geq h_d^s$$

$$x_p \in \mathbb{R}_+$$

Nevertheless, we keep the problem in form (18.47) bec
not apply to other protection/restoration mechanism
when extensions, as for example modular link capaci
In fact, as shown by the following N-L formulati
nomial for the single link failure scenario, and $\mathcal{NP}$-
scenarios (the last statement is proven in [70]).

$$\textbf{PD(N-L):} \quad \min\ F(x) = \sum_{e \in \mathcal{E}} \xi_e y_e$$

$$\sum_{e \in \delta^+(s(d))} x_{ed} - \sum_{e \in \delta^-(s(d))} x_{ed} = X_d$$

$$\sum_{e \in \delta^+(v)} x_{ed} - \sum_{e \in \delta^-(v)} x_{ed} = 0$$

$$\sum_{d \in \mathcal{D}} x_{ed} \leq y_e$$

$$X_d - x_{ed} \geq h_d^s$$

$$x_{ed} \in \mathbb{R}_+$$

$$X_d \in \mathbb{R}$$

Following [71] we now derive the pricing problem
will first derive the dual to (18.47). Certainly, for t
formulas for the LP dual that can be found in any han
(see Section 15.1.4 in Chapter 15, and for example [1

particular case we will derive the dual from scratch to illustrate the technique for the readers who are not familiar with dualization. In our derivation, analogously to (18.48), we get rid of constraint (18.47c) by substituting $\sum_{d \in \mathcal{D}} \sum_{p \in \mathcal{P}_{ed}} x_{dp}$ for y_e in the objective function (18.47a). Then we form the Lagrangean function $\mathcal{L}(x; \lambda)$ by dualizing constraints (18.47c) using dual variables $\lambda = (\lambda_d^s \geq 0 : d \in \mathcal{D}, s \in \mathcal{S})$:

$$\mathcal{L}(x; \lambda) = \sum_{e \in \mathcal{E}} \xi_e \left(\sum_{d \in \mathcal{D}} \sum_{p \in \mathcal{P}_{ed}} x_{dp} \right) + \sum_{d \in \mathcal{D}} \sum_{s \in \mathcal{S}} \lambda_d^s \left(h_d^s - \sum_{p \in \mathcal{P}_d^s} x_{dp} \right). \qquad (18.50)$$

The dual function $W(\lambda)$ is defined as

$$W(\lambda) = \min_{x \geq 0} \mathcal{L}(x; \lambda) \qquad (18.51)$$

and to express this function for a given $\lambda \geq 0$ we transform 18.50 into a suitable form with the primal variables separated as follows:

$$\mathcal{L}(x; \lambda) = \sum_{d \in \mathcal{D}} \sum_{s \in \mathcal{S}} h_d^s \lambda_d^s + \sum_{d \in \mathcal{D}} \sum_{p \in \mathcal{P}_d} \left(\sum_{e \in p} \xi_e - \sum_{s \in \mathcal{S}_p} \lambda_d^s \right) x_{dp}. \qquad (18.52)$$

Then we easily see that

$$W(\lambda) = \sum_{d \in \mathcal{D}} \sum_{s \in \mathcal{S}} h_d^s \lambda_d^s$$

when

$$\sum_{s \in \mathcal{S}_p} \lambda_d^s \leq \sum_{e \in p} \xi_e$$

for all $d \in \mathcal{D}$ and $p \in \mathcal{P}_d$, and $W(\lambda) = -\infty$ otherwise. Hence the dual problem is given by

$$\max \left\{ \sum_{d \in \mathcal{D}} \sum_{s \in \mathcal{S}} h_d^s \lambda_d^s : \lambda \geq 0, \ \sum_{s \in \mathcal{S}_p} \lambda_d^s \leq \sum_{e \in p} \xi_e, \ d \in \mathcal{D}, p \in \mathcal{P}_d \right\}. \qquad (18.53)$$

Note that problem (18.53) is always feasible ($\lambda = 0$ is a feasible solution). Introducing auxiliary dual variables $\Lambda_d, d \in \mathcal{D}$ we obtain a more handy form of the dual:

$$\textbf{PD(L-P)-D:} \quad \max W(\lambda) = \sum_{d \in \mathcal{D}} \sum_{s \in \mathcal{S}} h_d^s \lambda_d^s \qquad (18.54a)$$

$$\Lambda_d = \sum_{s \in \mathcal{S}} \lambda_d^s \qquad\qquad d \in \mathcal{D} \qquad (18.54b)$$

$$\Lambda_d \leq \sum_{e \in p} \xi_e + \sum_{s \in \bar{\mathcal{S}}_p} \lambda_d^s \qquad d \in \mathcal{D}, \ p \in \mathcal{P}_d \qquad (18.54c)$$

$$\lambda_d^s \in \mathbb{R}_+ \qquad\qquad\qquad d \in \mathcal{D}, \ s \in \mathcal{S}. \qquad (18.54d)$$

Given an optimal dual solution λ^* with respect to paths, the goal of the pricing problem for demand $d \in$ outside the current list of admissible paths $\mathcal{P}_d$ which (18.54c), i.e., which satisfies

$$\sum_{e \in p} \xi_e + \sum_{s \in \bar{\mathcal{S}}_p} \lambda_d^{s*} < \Lambda_d^*.$$

As explained in Subsection 18.3.1, if added to the li current formulation of PD(L-P), such a path $p(d)$ may function. Hence, using the pricing problem (18.55) in (18.47) using Algorithm PG given in Subsection 18.3

Under a single link failure scenario $\mathcal{S} \subseteq \{\{e\} : e \in$ for PD(L-P) can be solved in polynomial time, as obs note that with single link failures only, condition (18.5

$$\sum_{e \in p} (\xi_e + \lambda_d^{e*}) < \Lambda_d^*$$

where $\lambda_d^{e*} = 0$ if $\{e\} \notin \mathcal{S}$. The right-hand side of demand, and the link weights on the left-hand side a each demand $d \in \mathcal{D}$, violation of the dual constrair searching for a shortest path $p(d)$ between the end-n demand-dependent non-negative link weights $\gamma_e(d) =$ the Dijkstra algorithm, and comparing its length to th path $p(d)$ fulfills condition (18.56), then adding pat corresponding constraint $\Lambda_d \leq \sum_{e \in p(d)} (\xi_e + \lambda_d^{\{e\}})$ to can potentially improve the primal objective value. demand violates the dual constraints for the current s

As already mentioned, PD is $\mathcal{NP}$-hard for multipl a set s of links failing together is called the shared risk This in particular means that then the pricing problem Subsection 18.3.2 and [27]). In this case the pricing can be written as the following MIP:

$$\min \sum_{e \in \mathcal{E}} \xi_e + \sum_{s \in \mathcal{S}} \lambda_d^{s*} Y^s$$

$$\sum_{e \in \delta^+(v)} x_e - \sum_{e \in \delta^-(v)} x_e = \left\{ \begin{array}{ll} 0, & v \in \mathcal{V} \setminus \\ 1, & v = s(d) \\ -1, & v = t(d) \end{array} \right.$$

$$x_e \leq Y^s \leq 1 \qquad s \in \mathcal{S}, \ e \in s$$

$$x_e \in \{0, 1\} \qquad e \in \mathcal{E}.$$

An analogous formulation can be written for the ur formulation (18.57) according to (18.3) (see [71]).

Let (x^*, Y^*) be an optimal solution of (18.57). Then constraints (18.57b) together with the binary requirement (18.57d) will ensure that the flows x_e^* equal to 1 specify a directed single-path flow of value 1 from $s(d)$ to $t(d)$. Finally, variables Y^{s*} (identifying the failure states in which the so-specified flow fails) will also be binary when this matters (i.e., when $\lambda_d^{s*} > 0$) because they are minimized and restricted from below by the binary values.

More on path generation for problem PD(N-L) can be found in [71].

Hot-Standby

Perhaps the most common protection mechanism used in communication networks is hot-standby (HS), known also as 1+1 protection. In HS, the entire volume of every demand $d \in \mathcal{D}$ is realized on a single path and protected also by a single, dedicated failure disjoint path. HS assures 100% protection of the demands for the single-link failure scenarios. The related optimization problem is as follows.

HS(L-P): $\quad \min F(y) = \sum_{e \in \mathcal{E}} \xi_e y_e$ \hfill (18.58a)

$$\sum_{r \in \mathcal{R}_d} u_{dr} = 1 \qquad\qquad d \in \mathcal{D} \qquad (18.58b)$$

$$\sum_{d \in \mathcal{D}} \left(\sum_{r \in \mathcal{R}_{ed}^1} u_{dr} + \sum_{r \in \mathcal{R}_{ed}^2} u_{dr} \right) h_d \leq y_e \quad e \in \mathcal{E} \qquad (18.58c)$$

$$u_{dr} \in \{0,1\} \qquad\qquad d \in \mathcal{D}, \ r \in \mathcal{R}_d \quad (18.58d)$$

$$y_e \in \mathbb{R} \qquad\qquad e \in \mathcal{E}. \qquad (18.58e)$$

Note that reliable links can appear in both the primary and the backup path used for a demand; in such a case the link is loaded twice by the demand's volume.

The above problem is in fact very similar to the simple design problem SDP(L-P) (18.10) discussed in Section 18.2.3. As in SDP(L-P), the continuous link capacities are split among the demands and the resulting parts are dedicated to individual demands. Therefore, the problem can be split and solved separately for each demand. As for SDP(L-P), each of such separate formulations is totally unimodular (see Subsection 18.57) and will yield a binary optimal vertex solution even if the integrality condition (18.58d) is relaxed to $0 \leq u_{dr} \leq 1$. For a fixed $d \in \mathcal{D}$, such an optimal vertex solution will assign $u_{dr} = 1$ to a ξ-shortest pair $r \in \mathcal{R}_d$, i.e., to a pair $r = (p,q)$ with the minimum cost

$$\langle r \rangle = \sum_{e \in p} \xi_e + \sum_{e \in q} \xi_e. \qquad (18.59)$$

Note that by substituting the right-hand side of (18.58c) for y_e in objective (18.58a) we can get rid of constraint (18.58c) and then deriving the problem dual to the relaxation of HS(L-P) becomes straightforward. Using the dual, it can be easily

shown that column generation consists in finding, for
of failure-disjoint paths $r = (p, q)$ minimizing its cost

Hence, in this particular case, the primal problem
lem (18.59) are equivalent. With single link failures, a
minimizing (18.59) can be found in polynomial tim
rithm [73] or its modification described in [74]. For the
failures, finding a minimum-cost disjoint pair of path

We end the discussion of HS by a remark on ap
position. We simply notice that BD remains essenti
described in Subsection 18.3.1, with the only differenc
the feasibility test of Step 2 of Algorithm BD becomd

$$\lambda_d \leq \sum_{e \in p} \pi_e + \sum_{e \in q} \pi_e \qquad d \in \mathcal{D}, \ r =$$

Certainly, modular capacities can be considered in the

18.4.2   Restoration Models

Restoration mechanisms are dynamic in the sense t
are capable of restoring the failed flows on nonaffect
reconfigure the flow pattern in real-time.

Unrestricted Reconfiguration

We start our presentation with conceptually the simpl
ferred to as unrestricted reconfiguration (UR). When a
anism disconnects all the flows and realizes from scra
dependent flow pattern within the surviving link c
optimization problem reads:

$$\textbf{UR(L-P):} \quad \min \ F(y) = \sum_{e \in \mathcal{E}} \xi_e y_e$$

$$[\lambda_d^s] \quad \sum_{p \in \mathcal{P}_d^s} x_{dp}^s = h_d^s \qquad\qquad d \in \mathcal{D},$$

$$[\pi_e^s] \quad \sum_{d \in \mathcal{D}} \sum_{p \in \mathcal{P}_{ed}^s} x_{dp}^s \leq y_e \qquad\qquad e \in \mathcal{E},$$

$$x_{dp}^s \in \mathbb{R}_+ \qquad\qquad d \in \mathcal{D},$$

$$y_e \in \mathbb{R} \qquad\qquad e \in \mathcal{E}.$$

The problem dual to UR(L-P) can be written as follows:

$$\max \ W(\lambda, \pi) = \sum_{d \in \mathcal{D}} \sum_{s \in \mathcal{S}} h_d^s \lambda_d^s \tag{18.62a}$$

$$\lambda_d^s \leq \sum_{e \in p} \pi_e^s \qquad\qquad d \in \mathcal{D}, \ s \in \mathcal{S}, \ p \in \mathcal{P}_d^s \tag{18.62b}$$

$$\sum_{s \in \mathcal{S}_e} \pi_e^s = \xi_e \qquad\qquad e \in \mathcal{E} \tag{18.62c}$$

$$\lambda_d^s \in \mathbb{R} \qquad\qquad d \in \mathcal{D}, \ s \in \mathcal{S} \tag{18.62d}$$

$$\pi_e^s \in \mathbb{R}_+ \qquad\qquad e \in \mathcal{E}, \ s \in \mathcal{S}. \tag{18.62e}$$

In any optimal solution λ^*, π^* of the dual, the value λ_d^{s*} is the length of the shortest path $p \in \mathcal{P}_d^s$ with respect to link metrics $\pi^{s*} = (\pi_e^{s*}, \ e \in \mathcal{E}_s)$. Hence, for any failure scenario (both single and multiple link failure), the pricing problem reduces, for each network state $s \in \mathcal{S}$, to the separate shortest path problem for every demand $d \in \mathcal{D}$ for the link metrics π^{s*}. The pricing problem is thus solvable in polynomial time in the number of nodes, links, and network states. This means that UR is polynomial also when all elementary paths are admissible. Hence, it is not surprising that it can be expressed in a compact form using node-link formulation for any, even multiple, link failure scenario $\mathcal{S}$.

Also Benders decomposition (examined in detail in [76], see also [2]) for UR is quite simple, still instructive to perform (Exercise 18.6.11). In Step 2 of Algorithm BD we perform feasibility tests separately for each state $s \in \mathcal{S}$. Each such test is essentially the same as (18.20); for a fixed $s \in \mathcal{S}$ looks as follows:

$$\max \ W(\lambda, \pi) = \sum_{d \in \mathcal{D}} h_d^s \lambda_d - \sum_{e \in \mathcal{E}_s} y_e^* \pi_e \tag{18.63a}$$

$$\lambda_d \leq \sum_{e \in p} \pi_e \qquad\qquad d \in \mathcal{D}, \ p \in \mathcal{P}_d^s \tag{18.63b}$$

$$\sum_{e \in \mathcal{E}_s} \pi_e = 1 \tag{18.63c}$$

$$\pi_e \geq 0 \qquad\qquad e \in \mathcal{E}_s. \tag{18.63d}$$

For UR (and also for the problem RR consider below), Benders inequalities can be used to deduce strong valid inequalities for the incremental modular link capacity model [12, 38] (the incremental modular link capacity model is described in Subsection 18.2.3).

Certainly, UR would be highly impractical to control by a central unit, as for example in transmission networks. However, in the higher network layers (called the traffic layers) the routing protocols can sometimes realize UR in a distributed, automatic way, although with a considerable rerouting time. This is the case for example when dynamic call routing is used in an IDN (contemporary telephone network) or in an autonomous system using the OSPF protocol [2]. In any case,

the optimization problem related to UR is important
on the cost of the protected network since UR is the
network design problem we can think of.

Restricted Reconfiguration

A more realistic reconfiguration mechanism is referr
ration (RR). With RR, the flows that are not affecte
nected, i.e., nonfailing flows are preserved and only tl
In the balance of this paragraph, we assume that for
$s \in \mathcal{S}$. Although we could perform all the following d
restoration assumption, we adopt it not in order to
We will distinguish two cases of RR: with, and with
release, the capacity on the surviving parts (stubs) of
reused for backup flows; without stub-release, this c
restoration as it is reserved for the normal network s
   In the following problem RR/SR, we consider sit
of failed flows utilizing stub-release, i.e., the case w
the surviving stubs of failing paths can be reused for

$$\textbf{RR/SR(L-P):} \quad \min \ F(y) = \sum_{e \in \mathcal{E}} \xi_e y_e$$

$$[\lambda_d^{\emptyset}] \quad \sum_{p \in \mathcal{P}_d} x_{dp} = h_d \qquad\qquad d \in \mathcal{D}$$

$$[\lambda_d^{s}] \quad \sum_{q \in \mathcal{P}_d^{s}} x_{dq}^{s} = \sum_{p \in \bar{\mathcal{P}}_d^{s}} x_{dp} \qquad\qquad d \in \mathcal{D}$$

$$[\pi_e^{\emptyset}] \quad \sum_{d \in \mathcal{D}} \sum_{p \in \mathcal{P}_{ed}} x_{dp} \le y_e \qquad\qquad e \in \mathcal{E}$$

$$[\pi_e^{s}] \quad \sum_{d \in \mathcal{D}} \sum_{p \in \mathcal{P}_{ed}^{s}} x_{dp} + \sum_{d \in \mathcal{D}} \sum_{q \in \mathcal{P}_{ed}^{s}} x_{dq}^{s} \le y_e \qquad e \in \mathcal{E}$$

$$x_{dp} \in \mathbb{R}_+ \qquad\qquad d \in \mathcal{D}$$

$$x_{dp}^{s} \in \mathbb{R}_+ \qquad\qquad d \in \mathcal{D}$$

$$y_e \in \mathbb{R} \qquad\qquad e \in \mathcal{E}$$

In RR/SR(L-P), for every $d \in \mathcal{D}$, x_{dp} denote the nor
volume h_d (constraint (18.64b)), while x_{dp}^{s} are the re
in a failure state $s \in \mathcal{S} \setminus \{\emptyset\}$ to restore the failing n
(18.64c)). Constraints (18.64d) and (18.64e), in turn,
reservations are sufficient in the nominal state and in
   No compact LP formulation for problem RR/SR
link failure scenario, and the problem is most likely

failures. Hence, not surprisingly, path generation for formulation (18.64) is $\mathcal{NP}$-hard, as will be demonstrated now. The problem dual to RR/SR(L-P) reads:

$$\max \ W(\lambda) = \sum_{d \in \mathcal{D}} h_d \lambda_d^{\emptyset} \tag{18.65a}$$

$$\lambda_d^s \leq \sum_{e \in p} \pi_e^s \qquad\qquad d \in \mathcal{D}, \ s \in \mathcal{S} \setminus \{\emptyset\}, \ p \in \mathcal{P}_d^s \tag{18.65b}$$

$$\lambda_d^{\emptyset} \leq \sum_{e \in p} (\sum_{s \in \mathcal{S}_p} \pi_e^s) + \sum_{s \in \bar{\mathcal{S}}_p} \lambda_d^s \qquad d \in \mathcal{D}, \ p \in \mathcal{P}_d^s \tag{18.65c}$$

$$\sum_{e \in \mathcal{S}_e} \pi_e^s = \xi_e \qquad\qquad e \in \mathcal{E} \tag{18.65d}$$

$$\lambda_d^s \in \mathbb{R} \qquad\qquad d \in \mathcal{D}, \ s \in \mathcal{S} \tag{18.65e}$$

$$\pi_e^s \in \mathbb{R}_+ \qquad\qquad e \in \mathcal{E}, \ s \in \mathcal{S}_e. \tag{18.65f}$$

Denote the optimal dual variables solving (18.65) by λ^*, π^*. Similarly to the case of UR, it is easy to find an improving protection path $q(d)$ for every demand $d \in \mathcal{D}$ in any failure situation $s \in \mathcal{S} \setminus \{\emptyset\}$ by solving the shortest path problem between the end-nodes of d in the surviving network with respect to the link weights π^{s*}.

On the other hand, finding improving primary paths is $\mathcal{NP}$-hard already in the case of single link failures, as suggested in [12, 77], and shown later in [78] and [79]. To solve the pricing problem for a fixed demand $d \in \mathcal{D}$ we have to find a path $p(d)$ from $s(d)$ to $t(d)$ minimizing the quantity

$$\langle p \rangle = \sum_{e \in p} (\sum_{s \in \mathcal{S}_p} \pi_e^{s*}) + \sum_{s \in \bar{\mathcal{S}}_p} \lambda_d^{s*}. \tag{18.66}$$

The difficulty in minimizing the sum on the right-hand side stems from the term $\sum_{e \in p} \sum_{s \in \mathcal{S}_p} \pi_e^{s*}$. Because of that, the length of a path in the pricing problem is not a sum of independent link weights like in usual shortest-path problems. Instead, the contribution $\sum_{s \in \mathcal{S}_p} \pi_e^{s*}$ of a link to the path length depends on the set of failure situations in which the path survives, and thus on the whole path. Under a full single link failure scenario, this path length reduces to

$$\langle p \rangle = \sum_{e \in p} (\sum_{f \notin p} \pi_e^{f*}) + \sum_{s \in \bar{\mathcal{S}}_p} \lambda_d^{\{e\}*}. \tag{18.67}$$

It can be shown by reduction to the Hamilton path problem [79] or to the max-cut problem [78] that already minimizing $\sum_{e \in \mathcal{E}} (\sum_{f \notin p} \pi_e^{f*})$ is $\mathcal{NP}$-hard, which implies the $\mathcal{NP}$-hardness of minimizing (18.67). Observe that pricing problem (18.66) can be formulated as a MIP problem in the N-L notation using binary variables analogously to (18.57).

Problem RR/SR can be treated by Benders decomposition. It this case, however, the master problem (see Algorithm BD in Subsection 18.3.1) has to include variables $x = (x_{dp}, \ d \in \mathcal{D}, p \in \mathcal{P}_d)$ on top of the link y since both sets of variables influence all state $s \in \mathcal{S} \backslash$ program becomes:

$$\min \ F(y) = \sum_{e \in \mathcal{E}} \xi_e y_e$$

$$\sum_{p \in \mathcal{P}_d} x_{dp} = h_d$$

$$\sum_{d \in \mathcal{D}} \sum_{p \in \mathcal{P}_{ed}} x_{dp} \leq y_e$$

$$(x, y) \in \mathcal{Y}(\Omega)$$

and the feasibility test for state $s \in \mathcal{S} \setminus \{\emptyset\}$ follows fr

$$\max \ W(\lambda, \pi) = \sum_{d \in \mathcal{D}} (h_d - \sum_{p \in \mathcal{P}_d^s} x_{dp}^*) \lambda_d - \sum_{e \in \mathcal{E}_s} (y_e^* -$$

$$\lambda_d \leq \sum_{e \in p} \pi_e \qquad\qquad d \in \mathcal{D}, \ p \in \mathcal{P}_d^s$$

$$\sum_{e \in \mathcal{E}_s} \pi_e = 1$$

$$\pi_e \geq 0 \qquad\qquad e \in \mathcal{E}_s.$$

Note that $h_d - \sum_{p \in \mathcal{P}_d^s} x_{dp}^*$ is the actual demand volum restored in state s using the available link capacity y_e^* In Step 3 of the BD algorithm, we add an inequalit$\,$ with $W(\lambda^*, \pi^*) > 0$:

$$\Omega := \ \Omega \cup \{\sum_{e \in \mathcal{E}} \pi_e^* (y_e - \sum_{d \in \mathcal{D}} \sum_{p \in \mathcal{P}_{ed}^s} x_{dp}) \geq \sum_{d \in \mathcal{D}}$$

The counterpart for RR/SR that does not utilize $\,$

RR/NSR(L-P): $\min \ F(y) = \displaystyle\sum_{e \in \mathcal{E}} \xi_e y_e$

$$[\lambda_d^\emptyset] \quad \sum_{p \in \mathcal{P}_d} x_{dp} = h_d \qquad\qquad d \in$$

$$[\lambda_d^s] \quad \sum_{q \in \mathcal{P}_d^s} x_{dq}^s = \sum_{p \in \bar{\mathcal{P}}_d^s} x_{dp} \qquad\qquad d \in$$

$$[\pi_e^s] \quad \sum_{d \in \mathcal{D}} \sum_{p \in \mathcal{P}_{ed}} x_{dp} + \sum_{d \in \mathcal{D}} \sum_{q \in \mathcal{P}_{ed}^s} x_{dq}^s \leq y_e \qquad\qquad e \in$$

$$x_{dp} \in \mathbb{R}_+ \qquad\qquad d \in \mathcal{D}, \ p \in \mathcal{P}_d \qquad\qquad (18.70\text{e})$$

$$x_{dp}^s \in \mathbb{R}_+ \qquad\qquad d \in \mathcal{D}, \ s \in \mathcal{S}, \ p \in \mathcal{P}_d^s \qquad\qquad (18.70\text{f})$$

$$y_e \in \mathbb{R} \qquad\qquad e \in \mathcal{E}. \qquad\qquad (18.70\text{g})$$

The main difference between RR/NSR and RR/SR is in constraints (18.70d) and (18.64e), respectively. According to (18.70d) (RR/NSR), the surviving link in state s is loaded by all nominal flows, whether or not they survive in s, while according to (18.70d) (RR/SR) the link is loaded only by surviving nominal flows.

Omitting the formulation of the dual (Exercise 18.6.12), we directly proceed to the pricing problems, assuming the optimal dual variables λ^*, π^*. For the backup paths, the pricing problem consists in finding, for each state $s \in \mathcal{S} \setminus \{\emptyset\}$ and each demand $d \in \mathcal{D}$, a shortest path $q(d, s)$ between $s(d)$ and $t(d)$ in the graph composed of the nonfailing links E_s, with respect to the weights π^{s*}. For each such pair (d, s), if the path $q(d, s)$ is strictly shorter than λ_d^{s*} then it is added to the current path list $\mathcal{P}_d^s$. For the primary paths, the pricing problem is to find, for each demand $d \in \mathcal{D}$, a shortest path $p(d)$ with respect to the length defined by

$$\langle p \rangle = \sum_{e \in p} \xi_e + \sum_{s \in \bar{\mathcal{S}}_p} \lambda_d^{s*}. \qquad\qquad (18.71)$$

The above problem is identical to the pricing problem (18.56) for PD so all the observations made for (18.56) apply. As PD, problem RR/NSR is $\mathcal{NP}$-hard for multiple link failure scenarios [71].

The way of applying BD to RR/NSR is left to the reader as Exercise 18.6.13.

Situation-Independent Restoration

In restricted reconfiguration RR (and for that matter also in UR) the failed nominal flows are restored in a situation-dependent manner so in effect different backup paths can be used in different failure situations to restore the primary flow on path p. In situation-independent restoration, the failed flow on a primary path p is restored always using the same backup path q, no matter what particular failure situation affects the flow on path p. Certainly, for that we need to require that paths p and q are failure disjoint. For this type of restoration it is natural not to utilize stub-release, that is, the surviving but unused working link capacity is not reused for backup flows. Thus we concentrate on the case with no stub-release (the case with stub-release is presented in [71]). Using variables z_{dpq} to denote the flow on backup path $q \in \mathcal{Q}_p$ of working path $p \in \mathcal{P}_d$, the corresponding design problem can be formulated as follows:

SI/NSR(L-P): $\quad \min \ F(y) = \sum_{e \in \mathcal{E}} \xi_e y_e \qquad\qquad (18.72\text{a})$

$$[\lambda_d] \quad \sum_{p \in \mathcal{P}_d} x_{dp} = h_d \qquad\qquad d \in \mathcal{D} \qquad\qquad (18.72\text{b})$$

$$\sum_{q \in \mathcal{Q}_p} z_{dpq} = x_{dp} \qquad\qquad\qquad d$$

$$[\pi_e^s] \quad \sum_{d \in \mathcal{D}} \sum_{p \in \mathcal{P}_{ed}} x_{dp} + \sum_{d \in \mathcal{D}} \sum_{p \in \bar{\mathcal{P}}_d^s} \sum_{q \in \mathcal{Q}_{ep}} z_{dpq} \le y_e \qquad e$$

$$x_{dp} \in \mathbb{R}_+ \qquad\qquad\qquad\qquad d$$

$$z_{dpq} \in \mathbb{R}_+ \qquad d \in \mathcal{D}, \; p \in \mathcal{P}_d, \; q \in \mathcal{Q}_{dp}$$

$$y_e \in \mathbb{R} \qquad\qquad\qquad\qquad\qquad e$$

As far as PG for SI/NSR is concerned, note that ϵ
to primal constraints (18.72c) are not required as
(18.72c) by substituting $\sum_{q \in \mathcal{Q}_p} z_{dpq}$ for x_{dp} in (18.7
omit formulating dual to SI/NSR(L-P) (the dual pr
and proceed to the pricing problem. Assuming the c
the pricing problem consists of finding, for each dem
disjoint paths $r(d) = (p(d), q(d))$ from $s(d)$ to $t(d)$ mi

$$\langle r \rangle = \sum_{e \in p} \xi_e + \sum_{e \in q} \Big(\sum_{s \in \bar{\mathcal{S}}_p} \pi_e^{s*}$$

For each demand d, if the value $\langle r(d) \rangle$ is strictly sma
the admissible sets: $\mathcal{P}_d := \mathcal{P}_d \cup \{p(d)\}$ and $\mathcal{Q}_{p(d)} :=$
may already contain $p(d)$ or $\mathcal{Q}_{p(d)}$ may already conta
  The pricing problem (18.73) is $\mathcal{NP}$-hard already fo
as shown in [80]. In fact, it is demonstrated in [81] for
SI/NSR itself is $\mathcal{NP}$-hard (already for single link fai
using a reduction from Two Diverse Paths problem
$\mathcal{NP}$-hard Disjoint Connecting Paths problem [10] – c
disjoint paths between two distinct pairs of nodes.
  Application of Benders decomposition to SI/NSR t
as in the BD algorithm presented in Section 18.3.1, a
form the following version of FAP (see (18.16):

$$\min \; z$$

$$[\lambda_d] \quad \sum_{p \in \mathcal{P}_d} \sum_{q \in \mathcal{Q}_p} z_{dpq} = h_d$$

$$[\pi_e^s] \quad \sum_{d \in \mathcal{D}} \sum_{p \in \mathcal{P}_{ed}} \sum_{q \in \mathcal{Q}_p} z_{dpq} + \sum_{d \in \mathcal{D}} \sum_{p \in \bar{\mathcal{P}}_d^s} \sum_{q \in \mathcal{Q}_{ep}} z_{dpq} \le y_e^* \; +$$

$$z_{dpq} \in \mathbb{R}_+$$

where y^* is the solution of the BD master problem. Problem dual to (18.74) is as follows:

$$\max\ W(\lambda, \pi) = \sum_{d \in \mathcal{D}} h_d \lambda_d - \sum_{e \in \mathcal{E}} \left(\sum_{s \in \mathcal{S}_e \setminus \{\emptyset\}} \pi_e^s \right) y_e^* \tag{18.75a}$$

$$\lambda_d \le \sum_{e \in p} \left(\sum_{s \in \mathcal{S}_e \setminus \{\emptyset\}} \pi_e^s \right) + \sum_{e \in q} \left(\sum_{s \in \bar{\mathcal{S}}_p} \pi_e^s \right) \qquad d \in \mathcal{D},\ p \in \mathcal{P}_d,\ q \in \mathcal{Q}_{dp} \tag{18.75b}$$

$$\sum_{e \in \mathcal{E}} \sum_{s \in \mathcal{S}_e \setminus \{\emptyset\}} \pi_e^s = 1 \tag{18.75c}$$

$$\pi_e^s \ge 0 \qquad\qquad e \in \mathcal{E},\ s \in \mathcal{S}_e \setminus \{\emptyset\}. \tag{18.75d}$$

Therefore, if for optimal solution λ^*, π^* of the above dual it turns out that $W(\lambda^*, \pi^*) > 0$ then the BD inequality

$$\sum_{e \in \mathcal{E}} \left(\sum_{s \in \mathcal{S}_e \setminus \{\emptyset\}} \pi_e^{s*} \right) y_e \ge \sum_{d \in \mathcal{D}} h_d \lambda_d^*. \tag{18.76}$$

is added to the set Ω in Step 3 of the BD algorithm.

Note that using BD for SI/NSR can be beneficial (especially when modular link capacities are considered) because of a large number of the flow variables z required in the primal formulation of the problem.

Situation-Independent Restoration with Nonbifurcated Flows

Let us now consider the nonbifurcated version of SI/NSR which assumes that the entire volume h_d of each demand d is realized on a single primary path $p \in \mathcal{P}_d$ protected by a single failure-disjoint path $q \in \mathcal{Q}_p$. This version of SI/NSR is a backup capacity sharing counterpart of the problem (18.58) with the hot-standby 1+1 protection where backup capacity is dedicated to individual primary paths. The corresponding mixed-binary programming (MBP) problem reads:

$$\mathbf{SI/NBR(L\text{-}P):} \quad \min\ F(y) = \sum_{e \in \mathcal{E}} \xi_e (y_e^1 + y_e^2) \tag{18.77a}$$

$$\sum_{r \in \mathcal{R}_d} u_{dr} = 1 \qquad\qquad d \in \mathcal{D} \tag{18.77b}$$

$$\sum_{d \in \mathcal{D}} \sum_{r \in \mathcal{R}_{ed}^1} h_d u_{dr} \le y_e^1 \qquad\qquad e \in \mathcal{E} \tag{18.77c}$$

$$\sum_{d \in \mathcal{D}} \sum_{r = (p,q) \in \mathcal{R}_{ed}^2 :\ s \in \bar{\mathcal{S}}_p} h_d u_{dr} \le y_e^2 \qquad e \in \mathcal{E},\ s \in \mathcal{S}_e \setminus \{\emptyset\} \tag{18.77d}$$

$$u_{dr} \in \{0, 1\} \qquad\qquad d \in \mathcal{D},\ r \in \mathcal{R}_d \tag{18.77e}$$

$$y_e^1, y_e^2 \in \mathbb{R} \qquad\qquad e \in \mathcal{E}. \tag{18.77f}$$

Above, y_e^1 and y_e^2 denote, respectively, the primary an
vation on link $e \in \mathcal{E}$. The primary capacity y^1 is re
primary) flows, while the backup capacity y^2 is reserv
tainly, contrary to HS (see (18.58)), backup capacity i
flows in different failure situations $s \in \mathcal{S} \setminus \{\emptyset\}$.

Although looking quite different at first glance, t
(18.77) is equivalent to problem SI/NSR (18.72). Th
$r \in \mathcal{R}_d$ can have the same primary path $p \in \mathcal{P}_d$ an
sponding paths $q \in \mathcal{Q}_p$ are used to protect path p ir
problem dual to the relaxation of (18.77) is identical

Finally, observe that we can write down a compac
of SI/NBR, although for the LP relaxation of the pro
does not exist. This MIP formulation is as follows:

SI/NBR(N-L):   min $F(y) = \sum_{e \in \mathcal{E}} \xi_e (y_e^1 + y_e^2)$

$$\sum_{e \in \delta^+(v)} x_{ed} - \sum_{e \in \delta^-(v)} x_{ed} = \begin{cases} 0, & v \in \mathcal{V} \setminus \{s(c \\ 1, & v = s(d) \end{cases}$$

$$\sum_{e \in \delta^+(v)} z_{ed} - \sum_{e \in \delta^-(v)} z_{ed} = \begin{cases} 0, & v \in \mathcal{V} \setminus \{s(d \\ 1, & v = s(d) \end{cases}$$

$$x_{ed} + z_{ed} \leq 1 \quad s \in \mathcal{S}, \ e \in s, \ d \in \mathcal{D}$$

$$\sum_{d \in \mathcal{D}} h_d x_{ed} \leq y_e^1 \quad e \in \mathcal{E}$$

$$x_{ed} \leq W_{ds} \leq \sum_{e \in s} x_{ed} \quad s \in \mathcal{S} \setminus \{\emptyset\}, \ e \in s, \ d \in$$

$$Y_{eds} \geq z_{ed} - W_{ds} \quad e \in \mathcal{E}, \ d \in \mathcal{D}, \ s \in \mathcal{S} \setminus \{\emptyset\}$$

$$\sum_{d \in \mathcal{D}} h_d Y_{eds} \leq y_e^2 \quad e \in \mathcal{E}, \ s \in \mathcal{S} \setminus \{\emptyset\}$$

$$x_{ed}, z_{ed} \in \{0, 1\} \quad e \in \mathcal{E}, \ d \in \mathcal{D}$$

$$0 \leq Y_{eds} \leq 1 \quad e \in \mathcal{E}, \ d \in \mathcal{D}, \ s \in \mathcal{S} \setminus \{\emptyset\}.$$

In the above formulation, binary variables $x_{ed}, e \in$
$p(d) = \{e \in \mathcal{E} : x_{ed} = 1\}$ for demand $d \in \mathcal{D}$, and bin
backup path $q(d) = \{e \in \mathcal{E} : z_{ed} = 1\}$. Note that due
$q(d)$ are failure disjoint. Variables $W_{ds}, d \in \mathcal{D}, \ s \in$
continuous, can assume only binary values and indica
$p(d)$ works in failure situation s ($W_{ds} = 1$) or it is
variables $Y_{eds}, e \in \mathcal{E}, \ d \in \mathcal{D}, \ s \in \mathcal{S} \setminus \{\emptyset\}$ define the loa
path $q(d)$ on link e in situation s. In particular, Y_{eds}
link e in situation s because $e \in q(d)$ and $p(d)$ fai
constraints (18.78g) and (18.78j).

Link Protection

The last problem considered in this subsection is related to a restoration mechanism called link protection (LPR). LPR assumes a single link failure scenario, and contrary to the previous mechanisms, restores the failed capacity of link rather than end-to-end flows that use a failed link. This mechanism is applicable to facility networks like SDH/SONET [2]. The related NDP is as follows.

$$\textbf{LPR(L-P):} \quad \min \ F(y) = \sum_{e \in \mathcal{E}} \xi_e(y_e^1 + y_e^2) \tag{18.79a}$$

$$\sum_{p \in \mathcal{P}_d} x_{dp} = h_d \qquad d \in \mathcal{D} \tag{18.79b}$$

$$\sum_{d \in \mathcal{D}} \sum_{p \in \mathcal{P}_{ed}} x_{dp} \le y_e^1 \qquad e \in \mathcal{E} \tag{18.79c}$$

$$\sum_{q \in \mathcal{L}_e} u_{eq} = 1 \qquad e \in \mathcal{E} \tag{18.79d}$$

$$\sum_{q \in \mathcal{L}_{fe}} y_e^1 u_{eq} \le y_f^2 \qquad e \in \mathcal{E}, \ f \in \mathcal{E}, \ e \ne f \tag{18.79e}$$

$$x_{dp} \in \mathbb{R}_+ \qquad d \in \mathcal{D}, \ p \in \mathcal{P}_d \tag{18.79f}$$

$$u_{eq} \in \{0, 1\} \qquad e \in \mathcal{E}, \ q \in \mathcal{L}_e \tag{18.79g}$$

$$y_e^1, y_e^2 \in \mathbb{R} \qquad e \in \mathcal{E}. \tag{18.79h}$$

Above, $\mathcal{L}_e$ denotes the set of paths between the end nodes of link e that are admissible for restoring the primary capacity y_e^1, and $\mathcal{L}_{fe}$ denotes the set of paths from $\mathcal{L}_e$ that contain link f. Primary capacity is restored in a nonbifurcated way on a single path from $\mathcal{L}_e$—this is assured by constraint (18.79d). The restoration flow of each (individually) failed link e uses the protection capacity y_f^2, $f \in \mathcal{E} \setminus \{e\}$ of the other links. Note that the above formulation is bilinear (and hence not a MIP formulation) as it contains multiplication of variables in (18.79e). This bilinearity can be removed using standard methods (see [83]). In the LP relaxation, this bilinearity is not any issue, as we can use absolute continuous flows z_{eq} instead of binary flows u_{eq}, and substitute constraints (18.79d) and (18.79e) with

$$\sum_{q \in \mathcal{L}_e} z_{eq} = y_e^1 \qquad e \in \mathcal{E} \tag{18.80a}$$

$$\sum_{q \in \mathcal{L}_{fe}} z_{eq} \le y_f^2 \qquad e \in \mathcal{E}, \ f \in \mathcal{E}, \ e \ne f. \tag{18.80b}$$

Path generation for the LP relaxation of LPR(L-P) is polynomial and hence LPR can be approached with BP (BPC). LPR can also be written in a compact N-L notation and approached with BB (CB, BC).

18.4.3   Multihour and Multiperiod De

Multihour design refers to the situation when we are
many nonsimultaneous traffic matrices. This is for e
communication networks with noncoincident traffic
uncertain traffic when we have a large set of possible
know which of them will actually be realized. Thus w
i.e., a set $\mathcal{T}$ of the traffic states $\tau \in \mathcal{T}$. In this case, f
all states $\tau \in \mathcal{T}$ all links are available) and the states
vectors h^τ, $\tau \in \mathcal{T}$. In a demand scenario, we do not
    Consider the following multistate NDP:

$$\textbf{MH(L-P):} \quad \min \ F(y) = \sum_{e \in \mathcal{E}} \xi_e y_e$$

$$[\lambda_d^\tau] \quad \sum_{p \in \mathcal{P}_d^\tau} x_{dp}^\tau = h_d^\tau \qquad\qquad d \in \mathcal{D},$$

$$[\pi_e^\tau] \quad \sum_{d \in \mathcal{D}} \sum_{p \in \mathcal{P}_{ed}^\tau} x_{dp}^\tau \le y_e \qquad\qquad e \in \mathcal{E},$$

$$x_{dp}^\tau \in \mathbb{R}_+ \qquad\qquad d \in \mathcal{D},$$

$$y_e \in \mathbb{R} \qquad\qquad e \in \mathcal{E}.$$

The problem assumes state-dependent routing and in
problem UR (18.61) considered in Subsection 18.4.2
(examined in detail in [84]) are very very similar to th
can be easily written in the compact N-L notation (
the aggregated N-L notation).
    Formulation (18.81) assumes state-dependent (dy
is a correct assumption for different traffic hours, it be
we deal with traffic uncertainty. For the latter case
static state-independent routing (called also oblivious
this:

$$\textbf{UT(L-P):} \quad \min \ F(y) = \sum_{e \in \mathcal{E}} \xi_e y_e$$

$$[\lambda_d] \quad \sum_{p \in \mathcal{P}_d} x_{dp} = 1 \qquad\qquad d$$

$$[\pi_e^\tau] \quad \sum_{d \in \mathcal{D}} \sum_{p \in \mathcal{P}_{ed}} h_d^\tau x_{dp} \le y_e \qquad\qquad e$$

$$x_{dp} \in \mathbb{R}_+ \qquad\qquad d$$

$$y_e \in \mathbb{R} \qquad\qquad e$$

The nature of UT is different than that of MH. In part
PG is different, and BD is not applicable to UT. More

an N-L formulation, it does not possess an aggregated N-L formulation (Exercise 18.6.14).

To end the presentation of NDPs with nonsimultaneous traffic matrices we formulate, for completeness, the so-called multiperiod design problem. In this case the states $\tau \in \mathcal{T} = \{1, 2, \ldots, T\}$ correspond to consecutive periods of network operation, say, to consecutive years. At the beginning of each period the capacity of the network is extended (see Chapter 11 in [2]).

$$\textbf{MP(L-P):} \quad \min \ F(y) = \sum_{\tau=1}^{T} \sum_{e \in \mathcal{E}} \xi_e^\tau y_e^\tau \tag{18.83a}$$

$$[\lambda_d^\tau] \quad \sum_{p \in \mathcal{P}_d^\tau} x_{dp}^\tau = h_d^\tau \qquad\qquad d \in \mathcal{D},\ \tau \in \mathcal{T} \tag{18.83b}$$

$$[\pi_e^\tau] \quad \sum_{d \in \mathcal{D}} \sum_{p \in \mathcal{P}_{ed}^\tau} x_{dp}^\tau \le \sum_{t=1}^{\tau} y_e^t \qquad e \in \mathcal{E},\ \tau \in \mathcal{T} \tag{18.83c}$$

$$x_{dp}^\tau \in \mathbb{R}_+ \qquad\qquad d \in \mathcal{D},\ \tau \in \mathcal{T},\ p \in \mathcal{P}_d^\tau \tag{18.83d}$$

$$y_e^\tau \in \mathbb{R} \qquad\qquad e \in \mathcal{E},\ \tau \in \mathcal{T}. \tag{18.83e}$$

This problem also possesses an N-L formulation and an aggregated N-L formulation. It can also be easily treated by PG and BD.

18.5 Concluding Remarks

In this chapter, we have discussed optimization models for communication networks design and planning. An optimization model is understood as an appropriate mathematical problem formulation expressed in the language of multicommodity flow networks and mixed-integer programming, together with a set of algorithms that can be applied for finding the problem's optimal or near-optimal solutions. Section 18.2 is devoted to problem formulations. Consequently, it introduces the notation and discusses two basic types of network design problem (NDP) formulations: dimensioning problems (such as SDP) involving simultaneous optimization of link capacity reservations and flow patterns, and flow allocation problems (such as FAP) involving only flow patterns optimization. For both types of problems, we have used link-path notation leading to noncompact formulations, and node-link notation (and its variant – aggregated notation) leading to compact formulations. We have also discussed different variants of SDP and FAP including modular dimensioning and several routing requirements. Finally, we have discussed concave and convex link dimensioning models and their MIP/LP approximations.

In Section 18.3, we have surveyed basic optimization methods and algorithms for solving the considered multicommodity flow problem formulations, with an emphasis put on exact methods of mixed-integer programming. Here, such methods

as path generation, Benders decomposition, cutting p
its variants have been discussed.

Section 18.4 presents a number of advanced opti
multistate design of communication networks, thus en
networks, multihour design and multiperiod design. I
integer programming techniques presented in Sectio
presented problems. In particular, path generation i
noncompact LP models like PD, RR, and SI. Bran
modular dimensioning and for the single-path routing
the latter case, branch-and-price and branch-and-pric
when noncompact formulations are considered. Fin
may become a helpful means for improving the opti
creases the number of variables in the master proble
variables.

The potential number of valid NDPs combining c
routing, protection/restotration, and multihour and
mous. Certainly, in a chapter of this size, we have
aspects of NDPs and had to omit certain models. In
the case of multilayer networks (a communication net
several resource layers, as for example IDN-over-SO
IP-over-WDM networks), radio networks (for exampl
as wireless mesh networks or cellular networks), IP
routing (of the OSPF type), and others. These aspec
tant and valid NDPs even more. We hope, however, t
this chapter forms a critical mass for understanding t
ing and integer programming techniques, and will pre
the omitted models as well. For more reading, we re
books on network optimization, design, and planning
cited in the main body of this chapter, and, eventually.
valuable papers on network design and multicommodi
communications and operations research journals and

Finally, we should emphasize that the mixed-inte
can in many cases help to effectively solve the conside
to optimality or near-optimality. However, most of
network design are $\mathcal{NP}$-hard, and because of that we
approaches will always be effective, at least for lar
practical guideline, we may roughly say that in gene
a weak LP relaxation and/or with only weak cuttin
be efficiently solvable even with the best MIP solvers
(compact) linear programming formulations that gua
reasonable time. The reader is encouraged to check t
problem instances (and their LP relaxations) formula
state-of-the-art MIP solver for network examples from
described in [85].

18.6 Exercises

Exercise 18.6.1. Give an example of a simple network configuration with more than one path-flow pattern realizing a given link-flow pattern.

Exercise 18.6.2. Prove that the SPA rule leads to an optimal solution of SDP. Give an example when an optimal solution assigns non-zero flows to more than one path of a demand.

Exercise 18.6.3. Write down the incremental version of (18.12).

Exercise 18.6.4. Find a piecewise linear approximation of $f(z) = \sqrt{z}$ in $[0, +\infty)$ with $K = 4$ linear pieces and exact values for $z = 0, 1, 4, 9, 16$. Write down SDP/CV for this particular approximation.

Exercise 18.6.5. Find a piecewise linear approximation of $f(z) = z^2$ in $[0, +\infty)$ with $K = 4$ linear pieces and exact values for $z = 0, 1, 2, 3, 4$. Write down SDP/CX for this particular approximation.

Exercise 18.6.6. Consider a triangle symmetrical undirected network with 3 nodes ($\mathcal{V} = \{v_1, v_2, v_3\}$), 3 links ($\mathcal{E} = \{e_1 = \{v_1, v_2\}, e_2 = \{v_1, v_3\}, e_3 = \{v_2, v_3\}\}$), and 3 demands ($\mathcal{D} = \{d_1 = \{v_1, v_2\}, d_2 = \{v_1, v_3\}, d_3 = \{v_2, v_3\}\}$). Assume $c_1 = 10, c_2 = 20, c_3 = 20$ and $h_1 = 20, h_2 = 10, h_3 = 10$. Suppose that the initial lists of admissible paths for the demands contain just the direct paths ($\mathcal{P}_{d_i} = \mathcal{P}_i = \{p_i = \{e_i\}\}$, $i = 1, 2, 3$). Apply the PG algorithm to this setting. Then consider objective (18.1a) instead of (18.16a) and see what metrics will then be used in PG. (Hint: Write down the dual problem for (18.1).)

Exercise 18.6.7. Consider a fully connected network with the set of nodes $\mathcal{V} = \{v_1, v_2, \ldots, v_V\}$. The set of demands corresponds to all pairs of nodes. Suppose that all but one demand values are equal to 1 and that one demand value (for, say, demand $d_1 = \{v_1, v_2\}$) is equal to $1 + \varepsilon$. Suppose also that all capacity reservations c_e are equal to 1, except for the links $\{v_1, v_3\}, \{v_3, v_4\}, \ldots, \{v_{V-1}, v_V\}, \{v_V, v_2\}$ for which the capacity reservation is equal to $1 + \varepsilon$. What is the optimal solution of (18.16)? Suppose we start the PG process with the single direct paths on the demand admissible path-lists and apply the PG algorithm. Will it take a long time to get the final optimal solution with $z^* = 0$ (think of large $V = |\mathcal{V}|$)?

Exercise 18.6.8. Apply BD to the network from Example 1. What are the cuts corresponding to the generated inequalities. Write down the dual test (and the generated metric inequality) corresponding to (18.16) written in the N-L formulation.

Exercise 18.6.9. Write down the BB tree corresponding to the BB process executed for (18.31) and for (18.33). Compare the trees.

Exercise 18.6.10. Derive the problem dual to LP(Ω), a BB subproblem for (18.40), and in this way verify the PG rule.

684

Exercise 18.6.11. Execute the BD algorithm for nodes, one demand between them, and two parallel nodes, with the unit costs $\xi_{e_1} = 1, \xi_{e_2} = 2$. The nom while the demand volumes in the two considered si equal to $h^s = 1$. What is the final set of inequalities in terms of y? Give graphical interpretation of the ob by Benders inequalities.

Exercise 18.6.12. Derive the problem dual to RR, variables specified in formulation (18.70).

Exercise 18.6.13. Derive the BD feasibility tests a equalities for RR/NSR (18.70).

Exercise 18.6.14. Derive the pricing formulas for (1 where they differ? Why BD makes no sense for U notation is not applicable to UT?

References

[1] J. Vasseur, M. Pickavet, and P. Demeester, *Netw Restoration.* San Francisco: Morgan Kaufmann

[2] M. Pióro and D. Medhi, *Routing, Flow, and Cap tion and Computer Networks.* San Francisco: N

[3] W. D. Grover, *Mesh-Based Survivable Networks Optical, MPLS, SONET and ATM Networking.* Prentice Hall, 2003.

[4] L. R. Ford and D. R. Fulkerson, *Flows in Networ* University Press, 1962.

[5] L. Gouveia, P. Patricio, and A. Sousa, "Optim small number of hops," in *Telecommunications nology*, B. G. S. Raghvan and E. Wasil, Eds. S

[6] M. Belaidouni and W. Ben-Ameur, "On the m unsplittable flow problem," *RAIRO – Operations* 253–273, 2007.

[7] Y. Dinitz, N. Garg, and M. Goemans, "On the si problem," in *Proceedings 39th Annual Symposi puter Science*, 1998, pp. 290–299.

[8] J. Kleinberg, "Approximation algorithms for dis dissertation, 1996.

[9] A. Schrijver, P. Seymour, and P. Winkler, "The ring loading problem," *SIAM Journal of Discrete Mathematics*, vol. 11, no. 1, pp. 1–14, 1998.

[10] M. R. Garey and D. R. Johnson, *Computers and Intractability: A Guide to the Theory of NP-Completeness.* New York: W. H. Freeman & Co., 1979.

[11] M. Minoux, "Discrete cost multicommodity network optimization problems and exact solution methods," in *Annals of Operations Research*, 2001, vol. 106, pp. 19–46.

[12] R. Wessäly, "DImensioning Survivable Capacitated NETworks," Ph.D. dissertation, Technische Universität Berlin, April 2000.

[13] M. Minoux, "Network synthesis and optimum network design problems: Models, solution methods and applications," *Networks*, vol. 19, pp. 313–360, 1989.

[14] B. Yaged, "Minimum cost routing for static network models," *Networks*, vol. 1, pp. 139–172, 1971.

[15] M. Minoux, *Mathematical Programming: Theory and Algorithms.* New York: John Wiley & Sons, 1986.

[16] J. B. Rosen, "The gradient projection method for nonlinear programming: Part I, linear constraints," *SIAM Journal*, vol. 8, pp. 181–217, 1960.

[17] CPLEX, *ILOG CPLEX 11.0 User's Manual.* ILOG, 2007.

[18] L. Lasdon, *Optimization Theory for Large Systems.* New York: Macmillan, 1970.

[19] A. Schrijver, *Theory of Linear and Integer Programming.* New York: John Wiley & Sons, 1986.

[20] N. Garg and J. Könemann, "Faster and simpler algorithms for multicommodity flow and other fractional packing problems," *SIAM Journal on Computing*, vol. 37, no. 2, pp. 630–652, 2007.

[21] M. Bárász, Z. Fekete, A. Jüttner, M. Makai, and J. Szabó, "Qos aware and fair resource allocation scheme in transport networks," in *International Conference on Transport Optical Networks (ICTON)*, June 2006, pp. 239–242.

[22] J. F. Benders, "Partitioning procedures for solving mixed variable programming problems," *Numerische Mathematik*, vol. 4, pp. 238–252, 1962.

[23] H. Okamura and P. Seymour, "Multicommodity flows in planar graphs," *Journal of Combinatorial Theory*, vol. 31, no. 1, pp. 75–81, 1981.

[24] M. Padberg, "Classical cuts for mixed-integer programming and branch-and-cut," *Annals of Operations Research*, vol. 139, pp. 321–352, 2005.

686

[25] M. Padberg, "Cutting plane methods for mixed-
site tutorial.

[26] L. A. Wolsey, *Integer Programming.* New York

[27] G. L. Nemhauser and L. A. Wolsey, *Integer and*
New York: John Wiley & Sons, 1988.

[28] M. Grötschel, L. Lovász, and A. Schrijver, "T
consequences in combinatorial optimization," (
pp. 169–197, 1981.

[29] R. K. Ahuja, T. L. Magnanti, and J. B. Orlin, *N*
rithms, and Applications. Englewood Cliffs, NJ

[30] R. E. Gomory, *An Algorithm for the Mixed In*
Memoranda. RAND Corporation, 1960, no. RM

[31] R. E. Gomory, "An algorithm for integer solutic
Recent Advances in Mathematical Programming
Eds., 1963, pp. 269–302.

[32] A. Balakrishnan, "A dual-ascent oricedure for la
work design," *Operations Research*, vol. 37, no. 5

[33] F. Cruz, G. Mateus, and J. M. Smith, "A bra:
slve a multi-level network optimization problen
Modelling and Algorithms 2, pp. 37–56, 2003.

[34] A. D. Jongh, M. Gendreau, and M. Labbe, "Findi
munications networks with two technologies," *Op*
81–92, January 1999.

[35] E. Gourdin, M. Labbe, and H. Yaman, "Teleco
in *Facility Location: Applications and Theory*, Z
Eds. Berlin, Heidelberg, New York: Springer-V

[36] A. Jüttner, B. Szviatovszki, I. Mécs, and Z. Rajkó
method for the QoS routing problem," in *Proceed*
2001, pp. 859–868.

[37] M. Grötschel, C. Monma, and M. Stoer, *Netw*
in Operations Research and Management Science
Amsterdam, 1995, vol. 7, ch. 10 : Design of Survi
m.O. Ball, T.L. Magnanti, C.L. Monma, G.L. Ne

[38] G. Dahl and M. Stoer, "A cutting plane algorith
vivable network design problems," *INFORMS Jo*
no. 1, pp. 1–11, 1998.

[39] D. Alevras, M. Grötschel, and R. Wessäly, "Cost-efficient network synthesis from leased lines," *Annals of Operations Research*, vol. 76, pp. 1–20, 1998.

[40] V. Gabrel, A. Knippel, and M. Minoux, "Exact solution of multicommodity network optimization problems with general step cost functions," *Operations Research Letters*, vol. 25, pp. 15–23, 1999.

[41] C. Bienstock and G. Muratore, "Strong inequalities for capacitated survivable network design problems," *Mathematical Programming*, vol. A89, pp. 127–147, 2000.

[42] C. Barnhart, E. Johnson, G. Nemhauser, G. Savelsbergh, and P. Vance, "Branch-and-price: column generation for solving huge integer programs," *Operations Research*, vol. 46, no. 3, pp. 316–329, 1998.

[43] J. Geffard, "A solving method for singly routing traffic demand in telecommunication networks," *Annales des Telecommunicationes*, vol. 56, no. 3-4, pp. 140–149, 2001.

[44] M. Belaidouni and W. Ben-Ameur, "A superadditive approach to solve the minimum cost single path routing problem: preliminary results," In Proceedings of the 1st International Network Optimization Conference (INOC 2003), Paris, France, 2003, pp. 67–71.

[45] K. Hoffman and M. Padberg, "LP-based combinatorial problem solving," *Annals of Operations Research*, vol. 4, pp. 145–194, 1985.

[46] E. Balas, S. Ceria, G. Cornuéjols, and N. Natraj, "Gomory cuts revisited," *Operations Research Letters*, vol. 19, pp. 1–9, 1996.

[47] L. Wolsey, "Strong formulations for mixed integer programming: A survey," *Mathematical Programming*, no. 45, pp. 173–191, 1989.

[48] E. Balas, S. Ceria, and G. Cornuejols, "Mixed 0-1 programming by lift-and-project in a branch-and-cut framework," *Management Science*, no. 42, pp. 1229–1246, September 1996.

[49] O. Günlük, "A branch-and-cut algorithm for capacitated network design," *Mathematical Programming*, vol. A, no. 86, pp. 17–39, 1999.

[50] C. Barnhart, C. Hane, E. Johnson, and G. Sigismondi, "A column generation and partitioning approach for multi-commodity flow problems," *Telecommunication Systems*, vol. 3, pp. 239–258, 1995.

[51] J. Desrosiers, Y. Dumas, M. Solomon, and F. Soumis, "Time constrained routing and scheduling," in *Operations Research and Management Science*, M. Ball, T. Magnanti, C. Monma, and G. Nemhauser, Eds. Amsterdam: Elsevier Science, B. V., 1995, vol. 8, pp. 35–139.

[52] F. Vanderbeck and L. Wolsey, "An exact algorith
Operations Research Letters, vol. 19, pp. 151–15

[53] C. Barnhart, C. Hane, and P. Vance, "Using
solve origin-destination integer multicommodity
Research, vol. 48, no. 2, pp. 318–326, 2000.

[54] Z. Gu, G. L. Nemhauser, and M. W. P. Savelsber
for 0–1 integer programs: computation," INFO
vol. 10, pp. 427–438, 1998.

[55] S. Kirkpatrick, C. D. Gelatt, and M. P. Vecchi,
annealing," Science, vol. 220, no. 4598, pp. 671–

[56] D. S. Johnson, C. R. Aragon, L. A. McGeoch, an
by simulated annealing: An experimental evalu
vol. 39, no. 1, 1991.

[57] J. Korst, E. Aarts, and A. Korst, Simulated A
chines: A Stochastic Approach to Combinator
Computing. New York: John Wiley & Sons, 19

[58] D. Goldberg, Genetic Algorithms in Search,
Learning. Reading, Massachusetts: Addison-W

[59] Z. Michalewicz, Genetic Algorithms + Data Stru
Berlin, Heidelberg, New York: Springer, III editi

[60] E. Mulyana and U. Killat, "Load balancing in IF
weights," in Proc. 2nd Polish-German Teletraf
2000, Gdansk, Poland.

[61] F. Glover, "Tabu search fundamentals and uses
Boulder, Tech. Rep., 1994.

[62] M. Laguna and F. Glover, "Bandwidth packing
Management Science, vol. 39, pp. 492–500, 1993.

[63] P. Gajowniczek and M. Pióro, "Solving an O
simulated allocation," in The first Polish-Ger
(PGTS'2000), 2000, Dresden, Germany.

[64] P. Zhou, P.-H. Yuh, and S. Sapatnekar, "Applica
chip design using simulated allocation," in De
(ASP-DAC), January 2010, pp. 517–522.

[65] M. Fischetti, F. Glover, and A. Lodi, "The feas
Programming, vol. 104, no. 1, pp. 91–104, 2005.

[66] P. Raghavan and C. Thompson, "Randomized rounding: A technique for provably good algorithms and algorithmic proofs," *Combinatorica*, vol. 7, no. 4, pp. 365–374, 1987.

[67] A. Koster, A. Zymolka, M. Jäger, and R. Hülsermann, "Demand-wise shared protection for meshed optical networks," *Journal of Network and Systems Management*, vol. 13, no. 1, pp. 35–55, March 2005.

[68] R. Wessäly, S. Orlowski, A. Zymolka, A. Koster, and C. Gruber, "Demand-wise shared protection revisited: A new model for survivable network design," in *Proceedings of the 2nd International Network Optimization Conference (INOC 2005), Lisbon, Portugal*, March 2005, pp. 100–105.

[69] A. Koster and A. Zymolka, "Demand-wise shared protection and multiple failures," in *Proceedings of the 3rd International Network Optimization Conference (INOC 2007), Spa, Belgium*, April 2007.

[70] A. Tomaszewski, M. Pióro, and M. Żotkiewicz, "On the complexity of resilient network design," *Networks: an International Journal*, vol. 55, no. 2, pp. 109–118, 2010.

[71] S. Orlowski and M. Pióro, "On the complexity of column generation in network design with path-based survivability mechanisms," *Networks*, to appear in 2011.

[72] J. Strand, A. L. Chiu, and R. Tkach, "Issues for routing in the optical layer," *IEEE Communications Magazine*, pp. 81–87, 2001.

[73] J. W. Suurballe, "Disjoint paths in a network," *Networks*, vol. 4, pp. 125–145, 1974.

[74] R. Bhandari, *Survivable Networks – Algorithms for Diverse Routing*. Norwell, Massachusetts: Kluwer, 1999.

[75] J. Hu, "Diverse routing in optical mesh networks," *IEEE Trans. Com.*, vol. 51, no. 3, pp. 489–494, 2003.

[76] M. Minoux and J.-Y. Serreault, "Synthese optimal d'un reseau de telecommunication avec contraintes de securite," *Annales des Telecommunications*, vol. 36, no. 3-4, pp. 211–230, 1981.

[77] K. Murakami and H. Kim, "Optimal capacity and flow assignment for self-healing ATM networks based on line and end-to-end restoration," IEEE/ACM Trans. on Networking, vol. 6, no. 2, pp. 207–221, April 1998.

[78] S. Orlowski, "Local and global restoration of node and link failures in telecommunication networks," M.Sc. thesis, Technische Universität Berlin, February 2003, http://www.zib.de/orlowski/.

690

[79] J.-F. Maurras and S. Vanier, "Network synthe straints," *4OR – A Quarterly Journal of Opera* pp. 53–67, March 2004.

[80] T. Stidsen, B. Petersen, K. Rasmussen, S. Sp F. Rambach, and M. Kiese, "Optimal routing v tection," in *Proceedings of the 3rd International ference (INOC 2007), Spa, Belgium*, 2007.

[81] M. Żotkiewicz, M. Pióro, and A. Tomaszewski, work optimization," *European Transactions on* no. 7, pp. 701–709, 2009.

[82] S. Fortune, J. Hopcroft, and J. Wyllie, "The d phism problem," Ithaca, NY, USA, Tech. Rep.,

[83] H. P. Williams, *Model Building in Mathematical Revised.* Chichester, England: John Wiley & S

[84] M. Minoux, "Optimal synthesis of a network with modity flow requirements," in *Studies in Graph* P. Hansen, Ed. Amsterdam: North-Holland, 19

[85] S. Orlowski, R. Wessäly, M. Pióro, and A. Tom vivable network design library," *Networks*, vol. 5

Chapter 19 Game Theory

Erik G. Larsson[‡] and Eduard Jorswieck[♯]
‡Linköping University, Sweden
♯Technical University of Dresden, Germany

19.1 Introduction

Game theory is about optimization with multiple, conflicting objective functions. In conventional optimization, there is a single objective function that usually has a well-defined maximum or minimum. Finding this optimum point is then a matter of applying an appropriate numerical method. In game theory, the notion of optimality is not defined in terms of the maximum or minimum of a single cost function. Rather, the typical objective is to maximize two (or more) functions *jointly*, where the functions are coupled in such a way that increasing one of them necessarily means that the other must decrease.

Game theory as a scientific discipline mostly evolved from work in economics during the 20th century. Economics continues to be an important application area of game theory, but more recently the theory has been successfully used in other fields as well, such as resource allocation in engineering problems (especially, in communication systems [1, 2]). The goal of this chapter is to expose the most important key concepts of game theory in a manner accessible to graduate students in electrical engineering. Our writing is intentionally concise, and we have therefore been forced to omit many generalizations and proofs, while being as accurate as possible. Throughout the chapter, we provide exercises (some of which with solutions) that we highly recommend for understanding the material. As game theory is a fairly mature subject there exist a fair number of textbooks that treat the subject in various levels of detail [3–11]. Many of these books, particularly [4–6], have provided us with much inspiration in writing this chapter.

This chapter is organized in two parts: Sections 19.2–19.5 deal with basic theory and Sections 19.6–19.10 contain more advanced material. Finally, Section 19.11 provides pointers to the literature on applications of game theory in signal processing and communications.

19.2   Utility Theory

A basic assumption that underpins the entire field o▮
that enter a game can define an order of preference a▮
More precisely, if A and B are possible events, then ea▮
A over B (we write: $A > B$), B over A (we write:
the choice between A and B (we write: $A = B$). T▮
must be a weak linear, transitive order relation, whic▮
if $A = B$ and $B = C$, then $A = C$; (ii) if $A > B$ and
$A > B$, $B = C$ then $A > C$; and (iv) if $A = B$ and B▮

The preference relation tells us which one among s▮
a player favors, but it does not tell us *how much* more
over another. This is important to quantify because l
ations with randomness and we will then need to be a▮
events are preferable over others. This is ultimately
pected value, but this connection is not as simple as i▮
for example, Exercise 19.13.1. *Utility theory* makes t▮

The main result of utility theory essentially state
lation that constitutes a weak linear ordering as defin▮
tional technical axiom that we explain shortly below,
maps each event A, B, C, ..., onto a real number and▮
two conditions:

(a) $u(A) > u(B)$ whenever $A > B$.

(b) If R denotes the random event that "A occurs wit▮
     with probability $1 - p$", then

$$u(R) = pu(A) + (1 - p)u$$

This function, whenever it exists, is called the *utility f*▮
tion (a) represents a fairly natural requirement: the r▮
utility function should be ordered in the same way as
Condition (b) is much stronger: it essentially states
event is equal to a linear combination of the corresp▮
function. Events such as "A with probability p and B▮
often called "lotteries."

What are the conditions required for a utility funct▮
weak linear ordering of preferences, only a technical ▮
stating that if A, B and C are events such that $A >$▮
real number p, $0 < p < 1$ such that the player is in▮
the event "A with probability p and B with probabili▮
hard to justify.

Whenever a utility function $u(\cdot)$ exists, one can s▮
$\alpha u(\cdot) + \beta$ (for $\alpha > 0$) satisfies the requirements of a ▮

means that "utility" is necessarily invariant to scaling, and invariant to translation. The interpretation is that once a utility function is defined, it does not matter what units are used to measure the utility, nor does it matter where the "origin" is located. If utility is money, for example, then it is immaterial what currency is used to measure the amount of money owned.

Taken together, utility theory essentially allows us to reduce the notion of "preferences" between outcomes to a real number. The power of the theory is that we can deal with randomness by taking the expected value of this number. This means that throughout the chapter we can safely assume that we can describe outcomes of games using real numbers, and that the expected value of the utility function has a well-defined operational meaning.

In many engineering applications, a utility function is straightforward to define and write up. In real life, it is not necessarily so. Consider, for example, Exercise 19.13.2 which illustrates that coming up with a utility function is a nontrivial matter. The difficulty is, of course, that one cannot always consistently rank preferences in an order relation. This also explains why the application of game theory to situations that involve very large payoffs and very large risks (such as the loss of life) is not an easy matter. Indeed, the argument can be made that utility must be *bounded* (from below and from above) in the sense that there are events that will never be preferred over others. However, when the game played involves only small deviations around a status quo point, then meaningful utility functions can be normally defined, just like a nonlinear system can be linearized around an operating point. Exercise 19.13.3 illustrates this point.

19.3 Games on the Normal Form

There are many types of games and they can be represented in many forms. In colloquial language, a game usually refers to a match played between two individuals, such as poker or chess. In the engineering context, which is our focus, a game models a conflict about some resource. Since there are many, widely different types of games, it should come as no surprise that there are many different mathematical representations, each of which is useful in different circumstances.

Common for all games is that there are players, moves, strategies, and payoffs. Generally throughout the chapter, we will assume that there are K players. In Sections 19.3–19.9 though, K will be limited to $K = 2$ and we can simply call the players A and B (as in Alice and Bob). In the simplest setting, a game is only played once and when it is played, Alice and Bob simultaneously make a move. Depending on the moves they make, each of them will receive a payoff represented by a utility. To make this more formal, let α be the move made by A and let β be the move made by B, where $\alpha \in \{\alpha_1, \ldots, \alpha_m\}$ and $\beta \in \{\beta_1, \ldots, \beta_n\}$. The integers m and n represent the number of possible moves for Alice and Bob, respectively. We define $U_{i,j}$ to be the payoff that Alice receives and $V_{i,j}$ to be the payoff that Bob gets, for the moves $\alpha = \alpha_i$ and $\beta = \beta_j$. This terminology can easily be

Table 19.1: Normal form representation of

		β_1 (with prob. q_1)	(with
A's move	α_1 (with prob. p_1)	$(U_{1,1}, V_{1,1})$	$(U_{1,}$
	α_2 (with prob. p_2)	$(U_{2,1}, V_{2,1})$	
	$\vdots$	$\ldots$	
	α_m (with prob. p_m)	$(U_{m,1}, V_{m,1})$	$(U_{m,}$

summarized in terms of a *game matrix*, see Table 19
q_i in the table will be explained later in the text.) Th
is called *normal form*.

The way A and B decide what moves to make is de
strategy. Most simply, A and B would just decide be
manner, what move to make. If this is the case, we
strategies. This does not result in a particularly int
more interesting scenario is when A and B decide at r
To formalize this notion, let $\boldsymbol{p}$ and $\boldsymbol{q}$ be two discrete
describe how likely it is that the players choose differ
probability that A chooses the move α_i and q_j is th
the move β_j. In this scenario, we say that A and 1
interpretation is that A and B simultaneously throu
order to decide what moves to make. Clearly,

$$\sum_i p_i = \sum_j q_j = 1$$

$$0 \le p_i \le 1 \qquad 0 \le q_j \le 1$$

for otherwise, $\boldsymbol{p}$ and $\boldsymbol{q}$ are not valid probability distri
we shall assume that A and B make their moves indep
events associated with the distributions $\boldsymbol{p}$ and $\boldsymbol{q}$ are i
and in our text, $\boldsymbol{p}$ and $\boldsymbol{q}$ are called the *strategy profil*
elements of $\boldsymbol{p}$ (or $\boldsymbol{q}$) are zero, and the remaining elem
have the special case of pure strategies.

For given strategy profiles, the expected payoffs fo

$$u = E[U_{i,j}] = \sum_i \sum_j p_i q_j U$$

and

$$v = E[V_{i,j}] = \sum_i \sum_j p_i q_j V_i$$

Equations (19.3)–(19.4) can be conveniently written on matrix form as

$$u = \boldsymbol{p}^T \boldsymbol{U} \boldsymbol{q} \qquad \text{and} \qquad v = \boldsymbol{p}^T \boldsymbol{V} \boldsymbol{q}. \tag{19.5}$$

Of course, the pair (u, v) will depend on $\boldsymbol{p}$ and $\boldsymbol{q}$. More generally, we can view (u, v) as a function of $\boldsymbol{p}, \boldsymbol{q}$.

The region of possible payoffs (u, v) that can be obtained by valid probability distributions $\boldsymbol{p}, \boldsymbol{q}$ (see (19.2)) is called payoff region or *utility region*. See Figure 19.1. There are several points in the utility region that deserve special attention and which have received special names. A utility pair (u, v) is said to be *Pareto optimal* (after the Italian economist, Vilfredo Pareto) if it is impossible to increase u without simultaneously decreasing v, and vice versa. The set of all Pareto optimal points is called the *Pareto boundary*, and this is the northeast part of the boundary of the utility region.[1] The *egalitarian point* (u_e, v_e) is the unique point on the Pareto boundary where $u = v$, and it is found by intersecting the boundary with a straight line that starts at the origin and has a slope of $+1$. The utilitarian point (u_u, v_u) is the point where $u + v$ is as large as possible, and it is the (not necessarily unique) point where a line of slope -1 osculates the Pareto boundary.

Depending on the context, and on the philosophy adopted, one may want to operate at specific points inside the region or on its boundary, for example (u_e, v_e) or (u_u, v_u). However, there is no reason that any of the players would voluntarily use a strategy that reaches any such point. Rather, we will assume that each player is rational in the sense that she wants to maximize her outcome (u and v respectively) with no consideration of what happens to the other player. This conflict is precisely the one that game theory aims to model mathematically.

Throughout the discussion that follows, we shall assume that A does not know $\boldsymbol{q}$ and that B does not know $\boldsymbol{p}$. If they did, then the problem reduces to a classical decision problem, rather than a game in the sense defined here. Indeed, if A knew $\boldsymbol{q}$, then it is easy for her to compute the best response against this strategy of B. This is simply the strategy $\boldsymbol{p}$ that yields the largest possible payoff for a fixed $\boldsymbol{q}$:

$$\boldsymbol{p} = \arg \max \boldsymbol{p} \quad \boldsymbol{p}^T \boldsymbol{U} \boldsymbol{q} \tag{19.6}$$

$$\sum_i p_i = 1$$

$$p_i \geq 0$$

Equation (19.6) is a linear program. Its structure is similar to that of (19.16) in Section 19.3.1 below and it can be solved using the techniques that will be discussed there.

---

[1] A point is called *weakly* Pareto optimal if it is impossible to increase u and v at the same time, and vice versa.

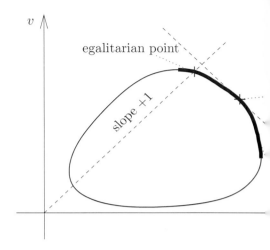

Figure 19.1: Utility region. The Pareto boundary is t the region.

Table 19.2: Game matrix (from A's point of view) for

		B's mo	
		rock	paper
A's move	rock	0	-1
	paper	+1	0
	scissors	-1	+1

19.3.1   Zero-Sum Games

In some games, the payoff for player A equals precise for player B, i.e., $V_{i,j} = -U_{i,j}$ so that

$$U_{i,j} + V_{i,j} = 0$$

The interpretation is that whatever A wins, B must the case, we say that the game is *zero sum*. As an ex payoff matrix for the classical rock-paper-scissors gam analyzed in Exercise 19.13.5, using the methodology

Zero-sum games are of interest because a rather str with relatively elementary mathematics. The result theorem due to von Neumann more than six decades will explain its significance and interpretation, but at

Consider first the game from A's point of view. achieves the payoff u in (19.3). Now suppose that E

any given strategy $\boldsymbol{p}$ used by A, he chooses the strategy $\boldsymbol{q}$ that makes A win as little as possible. Then for a given $\boldsymbol{p}$, the worst possible outcome for A is

$$f(\boldsymbol{p}) \triangleq \min_{\boldsymbol{q}} \boldsymbol{p}^T \boldsymbol{U} \boldsymbol{q}. \tag{19.8}$$

It is reasonable to assume that A would like to choose her strategy $\boldsymbol{p}$ in order to minimize her worst-case loss, i.e., to maximize $f(\boldsymbol{p})$. One can think of this as a way of minimizing the regret that A may ever feel, should she engage in this game. Minimizing the regret results in the optimization problem

$$\boldsymbol{p}^* = \arg\max_{\boldsymbol{p}} \min_{\boldsymbol{q}} \boldsymbol{p}^T \boldsymbol{U} \boldsymbol{q}, \tag{19.9}$$

where we let $(\cdot)^*$ stand for optimality. If A uses the strategy $\boldsymbol{p}^*$, she can ensure that her payoff will in no case be less than

$$U \triangleq f(\boldsymbol{p}^*) = \max_{\boldsymbol{p}} \min_{\boldsymbol{q}} \boldsymbol{p}^T \boldsymbol{U} \boldsymbol{q}. \tag{19.10}$$

U is called the value of the game, from A's point of view.

Conversely, let us consider the game from B's perspective. By symmetry, by choosing $\boldsymbol{q}$ according to

$$\boldsymbol{q}^* = \arg\min_{\boldsymbol{q}} \max_{\boldsymbol{p}} \boldsymbol{p}^T \boldsymbol{U} \boldsymbol{q} \tag{19.11}$$

he can force a payoff (value) for A that is equal to or smaller than

$$V \triangleq \min_{\boldsymbol{q}} \max_{\boldsymbol{p}} \boldsymbol{p}^T \boldsymbol{U} \boldsymbol{q}. \tag{19.12}$$

von Neumann's theorem [12] states that the value of the game from A's perspective is always equal to the value of the game from B's perspective. That is,

$$U = \max_{\boldsymbol{p}} \min_{\boldsymbol{q}} \boldsymbol{p}^T \boldsymbol{U} \boldsymbol{q} = \min_{\boldsymbol{q}} \max_{\boldsymbol{p}} \boldsymbol{p}^T \boldsymbol{U} \boldsymbol{q} = V \tag{19.13}$$

and there exists at least one pair $(\boldsymbol{p}, \boldsymbol{q})$ for which the equality is satisfied. It is natural to take $U = V$ to be the *value of the game*. This value and the associated strategies represent an equilibrium point at which the game is likely to end up if played many independent rounds. This equilibrium point is also called the *minimax solution* of the game.

The von Neumann equilibrium point can be computed in closed form or even found by inspection for many small, simple games [4, 5]. Special procedures also exist to solve games with specific structures. Moreover, games can sometimes be reduced in size. For example, if one row is dominated by all others in the sense that there exists an i_0 such that $U_{i,j} > U_{i_0,j}$ for all i and j, $i \neq i_0$, then there is no point for player A to ever use a mixed strategy that involves α_{i_0}; hence, at the equilibrium we will have $p_{i_0} = 0$. Similarly, columns that are dominated

can also be excluded. Furthermore, without much ϵ
game has a saddle point so that $U_{i,j}$ is the largest ϵ
smallest element in row i, then the pure strategy $\{\alpha_i,$
equilibrium, $p_i = 1$, $q_j = 1$ and all other elements of $\boldsymbol{p}$
as systematic, numerical solutions are of interest, the
are less interesting because as we show next, the von
zero-sum game can be efficiently computed by using

Solving Zero-Sum Games via Linear Programm

The procedure to solve zero-sum games via linear pr
show how to find the equilibrium by taking player
objective is to maximize the worst-case payoff, i.e., to

$$\max_{\boldsymbol{p}} \min_{\boldsymbol{q}} \boldsymbol{p}^T \boldsymbol{U} \boldsymbol{q}$$

Clearly, for fixed $\boldsymbol{p}$, the minimizer of $\boldsymbol{p}^T \boldsymbol{U} \boldsymbol{q}$ with re
for which one element is equal to unity and all the re
the element of the optimal $\boldsymbol{q}$ will have its unity elem
corresponding to the smallest element of the vector
equivalently be written as

$$\max_{\boldsymbol{p}} \min_{j} (\boldsymbol{p}^T \boldsymbol{U})_j$$

The task of finding $\boldsymbol{p}$ in (19.15) can be cast as a linea
slack variable t, we have

$$\max \ t$$

$$(\boldsymbol{p}^T \boldsymbol{U})_j = \sum_i p_i U_{i,j} \geq t,$$

$$\sum_i p_i = 1$$

$$p_i \geq 0, \qquad \forall i$$

The usefulness of the formulation in (19.15) shoul
ear programming is a mature technology and there
(including free implementations), such as the GLPK [
that can handle problems with up to tens of thousan
promising numerical stability. Hence, very large zer
very fast and without approximations.

Note that choosing player A as a reference here
owing to von Neumann's theorem, the same solutio
chose to start with B. (The computational complexity
be different, however.)

Table 19.3: Game matrix for the classical "prisoner's dilemma" game.

		Bob	
		Honest	Cheater
Alice	Honest	(1,1)	(-1,2)
	Cheater	(2,-1)	(0,0)

19.3.2   Non-Zero-Sum Games

So far we have assumed that the players strictly compete with one another. At best, if the players would cooperate, they could achieve a higher sum-payoff $U+V$ than if they did not cooperate. The excess gain could then be split among them in some manner so as to stimulate and reinforce the cooperation. Since in zero-sum games we have $U_{ij} + V_{ij} = 0$, there is no point for players to cooperate, and the analysis conducted in Section 19.3.1 should be sufficient for all conceivable purposes.

The story is different in games that are not zero-sum. Such games, called non-zero-sum here, are very common in reality. In what follows, we shall assume that the games we study are non-zero-sum. In non-zero-sum games, players can potentially benefit from cooperating because $U+V$ may be strictly larger than zero. Therefore, to analyze the games mathematically we need to distinguish between settings where the players are strictly competitive (the associated game is said to be noncooperative) and settings where the players cooperate (we then speak of cooperative games). The noncooperative case will be dealt with in Section 19.4. We will discuss cooperative games in Section 19.5.

19.4   Noncooperative Games and the Nash Equilibrium

Some basic complications that arise when the game is not zero-sum are illustrated by the example in Table 19.3. This game is called "prisoner's dilemma" in much of the literature, referring to a story with two arrested suspects who would mutually benefit if both confess, but where each suspect has no incentive to confess unless the other does so too. In the game of Table 19.3, we think of Alice and Bob engaging in a secret trade exchanging some merchandise for a bag of money. A leaves the merchandise at a pre-agreed but secret place and simultaneously, B leaves the money at another pre-agreed place. Now A and B may either act honestly, that is to drop the money or goods as agreed, or they may cheat the other by leaving an empty bag. In the payoff matrix of Table 19.3 we assign the value 1 for both A and B in the event of a successful deal. In the event that one of them cheats, we assign a value of 2 to the cheater and a value of -1 to the one being deceived. If both cheat, we assign a value of zero to both A and B.

Is there an equilibrium point for playing the game in Table 19.3? Clearly, if

A and B would agree beforehand to be honest, each
of one unit and it is easy to see that this outcome is
non-cooperative game theory, such deals cannot be s
can trust that the other will be honest. If one of the
be very bad off (he receives a payoff of -1). By contra
A and B cheating. In this case, none of them can imp
changing strategy. That is, none of them can gain
reasoning we infer that the pure strategy pair in whi
a form of equilibrium for this game. The notion of a
next, formalizes this reasoning.

Suppose that the sum-payoff is not necessarily zer
players are not allowed to cooperate so that the ra
p and q are independent. We are looking for strate
players operate at (p^*, q^*), then none of them has
deviate from this strategy. This means that p^* and q

$$p^T U q^* \leq p^{*T} U q^*, \qquad \forall$$

$$p^{*T} V q \leq p^{*T} V q^*, \qquad \forall$$

Under what conditions do strategies p^*, q^* that satisf
Nash [16] proved that for the games considered here
always exists, and it is called *Nash equilibrium*. Howe
fact, most non-zero sum games have multiple Nash eq
of Nash equilibrium is nontrivial and requires the use
There is free software available that can find the ec
games fast [17].

At the Nash equilibrium, none of the players has
strategy vector, provided that the other player does
guably, if a non-zero-sum game is played and the play
ing cooperation agreements by correlating their move
the only reasonable operating point at which the gar
In the example of Table 19.3, there is a unique equilib
as we found using intuitive reasoning above.

The payoffs at a Nash equilibrium are generally n
the players to cooperate is essential in order to reach
This is clear already from the example of Table 19
Nash equilibrium and the Pareto boundary can be ch
called the *price of anarchy* [18]. Of course, if the
with "memory" in the sense that the players can act
previous rounds, then things will be entirely different
up trust in one another and the incentive to cheat can
effect will be discussed in more detail in Section 19.8.

19.5   Cooperative Games

So far we have considered games where the players strictly compete. This has meant two things: First, Alice's and Bob's moves are chosen independently of one another. Hence, it is natural to assume that they will use strategies that maximize the worst-case payoff. The optimal strategies, in this sense, are given by the von Neumann (for zero-sum games) and Nash equilibria (for non-zero-sum games). Second, we have assumed that once A and B receive the payoffs $U_{i,j}, V_{i,j}$, there is no mechanism that allows them to share the sum-payoff $U_{i,j} + V_{i,j}$. Clearly, the possibility to jointly decide on the strategies, and the possibility to share the sum-payoff, according to some rule that makes sense to them both, would fundamentally change the way they play the game. This leads to the area of cooperative games. This section will summarize some key concepts from this field. Of course, in cooperative games the players must be able to communicate with one another to agree on the mode of cooperation, and there must be something or somebody who assures that the players stick to what they have negotiated.

For the noncooperative case, the Nash equilibrium strategies defined by (19.17) and (19.18) represented fairly natural models for how A and B would behave. When the possibility exists for the players to cooperate, it is less obvious how to model the players' behavior. One can ask, whether for a given game there is any way of finding a unique point of agreement (or *bargaining solution*) at which the game is likely to be resolved. More specifically, one can ask whether there is any sensible set of rules (axioms) under which such an agreement point would be uniquely defined. Interestingly, it turns out that one can find such sets of rules that are arguably compatible with the way many humans typically reason. The set of axioms due to John Nash [19] yields the most well-known bargaining theory, but there are other possibilities too. We will discuss the Nash axioms in Section 19.5.1. This way of reasoning is often called "axiomatic" game theory, because once the axioms are given, the agreement point can be computed by solving a set of mathematical equations. In axiomatic game theory, the problem of modeling the players' behavior is reduced to defining the set of rules (axioms) that eventually yield a unique agreement point.

Typically, in cooperative game theory, the model includes a threat point, which is the outcome that results if the cooperation breaks up. We denote this threat point by $(\bar{u}, \bar{v})$. The threat point may be located at the origin ($\bar{u} = \bar{v} = 0$), meaning that if the players withdraw from the cooperation, then both of them would receive a payoff of zero. More commonly, the threat point is taken to be the Nash equilibrium point, corresponding to the most likely outcome should the cooperation contract not work out. Clearly, under no circumstances the threat point can be outside the utility region.

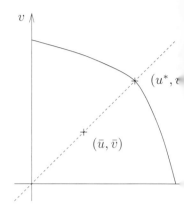

Figure 19.2: The symmetry axiom (A3) in

19.5.1   Bargaining without Transferab

We shall first assume that there is no mechanism that
another compensation for choosing a specific strategy
the theory of transferable utility and is briefly treate

John Nash proved in 1950 [19] that under a certa
below), there is a uniquely defined function $f(\mathcal{U}, \bar{u}, \bar{v})$
and the threat point onto a bargaining solution (u^*, v^*)
the bargaining solution can be uniquely determined a

A1. *Individual rationality:* $(u^*, v^*) \geq (\bar{u}, \bar{v})$. This
outcome of a bargain cannot be worse for any of
if no bargaining occurs (that is, the threat poin

A2. *Pareto-optimal:* $(u, v) \in \mathcal{U}$ and $(u, v) \geq (u^*, v^*$
condition simply states that the bargaining ou
Pareto boundary of the utility region. Clearly,
there would be another outcome which is bet
reasonable bargaining scheme would choose $(u,$

A3. *Symmetry (Figure 19.2):* If $\mathcal{U}$ is symmetric arou
v^*. This means that if the utility region is symm
through the origin and has a slope of $+1$, then
lie on this line of symmetry. This axiom can b
basic notion of fairness.

A4. *Independence of irrelevant alternatives (IIA)*
$\mathcal{U}' \subset \mathcal{U}$ and $(u^*, v^*) = f(\mathcal{U}, \bar{u}, \bar{v})$, then (u^*, v^*)

---

[2]Here $\geq$ means componentwise inequality.

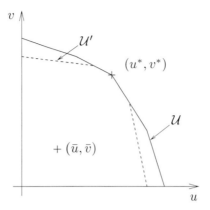

Figure 19.3: The independence of irrelevant alternatives (A4) axiom.

says that if bargaining in the utility region $\mathcal{U}$ results in a solution (u^*, v^*) which lies in a subset of $\mathcal{U}'$ of $\mathcal{U}$, then a hypothetical bargaining in the region $\mathcal{U}'$ would have resulted in the same solution.

A5. *Invariance to linear transformation:* Let $a_1, a_2, b_1, b_2 \in \mathbb{R}, a_1 > 0, a_2 > 0$ be arbitrary. Then this axiom states that if

$$(a_1\bar{u} + b_1, a_2\bar{v} + b_2) \in \mathcal{U} \tag{19.19}$$

then

$$f(\mathcal{U}, a_1\bar{u} + b_1, a_2\bar{v} + b_2) = [a_1, a_2]f(\mathcal{U}, \bar{u}, \bar{v}) + [b_1, b_2]. \tag{19.20}$$

This means that if the utility region and the threat point are scaled and translated, then the bargaining solution scales and translates in the same way. The operational meaning of the scaling invariance is that the units in which utilities are measured should not affect the result. The operational meaning of the translation invariance is that a utility increase from 0 to 1 should be worth as much as an increase from 100 to 101, and so forth. This axiom reflects a certain notion of "fairness" too. It is also highly natural in view of what we know about utility functions (see Section 19.2).

The function that determines the Nash bargaining solution is uniquely defined under axioms A1–A5 and is given by

$$(u^*, v^*) = f(\mathcal{U}, \bar{u}, \bar{v}) = \max_{(u,v) \in \mathcal{U}} (u - \bar{u})(v - \bar{v}). \tag{19.21}$$

The solution of (19.21) can be easily found numerically. It is the point where the Pareto boundary has a unique intersection with a hyperbola parametrized by $(u - \bar{u})(v - \bar{v}) = $ constant, see Figure 19.5.

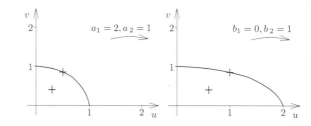

Figure 19.4: The invariance to translation and scali gaining.

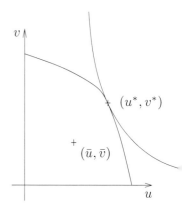

Figure 19.5: Nash bargaini

A fundamental point is that a player can be coo same time. That is, being cooperative does not me altruistic. The point is that even if players are eventu their own outcome, they may be willing to accept a found good enough for both. Note that Nash bargai with the Nash equilibrium, as the latter applies only which bargaining is not possible.

Bargaining is well defined for utility regions $\mathcal{U}$ tha ability mixture of outcomes as described in Section for more general problems. All that is required is a c threat point. Consider the following classical example one poor, who meet a genie on the street. The geni *provided that they can agree on how to split the mone* bargaining theory predict?

To solve this question, assume that the rich man

owns x_p and that utility of money is logarithmic in the amount owned so that $u_r = \log(x_r)$ and $u_p = \log(x_p)$ (this is a common assumption by economists). Also, let us assume that the rich man (R) is near infinitely rich ($x_r = \$10^{10}$) but that the poor man (P) owns only $x_p = \$10$ in total. Let x be the amount R gets in the bargain. Hence, after bargaining the utility for R is

$$\bar{u}_r = \log(10^{10} + x) \tag{19.22}$$

and the utility for P is

$$\bar{u}_p = \log(10 + (100 - x)). \tag{19.23}$$

Let us take the threat point to be $(\bar{u}_r, \bar{u}_p) = (\log(x_r), \log(x_p))$. This means that if no bargain occurs, both R and P will leave with exactly the amount they initially owned. The Nash bargaining solution follows by solving

$$\max_{u_r, u_p, x \in [0,100]} (u_r - \bar{u}_r)(u_p - \bar{u}_p) \tag{19.24}$$

and is given by $x \approx \$66$. Evidently, the bargaining outcome favors the rich man, who gets the larger share of the money. For comparison, if instead P had initially owned only $x_p = \$0.1 = 10$ cents (R has still $x_r = \$10^{10}$), then the Nash bargaining solution would be $x \approx \$84$ and the outcome would be even more unbalanced. The reason is that R has much more bargaining power. In fact, he can dictate a "my way or no way" outcome by threatening to walk away without a deal if he does not get most of the money. Especially he knows that not being able to reach a deal will hurt P more than R so that P will be more willing to accept a bad deal than no deal at all. In the special cases when $x_p \to 0$ the bargaining solution $x \to \$100$ and when $x_p \to x_r$ the solution approaches $x \to \$50$.

This example illustrates well the fundamental difficulty in objectively defining the notion of "fairness." Clearly, two of the axioms that underpin the Nash theory (symmetry and invariance to linear transformations) reflect a certain notion of fairness, but many would object to the outcome of the \$100 question being fair. In this context, Nash bargaining theory should be seen as a mathematical model for the fact that a stronger part in a conflict always has a larger power of negotiation and therefore will achieve a better outcome. See, [20] for an interesting discussion on the implications of this.

We end this section with a remark on a controversy that is associated with the bargaining solution obtained under axioms A1–A5 presented above. Namely, one can find examples where increasing the size of $\mathcal{U}$ will increase the bargaining outcome for one of the players, but decrease it for the other. More precisely, suppose $\mathcal{U} \subset \mathcal{U}'$ and let $(u_1^*, u_2^*) = f(\mathcal{U}, \bar{u}, \bar{v})$ and $(u_1'^*, u_2'^*) = f(\mathcal{U}', \bar{u}, \bar{v})$. Then we can construct cases where $u_1^* \geq u_1'^*$ but $u_2'^* < u_2^*$. Figure 19.6 exemplifies this point. The solid curve is the Pareto boundary of the original region $\mathcal{U}$. The dashed curve is the Pareto boundary of the expanded region $\mathcal{U}'$. The Nash bargaining solutions occur when the Pareto boundaries intersect the corresponding hyperbolas defined by (19.21). Clearly, expanding the region improves one of the utilities but not

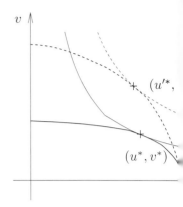

Figure 19.6: Controversy of Nash bargaining. Wh
the Nash bargaining solution does not necessarily in
players, so that $\bar{v}' > v^*$ but $u'^* < u^*$.

both. This illustrates that any requirement that $\mathcal{U}$
would be incompatible with axioms A1–A5 given abc

19.5.2   Bargaining with Transferable U

What if there is a way for Alice and Bob to share th
the possibility to transfer utilities between them, an
theory is called "transferable utility" (TU) game theo
utilities often exists when $U_{i,j}$ and $V_{i,j}$ represent son
it may not be possible, or allowed, if $U_{i,j}$ and $V_{i,j}$ rep
is not so easily transferred.

Suppose that for a given outcome, Alice achieves ι
be the amount that Alice pays to Bob as a compensat
The resulting utility region will then be the set of po
moving in the southeast or northwest direction from a
we get the region:

$$\mathcal{U}_{\text{tu}} = \{(u,v) + t \cdot (1,-1) : (u,v) \in \ell$$

If we draw a straight line, say $\mathcal{L}$ with slope -1 that os
at precisely one point, say (u',v'), then $\mathcal{U}_{\text{tu}}$ will be
side of this line. This is illustrated in Figure 19.7.
now be easily found by simple geometry. Consider ε
originates at the threat point $\bar{u}, \bar{v}$. The game is symm
bargaining solution must occur where $\mathcal{L}'$ intersects ℓ

for the straight lines $\mathcal{L}$ and $\mathcal{L}'$ in Figure 19.7, we have

$$
\begin{aligned}
v^* - u^* &= \bar{v} - \bar{u} \\
v^* + u^* &= u' + v'^{\cdot}
\end{aligned}
\tag{19.26}
$$

Solving for the bargaining solution we obtain

$$
\begin{aligned}
u^* &= -\frac{1}{2}(\bar{v} - \bar{u}) + \frac{1}{2}(u' + v') \\
v^* &= \frac{1}{2}(\bar{v} - \bar{u}) + \frac{1}{2}(u' + v')
\end{aligned}
\tag{19.27}
$$

Indeed, only the first three axioms (A1–A3: individual rationality, Pareto-optimality, and symmetry) of the Nash bargaining theory are needed to find this solution.

How would each player select its threat point in order to best serve its interest? One possibility is to reason in the same way as we did for zero-sum games and assume that A wants to maximize the bargaining outcome u^* that results if B selects its threat point $\bar{v}$ in the least favorable way A, and vice versa. To formulate this problem mathematically we note that $\bar{u}, \bar{v}$ must lie in $\mathcal{U}$. It is not sufficient that $(\bar{u}, \bar{v}) \in \mathcal{U}_{tu}$, since the threat point must be achievable without cooperation. To deal with this, recall that $\mathcal{U}$ consists of the convex hull of $\{U_{i,j}\}$ and $\{V_{i,j}\}$ over the probability distributions $\boldsymbol{p}, \boldsymbol{q}$, and let $(\bar{\boldsymbol{p}}, \bar{\boldsymbol{q}})$ be the strategies associated with the threat point $(\bar{u}, \bar{v})$ so that $\bar{u} = \bar{\boldsymbol{p}}^T \boldsymbol{U} \bar{\boldsymbol{q}}$ and $\bar{v} = \bar{\boldsymbol{p}}^T \boldsymbol{V} \bar{\boldsymbol{q}}$. From A's perspective we have the maximin problem

$$
\max_{\bar{\boldsymbol{p}}} \min_{\bar{\boldsymbol{q}}} u^* \quad \Leftrightarrow \quad \max_{\bar{\boldsymbol{p}}} \min_{\bar{\boldsymbol{q}}} \bar{u} - \bar{v} \quad \Leftrightarrow \quad \max_{\boldsymbol{p}} \min_{\boldsymbol{q}} \bar{\boldsymbol{p}}^T (\boldsymbol{U} - \boldsymbol{V}) \bar{\boldsymbol{q}}.
\tag{19.28}
$$

Note that $(u' + v')/2$ is a constant that does not change the value of the optimal $\bar{u}$; nor does the scaling factor $1/2$ change the problem. Similarly, from B's perspective,

$$
\max_{\bar{\boldsymbol{q}}} \min_{\bar{\boldsymbol{p}}} v^* \quad \Leftrightarrow \quad \max_{\bar{\boldsymbol{q}}} \min_{\bar{\boldsymbol{p}}} \bar{v} - \bar{u} \quad \Leftrightarrow \quad \min_{\bar{\boldsymbol{q}}} \max_{\bar{\boldsymbol{p}}} \bar{u} - \bar{v} \quad \Leftrightarrow
$$
$$
\min_{\bar{\boldsymbol{q}}} \max_{\bar{\boldsymbol{p}}} \bar{\boldsymbol{p}}^T (\boldsymbol{U} - \boldsymbol{V}) \bar{\boldsymbol{q}}^{\cdot}
\tag{19.29}
$$

Equations (19.28)–(19.29) can be solved by using the techniques presented in Section 19.3.1. In fact, (19.28)–(19.29) are formally equivalent to finding the value of a zero-sum game with payoff matrix $\boldsymbol{U} - \boldsymbol{V}$.

19.6   Games with Incomplete Information

In many scenarios, the players do not possess complete information on the preference relations of the other players or their utilities. A knows his own preferences but not necessarily the preferences of B. Then the game cannot be described in

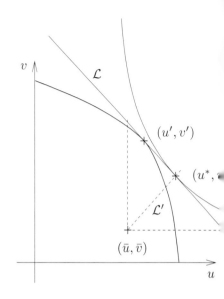

Figure 19.7: Bargaining with transfer

normal form, as the payoffs in Table 19.1 are not comm
concepts for normal games with complete information

In order to motivate the model for games with inco
for example a scenario in which a certain strategy
$c_A(a)$ and her payoff is computed as the difference
the case of companies producing goods, the costs are
Therefore, the other player B does not know $c_B(a)$
compute dominating strategies or the equilibrium solu
minmax or worst case solution can still be found.)
situation for player B is to estimate the costs of A b
and experience. Player B builds a subjective probab
player A for different strategies a. Now player A is
for the costs of player B. Additionally, player A kr
subjective probability for his costs. Player B knows tl
so on. This recursion can be resolved by Harsanyi's a

A private *type* t_k from a set of possible types T_k is
k. The type contains all information necessary to de
user. These types can be determined by nature or by
the actual game starts. They are private, i.e., player *A*
player builds a subjective probability on the types o
his own type, i.e., player A builds the conditional pro
a priori beliefs on the other player's types are collecte

A Bayesian game is described by the quintuple: th

strategy spaces, the set of types (T_A, T_B), the conditional probabilities π, and the utilities. The utilities depend on the strategies as well as on the types. It is assumed that this tuple is common knowledge. In addition, player A knows his own type t_A. This common prior assumption plays a central role in Bayesian games: There exists a common a priori probability distribution p, called *common prior*, if the conditional probabilities $p(t_B|t_A)$ and $p(t_A|t_B)$ are derived from some probability distribution p on $T_A \times T_B$ by

$$p(t_B|t_A) = \frac{p(t_A, t_B)}{\sum_{\tau_B \in T_B} p(t_A, \tau_B)} \quad \text{and} \quad p(t_A|t_B) = \frac{p(t_A, t_B)}{\sum_{\tau_A \in T_A} p(\tau_A, t_B)} \tag{19.30}$$

The assumption in (19.30) can be interpreted as follows: all players have a common "objective" belief on the probabilities which are used by nature to determine the types before the game starts.

The payoff for player A is computed based on his own type t_A, his strategy a and based on his belief on the type of player B (analogously for player B) as

$$\tilde{U}_a(a, t_A) = \sum_{t_B \in T_B} p(t_B|t_A) U^a(a, b(t_B), t_A) \quad \text{and}$$

$$\tilde{U}_b(b, t_B) = \sum_{t_A \in T_A} p(t_A|t_B) U^b(a(t_A), b, t_B) \tag{19.31}$$

where U^a is the instantaneous utility of player A evaluated for strategies a and b for type t_A and U^b for player B. $\tilde{U}_a(t_A)$ and $\tilde{U}_b(t_B)$ in (19.31) are the expected payoffs based on the common beliefs in (19.30). The strategy a of player A depends on his type, i.e., $a(t_A)$.

A strategy pair $a^*(t_A)$ and $b^*(t_B)$ is a *Bayesian equilibrium* [3, Section 6.4] if for all $t_A \in T_A$ and $t_B \in T_B$ it holds that

$$a^*(t_A) = \arg\max_{a \in A} \sum_{t_B \in T_B} p(t_B|t_A) U^a(a, b^*(t_B), t_A) \quad \text{and}$$

$$b^*(t_B) = \arg\max_{b \in B} \sum_{t_A \in T_A} p(t_A|t_B) U^b(a^*(t_A), b, t_B). \tag{19.32}$$

The definition of the Bayesian equilibrium in (19.32) is similar to the definition of the Nash equilibrium in (19.17) and (19.18). Therefore, the results on existence and uniqueness can be applied, too.

Consider the Cournot-Nash equilibrium Exercise 19.13.12 with incomplete information: Two firms A and B invest in a product and produce an amount x with costs $K(x)$. There is an upper bound on the production $x_i \le x_i^{max}$ and the price for the product depends on the market (supply and demand) as $p(x) = 100 - 2x$. Firm A obtains the revenue $G_A(x_A, x_B) = p(x_A + x_B)x_A - K_A(x_A)$ and firm B analogue. The costs K_A and K_B are known only by the firm itself. Assume that there are two types, high costs $K_{A1}(x_A) = 2x_A^2$ and low costs $K_{A2}(x_A) = x_A^2$, and

Table 19.4: Common priors for the Cournot-Nash information example.

	$t_B = 1$	$t_B = \ :$
$t_A = 1$	0.4	0.2
$t_A = 2$	0.1	0.3

analogously for firm B. Each firm builds a belief on t
cost of the other firm. The common prior is given in
     From Table 19.4, the subjective probabilities can
$\frac{p(t_{A2}, t_{B2})}{p(t_{A2})}$, and so forth. Then, the best responses for
can be computed. Finally, the intersection of the be:
is computed as the equilibrium outcome (see exercis
with incomplete information 19.13.12).

19.7   Extensive Form Games

In Section 19.3 the conflict situation was modeled a
normal form. The players decide before the game is
games are played in multiple rounds or stages, in v
the other. All sophisticated board games consist of n
are dynamic because the actions of a later stage dep
stages. They are also called *multistage games with*
extensive form model the sequential structure explicit
basis of the observed actions from former stages. At fi
setting is considered. However, games with incomplete
the last Section 19.6, can be easily modeled in extens
     In the model, the temporal flow of the game nec
set called *history* or path. All consecutive actions o
a sequence h. A complete play is described by a
sequence. The set of terminal sequences is denoted
is denoted with the empty set. Finally, we need to
some point in the play. Therefore, each sequence h is
a function P except the terminal sequences. Each ter
with a payoff vector.
     Consider the classical invader game illustrated in F
game with the invader (I) and the defender (D). The
to challenge (c) the defender or not (nc). If the de
decide to give up and quit (q) or to fight (f). If the
zero payoff. Otherwise, the one who gives up receiv
payoff two. There are three terminal sequences, nam

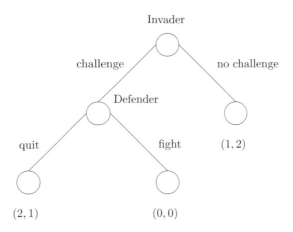

Figure 19.8: Invader game in extensive form.

The payoff vectors are $U(nc) = (1,2), U(c,q) = (2,1), U(c,f) = (0,0)$. Finally, the active player is $P(\emptyset) = I$ and $P(c) = D$. The game is represented in Figure 19.8 as a tree. The game starts at the root node. Each node corresponds to a turn of one player. Each edge corresponds to one action.

Thus the game in extensive form is described by the set of players, the set of terminal sequences, the assignment function, and a payoff function for each terminal sequence. If all terminal sequences have finite length, it is called *finite horizon*. If the number of terminal sequences is finite, too, the game is *finite*.

In order to completely model the game, we need to describe the strategies. A strategy of player A associates with each history h an action $a(h)$ if it is player A's turn, i.e., $P(h) = a$. This can be called *action plan* or strategy profile s. The terminal sequence which is induced by some strategy profile s is defined as $O(s)$. The payoff for user A for some strategy profile s is then given by $U^a(O(s))$.

For the invader game in Figure 19.8, a pure strategy of player I for $h = \emptyset$ could be c and a pure strategy of player D for $h = c$ could be f. The induced terminal sequence for this strategy profile is then $O(s) = (c, f)$ and the corresponding payoff vector is $U(O(s)) = (0,0)$.

19.7.1 Nash Equilibrium in Extensive Form Games

The concept of the Nash equilibrium for games in normal form can be carried over to games in extensive form. A strategy profile s^* is a Nash equilibrium for a game in extensive form with complete information if for all players i and all strategies

s_i of player i holds[3]

$$u_i(O(s^*)) \geq u_i(O(s^*_{-i}, s_i))$$

In order to compute Nash equilibria for extensive fo
is a canonical transform to a corresponding game i
strategies in the extensive form game are represent
form game. For the invader game in Figure 19.8, t
two-by-two matrix which is easily computed in Exerc

The disadvantage of the solution concept Nash equ
that it neglects the sequential structure of the extensi
game, the strategy pair (nc,f) is not possible becau
sequence since the game ends if the invader does not c
solution concept is needed.

19.7.2   Subgame-Perfect Equilibrium

The idea is to study the game starting at the last
$i = P(K)$ who is active in this stage assumes that so
and maximizes his payoff by choosing the action for t
$K - 1$, the active player $j = P(K - 1)$ is able to comp
the last step and thus can predict the action of playe
on his own actions. Hence, player j can maximize his
for stage $K - 1$. This *backward induction* works fine i
extended to infinite games.

The extensive form game is deconstructed into su
sequences h, we define a subgame which starts at h
tree as the original game. The Nash equilibrium fro
to all possible subgames. The terminal sequence $O_h($
s is obtained from the prefix h and the generated se
subgame-perfect equilibrium in an extensive form gam
is a strategy profile s^* if for all players i and all no
$P(h) = i$ it holds

$$u_i(O_h(s^*)) \geq u_i(O_h(s^*_{-i}, s_i))$$

for all strategies s_i of player i. It can be shown t
games with complete information have a subgame-pe
works with backward induction. The subgame-perfe
sarily unique. Clearly, all subgame-perfect equilibri
condition in (19.34) reduces the set of Nash equilibri

Reconsidering the invader game in Figure 19.8,
duction argument: Assume that the history is $h =$

---

[3]The term s_{-i} contains all but the i-th player strategies, i.e

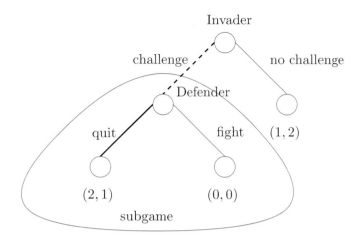

Figure 19.9: Invader game in extensive form with subgame.

dominating strategy which maximizes the payoff namely quit. Taking this knowledge into account, the invader will always challenge since this gives a payoff 2 instead of 1. The corresponding subgame is illustrated in Figure 19.9. The only subgame-perfect equilibrium is therefore (challenge, quit).

19.7.3 Incomplete Information in Extensive Form Games

If not all former actions can be observed by the active player, the uncertainty is collected in an *information set* . Every node of the tree is in exactly one information set. If there is no uncertainty, the information set is a singleton. If more nodes are in some information set, the player who has to choose a strategy at one of the nodes does not know which node he has achieved. He is uncertain about at which node from the information set he is. The nodes of one information set are connected by a dashed line in the tree representation.

Now it is possible to describe the classical prisoner's dilemma from Table 19.3 in extensive form with incomplete information. Let Alice begin. Then the nodes for Bob will be both in one information set because Bob does not know whether Alice cooperates or not. Figure 19.10 shows the corresponding tree.

Another interesting interpretation is the following: Normal form games with incomplete information from Section 19.6 can be described in extensive form with information sets. Consider that the nature moves first and determines the types of the players. The uncertainty about the types of the other players is modeled by corresponding information sets. Let us revise the Cournot-Nash-equilibrium example with incomplete information from Section 19.6. Two firms with two possible cost types play for maximizing their revenue. Assume that the production

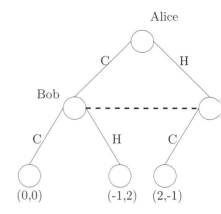

Figure 19.10: Prisoner's dilemma illustrated as t

amounts are chosen from a discrete set with two pos
one small production, say $x_h = 15$ and $x_l = 5$. The
is shown in Figure 19.11.

In Figure 19.11, the values on the edges from
spond to the types of players A and B. The followin
$(A_1, B_1), (A_1, B_2), (A_2, B_1), (A_2, B_2)$. The values on t
stage correspond to the two strategies of player A, cho
production. The dashed lines between the nodes in t
player A does not know the type of player B but on
uncertain whether he is in state (A_1, B_1) or (A_1, B_2)
on the edges from third to end stage correspond to t
The information sets for player B are indicated by a
and by a dashed-dotted line if his type is B_2. The
payoffs achieved for the two players. They are easil
price function $p(x) = 100 - 2x$, $K_1 = x^2$, $K_2 = 2x^2$
the strategy of the nature can be interpreted as a mi
probabilities in the common prior described in Table

19.8   Repeated Games and Evol

The simplest case of a dynamic game is the one witl
the very same game in normal form is played in one
repeated games is described. Two or more agents pl
several (possibly infinite many) times. Having devel
form games in the last section, it is easy to underst
the model is extended to include two or more indivic

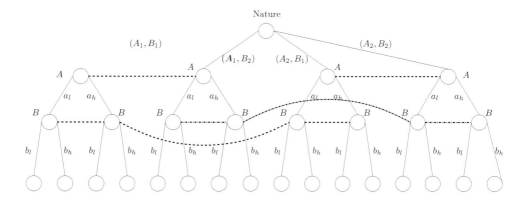

Figure 19.11: Cournot-Nash-equilibrium example with incomplete information and information sets.

of agents and including evolutionary concepts (e.g. survival of the fittest, inertia, myopia, and mutation) when updating the population, we arrive at evolutionary game theory.

19.8.1 Repeated Games

The idea of a *repeated game* is that in each period or stage, the same agents play the same stage or constituent game $\Gamma = \{N, s, u\}$ in normal form as discussed in Section 19.3. The complete game $\Gamma(T)$ is a version of the stage game repeated T times. If the game is repeated endlessly, it is called a *supergame*.

Consider first a finite repeated game with $T = 2$ stages of the simple Prisoner's dilemma game described in Table 19.3. If Alice and Bob know before the game starts that they will play precisely two rounds, a simple backward induction argument can be used to derive their selfish solutions. When they are in the last stage of the game, they will choose their dominating strategies, i.e., they will cheat. When they are in the round before the last one, knowing that they will cheat in the last round leads to the decision to cheat there, too.

In general, the solution for the stage game constitutes the solution of the finite repeated game as follows: Let $\mathbf{s}^*$ be the unique Nash equilibrium of the stage game $\Gamma = \{\mathcal{N}, s, u\}$. Then, the only subgame-perfect equilibrium of the finite repeated game $\Gamma(T)$ consists of T-times repeated play of the stage Nash equilibrium $\mathbf{s}^*$. From a social point of view, this result is disillusioning because even with punishment the repeated play of the stage strategy is optimal.

This pessimistic model looks entirely different if the agents do not know how many stages are played. One way to model this is to throw a (not necessarily fair) coin at the beginning of each stage. If the coin comes up head, then proceed with the game, otherwise stop the game. If the probability of head (continuing) is p and

each coin toss is independent, then the probability tha
is $p_k = p^k(1-p)$. The expected length of the game
backward induction argument from before cannot be

In order to model the objective of all players, a c
is introduced and it is assumed that all agents max
over all stages, e.g., $u_i(\mathbf{S}_i) = \sum_{j=0}^{\infty} \delta_i^j u_i(\mathbf{S}_{i,j})$ where
strategies of all players ($\mathbf{S}_{i,j}$ contains the strategy of
the utility of agent i. By choosing different forgettin
agents can be modeled, e.g., their patience or their n

Consider the following trigger strategy: All agen
achieves for all agents a better payoff than the Nash
$u_i(\mathbf{s}^*)$ for all $i \in \mathcal{N}$. For the two-player game descri
be the tuple $\mathbf{s}' = (honest, honest)$. If one agent de
order to achieve a short-run advantage), all other ag
their Nash strategy $\mathbf{s}^*$ until the game ends. If the ga
compute the overall payoff for the case in which all a

$$u_i(\mathbf{S}') = u_i(\mathbf{s}') \left(1 + \delta_i + \delta_i^2 + ...\right) = u_i(\mathbf{s}') \sum_{\ell=0}^{\infty}$$

On the other hand, if agent i deviates, the overall pay
Therefore, it is not beneficial to deviate if the followi

$$\delta_i > \frac{u_i(s_i^*, \mathbf{s}_{-i}') - u_i(\mathbf{s}')}{u_i(s_i^*, \mathbf{s}_{-i}') - u_i(\mathbf{s}^*)}.$$

It can be shown that applying these trigger strategies
repeated game to reach all feasible payoff vectors w
the Nash equilibrium. This result is known as one of
prisoners dilemma described in Table 19.3, for $\delta < 0$
equilibrium is the Nash equilibrium strategy, whereas
vectors better than the Nash outcome can be achi
to punish the deflecting agent by worst case strateg
larger set of feasible payoff vectors. Hence, there exi
see [10, Chapter 8]).

19.8.2   Evolutionary Game Theory

In [22], the author argues that one of game theory's g
insights it has provided in simpler biological systems
of the *evolutionary stable strategies*. However, the evo
restricted to the description of biological phenomena.
theory in which sensible agents are not always comple
decisions.

In the simplest and well-known version of the hawk-dove game, there is a population of animals in which always two randomly drawn animals meet and fight (i.e., play a two-player game in normal form). Another interpretation is that the same animal can play both strategies H and D and mixed strategies are possible. The animals fight on a resource with value V. A serious injury reduces the value by C. If a hawk meets a dove, hawk gets V and dove 0. If a dove meets a dove they divide V into two equal parts. If a hawk meets a hawk they divide after a fight and obtain $\frac{V-C}{2}$ each. It is an easy exercise to draw the matrix representation of the game. The central question is whether the hawk or the dove strategy will penetrate. For the hawk-dove game it was shown that the mixed strategy $\mathbf{x}^* = [p, 1-p]$ with $p = V/C$ is evolutionary stable in the following sense: If all members of a population play $\mathbf{x}^*$, there cannot exist a mutant which achieves a better fitness against $\mathbf{x}^*$ than $\mathbf{x}^*$ itself. Furthermore, $\mathbf{x}^*$ can invade any other population which plays another (possibly best response against $\mathbf{x}^*$) strategy than $\mathbf{x}^*$.

The payoffs from each round are interpreted as fitness of the individual and impact the number of offspring. Thereby, the strategy of the overall population is modified. The strategy adaptation happens according to the three criteria inertia, myopia, and mutation. In particular the mutation principle is important for evolutionary algorithms which could be interpreted as a nature analogue optimization approach. This technique is applied in order to approach difficult and complex programming problems.

The basic model of the strategy adaptation is as follows: There is a finite population with even number of individuals n. The adaptation works in discrete time. In each period two individuals randomly drawn meet and play the game in normal form. At the beginning of the period, the players choose their strategies which are constant during the period. The state $\mathbf{z}$ of the population in one period describes how many individuals choose a certain strategy σ_i. Based on one state $\mathbf{z}$, one player i can compute the empirical strategy distribution if he chooses strategy σ_k. Thereby, the average payoff of player i is easily computed $u_i(k, \mathbf{z})$. From this follows the best response of player i for given state $\mathbf{z}$. In order to include inertia and myopia, with a certain probability η the last strategy is kept irrespective of the next best response. Mutations are then finally included with certain probabilities.

The dynamics of the strategy adaptation is best described by a stochastic process. The state of the population forms a Markov chain and the transition probabilities can be computed based on the adaptation described above. In general it is difficult to find evolutionary stable strategies. A complete solution of the problem for all games by studying the stochastic adaptation process based on selection and mutation is not yet available.

19.9 Coalitional Form/Charact⋯
Form

Coalitional game theory mainly deals with the form
i.e., coalitions, that allow the cooperating players to
a given game. The two basic ingredients are the set ⋯
the function v which assigns a real number (value) to
A coalition $S \subseteq \mathcal{N}$ corresponds to an agreement bet⋯
S to cooperate and act as a single entity in the gam⋯
coalitions is an important task in many areas, inclu⋯
also in everyday life.

Examples of coalitions are the coalition with a sin⋯
This corresponds to a situation in which every age⋯
opposite is the grand coalition $S = \mathcal{N}$ in which all n⋯
coalition and a subset $S \subset \mathcal{N}$ are other possible coali⋯

The function v is called *characteristic function* ⋯
form if the value of the empty set is zero, i.e., $v(\emptyset) = $ ⋯
$S, T \subset \mathcal{N} : S \cap T = \emptyset$ it holds that the value of the jo⋯
equal to the sum of the values of the individual sets,
This property is called superadditivity. A special cla⋯
form are *connected games* in which for any partitio⋯
$v(\mathcal{N}) \geq \sum_{k=1}^{K} v(S_k)$.

In order to illustrate the properties of the charac⋯
simple example in which two persons $\mathcal{N} = \{1, 2\}$ are ⋯
One of them (1) owns a boat and the other one (2)
but does not own a boat. Both of them want to get t⋯
There are no bridges or ferries or other possibilities ⋯
single coalitions $\{1\}$ and $\{2\}$ they do not reach the ot⋯
is zero, i.e., $v(\{1\}) = v(\{2\}) = 0$. If they join forces ⋯
$\{1, 2\}$ they join the boat and paddles to traverse the⋯
value, say two, i.e., $v(\{1, 2\}) = 2$. Clearly, the functic⋯
a characteristic function of the game. In this simple
most likely solution for rational persons is to build th⋯

If a coalition S achieves a certain value $v(S) = c$,
lows how to split the value among the members of the ⋯
payoff of each member of the coalition. In a game wi⋯
straightforward to distribute the value among the coa⋯
in the following, we concentrate on cooperative gam⋯
In general, the value of the coalition S depends on th⋯
$k \in K$ who do not participate in the coalition $K = \mathcal{N}$⋯
computed based on the maxmin value of the coalition
tinguish between the payoffs which can be assured by t⋯
function) and the payoffs which cannot be prohibite⋯

by the others.

Before introducing solution concepts, we need the following simple definitions: A payoff vector **u** is called *imputation* if it is individual rational and Pareto-optimal. For games in characteristic form with transferable utilities, an imputation **u** satisfies $v(\mathcal{N}) = u(\mathcal{N})$ where $u(\mathcal{N}) = \sum_{k=1}^{N} v(\{k\})$ (efficiency) and $u_i \geq v(\{i\})$ (individual rational) for all $i \in \mathcal{N}$. The set of all imputations of a game $(\mathcal{N}, v)$ is $I(\mathcal{N}, v)$. A payoff vector **u**' *dominates* another payoff vector **u** in the coalition K if $u'_i \geq u_i$ for all $i \in K$ with at least one sharp inequality, and $\sum u'_i \leq v(K)$.

There exist various solution concepts for games in characteristic form. In the following, we present a subset of the concepts which are most often applied, namely the *core*, the *nucleolus*, and the *Shapley value*. An excellent introduction into coalitional games can be found in [23, Chapter 6].

19.9.1 The Core

The idea behind the definition of the core $C(\mathcal{N}, v)$ is similar to the Nash equilibrium for games in normal form. The core is the set of imputations under which no coalition has a value greater than the sum of its members' payoffs. In other words, the core is the set of all nondominated imputations. The formal definition is as follows

$$C(\mathcal{N}, v) = \{\mathbf{u} \in \mathbb{R}^N | u(\mathcal{N}) = v(\mathcal{N}), \quad u(S) \geq v(S) \text{ for all } S \subset \mathcal{N}\}. \quad (19.37)$$

The core is described by a set of inequalities, it is thus closed and compact. It is possible that the core contains more than one payoff vector. It might also happen that the core is empty. Therefore, other solution concepts are introduced, e.g., the ϵ-core or the least-core. The condition under which the core is not empty is characterized by the following result due to Bondareva and Shapley: A game in characteristic form with transferable utilities has a nonempty core if and only if it is *balanced*. The definition of balanced requires the *incidence vector* 1_S of a coalition S defined as $1_S(i) = 1$ if $i \in S$ otherwise $1_S(i) = 0$. A set of coalitions $\{S_1, ..., S_K\}$ is called balanced if for each player $i \in \mathcal{N}$ there are real-valued coefficients $0 \leq \alpha_j \leq 1$ with $j = 1...K$ such that $\sum_{j=1}^{K} \alpha_k 1_{S_j}(i) = 1$ for all S_j. Based on balanced sets of coalitions, a game in characteristic form is called balanced if for every balanced set of coalitions and corresponding coefficients hold that $\sum_{j=1}^{K} \alpha_j v(S_j) \leq v(\mathcal{N})$.

Before proceeding to the next solution concept, let us illustrate the core for one simple anecdotal example. Consider the following three-player coalitional game $(\mathcal{N}, v)$ with $\mathcal{N} = \{1, 2, 3\}$ and $v(\{i\}) = 0$ for all $i \in \mathcal{N}$, $v(\{1, 2\}) = 3$, $v(\{1, 3\}) = 13$, $v(\{2, 3\}) = 23$, and $v(\mathcal{N}) = 36$. The conditions for a payoff vector **u** to lie in the core are: $u_1 + u_2 + u_3 = 36$ (Pareto optimality), $u_1, u_2, u_3 \geq 0$ (individual rationality), and $u_1 + u_2 \geq 3$, $u_1 + u_3 \geq 13$, $u_2 + u_3 \geq 23$ (group rationality). The corresponding inequalities and the core are illustrated in Figure 19.12.

In order to understand the following set-based solution concepts, the *excess of a coalition* S with respect to payoff **u** is defined by $e(S, \mathbf{u}) = v(S) - u(S) =$

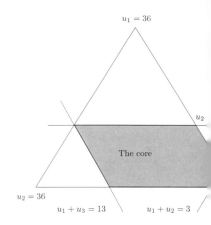

Figure 19.12: Illustration of the core of a three-player

$v(S) - \sum_{i \in S} u_i$. In the example above for the core, t
prefer the imputation $\mathbf{u} = [8, 18, 10]$ compared to $\mathbf{w} =$
$e(S, \mathbf{u}) = 23 - (18 + 10) = -5$ is smaller than the excess

19.9.2   The Kernel

The stability of a coalition depends on the *objectio*
against the coalition. The start point is an imputa
formulate an objection against another agent j to th
coalition S which includes i but not j with S-feasibl
all $k \in S$. This definition of objection means that
all agents in his favorite coalition S (which does not
the coalition, because they get a higher payoff $\mathbf{u}'$ than
The corresponding pair $(\mathbf{u}', S)$ is called an objection.

What the outsider j can do is to counter the obj
This counter argument is another pair of outcome vec
coalition contains agent j and not agent i, of course. I
l not in S to join t, it is sufficient to achieve a payof
agents in T and S should also achieve a higher payoff
$v_l \geq u'_l$ for $l \in T \cap S$.

Only those imputations are considered as a reali
characteristic form which are stable in the following
is the set of all imputations $\mathbf{u}$ for which for all obj
against j to $\mathbf{u}$ there exists a counterobjection $(\mathbf{v}, T)$

In order to define the kernel, we need to turn t
again with an imputation $\mathbf{u}$. An agent i raises an *obj*
a coalition S that excludes some agent j which has

single coalition $u_j > v(\{j\})$. The objection here is the coalition S. Agent j can counterobject by showing a coalition T which contains j but not i which has a larger excess at $\mathbf{u}$ than S, i.e., $e(T, \mathbf{u}) \geq e(S, \mathbf{u})$. The counterobjection to the objection S is here T. The *kernel* of a coalitional game is defined as the set of all imputations $\mathbf{u}$ for which for every objection S of some player i against j in $\mathbf{u}$ there is a counterobjection of j to S. The kernel always exists but is in general hard to compute.

19.9.3 The Nucleolus

If the core is empty, another approach is needed to find a reasonable solution concept. The nucleolus is one such solution concept introduced by Schmeidler in 1969.

Since there are multiple coalitions possible, and all need to be considered, there is a need for an order on the resulting vector of excesses. Let $\theta(\mathbf{u})$ be the vector with sorted components in nonincreasing order. The vector has 2^N components since there are in total 2^N coalitions $S \subseteq \mathcal{N}$ possible. The *vector of ordered excesses* can be written as

$$\theta(\mathbf{u}) = [\theta_1(\mathbf{u}), ..., \theta_{2^N}(\mathbf{u})] = [e(S_1, \mathbf{u}), ..., e(S_{2^N}, \mathbf{u})] \tag{19.38}$$

with $e(S_i, \mathbf{u}) \geq e(S_j, \mathbf{u})$ for $1 \leq i < j \leq 2^N$. For the example from above and the imputation $\mathbf{u} = [9, 15, 12]$, the vector of ordered excesses is given by $\theta(\mathbf{u}) = [0, 0, -4, -8, -9, -12, -15, -21]$. The first two zeros correspond to the grand and the zero coalition.

In order to compare different vectors of ordered excesses, we need an order to compare vectors. The lexicographic order is a total order and very intuitive. The vector $\mathbf{x} \in \mathbb{R}^n$ is lexicographically smaller than $\mathbf{y} \in \mathbb{R}^n$ if there is an index number $m \leq n$ such that $x_i = y_i$ for all $i = 1, ..., m - 1$ and $x_m < y_m$. This relation is written as $\mathbf{x} <_L \mathbf{y}$. Based on this order, the *nucleolus* of a game in characteristic form is defined as the set of all imputations which lexicographic minimize the vector of ordered excesses, i.e.,

$$\mathcal{H}(N, v) = \{\mathbf{u} \in I(N, v) | \theta(\mathbf{u}) \leq_L \theta(\mathbf{w}) \text{ for all } \mathbf{w} \in I(N, v)\}. \tag{19.39}$$

It can be shown that the nucleolus is a subset of the core if it is not empty. Furthermore, the nucleolus does always exist and consists of only a single payoff vector. There exist iterative algorithms which can compute the nucleolus by solving linear programs. An analytic solution for three-player games is computed in [24]. Thereby we can find the nucleolus for our example from above in Figure 19.12 as $\mathcal{H}(N, v) = [6.5, 11.5, 18]$.

19.9.4 The Shapley Value

In contrast to the preceding three set-based solution concepts, the *Shapley value* is a value-based solution concept for games in characteristic form with transferable

utilities. There exist two motivations or definitions fo
one follows the style of the Nash bargaining solution d
is an axiomatic definition. The second one follows t
kernel and defines the Shapley value by certain objec

The value ψ of a coalitional game with transfer:
feasible payoff vector. Feasible means that the sum ψ
to understand the axiomatic approach, we need to def
of player i to coalition S with $i \notin S$:

$$\Delta(i, S) = v(S \cup \{i\}) - v(S)$$

The following axioms define the Shapley value ψ:

SYM Symmetry: If i and j are interchangeable, i.e.,
coalition S that contains neither i nor j, then
i.e., $\psi(i, v) = \psi(j, v)$.

DUM Dummy agent: If the presence of an agent i d
the coalition by more than $v(\{i\})$ then i does n
is called a *dummy*. Then $\psi(i, v) = v(\{i\})$.

ADD Additivity: For any two characteristic function
payoff vectors $\mathbf{x}$ and $\mathbf{y}$, then they should get th
payoff vectors $\mathbf{x} + \mathbf{y}$ for the characteristic funct
$\psi(i, v) + \psi(i, w)$ for all $i \in \mathcal{N}$.

The Shapley value ψ is defined by the condition

$$\psi(i, v) = \sum_{i \in S \subset \mathcal{N}} \underbrace{\frac{|S|!(N - |S| - 1)!}{N!}}_{\alpha(|S|, N)}$$

Note that by the choice of $\alpha(|S|, N)$ in (19.41) the feas
$v(\mathcal{N})$. The Shapley value ψ is the only value which sat
It always exists and is unique. When the valuation f
Shapley value is individually rational, i.e., it is an i
function v is convex, the Shapley value is group-ration
are further values also known as power indices in the
Index, Deegan-Packel-Index, and the Public-Good-In

Another interpretation of the Shapley value define
us revisit the example from above in Figure 19.12 and
The power of agent one is computed by the gain of t
joins:

$$\Delta(1, \{2\}) = 3 - 0, \qquad \Delta(1, \{3\}) = 13 - 0, \qquad \text{a}$$

Each power value in (19.42) is weighted by the number of times it occurs if coalitions are drawn randomly and the Shapley values are computed as $\phi(1, v) = \frac{3}{6} + \frac{13}{6} + \frac{13}{3} = 7$, $\phi(2, v) = \frac{1}{2} + \frac{23}{6} + \frac{23}{3} = 12$ and $\phi(3, v) = \frac{13}{6} + \frac{23}{6} + \frac{33}{3} = 17$. Compare this result to the nucleolus $\mathcal{H}(N, v) = [6.5, 11.5, 18]$.

19.10 Mechanism Design and Implementation Theory

Having studied solution concepts for games in normal, extensive and characteristic form in this section, the following question of how to design the rules of a game such that a certain cooperative solution is implemented is answered.

There is one classical example to explain the idea of mechanism design. The problem is well known to anyone who has tried to divide some good fairly among two people. A typical setup is a mother who has one piece of cake that is to be shared between her two kids. The cake has not a perfect shape (circle or square) and it is not obvious how to divide the cake. Furthermore, the judgment of the two kids when the cake is divided which is the better, i.e., larger, piece is not clear. It is clear that the event that one kid has the (subjective) feeling that his part is smaller should be avoided by the mother. In the game theoretic terms introduced in Section 19.2 the problem can be described as follows: The two kids have private types (their judgment which is the better piece of the cake), which is not known to the mother before dividing the cake, which leads to a private preference relation. The task of the mother is to design a mechanism (a way to divide the cake) such that both agents (the kids) are happy. The solution to the problem is old but very elegant. The mechanism works as follows: One of the kids is dividing the cake. The other one decides on the distribution of the two pieces. This mechanism is optimal because the first kid will try to divide the cake completely fairly according to his preference relation because he cannot be sure which piece he will get. There is no incentive for the first kid to divide the cake unequally. The second kid will choose the piece that looks bigger to him and therefore has a clear incentive to make his private preference relation public. In the end, both kids are satisfied with the division. Interestingly, this mechanism only works for two participants. A similar efficient and simple solution is not yet available for three or more participants.

The example illustrates the main idea of mechanism design. The rules of the game should be chosen such that the outcome of the game has certain valuable properties. For engineers this method has some similarities with *reverse engineering*. In all previous sections of this chapter, game models are derived and then solution concepts developed to forecast the outcome of the game. In this section, we turn around and ask whether it is possible and how to design games such that the outcomes fulfill certain properties. Viewing the game and its outcome as a black box, first, the source code for the black box is reverse engineered and then modified to influence the outcome into a good direction.

There are fundamental limits in the possibilities to achieve certain outcomes.

In order to illustrate these limits, consider another c
with three candidates. Let us define the set of cand
the set of voters is $\mathcal{W} = \{1, 2, 3\}$. Each voter has a p
of candidates, e.g.,

$$A >_1 B >_1 C, \qquad B >_2 C >_2 A, \qquad (\quad$$

The inequalities in (19.43) say that voter 1 prefers A (
B over C over A and voter 3 prefers C over A over F
is used to decide the winner, this case leads to an ind
If there are more than two candidates, there has to be
in order to determine the social choice. This probl
Marquis des Condorcet in 1785.

19.10.1   Social Choice

Motivated by the two examples, we provide a desci
Every agent k has a preference relation which is dep
some set Θ_k. This type is private in general. During t
has a strategy s_k from some strategy space S_k. This
the one hand, a *social choice function* (SCF) f dete
based on all types (or preference relations) of the voter
On the other hand, an outcome rule g maps the str
outcome of the game: $g : S_1 \times S_2 \times \ldots \times S_K \to X$. In t
the SCF (the theory) coincides with the outcome of the
i.e.,

$$f(\theta_1, \theta_2, ..., \theta_K) = g(s_1(\theta_1), s_2(\theta_2), ...$$

The basic idea of mechanism design and a clear expla
is provided in [25]. Eric Maskin, Leonid Hurwicz, a
the 2007 Nobel Memorial Prize in Economics for ha
mechanism design theory.

In order to come back to the question whether a
plements a certain SCF, the properties of the outco
determined first. The approach is similar to the axi
described in Section 19.5.1. The following axioms are

1. A SCF fulfills *unanimity* if the outcome correspo
   of the agents.

2. An agent $k \in \mathcal{K}$ is called *dictator* if his prefere
   outcome irrespective of all other individual pref

3. The SCF is *independent of irrelevant alternat*
   between two alternatives a and b only depends (
   for these two.

Clearly, unanimity and independence of irrelevant alternatives are two favorable properties of a SCF. Additionally, a dictator should be avoided to have a fair SCF. Arrow showed in [26] the following impossibility results: Every SCF with more than two agents which fulfills unanimity and is independent of irrelevant alternatives implies a dictator. Arrow received the 1972 Nobel Memorial Prize in Economics with John Hicks.

19.10.2 Auctions

There are several ways out to arrive at possibility results. Either the axioms are weakend or other axioms are applied. Another way is to allow transferable utilities. One very important class of mechanisms with transferable utilities are *auctions*. The typical simple model for auctions is a set of alternatives $\mathcal{A}$, the set of bidders $\mathcal{K} = \{1, ..., K\}$. The preferences are described by functions corresponding to the value $v_i : \mathcal{A} \to \mathbb{R}$. The money the bidders have to pay is denoted by m. The utility of bidder i is given by $u_i = v_i - m$, i.e., the values of the object minus the money paid. The simplest case is an auction in which one good is auctioned between K bidders. Each player k has a scalar value w_k which he agrees to pay for the object. This is his private type. If a simple sealed first price auction is applied, the bidder j with highest bid wins and has to pay his bid p. His utility is $u_i = w_i - p$.

Here again the situation is similar to the cake problem from the beginning of this section. The bidders have private types $w_1, ..., w_K$ and the auctioneer aims for selling the good to the bidder who values it most. In the simple sealed, first-price auction, it is easy (and also rational) not to bid the true value. Imagine, one bidder k has value w_k, bids it and wins. Then, he has to pay $p = w_k$ and his utility is $u_k = w_k - p = w_k - w_k = 0$.

One auction, which has the property that bidding his correct value is the best (equilibrium) strategy, is the *Vickrey* auction [27]. It is a sealed, second-price auction in which the bidder with highest bid receives the good and pays the second highest bid. In order to explain the incentive-compatibility of the Vickrey auction, we study the two cases bidding more than the value and bidding less than the value separately. It is not advantageous to bid more - say v_1 - than the value w_1, i.e., $v_1 > w_1$ because if the bidder receives the bid, it can happen that the second bid is v_2 larger than the value of the highest bidder but of course smaller than the highest bid, i.e., $v_1 > v_2 > w_1$. Then, the highest bidder obtains a utility $u_1 = w_1 - v_2 < 0$ which is negative. But it is also not advantageous to bit less - say b_1 - than the value w_1, i.e., $b_1 < w_1$ because if another bidder bids b_2 between w_1 and b_1, i.e., $b_1 < b_2 < w_2$, then bidder two receives the good whereas bidder one values it more. Therefore, the optimal strategy is to bid the true value of the good and to apply an incentive compatible mechanism.

There are many different types of auctions proposed and performed in the literature and in practice. The four standard auctions are described as follows [28, Chapter 9]:

- First-Price, Sealed-Bid: Each bidder submits a
  highest bid wins and the bidder pays his bid fo

- Second-Price, Sealed-Bid: Each bidder submit
  The highest bid wins and pays the second-high

- Dutch Auction: The seller begins with a ver
  continuously. The first bidder who raises the h
  the current price.

- English Auction: The seller begins with a ver
  continuously. The bidder signals when he or s
  auction round. Once signed out, bidders canno
  is left, he pays the current price and gets the g

19.10.3   Direct Revelation Principle

In Subsection 19.10.1 it was assumed that the agen
based on their utility function which in turn depend
types. For the game designer who develops a mechai
SCF, the important question that arises is what typ
the agents is required.

Often there is one agent called principal which
other agents and makes a decision on some allocati
The direct revelation principle states that communica
the principal can be considered without loss of gener
types. In other words, the message space of the agen
After all agents have agreed to the proposed mechanisr
their types.

The direct revelation principle allows to separate
from the mechanism design. This separation result I
rium concepts and any general coordination mechar
mechanism and an incentive-compatible, individually
a direct-revelation mechanism in which truth telling
dividually rational, and which produces an identical
designer can restrict attention to direct-revelation m
assignments within those mechanisms.

19.10.4   Clarke-Groves and AGV-Arro

If transferable utilities can be used to alter the utilit
types are truthfully reported, two standard approach
was discovered by Groves and Clarke, the second by d'
and independently by Arrow [3, Section 7.4.3].

Let us model the types $\theta_1, ..., \theta_K$ of K agents drawn randomly from $\mathcal{A} \subseteq \mathbb{R}$ with certain probabilities $p_k(\theta_k)$. These probability distributions of the individual agents are common knowledge to all agents. The utility of agent k depends on the outcome x, his type θ_k and on his transfer $t_k \in \mathbb{R}$ in a quasi-linear manner: $u_k(x, \theta_k, t_k) = v_k(x, \theta_k) + t_k$. A mechanism is called budget balanced if $\sum_{k=1}^{K} t_k = 0$.

When developing the incentive-compatible mechanism, the design can either focus on the solution concept based on dominating strategies or Bayesian equilibrium. The first solution concept is stronger in the sense that the truth-telling strategy of one agent does not depend on the messages of the other agents. In contrast, the Bayesian equilibrium depends on the subjective probability of the agent on the types of the other agents. Based on this, the agent maximizes the average achievable payoff with respect to his reported type. The argument is then based on the equilibrium, i.e., if all agents report their values truthfully, there is no agent who can improve the expected payoff by deviating from reporting the true type.

Let us denote the function f which maps the reported types $\theta_1, ..., \theta_k$ to the outcome space $\mathcal{X}$, i.e., $f : \Theta_1 \times ... \times \Theta_K \to \mathcal{X}$ as in (19.44). Let the reported type of player k by $\hat{\theta}_k$. Then, the mechanism is implemented in dominating strategies if for each $\theta_k \neq \hat{\theta}_k, \theta_{-k}$ it holds

$$u_k(f(\theta_k, \theta_{-k}), \theta_k) \geq u_k(f(\hat{\theta}_k, \theta_{-k}), \theta_k). \tag{19.45}$$

The idea of the Groves mechanism is that the transfer of agent k is chosen such that his payoff is equal to the sum of all individual payoffs. The transfer for agent k is chosen as the sum payoff minus the payoff of agent k, i.e.,

$$t_k(\theta_k^*) = \sum_{l \neq k} v_l(f(\theta_k^*, \theta_{-k}), \theta_l).$$

In order to achieve budget balance, the transfer can be further modified to include an expected payoff (expected externality) for agent k when he announces the type $\hat{\theta}_k$.

The mechanism is implemented in Bayesian equilibrium if for each $\theta_k \neq \hat{\theta}_k, \theta_{-k}$ it holds

$$\mathbb{E}_{\theta_{-k}} \left[u_k(f(\theta_k, \theta_{-k}), \theta_k) \right] \geq \mathbb{E}_{\theta_{-k}} \left[u_k(f(\hat{\theta}_k, \theta_{-k}), \theta_k) \right]. \tag{19.46}$$

From (19.46) follows that if all agents $l \neq k$ report their true types θ_l, the expected payoff of agent k is maximized by reporting his true type θ_k. The idea of the AGV mechanism is similar to the Groves mechanism. The sum of the expected payoffs of the other agents is added to the utility of agent k, i.e.,

$$t_k(\hat{\theta}_k) = \mathbb{E}_{-\theta_k} \left[\sum_{l \neq k} v_l(f(\theta_k^*, \theta_{-k}), \theta_l) \right].$$

The maximization of the expected payoff of agent k
type $\hat{\theta}_k$ results in the true type θ_k.

19.11   Applications to Signal Pr
Communications

The application of game-theoretic tools to signal prc
started about 20 years ago. Most early efforts were (
game theory to communication networks [29]. Rece
theory, have been published by the *IEEE Journal on*
cations [30], the *IEEE Signal Processing Magazine* [1
on Advances in Signal Processing [31].

There are many recent results in the research 1
and signal processing available, in which game theore
resource allocation, transceiver and algorithm design
of references is by far not complete, and the references
the breadth of the field.

In multiple-antenna wireless communications rese
of transmit strategies for the interference channel is
sponse dynamics corresponds to the popular iterativ
der certain conditions, the global stability condition i
converges to the Nash equilibrium. In [33], the scer
unlicensed bands is first modeled as a noncooperative
the outcome is analyzed. Then, a repeated game app
the efficiency.

Another way to improve the efficiency is to allow cc
This situation can be modeled for the multiple-antenna
operative game. Axiomatic bargaining theory is appli
working point (the Nash bargaining solution) in [34
the Nash bargaining solution and the well-known pr
derived in [35] based on the interference function fran

In [36], an interference game where each link has in
the other player's channel conditions is studied. The l
and the ϵ-Nash equilibrium point is introduced. The s
spectrum utilization.

A complete framework of coalitional game theory
[37]. The solution concepts are categorized into three
games, coalition formation games, and coalitional gr
communication scenarios are developed.

In dynamic spectrum sharing scenarios, the next
lution concepts from mechanism design and implem
plied. The design of incentive compatible schemes is s

theoretical mechanism design methods are proposed. In [39], a repeated spectrum sharing game with cheat-proof strategies is proposed.

19.12 Acknowledgments

We would like to thank Johannes Lindblom for drawing several of the figures and for providing many useful comments on an early version of the manuscript. We thank Danyo Danev, Rami Mochaourab, and Fei Shen for reading and providing comments on an earlier version of the chapter.

19.13 Exercises

Exercise 19.13.1. What would you prefer?

(a) One million dollars guaranteed.

(b) Nothing with probability 50%, and ten million dollars with probability 50%.

Exercise 19.13.2. Consider again Exercise 19.13.1. Is the amount of money you own a valid utility function?

Exercise 19.13.3. Consider a big corporation, with a \$10B annual revenue and suppose that a deal is to be closed, which involves no higher stakes than a worth of \$1M. Is the amount of money the company makes in the deal a valid utility function?

Exercise 19.13.4 (Convexity of the utility region). Show that the utility region is compact and convex.

Exercise 19.13.5 (Rock-paper-scissors). Consider the classical rock-paper-scissors zero-sum game with payoffs in Table 19.2. Find the value of the game and the optimal strategies, and discuss its implications.

Exercise 19.13.6. Derive and sketch utility region for a zero-sum game.

Exercise 19.13.7 (Hazard game (inspired by discussion in [40])). Alice and Bob engage in a hazard game as follows. Alice tosses a coin, and Bob then attempts to guess how the coin comes up. The rules are as follows:

- If the coin comes up heads, and Bob guesses heads, nothing happens.

- If the coin comes up tails, and Bob correctly guesses tails, then he wins \$1 from Alice.

- If Bob's guess is incorrect, then he must pay Alice 50 cents.

Suppose that Alice's coin is designed to come up tails with probability $1-p$. Further, suppose Bob gue q and "tails" with probability $1-q$. Write down the ga Then use a linear programming toolbox of choice (or following questions:

(a) Suppose Alice's coin is fair, so that $p = 1/2$, and What is the optimal q, in the sense of maximizin

(b) Repeat (a) for $p = 1/4$.

(c) Suppose Alice's coin may be biased so that $p \neq$ know q. What is Bob's best strategy against A payoff (or loss)?

Exercise 19.13.8 (Monty-Hall game show (in [40])). We consider the classical Monty-Hall gamesh as the three-door problem) that goes as follows. Th are closed. Behind one of them, there is a car. Be nothing. The game is played between a game host (H the objective for P is to win the car. H placed the ca she knows where it is, but she does not reveal it. P c the car is. P is first asked to select one of the doors instead selects one of the other two doors, where sh *located.*

P may now choose between two option: either he had initially chosen, or he changes his mind to the c neither H nor P had pointed to).

Denote with (i, j) the moves that H have, where i placed the car ($i \in \{1, 2, 3\}$) and j is the door that $i \neq j$ by assumption, the moves available to H are $(1,$ and $(3, 2)$. Denote with 1ss, 1sc, 1cs, 1cc, s2s, s2c, the moves available to P, with meaning as follows: 1s 1, and stays regardless of which door H selects; 1sc stays if H selects door 2 but changes if H selects door door 1, and changes door regardless of which one H selects door 2, and stays regardless of which door H s

(a) Write up the payoff matrix U, assuming we as P winning the car and zero (0) otherwise.

(b) Show that the optimal strategy for host is to p ability $1/6$ and that the optimal strategy for the pa $cc3$ with probabilities $1/3$. Show that the expected pa

(c) Discuss the implications of the result in (b).

Exercise 19.13.9 (Chess). Consider chess. Argu uniquely defined, finite value.

Table 19.5: Game matrix for the motorway game.

		The other car	
		Keep left (L)	Keep right (R)
You	Keep left (L)	(-1,-1)	(-100,-100)
	Keep right (R)	(-100,-100)	(0,0)

Exercise 19.13.10 (Motorway driving). While driving on the motorway and meeting another car, you may either keep to the left (L) or the right (R). In many countries in the world, traffic rules prescribe that meeting cars on a motorway keep to the right, we assign a value of (0,0) to the event that both you and the other car keep to the right. If both you and the other car keep to the left, an accident is avoided but the memory of the event will persist, so we assign a payoff of (-1,-1). If you and the other car make a different choice, an accident will result and we assign (-100,-100). The game is summarized in Table 19.5.

Use Gambit [17] to find all Nash equilibria.

Exercise 19.13.11 (Cournot-Nash-Equilibrium). Two firms A and B invest in a product and produce an amount x with costs $K(x)$. There is an upper bound on the production $x_i \leq x_i^{max}$ and the price for the product depends on the market (supply and demand) as $p(x) = 100 - 2x$. Firm A obtains the revenue $G_A(x_A + x_B) = p(x_A + x_B)x_A - K_A(x_A)$ and firm B analogue. Assume $K_A(x_A) = x_A^2$ and $K_B(x_B) = x_B^2$. What is the optimal production x_A^* and x_B^*?

Exercise 19.13.12 (Cournot-Nash-Equilibrium with incomplete information). Consider the setting from the last Exercise 19.13.11 and the common prior in Table 19.4. The costs for type one are $K_{i1} = 2x_i^2$ whereas the costs for type two are $K_{i2} = x_i^2$. Compute the Bayesian Equilibrium for the game with incomplete information.

Exercise 19.13.13 (Normal form representation of invader game). Consider the invader game described in extensive form in Figure 19.8. Compute the two-by-two payoff matrix for the corresponding representation in normal form. Derive the two Nash equilibria.

Exercise 19.13.14 (Subgame-perfect equilibrium I). Compute the subgame-perfect equilibrium of the extensive-form game illustrated in Figure 19.13.

Exercise 19.13.15 (Subgame-perfect equilibrium II). Compute the subgame-perfect equilibrium of the extensive-form game illustrated in Figure 19.14.

Exercise 19.13.16 (Payoff for game in Figure 19.11). Compute the payoff for the outcomes of the game illustrated in Figure 19.11. There are two production strategies, one high and one low production, say $x_h = 15$ and $x_l = 5$. The price

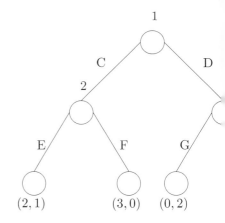

Figure 19.13: Extensive form game

function is $p(x) = 100 - 2x$, the cost types are $K_1 =$
revenue is given by $G_A(x_A, x_B) = p(x_A + x_B)x_A -$
$G_B(x_A, x_B) = p(x_A + x_B)x_B - K_B(x_B)$ for player B. C
equilibria.

Exercise 19.13.17 (The core 1). Consider the c
with $v(\mathcal{N}) = 1$, $v(S) = \alpha$ if $|S| = 2$ and $v(\{i\}) = 0$ for
not empty?

Exercise 19.13.18 (The core 2). An expedition
treasure in the mountains. Only four persons can carr
coalitional game for this situation is $v(S) = \lfloor |S|/4 \rfloor$.

Exercise 19.13.19 (Shapley value). Compute the
$\{(1, 2, 3, 4), v\}$ where $v((1, 2, 3, 4)) = 3$, $v(S) = 0$ if S c
$(1, 2, 3)$ and $v(S) = 2$ otherwise.

References

[1] E. Jorswieck, E. Larsson, M. Luise, and H. Poor,
cessing and communications [from the guest edito
Magazine, vol. 26, no. 5, pp. 17–132, Sept. 2009.

[2] A. MacKenzie and L. DaSilva, *Game Theory fo*
Rafael, CA: Morgan & Claypool publishers, 200(

[3] D. Fudenberg and J. Tirole, *Game theory*. Camb

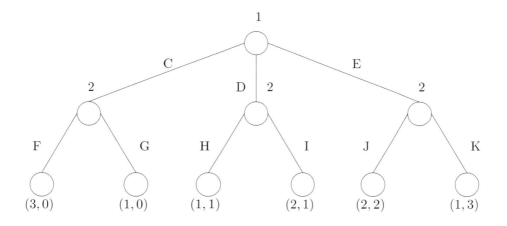

Figure 19.14: Extensive form game example 2.

[4] G. Owen, *Game Theory*, 3rd ed. Academic Press, 1995.

[5] T. S. Ferguson, "Game theory," http://www.math.ucla.edu/~tom/Game_Theory/Contents.html.

[6] T. W. Körner, *Naive Decision Making: Mathematics Applied to the Social World.* New York: Cambridge University Press, 2008.

[7] T. Basar and G. J. Olsder, *Dynamic Noncooperative Game Theory (Classics in Applied Mathematics, 23).* Harlow, Essex, UK: Soc. for Industrial & Applied Math (SIAM), December 1998.

[8] R. Gibbons, *A Primer in Game Theory.* Pearson Higher Education, June 1992.

[9] N. Nisan, T. Roughgarden, E. Tardos, and V. V. Vazirani, *Algorithmic Game Theory.* New York, NY, USA: Cambridge University Press, 2007.

[10] M. J. Osborne and A. Rubinstein, *A Course in Game Theory.* Cambridge, MA: MIT Press, 1999.

[11] H. Peters, *Axiomatic Bargaining Game Theory*, Boston: Kluwer Academic Publishers, 1992.

[12] J. von Neumann and O. Morgenstern, *Theory of Games and Economic Behavior.* Princeton, NJ: Princeton University Press, 1943.

[13] "Gnu linear programming kit," http://www.gnu.org/software/glpk/. August 2011.

734

[14] "lpsolve," http://sourceforge.net/projects/lpsolv

[15] "Cvx," http://cvxr.com/cvx/. August 2011.

[16] J. Nash, "Non-cooperative games," *The Annals c* pp. 286–295.

[17] "Gambit: software tools for game theory," htt gNU public license. August 2011.

[18] C. Papadimitriou, "Algorithms, games, and the i *ceedings of the thirty-third annual ACM symposi* New York, NY, USA: ACM, 2001, pp. 749–753.

[19] J. Nash, "The bargaining problem," *Econometric* April 1950. [Online]. Available: http://ideas v18y1950i2p155-162.html

[20] G. Rabow, "The social implications of nonzero-su *and Society Magazine*, vol. 7, no. 1, pp. 12–18, N

[21] J. C. Harsanyi, "Games with incomplete inform players," *Management Sciences*, vol. 14, pp. 159– 8.

[22] W. G. S. Hines, "Evolutionary stable strategies *Theoretical Population Biology*, vol. 31, pp. 195–

[23] M. J. Holler and G. Illing, *Einführung in die Spi* Springer-Verlag, 2000.

[24] M. Leng and M. Parlar, "Analytic solution for th cooperative game," *Naval Research Logistics (N* 672, 2010.

[25] D. Garg, Y. Narahari, and S. Gujar, "Foundati tutorial, part 1 key concepts and classical resul pp. 83–130, April 2008.

[26] K. J. Arrow, *Social Choice and Individual Valu* University Press; 2nd edition, 1951.

[27] V. Krishna, *Auction Theory*. London: Academi

[28] G. A. Jehle and P. J. Reny, *Advanced Microeco* Addison-Wesley, 2001.

[29] E. Altman, T. Boulogne, R. El-Azouzi, T. Jiméne on networking games in telecommunications," *C* no. 2, pp. 286–311, 2006.

[30] N. Mandayam, S. Wicker, J. Walrand, T. Basar, J. Huang, and D. Palomar, "Game theory in communication systems [guest editorial]," *IEEE Journal on Selected Areas in Communications*, vol. 26, no. 7, pp. 1042–1046, Sept. 2008.

[31] H. Boche, Z. Han, E. G. Larsson, and E. A. Jorswieck, "Game theory in signal processing and communications," *EURASIP Journal on Advances in Signal Processing*, vol. 2009, 2009.

[32] G. Scutari, D. Palomar, and S. Barbarossa, "Competitive design of multiuser MIMO systems based on game theory: A unified view," *IEEE Journal on Selected Areas in Communications*, vol. 26, no. 7, pp. 1089–1103, September 2008.

[33] R. Etkin, A. Parekh, and D. Tse, "Spectrum sharing for unlicensed bands," *IEEE Journal on Selected Areas in Communications*, vol. 25, no. 3, pp. 517–528, April 2007.

[34] E. G. Larsson, E. Jorswieck, J. Lindblom, and R. Mochaourab, "Game theory and the flat-fading Gaussian interference channel," *IEEE Signal Processing Magazine*, vol. 26, no. 5, pp. 18–27, Sept. 2009.

[35] H. Boche and M. Schubert, "Nash bargaining and proportional fairness for wireless systems," *IEEE/ACM Transactions on Networking*, vol. 17, no. 5, pp. 1453–1466, Oct. 2009.

[36] Y. Noam, A. Leshem, and H. Messer, "Competitive spectrum management with incomplete information," *IEEE Transactions on Signal Processing*, vol. 58, no. 12, pp. 6251–6265, Dec. 2010.

[37] W. Saad, Z. Han, M. Debbah, A. Hjorungnes, and T. Basar, "Coalitional game theory for communication networks," *IEEE Signal Processing Magazine*, vol. 26, no. 5, pp. 77–97, Sept. 2009.

[38] B. Wang, Y. Wu, Z. Ji, K. Liu, and T. Clancy, "Game theoretical mechanism design methods," *IEEE Signal Processing Magazine*, vol. 25, no. 6, pp. 74–84, Nov. 2008.

[39] Y. Wu, B. Wang, K. Liu, and T. Clancy, "Repeated open spectrum sharing game with cheat-proof strategies," *IEEE Transactions on Wireless Communications*, vol. 8, no. 4, pp. 1922–1933, Apr. 2009.

[40] O. Häggström, *Slumpens Skördar: Strövtåg i Sannolikhetsteorin.* Studentlitteratur, 2004, in Swedish.

Chapter 20 A Short Course on Frame Theory

Veniamin I. Morgenshtern and Helmut Bölcskei
ETH Zurich, Switzerland

Hilbert spaces [1, Def. 3.1-1] and the associated concept of orthonormal bases are of fundamental importance in signal processing, communications, control, and information theory. However, linear independence and orthonormality of the basis elements impose constraints that often make it difficult to have the basis elements satisfy additional desirable properties. This calls for a theory of signal decompositions that is flexible enough to accommodate decompositions into possibly nonorthogonal and redundant signal sets. The theory of frames provides such a tool.

This chapter is an introduction to the theory of frames, which was developed by Duffin and Schaeffer [2] and popularized mostly through [3–6]. Meanwhile frame theory, in particular the aspect of redundancy in signal expansions, has found numerous applications such as, e.g., denoising [7, 8], code division multiple access (CDMA) [9], orthogonal frequency division multiplexing (OFDM) systems [10], coding theory [11, 12], quantum information theory [13], analog-to-digital (A/D) converters [14–16], and compressive sensing [17–19]. A more extensive list of relevant references can be found in [20]. For a comprehensive treatment of frame theory we refer to the excellent textbook [21].

20.1 Examples of Signal Expansions

We start by considering some simple motivating examples.

Example 20.1.1 (Orthonormal basis in $\mathbb{R}^2$). Consider the orthonormal basis (ONB)

$$\mathbf{e}_1 = \begin{bmatrix} 1 \\ 0 \end{bmatrix}, \qquad \mathbf{e}_2 = \begin{bmatrix} 0 \\ 1 \end{bmatrix}$$

in $\mathbb{R}^2$ (see Figure 20.1). We can represent every signal $\mathbf{x} \in \mathbb{R}^2$ as the following linear combination of the basis vectors $\mathbf{e}_1$ and $\mathbf{e}_2$:

$$\mathbf{x} = \langle \mathbf{x} | \mathbf{e}_1 \rangle \, \mathbf{e}_1 + \langle \mathbf{x} | \mathbf{e}_2 \rangle \, \mathbf{e}_2. \tag{20.1}$$

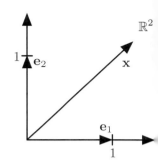

Figure 20.1: Orthonormal basis

To rewrite (20.1) in vector-matrix notation, we st
expansion coefficients as

$$\mathbf{c} = \begin{bmatrix} c_1 \\ c_2 \end{bmatrix} \triangleq \begin{bmatrix} \langle \mathbf{x}|\mathbf{e}_1 \rangle \\ \langle \mathbf{x}|\mathbf{e}_2 \rangle \end{bmatrix} = \begin{bmatrix} \mathbf{e}_1^\mathsf{T} \\ \mathbf{e}_2^\mathsf{T} \end{bmatrix} \mathbf{x} =$$

It is convenient to define the matrix

$$\mathbf{T} \triangleq \begin{bmatrix} \mathbf{e}_1^\mathsf{T} \\ \mathbf{e}_2^\mathsf{T} \end{bmatrix} = \begin{bmatrix} 1 & 0 \\ 0 & 1 \end{bmatrix}.$$

Henceforth we call $\mathbf{T}$ the *analysis matrix*; it multiplie
expansion coefficients

$$\mathbf{c} = \mathbf{T}\mathbf{x}.$$

Following (20.1), we can reconstruct the signal $\mathbf{x}$ f
according to

$$\mathbf{x} = \mathbf{T}^\mathsf{T}\mathbf{c} = \begin{bmatrix} \mathbf{e}_1 & \mathbf{e}_2 \end{bmatrix} \mathbf{c} = \begin{bmatrix} \mathbf{e}_1 & \mathbf{e}_2 \end{bmatrix} \begin{bmatrix} \langle \mathbf{x}|\mathbf{e}_1 \rangle \\ \langle \mathbf{x}|\mathbf{e}_2 \rangle \end{bmatrix} = \langle \mathbf{x}|\mathbf{e}$$

We call

$$\mathbf{T}^\mathsf{T} = \begin{bmatrix} \mathbf{e}_1 & \mathbf{e}_2 \end{bmatrix} = \begin{bmatrix} 1 & 0 \\ 0 & 1 \end{bmatrix}$$

the *synthesis matrix*; it multiplies the coefficient vect
It follows from (20.2) that (20.1) is equivalent to

$$\mathbf{x} = \mathbf{T}^\mathsf{T}\mathbf{T}\mathbf{x} = \begin{bmatrix} 1 & 0 \\ 0 & 1 \end{bmatrix} \begin{bmatrix} 1 & 0 \\ 0 & 1 \end{bmatrix}$$

The introduction of the analysis and the synthesis
may seem artificial and may appear as complicating

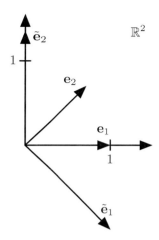

Figure 20.2: Biorthonormal bases in $\mathbb{R}^2$.

all, both $\mathbf{T}$ and $\mathbf{T}^\mathsf{T}$ are equal to the identity matrix in this special case. We will, however, see shortly that this notation paves the way to developing a unified framework for nonorthogonal and redundant signal expansions. Let us now look at a somewhat more interesting example.

Example 20.1.2 (Biorthonormal bases in $\mathbb{R}^2$). Consider two noncollinear unit norm vectors in $\mathbb{R}^2$. For concreteness, take (see Figure 20.2)

$$\mathbf{e}_1 = \begin{bmatrix} 1 \\ 0 \end{bmatrix}, \qquad \mathbf{e}_2 = \frac{1}{\sqrt{2}} \begin{bmatrix} 1 \\ 1 \end{bmatrix}.$$

For an arbitrary signal $\mathbf{x} \in \mathbb{R}^2$, we can compute the expansion coefficients

$$c_1 \triangleq \langle \mathbf{x} | \mathbf{e}_1 \rangle$$
$$c_2 \triangleq \langle \mathbf{x} | \mathbf{e}_2 \rangle.$$

As in Example 20.1.1 above, we stack the expansion coefficients into a vector so that

$$\mathbf{c} = \begin{bmatrix} c_1 \\ c_2 \end{bmatrix} = \begin{bmatrix} \langle \mathbf{x} | \mathbf{e}_1 \rangle \\ \langle \mathbf{x} | \mathbf{e}_2 \rangle \end{bmatrix} = \begin{bmatrix} \mathbf{e}_1^\mathsf{T} \\ \mathbf{e}_2^\mathsf{T} \end{bmatrix} \mathbf{x} = \begin{bmatrix} 1 & 0 \\ 1/\sqrt{2} & 1/\sqrt{2} \end{bmatrix} \mathbf{x}. \qquad (20.5)$$

Analogously to Example 20.1.1, we can define the analysis matrix

$$\mathbf{T} \triangleq \begin{bmatrix} \mathbf{e}_1^\mathsf{T} \\ \mathbf{e}_2^\mathsf{T} \end{bmatrix} = \begin{bmatrix} 1 & 0 \\ 1/\sqrt{2} & 1/\sqrt{2} \end{bmatrix}$$

and rewrite (20.5) as

$$\mathbf{c} = \mathbf{T}\mathbf{x}.$$

Now, obviously, the vectors $\mathbf{e}_1$ and $\mathbf{e}_2$ are not orthor
not unitary) so that we cannot write $\mathbf{x}$ in the form (2
to find a decomposition of $\mathbf{x}$ of the form

$$\mathbf{x} = \langle \mathbf{x}|\mathbf{e}_1 \rangle \, \tilde{\mathbf{e}}_1 + \langle \mathbf{x}|\mathbf{e}_2 \rangle \, \tilde{\mathbf{e}}_2$$

with $\tilde{\mathbf{e}}_1, \tilde{\mathbf{e}}_2 \in \mathbb{R}^2$. That this is, indeed, possible is ea
according to

$$\mathbf{x} = \begin{bmatrix} \tilde{\mathbf{e}}_1 & \tilde{\mathbf{e}}_2 \end{bmatrix} \mathbf{T} \mathbf{x}$$

and choosing the vectors $\tilde{\mathbf{e}}_1$ and $\tilde{\mathbf{e}}_2$ to be given by th
to

$$\begin{bmatrix} \tilde{\mathbf{e}}_1 & \tilde{\mathbf{e}}_2 \end{bmatrix} = \mathbf{T}^{-1}.$$

Note that $\mathbf{T}$ is invertible as a consequence of $\mathbf{e}_1$ anc
the specific example at hand we find

$$\begin{bmatrix} \tilde{\mathbf{e}}_1 & \tilde{\mathbf{e}}_2 \end{bmatrix} = \mathbf{T}^{-1} = \begin{bmatrix} 1 & 0 \\ -1 & \sqrt{2} \end{bmatrix}$$

and therefore (see Figure 20.2)

$$\tilde{\mathbf{e}}_1 = \begin{bmatrix} 1 \\ -1 \end{bmatrix}, \qquad \tilde{\mathbf{e}}_2 = \begin{bmatrix} 0 \\ \sqrt{2} \end{bmatrix}$$

Note that (20.8) implies that $\mathbf{T} \begin{bmatrix} \tilde{\mathbf{e}}_1 & \tilde{\mathbf{e}}_2 \end{bmatrix} = \mathbf{I}_2$, which i

$$\begin{bmatrix} \mathbf{e}_1^\mathsf{T} \\ \mathbf{e}_2^\mathsf{T} \end{bmatrix} \begin{bmatrix} \tilde{\mathbf{e}}_1 & \tilde{\mathbf{e}}_2 \end{bmatrix} = \mathbf{I}_2.$$

More directly the two sets of vectors $\{\mathbf{e}_1, \mathbf{e}_2\}$ and $\{\tilde{\mathbf{e}}$
mality" property according to

$$\langle \mathbf{e}_j | \tilde{\mathbf{e}}_k \rangle = \begin{cases} 1, & j = k \\ 0, & \text{else} \end{cases}, \qquad j, k$$

We say that $\{\mathbf{e}_1, \mathbf{e}_2\}$ and $\{\tilde{\mathbf{e}}_1, \tilde{\mathbf{e}}_2\}$ are biorthonormal
we can now define the synthesis matrix as follows:

$$\tilde{\mathbf{T}}^\mathsf{T} \triangleq \begin{bmatrix} \tilde{\mathbf{e}}_1 & \tilde{\mathbf{e}}_2 \end{bmatrix} = \begin{bmatrix} 1 & 0 \\ -1 & \sqrt{2} \end{bmatrix}$$

Our observations can be summarized according to

$$\mathbf{x} = \langle \mathbf{x}|\mathbf{e}_1 \rangle \, \tilde{\mathbf{e}}_1 + \langle \mathbf{x}|\mathbf{e}_2 \rangle \, \tilde{\mathbf{e}}_2$$
$$= \tilde{\mathbf{T}}^\mathsf{T} \mathbf{c} = \tilde{\mathbf{T}}^\mathsf{T} \mathbf{T} \mathbf{x}$$
$$= \begin{bmatrix} 1 & 0 \\ -1 & \sqrt{2} \end{bmatrix} \begin{bmatrix} 1 & 0 \\ 1/\sqrt{2} & 1/\sqrt{2} \end{bmatrix} \mathbf{x} =$$

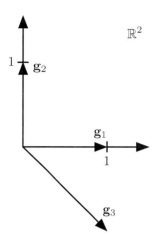

Figure 20.3: Overcomplete set of vectors in $\mathbb{R}^2$.

Comparing (20.9) to (20.4), we observe the following: To synthesize $\mathbf{x}$ from the expansion coefficients $\mathbf{c}$ corresponding to the nonorthogonal set $\{\mathbf{e}_1, \mathbf{e}_2\}$, we need to use the synthesis matrix $\tilde{\mathbf{T}}^{\mathsf{T}}$ obtained from the set $\{\tilde{\mathbf{e}}_1, \tilde{\mathbf{e}}_2\}$, which forms a biorthonormal pair with $\{\mathbf{e}_1, \mathbf{e}_2\}$. In Example 20.1.1, $\{\mathbf{e}_1, \mathbf{e}_2\}$ is an ONB and hence $\tilde{\mathbf{T}} = \mathbf{T}$, or, equivalently, $\{\mathbf{e}_1, \mathbf{e}_2\}$ forms a biorthonormal pair with itself.

As the vectors $\mathbf{e}_1$ and $\mathbf{e}_2$ are linearly independent, the 2×2 analysis matrix $\mathbf{T}$ has full rank and is hence invertible, i.e., there is a *unique* matrix $\mathbf{T}^{-1}$ that satisfies $\mathbf{T}^{-1}\mathbf{T} = \mathbf{I}_2$. According to (20.7) this means that for each analysis set $\{\mathbf{e}_1, \mathbf{e}_2\}$ there is precisely one synthesis set $\{\tilde{\mathbf{e}}_1, \tilde{\mathbf{e}}_2\}$ such that (20.6) is satisfied for all $\mathbf{x} \in \mathbb{R}^2$.

$\square$

So far we considered nonredundant signal expansions where the number of expansion coefficients was equal to the dimension of the vector space. Often, however, redundancy in the expansion is desirable.

Example 20.1.3 (Overcomplete expansion in $\mathbb{R}^2$, [20, Ex. 3.1]). Consider the following three vectors in $\mathbb{R}^2$ (see Figure 20.3):

$$\mathbf{g}_1 = \begin{bmatrix} 1 \\ 0 \end{bmatrix}, \quad \mathbf{g}_2 = \begin{bmatrix} 0 \\ 1 \end{bmatrix}, \quad \mathbf{g}_3 = \begin{bmatrix} 1 \\ -1 \end{bmatrix}.$$

Three vectors in a two-dimensional space are always linearly dependent. In particular, in this example we have $\mathbf{g}_3 = \mathbf{g}_1 - \mathbf{g}_2$. Let us compute the expansion coefficients $\mathbf{c}$ corresponding to $\{\mathbf{g}_1, \mathbf{g}_2, \mathbf{g}_3\}$:

$$\mathbf{c} = \begin{bmatrix} c_1 \\ c_2 \\ c_3 \end{bmatrix} \triangleq \begin{bmatrix} \langle \mathbf{x}|\mathbf{g}_1\rangle \\ \langle \mathbf{x}|\mathbf{g}_2\rangle \\ \langle \mathbf{x}|\mathbf{g}_3\rangle \end{bmatrix} = \begin{bmatrix} \mathbf{g}_1^{\mathsf{T}} \\ \mathbf{g}_2^{\mathsf{T}} \\ \mathbf{g}_3^{\mathsf{T}} \end{bmatrix} \mathbf{x} =$$

Following Examples 20.1.1 and 20.1.2, we define the

$$\mathbf{T} \triangleq \begin{bmatrix} \mathbf{g}_1^{\mathsf{T}} \\ \mathbf{g}_2^{\mathsf{T}} \\ \mathbf{g}_3^{\mathsf{T}} \end{bmatrix} = \begin{bmatrix} 1 & 0 \\ 0 & 1 \\ 1 & -1 \end{bmatrix}$$

and rewrite (20.10) as

$$\mathbf{c} = \mathbf{T}\mathbf{x}.$$

Note that here, unlike in Examples 20.1.1 and 20.1.2,
tion of $\mathbf{x}$ as we have *three* expansion coefficients for a
We next ask if $\mathbf{x}$ can be represented as a linear co

$$\mathbf{x} = \underbrace{\langle \mathbf{x}|\mathbf{g}_1\rangle}_{c_1} \tilde{\mathbf{g}}_1 + \underbrace{\langle \mathbf{x}|\mathbf{g}_2\rangle}_{c_2} \tilde{\mathbf{g}}_2 + \underbrace{\langle \mathbf{x}|}_{c}$$

with $\tilde{\mathbf{g}}_1, \tilde{\mathbf{g}}_2, \tilde{\mathbf{g}}_3 \in \mathbb{R}^2$? To answer this question (in the a
the vectors $\mathbf{g}_1, \mathbf{g}_2$ form an ONB for $\mathbb{R}^2$. We, therefor
true:

$$\mathbf{x} = \langle \mathbf{x}|\mathbf{g}_1\rangle\, \mathbf{g}_1 + \langle \mathbf{x}|\mathbf{g}_2\rangle\, \mathbf{g}_2$$

Setting

$$\tilde{\mathbf{g}}_1 = \mathbf{g}_1, \ \tilde{\mathbf{g}}_2 = \mathbf{g}_2, \ \tilde{\mathbf{g}}_3 = \mathbf{0}$$

obviously yields a representation of the form (20.11).
this representation is not unique and that an alter
form (20.11) can be obtained as follows. We start by
hand side of (20.12):

$$\mathbf{x} = \langle \mathbf{x}|\mathbf{g}_1\rangle\, \mathbf{g}_1 + \langle \mathbf{x}|\mathbf{g}_2\rangle\, \mathbf{g}_2 + \underbrace{\langle \mathbf{x}|\mathbf{g}_1 - \mathbf{g}}$$

Rearranging terms in this expression, we obtain

$$\mathbf{x} = \langle \mathbf{x}|\mathbf{g}_1\rangle\, 2\mathbf{g}_1 + \langle \mathbf{x}|\mathbf{g}_2\rangle\, (\mathbf{g}_2 - \mathbf{g}_1) - \langle \mathbf{x}|$$

We recognize that $\mathbf{g}_1 - \mathbf{g}_2 = \mathbf{g}_3$ and set

$$\tilde{\mathbf{g}}_1 = 2\mathbf{g}_1, \ \tilde{\mathbf{g}}_2 = \mathbf{g}_2 - \mathbf{g}_1, \ \tilde{\mathbf{g}}_3 =$$

This allows us to rewrite (20.13) as

$$\mathbf{x} = \langle \mathbf{x}|\mathbf{g}_1\rangle\, \tilde{\mathbf{g}}_1 + \langle \mathbf{x}|\mathbf{g}_2\rangle\, \tilde{\mathbf{g}}_2 + \langle \mathbf{x}|$$

The redundant set of vectors $\{\mathbf{g}_1, \mathbf{g}_2, \mathbf{g}_3\}$ is called a *frame*. The set $\{\tilde{\mathbf{g}}_1, \tilde{\mathbf{g}}_2, \tilde{\mathbf{g}}_3\}$ in (20.14) is called a *dual frame* to the frame $\{\mathbf{g}_1, \mathbf{g}_2, \mathbf{g}_3\}$. Obviously another dual frame is given by $\tilde{\mathbf{g}}_1 = \mathbf{g}_1$, $\tilde{\mathbf{g}}_2 = \mathbf{g}_2$, and $\tilde{\mathbf{g}}_3 = \mathbf{0}$. In fact, there are infinitely many dual frames. To see this, we first define the synthesis matrix corresponding to a dual frame $\{\tilde{\mathbf{g}}_1, \tilde{\mathbf{g}}_2, \tilde{\mathbf{g}}_3\}$ as

$$\tilde{\mathbf{T}}^{\mathsf{T}} \triangleq \begin{bmatrix} \tilde{\mathbf{g}}_1 & \tilde{\mathbf{g}}_2 & \tilde{\mathbf{g}}_3 \end{bmatrix}. \tag{20.15}$$

It then follows that we can write

$$\mathbf{x} = \langle \mathbf{x}|\mathbf{g}_1 \rangle\, \tilde{\mathbf{g}}_1 + \langle \mathbf{x}|\mathbf{g}_2 \rangle\, \tilde{\mathbf{g}}_2 + \langle \mathbf{x}|\mathbf{g}_3 \rangle\, \tilde{\mathbf{g}}_3$$
$$= \tilde{\mathbf{T}}^{\mathsf{T}}\mathbf{c} = \tilde{\mathbf{T}}^{\mathsf{T}}\mathbf{T}\mathbf{x},$$

which implies that setting $\tilde{\mathbf{T}}^{\mathsf{T}} = \begin{bmatrix} \tilde{\mathbf{g}}_1 & \tilde{\mathbf{g}}_2 & \tilde{\mathbf{g}}_3 \end{bmatrix}$ to be any left-inverse of $\mathbf{T}$ yields a valid dual frame. Since $\mathbf{T}$ is a 3×2 ("tall") matrix, its left-inverse is not unique. In fact, $\mathbf{T}$ has infinitely many left-inverses (two of them were found above). Every left-inverse of $\mathbf{T}$ leads to a dual frame according to (20.15).

Thanks to the redundancy of the frame $\{\mathbf{g}_1, \mathbf{g}_2, \mathbf{g}_3\}$, we obtain design freedom: In order to synthesize the signal $\mathbf{x}$ from its expansion coefficients $c_k = \langle \mathbf{x}|\mathbf{g}_k \rangle$, $k = 1, 2, 3$, in the frame $\{\mathbf{g}_1, \mathbf{g}_2, \mathbf{g}_3\}$, we can choose between infinitely many dual frames $\{\tilde{\mathbf{g}}_1, \tilde{\mathbf{g}}_2, \tilde{\mathbf{g}}_3\}$. In practice the particular choice of the dual frame is usually dictated by the requirements of the specific problem at hand. We shall discuss this issue in detail in the context of sampling theory in Section 20.4.2.

20.2 Signal Expansions in Finite-Dimensional Spaces

Motivated by the examples above, we now consider general signal expansions in finite-dimensional Hilbert spaces. As in the previous section, we first review the concept of an ONB, we then consider arbitrary (nonorthogonal) bases, and, finally, we discuss redundant vector sets — frames. While the discussion in this section is confined to the finite-dimensional case, we develop the general (possibly infinite-dimensional) case in Section 20.3.

20.2.1 Orthonormal Bases

We start by reviewing the concept of an ONB.

Definition 20.2.1. The set of vectors $\{\mathbf{e}_k\}_{k=1}^{M}$, $\mathbf{e}_k \in \mathbb{C}^M$, $k = 1, \dots, M$, is called an ONB for $\mathbb{C}^M$ if

1. $\operatorname{span}\{\mathbf{e}_k\}_{k=1}^{M} = \{c_1\mathbf{e}_1 + c_2\mathbf{e}_2 + \dots + c_M\mathbf{e}_M \,|\, c_1, c_2, \dots, c_M \in \mathbb{C}\} = \mathbb{C}^M$

2.

$$\langle \mathbf{e}_k | \mathbf{e}_j \rangle = \begin{cases} 1, & k = j \\ 0, & k \neq j \end{cases} \qquad k, j$$

When $\{\mathbf{e}_k\}_{k=1}^{M}$ is an ONB, thanks to the spanning every $\mathbf{x} \in \mathbb{C}^M$ can be decomposed as

$$\mathbf{x} = \sum_{k=1}^{M} c_k \mathbf{e}_k.$$

The expansion coefficients $\{c_k\}_{k=1}^{M}$ in (20.16) can be calculation:

$$\langle \mathbf{x} | \mathbf{e}_j \rangle = \left\langle \sum_{k=1}^{M} c_k \mathbf{e}_k \Big| \mathbf{e}_j \right\rangle = \sum_{k=1}^{M} c_k \langle \mathbf{e}$$

In summary, we have the decomposition

$$\mathbf{x} = \sum_{k=1}^{M} \langle \mathbf{x} | \mathbf{e}_k \rangle \, \mathbf{e}_k.$$

Just like in Example 20.1.1, in the previous section,

$$\mathbf{T} \triangleq \begin{bmatrix} \mathbf{e}_1^\dagger \\ \vdots \\ \mathbf{e}_M^\dagger \end{bmatrix}.$$

If we organize the inner products $\{\langle \mathbf{x} | \mathbf{e}_k \rangle\}_{k=1}^{M}$ into th

$$\mathbf{c} \triangleq \begin{bmatrix} \langle \mathbf{x} | \mathbf{e}_1 \rangle \\ \vdots \\ \langle \mathbf{x} | \mathbf{e}_M \rangle \end{bmatrix} = \mathbf{Tx} = \begin{bmatrix} \mathbf{e}_1^\dagger \\ \vdots \\ \mathbf{e}_M^\dagger \end{bmatrix}$$

Thanks to the orthonormality of the vectors $\mathbf{e}_1, \mathbf{e}_2, \dots$ i.e., $\mathbf{T}^\dagger = \mathbf{T}^{-1}$ and hence

$$\mathbf{T}\mathbf{T}^\dagger = \begin{bmatrix} \mathbf{e}_1^\dagger \\ \vdots \\ \mathbf{e}_M^\dagger \end{bmatrix} \begin{bmatrix} \mathbf{e}_1 & \cdots & \mathbf{e}_M \end{bmatrix} = \begin{bmatrix} \langle \mathbf{e}_1 | \mathbf{e}_1 \rangle & \cdots & \langle \mathbf{e} \\ \vdots & \ddots & \\ \langle \mathbf{e}_1 | \mathbf{e}_M \rangle & \cdots & \langle \mathbf{e} \end{bmatrix}$$

Thus, if we multiply the vector $\mathbf{c}$ by $\mathbf{T}^\dagger$, we synthesiz

$$\mathbf{T}^\dagger \mathbf{c} = \mathbf{T}^\dagger \mathbf{Tx} = \sum_{k=1}^{M} \langle \mathbf{x} | \mathbf{e}_k \rangle \, \mathbf{e}_k = \mathbf{I}$$

We shall therefore call the matrix $\mathbf{T}^\dagger$ the synthesis matrix, corresponding to the analysis matrix $\mathbf{T}$. In the ONB case considered here the synthesis matrix is simply the Hermitian adjoint of the analysis matrix.

20.2.2 General Bases

We next relax the orthonormality property, i.e., the second condition in Definition 20.2.1, and consider general bases.

Definition 20.2.2. The set of vectors $\{\mathbf{e}_k\}_{k=1}^M$, $\mathbf{e}_k \in \mathbb{C}^M$, $k = 1, \ldots, M$, is a basis for $\mathbb{C}^M$ if

1. $\operatorname{span}\{\mathbf{e}_k\}_{k=1}^M = \{c_1\mathbf{e}_1 + c_2\mathbf{e}_2 + \ldots + c_M\mathbf{e}_M \mid c_1, c_2, \ldots, c_M \in \mathbb{C}\} = \mathbb{C}^M$

2. $\{\mathbf{e}_k\}_{k=1}^M$ is a linearly independent set, i.e., if $\sum_{k=1}^M c_k\mathbf{e}_k = \mathbf{0}$ for some scalar coefficients $\{c_k\}_{k=1}^M$, then necessarily $c_k = 0$ for all $k = 1, \ldots, M$.

Now consider a signal $\mathbf{x} \in \mathbb{C}^M$ and compute the expansion coefficients

$$c_k \triangleq \langle \mathbf{x} | \mathbf{e}_k \rangle, \quad k = 1, \ldots, M. \tag{20.18}$$

Again, it is convenient to introduce the analysis matrix

$$\mathbf{T} \triangleq \begin{bmatrix} \mathbf{e}_1^\dagger \\ \vdots \\ \mathbf{e}_M^\dagger \end{bmatrix}$$

and to stack the coefficients $\{c_k\}_{k=1}^M$ in the vector $\mathbf{c}$. Then (20.18) can be written as

$$\mathbf{c} = \mathbf{T}\mathbf{x}.$$

Next, let us ask how we can find a set of vectors $\{\tilde{\mathbf{e}}_1, \ldots, \tilde{\mathbf{e}}_M\}$, $\tilde{\mathbf{e}}_k \in \mathbb{C}^M$, $k = 1, \ldots, M$, that is dual to the set $\{\mathbf{e}_1, \ldots, \mathbf{e}_M\}$ in the sense that

$$\mathbf{x} = \sum_{k=1}^M c_k \tilde{\mathbf{e}}_k = \sum_{k=1}^M \langle \mathbf{x} | \mathbf{e}_k \rangle \tilde{\mathbf{e}}_k \tag{20.19}$$

for all $\mathbf{x} \in \mathbb{C}^M$. If we introduce the synthesis matrix

$$\tilde{\mathbf{T}}^\dagger \triangleq [\tilde{\mathbf{e}}_1 \ \cdots \ \tilde{\mathbf{e}}_M],$$

we can rewrite (20.19) in vector-matrix notation as follows

$$\mathbf{x} = \tilde{\mathbf{T}}^\dagger \mathbf{c} = \tilde{\mathbf{T}}^\dagger \mathbf{T}\mathbf{x}.$$

This shows that finding vectors $\tilde{\mathbf{e}}_1, \ldots, \tilde{\mathbf{e}}_M$ that satisfy (20.19) is equivalent to finding the inverse of the analysis matrix $\mathbf{T}$ and setting $\tilde{\mathbf{T}}^\dagger = \mathbf{T}^{-1}$. Thanks to

the linear independence of the vectors $\{\mathbf{e}_k\}_{k=1}^{M}$, the n
therefore, invertible.

Summarizing our findings, we conclude that in the
analysis matrix and the synthesis matrix are inverse
$\mathbf{T}\tilde{\mathbf{T}}^{\dagger} = \mathbf{I}_M$. Recall that in the case of an orthonorm
matrix $\mathbf{T}$ is *unitary* and hence its inverse is simply
that in this case $\tilde{\mathbf{T}} = \mathbf{T}$.

Next, note that $\mathbf{T}\tilde{\mathbf{T}}^{\dagger} = \mathbf{I}_M$ is equivalent to

$$
\begin{bmatrix} \mathbf{e}_1^{\dagger} \\ \vdots \\ \mathbf{e}_M^{\dagger} \end{bmatrix}
\begin{bmatrix} \tilde{\mathbf{e}}_1 & \cdots & \tilde{\mathbf{e}}_M \end{bmatrix} =
\begin{bmatrix} \langle \tilde{\mathbf{e}}_1 | \mathbf{e}_1 \rangle & \cdots & \langle \tilde{\mathbf{e}} \\ \vdots & \ddots & \\ \langle \tilde{\mathbf{e}}_1 | \mathbf{e}_M \rangle & \cdots & \langle \tilde{\mathbf{e}} \end{bmatrix}
$$

or equivalently

$$
\langle \mathbf{e}_k | \tilde{\mathbf{e}}_j \rangle = \begin{cases} 1, & k = j \\ 0, & \text{else} \end{cases}, \qquad k, j =
$$

The sets $\{\mathbf{e}_k\}_{k=1}^{M}$ and $\{\tilde{\mathbf{e}}_k\}_{k=1}^{M}$ are biorthonormal bas
to themselves in this terminology, as already noted in
size that it is the fact that $\mathbf{T}$ and $\tilde{\mathbf{T}}^{\dagger}$ are square and
conclude that $\tilde{\mathbf{T}}^{\dagger}\mathbf{T} = \mathbf{I}_M$ implies $\mathbf{T}\tilde{\mathbf{T}}^{\dagger} = \mathbf{I}_M$ and he
holds. We shall see below that for redundant expan
$\tilde{\mathbf{T}}^{\dagger}\mathbf{T} \neq \mathbf{T}\tilde{\mathbf{T}}^{\dagger}$ ($\tilde{\mathbf{T}}^{\dagger}\mathbf{T}$ and $\mathbf{T}\tilde{\mathbf{T}}^{\dagger}$ have different dimensio
not be biorthonormal.

As $\mathbf{T}$ is a square matrix and of full rank, its inverse
for a given analysis set $\{\mathbf{e}_k\}_{k=1}^{M}$, the synthesis set $\{\tilde{\mathbf{e}}_k$
for a given synthesis set $\{\tilde{\mathbf{e}}_k\}_{k=1}^{M}$, there is a *unique*
uniqueness property is not always desirable. For exam
certain structural properties on the synthesis set $\{\tilde{\mathbf{e}}$
freedom in choosing the synthesis set as in Example

An important property of ONBs is that they are
of the coefficient vector $\mathbf{c}$ is equal to the norm of the
seen by noting that

$$
\|\mathbf{c}\|^2 = \mathbf{c}^{\dagger}\mathbf{c} = \mathbf{x}^{\dagger}\mathbf{T}^{\dagger}\mathbf{T}\mathbf{x} = \mathbf{x}^{\dagger}\mathbf{I}_M\mathbf{x}
$$

where we used (20.17). Biorthonormal bases are *not*
Rather, the equality in (20.21) is relaxed to a double-
the Rayleigh-Ritz theorem [22, Sec. 9.7.2.2] according

$$
\lambda_{\min}\left(\mathbf{T}^{\dagger}\mathbf{T}\right)\|\mathbf{x}\|^2 \leq \|\mathbf{c}\|^2 = \mathbf{x}^{\dagger}\mathbf{T}^{\dagger}\mathbf{T}\mathbf{x} \leq \lambda_{\max}(
$$

20.2.3 Redundant Signal Expansions

The signal expansions we considered so far are nonredundant in the sense that the number of expansion coefficients equals the dimension of the Hilbert space. Such signal expansions have a number of disadvantages. First, corruption or loss of expansion coefficients can result in significant reconstruction errors. Second, the reconstruction process is very rigid: As we have seen in Section 20.2.2, for each set of analysis vectors, there is a *unique* set of synthesis vectors. In practical applications it is often desirable to impose additional constraints on the reconstruction set, such as smoothness properties of the reconstruction functions or structural properties that allow for computationally efficient reconstruction.

Redundant expansions allow to overcome many of these problems as they offer design freedom and robustness to corruption or loss of expansion coefficients. We already saw in Example 20.1.3 that in the case of redundant expansions, for a given set of analysis vectors the set of synthesis vectors that allows perfect recovery of a signal from its expansion coefficients is not unique; in fact there are infinitely many sets of synthesis vectors, in general. This results in design freedom and provides robustness. Suppose that the expansion coefficient $c_3 = \langle \mathbf{x} | \mathbf{g}_3 \rangle$ in Example 20.1.3 is corrupted or even completely lost. We can still reconstruct $\mathbf{x}$ *exactly* from (20.12).

Now, let us turn to developing the general theory of redundant signal expansions in finite-dimensional Hilbert spaces. Consider a set of N vectors $\{\mathbf{g}_1, \ldots, \mathbf{g}_N\}$, $\mathbf{g}_k \in \mathbb{C}^M$, $k = 1, \ldots, N$, with $N \geq M$. Clearly, when N is *strictly* greater than M, the vectors $\mathbf{g}_1, \ldots, \mathbf{g}_N$ must be linearly dependent. Next, consider a signal $\mathbf{x} \in \mathbb{C}^M$ and compute the expansion coefficients

$$c_k = \langle \mathbf{x} | \mathbf{g}_k \rangle, \quad k = 1, \ldots, N. \tag{20.23}$$

Just as before, it is convenient to introduce the analysis matrix

$$\mathbf{T} \triangleq \begin{bmatrix} \mathbf{g}_1^\dagger \\ \vdots \\ \mathbf{g}_N^\dagger \end{bmatrix} \tag{20.24}$$

and to stack the coefficients $\{c_k\}_{k=1}^N$ in the vector $\mathbf{c}$. Then (20.23) can be written as

$$\mathbf{c} = \mathbf{Tx}. \tag{20.25}$$

Note that $\mathbf{c} \in \mathbb{C}^N$ and $\mathbf{x} \in \mathbb{C}^M$. Differently from ONBs and biorthonormal bases considered in Sections 20.2.1 and 20.2.2, respectively, in the case of redundant expansions, the signal $\mathbf{x}$ and the expansion coefficient vector $\mathbf{c}$ will, in general, belong to different Hilbert spaces.

The question now is how we can find a set of vectors $\{\tilde{\mathbf{g}}_1, \ldots, \tilde{\mathbf{g}}_N\}$, $\tilde{\mathbf{g}}_k \in \mathbb{C}^M$, $k = 1, \ldots, N$, such that

$$\mathbf{x} = \sum_{k=1}^N c_k \tilde{\mathbf{g}}_k = \sum_{k=1}^N \langle \mathbf{x} | \mathbf{g}_k \rangle \tilde{\mathbf{g}}_k \tag{20.26}$$

for all $\mathbf{x} \in \mathbb{C}^M$? If we introduce the synthesis matrix

$$\tilde{\mathbf{T}}^\dagger \triangleq [\tilde{\mathbf{g}}_1 \;\cdots\; \tilde{\mathbf{g}}_N],$$

we can rewrite (20.26) in vector-matrix notation as f

$$\mathbf{x} = \tilde{\mathbf{T}}^\dagger \mathbf{c} = \tilde{\mathbf{T}}^\dagger \mathbf{T} \mathbf{x}.$$

Finding vectors $\tilde{\mathbf{g}}_1, \ldots, \tilde{\mathbf{g}}_N$ that satisfy (20.26) for all
lent to finding a left-inverse $\tilde{\mathbf{T}}^\dagger$ of $\mathbf{T}$, i.e.,

$$\tilde{\mathbf{T}}^\dagger \mathbf{T} = \mathbf{I}_M.$$

First note that $\mathbf{T}$ is left-invertible if and only if $\mathbb{C}^M =$
if the set of vectors $\{\mathbf{g}_k\}_{k=1}^N$ spans $\mathbb{C}^M$. Next observe t
matrix $\mathbf{T}$ is a "tall" matrix, and therefore its left-inve
In fact, there are infinitely many left-inverses. The fo
2, Th. 1] provides a convenient parametrization of al

Theorem 20.2.1. *Let* $\mathbf{A} \in \mathbb{C}^{N \times M}$, $N \geq M$. *Assum*
$\mathbf{A}^+ \triangleq (\mathbf{A}^\dagger \mathbf{A})^{-1} \mathbf{A}^\dagger$ *is a left-inverse of* $\mathbf{A}$, *i.e.,* $\mathbf{A}^+ \mathbf{A} =$
solution $\mathbf{L} \in \mathbb{C}^{M \times N}$ *of the equation* $\mathbf{L}\mathbf{A} = \mathbf{I}_M$ *is give*

$$\mathbf{L} = \mathbf{A}^+ + \mathbf{M}\left(\mathbf{I}_N - \mathbf{A}\mathbf{A}^+\right.$$

where $\mathbf{M} \in \mathbb{C}^{M \times N}$ *is an arbitrary matrix.*

Proof: Since $\operatorname{rank}(\mathbf{A}) = M$, the matrix $\mathbf{A}^\dagger \mathbf{A}$ is inve
defined. Now, let us verify that $\mathbf{A}^+$ is, indeed, a left–

$$\mathbf{A}^+ \mathbf{A} = (\mathbf{A}^\dagger \mathbf{A})^{-1} \mathbf{A}^\dagger \mathbf{A} = \mathbf{I}$$

The matrix $\mathbf{A}^+$ is called the Moore-Penrose inverse c
Next, we show that every matrix $\mathbf{L}$ of the form
of $\mathbf{A}$:

$$\begin{aligned}
\mathbf{L}\mathbf{A} &= \left(\mathbf{A}^+ + \mathbf{M}(\mathbf{I}_N - \mathbf{A}\mathbf{A}^+\right. \\
&= \underbrace{\mathbf{A}^+ \mathbf{A}}_{\mathbf{I}_M} + \mathbf{M}\mathbf{A} - \mathbf{M}\mathbf{A}\underbrace{\mathbf{A}}_{} \\
&= \mathbf{I}_M + \mathbf{M}\mathbf{A} - \mathbf{M}\mathbf{A} = \mathbf{I}_M
\end{aligned}$$

where we used (20.29) twice.
Finally, assume that $\mathbf{L}$ is a valid left-inverse of $\mathbf{A}$
equation $\mathbf{L}\mathbf{A} = \mathbf{I}_M$. We show that $\mathbf{L}$ can be written
plying the equation $\mathbf{L}\mathbf{A} = \mathbf{I}_M$ by $\mathbf{A}^+$ from the right,

$$\mathbf{L}\mathbf{A}\mathbf{A}^+ = \mathbf{A}^+.$$

Adding $\mathbf{L}$ to both sides of this equation and rearranging terms yields

$$\mathbf{L} = \mathbf{A}^+ + \mathbf{L} - \mathbf{L}\mathbf{A}\mathbf{A}^+ = \mathbf{A}^+ + \mathbf{L}\left(\mathbf{I}_N - \mathbf{A}\mathbf{A}^+\right),$$

which shows that $\mathbf{L}$ can be written in the form (20.28) (with $\mathbf{M} = \mathbf{L}$), as required.

$\square$

We conclude that for each redundant set of vectors $\{\mathbf{g}_1, \ldots, \mathbf{g}_N\}$ that spans $\mathbb{C}^M$, there are infinitely many dual sets $\{\tilde{\mathbf{g}}_1, \ldots, \tilde{\mathbf{g}}_N\}$ such that the decomposition (20.26) holds for all $\mathbf{x} \in \mathbb{C}^M$. These dual sets are obtained by identifying $\{\tilde{\mathbf{g}}_1, \ldots, \tilde{\mathbf{g}}_N\}$ with the columns of $\mathbf{L}$ according to

$$[\tilde{\mathbf{g}}_1 \ \cdots \ \tilde{\mathbf{g}}_N] = \mathbf{L},$$

where $\mathbf{L}$ can be written as follows

$$\mathbf{L} = \mathbf{T}^+ + \mathbf{M}\left(\mathbf{I}_N - \mathbf{T}\mathbf{T}^+\right)$$

and $\mathbf{M} \in \mathbb{C}^{M \times N}$ is an arbitrary matrix.

The dual set $\{\tilde{\mathbf{g}}_1, \ldots, \tilde{\mathbf{g}}_N\}$ corresponding to the Moore-Penrose inverse $\mathbf{L} = \mathbf{T}^+$ of the matrix $\mathbf{T}$, i.e.,

$$[\tilde{\mathbf{g}}_1 \ \cdots \ \tilde{\mathbf{g}}_N] = \mathbf{T}^+ = (\mathbf{T}^\dagger\mathbf{T})^{-1}\mathbf{T}^\dagger$$

is called the *canonical dual* of $\{\mathbf{g}_1, \ldots, \mathbf{g}_N\}$. Using (20.24), we see that in this case

$$\tilde{\mathbf{g}}_k = (\mathbf{T}^\dagger\mathbf{T})^{-1}\mathbf{g}_k, \quad k = 1, \ldots, N. \tag{20.30}$$

Note that *unlike* in the case of a basis, the equation $\tilde{\mathbf{T}}^\dagger\mathbf{T} = \mathbf{I}_M$ *does not* imply that the sets $\{\tilde{\mathbf{g}}_k\}_{k=1}^N$ and $\{\mathbf{g}_k\}_{k=1}^N$ are biorthonormal. This is because the matrix $\mathbf{T}$ is *not* a square matrix, and thus, $\tilde{\mathbf{T}}^\dagger\mathbf{T} \neq \mathbf{T}\tilde{\mathbf{T}}^\dagger$ ($\tilde{\mathbf{T}}^\dagger\mathbf{T}$ and $\mathbf{T}\tilde{\mathbf{T}}^\dagger$ have different dimensions).

Similar to biorthonormal bases, redundant sets of vectors are, in general, not norm-preserving. Indeed, from (20.25) we see that

$$\|\mathbf{c}\|^2 = \mathbf{x}^\dagger\mathbf{T}^\dagger\mathbf{T}\mathbf{x}$$

and thus, by the Rayleigh-Ritz theorem [22, Sec. 9.7.2.2], we have

$$\lambda_{\min}(\mathbf{T}^\dagger\mathbf{T})\|\mathbf{x}\|^2 \leq \|\mathbf{c}\|^2 \leq \lambda_{\max}(\mathbf{T}^\dagger\mathbf{T})\|\mathbf{x}\|^2 \tag{20.31}$$

as in the case of biorthonormal bases.

We already saw some of the basic questions that a theory of orthonormal, biorthonormal, and redundant signal expansions should address. It should account for the signals and the expansion coefficient vectors belonging, potentially, to different Hilbert spaces; it should account for the fact that for a given analysis set, the synthesis set is not unique in the redundant case, it should prescribe how synthesis vectors can be obtained from the analysis vectors. Finally, it should apply not only to finite-dimensional Hilbert spaces, as considered so far, but also be applicable to infinite-dimensional Hilbert spaces. We now proceed to develop this general theory, known as the theory of frames.

20.3   Frames for General Hilber

Let $\{g_k\}_{k \in \mathcal{K}}$ ($\mathcal{K}$ is a countable set) be a set of ele
space $\mathcal{H}$. Note that this set need not be orthonormal

In developing a general theory of signal expansions
previous section, we start by noting that the central
the analysis matrix $\mathbf{T}$ associated to the (possibly n
set of vectors $\{\mathbf{g}_1, \ldots, \mathbf{g}_N\}$. Now matrices are not
finite-dimensional Hilbert spaces. In formulating fram
infinite-dimensional) Hilbert spaces, it is therefore se
operator $\mathbb{T}$ that assigns to each signal $x \in \mathcal{H}$ the
$\mathbb{T}x = \{\langle x | g_k \rangle\}_{k \in \mathcal{K}}$. Throughout this section, we assu
sequence, i.e., $\sum_{k \in \mathcal{K}} |\langle x | g_k \rangle|^2 < \infty$ for all $x \in \mathcal{H}$.

Definition 20.3.1. The linear operator $\mathbb{T}$ is define
the Hilbert space $\mathcal{H}$ into the space l^2 of square-sun
$\mathbb{T} : \mathcal{H} \to l^2$, by assigning to each signal $x \in \mathcal{H}$ the
$\langle x | g_k \rangle$ according to

$$\mathbb{T} : x \to \{\langle x | g_k \rangle\}_{k \in \mathcal{K}}.$$

Note that $\|\mathbb{T}x\|^2 = \sum_{k \in \mathcal{K}} |\langle x | g_k \rangle|^2$, i.e., the ene
pressed as

$$\|\mathbb{T}x\|^2 = \sum_{k \in \mathcal{K}} |\langle x | g_k \rangle|^2.$$

We shall next formulate the properties that the set $\{g_k$
$\mathbb{T}$ should satisfy if we have signal expansions in mind

1. The signal x can be perfectly reconstructed from
    This means that we want $\langle x | g_k \rangle = \langle y | g_k \rangle$, for a
    imply that $x = y$, for all $x, y \in \mathcal{H}$. In other
    to be left-invertible, which means that $\mathbb{T}$ is i
    $\mathcal{R}(\mathbb{T}) = \{y \in l^2 : y = \mathbb{T}x, \ x \in \mathcal{H}\}$.

    This requirement will clearly be satisfied if we
    constant $A > 0$ such that for all $x, y \in \mathcal{H}$ we ha

    $$A\|x - y\|^2 \le \|\mathbb{T}x - \mathbb{T}_l$$

    Setting $z = x - y$ and using the linearity of $\mathbb{T}$,
    equivalent to

    $$A\|z\|^2 \le \|\mathbb{T}z\|^2$$

    for all $z \in \mathcal{H}$ with $A > 0$.

---

[1]The fact that the range space of $\mathbb{T}$ is contained in l^2 is a
Bessel sequence.

2. The energy in the sequence of expansion coefficients $\mathbb{T}x = \{\langle x|g_k\rangle\}_{k\in\mathcal{K}}$ should be related to the energy in the signal x. For example, we saw in (20.21) that if $\{\mathbf{e}_k\}_{k=1}^M$ is an ONB for $\mathbb{C}^M$, then

$$\|\mathbf{T}\mathbf{x}\|^2 = \sum_{k=1}^M |\langle\mathbf{x}|\mathbf{e}_k\rangle|^2 = \|\mathbf{x}\|^2, \quad \text{for all} \quad \mathbf{x} \in \mathbb{C}^M. \tag{20.34}$$

This property is a consequence of the unitarity of $\mathbb{T} = \mathbf{T}$ and it is clear that it will not hold for general sets $\{g_k\}_{k\in\mathcal{K}}$ (see the discussion around (20.22) and (20.31)). Instead, we will relax (20.34) to demand that for all $x \in \mathcal{H}$ there exist a finite constant B such that[2]

$$\|\mathbb{T}x\|^2 = \sum_{k\in\mathcal{K}} |\langle x|g_k\rangle|^2 \leq B\|x\|^2. \tag{20.35}$$

Together with (20.33) this "sandwiches" the quantity $\|\mathbb{T}x\|^2$ according to

$$A\|x\|^2 \leq \|\mathbb{T}x\|^2 \leq B\|x\|^2.$$

We are now ready to formally define a frame for the Hilbert space $\mathcal{H}$.

Definition 20.3.2. A set of elements $\{g_k\}_{k\in\mathcal{K}}$, $g_k \in \mathcal{H}$, $k \in \mathcal{K}$, is called a frame for the Hilbert space $\mathcal{H}$ if

$$A\|x\|^2 \leq \sum_{k\in\mathcal{K}} |\langle x|g_k\rangle|^2 \leq B\|x\|^2, \quad \text{for all} \quad x \in \mathcal{H}, \tag{20.36}$$

with $A, B \in \mathbb{R}$ and $0 < A \leq B < \infty$. Valid constants A and B are called frame bounds. The largest valid constant A and the smallest valid constant B are called the (tightest possible) *frame bounds*.

Let us next consider some simple examples of frames.

Example 20.3.1 ([21]). Let $\{e_k\}_{k=1}^\infty$ be an ONB for an infinite-dimensional Hilbert space $\mathcal{H}$. By repeating each element in $\{e_k\}_{k=1}^\infty$ once, we obtain the redundant set

$$\{g_k\}_{k=1}^\infty = \{e_1, e_1, e_2, e_2, \ldots\}.$$

To see that this set is a frame for $\mathcal{H}$, we note that because $\{e_k\}_{k=1}^\infty$ is an ONB, for all $x \in \mathcal{H}$, we have

$$\sum_{k=1}^\infty |\langle x|e_k\rangle|^2 = \|x\|^2$$

and therefore

$$\sum_{k=1}^\infty |\langle x|g_k\rangle|^2 = \sum_{k=1}^\infty |\langle x|e_k\rangle|^2 + \sum_{k=1}^\infty |\langle x|e_k\rangle|^2 = 2\|x\|^2.$$

---

[2]Note that if (20.35) is satisfied with $B < \infty$, then $\{g_k\}_{k\in\mathcal{K}}$ is a Bessel sequence.

This verifies the frame condition (20.36) and shows th
bounds are given by $A = B = 2$.

Example 20.3.2 ([21]). Starting from the ONB
another redundant set as follows

$$\{g_k\}_{k=1}^\infty = \left\{ e_1, \frac{1}{\sqrt{2}}e_2, \frac{1}{\sqrt{2}}e_2, \frac{1}{\sqrt{3}}e_3, \frac{1}{\sqrt{3}} \right\}$$

To see that the set $\{g_k\}_{k=1}^\infty$ is a frame for $\mathcal{H}$, take a
that

$$\sum_{k=1}^\infty |\langle x|g_k\rangle|^2 = \sum_{k=1}^\infty k \left| \left\langle x \left| \frac{1}{\sqrt{k}} e_k \right. \right\rangle \right|^2 = \sum_{k=1}^\infty k \frac{1}{k} |\langle x|e_k$$

We conclude that $\{g_k\}_{k=1}^\infty$ is a frame with the frame

From (20.32) it follows that an equivalent formula

$$A\|x\|^2 \le \|\mathbb{T}x\|^2 \le B\|x\|^2, \quad \text{for al}$$

This means that the energy in the coefficient seque
below by bounds that are proportional to the signal en
frame bound $A > 0$ guarantees that the linear operato
first requirement above is satisfied. Besides that it a
of the set $\{g_k\}_{k \in \mathcal{K}}$ for $\mathcal{H}$, as we shall see next. To
following definition:

Definition 20.3.3. A set of elements $\{g_k\}_{k \in \mathcal{K}}$, $g_k \in$
Hilbert space $\mathcal{H}$ if $\langle x|g_k\rangle = 0$ for all $k \in \mathcal{K}$ and with a
only element in $\mathcal{H}$ that is orthogonal to all g_k, is $x =$

To see that the frame $\{g_k\}_{k \in \mathcal{K}}$ is complete for $\mathcal{H}$, ta
and assume that $\langle x|g_k\rangle = 0$ for all $k \in \mathcal{K}$. Due to th
bound $A > 0$ we have

$$A\|x\|^2 \le \sum_{k \in \mathcal{K}} |\langle x|g_k\rangle|^2 = 0$$

which implies $\|x\|^2 = 0$ and hence $x = 0$.

Finally, note that the existence of an upper fram
that $\mathbb{T}$ is a bounded linear operator[3] (see [1, Def. 2.7.-1
2.7.-9]), continuous[4] (see [1, Sec. 2.7]).

---

[3]Let $\mathcal{H}$ and $\mathcal{H}'$ be Hilbert spaces and $\mathbb{A} : \mathcal{H} \to \mathcal{H}'$ a linear
to be *bounded* if there exists a finite number c such that for all
[4]Let $\mathcal{H}$ and $\mathcal{H}'$ be Hilbert spaces and $\mathbb{A} : \mathcal{H} \to \mathcal{H}'$ a linear
to be *continuous* at a point $x_0 \in \mathcal{H}$ if for every $\epsilon > 0$ there is
satisfying $\|x - x_0\| < \delta$ it follows that $\|\mathbb{A}x - \mathbb{A}x_0\| < \epsilon$. The op
on $\mathcal{H}$, if it is continuous at every point $x_0 \in \mathcal{H}$.

Recall that we would like to find a general method to reconstruct a signal $x \in \mathcal{H}$ from its expansion coefficients $\{\langle x|g_k\rangle\}_{k \in \mathcal{K}}$. In Section 20.2.3, we saw that in the finite-dimensional case, this can be accomplished according to:

$$\mathbf{x} = \sum_{k=1}^{N} \langle \mathbf{x}|\mathbf{g}_k\rangle \, \tilde{\mathbf{g}}_k.$$

Here $\{\tilde{\mathbf{g}}_1, \ldots, \tilde{\mathbf{g}}_N\}$ can be chosen to be the canonical dual set to the set $\{\mathbf{g}_1, \ldots, \mathbf{g}_N\}$ and can be computed as follows: $\tilde{\mathbf{g}}_k = (\mathbf{T}^\dagger \mathbf{T})^{-1}\mathbf{g}_k$, $k = 1, \ldots, N$. We already know that $\mathbb{T}$ is the generalization of $\mathbf{T}$ to the infinite-dimensional setting. Which operator will then correspond to $\mathbf{T}^\dagger$? To answer this question we start with a definition.

Definition 20.3.4. The linear operator $\mathbb{T}^\times$ is defined as

$$\mathbb{T}^\times : l^2 \to \mathcal{H}$$
$$\mathbb{T}^\times : \{c_k\}_{k \in \mathcal{K}} \to \sum_{k \in \mathcal{K}} c_k g_k.$$

Next, we recall the definition of the adjoint of an operator.

Definition 20.3.5. Let $\mathbb{A} : \mathcal{H} \to \mathcal{H}'$ be a bounded linear operator between the Hilbert spaces $\mathcal{H}$ and $\mathcal{H}'$. The unique bounded linear operator $\mathbb{A}^* : \mathcal{H}' \to \mathcal{H}$ that satisfies

$$\langle \mathbb{A}x|y\rangle = \langle x|\mathbb{A}^*y\rangle \tag{20.37}$$

for all $x \in \mathcal{H}$ and all $y \in \mathcal{H}'$ is called the adjoint of $\mathbb{A}$.

Note that the concept of the adjoint of an operator directly generalizes that of the Hermitian transpose of a matrix: if $\mathbf{A} \in \mathbb{C}^{N \times M}$, $\mathbf{x} \in \mathbb{C}^M$, $\mathbf{y} \in \mathbb{C}^N$, then

$$\langle \mathbf{A}\mathbf{x}|\mathbf{y}\rangle = \mathbf{y}^\dagger \mathbf{A}\mathbf{x} = (\mathbf{A}^\dagger \mathbf{y})^\dagger \mathbf{x} = \langle \mathbf{x}|\mathbf{A}^\dagger \mathbf{y}\rangle,$$

which, comparing to (20.37), shows that $\mathbf{A}^\dagger$ corresponds to $\mathbb{A}^*$.

We shall next show that the operator $\mathbb{T}^\times$ defined above is nothing but the adjoint $\mathbb{T}^*$ of the operator $\mathbb{T}$. To see this consider an arbitrary sequence $\{c_k\}_{k \in \mathcal{K}} \in l^2$ and an arbitrary signal $x \in \mathcal{H}$. We have to prove that

$$\langle \mathbb{T}x|\{c_k\}_{k \in \mathcal{K}}\rangle = \langle x|\mathbb{T}^\times \{c_k\}_{k \in \mathcal{K}}\rangle.$$

This can be established by noting that

$$\langle \mathbb{T}x|\{c_k\}_{k \in \mathcal{K}}\rangle = \sum_{k \in \mathcal{K}} \langle x|g_k\rangle \, c_k^*$$

$$\langle x|\mathbb{T}^\times \{c_k\}_{k \in \mathcal{K}}\rangle = \left\langle x \left| \sum_{k \in \mathcal{K}} c_k g_k \right. \right\rangle = \sum_{k \in \mathcal{K}} c_k^* \, \langle x|g_k\rangle.$$

We therefore showed that the adjoint operator of $\mathbb{T}$ i

$$\mathbb{T}^\times = \mathbb{T}^*.$$

In what follows, we shall always write $\mathbb{T}^*$ instead o:
the concept of the adjoint of an operator generalizes
transpose of a matrix to the infinite-dimensional case
tion of $\mathbf{T}^\dagger$ to the infinite-dimensional setting.

20.3.1   The Frame Operator

Let us return to the discussion we had immediately
saw that in the finite-dimensional case, the canonical
set $\{\mathbf{g}_1, \ldots, \mathbf{g}_N\}$ can be computed as follows: $\tilde{\mathbf{g}}_k = ($
know that $\mathbb{T}$ is the generalization of $\mathbf{T}$ to the infinite-c
just seen that $\mathbb{T}^*$ is the generalization of $\mathbf{T}^\dagger$. It is no
$\mathbb{T}^*\mathbb{T}$ must correspond to $\mathbf{T}^\dagger\mathbf{T}$. The operator $\mathbb{T}^*\mathbb{T}$ is o
theory.

Definition 20.3.6. Let $\{g_k\}_{k \in \mathcal{K}}$ be a frame for the H
$\mathbb{S} : \mathcal{H} \to \mathcal{H}$ defined as

$$\mathbb{S} = \mathbb{T}^*\mathbb{T},$$

$$\mathbb{S}x = \sum_{k \in \mathcal{K}} \langle x | g_k \rangle \, g_k$$

is called the frame operator.

We note that

$$\sum_{k \in \mathcal{K}} |\langle x | g_k \rangle|^2 = \|\mathbb{T}x\|^2 = \langle \mathbb{T}x | \mathbb{T}x \rangle = \langle \mathbb{T}^*\mathbb{T}x |$$

We are now able to formulate the frame condition in t
by simply noting that (20.36) can be written as

$$A\|x\|^2 \leq \langle \mathbb{S}x | x \rangle \leq B\|x\|^2$$

We shall next discuss the properties of $\mathbb{S}$.

Theorem 20.3.1. *The frame operator $\mathbb{S}$ satisfies the*

   1. $\mathbb{S}$ is linear and bounded;

   2. $\mathbb{S}$ is self-adjoint, i.e., $\mathbb{S}^* = \mathbb{S}$;

   3. $\mathbb{S}$ is positive-definite, i.e., $\langle \mathbb{S}x | x \rangle > 0$ for all $x \in$

4. $\mathbb{S}$ has a unique self-adjoint positive-definite square root (denoted as $\mathbb{S}^{1/2}$).

Proof:

1. Linearity and boundedness of $\mathbb{S}$ follow from the fact that $\mathbb{S}$ is obtained by cascading a bounded linear operator and its adjoint (see (20.38)).

2. To see that $\mathbb{S}$ is self-adjoint simply note that

$$\mathbb{S}^* = (\mathbb{T}^*\mathbb{T})^* = \mathbb{T}^*\mathbb{T} = \mathbb{S}.$$

3. To see that $\mathbb{S}$ is positive-definite note that, with (20.40)

$$\langle \mathbb{S}x|x \rangle \geq A\|x\|^2 > 0$$

for all $x \in \mathcal{H}$, $x \neq 0$.

4. Recall the following basic fact from functional analysis [1, Th. 9.4-2].

   Lemma 20.3.1. Every self-adjoint positive-definite bounded operator $\mathbb{A} : \mathcal{H} \to \mathcal{H}$ has a unique self-adjoint positive-definite square root, i.e., there exists a unique self-adjoint positive-definite operator $\mathbb{B}$ such that $\mathbb{A} = \mathbb{B}\mathbb{B}$. The operator $\mathbb{B}$ commutes with $\mathbb{A}$, i.e., $\mathbb{B}\mathbb{A} = \mathbb{A}\mathbb{B}$.

   Property 4 now follows directly form Property 2, Property 3, and Lemma 20.3.1.

   $\square$

We next show that the tightest possible frame bounds A and B are given by the smallest and the largest spectral value [1, Def. 7.2-1] of the frame operator $\mathbb{S}$, respectively.

Theorem 20.3.2. *Let A and B be the tightest possible frame bounds for a frame with frame operator $\mathbb{S}$. Then*

$$A = \lambda_{\min} \quad and \quad B = \lambda_{\max}, \tag{20.41}$$

where $\lambda_{\min}$ and $\lambda_{\max}$ denote the smallest and the largest spectral value of $\mathbb{S}$, respectively.

Proof: By standard results on the spectrum of self-adjoint operators [1, Th. 9.2-1, Th. 9.2-3, Th. 9.2-4], we have

$$\lambda_{\min} = \inf_{x \in \mathcal{H}} \frac{\langle \mathbb{S}x|x \rangle}{\|x\|^2} \quad and \quad \lambda_{\max} = \sup_{x \in \mathcal{H}} \frac{\langle \mathbb{S}x|x \rangle}{\|x\|^2}. \tag{20.42}$$

This means that $\lambda_{\min}$ and $\lambda_{\max}$ are, respectively, the largest and the smallest constants such that

$$\lambda_{\min}\|x\|^2 \leq \langle \mathbb{S}x|x \rangle \leq \lambda_{\max}\|x\|^2 \tag{20.43}$$

is satisfied for every $x \in \mathcal{H}$. According to (20.40) thi
are the tightest possible frame bounds.

It is instructive to compare (20.43) to (20.31).
corresponds to the matrix $\mathbf{T}^\dagger \mathbf{T}$ in the finite-dimensi
tion 20.2.3. Thus, $\|\mathbf{c}\|^2 = \mathbf{x}^\dagger \mathbf{T}^\dagger \mathbf{T} \mathbf{x} = \langle \mathbf{S} \mathbf{x} | \mathbf{x} \rangle$, which
shows that (20.43) is simply a generalization of (20.3
case.

20.3.2   The Canonical Dual Frame

Recall that in the finite-dimensional case considered in
dual frame $\{\tilde{\mathbf{g}}_k\}_{k=1}^N$ of the frame $\{\mathbf{g}_k\}_{k=1}^N$ can be use
from the expansion coefficients $\{\langle \mathbf{x} | \mathbf{g}_k \rangle\}_{k=1}^N$ according

$$\mathbf{x} = \sum_{k=1}^N \langle \mathbf{x} | \mathbf{g}_k \rangle \, \tilde{\mathbf{g}}_k.$$

In (20.30) we saw that the canonical dual frame can

$$\tilde{\mathbf{g}}_k = (\mathbf{T}^\dagger \mathbf{T})^{-1} \mathbf{g}_k, \quad k = 1, \ldots$$

We already pointed out that the frame operator $\mathbb{S} = \mathbb{1}$
trix $\mathbf{T}^\dagger \mathbf{T}$ in the finite-dimensional case. The matrix $(\mathbf{T}$
to the operator $\mathbb{S}^{-1}$, which will be studied next.

From (20.41) it follows that $\lambda_{\min}$, the smallest
$\lambda_{\min} > 0$ if $\{g_k\}_{k \in \mathcal{K}}$ is a frame. This implies that ze
7.2-1] of $\mathbb{S}$ and hence $\mathbb{S}$ is invertible on $\mathcal{H}$, i.e., there
such that $\mathbb{S}\mathbb{S}^{-1} = \mathbb{S}^{-1}\mathbb{S} = \mathbb{I}_\mathcal{H}$. Next, we summarize th

Theorem 20.3.3. *The following properties hold:*

1. $\mathbb{S}^{-1}$ is self-adjoint, i.e., $(\mathbb{S}^{-1})^* = \mathbb{S}^{-1}$;

2. $\mathbb{S}^{-1}$ satisfies

$$\frac{1}{B} = \inf_{x \in \mathcal{H}} \frac{\langle \mathbb{S}^{-1} x | x \rangle}{\|x\|^2} \quad and \quad \frac{1}{A} =$$

where A and B are the tightest possible frame b

3. $\mathbb{S}^{-1}$ is positive-definite.

Proof:

1. To prove that $\mathbb{S}^{-1}$ is self-adjoint we write

$$(\mathbb{S}\mathbb{S}^{-1})^* = (\mathbb{S}^{-1})^* \mathbb{S}^* =$$

Since $\mathbb{S}$ is self-adjoint, i.e., $\mathbb{S} = \mathbb{S}^*$, we conclude that

$$(\mathbb{S}^{-1})^* \mathbb{S} = \mathbb{I}_{\mathcal{H}}.$$

Multiplying by $\mathbb{S}^{-1}$ from the right, we finally obtain

$$(\mathbb{S}^{-1})^* = \mathbb{S}^{-1}.$$

2. To prove the first equation in (20.45) we write

$$
\begin{aligned}
B &= \sup_{x \in \mathcal{H}} \frac{\langle \mathbb{S}x | x \rangle}{\|x\|^2} = \sup_{y \in \mathcal{H}} \frac{\langle \mathbb{S}\mathbb{S}^{1/2}\mathbb{S}^{-1}y | \mathbb{S}^{1/2}\mathbb{S}^{-1}y \rangle}{\langle \mathbb{S}^{1/2}\mathbb{S}^{-1}y | \mathbb{S}^{1/2}\mathbb{S}^{-1}y \rangle} \\
&= \sup_{y \in \mathcal{H}} \frac{\langle \mathbb{S}^{-1}\mathbb{S}^{1/2}\mathbb{S}\mathbb{S}^{1/2}\mathbb{S}^{-1}y | y \rangle}{\langle \mathbb{S}^{-1}\mathbb{S}^{1/2}\mathbb{S}^{1/2}\mathbb{S}^{-1}y | y \rangle} \\
&= \sup_{y \in \mathcal{H}} \frac{\langle y | y \rangle}{\langle \mathbb{S}^{-1}y | y \rangle},
\end{aligned}
\tag{20.46}
$$

where the first equality follows from (20.41) and (20.42); in the second equality we used the fact that the operator $\mathbb{S}^{1/2}\mathbb{S}^{-1}$ is one-to-one on $\mathcal{H}$ and we changed variables according to $x = \mathbb{S}^{1/2}\mathbb{S}^{-1}y$; in the third equality we used the fact that $\mathbb{S}^{1/2}$ and $\mathbb{S}^{-1}$ are self-adjoint, and in the fourth equality we used $\mathbb{S} = \mathbb{S}^{1/2}\mathbb{S}^{1/2}$. The first equation in (20.45) is now obtained by noting that (20.46) implies

$$\frac{1}{B} = 1 \bigg/ \left(\sup_{y \in \mathcal{H}} \frac{\langle y | y \rangle}{\langle \mathbb{S}^{-1}y | y \rangle} \right) = \inf_{y \in \mathcal{H}} \frac{\langle \mathbb{S}^{-1}y | y \rangle}{\langle y | y \rangle}.$$

The second equation in (20.45) is proved analogously.

3. Positive-definiteness of $\mathbb{S}^{-1}$ follows from the first equation in (20.45) and the fact that $B < \infty$ so that $1/B > 0$.

$\square$

We are now ready to generalize (20.44) and state the main result on canonical dual frames in the case of general (possibly infinite-dimensional) Hilbert spaces.

Theorem 20.3.4. *Let $\{g_k\}_{k \in \mathcal{K}}$ be a frame for the Hilbert space $\mathcal{H}$ with the frame bounds A and B, and let $\mathbb{S}$ be the corresponding frame operator. Then, the set $\{\tilde{g}_k\}_{k \in \mathcal{K}}$ given by*

$$\tilde{g}_k = \mathbb{S}^{-1} g_k, \quad k \in \mathcal{K}, \tag{20.47}$$

is a frame for $\mathcal{H}$ with the frame bounds $\tilde{A} = 1/B$ and $\tilde{B} = 1/A$.

The analysis operator associated to $\{\tilde{g}_k\}_{k \in \mathcal{K}}$ defined as

$$\tilde{\mathbb{T}} : \mathcal{H} \to l^2$$

$$\tilde{\mathbb{T}} : x \to \{\langle x | \tilde{g}_k \rangle\}_{k \in \mathcal{K}}$$

satisfies

$$\tilde{\mathbb{T}} = \mathbb{T}\mathbb{S}^{-1} = \mathbb{T}(\mathbb{T}^*\mathbb{T})^{-1}. \tag{20.48}$$

Proof: Recall that $\mathbb{S}^{-1}$ is self-adjoint. Hence, we
$\langle \mathbb{S}^{-1}x|g_k\rangle$ for all $x \in \mathcal{H}$. Thus, using (20.39), we obt

$$\sum_{k\in\mathcal{K}} |\langle x|\tilde{g}_k\rangle|^2 = \sum_{k\in\mathcal{K}} |\langle \mathbb{S}^{-1}x|g_k\rangle|^2$$

$$= \langle \mathbb{S}(\mathbb{S}^{-1}x)|\mathbb{S}^{-1}x\rangle = \langle x|\mathbb{S}^{-1}$$

Therefore, we conclude from (20.45) that

$$\frac{1}{B}\|x\|^2 \le \sum_{k\in\mathcal{K}} |\langle x|\tilde{g}_k\rangle|^2 \le \frac{1}{A}\|$$

i.e., the set $\{\tilde{g}_k\}_{k\in\mathcal{K}}$ constitutes a frame for $\mathcal{H}$ with
$\tilde{B} = 1/A$; moreover, it follows from (20.45) that $\tilde{A} =$
tightest possible frame bounds. It remains to show th

$$\tilde{\mathbb{T}}x = \{\langle x|\tilde{g}_k\rangle\}_{k\in\mathcal{K}} = \{\langle x|\mathbb{S}^{-1}g_k\rangle\}_{k\in\mathcal{K}} = \{\langle \mathbb{S}^{-1}$$

We call $\{\tilde{g}_k\}_{k\in\mathcal{K}}$ the *canonical dual frame* associa
is convenient to introduce the *canonical dual frame o*

Definition 20.3.7. The frame operator associated t

$$\tilde{\mathbb{S}} = \tilde{\mathbb{T}}^*\tilde{\mathbb{T}}, \quad \tilde{\mathbb{S}}x = \sum_{k\in\mathcal{K}} \langle x|\tilde{g}_k\rangle$$

is called the canonical dual frame operator.

Theorem 20.3.5. *The canonical dual frame operato*

Proof: For every $x \in \mathcal{H}$, we have

$$\tilde{\mathbb{S}}x = \sum_{k\in\mathcal{K}} \langle x|\tilde{g}_k\rangle \, \tilde{g}_k = \sum_{k\in\mathcal{K}} \langle x|\mathbb{S}^{-1}g_k\rangle \, \mathbb{S}$$

$$= \mathbb{S}^{-1} \sum_{k\in\mathcal{K}} \langle \mathbb{S}^{-1}x|g_k\rangle \, g_k = \mathbb{S}^{-1}\mathbb{S}\mathbb{S}^{-1}$$

where in the first equality we used (20.49), in the sec
third we made use of the fact that $\mathbb{S}^{-1}$ is self-adjoin
the definition of $\mathbb{S}$.

Note that canonical duality is a reciprocity relati
the canonical dual of the frame $\{g_k\}_{k\in\mathcal{K}}$, then $\{g_k\}_{k\in}$
frame $\{\tilde{g}_k\}_{k\in\mathcal{K}}$. This can be seen by noting that

$$\tilde{\mathbb{S}}^{-1}\tilde{g}_k = (\mathbb{S}^{-1})^{-1}\mathbb{S}^{-1}g_k = \mathbb{S}\mathbb{S}^{-1}g$$

20.3.3 Signal Expansions

The following theorem can be considered as one of the *central results in frame theory*. It states that every signal $x \in \mathcal{H}$ can be expanded into a frame. The expansion coefficients can be chosen as the inner products of x with the canonical dual frame elements.

Theorem 20.3.6. *Let $\{g_k\}_{k \in \mathcal{K}}$ and $\{\tilde{g}_k\}_{k \in \mathcal{K}}$ be canonical dual frames for the Hilbert space $\mathcal{H}$. Every signal $x \in \mathcal{H}$ can be decomposed as follows*

$$x = \mathbb{T}^* \tilde{\mathbb{T}} x = \sum_{k \in \mathcal{K}} \langle x | \tilde{g}_k \rangle \, g_k$$

$$x = \tilde{\mathbb{T}}^* \mathbb{T} x = \sum_{k \in \mathcal{K}} \langle x | g_k \rangle \, \tilde{g}_k. \tag{20.50}$$

Note that, equivalently, we have

$$\mathbb{T}^* \tilde{\mathbb{T}} = \tilde{\mathbb{T}}^* \mathbb{T} = \mathbb{I}_{\mathcal{H}}.$$

Proof: We have

$$\mathbb{T}^* \tilde{\mathbb{T}} x = \sum_{k \in \mathcal{K}} \langle x | \tilde{g}_k \rangle \, g_k = \sum_{k \in \mathcal{K}} \langle x | \mathbb{S}^{-1} g_k \rangle \, g_k$$

$$= \sum_{k \in \mathcal{K}} \langle \mathbb{S}^{-1} x | g_k \rangle \, g_k = \mathbb{S} \mathbb{S}^{-1} x = x.$$

This proves that $\mathbb{T}^* \tilde{\mathbb{T}} = \mathbb{I}_{\mathcal{H}}$. The proof of $\tilde{\mathbb{T}}^* \mathbb{T} = \mathbb{I}_{\mathcal{H}}$ is similar. $\square$

Note that (20.50) corresponds to the decomposition (20.26) we found in the finite-dimensional case.

It is now natural to ask whether reconstruction of x from the coefficients $\langle x | g_k \rangle$, $k \in \mathcal{K}$, according to (20.50) is the only way of recovering x from $\langle x | g_k \rangle$, $k \in \mathcal{K}$. Recall that we showed in the finite-dimensional case (see Section 20.2.3) that for each complete and redundant set of vectors $\{\mathbf{g}_1, \dots, \mathbf{g}_N\}$, there are infinitely many dual sets $\{\tilde{\mathbf{g}}_1, \dots, \tilde{\mathbf{g}}_N\}$ that can be used to reconstruct a signal $\mathbf{x}$ from the coefficients $\langle \mathbf{x} | \mathbf{g}_k \rangle$, $k = 1, \dots, N$, according to (20.26). These dual sets are obtained by identifying $\{\tilde{\mathbf{g}}_1, \dots, \tilde{\mathbf{g}}_N\}$ with the columns of $\mathbf{L}$, where $\mathbf{L}$ is a left-inverse of the analysis matrix $\mathbf{T}$. In the infinite-dimensional case the question of finding all dual frames for a given frame boils down to finding, for a given analysis operator $\mathbb{T}$, all linear operators $\mathbb{L}$ that satisfy

$$\mathbb{L} \mathbb{T} x = x$$

for all $x \in \mathcal{H}$. In other words, we want to identify all left-inverses $\mathbb{L}$ of the analysis operator $\mathbb{T}$. The answer to this question is the infinite-dimensional version of Theorem 20.2.1 that we state here without proof.

Theorem 20.3.7. *Let* $\mathbb{A} : \mathcal{H} \to l^2$ *be a bounded l*
$\mathbb{A}^*\mathbb{A} : \mathcal{H} \to \mathcal{H}$ *is invertible on* $\mathcal{H}$. *Then, the opera*
$\mathbb{A}^+ \triangleq (\mathbb{A}^*\mathbb{A})^{-1}\mathbb{A}^*$ *is a left-inverse of* $\mathbb{A}$, *i.e.,* $\mathbb{A}^+\mathbb{A} =$
operator on $\mathcal{H}$. *Moreover, the general solution* $\mathbb{L}$ *of t*
by

$$\mathbb{L} = \mathbb{A}^+ + \mathbb{M}(\mathbb{I}_{l^2} - \mathbb{A}\mathbb{A}^+)$$

where $\mathbb{M} : l^2 \to \mathcal{H}$ *is an arbitrary bounded linear op*
operator on l^2.

Applying this theorem to the operator $\mathbb{T}$ we see
be written as

$$\mathbb{L} = \mathbb{T}^+ + \mathbb{M}(\mathbb{I}_{l^2} - \mathbb{T}\mathbb{T}^+).$$

where $\mathbb{M} : l^2 \to \mathcal{H}$ is an arbitrary bounded linear ope

$$\mathbb{T}^+ = (\mathbb{T}^*\mathbb{T})^{-1}\mathbb{T}^*.$$

Now, using (20.48), we obtain the following importar

$$\mathbb{T}^+ = (\mathbb{T}^*\mathbb{T})^{-1}\mathbb{T}^* = \mathbb{S}^{-1}\mathbb{T}^* =$$

This shows that reconstruction according to (20.50),
tor $\tilde{\mathbb{T}}^*$ to the coefficient sequence $\mathbb{T}x = \{\langle x|g_k\rangle\}_{k\in\mathcal{K}}$
infinite-dimensional analog of the Moore-Penrose inv
already noted in the finite-dimensional case the exist
inverses of the operator $\mathbb{T}$ provides us with freedom i
We close this discussion with a geometric interp
tion (20.51). First observe the following.

Theorem 20.3.8. *The operator*

$$\mathbb{P} : l^2 \to \mathcal{R}(\mathbb{T}) \subseteq l^2$$

defined as

$$\mathbb{P} = \mathbb{T}\mathbb{S}^{-1}\mathbb{T}^*$$

satisfies the following properties:

1. $\mathbb{P}$ *is the identity operator* $\mathbb{I}_{l^2}$ *on* $\mathcal{R}(\mathbb{T})$.

2. $\mathbb{P}$ *is the zero operator on* $\mathcal{R}(\mathbb{T})^\perp$, *where* $\mathcal{R}(\mathbb{T})^\perp$
   plement of the space $\mathcal{R}(\mathbb{T})$.

In other words, $\mathbb{P}$ *is the orthogonal projection*
$\{\{c_k\}_{k\in\mathcal{K}} \mid \{c_k\}_{k\in\mathcal{K}} = \mathbb{T}x, x \in \mathcal{H}\}$, *the range space of*

Proof:

1. Take a sequence $\{c_k\}_{k \in \mathcal{K}} \in \mathcal{R}(\mathbb{T})$ and note that it can be written as $\{c_k\}_{k \in \mathcal{K}} = \mathbb{T}x$, where $x \in \mathcal{H}$. Then, we have

$$\mathbb{P}\{c_k\}_{k \in \mathcal{K}} = \mathbb{T}\mathbb{S}^{-1}\mathbb{T}^*\mathbb{T}x = \mathbb{T}\mathbb{S}^{-1}\mathbb{S}x = \mathbb{T}\mathbb{I}_{\mathcal{H}}x = \mathbb{T}x = \{c_k\}_{k \in \mathcal{K}}.$$

   This proves that $\mathbb{P}$ is the identity operator on $\mathcal{R}(\mathbb{T})$.

2. Next, take a sequence $\{c_k\}_{k \in \mathcal{K}} \in \mathcal{R}(\mathbb{T})^{\perp}$. As the orthogonal complement of the range space of an operator is the null space of its adjoint, we have $\mathbb{T}^*\{c_k\}_{k \in \mathcal{K}} = 0$ and therefore

$$\mathbb{P}\{c_k\}_{k \in \mathcal{K}} = \mathbb{T}\mathbb{S}^{-1}\mathbb{T}^*\{c_k\}_{k \in \mathcal{K}} = 0.$$

   This proves that $\mathbb{P}$ is the zero operator on $\mathcal{R}(\mathbb{T})^{\perp}$.

$\square$

Now using that $\mathbb{T}\mathbb{T}^+ = \mathbb{T}\mathbb{S}^{-1}\mathbb{T}^* = \mathbb{P}$ and $\mathbb{T}^+ = \mathbb{S}^{-1}\mathbb{T}^* = \mathbb{S}^{-1}\mathbb{S}\mathbb{S}^{-1}\mathbb{T}^* = \mathbb{S}^{-1}\mathbb{T}^*\mathbb{T}\mathbb{S}^{-1}\mathbb{T}^* = \tilde{\mathbb{T}}^*\mathbb{P}$, we can rewrite (20.51) as follows

$$\mathbb{L} = \tilde{\mathbb{T}}^*\mathbb{P} + \mathbb{M}(\mathbb{I}_{l^2} - \mathbb{P}). \tag{20.52}$$

Next, we show that $(\mathbb{I}_{l^2} - \mathbb{P}) : l^2 \to l^2$ is the orthogonal projection onto $\mathcal{R}(\mathbb{T})^{\perp}$. Indeed, we can directly verify the following: For every $\{c_k\}_{k \in \mathcal{K}} \in \mathcal{R}(\mathbb{T})^{\perp}$, we have $(\mathbb{I}_{l^2} - \mathbb{P})\{c_k\}_{k \in \mathcal{K}} = \mathbb{I}_{l^2}\{c_k\}_{k \in \mathcal{K}} - 0 = \{c_k\}_{k \in \mathcal{K}}$, i.e., $\mathbb{I}_{l^2} - \mathbb{P}$ is the identity operator on $\mathcal{R}(\mathbb{T})^{\perp}$; for every $\{c_k\}_{k \in \mathcal{K}} \in (\mathcal{R}(\mathbb{T})^{\perp})^{\perp} = \mathcal{R}(\mathbb{T})$, we have $(\mathbb{I}_{l^2} - \mathbb{P})\{c_k\}_{k \in \mathcal{K}} = \mathbb{I}_{l^2}\{c_k\}_{k \in \mathcal{K}} - \{c_k\}_{k \in \mathcal{K}} = 0$, i.e., $\mathbb{I}_{l^2} - \mathbb{P}$ is the zero operator on $(\mathcal{R}(\mathbb{T})^{\perp})^{\perp}$.

We are now ready to reinterpret (20.52) as follows. Every left-inverse $\mathbb{L}$ of $\mathbb{T}$ acts as $\tilde{\mathbb{T}}^*$ (the synthesis operator of the canonical dual frame) on the range space of the analysis operator $\mathbb{T}$, and can act in an arbitrary linear and bounded fashion on the orthogonal complement of the range space of the analysis operator $\mathbb{T}$.

20.3.4 Tight Frames

The frames considered in Examples 20.3.1 and 20.3.2 above have an interesting property: In both cases the tightest possible frame bounds A and B are equal. Frames with this property are called tight frames.

Definition 20.3.8. A frame $\{g_k\}_{k \in \mathcal{K}}$ with tightest possible frame bounds $A = B$ is called a tight frame.

Tight frames are of significant practical interest because of the following central fact.

Theorem 20.3.9. *Let $\{g_k\}_{k \in \mathcal{K}}$ be a frame for the Hilbert space $\mathcal{H}$. The frame $\{g_k\}_{k \in \mathcal{K}}$ is tight with frame bound A if and only if its corresponding frame operator satisfies $\mathbb{S} = A\mathbb{I}_{\mathcal{H}}$, or equivalently, if*

$$x = \frac{1}{A}\sum_{k \in \mathcal{K}} \langle x | g_k \rangle \, g_k \tag{20.53}$$

for all $x \in \mathcal{H}$.

Proof: First observe that $\mathbb{S} = A\mathbb{I}_\mathcal{H}$ is equivalent to $\mathbb{S}x$
which, in turn, is equivalent to (20.53) by definition

To prove that tightness of $\{g_k\}_{k \in \mathcal{K}}$ implies $\mathbb{S} =$
tion 20.3.8, using (20.40) we can write

$$\langle \mathbb{S}x|x \rangle = A \langle x|x \rangle, \text{ for all } x \in$$

Therefore

$$\langle (\mathbb{S} - A\mathbb{I}_\mathcal{H})x|x \rangle = 0, \text{ for all } x$$

which implies $\mathbb{S} = A\mathbb{I}_\mathcal{H}$.

To prove that $\mathbb{S} = A\mathbb{I}_\mathcal{H}$ implies tightness of $\{g_k\}_{k \in}$
with x on both sides of (20.53) to obtain

$$\langle x|x \rangle = \frac{1}{A} \sum_{k \in \mathcal{K}} \langle x|g_k \rangle \langle g_k|x$$

This is equivalent to

$$A\|x\|^2 = \sum_{k \in \mathcal{K}} |\langle x|g_k \rangle|^2 ,$$

which shows that $\{g_k\}_{k \in \mathcal{K}}$ is a tight frame for $\mathcal{H}$ with

The practical importance of tight frames lies in the
putation of the canonical dual frame, which in the ge
of an operator and application of this inverse to all
simple. Specifically, we have:

$$\tilde{g}_k = \mathbb{S}^{-1} g_k = \frac{1}{A} \mathbb{I}_\mathcal{H} g_k = \frac{1}{A}$$

A well-known example of a tight frame for $\mathbb{R}^2$ is t

Example 20.3.3 (The Mercedes-Benz frame [20]
(see Figure 20.4) is given by the following three vecto

$$\mathbf{g}_1 = \begin{bmatrix} 0 \\ 1 \end{bmatrix}, \quad \mathbf{g}_2 = \begin{bmatrix} -\sqrt{3}/2 \\ -1/2 \end{bmatrix}, \quad \mathbf{g}_3 =$$

To see that this frame is indeed tight, note that its an
the matrix

$$\mathbf{T} = \begin{bmatrix} 0 & 1 \\ -\sqrt{3}/2 & -1/2 \\ \sqrt{3}/2 & -1/2 \end{bmatrix}.$$

The adjoint $\mathbb{T}^*$ of the analysis operator is given by th

$$\mathbf{T}^\dagger = \begin{bmatrix} 0 & -\sqrt{3}/2 & \sqrt{3}/2 \\ 1 & -1/2 & -1/2 \end{bmatrix}$$

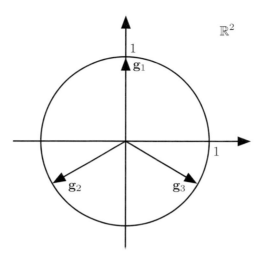

Figure 20.4: The Mercedes-Benz frame.

Therefore, the frame operator $\mathbb{S}$ is represented by the matrix

$$\mathbf{S} = \mathbf{T}^{\dagger}\mathbf{T} = \begin{bmatrix} 0 & -\sqrt{3}/2 & \sqrt{3}/2 \\ 1 & -1/2 & -1/2 \end{bmatrix} \begin{bmatrix} 0 & 1 \\ -\sqrt{3}/2 & -1/2 \\ \sqrt{3}/2 & -1/2 \end{bmatrix} = \frac{3}{2}\begin{bmatrix} 1 & 0 \\ 0 & 1 \end{bmatrix} = \frac{3}{2}\mathbf{I}_2,$$

and hence $\mathbb{S} = A\mathbb{I}_{\mathbb{R}^2}$ with $A = 3/2$, which implies, by Theorem 20.3.9, that $\{\mathbf{g}_1, \mathbf{g}_2, \mathbf{g}_3\}$ is a tight frame (for $\mathbb{R}^2$). $\qquad\square$

The design of tight frames is challenging in general. It is hence interesting to devise simple systematic methods for obtaining tight frames. The following theorem shows how we can obtain a tight frame from a given general frame.

Theorem 20.3.10. *Let $\{g_k\}_{k\in\mathcal{K}}$ be a frame for the Hilbert space $\mathcal{H}$ with frame operator $\mathbb{S}$. Denote the positive-definite square root of $\mathbb{S}^{-1}$ by $\mathbb{S}^{-1/2}$. Then $\{\mathbb{S}^{-1/2}g_k\}_{k\in\mathcal{K}}$ is a tight frame for $\mathcal{H}$ with frame bound $A = 1$, i.e.,*

$$x = \sum_{k\in\mathcal{K}} \left\langle x \middle| \mathbb{S}^{-1/2}g_k \right\rangle \mathbb{S}^{-1/2}g_k, \quad \text{for all } x \in \mathcal{H}.$$

Proof: Since $\mathbb{S}^{-1}$ is self-adjoint and positive-definite by Theorem 20.3.3, it has, by Lemma 20.3.1, a unique self-adjoint positive-definite square root $\mathbb{S}^{-1/2}$ that commutes with $\mathbb{S}^{-1}$. Moreover $\mathbb{S}^{-1/2}$ also commutes with $\mathbb{S}$, which can be seen as follows:

$$\mathbb{S}^{-1/2}\mathbb{S}^{-1} = \mathbb{S}^{-1}\mathbb{S}^{-1/2}$$

$$\mathbb{S}\mathbb{S}^{-1/2}\mathbb{S}^{-1} = \mathbb{S}^{-1/2}$$

$$\mathbb{S}\mathbb{S}^{-1/2} = \mathbb{S}^{-1/2}\mathbb{S}.$$

The proof is then effected by noting the following:

$$x = \mathbb{S}^{-1}\mathbb{S}x = \mathbb{S}^{-1/2}\mathbb{S}^{-1/2}\mathbb{S}x$$
$$= \mathbb{S}^{-1/2}\mathbb{S}\mathbb{S}^{-1/2}x$$
$$= \sum_{k \in \mathcal{K}} \left\langle \mathbb{S}^{-1/2}x \middle| g_k \right\rangle \mathbb{S}^{-1/2}$$
$$= \sum_{k \in \mathcal{K}} \left\langle x \middle| \mathbb{S}^{-1/2}g_k \right\rangle \mathbb{S}^{-1/2}$$

It is evident that every ONB is a tight frame w
that conversely a tight frame (even with $A = 1$) r
or orthogonal basis, as can be seen from Example 2
theorem shows, a tight frame with $A = 1$ and $\|g_k\| =$
an ONB.

Theorem 20.3.11. *A tight frame* $\{g_k\}_{k \in \mathcal{K}}$ *for the*
and $\|g_k\| = 1$, *for all* $k \in \mathcal{K}$, *is an ONB for* $\mathcal{H}$.

Proof: Combining

$$\langle \mathbb{S}g_k | g_k \rangle = A\|g_k\|^2 = \|g_k\|$$

with

$$\langle \mathbb{S}g_k | g_k \rangle = \sum_{j \in \mathcal{K}} |\langle g_k | g_j \rangle|^2 = \|g_k\|^4 + \sum_{j}$$

we obtain

$$\|g_k\|^4 + \sum_{j \neq k} |\langle g_k | g_j \rangle|^2 = \|g_k$$

Since $\|g_k\|^2 = 1$, for all $k \in \mathcal{K}$, it follows that $\sum_{j \neq k}$
This implies that the elements of $\{g_j\}_{j \in \mathcal{K}}$ are necessar

There is an elegant result that tells us that every t
$A = 1$ can be realized as an orthogonal projection of
larger dimension. This result is known as Naimark
the finite-dimensional version of this theorem, for the
see [24].

Theorem 20.3.12 (Naimark, [24, Prop. 1.1]). *L*
set $\{\mathbf{g}_1, \ldots, \mathbf{g}_N\}$, $\mathbf{g}_k \in \mathcal{H}$, $k = 1, \ldots, N$, *is a tight f*
Hilbert space $\mathcal{H}$ *with frame bound* $A = 1$. *Then, th*
Hilbert space $\mathcal{K} \supset \mathcal{H}$ *and an ONB* $\{\mathbf{e}_1, \ldots, \mathbf{e}_N\}$ *for*
$1, \ldots, N$, *where* $\mathbb{P} : \mathcal{K} \to \mathcal{K}$ *is the orthogonal projectio*

We omit the proof and illustrate the theorem by a

Example 20.3.4. Consider the Hilbert space $\mathcal{K} = \mathbb{R}^3$, and assume that $\mathcal{H} \subset \mathcal{K}$ is the plane spanned by the vectors $[1\ 0\ 0]^\mathsf{T}$ and $[0\ 1\ 0]^\mathsf{T}$, i.e.,

$$\mathcal{H} = \text{span}\{[1\ 0\ 0]^T, [0\ 1\ 0]^T\}.$$

We can construct a tight frame for $\mathcal{H}$ with three elements and frame bound $A = 1$ if we rescale the Mercedes-Benz frame from Example 20.3.3. Specifically, consider the vectors $\mathbf{g}_k$, $k = 1, 2, 3$, defined in (20.54) and let $\mathbf{g}_k' \triangleq \sqrt{2/3}\,\mathbf{g}_k$, $k = 1, 2, 3$. In the following, we think about the two-dimensional vectors $\mathbf{g}_k'$ as being embedded into the three-dimensional space $\mathcal{K}$ with the third coordinate (in the standard basis of $\mathcal{K}$) being equal to zero. Clearly, $\{\mathbf{g}_k'\}_{k=1}^3$ is a tight frame for $\mathcal{H}$ with frame bound $A = 1$. Now consider the following three vectors in $\mathcal{K}$:

$$\mathbf{e}_1 = \begin{bmatrix} 0 \\ \sqrt{2/3} \\ -1/\sqrt{3} \end{bmatrix}, \quad \mathbf{e}_2 = \begin{bmatrix} -1/\sqrt{2} \\ -1/\sqrt{6} \\ -1/\sqrt{3} \end{bmatrix}, \quad \mathbf{e}_3 = \begin{bmatrix} 1/\sqrt{2} \\ -1/\sqrt{6} \\ -1/\sqrt{3} \end{bmatrix}.$$

Direct calculation reveals that $\{\mathbf{e}_k\}_{k=1}^3$ is an ONB for $\mathcal{K}$. Observe that the frame vectors $\mathbf{g}_k'$, $k = 1, 2, 3$, can be obtained from the ONB vectors $\mathbf{e}_k$, $k = 1, 2, 3$, by applying the orthogonal projection from $\mathcal{K}$ onto $\mathcal{H}$:

$$\mathbf{P} \triangleq \begin{bmatrix} 1 & 0 & 0 \\ 0 & 1 & 0 \\ 0 & 0 & 0 \end{bmatrix},$$

according to $\mathbf{g}_k' = \mathbf{P}\mathbf{e}_k$, $k = 1, 2, 3$. This illustrates Naimark's theorem. □

20.3.5 Exact Frames and Biorthonormality

In Section 20.2.2 we studied expansions of signals in $\mathbb{C}^M$ into (not necessarily orthogonal) bases. The main results we established in this context can be summarized as follows:

1. The number of vectors in a basis is always equal to the dimension of the Hilbert space under consideration. Every set of vectors that spans $\mathbb{C}^M$ and has more than M vectors is necessarily redundant, i.e., the vectors in this set are linearly dependent. Removal of an arbitrary vector from a basis for $\mathbb{C}^M$ leaves a set that no longer spans $\mathbb{C}^M$.

2. For a given basis $\{\mathbf{e}_k\}_{k=1}^M$ every signal $\mathbf{x} \in \mathbb{C}^M$ has a *unique* representation according to

$$\mathbf{x} = \sum_{k=1}^M \langle \mathbf{x} | \mathbf{e}_k \rangle\, \tilde{\mathbf{e}}_k. \tag{20.55}$$

The basis $\{\mathbf{e}_k\}_{k=1}^M$ and its dual basis $\{\tilde{\mathbf{e}}_k\}_{k=1}^M$ satisfy the biorthonormality relation (20.20).

The theory of ONBs in infinite-dimensional spac
section, we ask how the concept of general (i.e., not
can be extended to infinite-dimensional spaces. Clear
case, we cannot simply say that the number of eleme
to the dimension of the Hilbert space. However, w
removing an element from a basis leaves us with an
motivate the following definition.

Definition 20.3.9. Let $\{g_k\}_{k\in\mathcal{K}}$ be a frame for the
frame $\{g_k\}_{k\in\mathcal{K}}$ *exact* if, for all $m \in \mathcal{K}$, the set $\{g_k\}_{k\neq r}$
the frame $\{g_k\}_{k\in\mathcal{K}}$ *inexact* if there is at least one elen.
from the frame, so that the set $\{g_k\}_{k\neq m}$ is again a fr

There are two more properties of general bases in f
carry over to the infinite-dimensional case, namely un
the sense of (20.55) and biorthonormality between
dual. To show that representation of a signal in ar
that an exact frame is biorthonormal to its canonical
following two lemmas.

Let $\{g_k\}_{k\in\mathcal{K}}$ and $\{\tilde{g}_k\}_{k\in\mathcal{K}}$ be canonical dual frar
states that for a fixed $x \in \mathcal{H}$, among all possible ex
$\{c_k\}_{k\in\mathcal{K}}$ satisfying $x = \sum_{k\in\mathcal{K}} c_k g_k$, the coefficients
l^2-norm.

Lemma 20.3.2 ([5]). Let $\{g_k\}_{k\in\mathcal{K}}$ be a frame fo
$\{\tilde{g}_k\}_{k\in\mathcal{K}}$ its canonical dual frame. For a fixed $x \in$
$x = \sum_{k\in\mathcal{K}} c_k g_k$. If it is possible to find scalars $\{a$
$x = \sum_{k\in\mathcal{K}} a_k g_k$, then we must have

$$\sum_{k\in\mathcal{K}} |a_k|^2 = \sum_{k\in\mathcal{K}} |c_k|^2 + \sum_{k\in\mathcal{K}} |c_k -$$

Proof: We have

$$c_k = \langle x|\tilde{g}_k\rangle = \langle x|\mathbb{S}^{-1}g_k\rangle = \langle \mathbb{S}^{-1}x|g$$

with $\tilde{x} = \mathbb{S}^{-1}x$. Therefore,

$$\langle x|\tilde{x}\rangle = \left\langle \sum_{k\in\mathcal{K}} c_k g_k \middle| \tilde{x}\right\rangle = \sum_{k\in\mathcal{K}} c_k \langle g_k|\tilde{x}\rangle = \sum_{k\in}$$

and

$$\langle x|\tilde{x}\rangle = \left\langle \sum_{k\in\mathcal{K}} a_k g_k \middle| \tilde{x}\right\rangle = \sum_{k\in\mathcal{K}} a_k \langle g_k|\tilde{x})$$

We can therefore conclude that

$$\sum_{k\in\mathcal{K}} |c_k|^2 = \sum_{k\in\mathcal{K}} a_k c_k^* = \sum_{k\in\mathcal{K}} a$$

Hence,

$$\sum_{k \in \mathcal{K}} |c_k|^2 + \sum_{k \in \mathcal{K}} |c_k - a_k|^2 = \sum_{k \in \mathcal{K}} |c_k|^2 + \sum_{k \in \mathcal{K}} (c_k - a_k)(c_k^* - a_k^*)$$

$$= \sum_{k \in \mathcal{K}} |c_k|^2 + \sum_{k \in \mathcal{K}} |c_k|^2 - \sum_{k \in \mathcal{K}} c_k a_k^* - \sum_{k \in \mathcal{K}} c_k^* a_k + \sum_{k \in \mathcal{K}} |a_k|^2 .$$

Using (20.57), we get

$$\sum_{k \in \mathcal{K}} |c_k|^2 + \sum_{k \in \mathcal{K}} |c_k - a_k|^2 = \sum_{k \in \mathcal{K}} |a_k|^2 .$$

$\square$

Note that this lemma implies $\sum_{k \in \mathcal{K}} |a_k|^2 > \sum_{k \in \mathcal{K}} |c_k|^2$, i.e., the coefficient sequence $\{a_k\}_{k \in \mathcal{K}}$ has larger l^2-norm than the coefficient sequence $\{c_k = \langle x | \tilde{g}_k \rangle\}_{k \in \mathcal{K}}$.

Lemma 20.3.3 ([5]). Let $\{g_k\}_{k \in \mathcal{K}}$ be a frame for the Hilbert space $\mathcal{H}$ and $\{\tilde{g}_k\}_{k \in \mathcal{K}}$ its canonical dual frame. Then for each $m \in \mathcal{K}$, we have

$$\sum_{k \neq m} |\langle g_m | \tilde{g}_k \rangle|^2 = \frac{1 - |\langle g_m | \tilde{g}_m \rangle|^2 - |1 - \langle g_m | \tilde{g}_m \rangle|^2}{2} .$$

Proof: We can represent g_m in two different ways. Obviously $g_m = \sum_{k \in \mathcal{K}} a_k g_k$ with $a_m = 1$ and $a_k = 0$ for $k \neq m$, so that $\sum_{k \in \mathcal{K}} |a_k|^2 = 1$. Furthermore, we can write $g_m = \sum_{k \in \mathcal{K}} c_k g_k$ with $c_k = \langle g_m | \tilde{g}_k \rangle$. From (20.56) it then follows that

$$1 = \sum_{k \in \mathcal{K}} |a_k|^2 = \sum_{k \in \mathcal{K}} |c_k|^2 + \sum_{k \in \mathcal{K}} |c_k - a_k|^2$$

$$= \sum_{k \in \mathcal{K}} |c_k|^2 + |c_m - a_m|^2 + \sum_{k \neq m} |c_k - a_k|^2$$

$$= \sum_{k \in \mathcal{K}} |\langle g_m | \tilde{g}_k \rangle|^2 + |\langle g_m | \tilde{g}_m \rangle - 1|^2 + \sum_{k \neq m} |\langle g_m | \tilde{g}_k \rangle|^2$$

$$= 2 \sum_{k \neq m} |\langle g_m | \tilde{g}_k \rangle|^2 + |\langle g_m | \tilde{g}_m \rangle|^2 + |1 - \langle g_m | \tilde{g}_m \rangle|^2$$

and hence

$$\sum_{k \neq m} |\langle g_m | \tilde{g}_k \rangle|^2 = \frac{1 - |\langle g_m | \tilde{g}_m \rangle|^2 - |1 - \langle g_m | \tilde{g}_m \rangle|^2}{2} .$$

$\square$

We are now able to formulate an equivalent condition for a frame to be exact.

Theorem 20.3.13 ([5]). *Let* $\{g_k\}_{k \in \mathcal{K}}$ *be a frame for the Hilbert space* $\mathcal{H}$ *and* $\{\tilde{g}_k\}_{k \in \mathcal{K}}$ *its canonical dual frame. Then,*

1. $\{g_k\}_{k\in\mathcal{K}}$ *is exact if and only if* $\langle g_m|\tilde{g}_m\rangle = 1$ *for*

2. $\{g_k\}_{k\in\mathcal{K}}$ *is inexact if and only if there exists a*
   $\langle g_m|\tilde{g}_m\rangle \neq 1$.

Proof: We first show that if $\langle g_m|\tilde{g}_m\rangle = 1$ for all m
plete for $\mathcal{H}$ (for all $m \in \mathcal{K}$) and hence $\{g_k\}_{k\in\mathcal{K}}$ is ar
fix an arbitrary $m \in \mathcal{K}$. From Lemma 20.3.3 we have

$$\sum_{k\neq m} |\langle g_m|\tilde{g}_k\rangle|^2 = \frac{1 - |\langle g_m|\tilde{g}_m\rangle|^2 - |1}{2}$$

Since $\langle g_m|\tilde{g}_m\rangle = 1$, we have $\sum_{k\neq m} |\langle g_m|\tilde{g}_k\rangle|^2 = 0$ so
for all $k \neq m$. But $\tilde{g}_m \neq 0$ since $\langle g_m|\tilde{g}_m\rangle = 1$. There
for $\mathcal{H}$, because $\tilde{g}_m \neq 0$ is orthogonal to all elements c

Next, we show that if there exists at least one m
then $\{g_k\}_{k\in\mathcal{K}}$ is inexact. More specifically, we will
frame for $\mathcal{H}$ if $\langle g_m|\tilde{g}_m\rangle \neq 1$. We start by noting that

$$g_m = \sum_{k\in\mathcal{K}} \langle g_m, \tilde{g}_k\rangle g_k = \langle g_m, \tilde{g}_m\rangle g_m + \sum_k$$

If $\langle g_m|\tilde{g}_m\rangle \neq 1$, (20.58) can be rewritten as

$$g_m = \frac{1}{1 - \langle g_m|\tilde{g}_m\rangle} \sum_{k\neq m} \langle g_m|\tilde{g}_k$$

and for every $x \in \mathcal{H}$ we have

$$|\langle x|g_m\rangle|^2 = \left|\frac{1}{1 - \langle g_m|\tilde{g}_m\rangle}\right|^2 \left|\sum_{k\neq m} \langle g_m|\tilde{g}_k\rangle \langle x|\right.$$

$$\leq \frac{1}{|1 - \langle g_m|\tilde{g}_m\rangle|^2} \left[\sum_{k\neq m} |\langle g_m|\tilde{g}_k\rangle|^2\right.$$

Therefore

$$\sum_{k\in\mathcal{K}} |\langle x|g_k\rangle|^2 = |\langle x|g_m\rangle|^2 + \sum_{k\neq m} |\langle x|g_k\rangle|^2$$

$$\leq \frac{1}{\left|1 - \left\langle g_m\middle|\tilde{g}_m\right\rangle\right|^2} \left[\sum_{k\neq m} |\langle g_m|\tilde{g}_k\rangle|^2\right]\left[\sum_{k}\right.$$

$$= \sum_{k \neq m} |\langle x | g_k \rangle|^2 \underbrace{\left[1 + \frac{1}{|1 - \langle g_m | \tilde{g}_m \rangle|^2} \sum_{k \neq m} |\langle g_m | \tilde{g}_k \rangle|^2 \right]}_{C}$$

$$= C \sum_{k \neq m} |\langle x | g_k \rangle|^2$$

or equivalently

$$\frac{1}{C} \sum_{k \in \mathcal{K}} |\langle x | g_k \rangle|^2 \leq \sum_{k \neq m} |\langle x | g_k \rangle|^2 .$$

With (20.36) it follows that

$$\frac{A}{C} \|x\|^2 \leq \frac{1}{C} \sum_{k \in \mathcal{K}} |\langle x | g_k \rangle|^2 \leq \sum_{k \neq m} |\langle x | g_k \rangle|^2 \leq \sum_{k \in \mathcal{K}} |\langle x | g_k \rangle|^2 \leq B \|x\|^2, \qquad (20.59)$$

where A and B are the frame bounds of the frame $\{g_k\}_{k \in \mathcal{K}}$. Note that (trivially) $C > 0$; moreover $C < \infty$ since $\langle g_m | \tilde{g}_m \rangle \neq 1$ and $\sum_{k \neq m} |\langle g_m | \tilde{g}_k \rangle|^2 < \infty$ as a consequence of $\{\tilde{g}_k\}_{k \in \mathcal{K}}$ being a frame for $\mathcal{H}$. This implies that $A/C > 0$, and, therefore, (20.59) shows that $\{g_k\}_{k \neq m}$ is a frame with frame bounds A/C and B.

To see that, conversely, exactness of $\{g_k\}_{k \in \mathcal{K}}$ implies that $\langle g_m | \tilde{g}_m \rangle = 1$ for all $m \in \mathcal{K}$, we suppose that $\{g_k\}_{k \in \mathcal{K}}$ is exact and $\langle g_m | \tilde{g}_m \rangle \neq 1$ for at least one $m \in \mathcal{K}$. But the condition $\langle g_m | \tilde{g}_m \rangle \neq 1$ for at least one $m \in \mathcal{K}$ implies that $\{g_k\}_{k \in \mathcal{K}}$ is inexact, which results in a contradiction. It remains to show that $\{g_k\}_{k \in \mathcal{K}}$ inexact implies $\langle g_m | \tilde{g}_m \rangle \neq 1$ for at least one $m \in \mathcal{K}$. Suppose that $\{g_k\}_{k \in \mathcal{K}}$ is inexact and $\langle g_m | \tilde{g}_m \rangle = 1$ for all $m \in \mathcal{K}$. But the condition $\langle g_m | \tilde{g}_m \rangle = 1$ for all $m \in \mathcal{K}$ implies that $\{g_k\}_{k \in \mathcal{K}}$ is exact, which again results in a contradiction. □

Now we are ready to state the two main results of this section. The first result generalizes the biorthonormality relation (20.20) to the infinite-dimensional setting.

Corollary 20.3.1 ([5]). Let $\{g_k\}_{k \in \mathcal{K}}$ be a frame for the Hilbert space $\mathcal{H}$. If $\{g_k\}_{k \in \mathcal{K}}$ is exact, then $\{g_k\}_{k \in \mathcal{K}}$ and its canonical dual $\{\tilde{g}_k\}_{k \in \mathcal{K}}$ are biorthonormal, i.e.,

$$\langle g_m | \tilde{g}_k \rangle = \begin{cases} 1, & \text{if } k = m \\ 0, & \text{if } k \neq m. \end{cases}$$

Conversely, if $\{g_k\}_{k \in \mathcal{K}}$ and $\{\tilde{g}_k\}_{k \in \mathcal{K}}$ are biorthonormal, then $\{g_k\}_{k \in \mathcal{K}}$ is exact.

Proof: If $\{g_k\}_{k \in \mathcal{K}}$ is exact, then biorthonormality follows by noting that Theorem 20.3.13 implies $\langle g_m | \tilde{g}_m \rangle = 1$ for all $m \in \mathcal{K}$, and Lemma 20.3.3 implies $\sum_{k \neq m} |\langle g_m | \tilde{g}_k \rangle|^2 = 0$ for all $m \in \mathcal{K}$ and thus $\langle g_m | \tilde{g}_k \rangle = 0$ for all $k \neq m$. To

show that, conversely, biorthonormality of $\{g_k\}_{k \in \mathcal{K}}$ a
frame $\{g_k\}_{k \in \mathcal{K}}$ is exact, we simply note that $\langle g_m |$
Theorem 20.3.13, implies that $\{g_k\}_{k \in \mathcal{K}}$ is exact.

The second main result in this section states that
frame is unique and, therefore, the concept of an exac
basis to infinite-dimensional spaces.

Theorem 20.3.14 ([5]). *If $\{g_k\}_{k \in \mathcal{K}}$ is an exact fr*
and $x = \sum_{k \in \mathcal{K}} c_k g_k$ with $x \in \mathcal{H}$, then the coefficients
given by

$$c_k = \langle x | \tilde{g}_k \rangle,$$

where $\{\tilde{g}_k\}_{k \in \mathcal{K}}$ is the canonical dual frame to $\{g_k\}_{k \in}$

Proof: We know from (20.50) that x can be written
assume that there is another set of coefficients $\{c_k\}_{k \in}$

$$x = \sum_{k \in \mathcal{K}} c_k g_k.$$

Taking the inner product of both sides of (20.60) with
mality relation

$$\langle g_k | \tilde{g}_m \rangle = \begin{cases} 1, & k = m \\ 0, & k \neq m \end{cases}$$

we obtain

$$\langle x | \tilde{g}_m \rangle = \sum_{k \in \mathcal{K}} c_k \langle g_k | \tilde{g}_m \rangle = c$$

Thus, $c_m = \langle x | \tilde{g}_m \rangle$ for all $m \in \mathcal{K}$ and the proof is con

20.4   The Sampling Theorem

We now discuss one of the most important results in
pling theorem. We will then show how the sampling
as a frame decomposition.

Consider a signal $x(t)$ in the space of square-integra
we cannot expect this signal to be uniquely specified
where T is the sampling period. The sampling theore
signal is strictly bandlimited, i.e., its Fourier transfor
finite interval, and if T is chosen small enough (relativ
then the samples $\{x(kT)\}_{k \in \mathbb{Z}}$ do uniquely specify the s
$x(t)$ from $\{x(kT)\}_{k \in \mathbb{Z}}$ perfectly. The process of obtair
from the continuous-time signal $x(t)$ is called A/D

---

[5]Strictly speaking A/D conversion also involves quantizatior

reconstruction of the signal $x(t)$ from its samples is called digital-to-analog (D/A) conversion. We shall now formally state and prove the sampling theorem.

Let $\widehat{x}(f)$ denote the Fourier transform of the signal $x(t)$, i.e.,

$$\widehat{x}(f) = \int_{-\infty}^{\infty} x(t)e^{-\mathrm{i}2\pi tf}\,dt.$$

We say that $x(t)$ is bandlimited to B Hz if $\widehat{x}(f) = 0$ for $|f| > B$. Note that this implies that the total bandwidth of $x(t)$, counting positive and negative frequencies, is $2B$. The Hilbert space of $\mathcal{L}^2$ functions that are bandlimited to B Hz is denoted as $\mathcal{L}^2(B)$.

Next, consider the sequence of samples $\{x[k] \triangleq x(kT)\}_{k \in \mathbb{Z}}$ of the signal $x(t) \in \mathcal{L}^2(B)$ and compute its discrete-time Fourier transform (DTFT):

$$\widehat{x}_d(f) \triangleq \sum_{k=-\infty}^{\infty} x[k]e^{-\mathrm{i}2\pi kf}$$

$$= \sum_{k=-\infty}^{\infty} x(kT)e^{-\mathrm{i}2\pi kf}$$

$$= \frac{1}{T}\sum_{k=-\infty}^{\infty} \widehat{x}\left(\frac{f+k}{T}\right), \tag{20.61}$$

where in the last step we used the Poisson summation formula[6] [25, Cor. 2.6].

We can see that $\widehat{x}_d(f)$ is simply a periodized version of $\widehat{x}(f)$. Now, it follows that for $1/T \geq 2B$ there is no overlap between the shifted replica of $\widehat{x}(f/T)$, whereas for $1/T < 2B$, we do get the different shifted versions to overlap (see Figure 20.5). We can therefore conclude that for $1/T \geq 2B$, $\widehat{x}(f)$ can be recovered exactly from $\widehat{x}_d(f)$ by means of applying an ideal lowpass filter with gain T and cutoff frequency BT to $\widehat{x}_d(f)$. Specifically, we find that

$$\widehat{x}(f/T) = \widehat{x}_d(f)\,T\,\widehat{h}_{\mathrm{LP}}(f) \tag{20.62}$$

with

$$\widehat{h}_{\mathrm{LP}}(f) = \begin{cases} 1, & |f| \leq BT \\ 0, & \text{otherwise.} \end{cases} \tag{20.63}$$

From (20.62), using (20.61), we immediately see that we can recover the Fourier transform of $x(t)$ from the sequence of samples $\{x[k]\}_{k \in \mathbb{Z}}$ according to

$$\widehat{x}(f) = T\,\widehat{h}_{\mathrm{LP}}(fT) \sum_{k=-\infty}^{\infty} x[k]e^{-\mathrm{i}2\pi kfT}. \tag{20.64}$$

---

[6] Let $x(t) \in \mathcal{L}^2$ with Fourier transform $\widehat{x}(f) = \int_{-\infty}^{\infty} x(t)e^{-\mathrm{i}2\pi tf}\,dt$. The Poisson summation formula states that $\sum_{k=-\infty}^{\infty} x(k) = \sum_{k=-\infty}^{\infty} \widehat{x}(k)$.

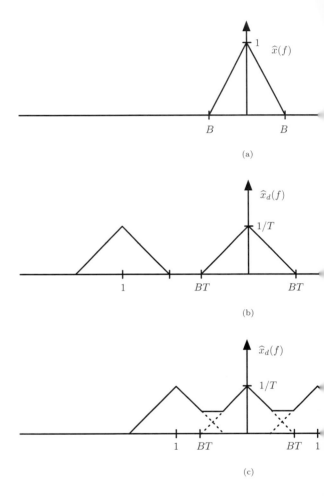

Figure 20.5: Sampling of a signal that is band-limite
the original signal; (b) spectrum of the sampled signal
of the sampled signal for $1/T < 2B$, where aliasing oc

We can therefore recover $x(t)$ as follows:

$$
\begin{aligned}
x(t) &= \int_{-\infty}^{\infty} \widehat{x}(f) e^{\mathrm{i}2\pi t f}\, df \\
&= \int_{-\infty}^{\infty} T \widehat{h}_{\mathrm{LP}}(fT) \sum_{k=-\infty}^{\infty} x[k] e^{-\mathrm{i}2\pi k fT} e^{\mathrm{i}2\pi f t}\, df \\
&= \sum_{k=-\infty}^{\infty} x[k] \int_{-\infty}^{\infty} \widehat{h}_{\mathrm{LP}}(fT) e^{\mathrm{i}2\pi fT(t/T-k)}\, d(fT) \\
&= \sum_{k=-\infty}^{\infty} x[k] h_{\mathrm{LP}}\left(\frac{t}{T} - k\right) \\
&= 2BT \sum_{k=-\infty}^{\infty} x[k]\, \mathrm{sinc}(2B(t - kT)),
\end{aligned}
\tag{20.65}
$$

where $h_{\mathrm{LP}}(t)$ is the inverse Fourier transform of $\widehat{h}_{\mathrm{LP}}(f)$, i.e,

$$
h_{\mathrm{LP}}(t) = \int_{-\infty}^{\infty} \widehat{h}_{\mathrm{LP}}(f) e^{\mathrm{i}2\pi t f}\, df,
$$

and

$$
\mathrm{sinc}(x) \triangleq \frac{\sin(\pi x)}{\pi x}.
$$

Summarizing our findings, we obtain the following theorem.

Theorem 20.4.1 (Sampling theorem [26, Sec. 7.2]). *Let $x(t) \in \mathcal{L}^2(B)$. Then $x(t)$ is uniquely specified by its samples $x(kT)$, $k \in \mathbb{Z}$, if $1/T \geq 2B$. Specifically, we can reconstruct $x(t)$ from $x(kT)$, $k \in \mathbb{Z}$, according to*

$$
x(t) = 2BT \sum_{k=-\infty}^{\infty} x(kT)\, \mathrm{sinc}(2B(t - kT)).
\tag{20.66}
$$

20.4.1 Sampling Theorem as a Frame Expansion

We shall next show how the representation (20.66) can be interpreted as a frame expansion. The samples $x(kT)$ can be written as the inner product of the signal $x(t)$ with the functions

$$
g_k(t) = 2B\, \mathrm{sinc}(2B(t - kT)), \quad k \in \mathbb{Z}.
\tag{20.67}
$$

Indeed, using the fact that the signal $x(t)$ is band-limited to B Hz, we get

$$
x(kT) = \int_{-B}^{B} \widehat{x}(f) e^{\mathrm{i}2\pi k fT}\, df = \langle \widehat{x} | \widehat{g}_k \rangle,
$$

where

$$\widehat{g}_k(f) = \begin{cases} e^{-\mathrm{i}2\pi k f T}, & |f| \leq E \\ 0, & \text{otherwise} \end{cases}$$

is the Fourier transform of $g_k(t)$. We can thus rewrit

$$x(t) = T \sum_{k=-\infty}^{\infty} \langle x|g_k \rangle \, g_k(t$$

Therefore, the interpolation of an analog signal from it interpreted as the reconstruction of $x(t)$ from its ex $\langle x|g_k \rangle$ in the function set $\{g_k(t)\}_{k \in \mathbb{Z}}$. We shall nex frame for the space $\mathcal{L}^2(B)$. Simply note that for $x(t)$

$$\|x\|^2 = \langle x|x \rangle = \left\langle T \sum_{k=-\infty}^{\infty} \langle x|g_k \rangle \, g_k(t) \middle| x \right\rangle =$$

and therefore

$$\frac{1}{T}\|x\|^2 = \sum_{k=-\infty}^{\infty} |\langle x|g_k \rangle|^2$$

This shows that $\{g_k(t)\}_{k \in \mathbb{Z}}$ is, in fact, a tight frame $A = 1/T$. We emphasize that the frame is tight irres (of course, as long as $1/T > 2B$).

The analysis operator corresponding to this frame as

$$\mathbb{T} : x \to \{\langle x|g_k \rangle\}_{k \in \mathbb{Z}},$$

i.e., $\mathbb{T}$ maps the signal $x(t)$ to the sequence of sample

The action of the adjoint of the analysis operator $\mathbb{T}$ interpolation according to

$$\mathbb{T}^* : \{c_k\}_{k \in \mathbb{Z}} \to \sum_{k=-\infty}^{\infty} c_k g_k$$

The frame operator $\mathbb{S} : \mathcal{L}^2(B) \to \mathcal{L}^2(B)$ is given by $\mathbb{S}$

$$\mathbb{S} : x(t) \to \sum_{k=-\infty}^{\infty} \langle x|g_k \rangle \, g_k(t$$

Since $\{g_k(t)\}_{k \in \mathbb{Z}}$ is a tight frame for $\mathcal{L}^2(B)$ with frame shown, it follows that $\mathbb{S} = (1/T)\mathbb{I}_{\mathcal{L}^2(B)}$.

The canonical dual frame can be computed easil the frame operator to the frame functions $\{g_k(t)\}_{k \in \mathbb{Z}}$

$$\tilde{g}_k(t) = \mathbb{S}^{-1} g_k(t) = T\mathbb{I}_{\mathcal{L}^2(B)} g_k(t) = Tg_k$$

Recall that exact frames have a minimality property in the following sense: If we remove any one element from an exact frame, the resulting set will be incomplete. In the case of sampling, we have an analogous situation: In the proof of the sampling theorem we saw that if we sample at a rate smaller than the *critical sampling rate* $1/T = 2B$, we cannot recover the signal $x(t)$ from its samples $\{x(kT)\}_{k\in\mathbb{Z}}$. In other words, the set $\{g_k(t)\}_{k\in\mathbb{Z}}$ in (20.67) is *not* complete for $\mathcal{L}^2(B)$ when $1/T < 2B$. This suggests that critical sampling $1/T = 2B$ could implement an exact frame decomposition. We show now that this is, indeed, the case. Simply note that

$$\langle g_k|\tilde{g}_k\rangle = T\,\langle g_k|g_k\rangle = T\|g_k\|^2 = T\|\widehat{g}_k\|^2 = 2BT, \quad \text{for all } k \in \mathbb{Z}.$$

For critical sampling $2BT = 1$ and, hence, $\langle g_k|\tilde{g}_k\rangle = 1$, for all $k \in \mathbb{Z}$. Theorem 20.3.13 therefore allows us to conclude that $\{g_k(t)\}_{k\in\mathbb{Z}}$ is an exact frame for $\mathcal{L}^2(B)$.

Next, we show that $\{g_k(t)\}_{k\in\mathbb{Z}}$ is not only an exact frame, but, when properly normalized, even an ONB for $\mathcal{L}^2(B)$, a fact well known in sampling theory. To this end, we first renormalize the frame functions $g_k(t)$ according to

$$g'_k(t) = \sqrt{T}g_k(t)$$

so that

$$x(t) = \sum_{k=-\infty}^{\infty} \langle x|g'_k\rangle\, g'_k(t).$$

We see that $\{g'_k(t)\}_{k\in\mathbb{Z}}$ is a tight frame for $\mathcal{L}^2(B)$ with $A = 1$. Moreover, we have

$$\|g'_k\|^2 = T\|g_k\|^2 = 2BT.$$

Thus, in the case of critical sampling, $\|g'_k\|^2 = 1$, for all $k \in \mathbb{Z}$, and Theorem 20.3.11 allows us to conclude that $\{g'_k(t)\}_{k\in\mathbb{Z}}$ is an ONB for $\mathcal{L}^2(B)$.

In contrast to exact frames, inexact frames are redundant, in the sense that there is at least one element that can be removed with the resulting set still being complete. The situation is similar in the *oversampled* case, i.e., when the sampling rate satisfies $1/T > 2B$. In this case, we collect more samples than actually needed for perfect reconstruction of $x(t)$ from its samples. This suggests that $\{g_k(t)\}_{k\in\mathbb{Z}}$ could be an inexact frame for $\mathcal{L}^2(B)$ in the oversampled case. Indeed, according to Theorem 20.3.13 the condition

$$\langle g_m|\tilde{g}_m\rangle = 2BT < 1, \quad \text{for all } m \in \mathbb{Z}, \tag{20.69}$$

implies that the frame $\{g_k(t)\}_{k\in\mathbb{Z}}$ is inexact for $1/T > 2B$. In fact, as can be seen from the proof of Theorem 20.3.13, (20.69) guarantees even more: for *every* $m \in \mathbb{Z}$, the set $\{g_k(t)\}_{k\neq m}$ is complete for $\mathcal{L}^2(B)$. Hence, the removal of *any* sample $x(mT)$ from the set of samples $\{x(kT)\}_{k\in\mathbb{Z}}$ still leaves us with a frame decomposition so that $x(t)$ can, in theory, be recovered from the samples $\{x(kT)\}_{k\neq m}$. The resulting frame $\{g_k(t)\}_{k\neq m}$ will, however, no longer be tight, which makes the computation of the canonical dual frame complicated, in general.

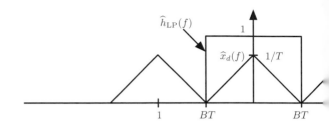

Figure 20.6: Reconstruction filter in the crit

20.4.2   Design Freedom in Oversample

In the critically sampled case, $1/T = 2B$, the ideal lov
with the transfer function specified in (20.63) is the on
reconstruction of the spectrum $\widehat{x}(f)$ of $x(t)$ according
In the oversampled case, there is, in general, an infin
filters that provide perfect reconstruction. The only re
filter has to satisfy is that its transfer function be co
range $-BT \leq f \leq BT$ (see Figure 20.7). Therefore,
has more freedom in designing the reconstruction filte
this design freedom is exploited to design reconstructi
characteristics, like, e.g., rolloff in the transfer functio
Specifically, repeating the steps leading from (20.6

$$x(t) = \sum_{k=-\infty}^{\infty} x[k] h\left(\frac{t}{T} - k\right.$$

where the Fourier transform of $h(t)$ is given by

$$\widehat{h}(f) = \begin{cases} 1, & |f| \leq BT \\ \mathrm{arb}(f), & BT < |f| \\ 0, & |f| > \frac{1}{2} \end{cases}$$

Here and in what follows arb($\cdot$) denotes an arbitrary
words, every set $\{h(t/T - k)\}_{k \in \mathbb{Z}}$ with the Fourie
ing (20.71) is a valid dual frame for the frame $\{g_k(t$
Obviously, there are infinitely many dual frames in th
We next show how the freedom in the design of t
transfer function specified in (20.71) can be interprete
choosing the left-inverse $\mathbb{L}$ of the analysis operator $\mathbb{T}$ a
Recall the parametrization (20.52) of all left-inverses

$$\mathbb{L} = \widetilde{\mathbb{T}}^* \mathbb{P} + \mathbb{M}(\mathbb{I}_{l^2} - \mathbb{P}),$$

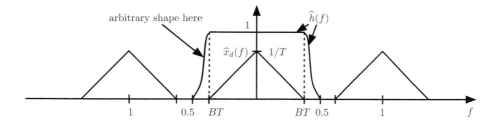

Figure 20.7: Freedom in the design of the reconstruction filter.

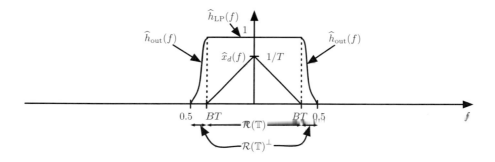

Figure 20.8: The reconstruction filter as a parametrized left-inverse of the analysis operator.

where $\mathbb{M} : l^2 \to \mathcal{H}$ is an arbitrary bounded linear operator and $\mathbb{P} : l^2 \to l^2$ is the orthogonal projection operator onto the range space of $\mathbb{T}$. In (20.61) we saw that the DTFT[7] of the sequence $\big\{x[k] = x(kT)\big\}_{k \in \mathbb{Z}}$ is compactly supported on the frequency interval $[-BT, BT]$ (see Figure 20.8). In other words, the range space of the analysis operator $\mathbb{T}$ defined in (20.68) is the space of l^2-sequences with DTFT supported on the interval $[-BT, BT]$ (see Figure 20.8). It is left as an exercise to the reader to verify (see Exercise 20.6.7), using Parseval's theorem[8], that the orthogonal complement of the range space of $\mathbb{T}$ is the space of l^2-sequences with DTFT supported on the set $[-1/2, -BT] \cup [BT, 1/2]$ (see Figure 20.8). Thus, in the case of oversampled A/D conversion, the operator $\mathbb{P} : l^2 \to l^2$ is the orthogonal

---

[7]The DTFT is a periodic function with period one. From here on, we consider the DTFT as a function supported on its fundamental period $[-1/2, 1/2]$.

[8] Consider the sequences $\{a_k\}_{k \in \mathbb{Z}}, \{b_k\}_{k \in \mathbb{Z}} \in l^2$ with DTFT $\widehat{a}(f) = \sum_{k=-\infty}^{\infty} a_k e^{-i2\pi k f}$ and $\widehat{b}(f) = \sum_{k=-\infty}^{\infty} b_k e^{-i2\pi k f}$, respectively. Parseval's theorem states the following result: $\sum_{k=-\infty}^{\infty} a_k b_k^* = \int_{-1/2}^{1/2} \widehat{a}(f) \widehat{b}^*(f) df$. In particular, $\sum_{k=-\infty}^{\infty} |a_k|^2 = \int_{-1/2}^{1/2} |\widehat{a}(f)|^2 df$.

projection operator onto the subspace of l^2-sequence
the interval $[-BT, BT]$; the operator $(\mathbb{I}_{l^2} - \mathbb{P}) : l^2 \to$
tion operator onto the subspace of l^2-sequences with
$[-1/2, -BT] \cup [BT, 1/2]$.

To see the parallels between (20.70) and (20.72
follows (see Figure 20.8)

$$h(t) = h_{\text{LP}}(t) + h_{\text{out}}(t),$$

where the Fourier transform of $h_{\text{LP}}(t)$ is given by (20.6
of $h_{\text{out}}(t)$ is

$$\widehat{h}_{\text{out}}(f) = \begin{cases} \text{arb}(f), & BT \leq \\ 0, & \text{otherwise.} \end{cases}$$

Now it is clear, and it is left to the reader to verify fo
that the operator $\mathbb{A} : l^2 \to \mathcal{L}^2(B)$ defined as

$$\mathbb{A} : \{c_k\}_{k \in \mathbb{Z}} \to \sum_{k=-\infty}^{\infty} c_k h_{\text{LP}}\left(\frac{t}{T}\right)$$

acts by first projecting the sequence $\{c_k\}_{k \in \mathbb{Z}}$ onto t
with DTFT supported on the interval $[-BT, BT]$ and
using the canonical dual frame elements $\tilde{g}_k(t) = h_{\text{L}}$
$\mathbb{A} = \tilde{\mathbb{T}}^*\mathbb{P}$. Similarly, it is left to the reader to verify fo
that the operator $\mathbb{B} : l^2 \to \mathcal{L}^2(B)$ defined as

$$\mathbb{B} : \{c_k\}_{k \in \mathbb{Z}} \to \sum_{k=-\infty}^{\infty} c_k h_{\text{out}}\left(\frac{t}{T}\right)$$

can be written as $\mathbb{B} = \mathbb{M}(\mathbb{I}_{l^2} - \mathbb{P})$. Here, $(\mathbb{I}_{l^2} - \mathbb{P}) : l^2 \to$
onto the subspace of l^2-sequences with DTFT support
$[BT, 1/2]$; the operator $\mathbb{M} : l^2 \to \mathcal{L}^2$ is defined as

$$\mathbb{M} : \{c_k\}_{k \in \mathbb{Z}} \to \sum_{k=-\infty}^{\infty} c_k h_M\left(\frac{t}{T}\right)$$

with the Fourier transform of $h_M(t)$ given by

$$\widehat{h}_M(f) = \begin{cases} \text{arb}_2(f), & -\frac{1}{2} \leq |f \\ 0, & \text{otherwis} \end{cases}$$

where $\text{arb}_2(f)$ is an arbitrary bounded function that e
$\frac{1}{2}$. To summarize, we note that the operator $\mathbb{B}$ corres
the right-hand side of (20.72).

We can therefore write the decomposition (20.70) as

$$
x(t) = \sum_{k=-\infty}^{\infty} x[k] h\left(\frac{t}{T} - k\right)
$$

$$
= \underbrace{\sum_{k=-\infty}^{\infty} x[k] h_{\mathrm{LP}}\left(\frac{t}{T} - k\right)}_{\tilde{\mathbb{T}}^* \mathbb{P} \mathbb{T} x(t)} + \underbrace{\sum_{k=-\infty}^{\infty} x[k] h_{\mathrm{out}}\left(\frac{t}{T} - k\right)}_{\mathbb{M}(\mathbb{I}_{l2} - \mathbb{P}) \mathbb{T} x(t)}
$$

$$
= \mathbb{L} \mathbb{T} x(t).
$$

20.4.3   Noise Reduction in Oversampled A/D Conversion

Consider again the bandlimited signal $x(t) \in \mathcal{L}^2(B)$. Assume, as before, that the signal is sampled at a rate $1/T \geq 2B$. Now assume that the corresponding samples $x[k] = x(kT)$, $k \in \mathbb{Z}$, are subject to noise, i.e., we observe

$$
x'[k] = x[k] + w[k], \ k \in \mathbb{Z},
$$

where the $w[k]$ are independent identically distributed zero-mean random variables, with variance $\mathbb{E}\,|w[k]|^2 = \sigma^2$. Assume that reconstruction is performed from the noisy samples $x'[k]$, $k \in \mathbb{Z}$, using the ideal lowpass filter with transfer function $\hat{h}_{\mathrm{LP}}(f)$ of bandwidth BT specified in (20.63), i.e., we reconstruct using the canonical dual frame according to

$$
x'(t) = \sum_{k=-\infty}^{\infty} x'[k] h_{\mathrm{LP}}\left(\frac{t}{T} - k\right).
$$

Obviously, the presence of noise precludes perfect reconstruction. It is, however, interesting to assess the impact of oversampling on the variance of the reconstruction error defined as

$$
\sigma_{\mathrm{oversampling}}^2 \triangleq \mathbb{E}_w\,|x(t) - x'(t)|^2, \tag{20.79}
$$

where the expectation is with respect to the random variables $w[k]$, $k \in \mathbb{Z}$, and the right-hand side of (20.79) does not depend on t, as we shall see below. If we decompose $x(t)$ as in (20.65), we see that

$$
\sigma_{\mathrm{oversampling}}^2 = \mathbb{E}_w\,|x(t) - x'(t)|^2 \tag{20.80}
$$

$$
= \mathbb{E}_w \left| \sum_{k=-\infty}^{\infty} w[k] h_{\mathrm{LP}}\left(\frac{t}{T} - k\right) \right|^2
$$

$$
= \sum_{k=-\infty}^{\infty} \sum_{k'=-\infty}^{\infty} \mathbb{E}_w\{w[k] w^*[k']\} h_{\mathrm{LP}}\left(\frac{t}{T} - k\right) h_{\mathrm{LP}}^*\left(\frac{t}{T} - k'\right)
$$

$$
= \sigma^2 \sum_{k=-\infty}^{\infty} \left| h_{\mathrm{LP}}\left(\frac{t}{T} - k\right) \right|^2. \tag{20.81}
$$

Next applying the Poisson summation formula (as
chapter) to the function $l(t') \triangleq h_{\mathrm{LP}}\left(\frac{t}{T} - t'\right) e^{-2\pi i t' f}$ w
$\widehat{h}_{\mathrm{LP}}(-f - f')e^{-2\pi i(t/T)(f+f')}$, we have

$$\sum_{k=-\infty}^{\infty} h_{\mathrm{LP}}\left(\frac{t}{T} - k\right) e^{-2\pi i k f} = \sum_{k=-\infty}^{\infty} l(k)$$

$$= \sum_{k=-\infty}^{\infty} \widehat{l}(k) = \sum_{k=-\infty}^{\infty} \widehat{h}$$

Since $\widehat{h}_{\mathrm{LP}}(f)$ is zero outside the interval $-1/2 \le f \le$

$$\sum_{k=-\infty}^{\infty} \widehat{h}_{\mathrm{LP}}(-f - k)e^{-2\pi i(t/T)(f+k)} = \widehat{h}_{\mathrm{LP}}(-f)e^{-2\pi i(}$$

We conclude from (20.82) and (20.83) that the DT
$h_{\mathrm{LP}}(t/T - k)\}_{k\in\mathbb{Z}}$ is given (in the fundamental in
$\widehat{h}_{\mathrm{LP}}(-f)e^{-2\pi i(t/T)f}$ and hence we can apply Parseval'
note 8 in this chapter) and rewrite (20.81) according

$$\sigma_{\mathrm{oversampling}}^2 = \sigma^2 \sum_{k=-\infty}^{\infty} \left|h_{\mathrm{LP}}\left(\frac{t}{T} - k\right)\right|^2 = \sigma^2 \int_{-1/}^{1/2}$$

$$= \sigma^2 \int_{-1/2}^{1/2} \left|\widehat{h}_{\mathrm{LP}}(f)\right|^2 df = \sigma^2 2BT.$$

We see that the average mean squared reconstruction ϵ
to the oversampling factor $1/(2BT)$. Therefore, each
factor decreases the mean squared error by 3 dB.
    Consider now reconstruction performed using a ge
fect reconstruction in the noiseless case. Specifically,

$$x'(t) = \sum_{k=-\infty}^{\infty} x'[k]h\left(\frac{t}{T} - k\right.$$

where $h(t)$ is given by (20.73). In this case, the avera
tion error can be computed repeating the steps leading
is given by

$$\sigma_{\mathrm{oversampling}}^2 = \sigma^2 \int_{-1/2}^{1/2} \left|\widehat{h}(f)\right|$$

where $\widehat{h}(f)$ is the Fourier transform of $h(t)$ and i
ing (20.73), we can now decompose $\sigma_{\mathrm{oversampling}}^2$ in (20.

to

$$\sigma^2_{\text{oversampling}} = \sigma^2 \underbrace{\int_{-BT}^{BT} \left| \widehat{h}_{\text{LP}}(f) \right|^2 df}_{2BT} + \sigma^2 \int_{BT \leq |f| \leq 1/2} \left| \widehat{h}_{\text{out}}(f) \right|^2 df. \qquad (20.86)$$

We see that two components contribute to the reconstruction error. Comparing (20.86) to (20.84), we conclude that the first term in (20.86) corresponds to the error due to noise in the signal-band $|f| \leq BT$ picked up by the ideal low-pass filter with transfer function $\widehat{h}_{\text{LP}}(f)$. The second term in (20.86) is due to the fact that a generalized inverse passes some of the noise in the out-of-band region $BT \leq |f| \leq 1/2$. The amount of additional noise in the reconstructed signal is determined by the bandwidth and the shape of the reconstruction filter's transfer function in the out-of-band region. In this sense, there exists a tradeoff between noise reduction and design freedom in oversampled A/D conversion. Practically desirable (or realizable) reconstruction filters (i.e., filters with rolloff) lead to additional reconstruction error.

20.5 Important Classes of Frames

There are two important classes of structured signal expansions that have found widespread use in practical applications, namely Weyl-Heisenberg (or Gabor) expansions and affine (or wavelet) expansions. Weyl-Heisenberg expansions provide a decomposition into time-shifted and modulated versions of a "window function" $g(t)$. Wavelet expansions realize decompositions into time-shifted and dilated versions of a mother wavelet $g(t)$. Thanks to the strong structural properties of Weyl-Heisenberg and wavelet expansions, there are efficient algorithms for applying the corresponding analysis and synthesis operators. Weyl-Heisenberg and wavelet expansions have been successfully used in signal detection, image representation, object recognition, and wireless communications. We shall next show that these signal expansions can be cast into the language of frame theory. For a detailed analysis of these classes of frames, we refer the interested reader to [4].

20.5.1 Weyl-Heisenberg Frames

We start by defining a linear operator that realizes time-frequency shifts when applied to a given function.

Definition 20.5.1. The Weyl operator $\mathbb{W}_{m,n}^{(T,F)} : \mathcal{L}^2 \to \mathcal{L}^2$ is defined as

$$\mathbb{W}_{m,n}^{(T,F)} : x(t) \to e^{i2\pi nFt} x(t - mT),$$

where $m, n \in \mathbb{Z}$, and $T > 0$ and $F > 0$ are fixed time and frequency shift parameters, respectively.

Now consider some prototype (or window) funct
rameters $T > 0$ and $F > 0$. By shifting the wind
integer multiples of T and in frequency by integer mu
structured set of functions according to

$$g_{m,n}(t) \triangleq \mathbb{W}_{m,n}^{(T,F)} g(t) = e^{i2\pi nFt} g(t - mT),$$

The set $\left\{ g_{m,n}(t) = e^{i2\pi nFt} g(t - mT) \right\}_{m \in \mathbb{Z},\, n \in \mathbb{Z}}$ is refe
(WH) set and is denoted by (g, T, F). When the WH
$\mathcal{L}^2$, it is called a *WH frame* for $\mathcal{L}^2$.

Whether or not a WH-set (g, T, F) is a frame for
answer. The answer depends on the window functior
parameters T and F. Intuitively, if the parameters T
given window function $g(t)$, the WH set (g, T, F) can
is because a WH set (g, T, F) with "large" paramete
the time-frequency plane" or equivalently in the Hil
intuition is correct and the following fundamental res

Theorem 20.5.1 ([21, Thm. 8.3.1]). *Let $g(t) \in$
Then the following holds:*

- *If $TF > 1$, then (g, T, F) is not a frame for $\mathcal{L}^2$*

- *If (g, T, F) is a frame for $\mathcal{L}^2$, then (g, T, F) is*
  $TF = 1$.

We see that (g, T, F) can be a frame for $\mathcal{L}^2$ only if
parameters T and F are such that the grid they induce
is sufficiently dense. Whether or not a WH set $(g, T,$
for $\mathcal{L}^2$ depends on the window function $g(t)$ and on th
is an important special case where a simple answer ca

Example 20.5.1 (Gaussian, [21, Thm. 8.6.1]). I
e^{-t^2}. Then the Weyl-Heisenberg (WH) set

$$\left\{ \mathbb{W}_{m,n}^{(T,F)} g(t) \right\}_{m \in \mathbb{Z},\, n \in \mathbb{Z}}$$

is a frame for $\mathcal{L}^2$ if and only if $TF < 1$.

20.5.2   Wavelets

Both for wavelet frames and WH frames we deal wit
tained by letting a special class of parametrized opera
In the case of WH frames, the underlying operator
shifts. In the case of wavelets, the generating opera
scaling. Specifically, we have the following definition.

Definition 20.5.2. The operator $\mathbb{V}_{m,n}^{(T,S)} : \mathcal{L}^2 \to \mathcal{L}^2$ is defined as

$$\mathbb{V}_{m,n}^{(T,S)} : x(t) \to S^{n/2}x(S^nt - mT),$$

where $m, n \in \mathbb{Z}$, and $T > 0$ and $S > 0$ are fixed time and scaling parameters, respectively.

Now, just as in the case of WH expansions, consider a prototype function (or mother wavelet) $g(t) \in \mathcal{L}^2$. Fix the parameters $T > 0$ and $S > 0$ and consider the set of functions

$$g_{m,n}(t) \triangleq \mathbb{V}_{m,n}^{(T,S)}g(t) = S^{n/2}g(S^nt - mT), \quad m \in \mathbb{Z}, \; n \in \mathbb{Z}.$$

This set is referred to as a *wavelet set*. When the wavelet set $\left\{g_{m,n}(t) = S^{n/2}g(S^nt - mT)\right\}_{m\in\mathbb{Z}, \; n\in\mathbb{Z}}$ with parameters $T, S > 0$ is a frame for $\mathcal{L}^2$, it is called a *wavelet frame*.

Similar to the case of Weyl-Heisenberg sets it is hard to say, in general, whether a given wavelet set forms a frame for $\mathcal{L}^2$ or not. The answer depends on the window function $g(t)$ and on the parameters T and S and explicit results are known only in certain cases. We conclude this section by detailing such a case.

Example 20.5.2 (Mexican hat, [21, Ex. 11.2.7]). Take $S = 2$ and consider the mother wavelet

$$g(t) = \frac{2}{\sqrt{3}}\pi^{-1/4}(1 - t^2)e^{-\frac{1}{2}t^2}.$$

Due to its shape, $g(t)$ is called the Mexican hat function. It turns out that for each $T < 1.97$, the wavelet set

$$\left\{\mathbb{V}_{m,n}^{(T,S)}g(t)\right\}_{m\in\mathbb{Z}, \; n\in\mathbb{Z}}$$

is a frame for $\mathcal{L}^2$ [21, Ex. 11.2.7]. □

20.6 Exercises

Exercise 20.6.1 (Tight frames [27]).

1. Prove that if $K \in \mathbb{Z} \setminus \{0\}$, then the set of vectors

$$\left\{\mathbf{g}_k = \begin{bmatrix} 1 & e^{\mathrm{i}2\pi k/(KM)} & \cdots & e^{\mathrm{i}2\pi k(M-1)/(KM)} \end{bmatrix}^{\mathsf{T}}\right\}_{0 \le k < KM}$$

is a tight frame for $\mathbb{C}^M$. Compute the frame bound.

2. Fix $T \in \mathbb{R} \setminus \{0\}$ and define for every $k \in \mathbb{Z}$

$$g_k(t) = \begin{cases} e^{\mathrm{i}2\pi kt/T}, & t \in [0, T] \\ 0, & \text{otherwise.} \end{cases}$$

Prove that $\{g_k(t)\}_{k\in\mathbb{Z}}$ is a tight frame for $\mathcal{L}^2$ integrable functions supported on the interval bound.

Exercise 20.6.2 (Discrete Fourier Transform (
sion). The DFT of an M-point signal $x[k]$, $k = 0, \ldots$

$$\hat{x}[n] = \frac{1}{\sqrt{M}} \sum_{k=0}^{M-1} x[k] e^{-\mathrm{i}2\pi \frac{\cdot}{\cdot}}$$

Find the corresponding inverse transform and show preted as a frame expansion in $\mathbb{C}^M$. Compute the fram frame special?

Exercise 20.6.3 (Unitary transformation of a frame for the Hilbert space $\mathcal{H}$ with frame bounds A unitary operator. Show that the set $\{Ug_k\}_{k\in\mathcal{K}}$ is agair the corresponding frame bounds.

Exercise 20.6.4 (Redundancy of a frame). Let with $N > M$. Assume that the frame vectors are n 1, $k = 1, \ldots, N$. The ratio N/M is called the redund

1. Assume that $\{\mathbf{g}_k\}_{k=1}^N$ is a tight frame with f $A = N/M$.

2. Now assume that A and B are the frame bou $A \leq N/M \leq B$.

Exercise 20.6.5 (Frame bounds [21]). Prove th frame bound are unrelated: In an arbitrary Hilbert with an upper frame bound $B < \infty$ but with the $A = 0$; find another set $\{g_k\}_{k\in\mathcal{K}}$ with lower frame tightest upper frame bound $B = \infty$. Is it possible to in the finite-dimensional space $\mathbb{C}^M$?

Exercise 20.6.6 (Tight frame as an orthogonal Let $\{\mathbf{e}_k\}_{k=1}^N$ be an ONB for an N-dimensional Hilbert be an M-dimensional subspace of $\mathcal{H}$. Let $\mathbb{P} : \mathcal{H} \to \mathcal{H}$ onto $\mathcal{H}'$. Show that $\{\mathbb{P}\mathbf{e}_k\}_{k=1}^N$ is a tight frame for frame bound.

Exercise 20.6.7. Consider the space of l^2-sequences v interval $[f_1, f_2]$ with $-1/2 < f_1 < f_2 < 1/2$. Show that (in l^2) of this space is the space of l^2-sequences with $[-1/2, 1/2] \setminus [f_1, f_2]$. [Hint: Use the definition of the apply Parseval's theorem.]

Exercise 20.6.8. Refer to Section 20.4.2 and consider the operator $\mathbb{A}$ in (20.75) and the operator $\mathbb{T}$ in (20.68).

1. Consider a sequence $\{a_k\}_{k\in\mathbb{Z}} \in \mathcal{R}(\mathbb{T})$ and show that $\mathbb{A}\{a_k\}_{k\in\mathbb{Z}} = \tilde{\mathbb{T}}^*\{a_k\}_{k\in\mathbb{Z}}$.

2. Consider a sequence $\{b_k\}_{k\in\mathbb{Z}} \in \mathcal{R}(\mathbb{T})^\perp$ and show that $\mathbb{A}\{b_k\}_{k\in\mathbb{Z}} = 0$.

3. Using the fact that every sequence $\{c_k\}_{k\in\mathbb{Z}}$ can be decomposed as $\{c_k\}_{k\in\mathbb{Z}} = \{a_k\}_{k\in\mathbb{Z}} + \{b_k\}_{k\in\mathbb{Z}}$ with $\{a_k\}_{k\in\mathbb{Z}} \in \mathcal{R}(\mathbb{T})$ and $\{b_k\}_{k\in\mathbb{Z}} \in \mathcal{R}(\mathbb{T})^\perp$, show that $\mathbb{A} = \tilde{\mathbb{T}}^*\mathbb{P}$, where $\mathbb{P} : l^2 \to l^2$ is the orthogonal projection operator onto $\mathcal{R}(\mathbb{T})$.

[Hints: Use the fact that $\mathcal{R}(\mathbb{T})$ is the space of l^2-sequences with DTFT supported on the interval $[-BT, BT]$; use the characterization of $\mathcal{R}(\mathbb{T})^\perp$ developed in Exercise 20.6.7; work in the DTFT domain.]

Exercise 20.6.9. Refer to Section 20.4.2 and use the ideas from Exercise 20.6.8 to show that the operator $\mathbb{B}$ in (20.76) can be written as $\mathbb{B} = \mathbb{M}(\mathbb{I}_{l^2} - \mathbb{P})$, where $\mathbb{P} : l^2 \to l^2$ is the orthogonal projection operator onto $\mathcal{R}(\mathbb{T})$ with $\mathbb{T}$ defined in (20.68); and $\mathbb{M} : l^2 \to \mathcal{L}^2$ is the interpolation operator defined in (20.77).

Exercise 20.6.10 (Weyl operator [28]). Refer to Definition 20.5.1 and show the following properties of the Weyl operator.

1. The following equality holds:

$$\mathbb{W}_{m,n}^{(T,F)}\mathbb{W}_{k,l}^{(T,F)} = e^{-i2\pi mlTF}\mathbb{W}_{m+k,n+l}^{(T,F)}.$$

2. The adjoint operator of $\mathbb{W}_{m,n}^{(T,F)}$ is given by

$$\left(\mathbb{W}_{m,n}^{(T,F)}\right)^* = e^{-i2\pi mnTF}\mathbb{W}_{-m,-n}^{(T,F)}.$$

3. The Weyl operator is unitary on $\mathcal{L}^2$, i.e.,

$$\mathbb{W}_{m,n}^{(T,F)}\left(\mathbb{W}_{m,n}^{(T,F)}\right)^* = \left(\mathbb{W}_{m,n}^{(T,F)}\right)^*\mathbb{W}_{m,n}^{(T,F)} = \mathbb{I}_{\mathcal{L}^2}.$$

Exercise 20.6.11 (Dual WH frame [3]). Assume that the WH set $\left\{g_{m,n}(t) = \mathbb{W}_{m,n}^{(T,F)}g(t)\right\}_{m\in\mathbb{Z}, \, n\in\mathbb{Z}}$ is a frame for $\mathcal{L}^2$ with frame operator $\mathbb{S}$.

1. Show that the frame operator $\mathbb{S}$ and its inverse $\mathbb{S}^{-1}$ commute with the Weyl operators, i.e.,

$$\mathbb{W}_{m,n}^{(T,F)}\mathbb{S} = \mathbb{S}\mathbb{W}_{m,n}^{(T,F)}$$
$$\mathbb{W}_{m,n}^{(T,F)}\mathbb{S}^{-1} = \mathbb{S}^{-1}\mathbb{W}_{m,n}^{(T,F)}$$

for $m, n \in \mathbb{Z}$.

2. Show that the minimal dual frame $\{\tilde{g}_{m,n}(t) = ($
   with prototype function $\tilde{g}(t) = (\mathbb{S}^{-1}g)(t)$, i.e., t

$$\tilde{g}_{m,n}(t) = \mathbb{W}_{m,n}^{(T,F)}\tilde{g}(t$$

Exercise 20.6.12 (WH frames in finite dimen
exercise. The point of the exercise is to understand
the frame operator and the dual frame mean in linea:
   Consider the space $\mathbb{C}^M$. Take M to be a large nu
resemble continuous-time waveforms, but small enou
program works. Take a prototype vector $\mathbf{g} = [g[1]$
choose, for example, the $fir1(.)$ function in MATLA
continuous-time Gaussian waveform $e^{-x^2/2}$. Next, fix
$\mathbb{N}$ in such a way that $L \triangleq M/T \in \mathbb{N}$. Now define

$$g_{k,l}[n] \triangleq g[(n - lT) \bmod M]\, e^{\mathrm{i}2\pi kn/K},$$
$$k = 0, \ldots, K - 1,\ l = 0, \ldots, L - 1,$$

and construct a discrete-time WH set according to

$$\left\{\mathbf{g}_{k,l} = [g_{k,l}[0]\ \cdots\ g_{k,l}[M-1]]^{\mathsf{T}}\right\}_{k=0,\ldots,}$$

1. Show that the analysis operator $\mathbb{T} : \mathbb{C}^M \to \mathbb{C}^{K}$
   M)-dimensional matrix. Specify this matrix in

2. Show that the adjoint of the analysis operator $\mathbb{T}^*$
   as an $(M \times KL)$-dimensional matrix. Specify th
   and M.

3. Specify the matrix corresponding to the frame o
   and M. Call this matrix $\mathbf{S}$. Compute and s
   MATLAB.

4. Given the matrix $\mathbf{S}$, check, if the Weyl
   $\{\mathbf{g}_{k,l}\}_{k=0,\ldots,K-1,\, l=0,\ldots,L-1}$ you started from is
   can verify this.

5. Prove that for $K = M$ and $T = 1$ and for every
   set $\{\mathbf{g}_{k,l}\}_{k=0,\ldots,K-1,\, l=0,\ldots,L-1}$ is a frame for $\mathbb{C}^M$

6. For the prototype vector $\mathbf{g}$ you have chosen, find
   (T_1, K_1) and (T_2, K_2) such that $\{\mathbf{g}_{k,l}\}_{k=0,\ldots,K-1,}$
   T_1 and $K = K_1$ and is not a frame for $T = T_2$
   where $\{\mathbf{g}_{k,l}\}_{k=0,\ldots,K-1,\, l=0,\ldots,L-1}$ is a frame, con

7. Compute the dual prototype vector $\tilde{\mathbf{g}} = [\tilde{g}[1] \cdots \tilde{g}[M]]^{\mathsf{T}} = \mathbf{S}^{-1}\mathbf{g}$. Show that the dual frame $\{\tilde{\mathbf{g}}_{k,l}\}_{k=0,\ldots,K-1,\,l=0,\ldots,L-1}$ is given by time-frequency shifts of $\tilde{\mathbf{g}}$, i.e.,

$$\left\{ \tilde{\mathbf{g}}_{k,l} = [\tilde{g}_{k,l}[0] \;\; \cdots \;\; \tilde{g}_{k,l}[M-1]]^{\mathsf{T}} \right\}_{k=0,\ldots,K-1,\,l=0,\ldots,L-1}$$

with

$$\tilde{g}_{k,l}[n] \triangleq \tilde{g}[(n - lT) \bmod M]\, e^{\mathrm{i}2\pi kn/K},$$
$$k = 0,\ldots,K-1,\; l = 0,\ldots,L-1,\; n = 0,\ldots,M-1.$$

References

[1] E. Kreyszig, *Introductory Functional Analysis with Applications.* Wiley, 1989.

[2] R. J. Duffin and A. C. Schaeffer, "A class of nonharmonic Fourier series," *Trans. Amer. Math. Soc.*, vol. 73, pp. 341–366, 1952.

[3] I. Daubechies, "The wavelet transform, time-frequency localization and signal analysis," *IEEE Trans. Inf. Theory*, vol. 36, pp. 961–1005, Sept. 1990.

[4] ——, *Ten lectures on wavelets.* CBMS-NSF Regional Conference Series in Applied Mathematics, 1992.

[5] C. E. Heil and D. F. Walnut, "Continuous and discrete wavelet transforms," *SIAM Rev.*, vol. 31, pp. 628–666, Dec. 1989.

[6] R. M. Young, *An Introduction to Nonharmonic Fourier Series.* New York: Academic Press, 1980.

[7] D. L. Donoho, "De-noising by soft-thresholding," *IEEE Trans. Inf. Theory*, vol. 41, no. 3, pp. 613–627, Mar. 1995.

[8] D. L. Donoho and I. M. Johnstone, "Ideal spatial adaptation via wavelet shrinkage," *Biometrika*, vol. 81, no. 3, pp. 425–455, Aug. 1994.

[9] M. Rupf and J. L. Massey, "Optimum sequence multisets for synchronous code-division multiple-access channels," *IEEE Trans. Inf. Theory*, vol. 40, no. 4, pp. 1261–1266, July 1994.

[10] M. Sandell, "Design and analysis of estimators for multicarrier modulation and ultrasonic imaging," Ph.D. dissertation, Luleå Univ. Technol., Luleå, Sweden, Sept. 1996.

[11] R. W. Heath, Jr. and A. J. Paulraj, "Linear dispersion codes for MIMO systems based on frame theory," *IEEE Trans. Signal Process.*, vol. 50, no. 10, pp. 2429–2441, Oct. 2002.

788

[12] M. Rudelson and R. Vershynin, "Geometric appr(
and reconstruction of signals," *International M(*
no. 64, pp. 4019–4041, 2005.

[13] Y. C. Eldar and G. D. Forney, Jr., "Optimal tigl
surement," *IEEE Trans. Inf. Theory*, vol. 48, no

[14] H. Bölcskei and F. Hlawatsch, "Noise reduction
using predictive quantization," *IEEE Trans. Inf*
155–172, Jan. 2001.

[15] H. Bölcskei, "Oversampled filter banks and predi(
dissertation, Technische Universität Wien, Nov.

[16] J. J. Benedetto, A. M. Powell, and Ö. Yılmaz, "Si
and finite frames," *IEEE Trans. Inf. Theory*, v(
May 2006.

[17] D. L. Donoho, "Compressed sensing," *IEEE Tran*
pp. 1289–1306, Apr. 2006.

[18] E. J. Candès and T. Tao, "Near-optimal signal
jections: Universal encoding strategies?" *IEEE*
no. 12, pp. 5406–5425, Dec. 2006.

[19] D. L. Donoho and M. Elad, "Optimally spars
(nonorthogonal) dictionaries via l^1 minimizatioi
of Sciences of the US, vol. 100, no. 5, Mar. 2003.

[20] J. Kovačević and A. Chebira, *An Introduction*
and Trends in Signal Processing. NOW Publisl

[21] O. Christensen, *An Introduction to Frames and*
U.S.A.: Birkhäuser, 2003.

[22] H. Lütkepohl, *Handbook of Matrices*. Chicheste

[23] A. Ben-Israel and T. N. Greville, *Generalized In*
tions, 2nd ed. Canadian Mathematical Society,

[24] D. Han and D. R. Larson, "Frames, bases and gr
oirs of the American Mathematical Society, vol.

[25] E. M. Stein and G. Weiss, *Introduction to For*
Spaces. Princeton Univ. Press, 1971.

[26] A. W. Oppenheim, A. S. Willsky, and S. Hamid,
Prentice Hall, 1996.

[27] S. Mallat, *A Wavelet Tour of Signal Processing: The Sparse Way*, 3rd ed. Elsevier, 2009.

[28] I. Daubechies, A. Grossmann, and Y. Meyer, "Painless nonorthogonal expansions," *J. Math. Phys.*, vol. 27, no. 5, pp. 1271–1283, May 1986.

Index